T0234009

In memory of Paul

who sadly passed away on 28 January 2021

just before the completion of this book

Foreword

Thoughts, Memories, and Commemoration of Paul Crutzen the Man and His Work

Memories are awakened and come alive in three dimensions. Images and events merge with mine, do not stay isolated, are combined in an outstanding personality, a clever, wise person, an important scientist ahead of his time: Paul Crutzen.

These memories vividly bring back general conversations, scientific activities, and also socio-political actions. They are mingled with the results of scientific discoveries and cutting-edge research that never made Paul Crutzen, the man, complacent or vain. He always regarded his scientific contribution as the result of collegial collaborative work at the Max Planck Institute for Chemistry, the Scripps Institute in California, and countless scientific conferences. Paul Crutzen never punctuated these results with an exclamation mark or with the self-assurance of those who regard any scepticism about the results of their research as sacrilege. He himself always placed a question mark after these results, never an exclamation mark. For him science was a dynamic process, and his commitment to scientific endeavour meant he was more concerned with questioning his results than with verifying them. The question mark was dominant, not the exclamation mark.

Vivid Memories of the Flight around Mount Everest

This flight started from Kathmandu, the capital of Nepal, and was made for the purpose of conducting a scientific investigation into the particles in the atmosphere. Paul Crutzen and "Ram" Ramanathan were the scientific leaders on board. The scientists were responsible for the UNEP-initiated project "Atmospheric Brown Cloud" (ABC), linked to the major scientific research project "Indian Ocean Experiment" (INDOEX). As the Executive Director of UNEP at the time, I was there with them. The flight was as unforgettable for me as the subsequent scientific discussions to evaluate the results of this excursion.

The Flight around Mount Everest

We flew through the "Brown Cloud", composed of particles from all kinds of mass combustion processes. Lots of aerosols formed this "cloud".

The central scientific questions were: "How do these particle clouds, which are by no means limited to the Asian region, affect the climate and the Radiation Budget? Are they masking an even faster global warming process? Do they confirm the noticeable changes associated with the massive infiltration into the atmosphere of particles from gigantic volcanic eruptions such as Toba, which caused a volcanic winter over 75,000 years ago, and Tombora (in Indonesia), which erupted in 1815 and made 1816 the "year without a summer", triggering misery, hunger and mass emigrations as far away as Germany?

Or do these particle clouds have very different effects on the different regions of Planet Earth? Will the Arctic, in particular, be "heated up" more intensely with an accelerated thawing of the polar ice and the permafrost soils? Will climate change be further magnified by the additional methane emissions caused by this?

These were questions that would become an exclamation mark due to the topicality of the significantly above-average warming in the Arctic region.

For Crutzen, the graphic consequences posed challenging food for thought—hotly contested disputes about active human intervention in the Radiation Budget in the natural cycles of the Earth; Solar Radiation Management (SRM), such as Carbon Dioxide Removal (CDR); the importance of natural and artificial CO_2 sinks —in his essay, "Geology of Mankind" (*Nature*, 415, 23/2002).

This "Anthropocene article", which will be discussed later, ends with a reference to "internationally accepted, large-scale geo-engineering projects, for instance to 'optimize' climate. At this stage, however, we are still largely treading on terra incognita."

Again the question mark, not the exclamation mark.

Memories of the Joint Efforts in Scientific and Socio-Political Debate to Take Decisive Action Against the Expansion of the "Ozone Hole"

Crutzen's Research Results

Man-made emissions of CFC (chlorofluorocarbons) cause a "hole" in the ozone layer in the Antarctic region. Halogens and bromides are therefore a major threat to human health, especially in the southern hemisphere.

These scientific results were by no means accepted without opposition or criticism! There was opposition from science, opposition from commerce, and, initially, equally widespread lack of understanding in the community because these substances were very cheap and could be used in a variety of ways. The CFCs in spray cans became the identifying feature in this debate.

Paul Crutzen's scientific wake-up call, reinforced in parallel by the scientific knowledge and results of Mario Molina and Frank Sherwood Rowland, prompted public demand for government action against these substances. Co-operation between science, non-governmental organizations and, increasingly, politicians led to success.

As early as 1986, the Vienna Convention for the Protection of the Ozone Layer was adopted as the basis for international law. It was also very quickly possible to successfully negotiate the Montreal Protocol in 1987 as the basis for specific interventions within the framework of the United Nations through the UNEP. What was still missing was solid funding for global action. This massive gap was finally bridged in 1990 with the establishment of the Multilateral Fund.

Without Paul Crutzen's almost agitatory activities, this success would not have been possible. Again the refrain: Paul Crutzen was never satisfied with the scientific result; he always felt it was his duty to share responsibility for social problems and to actively work to overcome them. Paul Crutzen, the staunch and convincing activist. For these contributions, which are crucial for successfully combating the threat to the ozone layer, the scientist Paul Crutzen was awarded the Nobel Prize in Chemistry in 1995, together with Mario Molina and Frank S. Rowland. Paul Crutzen's scientific interchanges, especially with Mario Molina, continued to be intensively maintained afterwards.

"Remembering is Only Fruitful When It Reminds You of What Still Needs to be Done"

This duty of remembering, which the philosopher Ernst Bloch urges, has an almost imperative quality in reminiscences of Paul Crutzen. How could it be otherwise when his scientific work ends with a question mark? So I must once more return to "Geology of Mankind", and the last two words of this most-cited scientific article of recent years, which I have already quoted: "terra incognita". Has man replaced nature? Has the natural age of the Holocene already ended, and have we already arrived at the first man-made age, the Anthropocene? Paul Crutzen accepted this finding—but not without the question mark mentioned above. Faced with a global catastrophe—Paul Crutzen cited as examples a meteorite impact, a world war, a pandemic—would mankind remain "a major environmental force for many millennia"? He exhorted scientists and engineers to lead mankind through the Anthropocene, stating, "This will require appropriate human behaviour". Crutzen left the question open as to who should enforce this human behaviour in a population that is rapidly advancing towards nine billion people. Above all, what this "appropriate human behaviour" looks like and how it is determined remains unanswered. The ultimate question that Paul Crutzen asked constitutes a challenge for society: can this "appropriate behaviour" be realized in an open parliamentary democracy? Does the ever-deeper foray of humans into the decoding of nature and life lead to human action to control the expected and unexpected consequences of previous action; or does it lead to a lack of alternatives, to path dependency?

In the Encyclical, *Laudato Si'*, Pope Francis focuses this question on the "technocratic paradigm": Do people control technology or does technology control people?

These questions are part of Paul Crutzen's compelling legacy. Finding answers to them is essential to gain scientific understanding that is always mindful of the question marks over one's own research; this is part of the legacy that Paul Crutzen left behind as a task. Again and again, especially in the Anthropocene, people will be confronted with "terra incognita" and will have to cope with the unexpected consequences of previous actions.

Science cannot brush this legacy aside. Not least, the Coronavirus pandemic, a global catastrophe mentioned by Paul Crutzen, is a powerful reminder of this. So it is highly appropriate that the Max Planck Society has started to set up a new Institute for Geo-Anthropology. It will be a mark of respect for Paul Crutzen's incredible achievement when the development work led by Professor Jürgen Renn leads to success.

Remember what still needs to be done!

This is a general challenge for a science that is constantly questioning itself, a warning to society and politics not to misuse this dynamic of science as an excuse for postponing action.

To me, Paul Crutzen was, consciously and unconsciously, a teacher and a reminder:

Scientific knowledge is the basis for social and political action. It is indispensable. It is equally necessary to define "appropriate human behaviour" on the basis of normative, ethical action. For me, this link between science and politics has been repeatedly confirmed at the intersections of science and normative relationships in social politics, but rarely so clearly. Time and again these question marks will be an incentive for scientific research and political responsibility. That is the legacy I owe to Paul Crutzen.

Höxter, Germany   Prof. Dr. Klaus Töpfer
June 2021

Prof. Dr. Klaus Töpfer is the founding Director and until 30 September 2015 was Executive Director of the *Institute for Advanced Sustainability Studies* (IASS) based in Potsdam and Council Chair at Agora Energiewende. He is also the former Executive Director of the *United Nations Environment Programme* (UNEP) based in Nairobi and Under-Secretary-General of the United Nations (1998–2006). He graduated from Mainz, Frankfurt and Münster in 1964 with a degree in Economics. From 1965 to 1971 he was a Research Assistant at the Central Institute for Spatial Research and Planning at the University of Münster, where he graduated in 1968 with a Ph.D. in "Regional development and location decision". From 1971 to 1978 he was Head of Planning and Information in the State of Saarland, as well as a visiting Professor at the Academy of Administrative Sciences in Speyer. During this period he also served as a consultant on development policy for the following countries: Egypt, Malawi, Brazil and Jordan. From 1978 to 1979 he was Professor and Director of the Institute for Spatial Research and Planning at the University of Hanover. In 1985 he was appointed by the University of Mainz Economics Faculty as an Honorary Professor. Since 2007 he has been a Professor of Environment and Sustainable Development at Tongji University, Shanghai. He is also a visiting Professor at the Frank-Loeb Institute, University of Landau. Klaus Töpfer has been a member of the *Christian Democratic Union* (CDU) party in Germany since 1972. He is the former Minister for Environment and Health, Rheinland-Pfalz (1985–1987). He was Federal Minister for the Environment, Nature Conservation and Nuclear Safety from 1987 to 1994 and Federal Minister for Regional Planning, Housing and Urban Development from 1994 to 1998. He was also a member of the German Bundestag during the period 1990 to 1998. He has received numerous awards and honours, including the Federal Cross of Merit in 1986 and the German Sustainability Award in 2008 for his lifetime achievement in the field of sustainability. He received the German Award for Culture 2010 and the Wilhelmine von Bayreuth Award in 2012. In 2012 he was inducted into the "Kyoto Earth Hall of Fame".

Preface

Rethinking Science for the Anthropocene: The Perspective of Geoanthropology

This book offers a survey of the development of and current discussions about the Anthropocene concept introduced by Paul Crutzen in 2000. Sadly, Crutzen passed away just before the publication of this book, but it will nevertheless serve as a tribute to his lifetime achievements. He grew up in difficult times—in war-torn Amsterdam during the German occupation. His ambition to attend university and pursue an academic career was thwarted by economic conditions. True to his realist mindset, he decided to train as a civil engineer in order not to be a financial burden on his parents. He later applied for a position as a computer programmer at the Meteorological Institute of *Stockholm University* (MISU), an opportunity that would profoundly shape his future career. By the 1960s, environmental and political issues were already present in discussions of atmospheric chemistry, but Crutzen was primarily driven by his interests in basic science. He was first and foremost a passionate scientist, fascinated by intellectual challenges. But what is remarkable about his further intellectual pathway that led him to popularising the notion of the Anthropocene—and this is as a characteristic for his scientific persona as it is for the time in which it came to maturity—is the intertwinement between basic research and societal concerns.

Paul Crutzen had an acute sensibility for the implications of his research and pursued their political and practical consequences to the very end. His work in 1986 on the cause of the ozone hole in the atmosphere helped to save humanity from potentially catastrophic consequences of its interventions into nature. But the scope of his work was much larger. He also engaged with other entanglements between human activities and natural processes: biomass burning and biofuel production, nuclear winter, and geoengineering to mitigate climate change, to name just a few examples. Many of the phenomena Crutzen studied are, in fact, not separable into natural and human dimensions but are characteristic of the novel epoch that he spontaneously dubbed the "Anthropocene"—without being aware of its longer intellectual prehistory. He thus introduced not only a new geological epoch but also a new era in which humanity, in general, and science, in particular, seek a more sustainable relationship with their home and habitat: planet Earth.

Crutzen's adoption of the term "Anthropocene" in 2000 had far-reaching consequences. It stimulated geologists to start examining the stratigraphic evidence for a new geological time period, a task taken up by the interdisciplinary *Anthropocene Working Group* (AWG) just a few years later. It encouraged Earth-system scientists to explore even more deeply the multiple entanglements between human activities and the various components of the Earth system. It shifted attention and stirred up profound discussions among the various disciplines of the humanities and social sciences that were now confronted with the repositioning of their genuine subject of interest: humans. Thanks to pioneering efforts such as *The Anthropocene Project* at the *Haus der Kulturen der Welt*, to which Paul Crutzen served as patron, the notion of the Anthropocene has triggered a wealth of cultural experiments in the arts, design, public engagement, and education. Innumerable publications, scientific as well as popular, have been dedicated to the notion of the Anthropocene, making it a familiar dictum in newspapers, political forums, and private conversations all around the world. Paul Crutzen has thus, in a significant way, contributed to the self-awareness of humans about their collective role as a geological agent. He has reminded us of what has been repressed by unstoppable progress due to the industrial implementations of science and technologies, namely, the need to deal with the externalities of an ever more expansive and extractive economy on a planetary scale. With his passing, we have lost an extraordinary researcher, but his legacy will surely continue to inspire future generations of scientists concerned with and concerned about the new planetary predicament that he helped to detect, describe, and analyse.

The ongoing transition into a new state of the Earth system, the Anthropocene, is the result of the nexus of multiple, human-induced crises that basically affect all living beings: rapid climatic change, an accelerating loss of biological and genetic diversity as well as critical habitats, the profound alteration of biogeochemical cycles, the introduction and dispersion of a plethora of novel synthetic materials, new mineral species, environmental toxins, radioactive elements, and many more such impacts that present an abrupt end—a shock—to the relatively steady state of the Holocene. This self-reinforcing nexus of planetary-wide environmental deterioration poses a number of challenges both to our understanding of it and our reactions to it. How did we get into this predicament? How can we survive and live in it? Does the Anthropocene perhaps even represent an opportunity to rethink and reshape our position as one among many species on this planet?

A key insight connected with the Anthropocene concept is the systemic nature of the ongoing changes and hence the need to take this systemic character into account when considering our actions. Partial improvements in environmental policy, economic incentives, and sociotechnical transformations are not only inadequate to cope with this systemic nature but also with the fundamental scale of the dynamics of the Anthropocene crisis. Improvements at one end risk becoming aggravations at other ends. To reduce the risk of systemic collapse of basic ecological, economic, and societal functions, it is vital to better understand the configuration and dynamic evolution of the crucial nexus between human agency and the Earth system in its complex, highly interactive, interdependent, and time-critical nature.

The Anthropocene is therefore also a challenge to the transformation of our systems of knowledge. The geosciences bore Earth-system science as a unifying concept, conceiving our planet as a complex dynamic system with subsystems, such as the atmosphere, the hydrosphere, the lithosphere, and the biosphere. In the past two decades, the Anthropocene has likewise become a driver of a human-centred Earth-system science, coevolving with global environmental policy and governance, and central to the question of how to respond to emerging instabilities in Earth's metabolism. But avoiding global systemic catastrophes, lowering risks, and enabling a holistic transformation of human societies to sustainability requires an even deeper understanding of the human-Earth nexus.

It is particularly necessary to go beyond the notion of humans as part and parcel of the biosphere, especially as such a notion purports a picture of humans as a mere biological species with equally distributed powers and equally distributed effects on the global environment. Instead, it seems more appropriate to consider the technological empowerment, sociotechnical evolution, and high carbon intensity of *certain parts* of modern humanity as the decisive agent behind the dangerous dynamics now unleashed. In that sense, humans are a sociocultural and political species, but most importantly they are a "technological species," able to construct their own niche as well as internal power relations by using certain tools, transforming matter, and increasingly controlling the energy pathways that sustain them.

In recent times, these forms of human niche construction and social stratifications have resulted in the emergence of a qualitatively new planetary sphere, the technosphere: a human-created fabric of industrial technologies, infrastructures, harnessed energy sources, knowledge systems, social institutions, and powers that increasingly interact with and function on a magnitude equivalent to that of natural spheres like the biosphere or hydrosphere. The technosphere is both a catalyst of human powers and a new substrate upon which a vastly expanded human population now depends. Although the technosphere might turn out to be only a transient stage, it appears to be the most decisive addition to the Earth system since the emergence of the biosphere and represents thereby not just a quantitative but also a qualitative step change in both human and Earth-system history. While the pre-industrial *anthropos* was largely embedded within the biosphere, the emergence and rapid development of the technosphere has elevated human agency to its current, hugely outsized rank in the Earth system's architecture.

The intricate interactions among the different Earth spheres require entirely new forms of analysis, which have much to learn both from Earth system analysis, with its sophisticated and powerful methods of data collection, analysis, and simulation, and from the no less sophisticated and powerful traditional forms of historical investigation. But there can be little doubt that their integration will lead to something completely novel that is neither reducible to the principles and methods of the natural sciences as we know them, nor to those of the humanities. We have tentatively called this novel outlook, integrating the advancing tools of Earth system observation and analysis with innovative investigations of the technological, economic, and societal drivers of the Anthropocene, "geoanthropology."

The proposed approach of geoanthropology responds to the challenges of the multiple couplings between ecological, technological, and social evolution by merging an updated version of Earth system research (the *geo*, including the *bio*) with cultural theories, social science studies, and historical investigations of socio-material, energetic, and informational flows (the *anthropos*) by forming a new discipline (the *logos*). Through analytical and interpretative approaches, geoanthropology aims to study the various drivers, dynamics, and dilemmas that have led us onto an Anthropocenic path and applies these insights to cope with its further unfolding and rapid intensification. Cast into a research framework that studies the complex coevolution of natural and human systems, geoanthropology will investigate the concrete human-created conditions of ongoing Earth system and biosphere destabilisation, the limits of socio-ecological carrying capacities, possible system thresholds and tipping elements, as well as necessary socio-economic and cultural reaction times. All of this will be crucial to identify possible intervention points and novel geosocial opportunities in the Anthropocene.

Currently, the Max Planck Society is exploring the possibility of establishing an institute dedicated to geoanthropology as a new form of cross-disciplinary research between the natural and the human sciences—an ambitious approach through which the legacy of Paul Crutzen lives on. "I don't carry out 'pure science'," Paul remarked in an interview a few years ago, "although originally that was my goal. Until I discovered … this is more than science, since humans are involved."[1] The foundation of a new institute exploring the dynamics of the techno-Earth system with a focus on the human component would truly maintain the spirit of his work, both at the Max Planck Society and beyond, as an individual scholar and as a supporter of cross-disciplinary collaborations on the burning issues that he helped to identify.

This book brings together some of the milestones that mark the emergence of the current Anthropocene debate, with a focus on Paul Crutzen's contributions to it. In this sense, it is a fitting sequel to *Paul J. Crutzen: A Pioneer on Atmospheric Chemistry and Climate Change in the Anthropocene*, which offers a unique entry point into Crutzen's life and work.[2] I am convinced that the present volume will itself become a milestone publication and a most useful reference and guide in the debates on the Anthropocene to come.[3]

March 2021   Jürgen Renn
                                                   Max Planck Institute for the History of Science
                                                                                            Berlin, Germany

---

[1]Carsten Reinhardt and Gregor Lax: "Interview with Paul Crutzen", 17 November 2011 in Mainz.

[2]Paul J. Crutzen and Hans Günter Brauch (Eds.), 2016: *Paul J. Crutzen: A Pioneer on Atmospheric Chemistry and Climate Change in the Anthropocene* (Cham: Springer International Publishing).

[3]This text is partly based on an unpublished white paper on geoanthropology co-authored with Christoph Rosol.

Jürgen Renn born on 11th July 1956 in Moers, degree in physics at Freie Universität Berlin (1983), doctorate in mathematics at Technische Universität Berlin (1987), collaborator and co-editor of the Collected Papers of Albert Einstein (1986–1992), Assistant Professor (1989) and Associate Professor at Boston University (1993), Simon Silverman Visiting Professor of History of Science at Tel Aviv University (1993), Visiting Professor of Philosophy at ETH Zürich (1993–1994), Director and Scientific Member at the Max Planck Institute for the History of Science (since 1994), honorary professor for History of Science at Humboldt-Universität Berlin (since 1996). Member of the Deutsche Akademie der Naturforscher Leopoldina (the German National Academy of Sciences, since 2005), honorary professor for History of Science at Freie Universität Berlin (since 2006). *Address*: Prof. Dr. Jürgen Renn, Max Planck Institute for the History of Science, Boltzmannstr. 22, 14195 Berlin, Germany, +49 30 22667-101. *Email*: rennoffice@mpiwg-berlin.mpg.de.

Acknowledgements

We gratefully acknowledge stimulating discussions and exchange with numerous colleagues and friends from the scientific community, and we wish to express our gratitude to Astrid Kaltenbach, Tanja Pallien und Andreas Zimmer (all MPIC) for their help with the compilation and completion of this book.

We would also like to thank Dr. Christian Witschel, head of the Department of Earth Sciences, Geography and Environment, Ms. Marion Schneider and Ms. Birke Dalia with Springer Nature in Heidelberg (Germany) and Mr. Arulmurugan Venkatasalam and Ms. Priyadharshini Subramani with Springer Nature and Scientific Publishing Services Private Limited in Chennai, Tamil Nadu (India) who were responsible for the production of this book.

March 2021
Susanne Benner
Gregor Lax
Jos Lelieveld
Ulrich Pöschl
Hans Günter Brauch

Introduction: Aims and Approach of This Book

At an international scientific conference in the year 2000, Paul Crutzen realised and stated that we have progressed from the geological epoch of the Holocene into the Anthropocene, a new era of Earth history dominated by human influence. His rationale was that humankind has become a major force shaping planet Earth through globally pervasive effects of human activity on the composition and properties of the atmosphere, oceans and terrestrial environment.

While geologists still argue about the formal recognition and dating of the Anthropocene as a new geological epoch, the term is already widely accepted and used as a new approach and concept of describing the relationship between nature and humankind across the natural sciences, humanities, and arts. Over the past years, a large and rapidly growing number of studies, journal articles, books, films, and exhibitions have addressed the Anthropocene or are building on it.

Within two decades Paul Crutzen's concept of the 'Anthropocene' has triggered more than 10.000 publications according to the World Catalogue and more than 5000 peer-reviewed articles and in selected books that were indexed separately in the 'World of Science' and in 'Scopus' (2000–2020) with a primary focus on English language literature.

This book outlines the development and perspectives of the Anthropocene concept by Paul Crutzen and colleagues from its inception to its implications for sciences, humanities, society and politics. The main text consists primarily of articles from peer-reviewed scientific journals and other scholarly sources. It comprises most articles published by Paul Crutzen and colleagues with whom he collaborated closely on the Anthropocene.

For those texts the copyright holders did not grant a free licence for this Crutzen anthology the editors included the abstracts and links to the websites of the publishers from where these texts may be read, downloaded or purchased.

The introductory text by Gregor Lax (Chap. 1) portrays Paul Crutzen's scientific work and experience leading from investigations of human influence on atmospheric chemistry and physics, climate, and the Earth system, to the conception of the Anthropocene. The subsequent document is structured in two parts:

- Part I contains 19 texts by Paul J. Crutzen and his Co-authors on the Anthropocene;
- Part II includes texts by Abraham Horowitz with 'reflections on the Anthropocene' and a bibliometric analysis (2000–2020) by Hans Günter Brauch on the The Anthropocene Concept in the Natural and Social Sciences, in the Humanities and Law;

The book backmatter adds an appendix with additional materials on the Anthropocene and on Paul J. Crutzen such as a complete list of Paul Crutzen's scholarly publications and a brief professional biography.

March 2021   Susanne Benner
                                                                                          Gregor Lax
                                                                                          Jos Lelieveld
                                                                                          Ulrich Pöschl
                                                                                          Hans Günter Brauch

Contents

1 Paul J. Crutzen and the Path to the Anthropocene   1
Gregor Lax

Part I   Texts by Paul J. Crutzen and his Close Colleagues
         and Co-authors on the Anthropocene

2 The 'Anthropocene' (2000) 19
Paul J. Crutzen and Eugene F. Stoermer

3 Geology of Mankind (2002) 23
Paul J. Crutzen

4 The 'Anthropocene' (2002) 27
Paul J. Crutzen

5 Atmospheric Chemistry in the 'Anthropocene' (2002) 33
Paul J. Crutzen

6 How Long Have We Been in the Anthropocene Era? (2003) 39
Paul J. Crutzen and Will Steffen

7 Atmospheric Chemistry and Climate in the Anthropocene:
   Where Are We Heading? (2004) 47
Paul J. Crutzen and Veerabhadran Ramanathan

8 Earth System Dynamics in the Anthropocene (2004) 75
Will Steffen, Meinrat O. Andreae, Peter M. Cox, Paul J. Crutzen,
Ulrich Cubasch, Hermann Held, Nebosja Nakicenovic,
Liana Talaue-McManus, and Billie Lee Turner II

9   The Anthropocene: Are Humans Now Overwhelming the Great
     Forces of Nature? (2007)............................... 103
     Will P. Steffen, Paul J. Crutzen, and John R. McNeill

10   Atmospheric Chemistry and Climate in the Anthropocene (2007)... 123
     Paul J. Crutzen

11   Fate of Mountain Glaciers in the Anthropocene (2011).......... 129
     Ajai, Lennart Bengtsson, David Breashears, Paul J. Crutzen,
     Sandro Fuzzi, Wilfried Haeberli, Walter W. Immerzeel, Georg Kaser,
     Charles F. Kennel, Anil Kulkarni, Rajendra Pachauri,
     Thomas H. Painter, Jorge Rabassa, Veerabhadran Ramanathan,
     Alan Robock, Carlo Rubbia, Lynn M. Russell,
     Marcelo Sánchez Sorondo, Hans Joachim Schellnhuber,
     Soroosh Sorooshian, Thomas F. Stocker, Lonnie G. Thompson,
     Owen B. Toon, Durwood Zaelke, and Jürgen Mittelstraß

12   Living in the Anthropocene: Toward a New Global Ethos (2011) ... 141
     Paul J. Crutzen and C. Schwägerl

13   The Anthropocene: From Global Change to Planetary
     Stewardship (2011) 145
     Will Steffen, Åsa Persson, Lisa Deutsch, Jan Zalasiewicz,
     Mark Williams, Katherine Richardson, Carole Crumley,
     Paul J. Crutzen, Carl Folke, Line Gordon, Mario Molina,
     Veerabhadran Ramanathan, Johan Rockström, Martin Scheffer,
     Hans Joachim Schellnhuber, and Uno Svedin

14   Atmospheric Chemistry and Climate in the Anthropocene (2012)... 175
     Klaus Lorenz, Paul J. Crutzen, Rattan Lal, and Klaus Töpfer

15   Climate, Atmospheric Chemistry and Biogenic Processes
     in the Anthropocene (2012).............................. 193
     Paul J. Crutzen

16   The Palaeoanthropocene – The Beginnings of Anthropogenic
     Environmental Change (2013)............................ 203
     Stephen F. Foley, Detlef Gronenborn, Meinrat O. Andrea,
     Joachim W. Kadereit, Jan Esper, Denis Scholz, Ulrich Pöschl,
     Dorrit E. Jacob, Bernd R. Schöne, Rainer Schreg, Andreas Vött,
     David Jordan, Jos Lelieveld, Christine G. Weller, Kurt W. Alt,
     Sabine Gaudzinski-Windheuser, Kai-Christian Bruhn, Holger Tost,
     Frank Sirocko, and Paul J. Crutzen

17 **Stratigraphic and Earth System Approaches to Defining
    the Anthropocene (2016)** 217
    Will Steffen, Reinhold Leinfelder, Jan Zalasiewicz, Colin N. Waters,
    Mark Williams, Colin Summerhayes, Anthony D. Barnosky,
    Alejandro Cearreta, Paul Crutzen, Matt Edgeworth, Erle C. Ellis,
    Ian J. Fairchild, Agnieszka Galuszka, Jacques Grinevald,
    Alan Haywood, Juliana Ivar do Sul, Catherine Jeandel, J. R. McNeill,
    Eric Odada, Naomi Oreskes, Andrew Revkin, Daniel de B. Richter,
    James Syvitski, Davor Vidas, Michael Wagreich, Scott L. Wing,
    Alexander P. Wolfe, and H. J. Schellnhuber

18 **Was Breaking the Taboo on Research on Climate Engineering
    via Albedo Modification a Moral Hazard, or a Moral Imperative?
    (2016/2017)** 253
    M. Lawrence and Paul J. Crutzen

19 **Declaration of the Health of People, Health of Planet and Our
    Responsibility-Climate Change, Air Pollution and Health
    Workshop (2017)** 267
    Pontifical Academy of Sciences

20 **Transition to a Safe Anthropocene (2017), Foreword to Well
    Under 2 °C: Fast Action Policies to Protect People and the Planet
    from Extreme Climate Change** 275
    Paul J. Crutzen

**Part II   Reflections and Review of Global Debate
           on the Anthropocene**

21 **The Anthropocene – Reflections (January 2018)** 281
    Abraham Horowitz

22 **The Anthropocene Concept in the Natural and Social
    Sciences, the Humanities and Law – A Bibliometric Analysis
    and a Qualitative Interpretation (2000–2020)** 289
    Hans Günter Brauch

Appendices 439

Appendix 1: Vita of Paul Josef Crutzen 441

Appendix 2: Complete Bibliography of Paul J. Crutzen (1965–2020) . . . 453

**Appendix 3: Facsimile of P.J. Crutzen's First Powerpoint
              on the Anthropocene** 503

Appendix 4: Abstracts of Anthropocene Texts 509

Appendix 5: Websites on the Anthropocene with Hyperlinks (Compiled by H.G. Brauch based on google.com, 10 October 2020). 531

Appendix 6: Anthropocene Videos with Hyperlinks 537

About the Nobel Prize in Chemistry in 1995 559

About the Max Planck Institute for Chemistry 563

On the Major Author Paul J. Crutzen 565

On the Co-Editors 567

About the Affiliations of the Contributors 571

Index 581

Abbreviations

^{14}C	Carbon-14, or radiocarbon
A&HCI	Arts & Humanities Citation Index
ACE-2	Aerosol Characterization Experiment
ACE-ASIA	Asian Pacific Regional Aerosol Characterization Experiment
AD (A.D.)	Anno domini (after Christ)
AERONET	AErosol RObotic NETwork
AFES-PRESS	Peace Research and European Security Studies
AI	Artificial intelligence
AMPG	Archives of the Max Planck Society
ANU	Austalian National University (Canberra, Australia),
AOD	Aerosol optical depth
Ar	Argon
ASI	Amsterdam Sustainability Institute
ASPO	Association of the Study of Peak Oil and Gas
ATSR	Along-Track Scanning Radiometer
AWG	Anthropocene Working Group
BC	Black carbon
BIBLE	International field research programme
BKCI-S	Book Citation Index–Science
BKCI-SSH	Book Citation Index–Social Sciences & Humanities
BNF	Biological N fixation
BP	Before Present
C	Carbon
Ca	Calcium
CC	Climate Change
CCS	Carbon capture and storage
CDIAC	Carbon Dioxide Information Analysis Center
CDR	Carbondioxide removal
CE	Climate engineering
CF_2Cl_2	Dichlordifluoromethane

CFC	Chlorofluorocarbon
$CFCl_3$	Trichlorofluoromethane
CFCs	Chlorofluorocarbons
CH_3CCl_3	Methyl chloroform
CH_4	Methane
CIAP	Climate Impact Assessment Program
ClO radicals	Chlor radicals
ClO_x	Chloroxide
ClO_x/NO_x	Chloroxides and nitrogen oxides
Cm	Cambrian
CMIP5	Coupled Model Intercomparison Project Phase 5
CMIP6	Coupled Model Intercomparison Project Phase 6
CNRS	Centre national de la recherche scientifique (French National Centre for Scientific Research)
CO	Carbon monoxide
CO_2	Carbon dioxide
COP15	Conference of Parties [of UNFCCC], Copenhagen, 2009
COP21	[UNFCCC] Conference of Parties No. 21 in Paris, 2015
Core Collection	A curated collection of Web of Science
COS	Carbonyl sulfide
CPCI-S	Conference Proceedings Citation Index-Science
CPCI-SSH	Conference Proceedings Citation Index-Social Science & Humanities
CPECC	The Committee to Prevent Extreme Climate Change
CRED	Centre for Research on the Epidemiology of Disasters
CSAB	Content Selection & Advisory Board
CSIC	Consejo Superior de Investigaciones Científicas (Spanish National Research Council)
D/O events	Dansgaard–Oeschger events
DGVMs	Dynamic Global Vegetation Models
DKN	German Committee Future Earth
DMSP	US Defense Meteorological Satellite Program
DNA	Deoxyribonucleic acid
DOAS	Different optical absorption spectrometer
EIA	Energy Information Agency
EJ	Exajoules
EMICs	Models of Intermediate Complexity
ENUWAR	Environmental Consequences of Nuclear War
ENVISAT	Environment Satellite
EPICA	European Project for Ice Coring in Antarctica
ERA-lab	PIK's laboratory
ESA	Earth System Analysis
ESCI	Emerging Sources Citation Index
ESS	Earth System Science
ESSP	Earth System Science Partnership

FAO	United Nations Food and Agriculture Organisation
FAOSTAT	UN Food and Agriculture Organisation, Statistics
FRG	Federal Republic of Germany
Ga	Giga annum (Billion years ago)
GCMs	General Circulation Models
GDP	Gross Domestic Product
GE	Genetic engineering
GEC	Global Environmental Change
GeoMIP	Geoengineering Model Intercomparison Project
GHGs	Greenhouse gases
GOME	Global Ozone Monitoring Experiment
GS	Geological Society, *London*
GSA	Geological Survey of America
GSSA	Global Standard Stratigraphic Age
GSSP	Global boundary Stratotype Section and Point
GWP	Global warming potential
H	Humanities
HCFC 31	Chlorofluoromethane
HCl	Hydrogen chloride
HDI	Human Development Index
HELSUS	Helsinki Institute of Sustainability Science in Finland
HFCs	Hydrofluorocarbons
HGBS	Hans Günter Brauch Foundation on Peace and Ecology in the Anthropocene
HNO_3	Nitric acid
IAHS Press	International Association of Hydrological Sciences Pubisher
IAHS	International Association of Hydrological Sciences
ICOLD	International Commission on Large Dams
ICS	International Commission on Stratigraphy
ICSU	International Council for Scientific Unions
IEA	International Energy Agency
IEEE	International Symposium on Electronics and the Environment
IGBP	International Geosphere Biosphere Programme
IGSD	Institute for Governance and Sustainable Development
IHDP	International Human Dimensions Programme on Global Environment Change
IHOPE	Integrated History and future of People on Earth
IMPLICC	Implications and Risks of Novel Options to Limit Climate Change
INDOEX	Indian Ocean Experiment
INQUA	International Union of Quaternary Studies
IPAT	Human Impact = Population $\times$ Affluence $\times$ Technology
IPCC	Intergovernmental Panel on Climate Change

IRD, France	Institut de Recherche pour le Développement (Research Institute for Development in France)
IRRI	International Rice Research Institute
ISI	Institute for Scientific Information
ISSN	International Standard Serial Number is registered with the ISSN International Centre
ITU	International Telecommunications Union
IUGS	International Union of Geological Sciences
IVM	Institute for Environmental Studies
K	Potassium
L	Law
LLL	Language, linguistics, literature
MA	Millennium Ecosystem Assessment
Ma	Million years ago
MEA	Millennium Ecosystem Assessment
Mg	Magnesium
MISU	Meteorological Institute of Stockholm University
MPIC	Max Planck Institute for Chemistry, Mainz, Germany
mPWP	Mid-Piacenzian Warm Period
N	Nitrogen
N_2	Nitrogen (diatomic gas)
N_2O	Nitrous oxide
Na	Sodium
NAS	National Academy of Science
NASA	US National Space Agency
NATO	North Atlantic Treaty Organisation
NCAR	National Center for Atmospheric Research, Boulder, USA
NERC	National Environment Research Council (in UK)
NH_3	Ammonia
NHx	Ammonia gas (NH_3) plus particulate ammonium (p-NH_4)
NKGCF	German National Committee on Global Change
NO	Nitrogen oxide
NO_3^-	Nitrate
Non-OECD	Developing and transitional countries
NO_x	Nitrogen oxides
NPP	Net primary production
NRC	National Research Council
NRW	North Rhine Westphalia, federal state in Germany
NS	Natural Science
$O(^1D)$	High-energy oxygen atom
O_2	Oxygen
O_3	Ozone
OC	Organic carbon
OCLC	Online Computer Library Center

OECD	Organisation for Economic Cooperation and Development
OFDA/CRED	Office of Foreign Disaster Assistance (OFDA-USAID) **a**/ Centre for Research on the Epidemiology of Disasters
OH	OH (Hydroxyl)
P	Phosphorus
PAGES	Past Global Changes
PB	Planetary Boundaries
PCSHE	Personal, Cultural, Social, Health and Economic Education
PD	Privatdozent (German academic title without equivalence)
PETM	The Paleocene-Eocene Thermal Maximum
Pg	Petagram
PIK	Potsdam Institute of Climate Impact Research
ppm	Parts per million
PSC	Polar stratospheric clouds
R & D	Research and Development
RC	Resilience Centre in Stockholm
RIVM	[Rijksinstituut voor Volksgezondheid en Milieu] National Institute for Public Health and the Environment
S	Sulphur
SafMA	Southern African Millennium Assessment
SCI	Science Citation Index
SCOPE	Scientific Committee on Problems of the Environment
SDI	Strategic Defense Initiative
SKAD	A sociology of knowledge approach to discourse
SKC	Scientific Knowledge Community
SLA	Scientific Lead Authors
SLCPs	Short-lived climate-forcing pollutants
SO_2	Sulphur dioxide
SQS	Subcommission on Quaternary Stratigraphy
SRM	Solar radiation management
SS	Social Science(s)
SSA	Single scattering albedo
SSCI	Social Sciences Citation Index
SSTs	Super Sonic Transports
TARFOX	Tropospheric Aerosol Radiative Forcing Observational Experiment
Tg	Teragram
THC	Thermohaline circulation
TLA	Top Lead Authors
TOA	Top of the atmosphere
TTAPS Study	Surnames of the authors (R. P. Turco, O. B. Toon, T. P. Ackerman, J. B. Pollack, and Carl Sagan
UCLA	University of California in Los Angeles
UK	United Kingdom

UNAM	National Autonomous University of Mexico
UNDP	United Nations Development Programme
UNEP	United Nations Environment Programme
UNESCO	United Nations Educational, Scientific and Cultural Organization
UNFCCC	United Nations Framework Convention on Climate Change
UNFPA	United Nations Populations Fund
UNISDR	United Nations International Strategy for Disaster Reduction
US	United States
USDA	United States Department of Agriculture
USGS	United States Geological Service
USSR	Union of Socialist Soviet Republics
WBGU	Wissenschaftlicher Beirat Globale Umweltfragen (German Advisory Council on Global Change)
WCRP	World Climate Research Programme
WMO	World Meteorological Society
WorldCat	World Catalogue
WoS	Web of Science
WRI	World Resources Institute
WWF	World Wildlife Fund

List of Figures

Fig. 5.1   An example of the complexity of Earth System research.
            The figure shows the reactions that are needed to describe
            stratospheric ozone chemistry. Not shown are reactions
            involving Br compounds, which are responsible for
            additional ozone loss (reproduced with permission from
            Crutzen (1996) *Angew Chem Int Engl* 35:1758–1777   35
Fig. 6.1   The increasing rates of change in human activity since the
            beginning of the Industrial Revolution (Steffen et al. 2003).
            Significant increases in the rates of change occur around the
            1950s in each case and illustrate how the past 50 years have
            been a period of dramatic and unprecedented change in
            human history. (U.S. Bureau of the Census 2000; Nordhaus
            1997; World Bank 2002; World Commission on Dams 2000;
            Shiklomanov 1990; International Fertilizer Industry
            Association 2002; UN Centre for Human Settlements 2002;
            Pulp and Paper International 1993; MacDonalds 2002; UNEP
            2000; Canning 2001; World Tourism Organization 2002)   42
Fig. 6.2   Global-scale changes in the Earth System as a result of the
            dramatic increase in human activity (Steffen et al. 2003): (a)
            atmospheric CO_2 concentration (Etheridge et al. 1996); (b)
            atmospheric N_2O concentration (Machida et al., 1995); (c)
            atmospheric CH_4 concentration (Blunier et al. 1993); (d)
            percentage total column ozone loss over Antarctica, using the
            average annual total column ozone, 330, as a base
            (Image: J. D. Shanklin, British Antarctic Survey); (e)
            northern hemisphere average surface temperature anomalies
            (Mann et al. 1999); (f) natural disasters after 1900 resulting in
            more than ten people killed or more than 100 people affected
            (OFDA/CRED 2002); (g) percentage of global fisheries either
            fully exploited, overfished or collapsed (FAOSTAT 2002);
            (h) annual shrimp production as a proxy for coastal zone

alteration (WRI, 2003; FAOSTAT 2002); (i) model-
calculated partitioning of the human-induced nitrogen
perturbation fluxes in the global coastal margin for the period
since 1850 (Mackenzie et al. 2002); (j) loss of tropical
rainforest and woodland, as estimated for tropical Africa,
Latin America and South and Southeast Asia (Richards 1990;
WRI 1990); (k) amount of land converted to pasture
and cropland (Klein et al. 1997); and (l) mathematically
calculated rate of extinction (based on Wilson 1992)   43

Fig. 7.1   **a** The 'Keeling curve', which shows the steady increase in
atmospheric CO_2 concentration recorded at Mauna Loa in
Hawaii, 1958–1999 (adapted from Keeling/Whorf 2000). **b**
The Antarctic total ozone as reported by Joe Farman and
colleagues in 1985 (left-hand panel) and currently typical
vertical ozone profiles (right-hand panel) measured by
Hofmann et al. (1989). Since then, ozone loss has continued
(with an unexpected exception during the spring of 2002),
reaching levels as low as about 100 DU. Lowest
concentrations of ozone occur in the same height range,
14–22 km, in which naturally occurring maximum ozone
concentrations are found   49

Fig. 7.2   The distributions of tropospheric NO_2, O_3, and CH_2O
measured with the DOAS (different optical absorption
spectrometer) instrument on board of the ENVISAT satellite
during two consecutive September months. Also shown are
the fire counts. *Source* Burrows et al., pers. comm.; figure
used with kind permission of J. P. Burrows and his team from
the University of Bremen   55

Fig. 7.3   Comparison of anthropogenic aerosol forcing with
greenhouse forcing. **a** The Indo-Asian region. The
greenhouse forcing was estimated from the NCAR
community climate model with an uncertainty of $\pm$ 20%. **b**
Same as above, but for global and annual average conditions.
The global average values are a summary of published
estimates from Ramanathan et al. (2001b)   56

Fig. 7.4   The averaged optical properties of different types of
tropospheric aerosol retrieved from the worldwide
AERONET network of ground-based radiometers (Dubovik
et al. 2002). Urban industrial, biomass burning, and desert
dust aerosols are shown for $\tau_{ext}(440) = 0.7$. Oceanic aerosol is
shown for $\tau_{ext}(440) = 0.15$ since oceanic background aerosol
loading does not often exceed 0.15. Also, $\omega_0(\lambda)$ and the
refractive index n shown for Bahrain was obtained only for
the cases when $\alpha \leq 0.6$ (for higher α, $\omega_0(\lambda)$ and refractive

index n were very variable due to a significant presence of urban-industrial aerosol). However, we show the particle size distribution representing all observations in Bahrain (complete range of α). Ångstrom parameter α is estimated using optical thickness at two wavelengths: 440 and 870 nm   58

Fig. 7.5   Over the past decade, the atmospheric chemistry research community has identified and quantified the distributions of a number of radiatively active substances. However, especially regarding the cooling (parasol) effect by aerosol in the hydrological cycle, as shown in the blue field, the level of scientific understanding in the calculated radiative forcing of these various substances is still very low (IPCC 2001)   64

Fig. 7.6   High concentrations of ClO radicals and the simultaneous rapid ozone destruction occur in winter when the temperature becomes very low (< -80 °C). Measurements by Anderson et al. (1989) show ozone and ClO mixing ratios near 20 km during two ER-2 aircraft flights. ClO radicals, together with Cl the catalysts that destroy ozone, show sharp increases in the low-temperature section south of about 68°S, resulting in strong depletions during the first half of September   71

Fig. 8.1   Schematic showing the methodology used by the IPCC to produce model projections. Continuous arrows show the inputs and outputs to each model in the chain. Dotted line shows the missing impacts-to-socioeconomic responses loop   87

Fig. 9.1   The mix of fuels in energy systems at the global scale from 1850 to 2000. Note the rapid relative decrease in traditional renewable energy sources and the sharp rise in fossil fuel-based energy systems since the beginning of the Industrial Revolution, and particularly after 1950. By 2000 fossil fuel-based energy systems generated about 80% of the total energy used to power the global economy   110

Fig. 9.2   The change in the human enterprise from 1750 to 2000 (Steffen et al. 2004). The Great Acceleration is clearly shown in every component of the human enterprise included in the figure. Either the component was not present before 1950 (e. g., foreign direct investment) or its rate of change increased sharply after 1950 (e.g., population)   112

Fig. 9.3   Global terrestrial nitrogen budget for (a) 1890 and (b) 1990 in Tg N year^{-1} (Hibbard et al. 2006). The emissions to the NOy box from the coal reflect fossil fuel combustion. Those from the vegetation include agricultural and natural soil emissions and combustion of biofuel, biomass (savanna and forest) and agricultural waste. The NHx emissions from the cow and feedlot reflect emissions from animal wastes. The transfers to the fish box represent the lateral flow of dissolved inorganic nitrogen from terrestrial systems to the coastal seas. Note the enormous amount of N_2 converted to NH_3 in the 1990 panel compared to 1890. This represents human fixation of nitrogen through the Haber-Bosch process, made possible by the development of fossil-fuel based energy systems (Hibbard/Crutzen/Lambin et al. 2006) 113

Fig. 9.4   The observed trajectory from 1850 to 2005 of carbon emissions due to fossil fuel combustion. Note the acceleration in emissions since 2000. The gap between current emission rates and those required to stabilise atmospheric CO_2 concentration at various levels (450, 650, and 1000 ppm) is growing rapidly 118

Fig. 10.1   Variations of the Earth's surface temperature for the last 140 years. There are uncertainties in the annual data due to data gaps, random instrumental errors and uncertainties, uncertainties in bias corrections in the ocean surface temperature data and also in adjustments for urbanisation over the land. Over both the last 140 years and 100 years, the best estimate is that the global average surface temperature has increased by 0.6 ± 0.2 °C. *Source* International Panel on Climate Change, *Climate Change 2001: The Scientific Basis,* Technical Summary (Fig. 10.1) 124

Fig. 10.2   What do we know about global energy variations? Earth receives energy from the sun in the form of ultra violet radiations and emits energy back in the form of infrared radiations. Over an average year the budget of these energy exchanges usually leads to a state of equilibrium. A modification in this equilibrium, from either more incoming energy or more outgoing energy, is termed radiative forcing. This modification can result from a concentration of greenhouse gases, to variations in the activity of the sun, to land use, etc. Figure 10.2 shows that our current scientific knowledge is uneven: there is more knowledge on forcing due to greenhouse gases than on aerosols effects. *Source* International Panel on Climate Change: *Climate Change 2001: The Scientific Basis,* Technical Summary 125

Fig. 10.3   Possible evolutions of global temperatures. The IPCC uses state of the art scientific information to build models of how our different human behaviours can affect global temperatures. The two maps in the figure are projections based on two different scenarios. The A2 storyline (map 1) describe a very heterogeneous world. The underlying theme is self-reliance and preservation of local identities. Fertility patterns across regions converge very slowly, which results in a continuously increasing population. The B2 scenario (map 2) describes a world in which the emphasis is on local solutions to economic, social, and environmental sustainability. It is a world with continuously increasing global population, at a rate lower than in A2, intermediate levels of economic development, and less rapid and more diverse technological change than in the B1 and A1 storylines. While the scenario is also oriented towards environmental protection and social equity, it focuses on local and regional levels. The scenarios do not include any new global climate policies beyond what has been done so far. Scenario B2, where global warming is slower thus shows that changes in our behaviours can have a measurable effect on the climate. *Source* International Panel on Climate Change, *Climate Change 2001: The Scientific Basis,* Technical Summary 127

Fig. 13.1   **a** The increasing rates of change in human activity since the beginning of the Industrial Revolution to 2000. Significant increases in rates change occur around the 1950s in each case and illustrate how the past 50 years have been a period of dramatic and unprecedented change in human history (Steffen et al.2004, and references therein). In the following part figures, the parameters are disaggregated into OECD (wealthy) countries (blue) and non-OECD (developing) countries (red); **b** Population change from 1960 through 2009, in 1000 millions of people (World Bank 2010); **c** Increase in real GDP from 1969 through 2010, in trillions 2005 USD (USDA 2010); **d** Communication: increase intelephones (millions), both land-lines and mobilephones, from 1950 through 2009 (Canning 1998; Canning/Farahani 2007; ITU 2010). 151

Fig. 13.2   I = PAT identity at the global scale from 1900 to the present. Note the difference in volume between the 1990–1950 period and the 1950–2011 period, which represents the Great Acceleration (Kolbert 2011) 153

Fig. 13.3   Global-scale changes in the Earth System as a result of the dramatic increase in human activity: **a** atmospheric CO_2 concentration, **b** atmospheric N_2O concentration, **c** atmospheric CH_4 concentration, **d** percentage total column

ozone loss over Antarctica, using the average annual total
column ozone, 330, as a base, **e** northern hemisphere average
surface temperature anomalies, **f** natural disasters after 1900
resulting in more than 10 people killed or more than 100
people affected, **g** percentage of global fisheries either fully
exploited overfished or collapsed, **h** annual shrimp
production as a proxy for coastal zone alteration, **i** model-
calculated partitioning of the human-induced nitrogen
perturbation fluxes in the global coastal margin for the period
since 1850, **j** loss of tropical rainforest and woodland, as
estimated for tropical Africa, Latin America and South and
Southeast Asia, **k** amount of land converted to pasture and
cropland, and **l** mathematically calculated rate of extinction
(Steffen et al. 2004, and references therein) 154

Fig. 13.4   The human domination of land systems in the Anthropocene.
Irrigated landscape, USA (photo: Azote) 156

Fig. 13.5   Changes in global average surface temperature through Earth
history, from ca. 70 million years ago to the present (adapted
from Zalasiewicz and Williams 2009). **a** The most recent 70
million years, showing the long cooling trend to the present,
coincident with decreasing atmospheric CO_2 concentrations;
the Antarctic ice sheets formed about 34 million years ago
and the northern hemisphere ice sheets about 2.5 million
years ago. **b** The most recent 3 million years, encompassing
the Quaternary period. The late Quaternary, the time during
which Homo sapiens evolved, is characterized by ca. 100
000-year rhythmic oscillations between long, variable cold
periods and much shorter warm intervals. The oscillations are
triggered by subtle changes in the Earth's orbit but the
temperature changes are driven by the waxing and waning of
ice sheets and changes in greenhouse gas concentrations. **c**
The most recent 60 000 years of Earth history, showing the
transition from the most recent ice age into the much more
stable Holocene about 12 000 years ago. The most recent ice
age, which humans experienced, was characterized by
repeated, rapid, severe, and abrupt changes in northern
hemisphere climate (Dansgaard-Oeschger events), with
changes in oceanic circulation, periodic major ice sheet
collapses, 5–10 m scale sea-level changes, and regional
changes in aridity/humidity. **d** The most recent 16 000 years
of Earth history, showing the Holocene and the transition into
it from the most recent ice age........................ 157

Fig. 13.6   National Human Development Index and Ecological
               Footprint trajectories, 1980–2007, compared with goal levels.
               (Global Footprint Network 2011) (see flash video at: <http://
               www.footprintnetwork.org/en/index.php/GFN/page/fighting_
               poverty_our_human_development_initiative/>)   159
Fig. 13.7   The planetary boundary for climate change is designed to
               avoid significant loss of ice from the large polar ice sheets.
               Melting Greenland ice sheet (photo: Bent
               Christensen, Azote).   161
Fig. 13.8   FAO food price index, 1990–2010. *Source* FAO (2011)   162
Fig. 13.9   The inner green shading represents the proposed safe
               operating space for nine planetary systems. The *red wedges*
               represent an estimate of the current position for each variable.
               The boundaries in three systems (rate of biodiversity
               loss, climate change and human interference with the
               nitrogen cycle) have already been exceeded (Rockström et al.
               2009a).   166
Fig. 13.10   A stability landscape with two stable states. The valleys, or
               basins of attraction, in the landscape represent the stable
               states at several different conditions, while the hilltops
               represent unstable conditions as the system transitions from
               one state to another. If the size of the basin of attraction is
               small, resilience is small, and even a moderate perturbation
               may bring the system into the alternative basin of attraction.
               Source Scheffer (2009)   167
Fig. 15.1   Photograph of an Arctic polar stratospheric cloud (PSC)
               taken at Kiruna, Sweden, on 27 January 2000. *Credit* Ross
               Salawitch. NASA source.   194
Fig. 15.2   Decrease of total ozone over the Halley Bay Station
               of the British Antarctic Survey   194
Fig. 15.3   Abrupt change of ozone concentration from maximum
               to the 'ozone hole' during just two months   195
Fig. 15.4   The 'Great Acceleration'.   196
Fig. 15.5   Changes in Greenhouse Gases   197
Fig. 15.6   Changes in temperature, sea level and northern hemisphere
               snow cover   197
Fig. 15.7   Global mean radiative forcing of the climate system
               for the year 2000, relative to 1750   198
Fig. 15.8   Sudden thawing, here in the Noatak National Preserve,
               Alaska.   199
Fig. 15.9   Greenhouse gas forcing and Vostok temperature   200
Fig. 16.1   Plotting the history of the Earth on a logarithmic scale gives
               three approximately equal sections for the Anthropocene, the
               Palaeoanthropocene and anthropogenically unaffected Earth

processes. The Palaeoanthropocene is a period of small and
regional effects that are more difficult to define and are
currently hotly debated. It is also a time for which the
research tools of several scientific disciplines overlap: the
integration of results from all these disciplines will be
essential to improve our understanding of processes in the
Palaeoanthropocene. 206

Fig. 16.2   Map showing the distribution of the first farming societies in
western Eurasia together with dates for major human
migration episodes (in years calibrated B.C.) beginning from
the Fertile Crescent, which had a much more temperate
climate at the time (after Gronenborn 2014). Greater temporal
resolution and precision and further understanding of the
reasons for migration will require input from many scientific
disciplines to assess the relative importance of short-scale
climate, environmental and species distributions. 208

Fig. 17.1   Sampling-standardised Phanerozoic marine diversity curve
(Alroy 2010), expressed as summed curves for constituent
groups. Regime shifts in the Earth System are reflected in the
transition from typical Cambrian (Cm) to Paleozoic to
Modern marine faunas, and at mass extinction events
(arrows). In this context 'Cambrian', 'Paleozoic', and
'Modern' do not refer to the respective time periods of the
same name, but instead to evolutionary stages of the biota.
Major alteration in the trajectory of evolution occurred at
each of the mass extinctions, recognizable by the estimated
loss of at least 75% of commonly fossilised marine species,
after which previously uncommon clades became dominant
(Barnosky et al. 2011) The dark gray area at top represents
genera not assigned to one of the three evolutionary faunas.
Ma = million years ago 224

Fig. 17.2   Global climate variation at six different timescales (modified
from Zalasiewicz and Williams, 2016 and references therein).
On the left side of the figure, the letter 'T' denotes relative
temperature, which can be taken as mean surface temperature
for panels a, b, and f, while panels c-e are predicted on a
reading of 'T' derived from the $\delta^{18}O$ of benthic marine
foraminifera for different time frames of the Cenozoic, which
for the intervals with permanent polar ice (within the
Oligocene and younger) will record a combination of ice
volume and ocean-floor temperature change. The
hyperthermals of the Mesozoic (e.g., the Turonian) are not
plotted. Ga = billion years ago; Ma = million years ago;
ka = thousand years ago 225

Fig. 17.3   Time line of geosphere-biosphere coevolution on Earth. Here
             the geosphere is defined as the atmosphere, hydrosphere,
             cryosphere, and upper part of the lithosphere. The biosphere
             is defined as the sum of all biota living at any one time and
             their interactions, including interactions and feedbacks with
             the geosphere. The time line runs from the bottom to top,
             starting with the accretion of planet Earth and ending at the
             present. Numbers indicate ages in billions of years ago (Ga).
             The major geological eons are indicated in the scale on the
             right. Left of the time line are major features of and changes
             in the state of the geosphere, including some perturbations
             from outside the system. Right of time line is the major
             transitions in the evolution of the biosphere, plus some other
             significant appearances. The major transitions in evolution
             are given abbreviated descriptions. The arrows crossing the
             two spheres depict patterns of coevolution and the fact that
             they are a single system. Eusocial behaviour has evolved in
             several organism groups including arthropods and mammals,
             perhaps first in the Mesozoic, but possibly much earlier.
             Based on a concept from Lenton et al. (2004)   229
Fig. 17.4   A ball-and-cup depiction of the Earth System definition of the
             Anthropocene, showing the Holocene envelope of natural
             variability and basin of attraction. The basin of attraction is
             more difficult to define than the envelope of variability and so
             its position is represented here with a higher degree of
             uncertainty   235
Fig. 17.5   A ball-and-cup depiction of a regime shift. The cup on the
             right represents a stable basin of attraction (the Holocene) and
             the orange ball, the state of the Earth System. The cup on the
             left and the pink ball represent a potential state (the
             Anthropocene) of the Earth System. Under gradual
             anthropogenic forcing, the cup becomes shallower and finally
             disappears (a threshold, ca. 1950), causing the ball to roll to
             the left (the regime shift) into the trajectory of the
             Anthropocene toward a potential future basin of attraction.
             The symbol τ represents the response time of the system to
             small perturbations. Adapted from Lenton et al. (2008)   236
Fig. 17.6   Two of the many possible scenarios for the Anthropocene,
             relative to the Holocene   238
Fig. 18.1   Trends in scientific publications on climate engineering
             (number of publications per year, indexed in *Web of Science*).
             Source Oldham et al. (2014).   255

Fig. 22.1   Anthropocene Texts Listed in the *Web of Science* Core
             Collection (2001–2020): 5530 on 1 February 2021.
             SourceWeb of Science. 314
Fig. 22.2   Anthropocene Texts Listed on *Scopus* (2001–2000): 5311
             on 1 February 2021. *Source Scopus* website, at: https://www.
             scopus.com (1 February 2021) 315
Fig. 22.3   Number of titles on the Anthropocene in biology that were
             recorded on the *Web of Science* from 2001 to 2020 332
Fig. 22.4   Citation of Paul J. Crutzen's 16 reported documents on the
             Anthropocene on the *Web of Science* (2004–2020).
             Web of Science database (13 February 2021) 364
Fig. 22.5   Citation of Will Steffen's 41 reported documents on the
             Anthropocene on the *Web of Science* (2004–2021).
             Web of Science database (13 February 2021) 365
Fig. 22.6   Citation of Jan Zalasiewicz's 43 reported documents on the
             Anthropocene on the *Web of Science* (2010–2021).
             Web of Science database (13 February 2021) 365
Fig. 22.7   Citations of all 5,668 documents on the Anthropocene from
             2000 to 4 March 2021 listed in the *Web of Science* (4 March
             2021) 366

List of Tables

Table 4.1   A partial record of the growths and impacts of human
             activities during the 20th century   28
Table 4.2   Composition of dry air at ground level in remote continental
             areas.   30
Table 7.1   Global source strength, atmospheric burden,
             and optical extinction due to the various types
             of aerosols (for the 1990s)   59
Table 7.2   Partial record of the growths and impacts of human
             activities during the twentieth century   68
Table 9.1   Atmospheric CO_2 concentration during the existence
             of fully modern humans on Earth. References given
             in notes below   111
Table 13.1   The planetary boundaries   165
Table 22.1   Anthropocene texts by year in the World Catalogue,
             the *Web of Science*, and *Scopus* (2000–2020).   313
Table 22.2   AWG members as lead authors on the Anthropocene in
             different scientific disciplines, according to the *Web of
             Science* and *Scopus*, based on data from 30 September 2020
             and 13 February 2021   360
Table 22.3   Top Lead Authors of Publications on the Anthropocene
             (excluding members of the *Anthropocene Working Group*).
             Source Compiled by the author based on information from
             Web of Science and *Scopus* databases on 12–14 November
             2020.   369
Table 22.4   Anthropocene texts by research areas and scientific
             disciplines in *the Web of Science* and in *Scopus*
             (2001–2021).   381
Table 22.5   Anthropocene texts by countries and languages in the *Web
             of Science* and in *Scopus* (2001–2021).   382
Table 22.6   Anthropocene texts by languages in the *World Cat*, the *Web
             of Science* and in *Scopus* (2001–2021).   382

Table 22.7   Anthropocene texts by type in the *Web of Science* and in
             Scopus (2001–2021). *Sources* Websites of *Web of Science*
             and *Scopus* (30 September 2020) 383
Table 22.8   Anthropocene texts by journals in the *Web of Science*
             and in *Scopus* (2001–2021) 384
Table 22.9   The first fifty Anthropocene texts by peer-reviewed book
             series in the *Web of Science* (2001–2021) 385
Table 22.10   Anthropocene texts by authors in the *Web of Science*
             and in *Scopus* (2001–2021) 386
Table 22.11   Fifty major editors of books in the Anthropocene in the *Web
             of Science* 387
Table 22.12   Anthropocene texts by universities and organisations in the
             Web of Science and in *Scopus* (2001–2021). *Sources*
             Websites of *Web of Science* and *Scopus* (30 September
             2020) 389
Table 22.13   Members of the Working Group on the Anthropocene
             (AWG), *Subcommission on Quaternary Stratigraphy* in
             2009 and in 2019 390

List of Boxes

Box 9.1:   Global Change and the Earth System. *Source* The authors   104
Box 9.2:   The Anthropocene System. *Source* The authors.   105
Box 17.1:   Two Contrasting Trajectories for the Anthropocene.   238

Chapter 1
Paul J. Crutzen and the Path to the Anthropocene

Gregor Lax

"Stop it! We are no longer in the Holocene, we are in the Anthropocene," with these words, Paul Crutzen remembers in retrospect a meeting of the IGBP in Cuernavaca (Mexico) in 2000. He addressed a speaker there in whose talk the term 'Holocene', at that time common for the current climatic era, was used repeatedly.[1] Crutzen could not have known then that with this interjection he would give a name to a potentially new geological era in which humans will be the decisive climatic factor: the Anthropocene. In 2000, together with Eugene Stoermer, Crutzen published an article barely two pages in length that outlined in greater detail the 'Anthropocene' – still presented in quotes at the time. As indicators of this era, the authors pointed to the global accumulation of human population and urbanisation, the consumption of fossil fuels and water, the increase of synthetic chemicals, overfishing, and the loss of biodiversity most notably in tropical regions, the massive increase of greenhouse gases and other harmful gaseous substances (esp. CO_2, CH_4, NO and SO_2), the long-term evidence of anthropogenic influences in lakes and the increase of natural disasters as well as the risk of manmade disasters, in particular the 'Nuclear Winter',[2] which will be discussed below.

In 2009, the *International Union of Geological Sciences* (IUGS) reacted to the Anthropocene proposal and deployed the 'Anthropocene Working Group'. This group was established within the Sub commission for Quaternary Stratigraphy, which itself was a part of the *International Commission on Stratigraphy* (ICS) at the IUGS.

Dr. Gregor Lax, Research Fellow, Max Planck Institute for the History of Science, Boltzmannstraße 22, 14195 Berlin, glax@mpiwg@mpiwg-berlin.mpg.de. Parts of this article were first published here: Gregor Lax: *From Atmospheric Chemistry to Earth System Science. Contributions to the recent history of the Max Planck Institute for Chemistry (Otto-Hahn-Institute), 1959–2000* (Berlin: GNT, 2018). Reprinted with permission.

[1] See Carsten Reinhardt/Gregor Lax: Interview with Paul Crutzen, 17 November 2011.

[2] Crutzen and Stoermer (2000) [see chap. 2 below]. The term became more popular following the often quoted essay: Crutzen (2002) [see chap. 3 below].

S. Benner et al. (eds.), *Paul J. Crutzen and the Anthropocene: A New Epoch in Earth's History*, The Anthropocene: Politik–Economics–Society–Science 1, https://doi.org/10.1007/978-3-030-82202-6_1

The objectives formulated by the working group were to examine the status, hierarchical level and definition of the Anthropocene as a potential new formal division of the Geological Time Scale.[3] The commission, however, includes solely proponents of an Anthropocene - including the chairman Jan Zalasiewicz. The scientific spectrum of the members is concentrated in climate and Earth System sciences, but occasionally also with other research areas, e. g. the history of sciences, which are represented by Naomi Oreskes (born 1958).[4] Against this background it is unsurprising that the working group concluded that the Anthropocene has been reached. But it has not been finally clarified whether it started after the Holocene or as part of the Holocene.[5]

The idea of a new Earth era widely triggered debates not only in geo and atmospheric sciences but also in social sciences and humanities, which have increasingly participated in the discussion since the 2010s. The latter not only made a valid joint claim as to the definition and conceptualisation of the new era, but also asked about the basic consequences of the Anthropocene. For example, to what extent does the concept of the Anthropos need to be revised (Palsson et al. 2013), whether the relationship between humans and nature ought to or could be rethought (Dalby et al. 2014) and whether the term Anthropocene is perhaps misleading because it could suggest that humans, as the main factor influencing the climate overall, do control, could control, or should control the Earth System (Malm et al. 2014).

By now, the Anthropocene was anticipated in the media as well. This included public mass media that had developed a growing interest over the past years and in which some science journalists had already shown an interest in the Anthropocene.[6] In addition, the Anthropocene had become a fixed component of exhibition and educational institutions, some highly prominent, for example in the context of the 'Haus der Kulturen der Welt' ('House of the World's Cultures') exhibition in Berlin, or in the Deutsches Museum in Munich. The former initiated the 'Anthropocene project' in 2013/14 with the participation of the Berlin Max Planck Institute for the History of Science among others. In 2016, the Deutsches Museum launched the

---

[3] Anthropocene Working Group of the Subcommission on Quarternary Stratigraphy (International Commission on Stratigraphy), Newsletter No. 1 (2009): 1; see: http://quaternary.stratigraphy.org/workinggroups/anthropocene/ (23 May 2018).

[4] See the list of the Commission members; at: http://quaternary.stratigraphy.org/workinggroups/anthropocene/ (10 October 2018).

[5] See online article of the committee member Erle Ellis (2013): 'Anthropocene', at: https://editors.eol.org/eoearth/wiki/Anthropocene (10 October 2018).

[6] For an example, see the work of Bojanowski and Schwägerl. To name only some readily accessible examples: Axel Bojanowski/Christian Schwägerl: "Debatte um neues Erdzeitalter: was vom Menschen übrig bleibt", in: *Spiegel Online*, 4 July 2011; at: https://www.spiegel.de/wissenschaft/natur/debatte-um-neues-erdzeitalter-was-vom-menschen-uebrig-bleibt-a-769581.html (29 September 2021); Christian Schwägerl: "Planet der Menschen", in: *Zeit Online*, 18 February 2014; at: https://www.zeit.de/zeit-wissen/2014/02/anthropozaen-planet-der-menschen (29 September 2021); Axel Bojanowski: "Debatte über Anthropozän: Forscher präsentieren Beweise für neues Menschenzeitalter", in: *Spiegel online*, 25 August 2014; at: http://www.spiegel.de/wissenschaft/natur/a987349.html (23 May 2018).

exhibition *Welcome to the Anthropocene. Our Responsibility for the Future of the Earth.*

The concept of the 'Anthropocene' marks the culmination of an occupation with the influence of humans on climate and environment that had extended throughout Crutzen's scientific career. This was reflected in the establishment of several research focuses in the Department for Atmospheric Chemistry at the MPIC. Under Crutzen's leadership, anthropogenic influences had played a greater role than ever before in the institute, and he, at an early stage saw it as his task to influence political decision making processes as an active voice speaking on the basis of scientific knowledge. In 2011, Crutzen stated in retrospect:

> I don't carry out 'pure science', although originally that was my goal. Until I discovered ... this is more than science, since humans are involved. And this was an important part of my research ... I discovered then that nitrogen oxides influence the ozone and even the climate.[7]

In fact, an interesting pattern can be seen in Crutzen's work over long stretches from the beginning of his scientific career: Crutzen's basic research was directly carried over to specific sociopolitical or environmental issues - frequently by himself. These issues in turn were often actively introduced by him in the sense of a compass in the contexts of political decision making. Several of Crutzen's key research areas during and before his time at the *Max Planck Institute for Chemistry* (MPIC) dealt in different ways with the anthropogenic role in climate and the Earth System. In 1968, he had shown in his dissertation that the status of the theory at that time could not adequately explain the ozone distribution in the stratosphere; this applied in particular at heights of 30–35 km.[8] He worked to determine numeric parameters that would allow more precise statements regarding O_3 distribution and soon after discovered the eminent importance of *nitrogen oxides* (NO_x) in the degradation of stratospheric ozone by catalysis (Crutzen 1970). Early on, this work directed his interest towards the global effects of anthropogenic influences which at the time were receiving little attention. In the dedication to his wife in one of his first papers on NO_x, he had anticipatorily written "I hope this will not disturb our lives too much" (Crutzen 1970).

In the following sections, I will sketch Crutzen's biography and thereafter describe a few of his most important research topics relating to anthropogenic influences in more detail.

1.1   Paul Crutzen: Biographical Notes

Paul Josef Crutzen's academic history is rather atypical for a scientific career. According to his own statement, he had wanted to study astronomy after high school

---

[7] Carsten Reinhardt/Gregor Lax: "Interview with Paul Crutzen", 17 November 2011.

[8] See Crutzen (1969); both the approach of Chapman (1929), already deemed 'classical theory', as well as Hampson's work from 1965/66 were the focus.

and go on to work as a scientist but his grade point average was not good enough for direct admission to study at Amsterdam University, which meant that he would have had to wait several semesters. Instead, he registered for a civil engineer programme at the 'Middelbare Technische School' in Amsterdam and, after graduating in 1954, he worked for the Amsterdam Bridge Construction Office for four years - with a temporary interruption for compulsory military service. In 1956, he met Finn Terttu Soininen. They married in 1958 and moved to Gaevle in Sweden, where Crutzen took a position as an engineer.[9] However, he was not particularly satisfied with his work and his desire to go into science remained. The opportunity finally arose in 1959, when the *Meteorological Institute of Stockholm University* (MISU), which was already highly renowned, advertised a position for a programmer, at that time a novel profession the details of which were unknown to Crutzen.[10] Nevertheless, he applied successfully for the position and met Bert Bolin, the head of MISU. Bolin was a pioneer of modern atmospheric research who had called for a more extensive examination of the cycles among the Earth's spheres since the late 1950's. Crutzen started to study meteorology, statistics and mathematics in addition to his employment and had excellent opportunities to work together with atmospheric scientists.[11] As a programmer at MISU, he was involved in the early development of computer based models especially to determine the ozone distribution in the atmosphere. In the first half of the 1960s, he worked for example with James R. Blankenship (Blankenship/Crutzen 1966), an officer from the U. S. Air Force, who at that time was working on his dissertation at MISU and later on played an important role in the context of the weather satellite based *US Defense Meteorological Satellite Program* (DMSP) (Hall 2001).

After completing his doctoral work in 1968, Crutzen moved to Oxford University on a scholarship until 1971, then returned to MISU and finally moved to the *National Center for Atmospheric Research* (NCAR) in Boulder to work on the 'Upper Atmospheric Programme' (Kant et al. 2012). At this point, he also came into closer contact to Christian Junge, who was the director of the Max Planck Institute for Chemistry in Mainz, since his work on the role of nitrogen oxides in ozone degradation in the atmosphere was directly related to the research interests of the working groups in Mainz, in particular the trace gas research of Junge and his employee Wolfgang Seiler, who became director of the Fraunhofer-Institute for Environmental Research in Garmisch-Partenkirchen in 1986. In 1975, Crutzen wrote an article on the possible effects of *carbonyl sulfide* (COS) on the atmosphere in interaction with the sulfur aerosol layer discovered by Junge (the Junge layer) (Crutzen 1976); he also worked on the importance of laughing gas (nitrous oxide, N_2O) for ozone degradation that time. In 1978, Crutzen was appointed as the successor to Junge as head of the Department for Atmospheric Chemistry.[12] He assumed the position in July 1980 and stayed

---

[9] Carsten Reinhardt/Gregor Lax: "Interview with Paul Crutzen", 17 November 2011.

[10] *Ibid.*

[11] *Ibid.*

[12] Protokoll der 90. Senatssitzung der MPG, 15 June 1978, in: AMPG, II. Abt., Rep. 60, No. 90. SP.Rep. 60, No. 90.SP.

there until his retirement in 2000. During these two decades Crutzen worked on a number of 'hot' topics like the 'Nuclear Winter'-thesis, the ozone layer, the effects of natural and anthropogenic induced biomassburning, and the 'Greenhouse-effect', to name but a few. At the same time Crutzen played a crucial role in the building and networking process in the atmospheric sciences in Germany but also world-wide. Amongst others, he was involved in several central funding-programmes of the Federal Government, and also wrote the first draft of the EUROTRAC-Proposal, a 1986 initiated 9-year funding-programme of the *European Community* (EC) to estab-lish a coordinated European network in atmospheric sciences. Moreover, Crutzen was deeply involved in the organisation of the *International Geosphere Biosphere Programme* (IGBP) under the aegis of the *World Meteorological Society* (WMO) and the 1987 initiated *International Council for Scientific Unions* (ICSU). This programme was key to the formation process of the Earth System Sciences on a global level.

Many of Crutzen's research topics are directly or indirectly connected to the Anthropocene, which was proposed in 2000 by himself, at first as a catchphrase and then as a scheme, which became an iconic catchphrase in the last 19 years and which finally advanced to a designation for a new geological epoch. In the following I will sketch some of the most important research fields, which influenced Crutzen allover his scientific career and which *vice versa* became influenced by Crutzen's work. We start approximately one decade prior to Crutzen's accession to office, indeed at the beginning of his scientific career, when he began to address the effect of NO_x in the atmosphere and the influence of emissions from airplanes.

1.2 NO_x, Supersonic Transports and Stratospheric Ozone

Among Crutzen's earlier areas of research was an interest in the possible effects of air traffic on the atmosphere, something he had studied since the early 1970s. This topic was significant in particular in the context of supersonic flights that had attracted the public's attention since the end of the 1960s; the medium-term goal was to introduce such flights in civil aviation as well. The first successful test flight of a supersonic passenger plane involved a Soviet Tupolev TU144 at the end of 1968; this plane was sometimes referred to as the 'Concordski' by Western media due to its striking similarity to the French Concorde. It took off before the Concorde test flight that was carried out a couple of weeks later.[13] The *Super Sonic Transports* (SSTs) were associated with a number of different challenges that gave rise to repeated criticism in the following decades: they were very expensive, very loud and there was reason to believe that the emission of NO_x into the stratosphere could dramatically affect the

---

[13] See the description of the exhibited original copy of the TU144 as part of the web site of the technical museums Sinsheim and Speyer https://sinsheim.technik-museum.de/de/tupolev-tu-144 (23 May 2018).

ozone layer. Several organisations against SSTs were founded, including the 'Anti-Concorde Project' in England, the 'Citizen's League against the Sonic boom', the 'Coalition against the SST' in the US and the 'Europäische Vereinigung gegen die schädlichen Auswirkungen des Luftverkehrs' ('European Association Against the Harmful Effects of Air Traffic') located in Frankfurt am Main in Germany. Several countries had already announced that they would not open their airspace to supersonic flights, including Canada, large parts of Scandinavia, the Netherlands and Switzerland (Crutzen 1972). In 1971, Crutzen was involved in the first conference of the *Climate Impact Assessment Program* (CIAP) Bureau of the U.S. Department of Transportation that had been established in connection with the SST issue. His contribution addressed the possible effects of NO_x emitted by SSTs on the stratosphere (Crutzen 1972a). In his opinion it was clear that insufficient attention was being given to environmental problems in general. In 1972, he published an article in which he stated that the criticism of SSTs until then had been dominated primarily by economic and political objections that have already been briefly outlined, while a potentially alarming impact on the ozone layer had only been discussed on the side (Crutzen 1972). The article explained in detail that the SSTs planned by Great Britain, the US and the USSR in the middle tropopause could cause a substantial increase of NO_x in the stratosphere. Below 40 km, where the ozone layer is densest, these flights would result in massive destruction of O_3 molecules by catalytic reactions (Crutzen 1972: 42,46). It was feared that a fleet of 500 regularly scheduled supersonic planes could reduce the ozone layer by half, or perhaps even completely (Crutzen 1972: 41). However, precise estimations were actually not possible at the time, since research on ozone degradation processes in the atmosphere was still in its.[14] Thus, Crutzen's prognosis was a consequence of the contemporary state of knowledge. In light of technical developments in aircraft construction and other innovations, more recent estimates show approximately 1% damage (Houghton/Filho/Callander et al. 1996: 96).

Ultimately, the criticism at the time did not prevent the use of SSTs in civil aviation. In 1976, the Concorde was finally placed into service as the first ultrasonic passenger plane and remained in service until 2003. For economic reasons, however, the number of jets remained very limited and in 2003, for safety reasons, the Concorde suffered the same fate as the Tupolev in 1978: the crash of Paris in July 2000, which was covered in detail by the media, resulted in the termination of scheduled Concorde flights.[15] Nevertheless, investigations of the effects of aviation remained a topic in atmospheric research as a whole. In the 1980s, as director at the MPIC, Crutzen

---

[14] A summary of the research that had intensified until the 1990s can be found in Stolarski et al.: 1995 Scientific Assessment.

[15] See Spiegel cover story: Pott et al., "Richtung Zukunft", in: *Der Spiegel*, 31 July 2000, 31: 112–126; see also: Deckstein: "Think small"; Michael Klaesgen: "Wilde Jagd am Himmel. In Paris zerschellte der Mythos der Sicherheit", in: *Zeit*, No. 31/2000; here from: *Zeit online archive*; at: https://www.zeit.de/2000/31/Wilde_Jagd_am_Himmel (23 May 2018).

initially focused on other fields and it was not until the early 1990s that he once again addressed questions about the impacts of aviation on the atmosphere.[16]

1.3   NO$_x$, CFCs and the Discovery of the Ozone Hole

Another major field of research in the 1970s was the effects of natural and artificial propellants on the ozone layer. In 1974, Mario Molina (born 1943) and Frank Sherwood Rowland (1927–2012) suggested that industrially produced *chlorofluorocarbons* (CFCs), which do not exist naturally, could play a significant role in ozone degradation. In their well-known article in the journal 'Nature', they pointed out the dangers of chlorofluoromethane (HCFC 31), which first, was being used to an ever-greater extent and thus would reach the atmosphere and second, had an expected residence time in the atmosphere of between 40 and 150 years (Rowland/Molina 1974: 810). Molina and Rowland went on to note that as the proportion of HCFC 31 grew, the proportion of chlorine atoms in the stratosphere would also increase sharply. The catalytic reaction between chlorine and O$_3$ molecules would result in rapid degradation of the ozone layer. Molina and Rowland listed NO$_x$ as a comparison for this catalytic reaction chain and cited amongst others Crutzen's work from 1971 (Rowland/Molina 1974). Also in 1974, Crutzen himself pointed out a possible ozone reduction by CFCs, in particular by CF$_2$Cl$_2$ and CFCl$_3$ (Crutzen 1974), and wrote an article in the same year that summarised the potential anthropogenic influences on the degradation of atmospheric ozone and estimated an overall rate that until then had been debated. CFCs, SSTs and the global production of NO$_x$ were all included (Crutzen 1974: 201). NO$_x$ emissions in the event of a possible nuclear war were also alluded to.[17] In September 1975, Rowland and Crutzen gave talks on the subject of CFCs at the conference of the *World Meteorological Organisation* (WMO).[18]

The political explosiveness of the work by Molina, Rowland, Crutzen, and others, was based on the fact that CFCs were almost irreplaceable substances at the time, with industrial applications ranging from coolant production to spray can propellants. Thus, these substances were of central importance to the largescale chemical industry around the world.[19] Public discussion of the possibility of restricting CFC production was initially more reserved in the FRG than in the United States, for example, where heated debate had started at a relatively early stage. The affected industry in the Federal Republic of Germany initially argued that no solid evidence

---

[16] Amongst others: Brühl/Crutzen (1991); Fischer et al. (1997). In the context of international events for example: Brühl/Crutzen (1990a); Grooß et al. (1994).

[17] See Crutzen (1974: 206 f.). This hypothesis was proposed by John Hampson (1974) in the same year.

[18] See AMPG, III. Abt., ZA 125, No. 9.

[19] In scientific research, several studies have been presented on the social debates around CFCs, in particular for the FRG and the US; see here: Böschen (2000); Grundmann (2001).

could be produced to support the hypothesis of ozone degradation by CFCs at such an alarming magnitude; they also highlighted the elimination of innumerable jobs if production were to stop (Brüggemann 2015: 175).

Even so, although somewhat later than in the US, a more profound change in policy occurred with the discovery of the Antarctic ozone hole in 1985, which became famous by the publication *Large Losses of Total Ozone in Antarctica Reveal Seasonal ClOₓ/NOₓInteraction* by Joe Farman, Brian Gardiner and Jonathan Shanklin (1985). But it were Paul Crutzen and Frank Arnold from the MPI for Nuclear Physics in Heidelberg who provided the underlying explanation shortly after this discovery (Crutzen/Arnold 1986): in the darkness of the polar winter, a cold air vortex is formed that supports the formation of *polar stratospheric clouds* (PSCs). These clouds consist to a large extent of acid molecules (primarily nitric acid) and are formed in the aerosol veil that had been discovered by Christian Junge at the beginning of the 1960s (Junge layer). Chlorine and bromine molecules are deposited on the surface of the particles in the PSCs.[20] At the end of the polar winter, increasing solar radiation means that the deposited molecules are activated in photochemical reactions. Then a catalytic reaction occurs, in which O_3 molecules are degraded so rapidly that a veritable hole is created in the ozone layer.

In Crutzen's Department for Atmospheric Chemistry in Mainz, anthropogenic influences on the ozone layer remained a fixed component of research even in the years that followed.[21] Crutzen's employee Christoph Brühl was particularly involved in the calculations of the ozone hole that were carried out by the MPIC until well into the 1990s.[22] Brühl had studied meteorology at Mainz University, worked at the MPIC, and graduated in 1987. In the meantime, he worked at the *National Center for Atmospheric Research* (NCAR) in Boulder and after finishing his doctorate he entered the working group for computer modeling in Crutzen's department.[23]

With the discovery of and explanation for the ozone hole, CFCs entered public awareness as a manmade "ozone killer" and international response on both economic and political levels followed, although at times delayed depending on the region. Whereas CFC producer DuPont in the US initiated a relatively quick response and made considerable R & D efforts towards the development of alternative substances, a reorientation at Hoechst AG and at Kali Chemie AG (which closed in 2011) was not carried out until 1986/87 (Brüggemann 2015: 184 ff.). Nevertheless, the ozone hole had caused a sharp increase in public interest in climate related topics in general and in the production of CFCs in particular. The position of proponents for regulation of CFC production was also strengthened in the framework of federal policy in Western Germany. Despite the somewhat delayed reaction by the German industry compared

---

[20] The features of chlorine and bromine that destroy the ozone layer were a focus of Crutzen's department from the beginning. Amongst others: Berg et al. (1980); first measurements: Gidel et al. (1983).

[21] See: Crutzen/Arnold (1986); Barrie et al. (1988); Crutzen et al. (1988)

[22] For example: Brühl/Crutzen (1988); Crutzen et al. (1988); Brühl/Crutzen (1990); Brühl et al. (1991); Crutzen et al. (1992).

[23] Email Christoph Brühl to Gregor Lax, 9 November 2015.

to the US, on a political level the FRG in 1987 finally became one of the driving forces in Europe advocating the establishment of the Montreal Protocol that restricts the production of CFCs on a global scale (Brüggemann 2015: 181).

For their pioneering work on NO_x and CFCs and later studies on the ozone hole, Crutzen, Molina and Rowland jointly won the first Nobel Prize for Chemistry in 1995 that was decidedly awarded for research on atmospheric chemistry.[24] Jaenicke considers this a final acknowledgment of this department that had been introduced by Christian Junge at the MPIC in 1968 (Jaenicke 2012: 187).

1.4 The 'Nuclear Winter'-Hypothesis

The first work on the hypothesis of a 'Nuclear Winter' arose at the beginning of Crutzen's term as director at the MPIC; this topic played a significant role in both scientific and public discourse of the 1980s (Badash 2009). The hypothesis predicted a long-term obfuscation of the Earth's Atmosphere caused by the extreme dust formation expected in the event of a nuclear war. Darkness, cold and subsequent poor harvests and famine around the globe would result. Thirty years later, Crutzen retrospectively referred to this idea as "probably ... the most important I ever had".[25]

Potential aftereffects of the use of nuclear weapons on the Earth's Atmosphere had already been discussed in the mid 1970s. Atmospheric researcher John Hampson, a professor at Laval University in Quebec, Canada at the time, had pointed out the danger of massive release of NO_x in the event of nuclear weapons use. At the time, Hampson felt that people were far from understanding the atmosphere, but a nuclear war could result in a massive thinning of the ozone layer through the well-known photochemical reaction of NO_x with O_3 (Hampson 1974). Initially, there was little focus on the question as to whether the use of nuclear weapons could also cause long term climatic cooling due to the ensuing formation of dust, resulting from bombshells and the following fires. This changed in 1982, when Crutzen and John William Birks published the article "The atmosphere after a nuclear war: Twilight at noon" in *Ambio* (Crutzen/Birks 1982). In addition to the expected high rate of NO_2 emission into the atmosphere, the authors emphasised the significance of the formation of smoke from extensive fires expected in cities, forests and oil and gas fields following a nuclear exchange. Thus, the focus of the approach was the consequence of the burning of biomass and material, which Crutzen had already started to work on before his time as MPI director. Like his earlier work on NO_x, these approaches were 'applied' to a certain extent in the context of the Nuclear Winter, which, against the background of the Cold War, was a highly politically charged scenario. That was all the more the case as the essay was published at the time of the rearmament debate, which

---

[24] See Press release of the Nobel Prize Organization, "The Nobel Prize in Chemistry 1995"; at: https://www.nobelprize.org/prizes/chemistry/1995/press-release/ (29 September 2021).

[25] Carsten Reinhardt/Gregor Lax: Interview with Paul Crutzen, November 17, 2011. See also the published extract of the interview in: Crutzen/Lax/Reinhardt, Paul Crutzen, 49, 2013.

burgeoned following the NATO Double Track Decision of December 12, 1979, and reached its peak in the mid 1980s. The decision combined the planned stationing of 108 Pershing II missiles and 464 cruise missiles in Europe with the offer to negotiate with the USSR on mutual disarmament of nuclear weapons.[26]

The primary feature of the Nuclear Winter hypothesis was that the approach was based on a computer simulation that fortunately has never been put to the test in reality until today. The authority of the hypothesis was underpinned by additional model studies whose results were not considered entirely reliable but nevertheless point in a similar direction: a massive use of nuclear weapons would overall result in the consequences outlined by Crutzen and Birks in 1982. Major contributions in this context came from Richard Peter Turco (born 1943) of the University of California, scientist and journalist Carl Sagan (1934–1996) and others. The model-based work of 1983, later referred to as the 'TTAPS Study' based on the surnames of the authors (R.P. Turco, O.B. Toon, T.P. Ackerman, J.B. Pollack, and Carl Sagan), concluded that a nuclear war would first obscure the Earth's surface, second, temperatures below freezing would be expected over several months, and third, substantial weather modifications would occur on a local level. This scenario was referred to as the 'Nuclear Winter'.[27] The TTAPS team later drew attention to the combustion potential of woodland and construction timber, crude and refined oil, plastic and polymer substances, asphalt surfaces and vegetation as specific central factors.[28]

The work of Crutzen, Birks, Turco, Sagan and others brought with its lasting consequences in international science as well as in public debates and political action. Organisational structures were also created on an international scale, such as the *Environmental Consequences of Nuclear War* (ENUWAR), a committee established in 1982 in the scope of the *Scientific Committee on problems of the environment* (SCOPE) of *the International Council of Scientific Unions* (ICSU), which included more than 300 scientists from around the world (Pittrock et al. 1986).

The snappy term Nuclear Winter resonated well with the public (Badash 2009: 4). The American science journal 'Discover' voted Crutzen as Scientist of the Year 1984 for his pioneering work on the Nuclear Winter (Overbye 1985), which was also picked up by the German media.[29] Crutzen's legacy reveals that many readers subsequently contacted him directly by post with a wide range of comments and requests relating to the Nuclear Winter. It is possible that this is what finally encouraged him to

---

[26] The field of history has four interpretations of the NATO Double Track Decision/ the political aims of NATO in the Cold War. While the approaches involving NATO's desire for a consensus in security policy and détente policy revisionism can themselves fundamentally be regarded as part of the contemporary debate, the approaches of synthesis of the history of society and of internationalism are interested above all in answering historical questions with various prioritizations. For extensive insight into the debate, see the anthology: Gassert et al., Zweiter Kalter Krieg.

[27] Turco et al., Nuclear Winter, 1290, 1983.

[28] See Turco et al., Climate and Smoke, 1, 1990.

[29] See for example: the cover of the *Jülicher Zeitung*, 29 December 1984; *Rhein Zeitung*, No. 300 (27 December 1984). A large dossier on the nuclear winter was published at the beginning of 1985 in: *Die Zeit*, No. 3 (1985): 10.

work journalistically as well to make the Nuclear Winter accessible to a broader audience. A good example is the anthology "Schwarzer Himmel – Auswirkungen eines Atomkriegs auf Klima und globale Umwelt" [Black Sky—Effects of a Nuclear War on Climate and the Global Environment] (Crutzen/ Hahn 1986), which he published in 1986 together with Jürgen Hahn, who had worked under Junge at the MPIC. In addition, Crutzen engaged himself politically. For example, he signed the declaration "Wir warnen vor der strategischen Verteidigungsinitiative" [We Warn of the Strategic Defense Initiative], which was directed against the *Strategic Defense Initiative* (hereafter referred to as the SDI programme) introduced into the discussion by the Reagan administration and sent to the Federal Chancellor and several ministries in mid-1985.[30] In October, the initiative was again expanded, when several political and cultural leaders joined in, including well known personalities such as then Minister President of NRW and later Federal President Johannes Rau (1931–2006), journalist, publicist and feminist Alice Schwarzer (born 1942) and leftwing intellectual songwriter Hannes Wader (born 1942).[31] The White House initially used the Nuclear Winter as a justification for the SDI programme: if it were possible to face a Soviet nuclear strike in the air, the dust other wise resulting from the impacts would fail to appear. The technical possibilities, however, would not have allowed the numerous Russian nuclear missiles to be repelled thoroughly enough for this programme to be able to prevent the scenario of a Nuclear Winter (Robock 1989: 360).

On the level of international politics, an important contribution in the context of nuclear disarmament during the final phase of the Cold War in the 1980s can to a certain extent be attributed to the 'Nuclear Winter'. In 1986, the Pentagon did not consider the effects of a nuclear exchange to be as high as estimated by the studies from the beginning of the 1980s, but nonetheless it perceived them as sufficiently threatening (Badash 2009: 165). Mikhail Gorbachev himself admitted that the theory had a certain influence on his own political stance (Robock 2010: 425). In 1988, the agreement on *Intermediate Range Nuclear Forces* (INF) that had been signed by Ronald Reagan and Gorbachev in the previous year came into force, initiating nuclear disarmament in the US and the USSR (Robock 1989: 361). In the same year, the UN recognised the Nuclear Winter as a scientifically established theory.[32] The number of potential ready to use nuclear weapons around the world dropped dramatically from approximately 70,000 warheads in the 1980s to around 8,500 in the 2000s. At the same time, this supply is still more than enough to cause a Nuclear Winter. Moreover, the number of warheads that could hypothetically be put to use again is estimated at 15,000 for Russia and the United States alone (Robock 2010: 419).

---

[30] Staudinger to Crutzen, 15 August 1985, in: AMPG, III. Abt., ZA 125, No. 4I; Appell gegen Waffen im Weltraum (1985), AMPG, III. Abt., ZA 125, No. 4I.

[31] See Starlinger to Crutzen, 10 October 1985, in: AMPG, III. Abt., ZA 125, No. 4II.

[32] See archive material provided online on the website of the UN: http://www.un.org/en/ga/search/view_doc.asp?symbol=A/43/351 (23 May 2018; see also: Robock (1989: 360).

1.5   Geo-Engineering: A Tricky Scientific Debate to Handle Responsibility in the Anthropocene?

The work on the influence of humans on the atmosphere and the global climate is a recurring theme throughout Crutzen's biography: from studies on the influence of aviation via CFCs, the ozone hole, greenhouse gases, (anthropogenic) biomass combustion and the Nuclear Winter scenario, to topics not described in detail here, such as methane emissions from mass animal breeding (Schade/Crutzen 1995; Berges/Crutzen 1996) and the emission of nitrous oxide from land treated with fertilisers (Crutzen/Ehhalt 1977). As already mentioned, in 2000 he finally proposed the term Anthropocene, a designation for a new geological era that is currently gaining increasing appeal in scientific and public discussions.

Crutzen himself clearly connects the concept with a call for action involving solutions that are both political as well as from the non-political sphere. The Montreal Protocol regarding the restriction of CFCs is without doubt an example of a relatively successful response on an international political level. In other areas, such successes have largely failed to appear, for example in the case of restricting CO_2 production. For Crutzen, therefore, the question of the responsibility of human kind remains prominent to this day.[33] In 2006, frustrated by the failure of political instruments, in particular in the case of reducing greenhouse gases,[34] Crutzen, in the search for alternatives that did not involve political decision making processes, triggered a heated debate. His proposal at the time was to intervene in the Earth's system with scientific methods to prevent a climate-based collapse or in other words: we should consider targeted geoengineering (Crutzen 2006).

In essence, the idea of geoengineering was not new and serious efforts to manipulate weather on a local scale had been made since the late 1940s, certainly advanced by the upcoming Cold War[35] and even promoted in the FRG (Achermann 2013). Geoengineering, however, reached a new dimension with Crutzen's 2006 proposal to manipulate global *climate*, a cross generational project. His first and subsequently highly controversial idea for such an intervention was to investigate whether the release of sulfur compounds into the atmosphere could significantly increase the Earth's potential to reflect sunlight into the space. Crutzen's suggestion explicitly did not relate to a practical implementation of this or a similar approach, but instead expressed that geoengineering should not be treated as taboo and that it is imperative to discuss any options and to develop and research suitable approaches. Initially, this proposal was met with some very heated reactions, based not only on justified criticism of the feasibility and potentially disastrous consequences of injecting sulfur into the atmosphere, but above all because of the basic idea of actively intervening with nature on a large scale.[36]

---

[33] Carsten Reinhardt/Gregor Lax: Interview with Paul Crutzen, 17 November 2011.

[34] Ibid.

[35] See Bonnheim 2010: 893; see also in particular for the USSR: Oldfield, Climate modification, 2013.

[36] Carsten Reinhardt/Gregor Lax: Interview with Paul Crutzen, 17 November 2011.

In the meantime, a serious scientific debate about geoengineering had been launched and some suggestions for operationalisation have been made that recently have even been addressed in policy in the FRG, in particular by the Federal Ministry for the Environment.[37] The proposals range from injecting aerosols into the atmosphere to solar sails in space to redirect a portion of sunlight (Low et al. 2013). In context of the geo-engineering debate, the Anthropocene is both a serious proposal for a new geological epoch and a call to mankind to deal with all possible – for someone impossible – measures and responsibilities concerning the future of the planet. In regard to recent developments, including the world-wide climate-movement of younger generations, originating in Sweden, the discussion on the responsibility in the Anthropocene could not be more actual. The future will show if global change and global warming can be managed – with or without geo-engineering.

References

Achermann, Dania, 2013: "Die Eroberung der Atmosphäre. Wetterbeeinflussung in Süddeutschland zur Zeit des Kalten Krieges", in: *Technikgeschichte*, 80,3: 225–239.

Badash, Lawrence, 2009: *A Nuclear Winter's Tale. Science and Politics in the 1980s* (Cambridge, MA: MIT Press).

Barrie, L.A.; Bottenheimer, J. W.; Schnell, R.C. et al., 1988: "Ozone destruction and photochemical reactions at polar sunrise in the lower Arctic atmosphere", in: *Nature*, 334: 138–141.

Berg, W.W.; Crutzen, Paul J.; Grahek, F.E. et al., 1980: "First measurements of total chlorine and bromine in the lower stratosphere", in: *Geophysical Research Letters*, 7: 937–940.

Berges, M.G.M.; Crutzen, Paul J., 1996: "Estimates of global N_2O emissions from cattle, pig and chicken manure, including a discussion of CH_4 emissions", in: *Journal Atmospheric Chemistry*, 24: 241–269.

Blankenship, James R.; Crutzen, Paul J., 1966: "A photochemical model for the space-time variations of the oxygen allotropes in the 20 to 100km layer", in: *Tellus*, 2,18: 160–175.

Bonnheim, Norah B., 2010: "History of climate engineering", in: *WIRE's Climate Change*, 1: 891–897.

Böschen, Stefan, 2000: Risikogenese. Prozesse gesellschaftlicher Gefahrenwahrnehmung: FCKW, Dioxin, DDT und Ökologische Chemie (Opladen: Leske und Budrich).

Brüggemann, Julia, 2015: "Die Ozonschicht als Verhandlungsmasse. Die deutsche Chemieindustrie in der Diskussion um das FCKW-Verbot 1974 bis 1991", in: *Zeitschrift für Unternehmensgeschichte*, 2,60: 168–193.

Brühl, Christoph; Crutzen, Paul J., 1988: "Scenarios of possible changes in atmospheric temperatures and ozone concentrations due to man's activities as estimated with a one-dimensional coupled photochemical climate model", in: *Climate Dynamics*, 2: 173–203.

Brühl, Christoph; Crutzen, Paul J., 1990: "Ozone and climate changes in the light of the Montreal Protocol. A model study", in: *Ambio* 19: 293–301.

Brühl, Christoph; Crutzen, Paul J., 1990a: "The atmospheric chemical effects of aircraft operations", in: Schumann, Ulrich (ed.): *Air Traffic and the Environment—Background, Tendencies and Potential Global Atmospheric Effects. Proceedings of a DLR International Colloquium, Bonn, Germany, November 15–16* (Berlin: Springer): 96–106.

Brühl, Christoph; Peter, Th.; Crutzen, Paul J., 1991: "Increase in the PSC-formation probability caused by high-flying aircraft", in: *Geophysical Research Letters*, 18: 1465–1468.

---

[37] Umweltbundesamt, Geo-Engineering, 2011.

Chapman, Sydney, 1929: "A Theory of upper-atmospheric ozone", in: *Memoirs of the Royal Meteorological Society*, 3,26: 103–125.

Crutzen, Paul J., 1969: "Determination of parameters appearing in the "dry" and the "wet" photochemical theories for ozone in the stratosphere", in: *Tellus*, 3,21: 368–388.

Crutzen, Paul J., 1970: "The influence of nitrogen oxides on the atmospheric ozone content", in: *Quarterly Journal of the Royal Meteorological Society*, 96: 320–325.

Crutzen, Paul J., 1972: "SST's—A Threat to the Earth's Ozone Shield", in: *Ambio*, 2/1: 41–51.

Crutzen, Paul J., 1972a: *The photochemistry of the stratosphere with special attention given to the effects of NOx emitted by supersonic aircraft* (Washingon, D.C.: US Department of Transportation, First Conference on CIAP 1972): 880–888.

Crutzen, Paul J., 1974: "Estimates of possible future ozone reductions from continued use of fluorochloromethanes (CF_2Cl_2, $CFCL_3$)", in: *Geophysical Research Letters*, 1: 205–208.

Crutzen, Paul J., 1976: "The possible importance of CSO for the sulfate layer of the stratosphere", in: *Geophysical Research Letters*, 3: 73–76.

Crutzen, Paul J., 2006: "Albedo Enhancement by Stratospheric Sulfur Injections. A Contribution to Resolve a Policy Dilemma?", in: *Climatic Change*, 77: 211–219.

Crutzen, Paul J.; Arnold, F., 1986: "Nitric acid cloud formation in the cold Antarctic stratosphere. A major cause for the springtime 'ozone hole'", in: *Nature*, 324: 651–655.

Crutzen, Paul J.; Brühl, Christoph; Schmailzl, U.; Arnold, F., 1988: "Nitric acid haze formation in the lower stratosphere: a major contribution factor to the development of the Antarctic 'ozone hole'", in: McCormick, M. P.; Hobbs, P.V. (eds.): *Aerosols and Climate* (Hampton, VA: A Deepak): 287–304.

Crutzen, Paul J.; Ehhalt, Dieter Hans, 1977: "Effects of nitrogen fertilizers and combustion on the stratospheric ozone layer", in: *Ambio*, 1–3,6: 112–117.

Crutzen, Paul J.; Hahn, Jürgen (eds.), 1986: *Schwarzer Himmel – Auswirkungen eines Atomkrieges auf Klima und globale Umwelt* (Frankfurt am Main: Fischer).

Crutzen, Paul J.; Lax, Gregor; Reinhardt, Carsten: Paul Crutzen on the Ozone Hole, Nitrogen Oxides, and the Nobel Prize. *Angewandte Chemie*, International Edition, 52,1 (2013): 48–50.

Crutzen, Paul J.; Müller, R.; Brühl, Christoph; Peter, Th., 1992: "On the potential importance of the gas phase reaction $CH_3O_2 + ClO \rightarrow ClOO+CH_3O$ and the heterogeneous reaction $HOCl + HCl \rightarrow H_2O + Cl_2$ in 'ozone hole'", in: *Geophysical Research Letters*, 19: 1113–1116.

Crutzen, Paul J.; Stoermer, Eugene F., 2000: "The 'Anthropocene'", in: *Global Change Newsletter*, 41: 17–18 [see in this vol. chap. 2 below].

Crutzen, Paul; Birks, John, 1982: "The Atmosphere after a Nuclear War: Twilight at Noon", in: *Ambio*, 11: 114–125.

Dalby, Simon; Lehman, Jessi; Nelson, Sara et al., 2014: "After the Anthropocene. Politics and geographic inquiry for a new epoch", in: *Progress in Human Geography*, 38,3: 439–456.

Ellis, Erle, 2018: "Anthropocene", at: https://editors.eol.org/eoearth/wiki/Anthropocene (10 October 2018).

Farman, Joe C.; Gardiner, B. G.; Shanklin, J. D., 1985: "Large Losses of Total Ozone in Antarctica Reveal Seasonam ClOx/NOx Interaction". *Nature*, 315: 207–210.

Fischer, H., Waibel, A. E.; Welling, M. et al., 1997: "Observations of high concentration of total reactive nitrogen (NO_y) and nitric acid (HNO_3) in the lower Arctic stratosphere during the Stratosphere-Troposphere Experiment by Aircraft Measurements (STREAM) II campaign in February 1995", in: *Journal of Geophysical Research*, 102: 23559–23571.

Gassert, Phillipp; Geiger, Tim; Wentker, Hermann (eds.), 2011: *Zweiter Kalter Krieg und Friedensbewegung: Der NATO-Doppelbeschluss in deutsch-deutscher und internationaler Perspektive* (München: Oldenbourg 2011).

Gidel, Louis T.; Crutzen, Paul J.; Fishman, Jack, 1983: "A two-dimensional photo-chemical model of the atmosphere. 1: Chlorocarbon emissions and their effect on stratospheric ozone", in: *Journal of Geophysical Research*, 88: 6622–6640.

Grooß, J.U.; Peter, Th.; Brühl, C.; Crutzen, P.J., 1994: "The Influence of high flying aircraft on polar heterogeneous chemistry", in: Schuhmann, U.; Wurzel, D. (eds.): *Proceedings of an international scientific Colloquium on Impact of Emissions from Aircraft and Spacecraft upon the Atmosphere* (Köln: DLR Mitteilung 94-06): 229–234.

Grundmann, Reiner, 2001: *Transnational Environmental Policy. Reconstructing Ozone.* (London: Routledge).

Hall, R. Cargill, 2001: *A History of the Military Polar Orbiting Meteorological Satellite Program* (Chantilly: National Reconnaissance Office).

Hampson, John, 1966: "Chemiluminescent emissions observed in the stratosphere and meso-sphere", in: *Les problèmes météorologiques de la stratosphere et de la mésosphère.* (Paris: Presses Universitaires de France): 393–440.

Hampson, John, 1974: "Photochemical war on the atmosphere", in: *Nature*, 250: 189–191.

Houghton, John T.; Filho, Luiz Meira, G.; Callander, B. A. et al. (eds.), 1996: *Climate Change 1995. The Science of Climate Change. Contribution of Working Group I to the Second Assessment Report of the Intergouvern-mental Panel on Climate Change.* (Cambridge: Cambridge University Press).

Jaenicke, Ruprecht, 2012: "Die Erfindung der Luftchemie – Christian Junge", in: Kant, Horst; Reinhardt, Carsten (eds.): *100 Jahre Kaiser-Wilhelm-/Max-Planck-Institut für Chemie* (Berlin: Archiv der Max-Planck-Gesellschaft): 187–202.

Kant, Horst; Lax, Gregor; Reinhardt, Carsten, 2012: "Die Wissenschaftlichen Mitglieder des Kaiser-Wilhelm-/Max-Planck-Instituts für Chemie (Kurzbiographien)", in: Kant, Horst; Reinhardt, Carsten (eds.): *100 Jahre Kaiser-Wilhelm-/Max-Planck-Institut für Chemie* (Berlin: Archiv der Max-Planck-Gesellschaft): 307–367.

Low, Sean; Schäfer, Stefan; Maas, Achim, 2013: "Climate Engineering", in: *IASS Fact Sheet*, 1: 1–5.

Malm, Andreas; Hornburg, Alf, 2014: "The geology of mankind? A critique of the Anthropocene narrative", in: *The Anthropocene Review*, 1: 1–8.

Oldfield, Jonathan D.: "Climate modification and climate change debates among Soviet physical geographers, 1940s–1960s". in: *WIRE's Climate Change*, 4 (2013): 513–524.

Overbye, Dennis, 1985: "Prophet of the cold and dark", in: *Discover*, 6: 24–32.

Palsson, Gisli; Szerszynski, Bronislaw; Sörlin, Sverker et al., 2013: "Reconceptualizing the 'Anthropos' in the Anthropocene. Integrating the social sciences and humanities in global environmental change research", in: *Environmental Science & Policy*, 28: 3–13.

Pittrock, A. B.; Ackerman, Thomas P.; Crutzen, Paul J. et al. (eds.), 1986: *Environmental Consequences of Nuclear War* (New York: Wiley).

Reinhardt, Carsten; Lax, Gregor, 2011: Interview with Paul Crutzen, 17 November 2011. See also the published extract of the interview in: Crutzen/Lax/Reinhardt (2013).

Robock, Alan, 1989: "Policy Implications of Nuclear Winter and Ideas for Solutions", in: Ambio, 7,18: 360–366.

Robock, Alan, 2010: "Nuclear Winter", in: *WIREs Climate Change*, 1: 418–427.

Rowland, Frank Sherwood; Molina, Mario, 1974: "Stratospheric sink for chlorofluoro-methanes. Chlorine atom-catalysed destruction of ozone", in: *Nature*, 249,5460: 810–812.

Schade, Gunnar W.; Crutzen, Paul J., 1995: "Emission of aliphatic amines from animal husbandry and their reactions. Potential source of N_2O and HCN", in: *Journal Atmospheric Chemistry*, 22: 319–346.

Stolarski, Richard S., Baughcum, Steven L.; Brune, William H. et al.: The 1995 Scientific Assessment of the Atmospheric Effects of Stratospheric Aircraft. NASA Reference Publication 1381, 1995.

Turco, Richard P.; Toon, Owen B.; Ackerman, Thomas P. et al.,1983: "Nuclear Winter: Global Consequences of Multiple Nuclear Explosions", in: *Science,* 222,4630: 1283–1292.

Turco, Richard P.; Toon, Owen B.; Ackerman, Thomas P. et al., 1990: "Climate and Smoke: An Appraisal of Nuclear Winter", in: *Science*, 247,4939: 166–176.

Umweltbundesamt (ed.), 2011: *Geo-Engineering. Wirksamer Klimaschutz oder Größenwahn? Methoden – rechtliche Rahmenbedingungen – umweltpolitische Forderungen* (Berlin: Umwelt-gundesamt).

Part I
Texts by Paul J. Crutzen and his Close Colleagues and Co-authors on the Anthropocene

Chapter 2
The 'Anthropocene' (2000)

Paul J. Crutzen and Eugene F. Stoermer

The name Holocene ('Recent Whole') for the post-glacial geological epoch of the past ten to twelve thousand years seems to have been proposed for the first time by Sir Charles Lyell in 1833, and adopted by the International Geological Congress in Bologna in 1885 (Encyclopaedia Britannica 1976). During the Holocene mankind's activities gradually grew into a significant geological, morphological force, as recognised early on by a number of scientists. Thus, G. P. Marsh already in 1864 published a book with the title *Man and Nature*, more recently reprinted as *The Earth as Modified by Human Action* (Marsh 1965). Stoppani in 1873 rated mankind's activities as a "new telluric force which in power and universality may be compared to the greater forces of earth" [quoted from Clark 1986]. Stoppani already spoke of the anthropozoic era. Mankind has now inhabited or visited almost all places on Earth; he has even set foot on the moon.

The great Russian geologist V. I. Vernadsky (Vernadski 1998) in 1926 recognised the increasing power of mankind as part of the biosphere with the following excerpt "... the direction in which the processes of evolution must proceed, namely towards increasing consciousness and thought, and forms having greater and greater influence on their surroundings." He, the French Jesuit P. Teilhard de Chardin and E. Le Roy in 1924 coined the term 'noösphere', the world of thought, to mark the growing role played by mankind's brainpower and technological talents in shaping its own future and environment.

The expansion of mankind, both in numbers and per capita exploitation of Earth's resources has been astounding (Turner 1990). To give a few examples: During the past 3 centuries human population increased tenfold to 6000 million, accompanied e.g. by a growth in cattle population to 1400 million (Crutzen/Graedel 1986) (about one cow

This text was first published by Paul J. Crutzen and Eugene F. Stoermer: "The 'Anthropocene'", in: *Global Change Newsletter*, 41 (May 2000): 17-18. The journal is out of print and the text is in the public domain. The authors thank the many colleagues, especially the members of the IGBP Scientific Committee, for encouraging correspondence and advice.

per average size family). Urbanisation has even increased tenfold in the past century. In a few generations mankind is exhausting the fossil fuels that were generated over several hundred million years. The release of SO_2, globally about 160 Tg/year to the atmosphere by coal and oil burning, is at least two times larger than the sum of all natural emissions, occurring mainly as marine dimethyl-sulfide from the oceans (Watson 1990); from Vitousek et al. (1997) we learn that 30–50% of the land surface has been transformed by human action; more nitrogen is now fixed synthetically and applied as fertilizers in agriculture than fixed naturally in all terrestrial ecosystems; the escape into the atmosphere of NO from fossil fuel and biomass combustion likewise is larger than the natural inputs, giving rise to photochemical ozone ('smog') formation in extensive regions of the world; more than half of all accessible fresh water is used by mankind; human activity has increased the species extinction rate by thousand to ten thousand fold in the tropical rain forests (Wilson 1992) and several climatically important 'greenhouse' gases have substantially increased in the atmosphere: CO_2 by more than 30% and CH_4 by even more than 100%. Furthermore, mankind releases many toxic substances in the environment and even some, the chlorofluorocarbon gases, which are not toxic at all, but which nevertheless have led to the Antarctic 'ozone hole' and which would have destroyed much of the ozone layer if no international regulatory measures to end their production had been taken. Coastal wetlands are also affected by humans, having resulted in the loss of 50% of the world's mangroves. Finally, mechanised human predation ('fisheries') removes more than 25% of the primary production of the oceans in the upwelling regions and 35% in the temperate continental shelf regions (Pauly/Christensen 1995). Anthropogenic effects are also well illustrated by the history of biotic communities that leave remains in lake sediments. The effects documented include modification of the geochemical cycle in large freshwater systems and occur in systems remote from primary sources (Stoermer/Smol 1999; Schelske/Stoermer 1971; Douglas et al. 1994).

Considering these and many other major and still growing impacts of human activities on earth and atmosphere, and at all, including global, scales, it seems to us more than appropriate to emphasise the central role of mankind in geology and ecology by proposing to use the term 'anthropocene' for the current geological epoch. The impacts of current human activities will continue over long periods. According to a study by Berger and Loutre (1996), because of the anthropogenic emissions of CO_2, climate may depart significantly from natural behaviour over the next 50,000 years.

To assign a more specific date to the onset of the "Anthropocene" seems somewhat arbitrary, but we propose the latter part of the 18th century, although we are aware that alternative proposals can be made (some may even want to include the entire holocene). However, we choose this date because, during the past two centuries, the global effects of human activities have become clearly noticeable. This is the period when data retrieved from glacial ice cores show the beginning of a growth in the atmospheric concentrations of several "greenhouse gases", in particular CO_2 and CH_4 (Vitousek 1997). Such a starting date also coincides with James Watt's invention of the steam engine in 1784. About at that time, biotic assemblages in most lakes began to show large changes (Berger/Loutre 1996).

Without major catastrophes like an enormous volcanic eruption, an unexpected epidemic, a large-scale nuclear war, an asteroid impact, a new ice age, or continued plundering of Earth's resources by partially still primitive technology (the last four dangers can, however, be prevented in a real functioning noosphere) mankind will remain a major geological force for many millennia, maybe millions of years, to come. To develop a world-wide accepted strategy leading to sustainability of ecosystems against human induced stresses will be one of the great future tasks of mankind, requiring intensive research efforts and wise application of the knowledge thus acquired in the noosphere, better known as knowledge or information society. An exciting, but also difficult and daunting task lies ahead of the global research and engineering community to guide mankind towards global, sustainable, environmental management (Schellnhuber 1999).

References

Berger, A.; Loutre, M.-F., 1996, in: *C. R. Acad. Sci.* (Paris): 323, II A, 1–16.

Clark, W.C., 1986; in: Clark, W.C.; Munn, R.E. (eds.): *Sustainable Development of the Biosphere* (Cambridge: Cambridge University Press), chapt. 1.

Crutzen, P. J.; Graedel, T.E., 1986, in: Clark, W. C.; Munn R. E., (eds.), 1986: *Sustainable Development of the Biosphere* (Cambridge University Press, Cambridge). chapt. 9.

Douglas, M. S. V.; Smol, J. P.; Blake Jr., W., 1994, in: *Science*, 266.

Encyclopaedia Britannica, Micropaedia, IX (London: *Encyclopaedia Britannica*, 1976).

Marsh, G. P., 1965: *The Earth as Modified by Human Action* (Cambridge, MA: Belknap Press, Harvard University Press).

Pauly, D.; Christensen, V., 1995, in: *Nature,* 374: 255-257.

Schellnhuber, H. J., 1999, in: *Nature*, 402: C19-C23.

Schelske, C. L.; Stoermer, E. F., 1971, in: *Science,* 173.

Stoermer, E. F.; Smol, J.P. (eds.), 1999: *The Diatoms: Applications for the Environmental and Earth Sciences* (Cambridge: Cambridge University Press).

Turner II, B. L. et al., 1990: *The Earth as Transformed by Human Action* (Cambridge: Cambridge University Press).

Vernadski, V. I., 1998: *The Biosphere, translated and annotated version from the original of 1926* (New York: Copernicus, Springer).

Verschuren, D., et al., 1998, in: *J. Great Lakes Res.,* 24.

Vitousek, P. M. et al. 1997, in: *Science*, 277: 494.

Watson, R. T. et al., 1990, in: Houghton, T.; Jenkins, G.J.; Ephraums, J.J. (eds.): *Climate Change. The IPCC Scientific Assessment* (Cambridge: Cambridge University Press), chapt. 1.

Wilson, E. O., 1992: *The Diversity of Life* (London: Penguin Books).

Chapter 3
Geology of Mankind (2002)

Paul J. Crutzen

> *The Anthropocene could be said to have started in the late eighteenth century, when analyses of air trapped in polar ice showed the beginning of growing global concentrations of carbon dioxide and methane.*

For the past three centuries, the effects of humans on the global environment have escalated. Because of these anthropogenic emissions of carbon dioxide, global climate may depart significantly from natural behaviour for many millennia to come. It seems appropriate to assign the term 'Anthropocene' to the present, in many ways human-dominated, geological epoch, supplementing the Holocene—the warm period of the past 10–12 millennia. The Anthropocene could be said to have started in the latter part of the eighteenth century, when analyses of air trapped in polar ice showed the beginning of growing global concentrations of carbon dioxide and methane. This date also happens to coincide with James Watt's design of the steam engine in 1784.

Mankind's growing influence on the environment was recognised as long ago as 1873, when the Italian geologist Antonio Stoppani spoke about a "new telluric force which in power and universality may be compared to the greater forces of earth," referring to the "anthropozoic era". And in 1926, V. I. Vernadsky acknowledged the increasing impact of mankind: "The direction in which the processes of evolution must proceed, namely towards increasing consciousness and thought, and forms having greater and greater influence on their surroundings." Teilhard de Chardin and Vernadsky used the term 'noösphere'—the 'world of thought'—to mark the growing role of human brain-power in shaping its own future and environment.

The rapid expansion of mankind in numbers and per capita exploitation of Earth's resources has continued apace. During the past three centuries, the human population has increased tenfold to more than 6 billion and is expected to reach 10 billion in this century. The methane-producing cattle population has risen to 1.4 billion.

The text was first published as: Paul J. Crutzen: "Geology of mankind", in: *Nature*, 415 (2002): 23. The permission to republish this article was granted on 19 August 2015 by Ms. Claire Smith, Nature Publishing Group & Palgrave Macmillan, London, UK.

About 30–50% of the planet's land surface is exploited by humans. Tropical rain-forests disappear at a fast pace, releasing carbon dioxide and strongly increasing species extinction. Dam building and river diversion have become commonplace. More than half of all accessible fresh water is used by mankind. Fisheries remove more than 25% of the primary production in upwelling ocean regions and 35% in the temperate continental shelf. Energy use has grown 16-fold during the twentieth century, causing 160 million tonnes of atmospheric sulphur dioxide emissions per year, more than twice the sum of its natural emissions. More nitrogen fertilizer is applied in agriculture than is fixed naturally in all terrestrial ecosystems; nitric oxide production by the burning of fossil fuel and biomass also overrides natural emissions. Fossil-fuel burning and agriculture have caused substantial increases in the concentrations of 'greenhouse' gases—carbon dioxide by 30% and methane by more than 100%—reaching their highest levels over the past 400 millennia, with more to follow.

So far, these effects have largely been caused by only 25% of the world population. The consequences are, among others, acid precipitation, photochemical 'smog' and climate warming. Hence, according to the latest estimates by the *Intergovernmental Panel on Climate Change* (IPCC), the Earth will warm by 1.4–5.8 °C during this century.

Many toxic substances are released into the environment, even some that are not toxic at all but nevertheless have severely damaging effects, for example the chlorofluorocarbons that caused the Antarctic 'ozone hole' (and which are now regulated). Things could have become much worse: the ozone-destroying properties of the halogens have been studied since the mid-1970s. If it had turned out that chlorine behaved chemically like bromine, the ozone hole would by then have been a global, year-round phenomenon, not just an event of the Antarctic spring. More by luck than by wisdom, this catastrophic situation did not develop.

Unless there is a global catastrophe—a meteorite impact, a world war or a pandemic—mankind will remain a major environmental force for many millennia. A daunting task lies ahead for scientists and engineers to guide society towards environmentally sustainable management during the era of the Anthropocene. This will require appropriate human behaviour at all scales, and may well involve internationally accepted, large-scale geo-engineering projects, for instance to 'optimise' climate. At this stage, however, we are still largely treading on terra incognita.

Further Reading

Berger, A.; Loutre, M.-F., 1996: *C. R. Acad. Sci. Paris 323* (IIA), 1–16.
Clark, W. C.; Munn, R. E. (eds.), 1986: *Sustainable Development of the Biosphere* Ch. 1 (Cambridge: Cambridge Univ. Press).
Crutzen, P. J.; Stoermer, E.F., 2000: *IGBP Newsletter*, 41 (Stockholm: Royal Swedish Academy of Sciences).
Houghton, J. T. *et al.* (eds.), 2001: *Climate Change 2001: The Scientific Basis* (Cambridge: Cambridge Univ. Press).

Marsh, G. P.: *Man and Nature* (1864): (Reprinted in 1965 as: *The Earth as Modified by Human Action* (Cambridge, Massachusetts: Belknap Press)).

McNeill, J. R., 2000: *Something New Under the Sun: An Environmental History of the Twentieth-Century World* (New York: W. W. Norton).

Schellnhuber, H. J., 1999: *Nature,* 402: C19–C23.

Turner, B. L. et al., 1990: *The Earth as Transformed by Human Action* (Cambridge: Cambridge Univ. Press).

Vernadski, V. I., 1998: *The Biosphere* (translated and annotated version from the original of 1926) (New York: Springer).

Chapter 4
The 'Anthropocene' (2002)

Paul J. Crutzen

4.1 The Holocene

Holocene ('Recent Whole') is the name given to the post-glacial geological epoch of the past ten to twelve thousand years as agreed upon by the International Geological Congress in Bologna in 1885 (Encyclopaedia Britannica 1976). During the Holocene, accelerating in the industrial period, mankind's activities grew into a significant geological and morphological force, as recognised early by a number of scientists. Thus, in 1864, G.P. Marsh published a book with the title *Man and Nature*, more recently reprinted as *The Earth as Modified by Human Action* (Marsh 1965). Stoppani in 1873 rated mankind's activities as a "new telluric force which in power and universality may be compared to the greater forces of earth" (quoted from Clark 1986). Stoppani already spoke of the anthropozoic era. Mankind has now inhabited or visited all places on Earth; he has even set foot on the moon. The great Russian geologist and biologist V.I. Vernadsky (1998) in 1926 recognised the increasing power of mankind in the environment with the following excerpt "… the direction in which the processes of evolution must proceed, namely towards increasing consciousness and thought, and forms having greater and greater influence on their surroundings". He, the French Jesuit priest P. Teilhard de Chardin and E. Le Roy in 1924 coined the term 'noösphere', the world of thought, to mark the growing role played by mankind's brainpower and technological talents in shaping its own future and environment.

This text was first published by Paul J. Crutzen: "The 'anthropocene'", in: *J.J. Phys. IV France*, 12,10 (November 2002), (Les Ulis: EDP Sciences); at: DOI: 10.1051 /jp4:20020447. The permission was granted on 28 May 2018 by M. Dominique Zumpano, Executive Assistant, EDF.

4.2   The Anthropocene

Supported by great technological and medical advancements and access to plentiful natural resources, the expansion of mankind, both in numbers and per capita exploitation of Earth's resources has been astounding (Turner 1990). To give some major examples:

During the past 3 centuries human population increased tenfold to 6000 million, growing by a factor of four during the past century alone (McNeill 2000). This growth in human population was accompanied e.g. by a growth in cattle population to 1400 million (about one cow per average size family; McNeill 2000). Urbanisation has even increased 13 times in the past century. Similarly large were the increases in several other factors, such as world economy and energy use (see Table 4.1). Industrial output even grew forty times (McNeill 2000). More than half of all accessible fresh water is used by mankind. Fisheries remove more than 25% of the primary production of the oceans in the upwelling regions and 35% in the temperate continental shelf regions (Pauly/Christensen 1995).

Table 4.1 A partial record of the growths and impacts of human activities during the 20th century

Item	Increase factor, 1890s–1990s
World population	4
Total world urban population	13
World economy	14
Industrial output	40
Energy use	16
Coal production	7
Carbon dioxide emissions	17
Sulfur dioxide emissions	13
Lead emissions	~8
Water use	9
Marine fish catch	35
Cattle population	4
Pig population	9
Irrigated area	5
Cropland	2
Forest area	20% decrease
Blue whale population (Southern Ocean)	99.75% decrease
Fin whale population	97% decrease
Bird and mammal species	1% decrease

Source J. R. McNeill (2000)

In a few generations mankind is exhausting the fossil fuels that were generated over several hundred million years, resulting in large emissions of air pollutants. The release of SO_2, globally about 160 Tg/year to the atmosphere by coal and oil burning, is at least two times larger than the sum of all natural emissions, occurring mainly as marine dimethyl-sulfide from the oceans (Houghton et al. 1990; Houghton 1996; Houghton et al. 2001). The oxidation of SO_2 to sulphuric acid has led to acidification of precipitation and lakes, causing forest damage and fish death in biologically sensitive regions, such as Scandinavia and the northeast of North America. Due to substantial reduction in SO_2 emissions, the situation in these regions has improved in the meanwhile. However, the problem is getting worse in east Asia.

From Vitousek et al. (1997) we learn that 30–50 % of the world's land surface has been transformed by human action; the land under cropping has doubled during the past century at the expense of forests which declined by 20% over the same period (McNeill 2000). Coastal wetlands are also affected by humans, having resulted for instance in the loss of 50% of the world's mangroves.

More nitrogen is now fixed synthetically and applied as fertilizers in agriculture than fixed naturally in all terrestrial ecosystems. Overapplication of nitrogen fertilizers in agriculture and especially its concentration in domestic animal manure have led to eutrophication of surface waters and even groundwater in many locations around the world. They also lead to the microbiological production of N_2O, a greenhouse gas and a source of NO in the stratosphere where it is strongly involved in stratospheric ozone chemistry. The issue of more efficient use of N fertilizer in food and energy production has recently been summarised in a special publication of *Ambio* (31,2 [March 2002]).

The release of NO into the atmosphere from fossil fuel and biomass combustion likewise is larger than the natural inputs, giving rise to photochemical ozone ('smog') formation in extensive regions of the world.

Human activity has increased the species extinction rate by thousand to ten thousand fold in the tropical rain forests (Wilson 1992). As a result of increasing fossil fuel burning, agricultural activities, deforestation, and intensive animal husbandry, especially cattle holding, several climatically important "greenhouse" gases have substantially increased in the atmosphere over the past two centuries: CO_2 by more than 30% and CH_4 by even more than 100% (see Table 4.2), contributing substantially to the observed global average temperature increase by about 0.5 °C that has been observed during the past century. According to the reports by the Intergovernmental Panel of Climate Change in 1995 (Houghton et al. 1996): "The balance of evidence suggests a discernable human influence on global climate" and in 2001: "There is new and stronger evidence that most of the warming observed over the last 50 years is attributable to human activities." Depending on the scenarios of future energy use and model uncertainties, the increasing emissions and resulting growth in atmospheric concentrations of CO_2 are estimated to cause a rise in global average temperature by 1.4–5.8 °C during the present century, accompanied by sea level rise of 9–88 cm (and 0.5–10 m until the end of the current millennium). Major anthropogenic climate changes are thus still ahead.

Table 4.2 Composition of dry air at ground level in remote continental areas

Constituent	Formula	Concentrations 1998/pre-industrial	Growth (%/year) average (1990–1999)
Nitrogen	N_2	78.1%	
Oxygen	O_2	20.9%	
Argon	Ar	0.93%	
Carbon dioxide	0 0	365/280 ppmv	+ 0.4
Methane	CH_4	1.745/0.7 ppmv	+ 0.3–0.5
Ozone	O_3	10–100/20(?) nmol/mol	variable
Nitrous oxide	N_2O	314/270 nmol/mol	+ 0.25
CFC-11	$CFCl_3$	0.27/0 nmol/mol	< 0 (decline)
CFC-12	CF_2Cl_2	0.53/0 nmol/mol	< 0 (decline)
OH (HYDROXYL)	OH	$\sim 4 \times 10^{-14}$	?

Furthermore, mankind also releases many toxic substances in the environment and even some, the chlorofluorocarbon gases ($CFCl_3$ and CF_2Cl_2), which are not toxic at all, but which nevertheless have led to the Antarctic springtime 'ozone hole' and which would have destroyed much more of the ozone layer if no international regulatory measures to end their production by 1996 had been taken. Nevertheless, due to the long residence times of the CFCs, it will take at least another 4–5 decades before the ozone layer will have recovered.

Considering these and many other major and still growing impacts of human activities on earth and atmosphere, and at all, including global, scales, it thus is more than appropriate to emphasise the central role of mankind in geology and ecology by using the term 'anthropocene' for the current geological epoch. The impact of current human activities is projected to last over very long periods. According to Loutre and Berger (2000), because of past and future anthropogenic emissions of CO_2, climate may depart significantly from natural behaviour even over the next 50,000 years.

To assign a more specific date to the onset of the 'anthropocene' is somewhat arbitrary, but we propose the latter part of the 18th century, although we are aware that alternative proposals can be made. However, we choose this date because, during the past two centuries, the global effects of human activities have become clearly noticeable. This is the period when data retrieved from glacial ice cores show the beginning of a growth in the atmospheric concentrations of several 'greenhouse gases', in particular CO_2, CH_4 and N_2O (Houghton et al. 1990; Houghton 1996; Houghton 2001). Such a starting date also coincides with James Watt's invention of the steam engine in 1784.

Without major catastrophes like an enormous volcanic eruption, an unexpected epidemic, a large-scale nuclear war, an asteroid impact, a new ice age, or continued plundering of Earth's resources by partially still primitive technology (the last four dangers can, however, be prevented in a real functioning noösphere) mankind will remain a major geological force for many millennia, maybe millions of years, to come. To develop a world-wide accepted strategy leading to sustainability of ecosystems against human induced stresses will be one of the great future tasks of mankind, requiring intensive research efforts and wise application of the knowledge thus acquired in the noösphere, now better known as knowledge or information society.

Hopefully, in the future, the 'anthropocene' will not only be characterised by continued human plundering of Earth's resources and dumping of excessive amounts of waste products in the environment, but also by vastly improved technology and management, wise use of Earth's resources, control of human and domestic animal population, and overall careful manipulation and restoration of the natural environment. There are enormous technological opportunities. Worldwide energy use is only 0.03% of the solar radiation reaching the continents. Only 0.6% of the incoming visible solar radiation is converted to chemical energy by photosynthesis on land and 0.13% in the oceans. Of the former about 10% go into agricultural net primary production. Thus, despite the fact that humans appropriate 10–55% of terrestrial photosynthesis products (Rojstaczer et al. 2001), there are plenty of opportunities for energy savings, solar voltaic and maybe fusion energy production, materials' recycling, soil conservation, more efficient agricultural production, et cetera. The latter makes it even possible to revert extended areas now used for agricultural to their natural state.

There is little doubt in my mind that, as one of the characteristic features of the 'anthropocene', distant future generations of '*homo sapiens*' will do all they can to prevent a new ice-age from developing by adding powerful artificial greenhouse gases to the atmosphere. Similarly, any drop in CO_2 levels to too low concentrations, leading to reductions in photosynthesis and agricultural productivity would be combated by artificial releases of CO_2. With plate tectonics and volcanism declining, this is not a scenario devoid of any realism, but of course not urgent in any way. And likewise, far to the future, 'homo sapiens' will deflect meteorites and asteroids before they could hit the Earth (Lewis 1996). Humankind is bound to remain a noticeable geological force, as long as it is not removed by diseases, wars, or continued serious destruction of Earth's life support system, which is so generously provided by nature cost-free.

4.3 Conclusions

To conclude: exciting, but also difficult and daunting task lies ahead of the global research and engineering community to guide mankind towards global, sustainable, environmental management into the anthropocene (Schellnhuber 1999).

References

Clark, W. C., 1986, in: Clark, W. C.; Munn, R. E. (eds.): *Sustainable Development of the Biosphere* (Cambridge: Cambridge University Press), chap. 1.

Encyclopaedia Britannica, 1976: *Micropaedia*, IX (London: Encyclopaedia Britannica).

Houghton, J. T. et al., 1996: *Climate Change* 1995 (Cambridge: Cambridge University Press);

Houghton, J. T. et al. (eds.), 2001: *Climate Change 2001: The Scientific Basis* (Cambridge: Cambridge University Press).

Houghton, J. T.; Jenkins, G. J.; Ehpraums, J. J. (eds.), 1990: *Climate Change: The IPCC Scientific Assessment,* (Cambridge: Cambridge University Press).

Lewis, J. S., 1996: *Rain of Iron and Ice* (Readings, MA: Addison-Wesley).

Loutre, M. F.; Berger, A., 2000, in: *Climatic Change*, 46: 61–90.

Marsh, G. P., 1965: *The Earth as Modified by Human Action* (Cambridge, MA: Belknap Press, Harvard University Press).

McNeill, J. R., 2000: *Something New Under the Sun* (New York-London: W. H. Norton and Company).

Pauly, D.; Christensen, V., 1995, in: *Nature*, 374: 255–257.

Rojstaczer, S.; Sterling, S. M.; Moore, N. J., 2001, in: *Science*, 294: 2549–2552

Schellnhuber, H. J., 1999, in: *Nature*, 402: C19–C23.

Turner II et al., B. L., 1990: *The Earth as Transformed by Human Action* (Cambridge: Cambridge University Press).

Vemadski, V. I., 1998: *The Biosphere*, translated and annotated version from the original of 1926 (New York: Copernicus, Springer).

Vitousek, P. M. et al., 1997, in: *Science*, 277: 494.

Wilson, E. O., 1992: *The Diversity of Life* (London: Penguin Books).

Chapter 5
Atmospheric Chemistry
in the 'Anthropocene' (2002)

Paul J. Crutzen

Since the beginning of the agro-industrial period, mankind's use of Earth's resources has grown so much that it seems justified to denominate the past two centuries into the future as a new geological epoch: 'The Anthropocene'. This transition is also marked by major changes in the chemistry and chemical composition of the atmosphere, such as:

- The atmospheric concentrations of several climatologically important 'greenhouse gases' have grown substantially: CO_2 by 30%, CH_4 by more than 100%, and tropospheric ozone regionally in the troposphere by more than 100%. These changes presently exert an additional infrared climate-warming forcing of about 2.5 W m^{-2}, which to a large degree can have contributed to the observed average warming of the planet by 0.6 °C. The increases in the concentrations of these gases were caused by the combustion of fossil fuels as well as deforestation and agricultural activities. The average global surface warming caused by these changes has been estimated with climate models to be in the range 1.4–5.8 °C by the end of this century, the wide span of this 'prediction' reflecting both insufficient knowledge about the many positive and negative feedback processes in the complex physical/chemical/biological climate system, as well as insufficient knowledge about future developments in the energy and agricultural sectors and resulting 'greenhouse gas' emissions. Thus, the possibility exists that concentrations of atmospheric 'greenhouse gases' and resulting climate warming may reach levels that have not existed on Earth since the emergence of *Homo sapiens;*
- The release of SO_2 (about 160 Tg S year^{-1}; Tg = Teragram = 10^{12} g) by coal and oil burning is at least two times larger than the sum of all natural sulphur emissions.

This text was first published as Chap. 7 by Paul J. Crutzen: "Atmospheric Chemistry in the 'Anthropocene'" in: W. Steffen; J. Jäger; D.J. Carson; C. Bradshaw (eds.): *Challenges of a Changing Earth. Proceedings of the Global Change Open Science Conference, Amsterdam, The Netherlands, 10–13 July 2001* (Berlin Heidelberg: Springer-Verlag, 2002): 45–48. Permission was granted by Springer Verlag in Heidelberg.

Over industrialised regions, the increase has been more than an order of magnitude, causing acid precipitation, health effects, and regionally poor visibility due to light scattering by sulphate aerosol. Interestingly, this factor has also led to increased reflection of solar radiation to space, causing a cooling effect on climate, and thereby partially counteracting the warming by the 'greenhouse gases';

- Release of NO to the atmosphere from fossil fuel and biomass burning is larger than or comparable to natural emissions, over extended regions causing photochemical smog, including high surface ozone concentrations which are harmful to human health and plant productivity. It should be noted here that photochemical smog is not only a phenomenon of the urban/suburban industrial world, but also extends to rural regions in the developing world as a consequence of widespread biomass burning during the dry season in the tropics and subtropics;

- A special class of products from the chemical industry, the chlorofluorocarbons $CFCl_3$ and CF_2Cl_2, which were never produced in nature, have caused the greatest large-scale change in the atmosphere: the 'ozone hole', the rapid loss of almost all ozone in the lower stratosphere over Antarctica during spring, due to 'cold chemistry' involving surface reactions on particles. The 'ozone hole' develops over a time period of a few weeks from 12–22 km, that is in the height region in which maximum ozone concentrations were always found until about 20 years ago. Smaller, but nevertheless, significant ozone depletions are also observed in some years during late winter/spring over the Arctic. It is important to note that these large, chemical ozone losses were not predicted. It was in fact thought that ozone in the polar regions was chemically inert. This experience shows the enormous importance of observations and limitations of model predictions. Especially in these times of increasing pressures of mankind on the environment, this experience raises questions about the stability of the environment and climate system, or parts thereof. Clearly, special human action has led to an instability in stratospheric ozone chemistry, in a region that seemed the least likely. It should be asked whether there might be other breakpoints in the complex Earth System that we are not aware of. How well may models predict these in the extremely complex Earth System with its many positive and negative feedbacks? See Fig. 5.1, which only depicts the complexity of stratospheric chemistry.

Fortunately, international agreements have been reached to phase out the production of the CFCs and some other halogenated organic compounds since the beginning of 1996. The impact of these restrictions on tropospheric CFC concentrations is already noticeable. Although reductions in stratospheric CFC levels will soon follow, it will take a relatively long time, on the order of some 50 years, before CFC levels will have decreased so much that the 'ozone hole' will be filled up. There is, however, the possibility of further delays due to the cooling of the stratosphere by increases in the concentrations of CO_2 or continued increase in water vapour. The international regulations that were reached by the phasing-out of CFC production may be considered an environmental success story. Unfortunately, one can not feel optimistic that international agreements against CO_2 emissions will be reached similarly in a timely fashion.

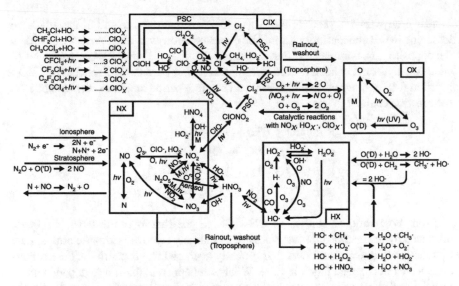

Fig. 5.1 An example of the complexity of Earth System research. The figure shows the reactions that are needed to describe stratospheric ozone chemistry. Not shown are reactions involving Br compounds, which are responsible for additional ozone loss (reproduced with permission from Crutzen (1996) *Angew Chem Int Engl* 35:1758–1777

We will next concentrate our discussions on a few issues involving atmospheric chemistry, which I see as particularly important: the self-cleansing activity of the atmosphere, and how it might be impacted by human activity, and the role of the tropics and subtropics in atmospheric chemistry and how it is disturbed by the large human population in this part of the world.

The self-cleansing propensity (oxidation efficiency) of the atmosphere. About three decades ago it was proposed that hydroxyl (OH) radicals play a key role in removing almost all gases that are emitted to the atmosphere by natural processes and human activities. Hydroxyl radicals are formed by the action of solar ultraviolet radiation on ozone, producing energetic $O(^1D)$ atoms, which possess enough energy to react with water vapour:

$$R1: O_3 + hv(< 420\,nm) \rightarrow O(^1D) + O_2$$
$$R2: O(^1D) + H_2O \rightarrow 2\,OH$$

The atmospheric concentrations of OH are very low, on average only 4 molecules per 10^{14} molecules of air. Nevertheless, for most gases, their atmospheric average residence time and spatial/temporal variability are determined by how fast they react with OH. For instance, CH_4 reacts only slowly with OH and shows much less variability than isoprene (C_5H_8), which is emitted by vegetation and can only be found in close vicinity to forests. Methane and carbon monoxide are the main gases with which hydroxyl reacts in most of the troposphere. Because both gases have been increasing by 0.5–1% year^{-1} (in the case of CO only up to the end of the 1980s), a lessening in OH concentrations could be expected. However, other processes work in the opposite way. For instance, due to anthropogenic emissions of NO_X ($NO + NO_2$) catalysts, hydroxyl concentrations can increase, first because NO reacts with HO_2, reaction R4, (or other peroxy radicals) and because ozone increases, for instance via the reaction chain:

$$R3: \quad CO + OH(+O_2) \rightarrow CO_2 + HO_2$$
$$R4: \quad HO_2 + NO \rightarrow OH + NO_2$$
$$R5: \quad NO_2 + h\upsilon(+O_2) \rightarrow NO + O_3$$
$$Net: \quad CO + 2O_2 \rightarrow CO_2 + O_3$$

From observational studies of CH_3CCl_3, a gas that is (until 1996 has been) produced by the chemical industry and that is largely removed from the atmosphere by reaction with OH, it follows that globally averaged tropospheric OH concentrations have not changed much in time. What will happen in the future depends much on what will happen in the tropical and subtropical regions of the world with the largest and growing fraction of the world population, because it is there that OH concentrations maximise due to high levels of water vapour and solar ultraviolet radiation.

The importance of the tropics and subtropics in atmospheric chemistry. Because of maximum abundance of hydroxyl radicals in the tropics and subtropics, maximum rainfall, and relatively low industrial activity, the tropics should be the cleanest part of our atmosphere. This is, however, not the case. Due to tropical deforestation activities and the burning of large amounts of biomass in various kinds of agricultural activities during the dry season, as well as lack of pollution controls, large amounts of pollutants are emitted into the atmosphere, including parts of the Southern Hemisphere. It is thus estimated that each year between some 2000–5000 Tg of biomass carbon is burned, releasing large amounts of photochemically active gases like CO, hydrocarbons and NO_X into the atmosphere, a mixture very similar to that which produces photochemical smog in the city plumes of the industrialised world. Thus high concentrations of ozone are observed in the dry season in many parts of the developed world, in Africa, South America and Asia. This affects the oxidation efficiency of the atmosphere; however, by how much, and in what direction, are not known. Besides the effects, just mentioned, there are other factors that influence the oxidising efficiency of the atmosphere. For instance, because of tropical deforestation, which is as high as 1% year^{-1} in some regions, reactive hydrocarbon emissions (such as isoprene) are declining. This will probably lead to an increase in OH concentrations both in the forested regions, where reactions with reactive hydrocarbons act as strong sinks for OH, and at larger scales where CO, an oxidation product of the hydrocarbons, will tend to decrease, and thus causing OH to increase.

A recently recognised, and maybe the greatest, impact on atmospheric chemistry and climate may well result from large anthropogenic emissions of particulate matter in both industrial and especially developing countries. Already now these are exerting a global average radiative (cooling) forcing of about 1.5 W m^{-2} on climate, mostly due to back-scattering of solar radiation to space, close to the current infrared warming forcing of about 2.5 W m^{-2} by the 'greenhouse gases'. Past studies were especially concerned with the effects of sulphate aerosol resulting from coal and oil burning in the developed world. Most of these studies have neglected the potentially very large contributions that can be caused by fossil fuel and biomass burning in the developing world. Already now, the emission of SO$_2$ to the atmosphere in the Asian countries is about equal to those from Europe and North America combined. In about 20 years, it will be 3 times larger. Because of lack of appropriate pollution controls, this is causing heavy air pollution in these countries. To the 'classic' air pollution emissions must be added those resulting from extensive biomass burning in primitive agriculture, and for cooking and heating purposes. As a consequence, especially during the dry season, high concentrations of smoke cover extensive regions of the developing world. During a field experiment (INDOEX [*Indian Ocean Experiment*]), which was conducted February-April, 1999 in India, the Maldives and the Indian Ocean, the observed particulate loading of the atmosphere was so high, it caused a reduction in solar radiation fluxes reaching the Indian Ocean surface north of the Intertropical Convergence Zone (the meteorological equator) by up to about 20 W m^{-2} (or about 10% of the solar irradiation), almost ten times larger than the global 'greenhouse top of the atmosphere forcing'. This may have a very strong impact on regional climate and the water cycle, an impact which will still strongly grow in coming decades. Contrary to the cooling forcing on climate resulting from fossil fuel derived sulphate, current emissions from fossil fuel and biomass burning in the developing world are characterised by large emissions of sunlight-absorbing black carbon (soot). Its impact on regional or global climate, thus has a quality which is quite distinct from that by the emissions in the developed world.

Future changes in atmospheric chemistry and climate will to a large extent depend on what is going to happen in the developing world. Especially, quantitative information about the production of the various kinds of particulate matter, their distribution and optical and microphysical properties is very uncertain, calling for strong enhancements in future research efforts.

Chapter 6
How Long Have We Been in the Anthropocene Era? (2003)

Paul J. Crutzen and Will Steffen

With great interest we have read Ruddiman's intriguing article which is in favour of placing the start of the Anthropocene at 5–8 millennia BP instead of the late quarter of the 18th century. He shows how land exploitation for agriculture and animal husbandry may have led to enhanced emissions of CO_2 and CH_4 to the atmosphere, thereby modifying the expected changes in the concentrations of these gases beyond those expected from variations in the Milankovich orbital parameters. Much of his argument depends on the correctness of their projected CH_4 concentration curve from 7,000 years BP to pre-industrial times showing a decline to about 425 ppb, according to Milankovich, instead of the measured 700 ppb. It appears, however, strange that in Ruddiman's analysis the proposed increase of CH_4 due to anthropogenic activities stopped at about 1000 years BP, because ice core data showed almost constant mixing ratios of CH_4 between 1000 years BP and about 200 years ago before the rapid rise of CH_4 in the industrial period (IPCC 2001). A major feature of Ruddiman's argument is that natural atmospheric CH_4 concentrations depend strongly on geological varying summer time insolations in the tropical northern hemisphere, controlling tropical wetlands and methane release from decaying organic matter under anaerobic conditions.

The choice of the start of the anthropocene remains rather arbitrary. The records of atmospheric CO_2, CH_4, and N_2O show a clear acceleration in trends since the end of the 18th century. For that reason, the start of the anthropocene was assigned to about that time, immediately following the invention of the steam engine in 1784 (Crutzen and Stoermer 2000; Crutzen 2002). The consequences of this innovation have been astounding, for instance, there has been a tenfold rise in human population to 6000 million, during the past three centuries, and a fourfold increase in the 20th

This text was first published as: Paul J. Crutzen and Will Steffen: "How Long Have We Been in the Anthropocene Era? An Editorial Comment", in: *Climatic Change*, 61: 251–257 (Heidelberg: Springer). The permission to republish this text was granted by Springer via the Copyright Clearance Center on 4 June 2018.

S. Benner et al. (eds.), *Paul J. Crutzen and the Anthropocene: A New Epoch in Earth's History*, The Anthropocene: Politik–Economics–Society–Science 1,
https://doi.org/10.1007/978-3-030-82202-6_6

century (Turner et al. 1990; McNeill 2000). This expansion was made possible by medical advances and a major growth in agriculture and animal husbandry leading for instance to a current cattle population of 1400 million (globally averaged about one cow per average size family). Let us give a few more examples.

In a few generations mankind is exhausting the fossil fuels that were generated over several hundred million years, resulting in large emissions of air pollutants. The release of SO_2, globally about 160 Tg/year to the atmosphere by coal and oil burning, is at least two times larger than the sum of all natural emissions, occurring mainly as marine dimethyl-sulfide from the oceans (IPCC 2001). The oxidation of SO_2 to sulphuric and NO_x to nitric acid has led to acidification of precipitation, causing forest damage and fish death in biologically sensitive lakes in regions, such as Scandinavia and the northeast of North America. Due to substantial reductions in SO_2 emissions, the situation has improved. However, the problem is now getting worse in Asia.

From Vitousek et al. (1997) we learn that 30–50% of the world's land surface has been transformed by human action, while the land under cropping has doubled during the past century at the expense of forests, which declined by 20% over the same period (McNeill 2000).

More nitrogen is now fixed synthetically and applied as fertilizers in agriculture than fixed naturally in all terrestrial ecosystems, 120 Tg/year vs 90 Tg/year (Galloway et al. 2002). The Haber-Bosch industrial process to produce ammonia from N_2 in the air made the human population explosion possible. (It is amazing to note the importance of this single invention for the evolution on our planet.) Only 20 Tg N/year is, however, contained in the food which is consumed by humans. Wasteful application of nitrogen fertilizers in agriculture and especially its concentration in domestic animal manure have led to eutrophication of surface waters and even groundwater in many locations around the world. Fossil fuel burning adds another 25 Tg N/year highly reactive NO_X to the atmosphere, causing photochemical ozone formation in extensive regions around the globe. The additional input of altogether 145 Tg N/year is almost twice as large as the global natural biological fixation on land. The disturbance of the N cycle also leads to the microbiological production of N_2O, a greenhouse gas and a source of NO in the stratosphere, where it is strongly involved in stratospheric ozone chemistry. Human disturbance of the nitrogen cycle has recently been treated in a special publication of Ambio (Galloway et al. 2002).

As a result of increasing fossil fuel burning, agricultural activities, deforestation, and intensive animal husbandry, especially cattle holding, several climatically important 'greenhouse' gases have substantially increased in the atmosphere over the past two centuries: CO_2 by more than 30% and CH_4 even by more than 100%, contributing substantially to the observed global average temperature increase by about 0.6 °C that has been observed during the past century. The Intergovernmental Panel of Climate Change (IPCC 1996) said in 1996: 'The balance of evidence suggests a discernable human influence on global climate' and in 2001: 'There is new and stronger evidence that most of the warming observed over the last 50 years is attributable to human activities' (IPCC 2001). Depending on the scenarios of future energy use and model uncertainties, further emissions of CO_2 and other greenhouse gases are estimated to cause a rise in global average temperature by 1.4–5.8 °C during the present century, accompanied by sea level rise of 9–88 cm (0.5–10 m until the end of this millennium). The largest anthropogenic climate changes are still ahead for future generations.

Furthermore, mankind also releases or have released many toxic substances in the environment and even some, the chlorofluorocarbon gases ($CFCl_3$ and CF_2Cl_2), which are not toxic at all, but which nevertheless have led to the Antarctic springtime 'ozone hole' and which would have destroyed much more of the ozone layer if no international regulatory measures to end their production by 1996 had been taken. Nevertheless, due to the long residence times of the CFC gases, it will take at least until the middle of this century before the ozone layer will have largely recovered.

The impact of humans on global economy and environment has undergone major stepwise expansions, especially during the second half of the past century. Figure 6.1 attempts to summarise the growing magnitude and changing nature of the Anthroposphere over the past 250 years in terms of changes in 12 key global-scale indicators. Note the distinct change in rate of increase of most indicators around 1950 and the sharp acceleration thereafter. The mid-20th century was a pivotal point of change in the relationship between humans and their life support system.

Figure 6.2 shows the impacts of the changing Anthroposphere on the functioning of the Earth System as a whole. In many ways, not just in climate, the human impress on the global environment is clearly discernable beyond natural variability. All components of the Earth System – atmosphere, land, ocean, coastal zone – are being significantly affected by human activities. The period of the Anthropocene since 1950 stands out as the one in which human activities rapidly changed from merely *influencing* the global environment *in some ways* to *dominating it in many ways:*

- Human impacts on Earth System structure (e.g., land cover, coastal zone structure) and functioning (e.g., biogeochemical cycling) now equal or exceed in magnitude many forces of nature at the global scale.
- The rates of human-driven change are almost always much greater than those of natural variability. For example, the current concentration of atmospheric CO_2 (about 90 ppmV higher that the pre-industrial level) has been reached at a rate at least 10 and possibly 100 times faster than natural increases in atmospheric CO_2 concentration during the previous 420,000 years at least (Falkowski et al. 2000).
- All of the changes to the Earth System depicted in Figures 6.1 and 6.2 are occurring simultaneously, and many are accelerating simultaneously.

In summary, we conclude that Earth is currently operating in a *no-analogue state.* In terms of key environmental parameters, the Earth System has recently moved well outside the range of natural variability exhibited over at least the last half million years. The nature of changes now occurring simultaneously in the Earth System, their magnitudes and rates of change are unprecedented and unsustainable.

We conclude that there may have been several distinct steps in the 'Anthropocene', the first, relatively modest, step can have been identified by Ruddiman, followed by a further major step from the end of the 18th century to 1950 and, from the perspective

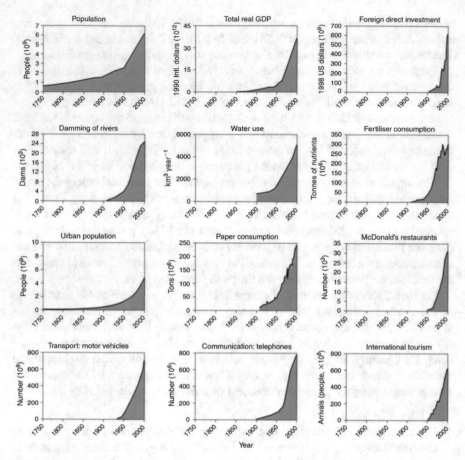

Fig. 6.1 The increasing rates of change in human activity since the beginning of the Industrial Revolution (Steffen et al. 2003). Significant increases in the rates of change occur around the 1950s in each case and illustrate how the past 50 years have been a period of dramatic and unprecedented change in human history. (U.S. Bureau of the Census 2000; Nordhaus 1997; World Bank 2002; World Commission on Dams 2000; Shiklomanov 1990; International Fertilizer Industry Association 2002; UN Centre for Human Settlements 2002; Pulp and Paper International 1993; MacDonalds 2002; UNEP 2000; Canning 2001; World Tourism Organization 2002)

of the functioning of the Earth System as a whole, the very significant acceleration since 1950.

Already in the 19th century, awareness of the upcoming human impact on the environment was identified among others by Marsh (1864) and further emphasised in particular by V. Vernadsky who wrote about 80 years ago: 'The surface of the earth has been transformed unrecognizably, and no doubt far greater changes will yet come...' '... We are confronted with a new form of biogenic migration resulting from the activity of the human reason'.

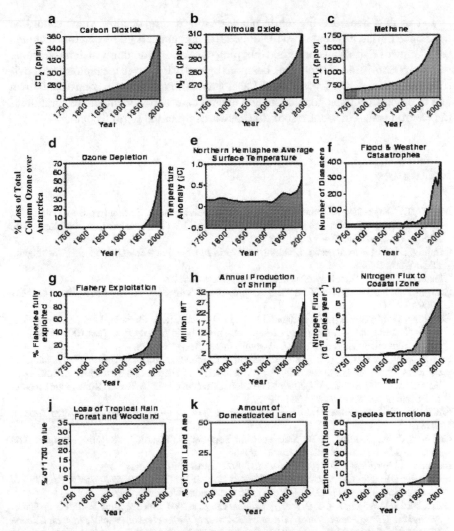

Fig. 6.2 Global-scale changes in the Earth System as a result of the dramatic increase in human activity (Steffen et al. 2003): (a) atmospheric CO_2 concentration (Etheridge et al. 1996); (b) atmospheric N_2O concentration (Machida et al., 1995); (c) atmospheric CH_4 concentration (Blunier et al. 1993); (d) percentage total column ozone loss over Antarctica, using the average annual total column ozone, 330, as a base (Image: J. D. Shanklin, British Antarctic Survey); (e) northern hemisphere average surface temperature anomalies (Mann et al. 1999); (f) natural disasters after 1900 resulting in more than ten people killed or more than 100 people affected (OFDA/CRED 2002); (g) percentage of global fisheries either fully exploited, overfished or collapsed (FAOSTAT 2002); (h) annual shrimp production as a proxy for coastal zone alteration (WRI, 2003; FAOSTAT 2002); (i) model-calculated partitioning of the human-induced nitrogen perturbation fluxes in the global coastal margin for the period since 1850 (Mackenzie et al. 2002); (j) loss of tropical rainforest and woodland, as estimated for tropical Africa, Latin America and South and Southeast Asia (Richards 1990; WRI 1990); (k) amount of land converted to pasture and cropland (Klein et al. 1997); and (l) mathematically calculated rate of extinction (based on Wilson 1992)

Let us thus hope that the fourth phase of the 'anthropocene', which should be developed during this century, will not be further characterised by continued human plundering of Earth's resources and dumping of excessive amounts of anthropogenic waste products in the environment, but more by vastly improved technology and environmental management, wise use of Earth's remaining resources, control of human and of domestic animal population, and overall careful treatment and restoration of the environment – in short, responsible stewardship of the Earth System.

References

Blunier, T.; Chappellaz, J.; Schwander, J.; Barnola, J.-M.; Desperts, T.; Stauffer, B.; Raynaud, D., 1993: "Atmospheric Methane Record from a Greenland Ice Core over the Last 1000 Years", in: *J. Geophys. Res.* 20: 2219–2222.

Canning, D., 2001: World Bank: *A Database of World Infrastructure Stocks, 1950–95* (Washington, D.C.: World Bank).

Crutzen, P. J.: 2002, "Geology of Mankind", in: *Nature* 415: 23.

Crutzen, P. J.; Stoermer, E. F., 2002: "The 'Anthropocene'", in: *Global Change Newsletter*, 41 (May 2000): 17–18.

Etheridge, D. M.; Steele, L. P.; Langenfelds, R. L.; Francey, R. J.; Barnola, J.-M.; Morgan, V. I., 1996: "Natural and Anthropogenic Changes in Atmospheric CO_2 over the Last 1000 Years from Air in Antarctic Ice and Firn", in: *J. Geophys. Res.* 101: 4115–4128.

Falkowski, P.; Scholes, R. J.; Boyle, E.; Canadell, J.; Canfield, D.; Elser, J.; Gruber, N.; Hibbard, K.; Hogberg, P.; Linder, S.; Mackenzie, F. T.; Moore III, B.; Pedersen, T.; Rosenthal, Y.; Seitzinger, S.; Smetacek, V.; Steffen, W., 2000: "The Global Carbon Cycle: A Test of Knowledge of Earth as a System", in: *Science,* 290: 291–296.

FAOSTAT, 2002: *Statistical Databases, Food and Agriculture Organization of the United Nations, Rome.*

Galloway, J. N.; Cowling, E. B.; Seitzinger, S.; Socolow, R. H.: 2002, "Reactive Nitrogen: Too Much of a Good Thing?", in: *Ambio,* 31: 60–63.

International Fertilizer Industry Association, 2002: "Fertilizer Indicators".

IPCC, 1996 [Houghton, J. T. et al. (eds.)]: *Climate Change: The IPCC Scientific Assessment, Climate Change 1995* (Cambridge, U.K. – New York: Cambridge University Press).

IPCC, 2001 [Houghton, J. T. et al. (eds.)]: *Climate Change 2001, The Scientific Basis, Contribution of Working Group 1 to the Third Assessment Request of the Intergovernmental Panel on Climate Change* (Cambridge, U.K. – New York: Cambridge University Press).

Klein Goldewijk, K.; Battjes, J. J., 1997: *One Hundred Year Database for Integrated Environmental Assessments* (Bilthoven, The Netherlands: National Institute for Public Health and the Environment [RIVM]).

Machida, T.; Nakazawa, T.; Fujii, Y.; Aoki, S.; Watanabe, O., 1995: "Increase in the Atmospheric Nitrous Oxide Concentration during the Last 250 Years", in: *Geophys. Res. Lett.,* 22: 2921–2924.

Mackenzie, F. T.; Ver, L. M.; Lerman, A., 2002: "Century-Scale Nitrogen and Phosphorus Controls of the Carbon Cycle", in: *Chem. Geol.,* 190: 13–32.

Mann, M. E.; Bradley, R. S.; Hughes, M. K., 1999: "Northern Hemisphere Temperatures during the Inferences, Uncertainties, and Limitations", in: *Geophys. Res. Lett.,* 26, 759–762.

Marsh, G. P., 1864: *Man and Nature,* reprinted in 1965 as: *The Earth as Modified by Human Action* (Cambridge, Massachusetts: Belknap Press).

McDonalds, 2002: Homepage.

McNeill, J. R, 2000: *Something New under the Sun* (New York-London: W. H. Norton and Company).

Nordhaus, 1997: "Do Real Wage and Output Series Capture Reality? The History of Lighting Suggests Not", in: Bresnahan, T.; Gordon, R. (eds.): *The Economics of New Goods,* (Chicago: University of Chicago Press).

OFDA/CRED, 2002: *Emergency Events Database (EM-DAT): The OFDA* (United States Office of Foreign Disaster Assistance)/CRED (Louvain: Catholic University of Louvain, Center for research on the Epidemiology of Disaster); international disaster database; at: http://www.cred.be

Pulp and Paper International, 1993: "PPI's International Fact and Price Book", in: *FAO Forest Product Yearbook 1960–1991* (Rome: Food and Agriculture Organization of the United Nations),

Richards, J., 1990: "Land Transformation", in: Turner II, B. L; Clark, W. C.; Kates, R. W.; Richards, J. F.; Mathews, J. T.; Meyer, W. B. (eds.): *The Earth as Transformed by Human Action: Global and Regional Changes in the Biosphere over the Past 300 Years,* (Cambridge: Cambridge University Press): 163–201.

Shiklomanov, I. A., 1990: "Global Water Resources", in: *Nature and Resources,* 26.

Steffen, W., Sanderson; A., Tyson, P.; Jager, J.; Matson, P.; Moore III, B.; Oldfield, F.; Richardson, K.; Schellnhuber, H.-J.; Turner II, B. L.; Wasson, R., 2003: *Global Change and the Earth System: A Planet under Pressure. IGBP Global Change Series* (Berlin, Heidelburg: New York Springer-Verlag).

Turner II, B. L. et al., 1990: *The Earth as Transformed by Human Action* (Cambridge: Cambridge University Press).

UN Center for Human Settlements, 2002: The state of the world's cities, 2001 (New York: United Nations).

UNEP, 2000: "Global Environmental Outlook 2000", in: Clarke, R. (ed.): United Nations Environment Programme (Nairobi: UNEP).

U.S. Bureau of the Census, 2000: "International Database"; at: http://www.census.gov/ipc/www/worldpop.htm (10 May 2000).

Vernadsky, V., 1986, *The Biosphere,* reprinted by Synergetic Press, Oracle AZ.

Vitousek, P. M. et al., 1997: "Human Domination of Earth's Ecosystems", in: *Science,* 277: 494–499.

Wilson, E. O., 1992: *The Diversity of Life (London:* Allen Lane – Harmondsworth: The Penguin Press).

World Bank, 2002: "Data and Statistics".

World Commission on Dams, 2000: Dams and Development: A New Framework for Decision Making, The Report of the World Commission on Dams (London - Sterling, VA: Earthscan Publications Ltd.).

World Tourism Organization, 2002: *Tourism Industry Trends, Industry Science Resources*; at: http://www.world-tourism.org (22 Oct. 2002).

WRI, 1990: "Forest and Rangelands", in: *A Guide to the Global Environment* (Washington, D.C.: World Resources Institute): 101–120.

WRI, 2003: *A Guide to World Resources 2002–2004: Decisions for the Earth.,* A joint Publication with UN Development Program, UN Environmental Program, World Bank and World Resources Institute (Washington, D.C.: WRI).

Chapter 7
Atmospheric Chemistry and Climate in the Anthropocene: Where Are We Heading? (2004)

Paul J. Crutzen and Veerabhadran Ramanathan

Abstract Humans are changing critical environmental conditions in many ways. Here, the important changes in atmospheric chemistry and climate are discussed. The most dramatic examples of major human impacts are the increase of the 'greenhouse' gases, especially carbon dioxide (CO_2), in the atmosphere and the unpredicted breakdown of much of the ozone in the lower stratosphere over Antarctica during the months of September to November, caused by theemissions of *chlorofluorocarbons* (CFCs). Other, more regional but ubiquitous examples include photochemical smog and acid rain. Industrial activities are not alone in causing air pollution and in changing the chemical composition of the atmosphere. Biomass burning, which takes place largely in the developing world, also contributes in major ways. In the future, climate warming due to CO_2 emissions will continue to increase over present levels and pose a major problem for humankind. Current radiative forcing by "greenhouse gases" can, to a substantial degree (up to half), be dampened by increased backscattering of solar radiation, either directly by aerosol particles or indirectly through their influence on cloud albedo, or also by cloud feedbacks independent of anthropogenic aerosols. Cloud and hydrological cycle feedbacks provide major challenges. It is unlikely and undesirable that aerosol emissions will continue to increase, as greater emphasis will be placed on air quality, also in the developing world. However, due to its long atmospheric lifetime and expected growth in global emissions, CO_2 will continue to accumulate, exacerbating climate warming and related problems in the future. Drastic measures are thus needed at the international level to reduce the emissions, in particular, of CO_2 through energy savings, alternative energy sources, and sequestration.

This text was first published as: Paul J. Crutzen and Veerabhadran Ramanathan: "Atmospheric Chemistry and Climate in the Anthropocene: Where Are We Heading?", in: Schellnhuber, H.J.; Crutzen, P J.; Clark, W.C. M.; Claussen, M.; Held, H. (Eds.): *Earth System Analysis for Sustainability*. Dahlem Workshop Report. pp., (Cambridge, MA: MIT Press): 265–292. This text is included here with the permission on 25 May 2018 by Ms. Pamela Quick of MIT Press. Paul J. Crutzen is associated with 'Max Planck Institute for Chemistry, 55020 Mainz, Germany; is affiliated with the Scripps Institution of Oceanography, La Jolla, CA 92037, U.S.A. Thanks go to Andi Andreae, Bert Bolin, Peter Cox, Ulrich Cubasch, Roland von Glasow, Jos Lelieveld, and Will Steffen for productive discussions and advice.

7.1   Introduction

The bulk of the Earth's mass, about 6×10^{27} g, is largely concentrated in the core
($\approx 30\%$) and mantle ($\approx 70\%$). The crust only contributes 0.4% to the total. The mass
of the biosphere is less than a millionth of that of the crust. Still, its influence on
shaping the surface of the Earth and biogeochemical cycles is profound. With about
10^{14} g of 'dry matter', humans constitute only 10^{-5} of the mass of the biosphere. Only
a minute fraction of this is human brains, the sites of the enormous collective thinking
power of the human race, unfortunately often badly used. Despite mass starvation,
epidemics, and wars, the human population has grown by a staggering factor of ten
over the past three centuries. Humankind's unique capacity to produce knowledge
and technology and transfer it to subsequent generations, its mastery of fire, as well
as the indention of agriculture through domestication of plants and animals has
increasingly impacted the environment over time, sharply out of proportion to the
relatively small brain mass of the human species. Generations of ambitious *Homo
sapiens* have played, and will continue to play, a major catalytic role in affecting the
basic properties of the atmosphere with impacts on climate, ecosystems, biological
diversity, and human health. Humankind's activities accelerated particularly during
the past few hundred years and precipitated the entry of a new geological era known
as the "Anthropocene" (Crutzen 2002; Crutzen/Steffen 2003; see also Appendix 7.l).

This was already foreseen eighty years ago by the Russian biologist/geologist
Vernadsky (1998, 1926) who wrote:

> Without life, the face of the Earth would become as motionless as the face of the moon ...
> the evolution of different forms of life throughout geological time increases the biogenic
> migration of elements in the biosphere.... In an insignificant time the biogenic migration
> has been increased by the use of man's skill to a degree far greater than that to be expected
> from the whole mass of living matter.... The surface of the Earth has been transformed
> unrecognizably, and no doubt far greater changes will yet come. We are confronted with a
> new form of biogenic migration resulting from the activity of the human reason.

Vernadsky could only see the beginning of the major changes in land use that
followed, but he could not foresee the changes in the chemical composition of the
atmosphere that were ahead. He recognised, however, the importance of ozone as a
cover against harmful solar radiation. The importance of carbon dioxide as a shield
against heat loss to space was discussed near theend of the nineteenth century especially by Svante Arrhenius, among others; however, he too could not possibly imagine
that less than a century later, climate warming together with ozone depletion would
become issues of great environmental concern requiring an international response
at the political level. In Fig. 7.1, we depict probably the most cited (here updated)
graphs in the history of the effects of human activities on the atmosphere: (a) Keeling's
Mauna Loa CO_2 growth curve (Keeling/Whorf 2000) and (b) the 'Ozone Hole' graph
by Farman et al. (1985). Also shown is a vertical ozone profile in the 'ozone hole'
by Hofmann et al. (1989).

Fossil-fuel and biomass burning, land-use changes, as well as agriculture have
caused increases in the atmospheric concentrations of the greenhouse gases: CO_2,

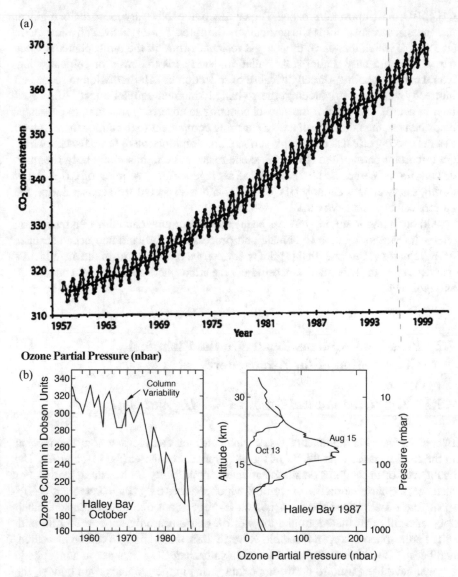

Fig. 7.1 a The 'Keeling curve', which shows the steady increase in atmospheric CO_2 concentration recorded at Mauna Loa in Hawaii, 1958–1999 (adapted from Keeling/Whorf 2000). **b** The Antarctic total ozone as reported by Joe Farman and colleagues in 1985 (left-hand panel) and currently typical vertical ozone profiles (right-hand panel) measured by Hofmann et al. (1989). Since then, ozone loss has continued (with an unexpected exception during the spring of 2002), reaching levels as low as about 100 DU. Lowest concentrations of ozone occur in the same height range, 14–22 km, in which naturally occurring maximum ozone concentrations are found

CH_4, N_2O, and tropospheric ozone. The same activities have also led to major regional increases in aerosol loadings, in particular in the optically most active sub micrometre size range, which also have the longest residence time in the atmosphere. If these particles would only scatter solar radiation, such as is the case for sulphate, they would cause a cooling, directly because of reflection of solar radiation to space and indirectly via cloud brightening, the so-called "indirect aerosol effect" (discussed later in detail). However, if the aerosol contains an absorbing material, in particular black carbon, then the aerosol adds a warming component to the atmosphere as well as a net cooling effect at the Earth's surface. The combination of these factors, which are important parts of the atmospheric chemistry system, has likely been the main driving force behind the observed global average surface warming of 0.6 ± 0.2 K during the twentieth century (IPCC 2001), and is expected to increase during the current century and beyond.

Although important progress has been made in our understanding of atmospheric chemistry and its relation to climate and processes at the land and ocean surface, much remains to be done. In this brief overview, we present some of the main factors that determine the chemical composition of the atmosphere and its impact on climate and biosphere.

7.2   Pressing Problems Related to the Chemical Composition of the Atmosphere

7.2.1   The 'Greenhouse' Gases CO_2, CH_4, and N_2O

Except for variable amounts of water vapour, which can increase to a few percent in the tropics, more than 99.9% of the atmosphere consists of N_2, O_2, and Ar. The abundance of these gases cannot be affected significantly by human activities. For instance, the current practice of burning, which releases 6 Pg (Pg = Petagram = 10^{15} g) of fossil-fuel carbon per year (i.e., on average about one ton per person, albeit very unevenly distributed around the world), consumes only about a 30 millionth of the atmospheric oxygen reservoir. Nevertheless, these changes can be measured, and because O_2 is much less soluble in seawater than CO_2, changes in atmospheric oxygen have been used to derive important terms in the global carbon budget, the uptake of CO_2 in the oceans and in terrestrial ecosystems (Keeling et al. 1996). A very important question to ask, however, is where the sinks are and to find out whether they are only temporarily active.

The most noticeable influence of humans on the global composition of the atmosphere and on Earth's climate is CO_2, the carbon source for photosynthesis. The volume mixing ratio of CO_2 is currently 370 ppmv (parts per million by volume), and it is growing at an annual rate of about 0.4%. This contrasts values of 280 ppmv during the interglacials and 180 ppmv during glacial periods (IPCC 2001). For the chemistry within the atmosphere, CO_2 does not play a primary role, except for a minor influence on cloud acidity: CO_2 alone would yield a pH of 5.6, but dissolution of other chemicals is normally more important.

In the 1990s, the estimated budget terms for CO_2, in units of Pg C per year, were as follows: emissions from fossil-fuel combustion $= 6.3 \pm 0.4$; growth rate in the atmosphere $= + 3.2 \pm 0.1$; uptake by the oceans $= 1.7 \pm 0.5$; uptake by land $= 1.4 \pm 0.7$ (IPCC 2001). The latter favourable condition may, however, not last. By the end of this century, the current terrestrial net carbon sink is expected to turn into a net source of 7 Pg C year^{-1}—larger than the oceanic net carbon sink which saturates at 5 Pg C year^{-1} (Jones et al. 2003)—and lead to a further increase in the atmospheric source of CO_2 in addition to the input of CO_2 from fossil-fuel burning. These results do not, however, agree with the findings of Melillo et al. (2002); their soil warming experiments in mid-latitude hardwood forests show only a small, short-lived release of CO_2 to the atmosphere.

Methane is globally the next most abundant greenhouse gas, with a direct *global warming potential* (GWP) of 23 over that of CO_2, integrated over a 100-year horizon (Ramaswamy et al. 2001). It is also chemically active on the global scale. In the troposphere, CH_4 is involved in reactions that determine the concentrations of ozone and hydroxyl radicals, and thus the oxidation power of the atmosphere. In the stratosphere, its oxidation is a significant source of water vapour. There it also serves partially as a sink for highly reactive ozone-destroying Cl and ClO radicals, by converting them to HCl, which does not react with ozone. Produced through the anaerobic decay of organic matter in wetlands, rice fields, in the rumen of cattle and in landfills, with further emissions coming from coal mines and natural gas leaks, the atmospheric methane content is strongly influenced by human activities. Its average atmospheric abundance has more than doubled since preindustrial times, starting from values around 0.7 ppmv (obtained from analysis of air trapped in ice cores) to a current level of about 1.75 ppmv. During glacial periods, its concentration was even lower, 0.4–0.5 ppmv. Recent observations have shown a slowing-down in the growth of atmospheric methane, which may indicate an approaching balance between sources and sinks, the latter largely due to reaction with hydroxyl radicals in the atmosphere (see below). The total source strength of CH_4, which can be derived from its model calculated loss by reaction with OH in troposphere and stratosphere, uptake at the surface, and present atmospheric growth rate, is about 600 million tons per year. If average OH concentrations have not changed much, as most model calculations suggest, then the preindustrial, natural methane source and sink was about 40% of its present value and stemmed primarily from wetlands. Although the total source and sink terms in the methane budget are rather well known, the individual, global emissions from wetlands and human activities, in particular from rice production, biomass burning, and fossil -fuel production, are much less certain. From the decline in the growth rate of CH_4 (Dlugokencky et al. 1998), an equilibrium between sources and sinks of CH_4 may soon become established. In fact, in the future it might even be possible to lower methane concentrations. To reach such a goal, it is important to have better quantitative information about the individual anthropogenic CH_4 sources. It is also imperative to monitor the releases of methane from natural tropical wetlands and high-latitude northern peat- and wetlands, which may occur under the climate warming regime of the Anthropocene, and which causes higher biological activity, a longer growing season, melting of permafrost, increased precipitation, and thus a possible expansion of the anaerobic zone. Similarly, although its current contribution is currently rather small, release of CH_4 from destabilizing methane hydrates may, in the distant future, become another significant source of CH_4.

With a mean residence time of about a century, N_2O is another significant greenhouse gas. It is present in the atmosphere at a mixing ratio of about 315 nmol/mol, increasing annually by 0.2–0.3%. N_2O is an intermediate product in the nitrogen cycle and is, therefore, among other factors, influenced by the application of nitrogen fertilizer in agriculture. It also plays a major role in stratospheric chemistry. Its reaction with $O(^1D)$ atoms produces NO, which, together with NO_2, catalytically destroys ozone. Thereby it controls the natural level of ozone in the stratosphere. Paradoxically, chemical interactions between ClO_x (Cl and ClO) and NO_x (NO and NO_2) radicals substantially reduce otherwise much larger ozone destruction.

As shown in the IPCC 2001 report, and as was also the case for CO_2, the total sinks of CH_4 and N_2O are now rather well known from atmospheric observations and knowledge of their chemistry. However, the contributions by individual sources are not.

A major question is whether the natural sources of CO_2, CH_4, and N_2O will increase as a result of climate change. The biogeochemical cycles of carbon and nitrogen, which are interconnected, have great uncertainties and require continued research.

7.2.2   Ozone, the Cleansing Effect of Hydroxyl in the Troposphere, and Acid Rain

In and downwind of those regions in which fossil-fuel and biomass burning take place, high ozone concentrations are generally observed in the lower troposphere, and this impacts human health and agricultural productivity. Ozone also acts as a greenhouse gas, whose anthropogenic increase is estimated to have contributed a positive radiative forcing of up to 0.5 Wm^{-2} since preindustrial times (IPCC 2001). Ozone is involved in the production of hydroxyl (OH) radicals, the atmospheric oxidizer, also dubbed the 'detergent of the atmosphere', which is present in the troposphere at a global average volume mixing ratio of only about 4×10^{-14}. Nevertheless, this ultra-minor constituent is responsible for the removal of almost all gases that are emitted to the atmosphere by natural processes and anthropogenic activity. OH is formed through photolysis of ozone by solar ultraviolet radiation yielding electronically excited $O(^1D)$ atoms, a fraction of which reacts with water vapour to produce OH radicals. Thus, while too much ultraviolet radiation and ozone can harm humans and the biosphere, OH radicals are also indispensable for cleansing the atmosphere. Whether the growth in ozone has led to an overall global increase in hydroxyl, and thus in the oxidizing power of the atmosphere, is not clear, since enhanced

OH production has been countered by destruction due to growing concentrations of carbon monoxide and methane, which both react with OH. Although major long-term changes in OH do not appear to have occurred, using methyl chloroform (CH_3CCI_3) as a chemical tracer to derive global average OH concentrations, the two most active research groups in this area (Prinn et al. 2001; Krol et al. 2003) have reached quite different conclusions about OH trends, especially for the early 1990s. During this period, CO and CH_4, the main gases with which OH reacts in the global troposphere, showed quite anomalous, unexplained behaviour. Since 1996, methyl chloroform has no longer been produced under the provisions of the Montreal Protocol and its amendments. Thus, CH_3CCI_3 concentrationsare decreasing, and other chemical tracers, such as HFCs or HCFCs (the CFC replacement products), have come into use as markers of change in OH. Accurate data on their release to the atmosphere are critical. Uncertainties in these have caused much of the disagreements between the two research groups, as mentioned above. Thus, Prinn et al. (2001) estimate a global, CH_3CCl_3 weighted increase by $15 \pm 20\%$ between 1979 and 1989, followed by a sharp decline to reach values in the year 2000 of $10 \pm 24\%$ below those in 1979. By contrast, Krol et al. (2003) derive an upward trend of $6.9 \pm 9\%$ from 1978 to 1993. One way to overcome this discrepancy may be to use dedicated tracers, solely fabricated for the purpose of deriving 'global average OH' and its trends.

Ozone is a driving force in atmospheric chemistry. It can be produced or destroyed in the troposphere, largely depending on the concentrations of NO. Its concentration distribution is highly variable in time and space. Unfortunately, measurements of ozone are still much too sparse to provide a satisfactory test for photochemical models and to derive the important terms in its budget and those of its precursors, such as hydrocarbons released from forests and from fossil-fuel and biomass burning, as well as NO emitted by soils and lightning discharges. Whereas the level of emissions of NO are quite well known (Galloway et al. 2002)—25 Tg N year^{-1} from fossil-fuel burning, including aviation—the emissions from soils, lightning, and biomass burning are very uncertain. Although downward transport from the stratosphere is no longer considered to be the main source of tropospheric ozone, it remains a major contribution to ozone in the upper troposphere, where its effect as a greenhouse gas is maximised.

Industrial emissions of SO_2 and NO, which are oxidised to sulfuric and nitric acid, have led to acidification of precipitation and caused the acidification of lakes and death of fish. This phenomenon was first reported in the Scandinavian countries and in the northeast section of the United States. Regulatory measures, especially against SO_2 release, have somewhat relaxed the situation, but acidity of the rain did not decrease as much as was hoped for, since the emissions of the neutralizing cations Ca, Na, Mg, and K from regional point-sources simultaneously decreased (Hedin et al. 1994). A special issue of the journal *Ambio* (Gunn et al. 2003) was devoted to the biological recovery from lake acidification. 'Acid rain' has grown into an environmental problem in several coal-burning regions in Asia.

7.2.3 Biomass Burning and the Consequences of Land-Use Change in the Developing World

Air pollution has traditionally been associated with fossil-fuel burning. Over the past 2–3 decades, however, biomass burning, which occurs mainly during the dry season in the poorer nations of the tropics and subtropics, has also been recognised as a major source of air pollution. It is estimated that annually 2–5 Pg of biomass carbon are burned in shifting cultivation, permanent deforestation (a net source for atmospheric CO_2), and savannas, as well as through the combustion of domestic and agricultural wastes and firewood. In the process, mostly recycled CO_2 is produced but so are also many chemically active gaseous air pollutants (e.g., CO, CH_4 and many other pure or partially oxidised hydrocarbons, as well as NO_x), thereby delivering both the "fuel" and the catalysts for photochemical ozone formation. During the dry season, high ozone concentrations are indeed widely observed in rural areas of the tropics and subtropics. Meteorological conditions are important. As shown in Figures 7.2a, b (Richter/Burrows 2002), for the tropical Southeast Asia/Australia region, inter-annual variability in atmospheric circulation strongly affects biomass burning and ozone chemistry, and caused concentrations of ozone and its precursors NO_2 and CH_2O that were higher during El Niño/1997 than during La Niña/1996 in September.

Land-use change, especially the removal of forest cover, can have a substantial influence on ozone and particulate matter. In an intact forest, NO_x, which is released from the soil, is largely recycled by uptake within the forest canopy, and ozone production is small. In fact, it is quite possible that reactions with isoprene and other reactive hydrocarbons, emitted by the vegetation, are a sink for OH, thereby suppressing regional photochemical activity. In a disturbed forest or savannas, however, NO_x can more easily escape into the troposphere. There, in the presence of natural hydrocarbons and NO_x as catalysts, ozone! can be produced, which in turn can harm the remaining vegetation. Note that the emissions of natural hydrocarbons (e.g., isoprene, C_5H_8) are very large (10^{15} g year^{-1} or ten times larger than their anthropogenic emissions; Guenther et al. 1995), so that availability of NO_x is the limiting factor for ozone production. This explains why high ozone concentrations are generally not found in the vicinity of tropical forests. In disturbed environments, especially where fires are a source of NO_x during the dry season, high ozone concentrations, approaching 100 nmol/mol, can be reached. This can also have consequences for aerosol production. Enhanced ozonolysis of mono- and sesquiterpenes and other higher terpenoid hydrocarbons can in supra-linear fashion lead to drastically enhanced production of secondary organic aerosols, which affect solar radiation scattering and which can also serve as cloud condensation nuclei (Kanakidou et al. 2000).

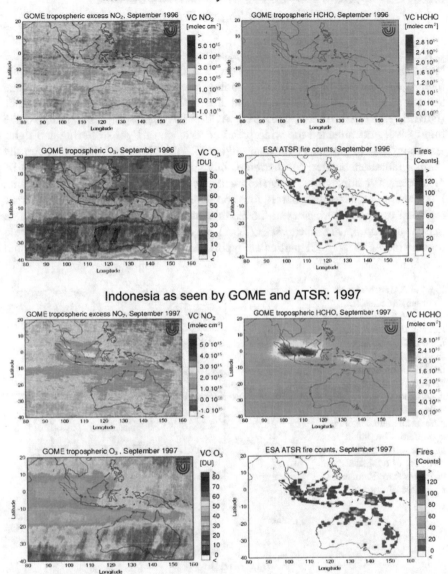

Fig. 7.2 The distributions of tropospheric NO_2, O_3, and CH_2O measured with the DOAS (different optical absorption spectrometer) instrument on board of the ENVISAT satellite during two consecutive September months. Also shown are the fire counts. *Source* Burrows et al., pers. comm.; figure used with kind permission of J. P. Burrows and his team from the University of Bremen

7.2.4 Light-Absorbing and Light-Scattering Particles and Regional Surface Climate Forcing

During the last decade, the role of aerosol particles in the radiation budget of the atmosphere and the hydrological cycle became a topic of great significance. Initially, interest was centred on the radiative cooling properties of nonadsorbing sulphate aerosol (Charlson et al. 1991). Recent international field research programmes (BIBLE, TARFOX, ACE-2, INDOEX, ACE-ASIA) and space observations have impressively documented the widespread occurrence of light-scattering and light-absorbing smoke particles over many regions qf the globe and their effect on the Earth's radiation budget. Furthermore, in particular, results from the INDOEX measurement campaign of 1999 (Ramanathan et al. 2001a) strongly suggest that much more emphasis than before must be given to the order of magnitude larger radiation energy disturbances at the Earth's surface and in the lower troposphere than at the *top of the atmosphere* (TOA) in- and downwind of heavily polluted regions (see Fig. 7.3). Of particular importance are the atmospheric heating and

Fig. 7.3 Comparison of anthropogenic aerosol forcing with greenhouse forcing. **a** The Indo-Asian region. The greenhouse forcing was estimated from the NCAR community climate model with an uncertainty of ± 20%. **b** Same as above, but for global and annual average conditions. The global average values are a summary of published estimates from Ramanathan et al. (2001b)

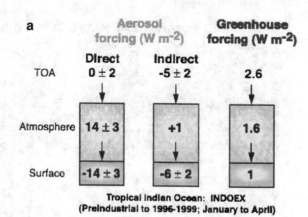

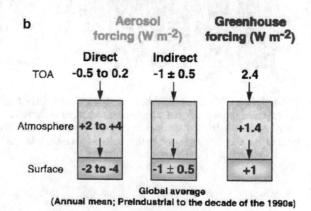

surface cooling effects that are caused by the absorption of solar radiation by the black carbon in the smoke, which, through thermal stabilisation of the boundary layer, reduces the strength of hydrological cycle. To date, model studies suggest that aerosols with *single scattering albedo* (SSA) greater than 0.95 will lead to surface cooling, whereas SSA values less than 0.85 will lead to surface warming. For SSA values in between, the sign of the net forcing will depend critically on cloud fraction, cloud type, and the reflectivity of the surface. High precision radiation measurements taken from space and from the surface during INDOEX have been used to Remonstrate that over the polluted Arabian Sea, radiation forcing at the surface was three to four times greater than at the TOA (Satheesh/Ramanathan 2000). Direct aerosol radiative forcing measurements similar to those obtained front INDOEX have been performed in other parts of the world (e.g., for the Mediterranean, see Markowicz et al. 2002; for other regions of the world, see Kaufman et al. these data should significantly help improve the accuracy of purely model-derived forcing estimates used in IPCC-type assessment studies. Worldwide SSA data from the AERONET network were published by Dubovik et al. (2002) and are reproduced in Fig. 7.4. A rough compilation of the sources, lifetimes, distributions, and optical properties of various kinds of aerosol in the atmosphere is given in Table 7.1.

Especially significant may be the possibilities for positive feedbacks. During the dry season, the residence time of aerosols is strongly enhanced by the lack of rain. With the addition of smoke from biomass burning, causing greater dynamic stability of the boundary layer, precipitation is suppressed, thereby enhancing the residence time of the particles, etc. In fact, even when clouds are formed, precipitation efficiency will decrease as a result of overseeding; this will bring about an increase in longevity, global cover, and reflectivity, and lead to a further cooling of the Earth's surface, less rainfall near populated regions, and more in remote areas.

In Asia, air pollution is attributed to a mixture of sources: biomass burning and fossil-fuel burning without, or with inadequate, emission controls. With the expected growth in population and industry, emissions from the Asian continent will become a major factor affecting not only the regional but most likely also global climates. Asia, however, is not the only part of the world with this kind of problem. Biomass burning is also heavily practiced in South America and Africa, causing high aerosol loadings, in particular over the savannas, as shown in Fig. 7.4.

Explanation to Table 7.1: This is an update of the compilation by Andreae (1995) with the following changes:

Desert and soil dust. From a range of 1000–3000 Tg year^{-1}, Andreae chose 1500 Tg and included this under natural sources. However, Tegen and Fung (1995) propose that 30–50% of the source might be derived from lands disturbed by human action. As this view is disputed, we adopt a range 0–40% as anthropogenic. The two values given for dust should not be interpreted as the range due to uncertainty; rather, the lower value (i.e., the 900 Tg year^{-1} estimate for primary dust) for the natural aerosol is based on the assumption that some of the dust (i.e., the 600 Tg year^{-1} value) may be anthropogenic.

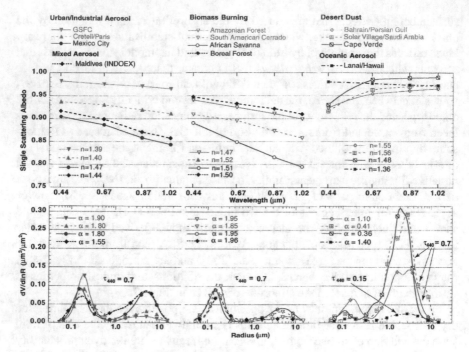

Fig. 7.4 The averaged optical properties of different types of tropospheric aerosol retrieved from the worldwide AERONET network of ground-based radiometers (Dubovik et al. 2002). Urban industrial, biomass burning, and desert dust aerosols are shown for $\tau_{ext}(440) = 0.7$. Oceanic aerosol is shown for $\tau_{ext}(440) = 0.15$ since oceanic background aerosol loading does not often exceed 0.15. Also, $\omega_0(\lambda)$ and the refractive index n shown for Bahrain was obtained only for the cases when $\alpha \leq 0.6$ (for higher α, $\omega_0(\lambda)$ and refractive index n were very variable due to a significant presence of urban-industrial aerosol). However, we show the particle size distribution representing all observations in Bahrain (complete range of α). Ångstrom parameter α is estimated using optical thickness at two wavelengths: 440 and 870 nm

Nitrate aerosol was not considered explicitly as much HNO_3 formed from NO_2 oxidation will be deposited on already existing particles. The fraction that forms new particles is highly uncertain. Published estimates of nitrate direct radiative forcing range from near zero to values similar to those for sulphates.

Sea salt. The given source strength is the average of Andreae (1995) and Penner et al. (2001). The scattering coefficient recommended by Andreae (1995) was about 0.4 m^2/g; but the inclusion of fine sea salt particles (radius smaller than 1 μm) has led to a substantial upward revision to a value of about 2.5 m^2/g at a relative humidity of about 80% (Haywood et al. 1999). We have adopted the average of these values.

Volcanic emissions. Ash is not considered because of the very short residence time of these larger particles. The sulphate flux of 20 Tg year^{-1} (Andreae 1995) to the troposphere derived from SO_2 oxidation is included. Also shown, but not included in the sum, is the contribution by the rare event of the Pinatubo eruption of 1991.

Table 7.1 Global source strength, atmospheric burden, and optical extinction due to the various types of aerosols (for the 1990s)

Source	Flux (Tg year^{-1})	Lifetime (days)	Column burden (mg m^{-2})	Specific scattering/absorption (m^2/g)	Optical depth (x 100) scattering/absorption
Natural					
Primary					
Dust (desert)	900–1500	4	19–33	0.6	1–2
Sea salt	2300	1	3	1.5	2
Biological debris	50	4	1	2	0.2
Secondary					
Sulfates from biogenic gases	70	5	2	8	1.6
Sulfates from volcanic SO$_2$ (troposphere)	20	10	1	8	0.8
Sulfates from Pinatubo (1991) (stratosphere)	(40)	(400)	(80)	(2)	(16)
Organic matter from biogenic hydrocarbons	20	5	0.6	8	0.5
Total natural	2400–3000		32–45		6–7 (± 3)
Anthropogenic					
Primary					
Dust (soil + desert)	0–600	4	0–13	0.7	0–0.9
Industrial dust	40	4	0.9	4	0.4
Black carbon (BC)	14	7	0.6	4/10	0.2/0.6
Organic carbon in smoke	54	6	1.8	6	1
Secondary					
Sulfates from SO$_2$	140	5	3.8	8	3
Organic hydrocarbons	20	7	0.8	8	0.6
Total anthropogenic	270–870		8–21		5–6 (± 3)/0.6 (± 3)
Total (natural and anthropogenic)					12 (± 4)/0.6 (± 3)

Black and organic carbon. We adopt 7 Tg for fossil-fuel burning (Penner et al. 2001) and 7 Tg for biomass burning (Penner et al. 2001; Haywood et al. 1999), with a total emission of 14 Tg.

Most aerosol source estimates, their lifetimes, and their optical effects are uncertain by at least a factor of 2, in particular sea salt and soil dust (see also Tegen et al. 2000). Despite the large uncertainties, this table indicates that the global average optical depths from natural and anthropogenic sources may be of similar magnitude with about 10% of the anthropogenic part due to black carbon. Note that, although the natural sources for aerosol are much larger than the anthropogenic inputs, the optical depths are very similar because the former produces fewer, but larger size, particles.

We adopted the following guidelines in arriving at the radiation parameters: The scattering coefficients account for the increase in scattering due to the hydration of the aerosols with a relative humidity of 80% in the first 1 km and 60% above 1 km. For sulfates, organics, and sea salt, we assume a scale height of 1 km for the vertical variation. Because of the relative humidity dependence, the sulfate scattering coefficient decreases from 8.5 m^2/g near the surface to about 7 m^2/g above 1 km. We assume that the uncertainties in *aerosol optical depth* (AOD) for natural and anthropogenic quantities are uncorrelated Also note that we do not cite the uncertainties in individual terms but the overall uncertainty in AOD. Although we separate the scattering and absorbing optical depth, the sum of the two yields the so-called extinction optical depth. When we sum the individual optical depths to estimate the total optical depth, we assume that the aerosols are externally mixed, i.e., the particles exist as chemically distinct species. If we make the other extreme assumption that all of the species are internally mixed, the scattering coefficient will be reduced and the absorption coefficient will increase (Jacobson 2001).

Traditional air pollution is on the decline in the industrialised world and now highest in the developing world. For many decades, developed nations were champions in CO_2 emissions on a per capita basis; now, however, emissions are rapidly growing, most noticeably in the Asian countries, even though traditional air pollution is declining due to health-related regulations. Thus, because of the long atmospheric residence time of CO_2 of more than a century, compared to only about a week for the aerosol particles, in the future the cumulative global climatic warming by CO_2 will outpace any cooling effect that comes from the backscattering of solar radiation into space by the aerosol particles.

7.3   Policy-Related Issues, Lessons to Be Learned

7.3.1   The 'Ozone Hole'

Human activity has caused major decreases in stratospheric ozone. Most dramatic has been the development of the stratospheric 'ozone hole', the complete destruction of ozone, which has been recurring almost annually during September–November, sometimes until midsummer, since the beginning of the 1980s over Antarctica in the 12–22 km height interval, precisely where maximum ozone concentrations had earlier always been observed (see Fig. 7.1 and Appendix 7.2). (The weak ozone hole of the spring of 2002 was a surprising excep-tion, most likely caused by unusual meteorological conditions.) Stratospheric ozone loss is caused by emissions of rela-tively small quantities of CFC gases, accumulating to a little over 3.5 nmol/mol levels of total chlorine, among which are the very reactive ozone-destroying Cl and ClO catalysts. Under the prevailing atmospheric conditions during late winter/early spring, production of these catalysts is strongly enhanced, leading to a chemical instability, especially over Antarctica and to a lesser extent over the Arctic. The development of the ozone hole takes place in a location and at a time of the year that were least expected. Conventional wisdom held that, especially over Antarctica, ozone should be chemically inert. This is indeed true for the natural stratosphere, but it does not hold in a stratosphere containing about six times more chlorine than under natural conditions. For the ozone hole to develop, five criteria must be fulfilled simultaneously (see Appendix 7.2 and accompanying figure), which explains why it was not predicted. In fact, when satellite measurements started to reveal a depletion of ozone, the data were first put aside and attributed to measurement errors. From this experience we should learn several lessons:

1. Do not release chemicals into the environment before their impacts are well studied. After their first release into the atmosphere, it took approximately four decades before Molina and Rowland (1974) could demonstrate that CFCs harm the ozone layer, and another two decades before their production was halted by international regulations. It will take up to an additional half century before the ozone hole will close again, which attests to the long timescales involved.
2. Although major strides are constantly being made in the development of compre-hensive climate models, we should not yet overly rely on models to make predictions. With their discrete space/time coordinate systems, which do not resolve important smaller-scale meteorological features, replacing them by so-called 'subgrid-scale parameterisations', models remain approximations of the complex, multidisciplinary feedback systems of the real world. Global climate models have particular difficulties with describing the various elements of the hydrological cycle and the physical-meteorological-chemical processes, which determine the distribution of aerosol and its impact on cloud properties and Earth's radiation budget—a topic to which we will return below. Numerical models are based on incomplete knowledge and have at best been partially tested against observations in the present world, which may not apply in the future, given the great uncertainty of societal responses.
3. Nevertheless, models are indispensable tools to combine available knowledge from several disciplines. Used in the research mode, they serve best when producing results that deviate from observations, thus indicating gaps in knowl-edge. Models can also be very valuable devices to identify potential low-probability/high-impact features in the complex environmental system. Under

certain circumstances, their results should at least be taken as warning signals for impending dangers.

4. Major changes may be triggered by relatively small disturbances. This was dramatically shown in the CFC case. Also, as we discuss below, in comparison with the average influx of 340 Wm^{-2} of solar radiation at the TOA, radiation flux disturbances due to human activities of a few Wm^{-2} are very small indeed but, nevertheless, very important.

5. If unexpected data (i.e., real data or calculated by models) appear, the so-called 'outliers' take them seriously, as they may be the first signals of shifts in environmental conditions (cf., e.g., the presence of 'negligibly' few automobiles on postcards from the beginning of the past century against pictures of today's ubiquitous traffic jams).

6. Do not assume that scientists always exaggerate; the ozone loss over Antarctica was much worse than originally thought.

7.3.2   How Well Do We Know the Science of Climate Change?

7.3.2.1   CFCs Compared to CO_2

The Mauna Loa CO_2 records of C.D. Keeling of the Scripps Institution of Oceanography and the ozone hole graph by Joe Farman and colleagues of the British Antarctic Survey constitute the most striking examples of human impacts on the global atmosphere (see Fig. 7.1). Following intensive research efforts, which clearly showed that the ozone loss was caused by catalytic reactions involving chlorine radicals, international agreements were set in force in 1996 to stop the production of CFCs as well as several other chlorine- and bromine-containing gases in the developed world. Due to the longevity of these products, however, it will take about half a century before the 'ozone hole' will largely close again. In the case of CO_2, regulations are much more difficult to achieve; it is estimated that it will take many more years before agreements on emissions can be reached to limit fossil-fuel use sufficiently. To demonstrate the magnitude of the task, consider the following: to prevent a further increase in levels of CO_2, the present level of worldwide CO_2 emissions from fossil fuels would need to be reduced by as much as 60%, a practically unattainable task, since without abundant renewable energy sources, much of the developing world can be expected to increase their fossil-fuel use. All efforts should be made to keep CO_2 growth at a climatically acceptable level. Energy savings and improved technology will have to make major contributions; however, technological fixes, such as carbon sequestrations, may become inevitable (Lackner 2003). Because of the long lifetime of atmospheric CO_2, it may take centuries before 'global warming' and its effects will peak; in terms of sea-level rise, this may take a millennium. One often cited reason for a delay in international regulations is that the science is so uncertain. Here, however, we must call attention to a potentially disturbing signal. The Earth's current radiation budget at the TOA for the period of1955 to 1996 was out of balance by 0.32 ± 0.15 Wm^{-2}, corresponding to the heat uptake in the oceans, plus lesser contributions from ice melting and heating of the atmosphere (Levitus et al. 2001). In comparison to the greenhouse gas climate forcing of 2.5–3.5 Wm^{-2} (see Fig. 7.5; IPCC 2001), this term may seem small, but it corresponds to the net heating of the Earth system, which accumulates from year to year. Note that this net global warming of 0.32 ± 0.15 Wm^{-2} cannot be confirmed by measurements at the TOA because of the lack of shortwave and infrared measurement capabilities with sufficient accuracy and precision. It is, however, close to the heat uptake in the oceans of 0.32 ± 0.15 Wm^{-2}, derived by Levitus et al. (2001). It also agrees rather well with the results from model calculations, which take into account the growth of 'greenhouse gases' and the direct cooling effect of anthropogenic sulphate particles during the same period (Barnett et al. 2001).

We next present in broad terms the changes in the Earth's radiative forcing between 1860 and the early 1990s. The increase in the infrared downward radiation flux forcing by greenhouse gases at the TOA is 2.3 ± 0.5 Wm^{-2}. From this flux we must subtract an upward radiation flux of 1 ± 0.3 Wm^{-2} due to the surface temperature rise of 0.6 ± 0.2 K, calculated for the case of a water vapour/surface temperature feedback (Ramanathan 1981). Because the atmosphere loses 0.3 Wm^{-2} as a result of heat uptake in the ocean (Levitus et al. 2001), the atmosphere would gain 1.0 ± 0.6 Wm^{-2} of energy, but it is unable to do so because of its small heat capacity. Consequently, this energy is backscattered as shortwave radiation, correspond-ding to $43 \pm 25\%$ of the greenhouse gas warming. We call this 'the parasol effect'. This energy loss must have occurred by enhanced reflection of solar radiation by aerosols and clouds. (We assume no change in surface albedo and neglect the variations in solar radiation by 0.3 ± 0.2 W m-^{2}). Remarkably, the range is also significantly caused by uncertainty in the trends in global average temperature.

In line with this analysis, there has been a 2% increase in cloud cover over the mid- to high-latitude land areas during the twentieth century (IPCC 2001). Enhanced albedo is also supported by the pan evaporation measurements of Roderick and Farquhar (2002) and cloud darkening observations in eastern Europe after the industrial cleanup as a result of the collapse of the communist regimes in Europe (Krüger/Graßl 2002). In addition, satellite-derived cloud and surface properties have shown that the Arctic has warmed, accompanied by increased cloudiness in all seasons, except spring (Wang/Key 2003), which has prevented even greater Arctic warming.

Without the water vapour/surface temperature feedback, 2 Wm^{-2} instead of 1 Wm^{-2} must be subtracted from the greenhouse forcing term (Ramanathan 1981). This no-feedback case is unlikely: Higher temperatures should allow for more evap-oration and a greater propensity of the atmosphere to hold water vapour, agreeing with what is shown by climate models and confirmed by observations, which show an increase by several percent in water vapour concentrations per decade over many regions of the Northern Hemisphere (IPCC 2001). Data are not available for the Southern Hemisphere.

Despite compensation of greenhouse warming by aerosols and clouds, due to the much shorter lifetimes of the aerosol, cumulative forcing by the greenhouse gases, especially CO_2, will in the future outpace the compensation provided by aerosols, implying accelerated climate warming. This was also pointed out by Anderson et al. (2003). The power of CO_2 and CH_4 as greenhouse gases has been highlighted in several recent studies, showing their importance] for climate change during the glacials and interglacials and other geological periods (e.g., Hinrichs et al. 2003; Barrett 2003; DeConto/Pollard 2003).

The global mean radiative forcing of the climate system for the year 2000, relative to 1750.

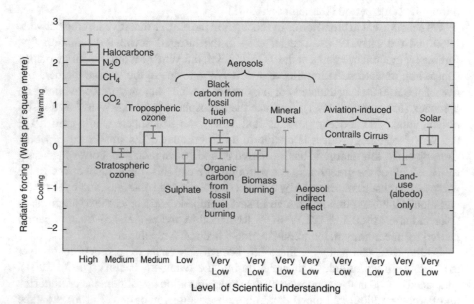

Fig. 7.5 Over the past decade, the atmospheric chemistry research community has identified and quantified the distributions of a number of radiatively active substances. However, especially regarding the cooling (parasol) effect by aerosol in the hydrological cycle, as shown in the blue field, the level of scientific understanding in the calculated radiative forcing of these various substances is still very low (IPCC 2001)

With an inevitable, continued growth in CO_2 concentrations and model predictions of an average global warming in the range of 1.4–5.8 K, there is great risk that humankind is rapidly moving into an uncertain and uncharged climate future with great risks of major climate warming and related problems, such as higher frequency of extreme events, sea-level rise, and regional lack of freshwater availability. The fourfold range in the estimated warming is due to a combination of uncertainties in scenarios, climate model feedbacks, and adopted future scenarios in greenhouse gas emissions and aerosol loadings. In all likelihood, climate change will represent— even more than now—a majorissue that humankind must face in this and, probably also, the next century.

7.3.2.2 Impacts of Particles on Human Health and Climate: A Dilemma?

Health problems associated with air pollution are substantial. The World Health Organization estimates that worldwide each year about 600,000 people fall victim to outdoor (Stone 2002) and 1.6 million to indoor air pollution (Smith 2002). Biomass burning in the households of the developing world for cooking and heating produces as much as 7–20% of the total emissions of CO, NO, and hydrocarbons (Ludwig et al. 2003). Thus, global, regional, local, and even! domestic air pollution effects intertwine. It is clear that for health reasons, priority should and will be given to the abatement of the heavy air pollution, which affects large tracts of land and people, especially in the developing world. As stated above, this can, however, have important consequences for regional and also global climate. Regionally, it will be advantageous as more solar radiation will again reach the surface with a positive effect onprecipitation. In addition, cloudiness may decrease via a lower second indirect aerosol effect and lead to a further increase in solar radiation. Black carbon is a special case. Its removal will cool the planet, but its budget is uncertain. Globally, however, if the cooling effect of the aerosol is greater than the heating effect by black carbon particles, the clean-up will add to global warming.

In summary, greenhouse gas emissions will increase surface temperature and rainfall. Aerosols counteract through surface cooling and drying. Since aerosol effects are concentrated regionally, global warming and regional cooling will happen simultaneously. Furthermore, as shown by Novakov et al. (2003), the aerosol effect itself fluctuates between warming and cooling, depending on how the various countries shift from inefficient technology (which produces more absorbing aerosols) to more efficient technology (which produces cooling aerosols). The opposite effects of aerosols and greenhouse gases on precipitation, on timescales of decades, may regionally lead intermittently to more frequent droughts and excess rainfall conditions.

7.4   Final Remark

Humankind is carrying out a grand experiment on its own planet with uncertain, but, most likely, major environmental consequences. This has been emphasised here, in part with some new arguments (as far as we know) on the climate side. Thus far, the political/economical actors have not been sufficiently impressed with the warnings by the climate research community (IPCC 2001), pointing to a major failure in communications.

Appendix 7.1: The 'Anthropocene'

Supported by great technological and medical advancements and access to plentiful natural resources, the expansion of humankind, both in numbers and per capita exploitation of Earth's resources, has been astounding (Turner et al. 1990). Several examples of the growth of human activities and economic factors impacting on the Environment during the twentieth century are given in Table 7.2.

During the past three centuries, human population increased tenfold to 6000 million, growing by a factor of four during the twentieth century alone (McNeill 2000). This growth in human population was accompanied, e.g., by a growth in the methane-producing cattle population to 1400 million (about one cow per average-size family).

In a few generations humankind will have exhausted the fossil fuels that were generated over several hundred million years, resulting in large emissions of air pollutants. The release of SO_2 (globally about 160 Tg year^{-1}) to the atmosphere by coal and oil burning is at least two times larger than the sum of all natural emissions, occurring mainly as marine dimethyl-sulfide from the oceans (Houghton et al. 1990, 1996, 2001). Oxidation of SO_2 to sulfuric and of NO_x to nitric acid has led to acidification of precipitation and lakes, causing forest damage and fish death in biologically sensitive lakes in regions such as Scandinavia and northeastern North America. Due to substantial reduction in SO_2 emissions, the situation has improved. However, the problem is now getting worse in Asia.

From Vitousek et al. (1997) we learn that 30–50% of the world's land surface has been transformed by human action; the land under cropping has doubled during the past century at the expense of forests, which declined by 20% (McNeill 2000) over the same period. More nitrogen is now fixed synthetically and applied as fertilizers in agriculture than fixed naturally in all terrestrial ecosystems (120 Tg year^{-1} vs. 90 Tg year^{-1}) (Galloway et al. 2002). The Haber–Bosch industrial process to fix N from N_2 in the air made human population explosion possible. It is remarkable to note the importance of a single invention for the evolution on our planet. Only 20 Tg N year^{-1} is contained in the food for human consumption. Wasteful application of nitrogen fertilizers in agriculture and especially its concentration in domestic animal manure have led to eutrophication of surface waters and even groundwater in many locations around the world. Fossil-fuel burning adds another 25 Tg N year^{-1} highly reactive NO_x to the atmosphere, causing photochemical ozone formation in extensive regions around the globe. Additional input of altogether 165 TgN year^{-1} is almost double as large as natural biological fixation. Disturbance of the N cycle also leads to the microbiological production of N_2O, a greenhouse gas and a source of NO in the stratosphere where it is strongly involved in stratospheric ozone chemistry. Human disturbance of the N cycle has recently been treated in a special publication of *Ambio* (vol. 31, March 2002). As a result of increasing fossil-fuel burning, agricultural activities, deforestation, and intensive animal husbandry, especially cattle holding, several climatically important 'greenhouse' gases have substantially increased in the atmosphere over the past two centuries: CO_2 by more than 30% and CH_4 by more than 100%, contributing substantially to the observed global average temperature increase by about 0.6 °C that has been observed during the past century. In 1995, IPCC stated: "The balance of evidence suggests a discernable human influence on global climate". In 2001: "There is new and stronger evidence that most of the warming observed over the last 50 years is attributable to human activities" (Houghton et al. 1990, 1996, 2001). Depending on the scenarios of future energy use and model uncertainties, increasing emissions and the resulting growth in atmospheric concentrations of CO_2 are estimated to cause a rise in global average temperature by 1.4–5.8 °C during the present century, accompanied by sea-level rise of 9–88 cm (and 0.5–10 m until the end of the current millennium). The largest anthropogenic climate changes are thus awaiting future generations.

Furthermore, humankind also releases many toxic substances in the environment and some, the chlorofluorocarbon gases ($CFCl_3$ and CF_2Cl_2), which are not toxic at all, have led to the Antarctic springtime 'ozone hole' and would have destroyed much more of the ozone layer if international regulatory measures to end their production by 1996 had not been taken. Nevertheless, due to the long residence times of CFCs, it will take at least until the middle of this century before the ozone layer will have largely recovered and the ozonehole will have disappeared.

Considering these and many other major and still growing impacts of human activities on Earth and atmosphere, at all scales, it is more than appropriate to emphasise the central role of humankind in geology and ecology by using the term 'Anthropocene' for the current geological epoch. The impact of current human activities is projected to last for very long periods. According to Loutre/Berger (2000), due to past and future anthropogenic emissions of CO_2, climate may depart significantly from natural behaviour even over the next 50,000 years.

To assign a more specific date to the onset of the 'Anthropocene' is somewhat arbitrary, but we propose the latter part of the eighteenth century, although we are aware that alternative proposals can be made. However, we choose this date because, during the past two centuries, the global effects of human activities have become clearly noticeable. This is the period when data retrieved from glacial ice cores show the beginning of a growth in the atmospheric concentrations of several 'greenhouse gases,'in particular CO_2, CH_4, and N_2O (Houghton et al. 1990, 1996, 2001). Such a starting date coincides with James Watt's invention of the steam engine in 1784.

Table 7.2 Partial record of the growths and impacts of human activities during the twentieth century

Item	Increase factor, 1890s–1990s
World population	4
Total world urban population	13
World economy	14
Industrial output	40
Energy use	16
Coal production	7
Carbon dioxide emissions	17
Sulfur dioxide emissions	13
Lead emissions	≈8
Water use	9
Marine fish catch	35
Cattle population	4
Pig population	9
Irrigated area	5
Cropland	2
Forest area	20% decrease
Blue whale population (Southern Ocean)	99.75% decrease
Fin whale population	97% decrease
Bird and mammal species	1% decrease [1]

Source McNeill (2000)

Without major catastrophes (e.g., enormous volcanic eruptions, an unexpected epidemic, a large-scale nuclear (or biological) war, an asteroid impact, a new ice age, or continued plundering of Earth's resources by wasteful technology), humankind will remain a major geological force for many millennia, maybe millions of years, to come. To develop a worldwide accepted strategy leading to sustainability of ecosystems against human-induced stresses will be one of the great tasks of human societies, requiring intensive research efforts and wise application of the knowledge thus acquired.

Exciting, but also difficult and daunting tasks lie ahead of the global research and engineering community to guide humankind toward global, sustainable, management of the ecosphere in the Anthropocene (Schellnhuber 1999).

Appendix 7.2: The Ozone Hole

Stratospheric ozone is formed through the photolysis of O_2 and recombination of the two resulting O atoms with O_2: $3O_2 \rightarrow 2O_3$. These reactions are clearly beyond human control. Reactions are also needed to reproduce O_2, otherwise within 10,000 years all oxygen would be converted to ozone. Because laboratory simulations of rate coefficients in the late 1960s had shown that the originally proposed reactions by Chapman in 1930 were too slow to balance ozone production, additional reactions involving several reactive radical species were postulated. The additional ozone destroying reaction chains can be written as:

$$X + O_3 \rightarrow XO + O_2$$
$$O_3 + h\nu \rightarrow O + O_2 \ (\lambda < 1140 \ nm)$$
$$O + XO \rightarrow X + O_2$$

net: $2O_3 \rightarrow 3O_2,$

where X stands for OH, NO, Cl, or Br, and XO correspondingly for HO_2, NO_2, ClO, and BrO. These catalysts are influenced by human activities, especially the emissions of industrial chlorine compounds which are transferred to the stratosphere, such as CCl_4, CH_3CCl_3, and most importantly, the chlorofluorocarbon ($CFCl_3$ and CF_2Cl_2) gases. The current content of chlorine in the stratosphere, about 3 nmol/mol, is about six times higher than what is naturally supplied by CH_3Cl.

For a long time it was believed that chemical loss of ozone by reactive chlorine would mostly take place in the 25–50 km height region and that at lower altitudes in the stratosphere, which contains most ozone, only relatively little loss would take place. The reason is that the NO_x and the ClO_x radicals, like two "Mafia" families, kill each other by forming $ClONO_2$ and HCl:

$$ClO + NO_2 + M \rightarrow ClONO_2 + M, \text{ and}$$
$$ClO + NO \rightarrow Cl + NO_2$$
$$Cl + CH_4 \rightarrow HCl + CH_3.$$

Most inorganic chlorine is normally present as HCl and $ClONO_2$, which react neither with each other nor with ozone in the gas phase, thus protecting ozone from otherwise much larger destruction.

This preferred situation does not always exist. In 1985, scientists from the British Antarctic Survey presented their observations showing total ozone depletions over the Antarctic by more than 50% during the late winter/springtime months (September to November), with ozone depletions taking place in the 14–22 km height region where normally maximum ozone concentrations are found; within a few weeks after polar sunrise, almost all ozone is destroyed, creating the 'ozone hole'. How was this possible? Nobody anticipated this; in fact, it was believed that at high latitudes, ozone in the lower stratosphere was largely chemically inert.

It only took some two years of research to identify the main processes that led to these large ozone depletions and to show that the CFCs were the culprits. The explanation involves each of five necessary conditions:

1. Low temperatures, below about $-80\ °C$, are needed to produce ice particles consisting of nitric acid and water (nitric acid trihydrate) or water molecules! In this process, NO_x catalysts are also removed from the stratosphere through the reactions:

$$NO + O_3 \rightarrow NO_2 + O_2$$
$$NO_2 + NO_3 + M \rightarrow N_2O_5 + M$$
$$N_2O_5 + H_2O \rightarrow 2HNO_3$$

thereby producing HNO_3 which is incorporated in the particles.

2. On the surface of the ice particles, HCl and $ClONO_2$ react with each other to produce Cl_2 and HNO_3:

$$HCl + ClONO_2 \rightarrow Cl_2 + HNO_3.$$

The latter is immediately incorporated in the particles.

3. After the return of daylight following the polar night, Cl_2 is photolysed to produce 2 Cl atoms:

$$Cl_2 + h\nu \rightarrow 2Cl.$$

4. The chlorine atoms start a catalytic chain of reactions, leading to the destruction of ozone:

$$Cl + O_3 \rightarrow ClO + O_2$$
$$Cl + O_3 \rightarrow ClO + O_2$$
$$ClO + ClO + M \rightarrow Cl_2O_2 + M$$
$$Cl_2O_2 + h\nu \rightarrow Cl + ClO_2 \rightarrow 2Cl + O_2$$

net: $2O_3 \rightarrow 3O_2.$

Note that the breakdown of ozone is proportional to the square of the ClO concentrations. As these grew for a long time by more than 4% per year, ozone loss increased by 8% from one year to the next. Also, because there is now about six times more chlorine, about 3 nmol/mol, in the stratosphere, compared to natural conditions when chlorine was solely provided by CH_3Cl the ozone depletion is now 36 times more powerful than prior to the 1930s when CFC production started. Earlier, under natural conditions, chlorine-catalysed ozone destruction was unimportant and it will be so again in 1–2 centuries.

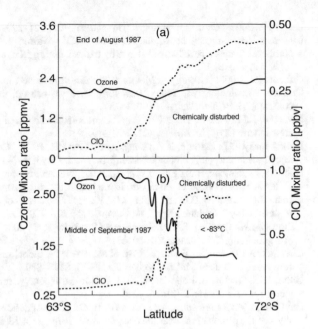

Fig. 7.6 High concentrations of ClO radicals and the simultaneous rapid ozone destruction occur in winter when the temperature becomes very low ($< -80\ °C$). Measurements by Anderson et al. (1989) show ozone and ClO mixing ratios near 20 km during two ER-2 aircraft flights. ClO radicals, together with Cl the catalysts that destroy ozone, show sharp increases in the low-temperature section south of about 68°S, resulting in strong depletions during the first half of September

5.   Enhanced inorganic chlorine (Cl, ClO, HCl, ClONO$_2$, Cl$_2$O$_2$) concentrations, produced by CFC photolysis above 25–30 km are brought down during winter into the lower stratosphere by downwind transport from the middle and upper stratosphere within a meteorologically stable vortex with the pole more or less at the center. This is important because at the higher altitudes more organic chlorine is converted to much more reactive inorganic chlorine gases, including the ozone-destroying catalysts Cl, ClO, and Cl$_2$O$_2$.

All five factors have to come together to cause the ozone hole (see Fig. 7.6). It is thus not surprising that the ozone hole was not predicted. This experience shows the critical importance of measurements. What other surprises lie ahead involving instabilities in other parts of the complex Earth system?

References

Anderson, J.G; Brune, W.H.; Proffitt, M.H., 1989: "Ozone destruction by chlorine radicals within the Antarctic vortex: The spatial and temporal evolution of ClO-O$_3$ anticorrelation based on *in situ* ER-2 data", in: *J. Geophys. Res.*94,13: 46—11: 479.

Anderson, T.L.; Charlson, R.J.; Schwartz, S.E., et al., 2003: "Climate forcing by aerosols—A hazy picture"; in: *Science* 300: 1103–1104.

Andreae, M.O., 1995: "Climatic effects of changing atmospheric aerosol levels", in: Henderson-Sellers, A. (ed.): *Future Climates of the World: World Survey of Climatology* (Amsterdam: Elsevier), 16: 341–392.

Barnett, T.P.; Pierce, D.W.; Schnur, R., 2001: "Detection of anthropogenic climate change in the world's oceans", in: *Science*, 292: 270–274

Barrett, P., 2003: "Cooling a continent", in: *Nature*, 421: 221–223.

Charlson, R.J.; Langner, J.; Rohde, H.; Leovy, C.B.; Warren, S.G., 1991: "Perturbation of the Northern Hemisphere radiative balance by backscattering from anthropogenic sulfate aerosols", in: *Tellus*, 43,AB: 152–163.

Crutzen, P.J., 2002: "Geology of Mankind: The Anthropocene", in: *Nature*, 415: 23.

Crutzen, P.J.; Steffen, W., 2003: "How long have we been in the Anthropocene: An editorial comment", in: *Clim. Change*, 61: 251–257.

DeConto, R.M.; Pollard, D., 2003: "Rapid Cenozoic Glaciation of Antarctic induced by declining atmospheric CO_2", in: *Nature*, 421: 245–249

Dlugokencky, E.J.; Maserie, K.A.; Lang, P.M.; Tans,P.P., 1998: "Continuing decline in the growth rate of the atmospheric methane burden", in: *Nature*, 393: 447–450.1

Dubovik, O.; Holben, B.N.; Eck, T.F. et al., 2002: "Variability of absorption and optical properties of key aerosol types observed in worldwide locations", in: *J. Atmos. Sci.*,59: 590–608

Farman, J.C.; Gardiner, B.G.; Shanklin, J.D., 1985: "Large losses of total Ozone in Antarctica reveal seasonal ClO_x/NO_x interaction", in: *Nature*, 315: 207–210.

Galloway, J.N.; Cowling, E.B.; Seitzinger, S.; Socolow, R.H., 2002: "Reactive nitrogen: Too much of a good thing?", in: *Ambio*, 31: 60–63.

Guenther, A.; Hewitt, D. Erickson. C.N. et al., 1995: "A global model of natural volatile organic compound emissions", in: *J. Geophys. Res.*, 100: 8873–8892.

Gunn, J.; S. Sandoy, B. Keller et al. 2003. Biological recovery from acidification: Northern lakes recovery study. *Ambio* 32(3): 161–248.

Haywood, J.M.; Ramaswamy, V.; Soden, B., 1999: "Tropospheric aerosol climate forcing in clear-sky satellite observations over the oceans", in: *Science*, 283: 1299–1303.

Hedin, L.O.; Granat, L.; Likens G.E. et al., 1994: "Steep declines in atmospheric base cations in regions of Europe and North America", in: *Nature*, 367: 351–354.

Hinrichs, K.U.; Hmelo, L.R.; Sylva, S.P., 2003: "Molecular fossil record of elevated methane levels in Late Pleistocene coastal waters", in: *Science*, 299: 1214–1217.

Hofmann, D.J.; Harder, J.W.; Rosen, J.M.; Hereford, J.; Carpenter, J.R., 1989: "Ozone profile measurements at McMurdo Station, Antarctica, during the spring of 1987", in: *J. Geophys. Res.*, 94: 16:527–16:536.

Houghton, J.T.; Jenkins, G.J.; Ephraums, J.J. (eds.), 1990: *Climate Change: The IPCC Scientific Assessment* (Cambridge: Cambridge Univ. Press).

Houghton, J.T.; Meiro Filho, L.G.; Callander B.A. et al., 1996: *Climate Change 1995* (Cambridge: Cambridge Univ. Press).

Houghton, J.T.; Ding, Y.; Griggs, D.J. et al. (eds.), 2001: *Climate Change 2001: The Scientific Basis* (Cambridge: Cambridge Univ. Press).l

Houghton, J.T.; Ding Y.; Griggs, D.J. et al. (eds.) [Intergovernmental Panel on Climate Change], 2001: *Climate Change 2001. The Scientific Basis. Working Group I Contribution, Third Assessment Report of the IPCC* (Cambridge: Cambridge Univ. Press).

Jacobson, M.Z., 2001: "Global direct radiative forcing due to multicomponent anthropogenic and natural aerosols", in: *J. Geophys. Res.*, 106: 1551—11568.

Jones, C.D.; Cox, P.M.; Essery, R.L.H.; Roberts, D.L.; Woodage, M.J., 2003: "Strong carbon cycle feedbacks in a climate model with interactive CO_2 and sulphate aerosols", in: *Geophys. Res. Lett.*, 30,9: 1479.

Kanakidou, M.; Tsigaridis, K.; Dentener, F.J.; Crutzen, P.J., 2000: "Human activity-enhanced formation of organic aerosols by biogenic hydrocarbon oxidation", in: *J. Geophys. Res.*, 105: 9243–9254.

Kaufman, Y.J.; Tanre, D.; Boucher, O., 2002: "A satellite view of aerosols in the climate system", in: *Nature*, 419: 215–223.

Keeling, C.D.; Whorf, T.P., 2000: "Atmospheric CO_2 records from sites in the SIO sampling network", in: *Trends: A Compendium of Data on Global Change* (Oak Ridge, TN: Carbon Dioxide Information Analysis Center, Oak Ridge Natl. Lab., U.S. Dept, of Energy).

Keeling, R.F.; Piper, S.C.; Heimann, M., 1996: "Global and hemispheric CO_2 sinks deduced from changes in atmospheric O_2 concentration", in: *Nature*, 381: 218–221

Krol, M.C.; Lelieveld, J.; Oram, D.E. et al., 2003: "Continuing emissions of methyl chloroform from Europe", in: *Nature*, 421: 131–135.

Krüger, O.; Graßl, H., 2002: "The indirect aerosol effect over Europe", in: *Geophys. Res. Lett.*, 29,19: 1925.

Lackner, K.S., 2003: "A guide to CO_2 sequestration", in: *Science*, 300: 1677–1678.

Levitus, S.; Antonov, J.I.; Wang, J. et al., 2001: "Anthropogenic warming of Earth's climate system", in: *Science*, 292: 267–270.

Loutre, M.F.; Berger, A., 2000: "Future climatic changes: Are we entering an exceptionally long interglacial?", in: *Clim. Change*, 46: 61–90.

Ludwig, J.; Marafu, L.T.; Huber, B.; Andreae, M.O.; Hebs, G., 2003. "Domestic combustion of biomass fuels in developing countries: A major source of atmospheric pollutants", in: *J. Atmos. Chem.*, 44: 23–37.

Markowicz, K.M.; Flatau, P.J.; Ramana, M.V.; Crutzen, P.J.; Ramanathan, V., 2002: "Absorbing Mediterranean aerosols lead to a large reduction in the solar radiation at the surface", in: *Geophys. Res. Lett.*, 29, 20: 1967.

McNeill, J.R., 2000: *Something New Under the Sun* (New York: Norton).

Melillo, J.M.; Steudler, P.A.; Aber, J.D. et al., 2002: "Soil warming and carbon-cycle feedbacks to the climate system", in: *Science*, 298: 2173–2176.

Molina, M.; Rowland, F.S., 1974: "Stratospheric sink for chlorofluoromethanes: Chlorine atom catalyzed destruction of ozone", in: *Nature*, 249: 810–812.

Novakov, T.; Ramanathan, V.; Hansen, J.E. et al., 2003: "Large historical changes of fossil-fuel black carbon aerosol", in: *Geophys. Res. Lett.*, 30,6: 1324.

Penner, J.E.; Andreae, M.O.; Annegam, H. et al., 2001: "Aerosols, their direct and indirect effects", in: Houghton, J.T.; Ding, Y.; Griggs, D. J. et al. (eds.): *Climate Change 2001. The Scientific Basis. Working Group I Contribution, Third Assessment Report of the IPCC* (Cambridge: Cambridge Univ. Press): 289–348.

Prinn, R.G; Huang, J.; Weiss, R.et al., 2001: "Evidence for substantial variations of atmospheric hydroxyl radicals in the past two decades", in: *Science*, 292: 1882–1888.

Ramanathan, V., 1981: "The role of ocean-atmosphere interactions in the CO_2 climate problem", in: *J. Atmos. Set.*, 38: 918–930

Ramanathan, V.;Crutzen P.J.; Lelieveld, J. et al., 2001a: "Indian Ocean Experiment: An integrated analysis of the climate forcing and effects of the great Indo-Asian haze", in: *J. Geophys. Res.*, 106: 28: 371–398.

Ramanathan, V.; Crutzen, P.J.; Kiehl, J.T.; Rosenfeld, D., 2001b: "Aerosols, climate, and the hydrological cycle", in: *Science*, 294: 2119–2124.

Ramaswamy, V.; Boucher O.; Haigh, J. et al., 2001: "Radiative forcing of climate change", in: Houghton, J.T.; Ding, Y.; Griggs, D.J. et al. (eds.): *Climate Change 2001. The Scientific Basis. Working Group I Contribution, Third Assessment Report of the IPCC* (Cambridge: Cambridge Univ. Press): 349–416.

Richter, A.; Burrows, J.P., 2002: "Retrieval of tropospheric NO_2 from GOME measurements", in: *Adv. Space Res.*, 29: 1673–1683.

Roderick, M.L.; Farquhar, G.D., 2002: "The cause of decreased pan evaporation over the past 50 years", in: *Science*, 298: 1410–1411.

Satheesh, S.K.; Ramanathan, V., 2000: "Large differences in tropical aerosol forcing at the top of the atmosphere and Earth's surface", in: *Nature*, 405: 60–63.

Schellnhuber, H.J., 1999: "'Earth system' analysis and the second Copernican revolution", in: *Nature*, 402: C19–C23.

Smith, K.R., 2002: "In praise of petroleum?", in: *Science*, 298: 1847.

Stone, R., 2002: "Counting the cost of London's killer smog", in: *Science*, 298: 2106–2107.

Tegen, I.D.; Fung, I., 1995: "Contribution to the atmospheric mineral aerosol load from land-surface modification", in: *J. Geophys Res.*, 100,D9: 18,707–726.

Tegen, I.D.; Koch, A.; Lacis, A.; Sato, M., 2000: "Trends in tropospheric aerosol loads and corresponding impact on direct radiative forcing between 1950 and 1990: A model study", in: *J. Geophys. Res.*, 105,D22, 26,971–989.

Turner, B.L., II; Clark, W.C.; Kates, R.W. et al. (eds.), 1990: *The Earth as Transformed by Human Action* (Cambridge: Cambridge Univ. Press).

Vernadsky, V.I., 1998/1926: *The Biosphere* (translated and annotated version from the original of 1926) (New York: Springer).

Vitousek, P.M.; Mooney, H.A.; Lubchenco, J.; Melillo, J.M., 1997: "Human domination of Earth's ecosystems", in: *Science,* 277: 494–499.

Wang, X.; Key, J.R., 2003: "Recent trends in Arctic surface, cloud, and radiation properties from space", in: *Science*, 299:1725–1728.

Chapter 8
Earth System Dynamics
in the Anthropocene (2004)

Will Steffen, Meinrat O. Andreae, Peter M. Cox, Paul J. Crutzen,
Ulrich Cubasch, Hermann Held, Nebosja Nakicenovic,
Liana Talaue-McManus, and Billie Lee Turner II

8.1   The Earth System in the Anthropocene

Human-driven changes to many features of the Earth system have become so ubiquitous and significant in magnitude that a new era for the planet—the 'Anthropocene'—
has been proposed (Crutzen and Stoermer 2001; Clark et al. 2001). Many of these
changes are large in magnitude at the planetary-scale, sometimes even exceeding
natural flows in major aspects of biogeochemical cycling. In addition, anthropogenic
changes invariably occur at rates that are much larger than those of natural variability,
often by an order of magnitude or more. The magnitudes and rates of these changes,
coupled with the fact that changes to a large number of Earth system processes and
compartments are occurring simultaneously, has led to the recognition that the Earth
is now operating in a 'no-analogue state' (Steffen et al. 2004).

The features of this no-analogue state present challenges to human responsiveness,
challenges that have not been experienced in coping with any previous environmental
changes, which have occurred at local and regional scales. These features include
the facts that:

- Large parts of the problem are global in scale, transcending any region, continent,
  or ocean basin on its own.
- Connectivity between different biophysical processes and geographical areas of
  the planet is much greater than previously thought, and the connectivity of human
  activities is increasing at a rapid rate.
- Human-driven changes to Earth system functioning operate on very long
  timescales, where the consequences of some human actions may be present

This text was first published as: Steffen, W.; Andreae, M.O.; Cox, P.M.; Crutzen, P.J.; Cubasch, U.;
Held, H.; Nakicenovic, N.; Talaue-McManus, L.; Turner II, B.L., 2004: *Earth System Dynamics in
the Anthropocene*, in: Schellnhuber, H.J.; Crutzen, P.J.; Clark, W.C.; Claussen, M.; Held, H. (Eds.):
Earth System Analysis for Sustainability (Cambridge, MA: MIT Press): 313–40. The permission to
republish this text was granted by MIT Press. We thank Bob Scholes for his written contributions
to this report.

for decades and centuries, and across very large space scales, where causes are spatially de-linked from consequences.

- The impacts of global change on human—environment systems can no longer be understood by simple cause-effect relationships, but rather in terms of cascading effects that result from multiple, interacting stresses.
- New forcing functions arising from human actions (e.g., synthetic chemicals) and rapid rates of change imply that the natural resilience of ecological systems may not be sufficient to cope with the change.

These features of the Anthropocene are already leading to discernible changes in the functioning of the Earth system and may well lead to accelerating change in many ways throughout this century (IPCC 2001; Steffen et al. 2004). Significant improvements in the ability to observe, understand, and simulate past, contemporary, and potential future change are essential to provide the knowledge base required to achieve global sustainability.

The aim of this report is to examine a few critical areas where human activities are having, or have the potential to have, significant impacts on the functioning of the Earth system. We begin by focusing more closely on the climate component of the Earth system before moving to broader considerations of geography and to a discussion of the potential for abrupt changes in several components of the Earth system. A brief discussion of the challenges for modelling and observation follows. We conclude by examining whether or not technical substitution or fix can address the issues raised earlier in the report.

8.2   Climate Sensitivity

How does the increasing understanding of the role of aerosols change our understanding of the climate sensitivity to greenhouse gases?

Climate has been changing during the last century. The IPCC (2001) has established that the global mean temperature has increased by 0.6 ± 0.2 °C. About half of this increase has occurred during the last 40 years or so. It is also clear that this change is unique as compared with the rest of the last millennium (Mann et al. 1999), although there are some uncertainties about the variations of the mean temperature during this period as deduced from the many different indicators that have been used.

The IPCC has also concluded that the change during the twentieth century cannot be explained without including the rote of increasing concentrations of greenhouse gases in the atmosphere during this period of time. Although few in the scientific community now contest this conclusion, there are still occasional claims that the recent warming is the result of internal natural variations of the climate system. To accept this as a plausible possibility, the following question must be answered: Why did a major change of the internal variability occur toward the end of the twentieth century? Further, such changes must also be associated with changes of the internal fluxes of energy through the atmosphere. A plausible analysis to support the idea that random internal variations would be the prime reason for the increase of the global mean temperature during the twentieth century has yet to be presented. The conclusion that human activities have led to a significant change of climate, characterised to first order by a global warming, is now beyond reasonable doubt.

The observed increase in the global mean surface temperature is the net result of opposing forcings and effects. Greenhouse gases exert a warming forcing that is modified by feedbacks, most prominently as a result of changes in atmospheric water vapour and clouds. Aerosols have direct, radiative cooling and warming effects as well as a series of indirect effects due to aerosol-induced changes in cloud properties, abundance, and dynamics. Large uncertainty is associated both with the magnitude of the enhancement of greenhouse gas forcing (warming), resulting from cloud feedbacks, and the (net cooling) aerosol effects, including the various aerosol-cloud effects. Consequently, the observed temperature increase could be explained by a large greenhouse gas effect (implying a large greenhouse gas-cloud feedback), which is opposed by a large aerosol-cloud effect, or alternatively by small greenhouse gas-cloud and aerosol-cloud effects. At present, these alternatives yield nearly the same solution for the interpretation of the observed climate history. The magnitude of the greenhouse gas-temperature-cloud feedback is a long-standing uncertainty. The magnitude of the cooling owing to aerosol-related effects is also difficult to determine directly because of the inhomogeneous distribution of aerosols, as well as clouds, and limited knowledge about their optical characteristics.

There is, however, a very important difference in the way these alternatives affect future climate change. 'Climate sensitivity' is defined as the amount of climate change per amount of radiative forcing (greenhouse gas, aerosol) added to the atmosphere (usually expressed as degrees of global temperature rise per doubling of CO_2 in an equilibrium climate model run). At present, different climate models predict very different climate sensitivities, mostly because they contain different ways of representing greenhouse gas-cloud feedbacks, and there appears to be no *a priori* way of deciding which of these models gives the 'better,' more accurate answer regarding climate sensitivity. The net effect of the present-day greenhouse gas and aerosol forcings (including delayed effects due to latency in the system) is simply the observed present-day climate change. Therefore, the potentially large aerosol effects (as summarised by Anderson et al. 2003) would imply that climate sensitivities are more consistent with the high end of the range presented in IPCC (2001).

A quantitative analysis, based on Crutzen and Ramanathan (this volume), of the current situation illustrates the qualitative point made above. The enhanced concentrations of greenhouse gases in the atmosphere have reduced the outgoing long-wave radiation by 2.7 ± 0.5 W m^{-2} (IPCC 2001). This energy flux must be balanced by (a) increases in outgoing radiation due to warming at the Earth's surface, (b) flux of heat into the Earth's surface, primarily the surface ocean, and (c) the sum of aerosol radiative effects and albedo feedbacks among aerosols, greenhouse gases, and clouds. We can deduce the approximate outgoing flux of energy from the observed warming at the surface of the Earth (Ramanathan 1981) to be 1.5 ± 0.5 W m^{-2}. The smaller value of the range $(1.0$ W m$^{-2})$ includes the role of feedback mechanisms, primarily due to water vapour, which reduce the long-wave radiation to space (water vapour feedback), whereas the large value (2.0 W m^2) is obtained for an atmosphere without a water vapour feedback. The increase of the temperature in the atmosphere is also driving a flux of heat into the oceans of about 0.32 ± 0.15 W m^{-2} (Levitus et al. 2000; Barnett et al. 2001). To achieve a balance of the energy budget, aerosol-greenhouse gas-cloud interaction processes must induce a net flux of radiation back to space, ranging from about 0 W m^2 in the case of no water vapour feedbacks, which is known to be unrealistic (IPCC 2001), to 1.9 ± 0.5 W m^{-2}, corresponding to 60% of the greenhouse gas forcing, if the feedback mechanisms due to water vapour, etc. are considered. In effect, the cooling that results from aerosol-related processes counteracts a significant increase of surface temperature that would have otherwise occurred in their absence. It is important to note that this quantitative analysis is valid for the situation around the year 2000 and does not scale into the future as both greenhouse gases and aerosol loadings in the atmosphere will change with time.

This result has a major consequence for estimates of future climate change. Atmospheric aerosol loadings will not likely increase strongly; they may even decline in the coining decades. Thus, their cooling effect will level off or decrease. On the other hand, greenhouse gases will continue to accumulate in the atmosphere. In the case of high climate sensitivity, this must lead to a considerably sharper increase in global temperatures than has been experienced so far and to temperature increases closer to the upper end of the range given by the IPCC Third Assessment Report (2001), even without extreme emission scenarios.

It is evident that great effort should be invested in resolving these issues. To make progress in this direction, we need (a) to improve the representation of cloud effects in *general circulation models* (GCMs), (b) to develop parameterisations of aerosol effects on clouds and find ways to incorporate them into GCMs. and (c) to use the analysis of parameters besides temperature (precipitation, heat fluxes, etc.) and of the spatiotemporal distribution of climate change to diagnose the relative contributions of greenhouse gas and cloud effects. In addition, a much more systematic, consistent, and continuous climate observing system is required to test models and to improve understanding of the climate system in general.

8.3   Earth System Geography in the Anthropocene

What are the important regions in the Earth System and in what ways? Do midlatitudes really matter compared to the tropics and high latitudes? Can we differentiate geographically between drivers and impacts of global changes?

The Earth's surface is highly heterogeneous, and the distribution of humans and our activities are highly skewed, features that have important implications for the functioning of the Earth system in the Anthropocene. The implications of heterogeneity vary, however, according to the question being asked. For example, very different subglobal patterns of important areas, or *hot spots* emerge from analyses of the physical climate system and of the socioeconomic sphere of the Earth system, respectively. For any aspect, however, understanding of the Earth system is only as good as its least understood region, implying that the scientific effort must be much better distributed around the globe than in the past.

In terms of *socioeconomic aspects,* an important feature is that the areas that are currently important as drivers of change (the midlatitudes) are not necessarily those that will largely bear the brunt of global change impacts (e.g., Shah 2002) and that maintain critical processes for Earth system functioning (the tropics, e.g., the role of tropical forests in heat and water vapour exchange). Yet understanding, either in a biophysical or a socioeconomic sense, is far less for the tropics than for the midlatitudes, and the disparity in research effort appears to be increasing. The implications of this for the quest for global sustainability are significant, as projections of demographic change suggest that by 2030 about 90% of the population will live in the tropics (UN 2001). The increasing connectivity of the global economy (e.g., through production-consumption chains of key commodities like food) are linking the tropics more tightly to the midlatitudes so that impacts in the tropics will reverberate further in the Earth system.

The impacts of human-driven change in the tropics for the functioning of the Earth system are equally less understood in comparison to the midlatitudes. Much contemporary land-cover change is occurring at local and regional scales in the tropics, usually involving conversion of forest to agriculture and pasture as well as secondary regrowth that is subsequently recut. Nearly all of the rapid change to the structure of the coastal zone ecosystems is also occurring in the tropics. Understanding of the implications of these changes for biogeochemical cycling, biodiversity, and the physical climate system considerably lags behind that for similar changes in the midlatitudes. One exception is the Large-scale Biosphere-Atmosphere Experiment in Amazonia (Nobre et al. 2002), where a decade-long, multinational study involving hundreds of researchers is rapidly building a better understanding of the dynamics and consequences of land-cover change in the Amazon Basin, from the local up to the global scale. Many other examples could be given, all pointing to the need for a significantly enhanced research effort in the tropics in all aspects of global change.

For the *physical climate system,* a few well-known hot spots have received considerable attention by the research community. Examples include El Niño research in the tropical Pacific Ocean and a rapidly increasing number of investigations of the potentially critical branch of the *thermohaline circulation* (THC) in the North Atlantic Ocean. The Southern Ocean, however, has been much less studied. Issues such as deepwater formation and its role in THC and the relative importance of the Southern Ocean in the marine carbon cycle demand increased attention.

A strong case can also be made for an enhanced effort in the high latitudes, which are now experiencing the most rapid rates of climate change and which also play an important role in the Earth system through the albedo-ice feedback, the taiga-tundra feedback, and the potentially large releases of carbon compounds from the terrestrial biosphere with increased warming. All of these potential feedbacks (e.g., possible disappearance of Arctic sea ice in summer, movement of boreal forests northward) are positive, that is, they enhance the warming that triggered them and could occur within a 50- to 100-year time frame. This suggests that a concerted effort is required to improve the knowledge base of high-latitude systems under rapid warming; the need is particularly acute for northern Eurasia, which is clearly a very important region from the perspective of Earth system dynamics and which currently suffers from decaying scientific infrastructure and a lack of adequate support for the large scientific community that has worked there through much of the previous century.

On a longer time frame (a few millennia), an intriguing question concerns the sensitivity of the area north of 65°, particularly in North America, as it is known to be the site of glacial inception and thus might be the starting point of the next ice age. Although solar output is currently increasing slightly, and thus is a minor contributor to the observed warming, solar insolation will reach a minimum during the next 200 to 500 years as a result of orbital variation. Model experiments suggest that in a 280 ppm CO_2 world, such insolation conditions could lead to a formation of an ice sheet, although this would almost surely require significant additional cooling due to, for example, injection of massive amounts of aerosols into the atmosphere from volcanoes or anthropogenic activities. However, given that the minimum in solar insolation is rather shallow and will not be much different from current insolation (e.g., compared to greenhouse gas forcing), it appears that the projected increased level of CO_2 over the next 200 to 300 years will more than compensate for the insolation minimum. In addition, another study (Loutre/Berger 2000) suggests that changes in orbital forcing alone, without any anthropogenic increase in greenhouse gas concentration in the atmosphere, will lead to a continuation of the present inter-glacial period for another 30,000–50,000 years. Suggestions of glacial inception notwithstanding, it must be clearly stated that the most important issue by far in the high latitudes in terms of immediacy and rate and magnitude of change, is the current strong warming and the potential for positive feedbacks to the climate system.

8.4 "Achilles' Heels" in the Earth System

What are the Achilles' Heels in the Earth System? Can abrupt changes in the operations of the Earth System be anticipated and predicted? Can those that are most susceptible to triggering by human actions be identified?

Earth's environment shows significant variability on virtually all time and space scales, and thus global change will never be linear or steady under any scenario. Of particular interest and importance are abrupt changes that can affect large regions of Earth. For example, the paleo-record gives unequivocal evidence of such abrupt climate change in the recent past, such as the *Dansgaard/Oeschger* (D/O) events that happened in the period 70,000–15,000 years ago (Grootes et al. 1993). Although presumably not having been of global scale, the significance of abrupt changes such as D/O events is that (a) they can involve a scale of change, up to 10 °C in a decade or so (see Rahmstorf/Sirocko 2004), which could devastate modern economies should such changes occur in these regions, (b) they have occurred during the time of human occupation of the planet, and (c) they have occurred in regions (western Europe and North America) now heavily populated. Abrupt changes cannot be dismissed as either implausible or irrelevant in terms of spatial or temporal scales.

Furthermore, one abrupt change of a different kind has already occurred. The formation of the ozone hole over Antarctica was the unexpected result of the release of human-made chemicals thought to be environmentally harmless (see Crutzen/Ramanathan 2004; Schneider et al. 1998). The event was one of chemical instability in the atmosphere rather than an abrupt change in the physical climate system. In addition, it occurred in a far distant part of the planet, well away from origin of the cause. In several ways, humankind was lucky in that the ozone hole could have been global and present through all seasons (Crutzen 1995).

Such evidence of instabilities in the chemical system in the stratosphere and in the THC in the North Atlantic (thought to underlie the D/O and Heinrich events seen in the Greenland ice core records; Ganopolski/Rahmstorf 2001; Clark et al. 2001) gives a warning that human activities could trigger similar or even as-yet unimagined instabilities in the Earth system, in its physical, chemical, or biological components or in coupled human-environment systems.

8.4.1   Abrupt Changes in the Physical Earth System

Initially it may appear an impossible task to anticipate abrupt changes in the Earth system (NRC 2002; cf. Schneider et al. 1998). Abrupt changes, by definition, occupy small regions of a potentially high-dimensional climate phase space, such that it is impractical to search for such changes with a comprehensive model (such as a GCM). However, the special nature of abrupt changes actually makes them amenable to analytical techniques. By 'abrupt' we mean changes that occur much more quickly than changes in anthropogenic forcing. In a typical setting, for this to occur requires the existence of multiple equilibria (or 'fixed points'), of which the 'current' Earth system equilibrium state becomes linearly unstable or even vanishes (such a point in the phase space/forcing diagram is called 'bifurcation'). Under these conditions an arbitrarily small perturbation to the formerly stable equilibrium can result in a transition to a different equilibrium state even in the absence of changes in forcing.

The classical example is the ocean's THC, which in its current state transports heat from equator to pole, helping to keep western Europe unusually warm for its latitude. Thermohaline circulation takes warm surface waters to the North Atlantic, where they cool (releasing their heat to the atmosphere), become denser, and sink to depth. Simple models of the THC exhibit both 'on' and 'off' states with the potential for rapid switching between these states based on the freshwater input to the North Atlantic. The current 'on' state can be destabilised by additional freshwater inputs to the North Atlantic, which freshen the surface waters, make them less dense, and inhibit sinking (Rahmstorf 2000). It is hypothesised that such a perturbation arose from the melting of the North American ice sheet, leading to a shutdown of the THC and a cooling of the Northern Hemisphere during the Younger Dryas event

12,000 years ago. Some comprehensive GCMs also suggest that the THC could be similarly shut down by increases in rainfall at high latitudes under greenhouse warming (thereby increasing the freshwater flow in Russian rivers to the Arctic Sea); however, this sensitivity is by no means common to all models.

Thermohaline circulation offers an excellent example of where a possible abrupt change in the Earth system has been anticipated using a combination of models and data. Although the precise timing of a THC shutdown cannot be predicted with any certainty, the topology of the THC phase space is sufficiently well known to inform attempts to monitor for signs of an impending switch to the off state, furthermore, the transition from one equilibrium state to another (triggered by a bifurcation) is typically preceded by enhanced variability in the THC (Kleinen et at. 2003), offering an additional warning of possible change.

Other aspects of Earth system dynamics are also believed to exhibit multiple equilibrium states and may therefore display abrupt transitions between these equilibria. These include evidence for a transition from a green to an arid Sahara in the mid-Holocene 5,500 years ago (Claussen et al. 1999; de Menocal et al. 2000) and model-derived results which suggest that Greenland can support both ice-covered and ice-free conditions under current CO_2 conditions. These subsystems display 'hysteresis' or path-dependence in their response to control variables. Thus, for example, under sustained increases in CCS level (equivalent to a 3 °C warming over millennia), the Greenland ice sheet is predicted to melt in an irreversible manner (IPCC 2001), such that much lower CO_2 values would be required before it would return.

The generic properties of multiple equilibria, linear instabilities, and bifurcations offer the possibility of cataloguing possible abrupt changes in the Earth system in a much more thorough way than has been achieved to date. In principle, Earth system equilibria can be defined by setting time derivatives to zero within current Earth system models. Linear stability theory requires that only linear terms are kept within the full nonlinear equations, significantly simplifying the analysis. Therefore, the initial cataloguing of possible Earth system instabilities can be based on well-founded analytical and semi-analytical mathematical techniques, potentially providing a map of hot spots in the Earth system where abrupt change is possible. Once an equilibrium has been found in a model, path-continuation numerics (Feudel/Jansen 1992) make it possible to derive automatically a bifurcation diagram. This technique is becoming increasingly feasible even for comprehensive models (Dijkstra 2000).

Another approach is to use the phenomenon of stochastically induced jumps between multiple equilibria. According to Kramer's rule (Gardiner 1994), an increase in noise in a complex system can trigger an abrupt shift from one state to another. The related timescale is determined by the potential well between the equilibrium states and the amplitude of the noise. The interplay between multiple equilibria and noise can amplify an existing periodic forcing ('stochastic resonance'; Gammaitoni et al. 1998) or may trigger an excitable cycle ('coherence resonance'; Pikovsky/Kurths 1997). Stochastic resonance occurs where the period of the forcing matches the time for transitions between alternative equilibrium states of the system. In analogy, coherence resonance occurs where the time for excitation by noise fulfils a certain matching condition with the period of the excited cycle. Stochastic resonance has been suggested as a contributing factor in D/O events (Ganopolski/Rahmstorf 2002).

Instabilities in the Earth system could be explored by subjecting Earth system models to a noise and systematically tuning this noise until a resonance is achieved (defined by a significant amplification of the variability in internal model variables at a characteristic frequency). The resonance would be indicative of multiple equilibrium states, which might yield abrupt changes under anthropogenic forcing, but it would also give insights into the magnitude of the abrupt change and the amount of noise needed to trigger such a state change. Related ideas have already been successfully applied to complex systems (Majda el al. 1999; Fischer et al. 2002). In the latter case, the metastable states of a molecular dynamical system were extracted from time series of the stochastically perturbed system. A similar approach has the potential to yield invaluable insights into abrupt transitions in the Earth system.

8.4.2   Complexity in the Chemistry of the Atmosphere

The stability of chemical systems in the atmosphere is of concern following the discovery of the ozone hole. Tropospheric chemistry is as complex as that in the stratosphere and is of high importance for the health and well-being of humans, as well as for the functioning of the Earth system. The troposphere is an oxidizing medium, removing compounds emitted naturally by the terrestrial and marine biospheres and pollutants emitted by human activities. It also affects climate in many ways, for example, through the destruction of the potent greenhouse gas methane, CH_4. Without this cleansing ability, a large range of natural and human-made compounds would accumulate in the atmosphere to very high concentrations. The most important of the oxidizing species in the atmosphere is the highly reactive hydroxyl radical, OH.

Because of its short lifetime, the concentration of OH shows large variations in space and time. Models indicate that the regions with the highest abundance of OH are located over the tropics. Therefore, most of the self-cleansing reactions of the troposphere occur in the tropical zone, and this region consequently plays a key role in the regulation of atmospheric composition. In spite of the well-established importance of the OH radical in atmospheric chemistry, measurements of this species are still very sparse. In particular, there are no measurements at all of OH over the tropical continents, where anthropogenic perturbations of the atmospheric oxidant cycle are likely to occur and where they may have the most pronounced effect. Such measurements are urgently needed as a test of our basic understanding of atmospheric photochemistry.

8.4.3 Ecological Complexity and Earth System Functioning

Major anthropogenic activities have manifested their impacts on the global biosphere. Overfishing and eutrophication due to human activities stand out as among the most serious issues threatening the marine biosphere worldwide. Myers and Worm (2003) recently reported that about 90% of the large predatory fish biomass has been removed from the world's oceans, with removal rates being highest with the onset of post-World War II industrial fisheries. Ecosystem impacts include intermediate results of compensation by no target fish populations. However, because of accelerated expansion of fishing in the 1980s, fishing pressure exceeded these compensatory mechanisms and has now led to unequivocal evidence of decline in most pelagic and ground fisheries of continental shelves. There is less evidence for oceanic fishing grounds. Given the importance of top-down controls on the dynamics of marine ecosystems, there is the possibility that such overfishing could lead to significant, abrupt changes to marine ecosystems (often called 'regime shifts'), which reverberate through to lower trophic levels such as zooplankton, phytoplankton, and bacteria.

Other anthropogenic pressures on the coastal zone have led to abrupt changes (from an Earth system perspective) in the functioning of marine ecosystems. For example, because of its ubiquity, human-dominated waste loading is altering coastal ecosystems on a global scale. This has led to a state of eutrophication, the latter being a biogeochemical response to heavy nutrient loading. Primary producers synthesise organic matter in addition to what is delivered as waste from populations and manufacturing systems. The excess organic matter undergoes oxygen-consuming degradation. From the 1970s to the 1990s, anthropogenic loads of dissolved inorganic nitrogen increased about sixfold to 13.3 Tg (1 Tg $= 10^{12}$g). Over the same period, dissolved inorganic nitrogen increased fourfold to 1.6 Tg.

There are secondary consequences of eutrophication. Hypoxic zones under certain conditions can release nitrous oxide to the atmosphere during the process of denitrification. This has been documented for the western shelf of India, which obtains dissolved inorganic nitrogen inputs both from seasonal upwelling and from horizontal delivery from land. In the Gulf of Mexico, hypoxia is a major summer feature, but denitrification has not been detected. Competing microbial pathways such as dissimilatory nitrate reduction to ammonium may keep the reactive substrate in the water column.

The ecosystem effects of eutrophication are just beginning to be studied. Jackson et al. (2001) argue that historical overfishing, including the removal of suspension feeders because of trawling and other top predators, has resulted in the simplification of trophic and other functional relationships and the microbialisation of coastal systems. Phase shifts include the shift from long-lived macrophytes to short-lived epiphytes and the increasing frequency of phytoplankton blooms and cyanobacteria. In sediments, shifts toward hetero-trophic microbial processes are evident.

It remains to be seen how overfishing and eutrophication will alter biogeochemical cycles and the resulting global inventories of carbon, nitrogen, phosphorus, and silica. There is, seemingly, consensus that the nearshore estuarine systems most proximal to human populations are carbon sources, being net heterotrophic and microbe dominated. In open shelf and oceanic domains, the systems remain as carbon sinks, being net autotrophic. Despite the apparent capacity of oceanic ecosystems to assimilate the impacts of waste loading and overfishing, governments should consider the imminent collapse of coastal ecosystems as symptoms that demand immediate mitigation.

In contrast to their marine equivalents, terrestrial ecosystems generally lost many of their top predators and underwent trophic pathway simplification several centuries ago. There has not been widespread ecosystem failure as a result. Terrestrial ecologists generally favour a more 'bottom-up' view of ecosystem regulation.

There is evidence (Tilman 1999; Loreau et al. 2001) of a relationship between terrestrial biodiversity and aspects of ecosystem functioning, particularly when the biodiversity is expressed in 'functional type' terms. However, it appears that quite modest levels of biodiversity are sufficient to maintain processes such as primary production and nutrient cycling at close to maximum levels, and there is no obvious threshold below which loss of ecosystem function or services suddenly occurs.

If such an effect does occur, it is most likely within the radically simplified agricultural systems. Widespread failure of these systems would have dire consequences for human welfare, but not for life on Earth. Agricultural systems not only replace more diverse natural and seminatural systems with a small group of domesticates, they also simplify the landscape when conducted at large scale, and within the agricultural species, the genetic base is progressively narrower.

The argument, largely unsupported by data, is that agricultural systems of low spatial and genetic diversity are more vulnerable to pest outbreaks and environmental change.

8.4.4   Pandemics

Critical breakpoints for the Earth system may also lie in the still very inadequately explored interactions of climate and environmental change, socioeconomic development, and human and animal health. The preeminent feature of the Anthropocene is that human activities have become a geophysical and biogeochemical force that rivals the 'natural,' nonhuman processes. This implies that major discontinuities in the socioeconomic domain may lead to corresponding disruptions in the biogeochemical/physical domain. An example of such a discontinuity may be the spread of a new disease vector resulting in a pandemic. High population densities in close contact with animal reservoirs of infectious disease make the rapid exchange of genetic material possible, and the resulting infectious agents can spread quickly through a worldwide contiguous, highly mobile human population with few barriers to transmission. Warmer and wetter conditions as a result of climate change may also facilitate the spread of diseases. Malnutrition, poverty, and inadequate public health systems in many developing countries provide large immune-compromised populations with few immunological and institutional defences against the infectious disease. An event similar to the 1918 Spanish Flu pandemic, which is thought to have cost 20–40 million lives worldwide at the time, may result in over 100 million deaths worldwide within a single year. Such a catastrophic event, which is not considered to be unlikely by the epidemiological community, might lead to rapid economic collapse in a world economy dependent on fast global exchange of goods and services. In a worst case this might lead to a drastic, and probably long-lasting, change in the way humans affect the Earth system.

8.4.5 Current Knowledge Base on Abrupt Changes

The preceding discussion of the '"Achilles" heels' of the Earth system can be summarised as follows:

- It is well established that abrupt changes in major features of Earth system functioning can occur and indeed have occurred. Prominent examples include the D/O events and the formation of the Antarctic ozone hole, which have been regional in scale but may trigger impacts at the global scale.
- It is further known where some of these abrupt changes can occur. In addition to the two examples given above, the switching of northern African vegetation between savannah and desert, the existence or not of Greenland ice cover, and the large regions of permafrost in northern Eurasia are further areas of instability where a part of the Earth system can change relatively rapidly from one well-defined state to another.
- Not all of the potential abrupt changes in all components of the Earth system (climate, chemical, biological, human and their coupling) are known, nor are they likely to be. However, promising techniques exist to identify more of them.
- Beyond knowing that a potential abrupt change might occur, it is more difficult to determine what triggers abrupt changes or how close a system may be to a threshold.
- Both the magnitude and rate of human forcing are important in determining whether an abrupt change is triggered in a system or not. In general, the probability of abrupt changes in complex systems increases with the magnitude and rate of forcing.
- The Earth system as a whole in the late Quaternary appears to exist in two stales (glacial and interglacial) with well-defined boundary conditions in atmospheric composition (CO_2, CH_4) and climate (inferred temperature) (Petit et al. 1999). The nature of the controls on the boundary conditions are not known (cf. Watson et al. 2004) nor are the consequences of the present large, ongoing, human-driven excursion beyond these boundaries (e.g., Keeling and Whorf 2000). Model-based exploration of Earth system phase space cannot yet find a third equilibrium state at a warmer, higher CO_2 level than the interglacial (Falkowski et al. 2000).

8.5   Systems of Models and Observations

What sort of models and data do we need to understand and anticipate Earth System change in the Anthropocene?

8.5.1   The Current State-of-the-Art in Climate Projection

Many critical Earth system characteristics are undergoing rapid change in the Anthropocene, but climate change is the most obvious example of where international research has been organised to address a policy-relevant question. The production of climate change projections for the twenty-first century, as embodied in the assessments of the IPCC, is multidisciplinary (see Fig. 8.1). The drivers of climate change (anthropogenic emissions of greenhouse gases and aerosols and land-use change) are derived using socioeconomic models, based on a range of 'storylines' regarding population growth, economic development, and technological change. High emissions scenarios assume major technological developments to permit extensive use of nonconventional oil and gas resources. We do not know how plausible such developments might be. The emissions are then used to drive atmospheric chemistry models, which produce corresponding scenarios of changes in the concentrations of greenhouse gases and aerosols for use within climate models. The resulting climate projections are used by impact modellers, who estimate the extent to which the projections

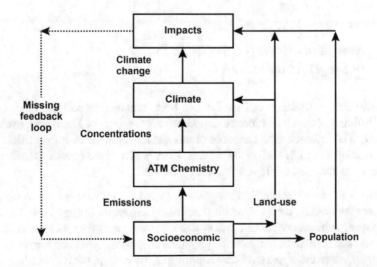

Fig. 8.1 Schematic showing the methodology used by the IPCC to produce model projections. Continuous arrows show the inputs and outputs to each model in the chain. Dotted line shows the missing impacts-to-socioeconomic responses loop

will affect humankind (e.g., through climate-driven changes in water and crops, as well as changing demands driven by population growth).

Each stage of this process involves models of some complexity and with widely differing structures. The socioeconomic models operate at large regional scales (e.g., North America, Europe), are not gridded, and are often based on optimisation assumptions under equilibrium conditions. By contrast, atmospheric chemistry models and GCMs use a grid (e.g., with boxes of equal size in latitude and longitude) to represent the Earth system and are based on deterministic differential equations. The computational cost of running these models is very high, which limits the resolution they can employ (i.e., the minimum size of the grid boxes) to about 250 km at present. On the other hand, impacts are generally felt at finer scales (e.g., at the scale of a river catchment for hydrology), so it is normally necessary to 'downscale' the outputs of the GCM before they can be used to drive impact models, either using statistical techniques or high-resolution regional climate models, which currently operate with grid boxes of about 50 km. The validity of these downscaling models has not yet been well tested.

At each stage of the IPCC modelling process there is a change in the way the Earth system is represented, which leads to difficulties at the interfaces, requiring downscaling, upscaling, or arbitrary definition of the outputs of one model in terms of the inputs to another. In addition, the modelling methodology is one-way in the sense that information flows bottom-to-top in Fig. 8.1 but not the reverse. This means that the subcomponents of the modelling system do not generally feedback on one another in the way in which the real Earth system operates, which of course is a principal deficiency.

8.5.2 New Tools Required to Guide Policy in the Anthropocene

The methodology outlined in Fig. 8.1 has been remarkably successful in coordinating different research disciplines to address a key aspect of Earth system change. However, this approach is not capable of answering some of the most critical questions posed by scientists and policy makers. Here we list these questions and suggest the new tools and methodologies that these demand.

1. *How will the coupled Earth system respond to anthropogenic forcing?* As noted previously, the existing climate modelling methodology lacks feedbacks between subsystems of the Earth system. Some recent attempts have been made to include feedbacks between the physical, biological, and chemical parts of the Earth system through a two-way coupling of the various subsystems (e.g., Jones et al. 2003; Johnson et al. 2001). Integrated assessment models also represent the feedbacks between the socioeconomic and natural parts of the Earth system, but they do this at the expense of drastic simplifications in the sub models (e.g., climate may be represented solely by global mean temperature). An intermediate

complexity approach is required in which the subcomponents are 'traceable' to more comprehensive models but which are sufficiently economical to enable exploration of additional feedbacks.

2.   *What are the impacts of climate change at the scale of communities?* This question is difficult to answer because climate models are currently too coarse-grained for regional impacts assessments. Furthermore, impacts are generally determined by climatic extremes (e.g., droughts or floods), and these are not well represented at low resolutions. Higher-resolution climate projections are therefore required, either through embedding regional climate models in GCMs or through basic enhancements in GCM resolution as computer power increases. The latter approach is typified by the Japanese Earth Simulator Centre, which has plans to run global climate projections with a resolution of 10 km, compared to 250 km in current GCMs. Note, however, that such significant increases in resolution may compromise the ability to include the full Earth system teed-backs, to explore abrupt change, and to assess the uncertainty in the projections (see the first and third questions in this section).

3.   *What are the uncertainties in the projections?* A key deficiency in climate modeming has been an inability to define 'error bars' for projections. Some qualitative measure of uncertainty is given by the spread in results from different GCMs: however, policy makers actually require a more meaningful estimate of the 'probability distribution function' (PDF) for future climate. The probability of certain critical thresholds being crossed (e.g. > 2K warming by 2050) is required for risk analysis. Attempts are now underway to define the climate PDF using 'physics ensembles', which are made up of structurally identical models each of which has different plausible sets of internal parameters. Climate projections are then weighted by their ability to reproduce key features of the current climate (with 'better models' receiving more weight). This approach is promising but requires many climate model runs (~hundreds) rather than a 'one shot' model. There is, therefore, a tension between greater model reso-lution and sufficient model speed to enable such estimation of uncertainty. A fundamental difficulty remains, however, in that the socioeconomic future of the global society cannot be predicted, since this system is indeed chaotic and in principle unpredictable, except for some overarching features and within some limited period of time. Furthermore, the results of projections of future devel-opments cannot be tested against real data, since in reality there will be only one experiment and we are in the midst of it. In addition, there are still considerable differences in the regional changes of climate as simulated by different models.

4.   *Can we reduce uncertainty?* The inclusion of additional feedbacks in the Earth system is likely to increase rather than reduce the spread among model projec-tions, since the additional components provide new ways for the models to differ. However, more complete Earth system models will provide a more realistic (and larger) estimate of the uncertainty in the behaviour of the real Earth system. The uncertainty is valuable information in its own right (e.g., for assessing the prob-ability of some abrupt change occurring); however, the fact that it appears to be growing is in danger of being misinterpreted by our paymasters (who may

wish to wait for less ambiguous results and conclude that since more money into model development increases uncertainty, less money might have the desired effect!).

Model development alone is unlikely to reduce uncertainty in the foreseeable future, and some uncertainties can never be eliminated since the climate system is chaotic. Still, additional data on changes in the Earth system can constrain models and thereby reduce uncertainty. Thus, there is an urgent need to maintain and develop the monitoring of the Earth system (e.g., through the Global Climate Observing System). A wide spectrum of Earth system quantities needs to be monitored, ranging from the maintenance of historical records (e.g., of riverflow) to the utilisation of new satellite data (e.g., CO_2 from space). There is also an urgent need for socioeconomic data of particular interest in the context of climate change (e.g., data on land use and land-use change).

8.5.3 Further Developments in Earth System Modelling

In addition to the developments outlined above, full Earth system models must consider other processes. Dynamics of the biological and human systems of the planet are relevant at the global scale and, through their interactions, must be included in simulations of the dynamics of the whole Earth system. At present, with the focus of many global models on climate change, the primary emphasis in terms of human activities has been on greenhouse gas emissions, land-use change, and aerosol emissions. The influence of biological systems is modelled mainly through their biogeochemical cycles.

There are, however, other important aspects of Earth system dynamics that are not climate related. For example, the growth of the world's population, evolution of technology, transformations of the economy or relevant changes in global lifestyles, and political ambitions occur with or without climate change. Such factors are becoming increasingly global in scale and character. These dynamics in the human part of the Earth system have significant, first-order effects upon the whole system, and future projections of biosphere-sociosphere interactions are undoubtedly crucially important in simulating the future evolution of the Earth system as a whole. Bio geochemically, the material basis of the human economy, which at the core concerns the distribution and redistribution of materials extracted from the physical and biological systems of the world under various constraints, can be treated as an extension of 'natural' biogeochemical flows to flows through human systems. As noted in the preceding section, these societal processes are presently included as 'given' scenarios external to the model itself and not included in the internal dynamics of the model.

Thus, three types of activities are currently cm the near-term agenda to develop more complete Earth system models:

1. Coupling full terrestrial biosphere models (*Dynamic Global Vegetation Models*, DGVMs) to climate models to capture not just primary feedbacks of the terrestrial biosphere to the climate system but also, in a consistent way, the effects of climate on the terrestrial biosphere (similar models, the so-called 'Green Ocean' models, are being developed for putting the marine biosphere into Earth system models).
2. Expansion of DGVMs to incorporate fully human land use, particularly agriculture and water use, including the development of parameters that allow quantification of ecosystem services to society.
3. Coupling of DGVMs through their land-use modules to economic models (including endogenous technology dynamics), themselves perhaps drawing upon models of lifestyle dynamics.

The technical challenges of such model development and coupling are considerable. For example, climate and biosphere models are time-step models, whereas economic models are mostly based on optimisation approaches under equilibrium conditions. Economic models are therefore not gridded but rather act upon 10–20 world regions, whereas climate and biosphere models are spatially explicit (similar differences occur in the data sets available for parameterizing and driving the models). Coupling requires considerable efforts in downscaling and development of software metastructures for fuzzy information exchange (hard-wired coupling may be less preferable than a 'mutual envelope' approach). With respect to the economic system, models of price dynamics have to be interpreted more consciously in terms of material and energy flows, including those that are not currently assigned monetary value (such as use of clean air). In the social sphere, formulating quantified scenarios of lifestyle dynamics seems an urgent task. Progress is being made in all of these fields, but many efforts are still at an early stage. The promise of enhanced understanding is great. For example, such integrated socioeconomic-biophysical models may be used to explore whether gradual changes in the biophysical realm of the Earth system can trigger abrupt changes in the socioeconomic sphere.

Observations designed to monitor the anthroposphere in the context of the Earth system involve the human subsystem as well. Various human dimension and related initiatives have yet to concur on a shortlist of high priority areas that require monitoring, in part because of the large variation in the ways in which different communities perceive the problems inherent in the anthroposphere. Focusing on the immediate or proximate factors that register humanity's demands on the Earth system and resources (e.g., Turner 2002), such a list would include: population variables (e.g., fertility, age structure, rural-urban mix), wealth and changes of behaviour associated with changes in wealth (e.g., diet), energy-material consumption and waste emissions by level of economic development location, efficacy of institutional controls on resource—environment issues, and critical land-use/land-cover trajectories. Visionary approaches to building an Earth-observing system focused on socioeconomics are embodied in such projects as the Geoscope.

8.5.4   Models of Biodiversity

The biological complexity of the planet also plays a role in the functioning of the Earth system. The capability to predict where, and to what degree, biodiversity is likely to be lost as a result of the combined impact of climate change, land-use change, direct use, and the impact of pollutants is an emerging field. In the climate change field, models have progressed from simple bioclimatic envelope approaches that are applied to whole biomes, to similar models applied to functional types and then individual species (including no climatic constraints), to fully dynamic models that track the movement of populations to determine if they can keep up with the rate of change. The next step will be 'ecosystem' models, which take into consideration the presence of competitors, mutualisms, food and predator species, as well as habitat structure.

On another track, integrated assessment models aim at expressing the complexity of biodiversity in synthetic index terms (macro-ecological indicator), and then relate changes in this index to various types and intensities of human activities. This allows scenarios to be developed, targets set. and performance to be monitored. Examples include the RIVM natural capital index (ten Brink 2000) and the SafMA biodiversity intactness index (in preparation).

8.6   Technological Fix and Substitution

How effective will technological substitution be in dealing with increasing impacts of the human-environment relationship on the Earth system?

The preceding sections make a strong case for the necessity of a societal response to global change. A business-as-usual approach to the future will not achieve global sustainability. Prominent among the proposed responses to global change are technological options, ranging from treatment of the fundamental causes of the problem, such as the development of no carbon-based energy systems, to treatment of the symptoms of the problem through highly controversial geo-engineering approaches.

Throughout history, society has responded in two principal ways to environmental vagaries, flux, hazards, and drawdown, including resource depletion: *move,* either through designed mobility as in pastoral nomadic systems or 'forced' relocation owing to environmental or resource degradation as exemplified in the salinisation-relocation pattern of irrigation in Mesopotamia (Adams 1965); and *change techno-managerial strategies,* as in the adoption of fossil-fuel energy or genomics (Grübler 1998). The first option has decreased in significance in an ever more crowded and politically controlled world. The second option—to modify or transform biophysical conditions in order to gain a measure of 'control' over some portion of the environment or to deliver a substitute for a depleted resource—is not only ancient but has become a defining element of our species (Diamond 1997; Redman 1999; Turner et al. 1990). Such responses are labelled technological fix and substitution.

Modem society has raised the bar in pursuit of techno-managerial solutions, with long-standing success in regard to deliveries of food, fuel, and fibre to increasingly larger and highly consumptive populations (Grübler 1998; Kasperson et al. 1995; Kinzig et al. 2004).

This approach to human-environment relations and ensuing problems is the cornerstone of the modern conditions of life, be it the Industrial Revolution or the Green Revolution. Society has become so reliant on ever-increasing advances in technology to overcome the next generation of problems that a disconnect or gap has emerged between the environmental consequences of production and consumption and the public consumer (e.g., Sack 1992). Technological solutions also offer a means to avoid the thorny issues involved in alternative solutions that are often perceived to affect lifestyles.

Science for sustainable development confronts technological fix and substitution in the face of natural and anthropogenic changes in the Earth system, culminating in global environment change. Technology constitutes one of a set of responses to deal with the problems inherent in changes in the Earth system. Indeed, some researchers believe that technological solutions will, in fact, liberate the Earth system of many of its current threats (Ausubel 2000, 2002). As noted above, however, these changes have no known analogues (Steffen et al. 2004), and some of them constitute qualitative shifts in the structure and functioning of the Earth system. These qualities raise a fundamental question and set of sub-questions about the efficacy of technological fix and substitution *alone* to cope with the problems: What evidence exists to indicate that the changes underway in the Earth system constitute a no-analogue situation, not only in the changes themselves but also in the sole use of technological fix and substitution to address these changes?

Furthermore, regarding technological fix and substitution, does the evidence suggest:

- Reduced effectiveness to address changes at the scale of the Earth system?
- Excessive cost to develop and deploy them compared to alternatives (e.g., societal changes or preservation of goods and services of the Earth system)?
- Temporal mismatch between the capacity of the potential fixes to become operative and the increasing environmental problems, with potential abrupt changes?

The antecedents and antiquity of global environmental change notwithstanding, the human-environment condition has entered a new phase that constitutes a qualitative or threshold shift: (a) The capacity of humankind to change directly the biogeochemical cycles that sustain the structure and functioning of ecosystems and the biosphere as a whole. Anthropogenic input into many of these cycles now exceeds nature's input (Steffen et al. 2004; Turner et al. 1990). For example, more nitrogen is now fixed from the atmosphere as a result of human activities than all natural nitrogen-fixing processes in the terrestrial biosphere combined. Technology also introduces new, synthetic compounds into the Earth system. The release of the well-known *chlorofluorocarbons* (CFCs) is only one of many examples; globally over

100,000 industrial chemicals-many of them unknown in the natural world—are in use today (Raskin et al. 1996). (b) The combination of these emissions has complex, systemic consequences, that is, numerous unforeseen feedbacks with far-ranging consequences and connections invariably leading to 'surprises'. Perhaps the most dramatic example to date was the formation of the ozone hole over Antarctica.

This qualitative shift in the human impacts on the Earth system generates at least three new conditions to confront technological responses:

1. Earth system changes are global in scale; climate and other environmental changes are taking place worldwide. Whereas these changes vary by region and locale, the historical societal response of moving the location of production and consumption to account for these changes or those of resource depletion or degradation appears to be attenuated. An increasingly occupied and crowded planet reduces new spaces in which to move and fosters more intensive uses of the spaces already occupied.
2. Systemic changes are inherently transboundary, and thus changes in one place affect places far away. For example, the burning of fossil fuels in North America and western Europe probably made a major contribution to drought and subsequent famine and starvation in the Sahel region in the 1970s and 1980s (Rotstayn/Lohmann 2002).
3. Many of the changes currently underway drive processes that operate over long timescales with impacts that will affect the functioning of the Earth system long after the forcing function is relaxed. Examples include the atmospheric emissions of CO_2, whose effects have a lifetime of 50–150 years; the closure of the ozone hole, which despite the reduction in CFC emissions following the Montreal Protocol, is expected to take at least four or five decades, perhaps more, to close fully; and the accumulation of reactive nitrogen compounds in terrestrial and marine ecosystems with consequences that will be played out over century timescales.

To date, humankind has directed technology to environmental problems focused primarily on resource extraction of food, fuel, and fibre, on the reduction in resource stocks (enlarging or changing), or on reducing the consequences of environmental hazards (e.g., drought to floods). The Earth system and the major societal activities affecting it have redirected these characteristics. The impacts of waste from production and consumption, such as CO_2 emissions, are equivalent to or exceed the consequences of resource extraction, including land-cover conversion, and the changes underway have shifted from resource stocks to functioning of ecosystems and the biosphere. It is highly improbable that ecosystems can be significantly altered and their many functions replaced technologically. It is even less probable that technological replacements can be found for the functioning the Earth system as a whole, especially its ability to absorb and process wastes.

The kind of environmental changes underway challenge the historical relationships between technology and environment. Other factors, however, affect this relationship as well.

The temporal dimension of the development and deployment of new technologies varies considerably by case, and the overall process may be accelerating through time (Grübler 1998). The Green Revolution, for example, transpired rapidly; it took no more than thirty years from the founding of research development centres for hybrid crops to dominate the world (Conway/Ruttan 1999). Regardless, changes currently underway in the Earth system are likely to play out over much longer timescales unless technologies of 'reversal' are developed (see examples quoted above). In addition, the Earth system could shift in ways that would change the very aims or goals of technological controls. For example, if an abrupt shutdown of the THC in the North Atlantic Ocean leads to no net warming or even cooling, societies in northern Europe would have to abandon plans to change their infrastructure to cope with strong warming.

These characteristics of the Earth system and changes underway indicate that there are few analogues regarding past technological fixes and substitutions. Also, the lock-in of significant growth in human population (Population Reference Bureau 2002) and the near-universal call for increases in per capita consumption within the developing world indicate a world in 2050 that will demand more, not less, from the Earth system (Kates et al. 2001). These conditions require new ways of approaching human-environment problems that deviate from 'business as usual' and are capable of provisioning (resources) and conserving (ecosystem-biosphere) more while degrading and changing less (Earth system). The 'precautionary principle', uncertainty and surprise, and the no-analogue conditions noted above suggest caution in a solution focused solely on technological fix and substitution and raise consideration of alternatives that address values, institutions, and other societal structures (Kinzig et al. 2004).

These nontechnical solutions need not be necessarily invented anew; various examples exist or are emerging, research on which provides clues for exploration. For example, comparative case study work indicates that sociopolitical structures which facilitate the flow of information among many stakeholders and decision makers tend to encourage learning in such forms as recognition of local and regional threats to environmental systems, a critical step toward any action taken (Kasperson et al. 1995; Social Learning Group 2001). Likewise, structures providing checks and balances on resources and environmental decisions tend to prevent potential threats to extant uses of local and regional ecosystems that might otherwise be inflicted from decisions made from afar. For example, absence of these checks and balances permitted the Soviet government to reduce the Aral Sea ecosystem to near-death conditions, despite local recognition of its demise (Micklin 1988; Kasperson et al. 1995). This observation, however, does not mean that structures promoting strong checks and balances necessarily lead to improved environmental conditions. Finally, it is important to recognise that global structures designed to provide some measures of checks and balances regarding environmental issues constitute a relatively new phenomenon (Young 1999, 2002).

These structures are emerging within a political-economic process labelled globalisation in which production, consumption, and information operate in worldwide

networks that connect virtually every place. This process is argued by some to atten-
uate the repercussions of environmental and resource disasters, for example, by
marshalling large amounts of food aid to famine areas (Kates/Parris 2003). Alternatively, others claim that it amplifies environmental problems by disconnecting more
than ever in human history, the location and impacts of production and consumption, which exacerbates environmental degradation in marginalised locations. The
large-scale destruction of the Indonesian forests for the international timber industry
(Brookfield et al. 1995; Dauvergne 1997) is a case in point. Less explored is
the concept that increasing globalisation potentially sets the stage for worldwide
collapses of social and environmental systems because the geographical, and in
some cases temporal, buffering of subsystems is diminished, in terms of technological substitution, globalisation could, in principle, increase the ability of new
technologies to diffuse and penetrate more rapidly from their point of development
to other regions of the world.

The challenges to technological fix and substitution notwithstanding, technology
will constitute part of the solutions directed to environmental problems—global and
local—in the future. Indeed, inasmuch as technology is responsible for some of
these problems, so can it help to alleviate them. Technological advances promise
increasing efficiencies in existing technologies whereas various emerging technologies will likely be critical in the future; these include genomics and biotechnology,
nanotechnology and information, as well as 'alternative' energy.

8.7 Research Challenges

Significant progress has been made over recent years in understanding the dynamics
of the Earth system in the Anthropocene. The complexities of atmospheric composition in influencing the climate system are increasingly well understood; the possibility of abrupt changes in the Earth system is apparent and promising approaches
for understanding and anticipating them are being developed; and a suite of Earth
system models of varying emphases and complexities is being developed to simulate
past, present, and future functioning of the planet. Such progress helps to sharpen
the focus of the near-term research effort and leads to a set of research questions to
help guide Earth system science over the next five to ten years.

8.7.1 Climate Sensitivity

- What is the quantitative importance of greenhouse gas-aerosol-cloud dynamics
  in enhancing or counteracting the direct radiative effects of greenhouse gases in
  the atmosphere?

- What are the radiative and chemical characteristics of aerosol particles, their emission/formation processes, regional and intercontinental dispersion, and deposition on a regional and global basis?
- Can the energy balance at the Earth's surface be closed at the regional and global scales for the Anthropocene? If so, what insight does that give about the climate sensitivity to greenhouse gases?

8.7.2   Earth System Geography

- What strategies are required to achieve a better balance of research and observation effort around the world?

8.7.3   Abrupt Changes

- What is the catalogue of possible abrupt changes in the Earth system resulting from a model-based, systematic exploration of Earth system phase space using equilibrium and stochastic resonance approaches?
- What research approaches can be developed to anticipate abrupt changes in the socioeconomic sphere of complex, coupled human-environment systems?

8.7.4   Models and Observations

- What spectrum of Earth system models is required to examine the wide range of questions associated with Earth system functioning, from exploring critical thresholds and abrupt change to high resolution impacts studies? How can we build a 'traceable' spectrum of Earth system models?
- What is the best strategy for developing models that incorporate the human dimension as a fully interactive component of the Earth system?
- How can data-model fusion be developed further to provide a more complete diagnosis of the Earth system? What critical parameters need to be observed routinely to monitor the 'vital signs' of Earth system functioning?
- What is the best strategy to test and improve Earth system models in the context of gradually evolving global change punctuated by extreme events in nature and society?

8.7.5   Technological Substitution

- What is the probability that technological change will be able to support the projected global population of 2050 at significantly higher average levels of consumption while reducing the emissions of CO_2, CH_4, and other gases and particles to the atmosphere and slowing down and ultimately stopping the degradation of marine and terrestrial ecosystems?
- Will technological fix and substitution directed to environmental concerns be offset by that directed to other concerns (e.g., economic growth)?
- Which institutional and organisational structures have proven most effective (including public acceptance) in enforcing environmental regulations under different human-environment conditions and different scales of governance (Kinzig et al. 2004)?
- What kinds of programmes and policies effectively support the conversion to and maintenance of consumption-production processes (industrial and agricultural) that are more environmentally benign (compared to extant or conventional processes) in both developed and developing countries?

References

Adams, R.M., 1965: *Land Behind Baghdad: A History of Settlement on the Diyala Plain* (Chicago: Univ. of Chicago Press).

Anderson, T.L.; Charlson, R.J.; Schwartz, S.E. et al., 2003: "Climate forcing by aerosols—A hazy picture", in: *Science*, 300: 1103–1104.

Ausubel, J., 2000: "The great reversal: Nature's chance to restore land and sea", in: *Technol. Soc.*, 22: 289–301.

Ausubel, J., 2002: "Maglevs and the vision of St Hubert, or the great restoration of nature: Why and how", in: Steffen, W.; Jäger, J.; Carson, D.; Bradshaw, C. (eds.): *Challenges of a Changing Earth: Proc. of the Global Change Open Science Conf., Amsterdam, 10–13 July 2001* (Berlin: Springer): 175–182.

Barnett, T.P.; Pierce, D.W.; Schnur, R., 2001: "Detection of anthropogenic climate change in the world's oceans", in: *Science*, 292: 270–274.

Brookfield, H.C.; Potter, L.; Byron, Y., 1995: *In Place of the Forest Environmental and Socioeconomic Transformation in Borneo and the Eastern Malay Peninsula* (Tokyo: United Nations Univ. Press).

Clark, P.U.; Marshall, S.J.; Clarke, G.K.C. et al., 2001: "Freshwater forcing of abrupt climate change during the last glaciation", in: *Science*, 293: 283–287.

Claussen, M.; Kubatzki, C.; Brovkin, V. et al., 1999: "Simulation of an abrupt change in Saharan vegetation at the end of the mid-Holocene", in: *Geaphys. Res. Lett.*, 24: 2037–2040.

Conway, G.; Ruttan, V.W., 1999: *The Doubly Green Revolution: Food for All in the Twenty-First Century* (Ithaca, NY: Comstock Publ.).

Crutzen, P.J., 1995: "My life with O_3, NO_x and other YZO_xs", in: *Les Prix Nobel - The Nobel Prizes* (Stockholm: Almqvist and Wiksell Intl.): 123–157.

Crutzen, P.J.; Stoermer, E., 2001: "The 'Anthropocene'", in: *Glob. Change Newsl.*, 41: 12–13.

Dauvergne, P., 1997: *Shadows in the Forest: Japan and the Politics of Timber in Southeast Asia.* (Cambridge, MA: MIT Press).

DcMenocal, P.B.; Ortiz, J.; Guilderson, T. et al., 2000: "Abrupt onset and termination of the African Humid Period: Rapid climate response to gradual insolation forcing", in: *Quat. Sci. Rev.*, 19: 347–361.

Diamond, J., 1997: *Guns, Germs, and Steel: The Fates of Human Societies* (New York: Norton).

Dijkstra, H.A., 2000: *Nonlinear Physical Oceanography*. Atmospheric and Oceanographic Library Series (Dordrecht: Kluwer Academic).

Falkowski, P.; Scholes, R.J.; Boyle, E. et al., 2000: "The global carbon cycle: A test of our knowledge of Earth as a system", in: *Science*, 290: 291–296.

Feudel, U.; Jansen, W., 1992: CANDYS/QA-a software system for the qualitative analysis of nonlinear dynamical systems", in: *Intl. J. Bifurc. & Chaos*, 2: 773–794.

Fischer, A.; Schutte, C.; Deuflhard, P.; Cordes, F., 2002: "Hierarchical uncoupling-coupling of metastable conformations, in: Schlick, T.; Gan, H.H. (eds.): *Computational Methods for Macromolecules: Challenges and Applications*, Proc. 3rd Intl. Workshop on Algorithms tor Macromolecular Modeling, Lecture Notes in Computational Science and Engineering Series, vol. 24 (Berlin: Springer): 235–259.

Gammaitoni, L.; Honggi, P.; Jung, P.; Marchesoni, F., 1998: "Stochastic resonance", in: *Rev. Mod. Physics*, 70: 223–287.

Ganopolski, A.; Rahmstorf, S., 2001: "Rapid changes of glacial climate simulated in a coupled climate model", in: *Nature*, 409: 153–158.

Ganopolski, A.; Rahmstorf, S., 2002: "Abrupt glacial climate changes due to stochastic resonance", in: *Phys. Rev. Lett.*, 88: 038501-1–038501-4.

Gardiner, C.W., 1994: *Handbook of Stochastic Methods*, 2nd ed. (Berlin: Springer)

Grootes, P.M.; Stuiver, M.; White, J.W.C.; Johnsen, S.; Jouzel, J., 1993: "Comparison of oxygen isotope records from the GISP2 and GRIP Greenland ice cores", in: *Nature* 366: 552–554.

Grübler, A., 1998: *Technology and Global Change* (Cambridge: Cambridge Univ. Press).

IPCC (Intergovernmental Panel on Climate Change), 2001: [Houghton, J.T.; Ding, Y.; Griggs, D.J. et al. (eds.)]: *Climate Change 2001: The Scientific Basis. Working Group 1. Contribution. Third Assessment Report of the IPCC* (Cambridge: Cambridge Univ. Press).

Jackson, J.B.C.; Kirby, M.X.; Berger, W.H. et al. 2001: "Historical overfishing and the recent collapse of coastal ecosystems", in: *Science*, 293: 629–637.

Johnson, C.E.; Stevenson, D.S.; Collins, W.J.; Derwent, R.G., 2001: "The role of climate feedback on methane and ozone studied with a coupled ocean-atmosphere-chemistry model", in: *Geophys. Res. Lett.*, 28: 1723–1726.

Jones, C.D.; Cox, P.M.; Essery, R.L.H. et al., 2003: "Strong carbon cycle feedbacks in a model with interactive CO_2 and sulphate aerosols", in: *Geophvs. Res. Lett.*, 30: 1479–1482.

Kasperson, J.X.; Kasperson, R.E.; Turner II, B.L. (eds.), 1995: *Regions at Risk: Comparisons of Threatened Environments* (Tokyo: United Nations Univ. Press).

Kates, R.W.; Clark, W.C.; Corell, R. et al., 2001: "Sustainability science", in: *Science*, 292: 641–642.

Kates, R.W.; Parris, T.M., 2003: "Long-term trends and a sustainability transition", in: *Prvc. Natl. Acad. Sci. USA*, 100: 8062–8067.

Keeling, C.D.; Whorf, T.P., 2000: "Atmospheric CO_2 records from sites in the SIO air sampling network", in: *Trends: A Compendium of Data on Global Change* (Carbon Dioxide Information Analysis Center, Oak Ridge Natl. Laboratory, Oak Ridge, TN: U.S. Dept. of Energy).

Kinzig, A.P.; Clark, W.C.; Edenhofer, O.; Gallopín, G.C.; Lucht, W.; Mitchell, R.B.; Romero Lankao, P.; Sreekesh, S.; Tickell, C.; Young, O.R., 2004: "Group Report: Sustainability", in: Schellnhuber, H.J.; Crutzen, P.J.; Clark, W.C.; Claussen, M.; Held, H. (eds.): *Earth System Analysis for Sustainability* (Cambridge, Mass.: The MIT Press): 409–434.

Kleinen, T.; Held, H.; Petschel-Held, G., 2003: "The potential role of spectral properties in detecting thresholds in the Earth system: Application to the thermohaline circulation", in: *Ocean Dyn.*, 53: 53–63.

Levitus, S.; Antonov, J.I.; Boyer, T.P.; Stephens, C., 2000: "Warming of the world ocean", in: *Science*, 287: 2225–2229.

Loreau, M.; Naeem, S.; Inchausti, P. et al., 2001: "Biodiversity and ecosystem functioning: Current knowledge and future challenges", in: *Science,* 294: 804–808.

Loutre, M.F.; Berger, A., 2000: "Future climatic changes: Are we entering an exceptionally long interglacial?", in: *Clim. Change*, 46: 61–90.

Majda, A.J.; Timofeyev, I.; Vanden Eijnden, E., 1999: "Models for stochastic climate prediction", in: *Proc. Natl. Acad. Sci. USA*, 96: 14: 687–691.

Mann, M.E.; Bradley, R.S.; Hughes, M.K., 1999: "Northern Hemisphere temperatures during the past millennium: Inferences, uncertainties, and limitations", in: *Geophys. Res- Lett.*, 26: 759–762.

Micklin, P. 1988: "Desiccation of the Aral Sea: A water management disaster in the Soviet Union, in: *Science*, 241: 1170–1176.

Myers, R. A.; Wonn, B., 2003: "Rapid worldwide depletion of predatory fish communities", in: *Nature*, 423: 280–283.

Nobre, C.A.; Artaxo, P.; Silva Dias, M.A.F. et al., 2002: "The Amazon Basin and land-cover change: A future in the balance?", in: Steffen, W.; Jager, J.; Carson, D.; Bradshaw, C. (eds.): *Challenges of a Changing Earth, Proc. of the Global Change Open Science Conf., Amsterdam, NL, 10–13 July 2001.* (Berlin: Springer): 137–141.

NRC (National Research Council), 2002: *Abrupt Climate Change: Inevitable Surprises* (Washington, D.C.: Natl. Academy Press).

Petit, J.R.; Jouzel, J.; Raynaud, D. et al., 1999: "Climate and atmospheric history of the past 420,000 years from the Vostok ice core, Antarctica", in: *Nature*, 399: 429–436.

Pikovsky, A.S.; Kurths, J., 1997: "Coherence resonance in a noise-driven excitable system", in: *Phys. Rev. Lett.*, 78: 775–778.

Population Reference Bureau, 2002: "2002 World Population Data Sheet".

Rahmstorf, S., 2000: "The thermohaline circulation: A system with dangerous thresholds?", in: *Clim. Change*, 46: 247–256.

Ramanathan, V., 1981: "The role of ocean-atmosphere interactions in the CO_2 climate problem", in: *J. Atmos. Sci.*, 38: 918–930.

Raskin, P.; Chadwick, M.; Jackson, T.; Leach, G., 1996: *The sustainability transition: Beyond conventional development*. Polestar Series 1 (Stockholm: Stockholm Environment Institute).

Redman, C.L., 1999: *Human Impact on Ancient Environments* (Tucson: Univ. of Arizona Press).

Rotstayn, L.D.; Lohmann, U., 2002: "Tropical rainfall trends and the indirect aerosol effect", in: *J. Climate*, 15: 2103–2116.

Sack, R.D, 1992: *Place, Modernity, and the Consumer's World* (Baltimore: Johns Hopkins Press).

Schneider, S.; Turner II., B.L.; Morehouse Garriga, H., 1998: "Imaginable surprise in global change science", in: *J. Risk Res.*, 1: 165–185.

Shah, M., 2002: "Food in the 21st century: Global climate of disparities", in: Steffen, W.; Jäger, J.; Carson, D.; Bradshaw, C. (eds.) *Challenges of a Changing Earth, Proc. Of the Global Change Open Science Conf., Amsterdam, 10–13 July 2001* (Berlin: Springer): 31–38.

Social Learning Group, 2001: *Learning to Manage Global Environmental Risks*, 2 vols. (Cambridge, MA: MIT Press).

Steffen, W.; Sanderson, A.; Tyson, P.D. et al., 2004: *Global Change and the Earth System: A Planet Under Pressure*. The IGBP Book Series (Berlin: Springer).

ten Brink, B.J.E., 2000: *Biodiversity indicators for the OECD Environmental Outlook and Strategy: A feasibility study*. Report 402001014 (Bilthoven: RIVM).

Tilman, D., 1999: "The ecological consequences of changes in biodiversity: A search for general principles. Robert H. MacArthur Award Lecture", in: *Ecology*, 80: 1455–1474.

Turner, B.L., II., 2002: "Toward integrated land-change science: Advances in 1.5 decades of sustained international research on land-use and land-cover change", in: Steffen, W.; Jäger, J.; Carson, D.; Bradshaw, C. (eds.): *Challenges of a Changing Earth, Proc. of the Global Change Open Science Conf. Amsterdam. 10–13 July 2001* (Berlin: Springer): 21–26.

Turner, B.L., II; Clark, W.C.; Kates, R.W. et al., 1990: *The Earth as Transformed by Human Action: Global and Regional Changes in the Biosphere over the Past 300 years* (Cambridge: Cambridge Univ. Press).

UN (United Nations), 2001: *World Population Monitoring 2001*. Population, Environment and Development ST/ESA/SER.A/203 (New York: UN Publ.).

Young, O.R., 1999: *Governance in World Affairs* (Ithaca: Cornell Univ. Press).

Young, O.R., 2002: "Can new institutions solve atmospheric problems? Confronting acid rain, ozone depletion and climate change", in: Steffen, W.; Jäger, J.; Carson, D.; Bradshaw, C. (eds.): *Challenges of a Changing Earth, Proc. of the Global Change Open Science Conf., Amsterdam, 10–13 July 2001* (Berlin: Springer): 87–91.

Chapter 9
The Anthropocene: Are Humans Now Overwhelming the Great Forces of Nature? (2007)

Will P. Steffen, Paul J. Crutzen, and John R. McNeill

We explore the development of the Anthropocene, the current epoch in which humans and our societies have become a global geophysical force. The Anthropocene began around 1800 with the onset of industrialisation, the central feature of which was the enormous expansion in the use of fossil fuels. We use atmospheric carbon dioxide concentration as a single, simple indicator to track the progression of the Anthropocene. From a preindustrial value of 270–275 ppm, atmospheric carbon dioxide had risen to about 310 ppm by 1950. Since then the human enterprise has experienced a remarkable explosion, the Great Acceleration, with significant consequences for Earth System functioning. Atmospheric CO_2 concentration has risen from 310 to 380 ppm since 1950, with about half of the total rise since the preindustrial era occurring in just the last 30 years. The Great Acceleration is reaching criticality. Whatever unfolds, the next few decades will surely be a tipping point in the evolution of the Anthropocene.

9.1 Introduction

Global warming and many other human-driven changes to the environment are raising concerns about the future of Earth's environment and its ability to provide the services required to maintain viable human civilizations. The consequences of this unintended experiment of humankind on its own life support system are hotly debated, but worst-case scenarios paint a gloomy picture for the future of contemporary societies.

Underlying global change (Box 9.1) are human-driven alterations of (*i*) the biological fabric of the Earth; (*ii*) the stocks and flows of major elements in the planetary machinery such as nitrogen, carbon, phosphorus, and silicon; and (*iii*) the energy

This text was first published as: Steffen, W.P.; Crutzen, P.C.; McNeill, J.R.: "The Anthropocene: Are Humans Now Overwhelming the Great Forces of Nature?", in: *Ambio*, 36,8 (December 2007): 614–621. The permission was granted by the Royal Swedish Academy of Sciences.

balance at the Earth's surface (Hansen et al. 2005: 1431–1435). The term *Anthropocene* (Box 9.2) suggests that the Earth has now left its natural geological epoch, the present interglacial state called the Holocene. Human activities have become so pervasive and profound that they rival the great forces of Nature and are pushing the Earth into planetary *terra incognita*. The Earth is rapidly moving into a less biologically diverse, less forested, much warmer, and probably wetter and stormier state.

The phenomenon of global change represents a profound shift in the relationship between humans and the rest of nature. Interest in this fundamental issue has escalated rapidly in the international research community, leading to innovative new research projects like *Integrated History and future of People on Earth* (IHOPE) (Costanza et al. 2006). The objective of this paper is to explore one aspect of the IHOPE research agenda—the evolution of humans and our societies from hunter-gatherers to a global geophysical force.

Box 9.1: Global Change and the Earth System. *Source* **The authors**

The term *Earth System* refers to the suite of interacting physical, chemical and biological global-scale cycles and energy fluxes that provide the life-support system for life at the surface of the planet (Oldfield/Steffen 2004: 7). This definition of the Earth System goes well beyond the notion that the geophysical processes encompassing the Earth's two great fluids—the ocean and the atmosphere—generate the planetary life-support system on their own. In our definition biological/ecological processes are an integral part of the functioning of the Earth System and not merely the recipient of changes in the coupled ocean-atmosphere part of the system. A second critical feature is that forcings and feedbacks *within* the Earth System are as important as external drivers of change, such as the flux of energy from the sun. Finally, the Earth System includes humans, our societies, and our activities; thus, humans are not an outside force perturbing an otherwise natural system but rather an integral and interacting part of the Earth System itself.

We use the term *global change* to mean both the biophysical and the socioeconomic changes that are altering the structure and the functioning of the Earth System. Global change includes alterations in a wide range of global-scale pheno-mena: land use and land cover, urbanisation, globalisation, coastal ecosystems, atmospheric composition, riverine flow, nitrogen cycle, carbon cycle, physical climate, marine food chains, biological diversity, population, economy, resource use, energy, transport, communication, and so on. Interactions and linkages between the various changes listed above are also part of global change and are just as important as the individual changes themselves. Many components of global change do not occur in linear fashion but rather show strong nonlinearities.

Box 9.2: The Anthropocene System. *Source* **The authors**

Holocene ("Recent Whole") is the name given to the postglacial geological epoch of the past ten to twelve thousand years as agreed upon by the International Geological Congress in Bologna in 1885 (Encyclopaedia Britannica 1976). During the Holocene, accelerating in the industrial period, humankind's activities became a growing geological and morphological force, as recognised early by a number of scientists. Thus, in 1864, Marsh published a book with the title *Man and Nature*, more recently reprinted as *The Earth as Modified by Human Action* (Marsh 1965). Stoppani in 1873 rated human activities as a "new telluric force which in power and universality may be compared to the greater forces of earth" (quoted from Clark [1986]). Stoppani already spoke of the anthropozoic era. Humankind has now inhabited or visited all places on Earth; he has even set foot on the moon. The great Russian geologist and biologist Vernadsky (1998) in 1926 recognised the increasing power of humankind in the environment with the following excerpt "… the direction in which the processes of evolution must proceed, namely towards increasing consciousness and thought, and forms having greater and greater influence on their surroundings." He, the French Jesuit priest P. Teilhard de Chardin and E. Le Roy in 1924 coined the term "noösphere," the world of thought, knowledge society, to mark the growing role played by humankind's brainpower and technological talents in shaping its own future and environment. A few years ago the term 'Anthropocene' has been introduced by one of the authors (Crutzen 2002) for the current geological epoch to emphasise the central role of humankind in geology and ecology. The impact of current human activities is projected to last over very long periods. For example, because of past and future anthropogenic emissions of CO_2, climate may depart significantly from natural behaviour over the next 50 000 years.

To address this objective, we examine the trajectory of the human enterprise through time, from the arrival of humans on Earth through the present and into the next centuries. Our analysis is based on a few critical questions:

- Is the imprint of human activity on the environment discernible at the global scale? How has this imprint evolved through time?
- How does the magnitude and rate of human impact compare with the natural variability of the Earth's environment? Are human effects similar to or greater than the great forces of nature in terms of their influence on Earth System functioning?
- What are the socioeconomic, cultural, political, and technological developments that change the relationship between human societies and the rest of nature and lead to accelerating impacts on the Earth System?

9.2   Pre-anthropocene Events

Before the advent of agriculture about 10 000–12000 years ago, humans lived in small groups as hunter-gatherers. In recent centuries, under the influence of noble savage myths, it was often thought that pre-agricultural humans lived in idyllic harmony with their environment. Recent research has painted a rather different picture, producing evidence of widespread human impact on the environment through predation and the modification of landscapes, often through use of fire (Pyne 1997). However, as the examples below show, the human imprint on environment may have been discernible at local, regional, and even continental scales, but preindustrial humans did not have the technological or organisational capability to match or dominate the great forces of nature.

The mastery of fire by our ancestors provided humankind with a powerful monopolistic tool unavailable to other species, that put us firmly on the long path towards the Anthropocene. Remnants of charcoal from human hearths indicate that the first use of fire by our bipedal ancestors, belonging to the genus *Homo erectus,* occurred a couple of million years ago. Use of fire followed the earlier development of stone tool and weapon making, another major step in the trajectory of the human enterprise.

Early humans used the considerable power of fire to their advantage (Pyne 1997). Fire kept dangerous animals at a respectful distance, especially during the night, and helped in hunting protein-rich, more easily digestible food. The diet of our ancestors changed from mainly vegetarian to omnivorous, a shift that led to enhanced physical and mental capabilities. Hominid brain size nearly tripled up to an average volume of about 1300 cm^3, and gave humans the largest ratio between brain and body size of any species (Tobias 1976). As a consequence, spoken and then, about 10 000 years ago, written language could begin to develop, promoting communication and transfer of knowledge within and between generations of humans, efficient accumulation of knowledge, and social learning over many thousands of years in an impressive catalytic process, involving many human brains and their discoveries and innovations. This power is minimal in other species.

Among the earliest impacts of humans on the Earth's biota are the late Pleistocene megafauna extinctions, a wave of extinctions during the last ice age extending from the woolly mammoth in northern Eurasia to giant wombats in Australia (Martin/Klein 1984; Alroy 2001; Roberts et al. 2001). A similar wave of extinctions was observed later in the Americas. Although there has been vigorous debate about the relative roles of climate variability and human predation in driving these extinctions, there is little doubt that humans played a significant role, given the strong correlation between the extinction events and human migration patterns. A later but even more profound impact of humans on fauna was the domestication of animals, beginning with the dog up to 100 000 years ago (Leach 2003) and continuing into the Holocene with horses, sheep, cattle, goats, and the other familiar farm animals. The concomitant domestication of plants during the early to mid-Holocene led to agriculture, which initially also developed through the use of fire for forest clearing and, somewhat later, irrigation (Smith 1995).

According to one hypothesis, early agricultural development, around the mid-Holocene, affected Earth System functioning so fundamentally that it prevented the onset of the next ice age (Ruddiman 2003). The argument proposes that clearing of forests for agriculture about 8000 years ago and irrigation of rice about 5000 years ago led to increases in atmospheric carbon dioxide (CO_2) and methane (CH_4) concentrations, reversing trends of concentration decreases established in the early Holocene. These rates of forest clearing, however, were small compared with the massive amount of land transformation that has taken place in the last 300 years (Lambin/Geist 2006). Nevertheless, deforestation and agricultural development in the 8000 to 5000 BP period may have led to small increases in CO_2 and CH_4 concentrations (maybe about 5–10 parts per million for CO_2) but increases that were perhaps large enough to stop the onset of glaciation in northeast Canada thousands of years ago. However, recent analyses of solar forcing in the late Quaternary (EPICA Community Members 2004) and of natural carbon cycle dynamics (Broecker/Stocker 2006; Joos et al. 2004) argue that natural processes can explain the observed pattern of atmospheric CO_2 variation through the Holocene. Thus, the hypothesis that the advent of agriculture thousands of years ago changed the course of glacial-interglacial dynamics remains an intriguing but unproven beginning of the Anthropocene.

The first significant use of fossil fuels in human history came in China during the Song Dynasty (960–1279) (Hartwell 1962, 1967). Coal mines in the north, notably Shanxi province, provided abundant coal for use in China's growing iron industry. At its height, in the late 11th century, China's coal production reached levels equal to all of Europe (not including Russia) in 1700. But China suffered many setbacks, such as epidemics and invasions, and the coal industry apparently went into a long decline. Meanwhile in England coal mines provided fuel for home heating, notably in London, from at least the 13th century (TeBrake 1975; Brimblecombe 1987). The first commission charged to investigate the evils of coal smoke began work in 1285 (Brimblecombe 1987). But as a concentrated fuel, coal had its advantages, especially when wood and charcoal grew dear, so by the late 1600s London depended heavily upon it and burned some 360 000 tons annually. The iron forges of Song China and the furnaces of medieval London were regional exceptions, however; most of the world burned wood or charcoal rather than resorting to fuel subsidies from the Carboniferous.

Preindustrial human societies indeed influenced their environment in many ways, from local to continental scales. Most of the changes they wrought were based on knowledge, probably gained from observation and trial-and-error, of natural ecosystem dynamics and its modification to ease the tasks of hunting, gathering, and eventually of farming. Preindustrial societies could and did modify coastal and terrestrial ecosystems but they did not have the numbers, social and economic organisation, or technologies needed to equal or dominate the great forces of Nature in magnitude or rate. Their impacts remained largely local and transitory, well within the bounds of the natural variability of the environment.

9.3   The Industrial Era (ca. 1800–1945): Stage 1
of the Anthropocene

One of the three or four most decisive transitions in the history of humankind, potentially of similar importance in the history of the Earth itself, was the onset of industrialisation. In the footsteps of the Enlightenment, the transition began in the 1700s in England and the Low Countries for reasons that remain in dispute among historians (Mokyr 1999). Some emphasise material factors such as wood shortages and abundant water power and coal in England, while others point to social and political structures that rewarded risk-taking and innovation, matters connected to legal regimes, a nascent banking system, and a market culture. Whatever its origins, the transition took off quickly and by 1850 had transformed England and was beginning to transform much of the rest of the world.

What made industrialisation central for the Earth System was the enormous expansion in the use of fossil fuels, first coal and then oil and gas as well. Hitherto humankind had relied on energy captured from ongoing flows in the form of wind, water, plants, and animals, and from the 100- or 200-year stocks held in trees. Fossil fuel use offered access to carbon stored from millions of years of photosynthesis: a massive energy subsidy from the deep past to modern society, upon which a great deal of our modern wealth depends.

Industrial societies as a rule use four or five times as much energy as did agrarian ones, which in turn used three or four times as much as did hunting and gathering societies (Sieferle 2001). Without this transition to a high-energy society it is inconceivable that global population could have risen from a billion around 1820 to more than six billion today, or that perhaps one billion of the more fortunate among us could lead lives of comfort unknown to any but kings and courtiers in centuries past.

Prior to the widespread use of fossil fuels, the energy harvest available to humankind was tightly constrained. Water and wind power were available only in favoured locations, and only in societies where the relevant technologies of watermills, sailing ships, and windmills had been developed or imported. Muscular energy derived from animals, and through them from plants, was limited by the area of suitable land for crops and forage, in many places by shortages of water, and everywhere by inescapable biological inefficiencies: plants photosynthesise less than a percent of the solar energy that falls on the Earth, and animals eating those plants retain only a tenth of the chemical energy stored in plants. All this amounted to a bottleneck upon human numbers, the global economy, and the ability of humankind to shape the rest of the biosphere and to influence the functioning of the Earth System.

The invention (some would say refinement) of the steam engine by James Watt in the 1770s and 1780s and the turn to fossil fuels shattered this bottleneck, opening an era of far looser constraints upon energy supply, upon human numbers, and upon the global economy. Between 1800 and 2000 population grew more than six-fold, the global economy about 50-fold, and energy use about 40-fold (McNeill 2001). It also opened an era of intensified and ever-mounting human influence upon the Earth System.

Fossil fuels and their associated technologies—steam engines, internal combustion engines—made many new activities possible and old ones more efficient. For example, with abundant energy it proved possible to synthesise ammonia from atmospheric nitrogen, in effect to make fertilizer out of air, a process pioneered by the German chemist Fritz Haber early in the 20th century. The Haber-Bosch synthesis, as it would become known (Carl Bosch was an industrialist) revolutionised agriculture and sharply increased crop yields all over the world, which, together with vastly improved medical provisions, made possible the surge in human population growth.

The imprint on the global environment of the industrial era was, in retrospect, clearly evident by the early to mid 20th century (Steffen et al. 2004). Deforestation and conversion to agriculture were extensive in the midlatitudes, particularly in the northern hemisphere. Only about 10% of the global terrestrial surface had been 'domesticated' at the beginning of the industrial era around 1800, but this figure rose significantly to about 25–30% by 1950 (Lambin/Geist 2006). Human transformation of the hydrological cycle was also evident in the accelerating number of large dams, particularly in Europe and North America (Vörösmarty et al. 1997). The flux of nitrogen compounds through the coastal zone had increased over 10-fold since 1800 (Mackenzie et al. 2002).

The global-scale transformation of the environment by industrialisation was, however, nowhere more evident than in the atmosphere. The concentrations of CH_4 and nitrous oxide (N_2O) had risen by 1950 to about 1250 and 288 ppbv, respectively, noticeably above their preindustrial values of about 850 and 272 ppbv (Blunier et al. 1993; Machida et al. 1995). By 1950 the atmospheric CO_2 concentration had pushed above 300 ppmv, above its preindustrial value of 270–275 ppmv, and was beginning to accelerate sharply (Etheridge et al. 1996).

Quantification of the human imprint on the Earth System can be most directly related to the advent and spread of fossil fuel-based energy systems (Fig. 9.1), the signature of which is the accumulation of CO_2 in the atmosphere roughly in proportion to the amount of fossil fuels that have been consumed. We propose that atmospheric CO_2 concentration can be used as a single, simple indicator to track the progression of the Anthropocene, to define its stages quantitatively, and to compare the human imprint on the Earth System with natural variability (Table 9.1).

Around 1850, near the beginning of Anthropocene Stage 1, the atmospheric CO_2 concentration was 285 ppm, within the range of natural variability for interglacial periods during the late Quaternary period. During the course of Stage 1 from 1800/50 to 1945, the CO_2 concentration rose by about 25 ppm, enough to surpass the upper limit of natural variation through the Holocene and thus provide the first indisputable evidence that human activities were affecting the environment at the global scale. We therefore assign the beginning of the Anthropocene to coincide with the beginning of the industrial era, in the 1800–1850 period. This first stage of the Anthropocene ended abruptly around 1945, when the most rapid and pervasive shift in the human-environment relationship began.

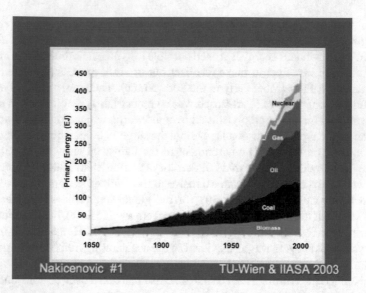

Fig. 9.1 The mix of fuels in energy systems at the global scale from 1850 to 2000. Note the rapid relative decrease in traditional renewable energy sources and the sharp rise in fossil fuel-based energy systems since the beginning of the Industrial Revolution, and particularly after 1950. By 2000 fossil fuel-based energy systems generated about 80% of the total energy used to power the global economy

9.4   The Great Acceleration (1945-ca. 2015): Stage 2 of the Anthropocene

The human enterprise suddenly accelerated after the end of the Second World War (McNeill 2001) (Fig. 9.2) Population doubled in just 50 years, to over 6 billion by the end of the 20th century, but the global economy increased by more than 15-fold. Petroleum consumption has grown by a factor of 3.5 since 1960, and the number of motor vehicles increased dramatically from about 40 million at the end of the War to nearly 700 million by 1996. From 1950 to 2000 the percentage of the world's population living in urban areas grew from 30 to 50% and continues to grow strongly. The interconnectedness of cultures is increasing rapidly with the explosion in electronic communication, international travel and the globalisation of economies.

The pressure on the global environment from this burgeoning human enterprise is intensifying sharply. Over the past 50 years, humans have changed the world's ecosystems more rapidly and extensively than in any other comparable period in human history (Pimm et al. 1995). The Earth is in its sixth great extinction event, with rates of species loss growing rapidly for both terrestrial and marine ecosystems (Intergovernmental Panel on Climate Change 2007). The atmospheric concentrations of several important greenhouse gases have increased substantially, and the Earth is warming rapidly (Galloway/Cowling 2002). More nitrogen is now converted from the atmosphere into reactive forms by fertilizer production and fossil fuel combustion

Table 9.1 Atmospheric CO_2 concentration during the existence of fully modern humans on Earth. References given in notes below

Year/Period	Atmospheric CO_2 concentration (ppmv)[1]
250 000–12 000 years BP[2]:	
Range during interglacial periods:	262–287
Minimum during glacial periods:	182
12 000–2 000 years BP:	260–285
Holocene (current interglacial)	
1000	279
1500	282
1600	276
1700	277
1750	277
1775	279
1800 (Anthropocene Stage I begins)	283
1825	284
1850	285
1875	289
1900	296
1925	305
1950 (Anthropocene Stage II begins)	311
1975	331
2000	369
2005	379

[1] The CO_2 concentration data were obtained from: (a) http://cdiac.ornl.gov/trends/trends.htm for the 250 000–12 000 BP period and for the 1000 AD-2005 AD period. More specifically, data were obtained from (Barnola et al. 2003; 250 000–12 000 BP), (Etheridge et al. 1998; 1000–1950 AD), and (Indermuhle et al. 1999; 1975–2000 AD). (b) CO_2 concentrations for the 12 000–2000 BP period (the Holocene) were obtained from (Millennium Ecosystem Assessment 2005).

[2] The period 250 000–12 000 years BP encompasses two interglacial periods prior to the current interglacial (the Holocene) and two glacial periods. The values listed in the table are the maximum and minimum CO_2 concentrations recorded during the two interglacial periods and the of fully modern humans was approximately 250 000 years BP.

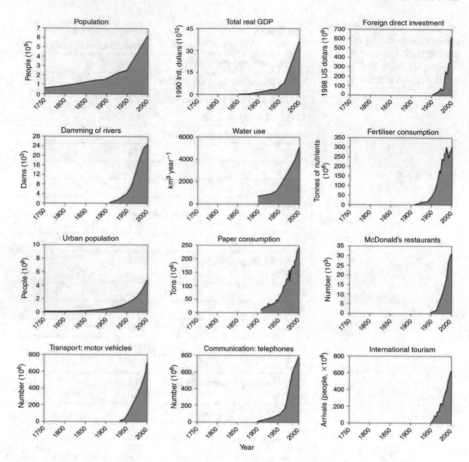

Fig. 9.2 The change in the human enterprise from 1750 to 2000 (Steffen et al. 2004). The Great Acceleration is clearly shown in every component of the human enterprise included in the figure. Either the component was not present before 1950 (e.g., foreign direct investment) or its rate of change increased sharply after 1950 (e.g., population)

than by all of the natural processes in terrestrial ecosystems put together (Fig. 9.3) (Hibbard et al. 2006).

The remarkable explosion of the human enterprise from the mid-20th century, and the associated global-scale impacts on many aspects of Earth System functioning, mark the second stage of the Anthropocene—the Great Acceleration (Keeling/Whorf 2005). In many respects the stage had been set for the Great Acceleration by 1890 or 1910. Population growth was proceeding faster than at any previous time in human history, as well as economic growth. Industrialisation had gathered irresistible momentum, and was spreading quickly in North America, Europe, Russia, and Japan. Automobiles and airplanes had appeared, and soon rapidly transformed mobility. The world economy was growing ever more tightly linked by mounting flows of migration, trade, and capital. The years 1870 to 1914 were, in fact, an age of globalisation

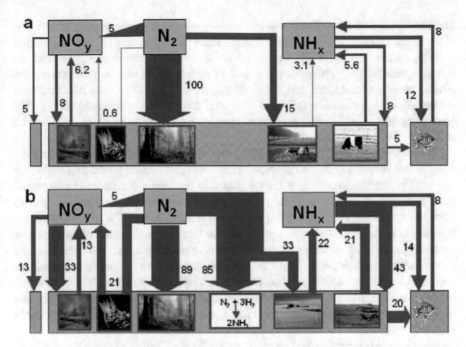

Fig. 9.3 Global terrestrial nitrogen budget for (a) 1890 and (b) 1990 in Tg N year[-1] (Hibbard et al. 2006). The emissions to the NOy box from the coal reflect fossil fuel combustion. Those from the vegetation include agricultural and natural soil emissions and combustion of biofuel, biomass (savanna and forest) and agricultural waste. The NHx emissions from the cow and feedlot reflect emissions from animal wastes. The transfers to the fish box represent the lateral flow of dissolved inorganic nitrogen from terrestrial systems to the coastal seas. Note the enormous amount of N_2 converted to NH_3 in the 1990 panel compared to 1890. This represents human fixation of nitrogen through the Haber-Bosch process, made possible by the development of fossil-fuel based energy systems (Hibbard/Crutzen/Lambin et al. 2006)

in the world economy. Mines and plantations in diverse lands such as Australia, South Africa, and Chile were opening or expanding in response to the emergence of growing markets for their products, especially in the cities of the industrialised world.

At the same time, cities burgeoned as public health efforts, such as checking waterborne disease through sanitation measures, for the first time in world history made it feasible for births consistently to outnumber deaths in urban environments. A major transition was underway in which the characteristic habitat of the human species, which for several millennia had been the village, now was becoming the city. (In 1890 perhaps 200 million people lived in cities worldwide, but by 2000 the figure had leapt to three billion, half of the human population). Cities had long been the seats of managerial and technological innovation and engines of economic growth, and in the Great Acceleration played that role with even greater effect.

However, the Great Acceleration truly began only after 1945. In the decades between 1914 and 1945 the Great Acceleration was stalled by changes in politics and the world economy. Three great wrenching events lay behind this: World War I, the Great Depression, and World War II. Taken together, they slowed population growth, checked—indeed temporarily reversed—the integration and growth of the world economy. They also briefly checked urbanisation, as city populations led the way in reducing their birth rates. Some European cities in the 1930s in effect went on reproduction strikes, so that (had they maintained this reluctance) they would have disappeared within decades. Paradoxically, however, these events also helped to initiate the Great Acceleration.

The lessons absorbed about the disasters of world wars and depression inspired a new regime of international institutions after 1945 that helped create conditions for resumed economic growth. The United States in particular championed more open trade and capital flows, reintegrating much of the world economy and helping growth rates reach their highest ever levels in the period from 1950 to 1973. At the same time, the pace of technological change surged. Out of World War II came a number of new technologies—many of which represented new applications for fossil fuels—and a commitment to subsidised research and development, often in the form of alliances among government, industry, and universities. This proved enormously effective and, in a climate of renewed prosperity, ensured unprecedented funding for science and technology, unprecedented recruitment into these fields, and unprecedented advances as well.

The Great Acceleration took place in an intellectual, cultural, political, and legal context in which the growing impacts upon the Earth System counted for very little in the calculations and decisions made in the world's ministries, boardrooms, laboratories, farmhouses, village huts, and, for that matter, bedrooms. This context was not new, but it too was a necessary condition for the Great Acceleration.

The exponential character of the Great Acceleration is obvious from our quantification of the human imprint on the Earth System, using atmospheric CO_2 concentration as the indicator (Table 9.1). Although by the Second World War the CO_2 concentration had clearly risen above the upper limit of the Holocene, its growth rate hit a take-off point around 1950. Nearly three-quarters of the anthropogenically driven rise in CO_2 concentration has occurred since 1950 (from about 310 to 380 ppm), and about half of the total rise (48 ppm) has occurred in just the last 30 years.

9.5 Stewards of the Earth System? (ca. 2015-?): Stage 3 of the Anthropocene

Humankind will remain a major geological force for many millennia, maybe millions of years, to come. To develop a universally accepted strategy to ensure the sustainability of Earth's life support system against human-induced stresses is one of the greatest research and policy challenges ever to confront humanity. Can humanity meet this challenge?

Signs abound to suggest that the intellectual, cultural, political and legal context that permitted the Great Acceleration after 1945 has shifted in ways that could curtail it (Keeling/Whorf 2005). Not surprisingly, some reflective people noted human impact upon the environment centuries and even millennia ago. However, as a major societal concern it dates from the 1960s with the rise of modern environmentalism. Observations showed incontrovertibly that the concentration of CO_2 in the atmosphere was rising markedly (Crutzen 1995). In the 1980s temperature measurements showed global warming was a reality, a fact that encountered political opposition because of its implications, but within 20 years was no longer in serious doubt (Galloway/Cowling 2002). Scientific observations showing the erosion of the Earth's stratospheric ozone layer led to international agreements reducing the production and use of CFCs (*chlorofluorocarbons*) (Schellnhuber 1998). On numerous ecological issues local, national, and international environmental policies were devised, and the environment routinely became a consideration, although rarely a dominant one, in political and economic calculations.

This process represents the beginning of the third stage of the Anthropocene, in which the recognition that human activities are indeed affecting the structure and functioning of the Earth System as a whole (as opposed to local- and regional-scale environmental issues) is filtering through to decision-making at many levels. The growing awareness of human influence on the Earth System has been aided by (*i*) rapid advances in research and understanding, the most innovative of which is interdisciplinary work on human-environment systems; (*ii*) the enormous power of the internet as a global, self-organizing information system; (*iii*) the spread of more free and open societies, supporting independent media; and (*iv*) the growth of democratic political systems, narrowing the scope for the exercise of arbitrary state power and strengthening the role of civil society. Humanity is, in one way or another, becoming a self-conscious, active agent in the operation of its own life support system (Lomborg 2001).

This process is still in train, and where it may lead remains quite uncertain. However, three broad philosophical approaches can be discerned in the growing debate about dealing with the changing global environment (Steffen et al. 2004; Lomborg 2001).

Business-as-usual. In this conceptualisation of the next stage of the Anthropocene, the institutions and economic system that have driven the Great Acceleration continue to dominate human affairs. This approach is based on several assumptions. First, global change will not be severe or rapid enough to cause major disruptions to the global economic system or to other important aspects of societies, such as human health. Second, the existing market-oriented economic system can deal autonomously with any adaptations that are required. This assumption is based on the fact that as societies have become wealthier, they have dealt effectively with some local and regional pollution problems (Rahmstorf 2007). Examples include the clean-up of major European rivers and the amelioration of the acid rain problem in western Europe and eastern North America. Third, resources required to mitigate global change proactively would be better spent on more pressing human needs.

The business-as-usual approach appears, on the surface, to be a safe and conservative way forward. However, it entails considerable risks. As the Earth System changes in response to human activities, it operates at a time scale that is mismatched with human decision-making or with the workings of the economic system. The long-term momentum built into the Earth System means that by the time humans realise that a business-as-usual approach may not work, the world will be committed to further decades or even centuries of environmental change. Collapse of modern, globalised society under uncontrollable environmental change is one possible outcome.

An example of this mis-match in time scales is the stability of the cryosphere, the ice on land and ocean and in the soil. Depending on the scenario and the model, the *Intergovernmental Panel on Climate Change* (IPCC) (Galloway/Cowling 2002) projected a global average warming of 1.1–6.4 °C for 2094–2099 relative to 1980–1999, accompanied by a projected sea-level rise of 0.18–0.59 m (excluding contributions from the dynamics of the large polar ice sheets). However, warming is projected to be more than twice as large as the global average in the polar regions, enhancing ice sheet instability and glacier melting. Recent observations of glacial dynamics suggest a higher degree of instability than estimated by current cryospheric models, which would lead to higher sea level rise through this century than estimated by the IPCC in 2001 (Schellnhuber et al. 2006). It is now conceivable that an irreversible threshold could be crossed in the next several decades, eventually (over centuries or a millennium) leading to the loss of the Greenland ice sheet and consequent sea-level rise of about 5 m.

Mitigation. An alternative pathway into the future is based on the recognition that the threat of further global change is serious enough that it must be dealt with proactively. The mitigation pathway attempts to take the human pressure off of the Earth System by vastly improved technology and management, wise use of Earth's resources, control of human and domestic animal population, and overall careful use and restoration of the natural environment. The ultimate goal is to reduce the human modification of the global environment to avoid dangerous or difficult-to-control levels and rates of change (Steffen 2002), and ultimately to allow the Earth System to function in a pre-Anthropocene way.

Technology must play a strong role in reducing the pressure on the Earth System (Haberl 2006). Over the past several decades rapid advances in transport, energy, agriculture, and other sectors have led to a trend of dematerialisation in several advanced economies. The amount and value of economic activity continue to grow but the amount of physical material flowing through the economy does not.

There are further technological opportunities. Worldwide energy use is equivalent to only 0.05% of the solar radiation reaching the continents. Only 0.4% of the incoming solar radiation, 1 W m^{-2}, is converted to chemical energy by photosynthesis on land. Human appropriation of net primary production is about 10%, including agriculture, fiber, and fisheries (Fischer et al. 2007). In addition to the many opportunities for energy conservation, numerous technologies—from solar thermal and photovoltaic through nuclear fission and fusion to wind power and biofuels from forests and crops—are available now or under development to replace fossil fuels.

Although improved technology is essential for mitigating global change, it may not be enough on its own. Changes in societal values and individual behaviour will likely be necessary (Rahmstorf et al. 2007). Some signs of these changes are now evident, but the Great Acceleration has considerable momentum and appears to be intensifying (Andreae et al. 2005). The critical question is whether the trends of dematerialisation and shifting societal values become strong enough to trigger a transition of our globalising society towards a much more sustainable one.

Geo-engineering options. The severity of global change, particularly changes to the climate system, may force societies to consider more drastic options. For example, the anthropogenic emission of aerosol particles (e.g., smoke, sulphate, dust, etc.) into the atmosphere leads to a net cooling effect because these particles and their influence on cloud properties enhance backscattering of incoming solar radiation. Thus, aerosols act in opposition to the greenhouse effect, masking some of the warming we would otherwise see now (Friedlingstein et al. 2006). Paradoxically, a clean-up of air pollution can thus increase greenhouse warming, perhaps leading to an additional 1 °C of warming and bringing the Earth closer to 'dangerous' levels of climate change. This and other amplifying effects, such as feedbacks from the carbon cycle as the Earth warms (Intergovernmental Panel on Climate Change 2005), could render mitigation efforts largely ineffectual. Just to stabilise the atmospheric concentration of CO_2, without taking into account these amplifying effects, requires a reduction in anthropogenic emissions by more than 60%—a herculean task considering that most people on Earth, in order to increase their standard of living, are in need of much additional energy. One engineering approach to reducing the amount of CO_2 in the atmosphere is its sequestration in underground reservoirs (The Royal Society 2005). This 'geosequestration' would not only alleviate the pressures on climate, but would also lessen the expected acidification of the ocean surface waters, which leads to dissolution of calcareous marine organisms (Schneider 2001).

In this situation some argue for geo-engineering solutions, a highly controversial topic. Geo-engineering involves purposeful manipulation by humans of global-scale Earth System processes with the intention of counteracting anthropogenically driven environmental change such as greenhouse warming (Crutzen 2006). One proposal is based on the cooling effect of aerosols noted in the previous paragraph (Raupach et al. 2007). The idea is to artificially enhance the Earth's albedo by releasing sunlight-reflective material, such as sulphate particles, in the stratosphere, where they remain for 1–2 years before settling in the troposphere. The sulphate particles would be produced by the oxidation of SO_2, just as happens during volcanic eruptions. In order to compensate for a doubling of CO_2, if this were to happen, the input of sulphur would have to be about 1–2 Tg S y^{-1} (compared to an input of about 10 Tg S by Mount Pinatubo in 1991). The sulphur injections would have to occur for as long as CO_2 levels remain high.

Looking more deeply into the evolution of the Anthropocene, future generations of *H. sapiens* will likely do all they can to prevent a new ice-age by adding powerful artificial greenhouse gases to the atmosphere. Similarly, any drop in CO_2 levels to low concentrations, causing strong reductions in photosynthesis and agricultural productivity, might be combated by artificial releases of CO_2, maybe from earlier

CO_2 sequestration. And likewise, far into the future, *H. sapiens* will deflect meteorites and asteroids before they could hit the Earth.

For the present, however, just the suggestion of geoengineering options can raise serious ethical questions and intense debate. In addition to fundamental ethical concerns, a critical issue is the possibility for unintended and unanticipated side effects that could have severe consequences. The cure could be worse than the disease. For the sulphate injection example described above, the residence time of the sulphate particles in the atmosphere is only a few years, so if serious side-effects occurred, the injections could be discontinued and the climate would relax to its former high CO_2 state within a decade.

The Great Acceleration is reaching criticality (Fig. 9.4). Enormous, immediate challenges confront humanity over the next few decades as it attempts to pass through a bottleneck of continued population growth, excessive resource use and environmental deterioration. In most parts of the world the demand for fossil fuels overwhelms the desire to significof ecosystem servicesantly reduce greenhouse gas emissions. About 60% are already degraded and will continue to degrade further unless significant societal changes in values and management occur (Pimm et al.

Fig. 9.4 The observed trajectory from 1850 to 2005 of carbon emissions due to fossil fuel combustion.[1] Note the acceleration in emissions since 2000. The gap between current emission rates and those required to stabilise atmospheric CO_2 concentration at various levels (450, 650, and 1000 ppm) is growing rapidly

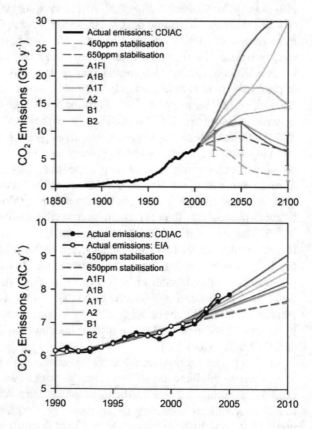

[1] This paper grew out of discussions at the 96th Dahlem Conference ("*Integrated History and future of People on Earth* [IHOPE]"), held in Berlin in June 2005. We are grateful to the many colleagues at the Conference who contributed to the stimulating discussions, and to Dr. Julia Lupp, the Dahlem Conference organiser, for permission to base this paper on these discussions.

1995). There is also evidence for radically different directions built around innovative, knowledge-based solutions. Whatever unfolds, the next few decades will surely be a tipping point in the evolution of the Anthropocene.

References

Alroy, J., 2001: "A multispecies overkill simulation of the End-Pleistocene Megafaunal mass extinction", in: *Science, 292:* 1893–1896.

Andreae, M.O.; Jones, C.D.; Cox, P.M., 2005: "Strong present day aerosol cooling implies a hot future", in: *Nature, 435:* 1187–1190.

Barnola, J.-M.; Raynaud, D.; Lorius, C.; Barkov, N.I., 2003: "Historical CO_2 record from the Vostok ice core", in: *Trends: A Compendium of Data on Global Change* (Oak Ridge, TN: U.S. Department of Energy, Oak Ridge National Laboratory, Carbon Dioxide Information Analysis Center).

Blunier, T.; Chappellaz, J.; Schwander, J.; Barnola, J.-M.; Desperts, T.; Stauffer, B.; Raynaud, D., 1993: "Atmospheric methane record from a Greenland ice core over the last 1000 years", in: *J. Geophys. Res., 20*: 2219–2222.

Brimblecombe, P., 1987: *The Big Smoke: A History of Air Pollution in London since Medieval Times* (London: Methuen).

Broecker, W.C.; Stocker, T.F., 2006: "The Holocene CO_2 rise: anthropogenic or natural?", in: *Eos, 87,3:* 27–29.

Clark, W.C., 1986: Chapter 1., in: Clark, W.C.; Munn, R.E. (eds.): *Sustainable Development of the Biosphere* (Cambridge, UK: Cambridge University Press).

Costanza, R.; Graumlich, L.; Steffen, W. (eds), 2006: *Integrated History and Future of People on Earth.* Dahlem Workshop Report 96 (Cambridge, MA: MIT Press).

Crutzen, P.J., 1995: "My life with O_3, NOx and other YZOxs", in: *Les Prix Nobel (The Nobel Prizes) 1995* (Stockholm: Almqvist & Wiksell International): 123–157.

Crutzen, P.J., 2002: "Geology of mankind: the Anthropocene", in: *Nature, 415:* 23.

Crutzen, P.J., 2006: "Albedo enhancement by stratospheric sulfur injections: A contribution to resolve a policy dilemma", in: *Clim. Chang., 77*: 211–220.

Encyclopaedia Britannica, 1976: *Micropadia*, IX (London: Encyclopaedia Britannica).

EPICA Community Members, 2004: "Eight glacial cycles from an Antarctic ice core", in: *Nature, 429:* 623–628.

Etheridge, D.M.; Steele, L.P.; Langenfelds, R.L.; Francey, R.J.; Barnola, J.-M.; Morgan, V.I., 1996: "Natural and anthropogenic changes in atmospheric CO_2 over the last 1000 years from air in Antarctic ice and firn", in: *J. Geophys. Res., 101:* 4115–4128.

Etheridge, D.M.; Steele, L.P.; Langenfelds, R.L.; Francey, R.J.; Barnola, J.-M.; Morgan, V.I., 1998: "Historical CO_2 records from the Law Dome DE08, DE08-2, and DSS ice cores", in: *Trends: A Compendium of Data on Global Change* (Oak Ridge, TN: U.S. Department of Energy, Oak Ridge National Laboratory, Carbon Dioxide Information Analysis Center).

Fischer, J.; Manning, A.D.; Steffen, W.; Rose, D.B.; Danielle, K.; Felton, A.; Garnett, S.; Gilna, B. et al., 2007: "Mind the sustainability gap", in: *Trends Ecol. Evol.*, in press.

Friedlingstein, P.; Cox, P.; Betts, R.; Bopp, L.; von Bloh, W.; Brovkin, V.; Doney, V.S.; Eby, M.I., et al., 2006: "Climate-carbon cycle feedback analysis, results from the C^4MIP model intercomparison", in: *J. Clim., 19:* 3337–3353.

Galloway, J.N.; Cowling, E.B., 2002: "Reactive nitrogen and the world: two hundred years of change", in: *Ambio, 31:* 64–71.

Haberl, H., 2006: "The energetic metabolism of the European Union and the United States, decadal energy inputs with an emphasis on biomass", in: *J. Ind. Ecol., 10:* 151–171.

Hansen, J.; Nazarenko, L.; Ruedy, R.; Sato, M.; Willis, J.; Del Genio, A.; Koch, D.; Lacis, A., et al. 2005: "Earth's energy imbalance: comfirmation and implications", in: *Science, 308:* 1431–1435.

Hartwell, R., 1962: "A revolution in the iron and coal industries during the Northern Sung", in: *J. Asian Stud., 21:* 153–162.

Hartwell, R., 1967: "A cycle of economic change in Imperial China: coal and iron in northeast China, 750–1350", in: *J. Soc. and Econ. Hist. Orient, 10:* 102–159.

Hibbard, K.A.; Crutzen, P.J.; Lambin, E.F.; Liverman, D.; Mantua, N.J.; McNeill, J.R.; Messerli, B.; Steffen, W., 2006: "Decadal interactions of humans and the environment", in: Costanza, R.; Graumlich, L.; Steffen, W. (eds): *Integrated History and Future of People on Earth.* Dahlem Workshop Report 96 (Cambridge, MA: MIT Press): 341–375.

Indermuhle, A.; Stocker, T.F.; Fischer, H.; Smith, H.J.; Joos, F.; Wahlen, M.: Deck, B.; Mastroianni, D., et al., 1999: "High-resolution Holocene CO_2-record from the Taylor Dame ice core (Antarctica)", in: *Nature, 398:* 121–126

Intergovernmental Panel on Climate Change (IPCC), 2005: *Carbon Dioxide Capture and Storage. A Special Report of Working Group III* (Geneva, Switzerland: Intergovernmental Panel on Climate Change).

Intergovernmental Panel on Climate Change (IPCC), 2007: *Climate Change 2007: The Physical Science Basis. Summary for Policymakers* (Geneva, Switzerland: Intergovernmental Panel on Climate Change).

Joos, F.; Gerber, S.; Prentice, I.C.; Otto-Bliesner, B.L.; Valdes, P.J., 2004: "Transient simulations of Holocene atmospheric carbon dioxide and terrestrial carbon since the Last Glacial Maximum", in: *Global Biogeochem. Cycles* 18, GB2002.

Keeling, C.D.; Whorf, T.P., 2005: "Atmospheric CO_2 records from sites in the SIO air sampling network", in: *Trends: A Compendium of Data on Global Change* (Oak Ridge, TN: U.S. Department of Energy, Oak Ridge National Laboratory, Carbon Dioxide Information Analysis Center).

Lambin, E.F.; Geist, H.J. (eds), 2006: *Land-Use and Land-Cover Change: Local Processes and Global Impacts.* The IGBP Global Change Series (Berlin, Heidelberg, New York: Springer-Verlag).

Leach, H.M., 2003: "Human domestication reconsidered", in: *Curr. Anthropol., 44:* 349–368.

Lomborg, B., 2001: *The Skeptical Environmentalist: Measuring the Real State of the World* (Cambridge, UK: Cambridge University Press).

Machida, T.; Nakazawa, T.; Fujii, Y.; Aoki, S.; Watanabe, O., 1995: "Increase in the atmospheric nitrous oxide concentration during the last 250 years", in: *Geophys. Res. Lett., 22:* 2921–2924.

Mackenzie, F.T.; Ver, L.M.; Lerman, A., 2002: "Century-scale nitrogen and phosphorus controls of the carbon cycle", in: *Chem. Geol., 190*: 13–32.

Marsh, G.P., 1965: *The Earth as Modified by Human Action* (Cambridge, MA: Belknap Press, Harvard University Press).

Martin, P.S.; Klein, R.G., 1984: *Quaternary Extinctions: A Prehistoric Revolution* (Tucson: University of Arizona Press).

McNeill, J.R., 2001: *Something New Under the Sun* (New York, London: W.W. Norton).

Millennium Ecosystem Assessment, 2005: *Ecosystems & Human Wellbeing: Synthesis* (Washington: Island Press).

Mokyr, J. (ed.), 1999: *The British Industrial Revolution: An Economic Perspective* (Boulder, CO: Westview Press).

Oldfield, F.; Steffen, W. 2004: "The earth system", in: *Global Change and the Earth System: A Planet Under Pressure (Heidelberg Berlin: Springer).*

Pimm, S.L.; Russell, G.J.; Gittleman, J.L.; Brooks, T.M., 1995: "The future of biodiversity", in: *Science, 269:* 347–350.

Pyne, S., 1997: *World Fire: The Culture of Fire on Earth* (Seattle: University of Washington Press).

Rahmstorf, S., 2007: "A semi-empirical approach to projecting future sea-level rise", in: *Science, 315*: 368–370.

Rahmstorf, S.; Cazenave, A.; Church, J.A.; Hansen, J.E.; Keeling, R.F.; Parker, D.E.; Somerville, R.C.J., et al. 2007: "Recent climate observations compared to projections", in: *Science, 316:* 709.

Raupach, M.R.; Marland, G.; Ciais, P.; Le Quere, C.; Canadell, J.G.; Klepper, G.; Field, C.B., 2007: "Global and regional drivers of accelerating CO_2 emissions", in: *Proc. Nat. Acad. Sci. USA*.

Roberts, R.G.; Flannery, T.F.; Ayliffe, L.K.; Yoshida, H.; Olley, J.M.; Prideaux, G.J.; Laslett, G.M.; Baynes, A., et al., 2001: "New ages for the last Australian Megafauna: continent-wide extinction about 46,000 years ago", in: *Science, 292:* 1888–1892.

Ruddiman, W.F., 2003: "The anthropogenic greenhouse era began thousands of years ago", in: *Climat. Chang., 61:* 261–293.

Schellnhuber, H.-J., 1998: "Discourse: Earth System analysis: the scope of the challenge", in: Schellnhuber, H.-J.; Wetzel, V. (eds.): *Earth System Analysis* (Berlin, Heidelberg, New York: Springer-Verlag): 3–195.

Schellnhuber, H.J.; Cramer, W.; Nakicenovic, N.; Wigley, T.; Yohe, G. (eds.), 2006: *Avoiding Dangerous Climate Change* (Cambridge, UK: Cambridge University Press).

Schneider, S.H., 2001: "Earth systems engineering and management", in: *Nature, 409:* 417–421.

Sieferle, R.-P., 2001: *Der Europäische Sonderweg: Ursachen und Faktoren* (Stuttgart: Breuninger Stiftung).

Smith, B.D., 1995: *The Emergence of Agriculture* (New York: Scientific American Library).

Steffen, W., 2002: "Will technology spare the planet?", in: Steffen, W.; Jäger, J.; Carson, D.; Bradshaw, C. (eds.): *Challenges of a Changing Earth: Proceedings of the Global Change Open Science Conference. Amsterdam, The Netherlands, 10–13 July 2001.* The IGBP Global Change Series (Berlin, Heidelberg, New York: Springer-Verlag): 189–191.

Steffen, W.; Sanderson, A.; Tyson, P.; Jäger, J.; Matson, P.; Moore, B. III; Oldfield, F., Richardson, K., et al. (eds.), 2004: *Global Change and the Earth System: A Planet Under Pressure.* The IGBP Global Change Series (Berlin, Heidelberg, New York: Springer-Verlag): 7.

Steffen, W.; Sanderson, A.; Tyson, P.D.; Jäger, J.; Matson, P.; Moore, B. III; Oldfield, F., Richardson, K., et al., 2004: *Global Change and the Earth System: A Planet Under Pressure.* The IGBP Global Change Series (Berlin, Heidelberg, New York: Springer-Verlag).

TeBrake, W.H., 1975: "Air pollution and fuel crisis in preindustrial London, 1250–1650", in: *Technol. Culture, 16:* 337–359.

The Royal Society, 2005: *Ocean Acidification Due to Increasing Atmospheric Carbon Dioxide.* Policy document 12/05 (London: The Royal Society, UK).

Tobias, P.V., 1976: "The brain in hominid evolution", in: *Encyclopaedia Britannica,* Macropaedia. Vol. 8 (London: Encyclopedia Britannica): 1032.

Vernadski, V.I., 1998: *The Biosphere (translated and annotated version from the original of 1926),* (New York: Copernicus, Springer).

Vörösmarty, C.J.; Sharma, K.; Fekete, B.; Copeland, A.H.; Holden, J.; Marble, J.; Lough, J.A., 1997: "The storage and aging of continental runoff in large reservoir systems of the world", in: *Ambio, 26:* 210–219.

Chapter 10
Atmospheric Chemistry and Climate in the Anthropocene (2007)

Paul J. Crutzen

In the following I shall canvass the changes that have taken place in global chemistry and climate, particularly from the beginning of the 19th century. Since that time, humankind's growing activities have plunged the Earth into a new geological era, which I call the 'era of humankind', or the Anthropocene. This is clearly visible in the effects of human activities on climate and also in the chemistry of the atmosphere, including the ozone hole.

Over the past three centuries, the human population has increased tenfold, to more than 6 billion, and in the 20th century by a factor of four. The cattle population has increased to 1.4 billion, so that on average there is now one cow per family supplying us with meat, cheese and milk. But cattle also produce methane gas, and more methane in the atmosphere means that temperatures will rise. During the last century, the cattle population of the world has gone up by a factor of four.

There are many other ways in which humankind is affecting the environment. Urbanisation grew more than tenfold in the last century, so that almost half of the population now lives in cities and megacities. Industrial output increased forty times during the same period and energy use sixteen times. Almost 50% of the land surface has been transformed by human activities.

Water use increased ninefold during the past century to about 800 cubic metres per capita per year—65% of which is used for irrigation, 25% in industry and 10% in households. The way we use water is not the most efficient; for instance, according to *New Scientist* (Pearce 2006), it takes about 20,000 l of water to grow one kilogram of coffee, 11,000 l of water to make a quarter-pound hamburger, and 5,000 l of water to make one kilogram of cheese. So it is no wonder that we have a problem with the water supply.

This text was first published as: Crutzen, P.J., 2007: "Atmospheric Chemistry and Climate in the Anthropocene", in: Bindé, J. (Ed.): *Making Peace with the Earth – What Future for the Human Species and the Planet?* (Oxford – New York: Berghahn Books - UNESCO Publishing): 113–120. The permission was granted on 20 August 2018 by Dr. Marion Berghahn, Publisher.

Human appropriation of terrestrial net primary productivity is now about 30%, although the number is rather uncertain. Fish catch increased forty times. The rise of *sulphur dioxide* (SO_2) to the atmosphere—amounting to some 100 million tons annually by coal and oil burning—is at least twice the sum of all natural emissions. And overland the increase has been sevenfold, causing acid rain, health effects, poor visibility and climate change due to sulphate aerosols. Releases *of nitrous dioxide* (NO_2) into the atmosphere from fossil fuel and biomass burning are larger than the natural inputs, causing high ozone levels during summertime over vast regions of the globe. Several climatically important greenhouse gases have substantially increased in the atmosphere, for instance, carbon dioxide (CO_2) by 30% and *methane* (CH_4) by over 100%.

Humanity is also responsible for the presence of many toxic substances in the environment, and also some that are not toxic at all but which have nevertheless led to the ozone hole. I am referring here, of course, to *chlorofluorocarbon* (CFC) gases. This year there was a record ozone hole over Antarctica, which was quite unexpected because we thought that, after the regulations passed against the use of CFC gases, the situation would improve. This situation is important because the ozone layer protects us from harmful ultra-violet radiation, causing an increase in skin cancer and other effects in the biosphere.

The use of nitrogen in agriculture and also as a by-product of industry and transport has more than doubled the amount of fixed nitrogen emissions in the environment. This, despite the fact that we directly use only 10–15% of that nitrogen in the food we eat—the rest being wasted. There is now very little doubt that temperatures on Earth is increasing; they have risen especially over the last twenty to thirty years.

We do not see any good reason for the rise in global mean temperatures other than human activities involving emissions of carbon dioxide and other greenhouse gases in the atmosphere. The warming of the atmosphere by greenhouse gases is a well-known phenomenon.

Variations of the Earth's surface temperature for the past 140 years

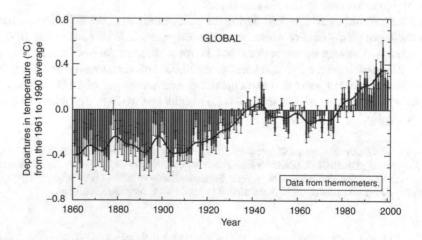

◀**Fig. 10.1** Variations of the Earth's surface temperature for the last 140 years. There are uncertainties in the annual data due to data gaps, random instrumental errors and uncertainties, uncertainties in bias corrections in the ocean surface temperature data and also in adjustments for urbanisation over the land. Over both the last 140 years and 100 years, the best estimate is that the global average surface temperature has increased by 0.6 ± 0.2 °C. *Source* International Panel on Climate Change, *Climate Change 2001: The Scientific Basis,* Technical Summary (Fig. 10.1)

But there are also ingredients that cool the atmosphere. These are aerosol particles, small particles in the atmosphere which reflect solar radiation back to space.

The Global Mean Radiative Forcing of the Climate System for the Year 2000, Relative to 1750

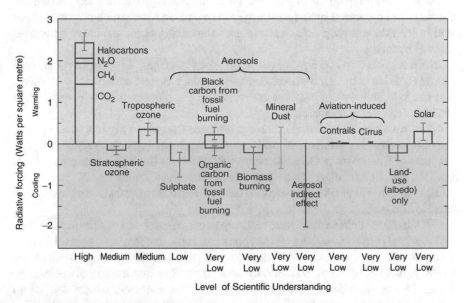

Fig. 10.2 What do we know about global energy variations? Earth receives energy from the sun in the form of ultra violet radiations and emits energy back in the form of infrared radiations. Over an average year the budget of these energy exchanges usually leads to a state of equilibrium. A modification in this equilibrium, from either more incoming energy or more outgoing energy, is termed radiative forcing. This modification can result from a concentration of greenhouse gases, to variations in the activity of the sun, to land use, etc. Figure 10.2 shows that our current scientific knowledge is uneven: there is more knowledge on forcing due to greenhouse gases than on aerosols effects. *Source* International Panel on Climate Change: *Climate Change 2001: The Scientific Basis,* Technical Summary

These serve as cloud condensation nuclei, which means that the clouds formed in the polluted atmosphere contain many more small particles and are thus much more effective in reflecting solar radiation back into space. So part of the greenhouse warming is actually neutralised by the presence of air pollution in the atmosphere. This is creating a dilemma for policy-makers, since we want to get rid of air pollution in the lower atmosphere, but when we do that, temperatures will go up.

What should we do to stabilise the amount of carbon dioxide in the atmosphere? We have to aim at reductions in emissions of more than 60% in the case of carbon dioxide. And that is in a world in which many people in developing countries are not even contributing in any real way to carbon dioxide emissions.

The increase in the atmosphere in concentrations of methane, another greenhouse gas, has stopped at the moment. That does not mean that it will always be the case, because by warming up the higher latitude permafrost regions, methane and carbon dioxide can be released again increasingly in the atmosphere, which may well mean that in the future atmospheric methane may start rising again and cause additional climate warming.

Nitrous oxide (N_2O) is another powerful greenhouse gas. To stabilise the amount of nitrous oxide in the atmosphere, we have to reduce current emissions by 70–80%. Yet that will be very difficult, because much of the nitrous oxide is coming from nitrogen fertilizer in agriculture.

Then we have CFC gases. Fortunately, we have a success story in this case, since they are no longer produced, or only in very small amounts. So, very slowly, the amounts of CFC gases in the atmosphere will decrease. But it will take a long time, until the end of this century, before they have been reduced enough for the ozone hole to disappear. It will take so long because the lifetime of these gases in the atmosphere is 50 to 100 years.

We should not believe that nature will help us out. Measurements of carbon dioxide and methane in the atmosphere and temperatures show that they are positively correlated. So when temperatures rise in the atmosphere in keeping with the Milankovitch cycle, namely in terms of Earth's orbit around the Sun, the heating effect is reinforced by the greenhouse gases, carbon dioxide and methane, which also go up when temperatures go up. So nature is not stabilising the climate (Fig. 10.3).

We know that the high-latitude regions are comparatively most affected. The doubling of the amount of CO_2 in the atmosphere at high latitudes can lead to temperature increases by up to 6-8 °C, which means that the permafrost regions may melt, leading to the release of carbon dioxide and methane into the atmosphere and further warming.

Possible evolutions of global temperatures

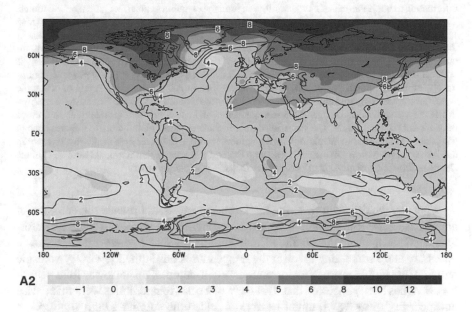

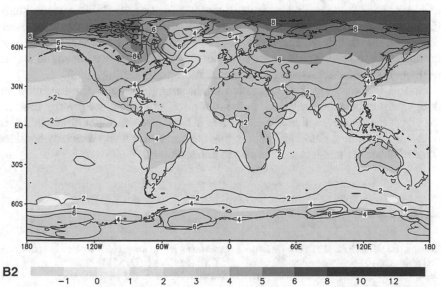

◄**Fig. 10.3** Possible evolutions of global temperatures. The IPCC uses state of the art scientific information to build models of how our different human behaviours can affect global temperatures. The two maps in the figure are projections based on two different scenarios. The A2 storyline (map 1) describe a very heterogeneous world. The underlying theme is self-reliance and preservation of local identities. Fertility patterns across regions converge very slowly, which results in a continuously increasing population. The B2 scenario (map 2) describes a world in which the emphasis is on local solutions to economic, social, and environmental sustainability. It is a world with continuously increasing global population, at a rate lower than in A2, intermediate levels of economic development, and less rapid and more diverse technological change than in the B1 and A1 storylines. While the scenario is also oriented towards environmental protection and social equity, it focuses on local and regional levels. The scenarios do not include any new global climate policies beyond what has been done so far. Scenario B2, where global warming is slower thus shows that changes in our behaviours can have a measurable effect on the climate. *Source* International Panel on Climate Change, *Climate Change 2001: The Scientific Basis,* Technical Summary

What can we do about this? In the first place, we should reduce the emission of greenhouse gases into the atmosphere. That is an absolute priority. Yet we are unfortunately far away from that condition, and therefore I wonder whether it might not be necessary to do something to lower the temperatures on Earth. After every volcanic eruption, the global average temperature is a half-degree Celsius lower for a year or two. So it has been suggested that if, in the worst case, temperatures were to increase unexpectedly quickly, we might be able to add some sulphur to the stratosphere and make sunlight-reflecting sulphate particles, as volcanoes do. In a recent study (Crutzen/Rasch 2006), Phil Rasch of the National Center for Atmospheric Research and I have shown that the addition of about 1 million tons of sulphur (as SO_2) into the stratosphere would cause a cooling effect that would largely compensate for the heating effect of a doubling of CO_2 concentrations in the atmosphere.

So this is something that humankind might attempt sometime in the future, if things were to become really bad. In the first place, of course, our priority must be to prevent the emissions of greenhouse gases in the atmosphere. But if humankind is unable to do so, then I am fairly certain that our descendants will seriously consider some unpleasant measures such as albedo enhancement.

References

Crutzen, P.J.; Rasch, P. J., 2006: "Albedo enhancement by stratospheric sulphur injections: A contribution to resolve a policy dilemma?", in: *Climatic Change,* 77: 211–220.
Pearce, F, 2006: "The parched planet", in: *New Scientist,* No. 2540 (25 February): 32.

Chapter 11
Fate of Mountain Glaciers
in the Anthropocene (2011)

A Report by the Working Group Commissioned by the Pontifical Academy of Sciences

Ajai, Lennart Bengtsson, David Breashears, Paul J. Crutzen, Sandro Fuzzi,
Wilfried Haeberli, Walter W. Immerzeel, Georg Kaser, Charles F. Kennel,
Anil Kulkarni, Rajendra Pachauri, Thomas H. Painter, Jorge Rabassa,
Veerabhadran Ramanathan, Alan Robock, Carlo Rubbia, Lynn M. Russell,
Marcelo Sánchez Sorondo, Hans Joachim Schellnhuber,
Soroosh Sorooshian, Thomas F. Stocker, Lonnie G. Thompson,
Owen B. Toon, Durwood Zaelke, and Jürgen Mittelstraß

Main Rongbuk Glacier, *Source*: Glacier Works

The working group consists of glaciologists, climate scientists, meteorologists, hydrologists, physicists, chemists, mountaineers, and lawyers organised by the Pontifical Academy of Sciences at the Vatican, to contemplate the observed retreat

This text was first published under this title at: http://www.casinapioiv.va/content/accademia/en/
events/2017/health/declaration.html.

of the mountain glaciers, its causes and consequences. This report resulted from a workshop in April 2011 at the Vatican.[1]

Declaration by the Working Group

We call on all people and nations to recognise the serious and potentially irreversible impacts of global warming caused by the anthropogenic emissions of greenhouse gases and other pollutants, and by changes in forests, wetlands, grasslands, and other land uses. We appeal to all nations to develop and implement, without delay, effective and fair policies to reduce the causes and impacts of climate change on communities and ecosystems, including mountain glaciers and their watersheds, aware that we all live in the same home. By acting now, in the spirit of common but differentiated responsibility, we accept our duty to one another and to the stewardship of a planet blessed with the gift of life.

We are committed to ensuring that all inhabitants of this planet receive their daily bread, fresh air to breathe and clean water to drink as we are aware that, if we want justice and peace, we must protect the habitat that sustains us. The believers among us ask God to grant us this wish.

Morteratsch glacier (Alps). Courtesy of J. Alean, SwissEduc.

---

[1] *Main Rongbuk Glacier, 11 May 2011;* Location: Mount Everest, 8848 m, Tibet Autonomous Region; Range: Mahalangur Himal, Eastern Himalayas; Coordinates: 27°59'15"N, 86°55'29"E; Elevation of Glacier: 5,060 - 6,462 m; Average Vertical Glacier Loss: 101 m, 1921 – 2008.

11.1   Summary

11.1.1   Receding Glaciers Require Urgent Responses

Kyetrak Glacier 1921. Location: Cho Oyu, 8201 m, Tibet Autonomous Region; Eastern Himalayas. Elevation of Glacier: 4,907 – 5,883 m. Courtesy of Royal Geographical Society

Kyetrak Glacier 2009; Photography 2009: Courtesy of Glacier Works

11.1.1.1   Anthropocene

Aggressive exploitation of fossil fuels and other natural resources has damaged the air we breathe, the water we drink, and the land we inhabit. To give one example, some 1000 billion tons of carbon dioxide and other climatically important 'greenhouse' gases have been pumped into the atmosphere. As a result, the concentration of carbon dioxide in the air now exceeds the highest levels of the last 800,000 years. The climatic and ecological impacts of this human interference with the Earth System are expected to last for many millennia, warranting a new name, *The Anthropocene*, for the new 'man-made' geologic epoch we are living in.

11.1.1.2   Glacier Retreat

Glaciers are shrinking in area worldwide, with the highest rates documented at lower elevations. The widespread loss of glaciers, ice, and snow on the mountains of tropical, temperate, and polar regions is some of the clearest evidence we have for a change in the climate system, which is taking place on a global scale at a rapid rate. Long-term measurement series indicate that the rate of mass loss has more than doubled since the turn of the century. Melting mountain glaciers and snows have contributed significantly to the sea level rise observed in the last century. Retreat of the glaciers in the European Alps has been observed since the end of the 'Little Ice Age' (first part of the 19th century), but the pace of retreat has been much faster since the 1980s. The Alpine glaciers have already lost more than 50% of their mass. Thousands of small glaciers in the Hindukush-Himalayan-Tibetan region continue to disintegrate, a threat to local communities and the many more people farther away who depend on mountain water resources. Robust scenario calculations clearly indicate that many mountain ranges worldwide could lose major parts of their glaciers within the coming decades.

The recent changes observed in glacial behaviour are due to a complex mix of causal factors that include greenhouse gas forcing together with large scale emissions of dark soot particles and dust in 'brown clouds', and the associated changes in regional atmospheric energy and moisture content, all of which result in significant warming at higher altitudes, not least in the Himalayas.

11.1.1.3   Perspective on Past Changes

In response to the argument that "since the Earth has experienced alternating cold periods (ice ages or glacials) and warm periods (interglacials) during the past, today's climate and ice cover changes are entirely natural events", we state:

> The primary triggers for ice ages and interglacials are well understood to be changes in the astronomical parameters related to the motion of our planet within the solar system and natural feedback processes in the climate system. The time scales between these triggers are in the range of 10,000 years or longer. By contrast, the observed human-induced changes in

carbon dioxide, other greenhouse gases, and soot concentrations are taking place on 10–100 year timescales—at least a hundred times as fast. It is particularly worrying that this release of global warming agents is occurring during an interglacial period when the Earth was already at a natural temperature maximum.

11.1.1.4   Three Recommended Measures

Human-caused changes in the composition of the air and air quality result in more than 2 million premature deaths worldwide every year and threaten water and food security—especially among those 'bottom 3 billion' people who are too poor to avail of the protections made possible by fossil fuel use and industrialisation. Since a sustainable future based on the continued extraction of coal, oil and gas in the 'business-as-usual mode' will not be possible because of both resource depletion and environmental damages (as caused, e.g., by dangerous sea level rise) we urge our societies to:

1. *Reduce worldwide carbon dioxide emissions without delay, using all means possible to meet ambitious international global warming targets and ensure the long-term stability of the climate system.* All nations must focus on a rapid transition to renewable energy sources and other strategies to reduce CO_2 emissions. Nations should also avoid removal of carbon sinks by stopping deforestation, and should strengthen carbon sinks by reforestation of degraded lands. They also need to develop and deploy technologies that draw down excess carbon dioxide in the atmosphere. These actions must be accomplished within a few decades.
2. *Reduce the concentrations of warming air pollutants (dark soot, methane, lower atmosphere ozone, and hydrofluorocarbons) by as much as 50%,* to slow down climate change during this century while preventing millions of premature deaths from respiratory disease and millions of tons of crop damages every year.
3. *Prepare to adapt to the climatic changes, both chronic and abrupt, that society will be unable to mitigate.* In particular, we call for a global capacity building initiative to assess the natural and social impacts of climate change in mountain systems and related watersheds.

The cost of the three recommended measures pales in comparison to the price the world will pay if we fail to act now.

11.2   Specific Findings and Recommendations

Mountain glaciers are lethally vulnerable to ongoing climate change

Qori Kalis outlet glacier (the largest outlet glacier from the Quelccaya Ice Cap in the southern Andes of Peru). Credit: Lonnie Thompson, Ohio State University

11.2.1 Anthropocene: A New Geological Epoch

The last two centuries have seen an unprecedented expansion of human population and exploitation of Earth's resources. This exploitation has caused increasingly negative impacts on many components of the Earth System—on the air we breathe, the water we drink, and the land we inhabit. Humanity is changing the climate system through its emissions of greenhouse gases and heat-absorbing particulate pollution. Today's atmospheric concentration of carbon dioxide, the principal greenhouse gas, exceeds all other maxima observed over the last 800,000 years. Vast transformations of the land surface, including loss of forests, grasslands, wetlands, and other ecosystems, are also causing climate change. In recognition of the fact that human activities are profoundly altering these components of the Earth System, Nobel Laureate Paul Crutzen has given the name 'Anthropocene' to the new geological epoch we have created for ourselves.

An expert group of scientists met under the auspices of the Pontifical Academy of Sciences at the Casina Pio IV in the Vatican from 2 to 4 April 2011 to discuss the fate of mountain glaciers in the Anthropocene and consider the responses required to stabilise the climate change affecting them. This group's consensus statement is a warning to humanity and a call for fast action—to mitigate global and regional warming, to protect mountain glaciers and other vulnerable ecosystems, to assess national and local climate risks, and to prepare to adapt to those climate impacts that cannot be mitigated. The group also notes that another major anthropogenic risk to the climate system is from the threat of nuclear war, which can be lessened by rapid and large reductions in global nuclear arsenals.

11.2.2 The Earth Is Warming and the Impacts of Climate Change Are Increasing

Warming of the Earth is unequivocal. Most of the observed increase in globally averaged temperature since the mid-20th century is 'very likely'—defined as more than 90% likely—to be the result of the observed increase in anthropogenic greenhouse gas concentrations. This warming is occurring in spite of masking by cooling aerosol particles—many of which are co-emitted by CO_2-producing processes.

Some of the current and anticipated impacts of climate change include losses of coral reefs, forests, wetlands, and other ecosystems; a rate of species extinction many times faster than the historic average; and water and food shortages for many vulnerable peoples. Increasing sea level rise and stronger storm surges threaten vulnerable ecosystems and peoples, especially those in low-lying islands and coastal nations. The loss of mountain glaciers discussed here threatens downstream populations, especially during the dry season when glacial runoff is most needed.

11.2.3 The Earth's Glaciers Are Retreating: Causes and Consequences

The widespread loss of ice and snow in the world's mountain glaciers is some of the clearest evidence we have for global changes in the climate system. The present losses of mountain glaciers cause more than 1 mm per year of sea level rise, or about one-third of the observed rate. In the most recent part of the Anthropocene, much of the reduction of glacier mass and length in tropical, temperate, and polar regions results from the observed increases in greenhouse gases and the increases in sunlight absorbing particles such as soot, from inefficient combustion processes, and dust, from land cover change.

As shown in the 2007 report of the Intergovernmental Panel on Climate Change, extrapolations from mass change studies carried out on about 400 selected glaciers worldwide indicate a present average annual thinning of about 0.7 m in water equivalent. The equilibrium line between accumulation and ablation area of a glacier has shifted upward by several hundred meters in most mountain ranges compared to the mid-1970s. For many glaciers in lower mountain ranges, the snow line at the end of summer is above the maximum altitude of the mountains, leaving them lethally vulnerable to ongoing climate change. Glacier fragmentation is occurring in most areas, leaving the resultant smaller glacier closer to disappearing altogether.

Glacier areas are observed to be shrinking worldwide, with the highest rates at lower elevations. Large glaciers lose their tongues, leaving unstable moraines and fragilely dammed lakes behind, such as Imja Lake in Nepal. Such fragile dams have been subject to failure, causing outburst floods that ravage the already fragile infrastructure of poor communities downstream.

In Western North America, human disturbance is increasing the dust load from the deserts of the Colorado Plateau and Great Basin, which darkens and thus shortens the snow season in the Colorado Rocky Mountains by 4–7 weeks. The dust particles also add to atmospheric warming by absorbing sunlight. Elsewhere the widespread 'brown clouds' of black carbon from inefficient combustion could have a large impact in regions such as the Himalayas. We have very limited—and in some cases no—energy and mass balance studies that quantify the black carbon effects on snow and ice in such remote mountain areas. The impacts that we do understand with detailed measurements in the Western North America provide insight into the snow and glacier responses in other similarly affected regions.

The amounts and rates of glacier mass loss differ by region, and so also do the associated impacts on seasonal water availability in close-by valleys and neighbouring lowlands. In regions with dry and warm seasons, such as Central Asia, mountains and their glaciers and winter snows are like 'water towers' that store water for millions of people. Their behaviour can be deceptive. Glacial mass loss can cause an initial temporary increase in runoff downstream from water that has been stored for a long time, as has been observed in several basins, but runoff inevitably decreases as the parent glaciers decrease further.

Mountain glaciers serve another critical function: they preserve detailed information on past climate and the ability of glaciers to respond to different climate variables. This makes glaciers powerful tools for understanding past and present climate dynamics. The full potential of mountain glaciers as climate research tools is just beginning to be realised. The additional research needed to reduce uncertainties, delineate governing processes, and quantify regional impacts could have a big payoff. It is time we pay more careful attention to mountain glaciers before their archives are lost forever.

11.2.4 Avoiding 'Dangerous Anthropogenic Interference' Requires Clear and Binding Climate Targets

The goal of a climate policy is to stabilise greenhouse gas emissions at a level that would prevent 'dangerous anthropogenic interference with the climate system' and 'allow ecosystems to adapt naturally to climate change, ensure that food production is not threatened and enable economic development to proceed in a sustainable manner', as laid down in Article 2 of the UN Framework Convention on Climate Change.

The temperature guardrail for avoiding 'dangerous anthropogenic interference' is now proposed to be at 2 °C warming (above the pre-industrial level), although many scientists argue and many nations agree that 1.5 °C is a safer upper limit. Scientific, political, and economic considerations have contributed to the identification of this threshold, which has been adopted by the international climate negotiations. The Earth has already warmed by 0.75 °C since 1900 AD, and might reach some 2 °C by the year 2100 AD, even if today's greenhouse gas concentrations are not increased further and air pollution is curbed for humanitarian reasons. There is a risk that the warming can well exceed 3 °C if emissions of greenhouse gases continue to increase at present rates. Thus exceeding the 2 °C climate target is a real and serious possibility.

11.2.5 Rapid Mitigation Is Required If Warming and Associated Impacts Are to Be Limited

Understanding the causes of climate change, as well as its current and projected impacts, presents society the opportunity to avoid unmanageable impacts through mitigation and to manage unavoidable impacts through adaptation. The time to act is now if society is to have a reasonable chance of staying below the 2 °C guardrail.

11.2.5.1 Possible Mitigation by Reducing Carbon Dioxide Emissions and Expansion of Carbon Sinks

CO_2 is the largest single contributor to greenhouse warming. While more than half of CO_2 is absorbed by ocean and terrestrial sinks within a century, approximately 20% remains in the atmosphere to cause warming for millennia. Every effort must be made to cut CO_2 direct emissions from fossil fuel burning, cut indirect emissions by avoiding deforestation, and expand forests and other sinks, as fast as possible to avoid the profoundly long warming and associated effects that CO_2 causes.

11.2.5.2 Possible Mitigation by Reducing the Emission of Non-CO_2 Short-Lived Drivers

The second part of an integrated mitigation strategy is to cut the climate forcers that have short atmospheric lifetimes. These include black carbon soot, tropospheric ozone and its precursor methane, and *hydrofluorocarbons* (HFCs). *Black carbon* (BC) and tropospheric ozone strongly impact regional as well as global warming. Cutting the short-lived climate forcers using existing technologies can reduce the rate of global warming significantly by the latter half of this century, and the rate of Arctic warming by two-thirds, provided CO_2 is also cut.

Reducing local air pollutants can save about 2 million lives each year, increase crop productivity, and repair the ability of plants to sequester carbon. Black carbon management should be part of an integrated aerosol management strategy, to ensure that BC warming is cut faster than the cooling from other aerosols. In many areas, there is a real potential to reduce the BC and dust loading that accelerates glacier and snow melt, by: reducing BC emissions from traditional cook stoves by replacing them with energy efficient and less polluting cook stoves, trapping BC from diesel combustion with filters, and restabilising desert surfaces and other soils to reduce their dust emissions.

HFCs are synthetic gases and are the fastest growing climate forcer in many countries. The production and use of HFCs can be phased down under the Montreal Protocol on Substances that Deplete the Ozone Layer, while leaving the downstream emissions of HFCs in the Kyoto Protocol. This would provide the equivalent of 100 giga tonnes of CO_2 in mitigation by 2050 or earlier. The Montreal Protocol is widely considered the world's best environmental treaty; it has already phased out 98% of nearly 100 chemicals that are similar to HFCs, for a net of 135 giga tonnes of climate mitigation between 1990 and 2010.

In sum, air pollution and climate change policies are still treated as if they were two separate problems, when they actually represent the same scourge. Emission sources for air pollutants and greenhouse gases coincide, and a combined policy strategy reduces the cost of counteracting both these threats to human health and the wellbeing of society. These mitigation strategies must be pursued simultaneously and as aggressively as the dictates of science demand. Together they have the potential to restore the climate system to a safe level, and reduce climate injustice. But time is short. Warming and associated effects in the Earth System caused by the cumulative CO_2 emissions that remain in the atmosphere for millennia may soon become unmanageable.

11.2.5.3   Adaptation Must Begin Now

Because of the time lag between mitigation action and climate response, vulnerable ecosystems and populations will face significant climate impacts and possibly unacceptable risks even with ultimately successful mitigation. Therefore, in addition to mitigation, adaptation must also start now and be pursued aggressively.

We cannot adapt to changes we cannot understand. Adaptation starts with assessment. An international initiative to observe and model mountain systems and their watersheds with high spatial resolution, realistic topography, and processes appropriate to high altitudes is a prerequisite to strengthening regional and local capacities to assess the natural and social impacts of climate change.

11.2.5.4   Glacier Measurements Need to Be Expanded and Improved

We need to characterise the critical climate and radiative forcings on mountain glaciers and their corresponding responses that are not yet sufficiently understood. Among these, we must improve our understanding of the regional differences in glacial response around the globe in terms of the regional changes in climate and in absorbing impurities. Our observations of the glacier volumes, precipitation, and respective changes in mountain catchments are severely limited. This limits our ability to create scenarios of future runoff. Our climate models cannot resolve the rough terrain of mountainous regions and therefore poorly represent precipitation, temperature variations, and capture of aerosol loading. Likewise, our modelling and monitoring of the connections between the changes in an upstream glacierised water catchment to water resources at the downstream basin scale are at the initial stages.

The remoteness and dangerous nature of work above 6000 m is one reason why we have few detailed measurements, other than of glacier length and size, in high mountain systems like the Himalayas and Andes. Current remote sensing technologies can detect changes in glacier and snow extent, but do not quantify relative forcings or provide important snow and ice properties, such as grain size, local impurities, and surface liquid water content. However, airborne and space-borne imaging spectrometers will soon allow us to make spatially comprehensive measurements of these surface properties. Put in context by more extensive observations from large-scale field campaigns, and in situ energy balance and mass balance measurements, imaging spectrometers will be used to construct and validate the next generation of high resolution glacier mass balance models. Quantitative observations are the key.

11.2.5.5 Geoengineering: Further Research and International Assessment Are Required

Geoengineering is no substitute for climate change mitigation. There are many questions that need to be answered about potential irreversibilities, and of the disparities in regional impacts, for example, before geoengineering could be responsibly considered. There has not been a dedicated international assessment of geoengineering. Geoengineering needs a broadly representative, multi-stakeholder assessment performed with the highest standards, based for example on the IPCC model. The foundation for such an assessment has to be much broader with deeper scientific study than there has been a chance to carry out thus far.

It may be prudent to consider geo-engineering if irreversible and catastrophic climate impacts cannot be managed with mitigation and adaptation. A governance system for balancing the risks and benefits of geoengineering, and a transparent, broadly consultative consensus decision-making process to determine what risks are acceptable must be developed before any action can be taken.

11.2.5.6 Individuals and Nations Have a Duty to Act Now

Humanity has created the *Anthropocene* era and must live with it. This requires a new awareness of the risks human actions are having on the Earth and its systems, including the mountain glaciers discussed here. It imposes a new duty to reduce these risks. Failure to mitigate climate change will violate our duty to the vulnerable of the Earth, including those dependent on the water supply of mountain glaciers, and those facing rising sea level and stronger storm surges. Our duty includes the duty to help vulnerable communities adapt to changes that cannot be mitigated. All nations must ensure that their actions are strong enough and prompt enough to address the increasing impacts and growing risk of climate change and to avoid catastrophic irreversible consequences.

We call on all people and nations to recognise the serious and potentially irreversible impacts of global warming caused by the anthropogenic emissions of greenhouse gases and other pollutants, and by changes in forests, wetlands, grasslands, and other land uses. We appeal to all nations to develop and implement, without delay, effective and fair policies to reduce the causes and impacts of climate change on communities and ecosystems, including mountain glaciers and their watersheds, aware that we all live in the same home. By acting now, in the spirit of common but differentiated responsibility, we accept our duty to one another and to the stewardship of a planet blessed with the gift of life. We are committed to ensuring that all inhabitants of this planet receive their daily bread, fresh air to breathe and clean water to drink, as we are aware that, if we want justice and peace, we must protect the habitat that sustains us.

Chapter 12
Living in the Anthropocene: Toward a New Global Ethos (2011)

Paul J. Crutzen and C. Schwägerl

A decade ago, Nobel Prize-winning scientist Paul Crutzen first suggested we were living in the 'Anthropocene', a new geological epoch in which humans had altered the planet. Now, in an article for *Yale Environment 360*, Crutzen and a coauthor explain why adopting this term could help transform the perception of our role as stewards of the Earth.

It's a pity we're still officially living in an age called the Holocene. The Anthropocene—human dominance of biological, chemical and geological processes on Earth—is already an undeniable reality. Evidence is mounting that the name change suggested by one of us more than ten years ago is overdue. It may still take some time for the scientific body in charge of naming big stretches of time in Earth's history, the International Commission on *Stratigraphy* (ICS), to make up its mind about this name change. But that shouldn't stop us from seeing and learning what it means to live in this new Anthropocene epoch, on a planet that is being anthroposised at high speed.

For millennia, humans have behaved as rebels against a superpower we call 'Nature'. In the 20th century, however, new technologies, fossil fuels, and a fast-growing population resulted in a 'Great Acceleration' of our own powers. Albeit clumsily, we are taking control of Nature's realm, from climate to DNA. We humans are becoming the dominant force for change on Earth. A long-held religious and philosophical idea—humans as the masters of planet Earth—has turned into a stark reality. What we do now already affects the planet of the year 3000 or even 50,000.

This text was first published as: Crutzen, P.J.; Schwägerl, C., 2011; "Living in the Anthropocene: Toward a New Global Ethos", in: *Yale Environment 360* (New Haven: Yale, Yale School of Forestry & Environmental Studies, 24 January). *Christian Schwägerl* is a Berlin-based journalist who writes for *GEO* magazine, the German newspaper *Frankfurter Allgemeine*, and other media outlets. He is co-founder of *Riff Reporter*, a freelance cooperative, and author of *The Anthropocene: The Human Era and How it Shapes Our Planet*.

© The Author(s), under exclusive license to Springer Nature Switzerland AG 2021   141
S. Benner et al. (eds.), *Paul J. Crutzen and the Anthropocene: A New Epoch in Earth's History*, The Anthropocene: Politik–Economics–Society–Science 1,
https://doi.org/10.1007/978-3-030-82202-6_12

Changing the climate for millennia to come is just one aspect. By cutting down rainforests, moving mountains to access coal deposits and acidifying coral reefs, we fundamentally change the biology and the geology of the planet. While driving uncountable numbers of species to extinction, we create new life forms through gene technology, and, soon, through synthetic biology.

Human population will approach ten billion within the century. We spread our man-made ecosystems, including 'mega-regions' with more than 100 million inhabitants, as landscapes characterised by heavy human use—degraded agricultural lands, industrial wastelands, and recreational landscapes—become characteristic of Earth's terrestrial surface. We infuse huge quantities of synthetic chemicals and persistent waste into Earth's metabolism. Where wilderness remains, it's often only because exploitation is still unprofitable. Conservation management turns wild animals into a new form of pets.

It's no longer us against 'Nature.' It's we who decide what nature is what it will be.

Geographers Erle Ellis and Navin Ramankutty argue we are no longer disturbing natural ecosystems. Instead, we now live in "human systems with natural ecosystems embedded within them". The long-held barriers between nature and culture are breaking down. It's no longer us against 'Nature'. Instead, it's we who decide what nature is and what it will be.To master this huge shift, we must change the way we perceive ourselves and our role in the world. Students in school are still taught that we are living in the Holocene, an era that began roughly 12,000 years ago at the end of the last Ice Age. But teaching students that we are living in the Anthropocene, the Age of Men, could be of great help. Rather than representing yet another sign of human hubris, this name change would stress the enormity of humanity's responsibility as stewards of the Earth. It would highlight the immense power of our intellect and our creativity, and the opportunities they offer for shaping the future.

If one looks at how technology and cultures have changed since 1911, it seems that almost anything is possible by the year 2111. We are confident that the young generation of today holds the key to transforming our energy and production systems from wasteful to renewable and to valuing life in its diverse forms. The awareness of living in the Age of Men could inject some desperately needed eco-optimism into our societies.

To accommodate the Western lifestyle for 9 billion people, we'd need several more planets.

What then does it mean to live up to the challenges of the Anthropocene? We'd like to suggest three avenues for consideration:

First, we must learn to grow in different ways than with our current hyper-consumption. What we now call economic 'growth' amounts too often to a Great Recession for the web of life we depend on. Gandhi pointed out that "the Earth provides enough to satisfy every man's needs, but not every man's greed." To accommodate the current Western lifestyle for 9 billion people, we'd need several more planets. With countries worldwide striving to attain the 'American Way of Life', citizens of the West should redefine it—and pioneer a modest, renewable, mindful, and less material lifestyle. That includes, first and foremost, cutting the consumption of industrially produced meat and changing from private vehicles to public transport.

Second, we must far surpass our current investments in science and technology. Our troubles will deepen exponentially if we fail to replace the wasteful fossil-fueled infrastructure of today with a system fueled by solar energy in its many forms, from artificial photosynthesis to fusion energy. We need bio-adaptive technologies to render "waste" a thing of the past, among them compostable cars and gadgets. We need innovations tailored to the needs of the poorest, for example new plant varieties that can withstand climate change and robust iPads packed with practical agricultural advice and market information for small-scale farmers. Global agriculture must become high-tech and organic at the same time, allowing farms to benefit from the health of natural habitats. We also need to develop technologies to recycle substances like phosphorus, a key element for fertilizers and therefore for food security.

To prevent conflicts over resources and to progress towards a durable 'bioeconomy' will require a collaborative mission that dwarfs the Apollo programme. Global military expenditure reached 1,531 billion U.S. dollars in 2009, an increase of 49 percent compared to 2000. We must invest at least as much in understanding, managing, and restoring our 'green security system'—the intricate network of climate, soil, and biodiversity. To reduce CO_2 concentrations in the atmosphere to safe levels, we need to move towards 'negative emissions,' e.g. by using plant residues in power plants with carbon capture and storage technologies. We also need to develop geoengineering capabilities in order to be prepared for worst-case scenarios. In addition to cutting industrial CO_2 emissions and protecting forests, large investments will be needed to maintain the huge carbon stocks in fertile soils, currently depleted by exploitative agricultural practices. For biodiversity, green remnants in a sea of destruction will not be enough—we need to build a 'green infrastructure', where organisms and genes can flow freely over vast areas and maintain biological functions.

We must build a culture that grows with Earth's biological wealth instead of depleting it.

Finally, we should adapt our culture to sustaining what can be called the 'world organism'. This phrase was not coined by an esoteric Gaia guru, but by eminent German scientist Alexander von Humboldt some 200 years ago. Humboldt wanted us to see how deeply interlinked our lives are with the richness of nature, hoping that we would grow our capacities as a part of this world organism, not at its cost. His message suggests we should shift our mission from crusade to management, so we can steer nature's course symbiotically instead of enslaving the formerly natural world.

Until now, our behaviours have defied the goals of a functioning and fruitful Anthropocene. But at the end of 2010, two United Nations environmental summits offered some hope for progress. In October, in Nagoya, Japan, 193 governments agreed on a strategic plan for global conservation that includes protecting an unprecedented proportion of Earth's ecosystems and removing ecologically harmful subsidies by 2020. And in December, in Cancun, countries agreed that Earth must not warm more than 2 degrees Celsius above the average temperature level before industrialisation. This level is already very risky—it implies higher temperature increases in polar regions and therefore greater chance of thawing in permafrost regions, which could release huge amounts of CO_2 and methane. But at least, Cancun and Nagoya turned out not to be *cul-de-sacs* for environmental policy. After years of stalemate and the infamous Copenhagen collapse, there is now at least a glimmer of hope that humanity can act together. Between now and 2020, however, the commitments on paper must be turned into real action.

Imagine our descendants in the year 2200 or 2500. They might liken us to aliens who have treated the Earth as if it were a mere stopover for refueling, or even worse, characterise us as barbarians who would ransack their own home. Living up to the Anthropocene means building a culture that grows with Earth's biological wealth instead of depleting it. Remember, in this new era, nature is us.

Chapter 13
The Anthropocene: From Global Change to Planetary Stewardship (2011)

Will Steffen, Åsa Persson, Lisa Deutsch, Jan Zalasiewicz, Mark Williams,
Katherine Richardson, Carole Crumley, Paul J. Crutzen, Carl Folke,
Line Gordon, Mario Molina, Veerabhadran Ramanathan, Johan Rockström,
Martin Scheffer, Hans Joachim Schellnhuber, and Uno Svedin

Abstract Over the past century, the total material wealth of humanity has been
enhanced. However, in the twenty-first century, we face scarcity in critical resources,
the degradation of ecosystem services, and the erosion of the planet's capability to
absorb our wastes. Equity issues remain stubbornly difficult to solve. This situation is
novel in its speed, its global scale and its threat to the resilience of the Earth System.
The advent of the Anthropence, the time interval in which human activities now
rival global geophysical processes, suggests that we need to fundamentally alter our
relationship with the planet we inhabit. Many approaches could be adopted, ranging
from geoengineering solutions that purposefully manipulate parts of the Earth System
to becoming active stewards of our own life support system. The Anthropocene is a
reminder that the Holocene, during which complex human societies have developed,
has been a stable, accommodating environment and is the only state of the Earth
System that we know for sure can support contemporary society. The need to achieve
effective planetary stewardship is urgent. As we go further into the Anthropocene,
we risk driving the Earth System onto a trajectory toward more hostile states from
which we cannot easily return.

Keywords Earth system · Anthropocence · Planetary stewardship · Ecosystem
services · Resilience

This text was first published as: Steffen, W.; Persson, Å.; Deutsch, L.; Zalasiewicz, J.; Williams,
M.; Richardson, K.; Crumley, C.; Crutzen, P.J.; Folke, C.; Gordon, L.; Molina, M.; Ramanathan,
V.; Rockström, J.; Scheffer, M.; Schellnhuber, H.J.; Svedin, U., 2011: "The Anthropocene: From
Global Change to Planetary Stewardship", in: *Ambio*, 40,7 (November): 739–761; https://doi.org/10.
1007/s13280-011-0185-x. The permission was granted by Springer Nature through the Copyright
Clearance Center.

13.1 People and the Planet: Humanity at a Crossroads in the Twenty-First Century

The twin challenges of 'peak oil'—decreasing petroleum resources and increasing demand—and climate change are redefining the pathways of human development in the twenty-first century (Sorrell et al. 2009; ASPO 2010; Richardson et al. 2011). Less well known is the potential shortage of the mineral phosphorus and the increasing competition for land—sometimes referred to as the 'land grab' in relation to Africa— as the new economic giants of Asia move to secure food resources in non-Asian territories. The pathways of development followed by today's wealthy countries after the Second World War—built on plentiful, cheap fossil fuel energy resources, an abundance of other material resources, and large expanses of productive land to be developed—cannot be followed by the 75–80% of the human population who are now at various stages of their trajectories out of poverty, and are beginning to compete with today's wealthy countries for increasingly scarce resources.

A large fraction of our population of nearly 7000 million people needs more access to food, water and energy to improve their material standard of living, and the prospect of an additional 2000 million by 2050 intensifies the need for basic resources. These challenges come at a time when the global environment shows clear signs of deterioration and, as a consequence, questions the continuing ability of the planet to provide the same accommodating environment that has facilitated human development over the past 10 000 years.

Climate change is a prominent sign of human-driven changes to the global environment. The evidence that the Earth is warming is unequivocal, and human emissions of greenhouse gases, most importantly carbon dioxide (CO_2), have been responsible for most of the warming since the middle of the twentieth century (IPCC 2007). The manmade greenhouse gases have already trapped enough infrared energy to warm the planet by more than 2 °C (Ramanathan/Feng 2008). Although many uncertainties still surround the risks associated with climate change, impacts are already observable at today's mean global surface temperature rise of about 0.8 °C since the mid nineteenth century. These risks, such as those associated with sea-level rise, extreme events, and shifts in rainfall patterns, rise sharply as the temperature climbs toward 2 °C above pre-industrial and quite possibly beyond (Richardson et al. 2011).

At least as disturbing as climate change, and far less well known and understood, is the erosion over the past two centuries of ecosystem services, those benefits derived from ecosystems that support and enhance human wellbeing. The *Millennium Ecosystem Assessment* (MA 2005) assessed 24 ecosystems services, from direct services such as food provision to more indirect services such as ecological control of pests and diseases, and found that 15 of them are being degraded or used unsustainably. Humanity now acquires more than the ongoing productivity of Earth's ecosystems can provide sustainably, and is thus living off the Earth's natural capital in addition to its productivity. This can lead to continued improvements to human well-being for some time, but cannot be sustained indefinitely.

The challenges of peak oil, peak phosphorus (where the demand for phosphorus may soon outstrip supply; Cordell et al. 2009; Sverdrup/Ragnarsdottir 2011) and climate change demonstrate the existence of limits to the rate or magnitude at which humanity can consume the planet's geophysical resources. Furthermore, climate change and the appearance of the ozone hole owing to man-made chemicals are strong evidence that humanity can overwhelm important chemical, physical, and biological processes that modulate the functioning of the Earth System. These unintended consequences on the global life support system that underpins the rapidly expanding human enterprise lie at the heart of the interconnected twenty-first century challenges.

The classification system developed to define ecosystem services (MA 2005) might be extended to include geophysical goods and services and expanded to the scale of the planet as a whole. These could be called Earth System goods and services. The classification, based on three of the four ecosystem services of the MA (2005), would include the following types.

13.1.1   Provisioning Goods and Services

Most commonly known as 'resources', these include the well-known ecosystem services of food, fibre, and fresh water (natural resources), but would now also include fossil fuels, phosphorus, metals, and other materials derived from Earth's geological resources. Many, but far from all, of these types of goods and services have market prices, which can regulate supply and demand to some extent.

13.1.2   Supporting Services

In the ecosystem framework, these include nutrient cycling, soil formation and primary production. All are necessary to support, for example, well-functioning agricultural systems. They are also sometimes called 'environmental resources'. Geophysical processes also provide supporting services that indirectly yield benefits for humanity. Examples include the long-term provision of fertile soils through glacial action, the upwelling branches of ocean circulation that bring nutrients from the deep ocean to support many of the marine ecosystems that provide protein-rich food, and the Himalayan glaciers that act as giant water storage facilities for the provision of water resources.

13.1.3   Regulating Services

Two of the most well-known of these are the ecological control of pests and diseases and regulation of the climate system through the uptake and storage of carbon by ecosystems. These regulating services, also sometimes considered environmental resources, help maintain an environment conducive for human life, rather than directly contributing to provisioning goods and services. Storage of carbon by ecosystems is a part of a larger, Earth System regulatory service that has a significant geophysical component—the dissolution of atmospheric CO_2 into the ocean. Other Earth System services include the set of chemical reactions in the stratosphere that continually form ozone, essential for filtering out biologically damaging ultraviolet radiation from the sun, and the role of the large polar ice sheets in regulating temperature. Regulating services are generally considered as 'free services' provided by nature.

The accelerating pressures on all three types of Earth System goods and services that connect people and the planet are coming together in the first decades of the twenty-first century to generate a global sustainability crisis. The concept of social-ecological systems is proving to be a powerful concept to deal with sustainability challenges arising from the complex interaction of people and environment at local and regional scales. It has been little applied yet to the global scale (Folke et al. 2011). However, the concept of a planetary-scale social-ecological- geophysical system is rapidly becoming a reality. Or, more simply, the human enterprise is now a fully coupled, interacting component of the Earth System itself (Steffen et al. 2004).

A human-inclusive Earth System implies that global-scale social and economic processes are now becoming significant features in the functioning of the System, like atmospheric and oceanic circulation. Prominent social processes are the globalisation of trade and finance and the rapid increase in communication, especially via the internet (Castells 2010). This level of social and economic connectivity is generating some instabilities in the human enterprise. The Global Financial Crisis is a good example, where an instability in one country—in the US sub-prime market—quickly propagated and amplified to drive a drop in US GDP of about 400 times the total value of the subprime market, and cascaded internationally to trigger a global recession, shrink the availability of credit, and increase the levels of poverty and unemployment in many countries around the world (Taylor 2009), with long-term effects for some but relatively rapid recovery for others.

When the hyper-connectivity of the human enterprise intersects with the pressures on Earth System goods and services, some concatenated global crises can propagate rapidly through the Earth System. The food price crisis of 2008 is a recent example. Global prices of staples such as rice and wheat rose sharply (wheat by 81% and rice by 255%) from 2004 to 2008, with most of the rise coming in the last 12 months (IRRI 2010), leading to food riots in some countries and affecting 100 million people worldwide. One analysis points to several interacting drivers as the cause—rising energy prices, pro-biofuel policies, and export restrictions by managers in middle- and low-income countries (Biggs et al. 2011), and speculative actions by strong players in the market may also have played a role. However, it now seems likely that energy price rises were the dominant global driver, overshadowing the other contributing factors. The connectivity provided by the Internet and mobile phones has also likely played a role in the unrest in North Africa in early 2011. Citizens there were able to see what others elsewhere in the world have while they face rising food prices, in part driven by a grain export ban in Russia in 2010 owing to a fire-related reduction in yields (Fraser/Rimas 2011).

As these twenty-first century problems become better understood, the focus turns toward finding solutions. One of the key developments in moving from problem definition to solution formulation is the concept of the Anthropocene (Crutzen 2002), which cuts through a mass of complexity and detail to place the evolution of the human enterprise in the context of a much longer Earth history. This analysis sharpens the focus on an overarching long-term goal for humanity—keeping the Earth's environment in a state conducive for further human development.

The Anthropocene implies that the human imprint on the global environment is now so large that the Earth has entered a new geological epoch; it is leaving the Holocene, the environment within which human societies themselves have developed. Humanity itself has become a global geophysical force, equal to some of the 'great forces of Nature' in terms of Earth System functioning (Williams et al. 2011). The term is still informal, but is being analysed by a working group of the International Commission on Stratigraphy as regards potential formalisation (Zalasiewicz et al. 2012).

The concept of the Anthropocene focuses the twenty-first century challenges for humanity away from resource constraints and environmental impacts toward more fundamental questions. What are the implications of the Anthropocene for the future of humanity in the twenty-first century and beyond? Can we become active, effective stewards of the Earth System, our own life support system (Schellnhuber 1999)?

13.2   The Anthropocene: From Hunter-Gatherers to a Global Geophysical Force

For well over 90% of its 160 000 year history, *Homo sapiens* have existed as hunter-gatherers only. During that time our ancestors had demonstrable impacts on their environment, even at scales approaching continental, through, for example, fire-stick farming and hunting of mega-fauna during the latest Pleistocene. However, these human impacts registered only slightly at the global scale, and the functioning of the Earth System continued more or less unchanged.

About 10 000 years ago, near the onset of the Holocene, agriculture was developed in four different parts of the world. This eventually led to a more sedentary lifestyle, the development of villages and cities, and the creation of complex civilizations that eventually spanned large regions. Land-clearing for agriculture affected large areas of the land surface but the rate of clearing was tightly constrained by the availability of energy; only human and animal power was available. These early agricultural activities may have had an appreciable effect on the functioning of the Earth System via an increase in atmospheric CO_2 concentration (Ruddiman 2003), but any increase was not enough to raise the CO_2 concentration beyond the envelope of natural variability (Steffen et al. 2007). The Earth System was still operating within the Holocene state, even with the influence of early agriculture.

Around 1800 AD, the industrial era began with greatly enhanced use of fossil fuels. Land-clearing occurred at a much greater rate, and land ecosystems were converted from mostly wild to mostly anthropogenic, passing the 50% mark early in the twentieth century (Ellis et al. 2010). The industrial fixation of nitrogen from the atmosphere, now possible with fossil fuels, produced large amounts of fertilizer, breaking a constraint on food production. Sanitation systems were improved, yielding great benefits for human health and improving urban environments, while passing the effluent to downstream ecosystems whose buffering capacities were able to purify the water until being overwhelmed in recent decades (Scheffer et al. 2001). Population grew more rapidly, with increases in life expectancy and well-being. Fossil fuel-based manufacturing systems enhanced the production of goods, and consumption began to grow with population. Unknown to human societies at the time (but see Arrhenius 1986), the rapid expansion of fossil fuel usage was slowly raising the CO_2 concentration in the atmosphere, and by the early twentieth century the CO_2 concentration was clearly above the upper limit of Holocene variability (Fig. 13.1).

The remarkable discontinuity in the human enterprise about the middle of the twentieth century defines the beginning of second stage of the Anthropocene. The speeding up of just about everything after the Second World War—sometimes called the Great Acceleration (Hibbard et al. 2006)—is shown in Fig. 13.1a. Human population has tripled, but the global economy and material consumption have grown many times faster. The connectivity of humanity has grown at an astounding rate since 1950, as seen in foreign direct investment, international tourism and the numbers of motor vehicles and telephones.

A simple way to estimate the overall impact of the Great Acceleration on the global environment is via the IPAT identity, where the impact is the aggregate of changes in population, affluence (an indicator for consumption) and technology. The volume of the box in Fig. 13.2 depicts the overall impact (I) and the three axes represent the three drivers (P, A, T). The enormous increase in the volume of the box from 1950 to 2011 relative to the 1900–1950 period shows the Great Acceleration. Also evident is the change in the relative importance of the factors. From 1900 to 1950 population, consumption and technology had roughly equal effects, while from 1950 to the present increases in consumption and technology have become the dominant factors driving environmental impact.

Figure 13.3, using the same time period as Fig. 13.1a, shows the corresponding changes in the structure and functioning of the Earth System. Human causation of the trends is obvious, indeed, by definition, in four of the six lower panels—exploitation of fisheries, conversion of mangrove forests to shrimp farms, tropical deforestation and increase in domesticated land. Of the other two, increasing nitrogen fluxes in the environment can be traced directly to human fixation of atmospheric nitrogen (Galloway/Cowling 2002), and the increasing loss of biodiversity (Barnosky et al. 2011) is undoubtedly caused by a number of human activities (MA 2005). Of the top six panels, increases in the three well-known greenhouse gases can be unequivocally linked to anthropogenic sources (IPCC 2007), and the role of human-made chemicals in the reduction of stratospheric ozone has been beyond doubt for

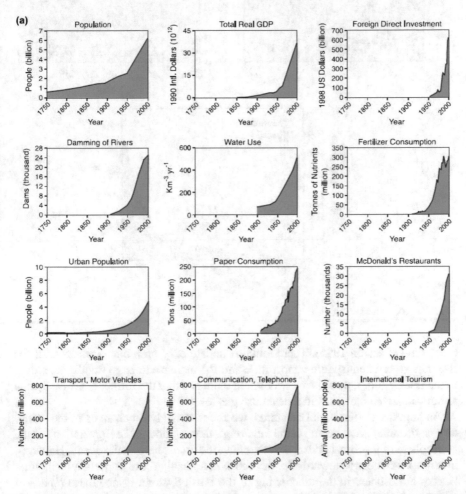

Fig. 13.1 a The increasing rates of change in human activity since the beginning of the Industrial Revolution to 2000. Significant increases in rates change occur around the 1950s in each case and illustrate how the past 50 years have been a period of dramatic and unprecedented change in human history (Steffen et al. 2004, and references therein). In the following part figures, the parameters are disaggregated into OECD (wealthy) countries (blue) and non-OECD (developing) countries (red); **b** Population change from 1960 through 2009, in 1000 millions of people (World Bank 2010); **c** Increase in real GDP from 1969 through 2010, in trillions 2005 USD (USDA 2010); **d** Communication: increase intelephones (millions), both land-lines and mobilephones, from 1950 through 2009 (Canning 1998; Canning/Farahani 2007; ITU 2010)

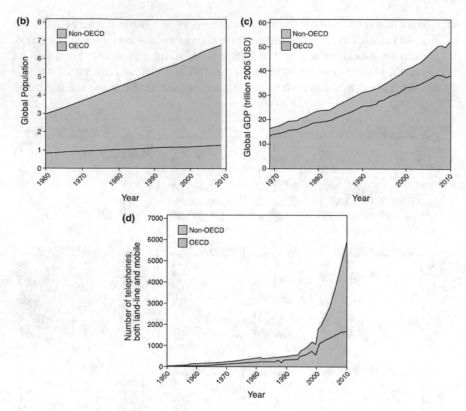

Fig. 13.1 (continued)

some time (Crutzen 1995). There remains debate only on human causation of the rise in northern hemisphere temperature and the increase in large floods; according to the IPCC (2007), there is overwhelming evidence that the former is primarily due to human-driven increases in greenhouse gas levels.

As humans are terrestrial creatures, we focus strongly on changes in the planetary environment that occur on the land (e.g., degradation and deforestation) or the atmosphere (e.g., climate change) rather than the cryosphere or the ocean. However, in terms of planetary stewardship, the ocean is arguably more important than either land or atmosphere in the functioning of the Earth System; it modulates modes of climate variability, provides the moisture for most of rainfall over land that supports agriculture and cities, and stores much more carbon than the land and atmosphere combined

The concept of Earth System goods and services is an effective framework to explore the important roles of the ocean. Provisioning services include food, medicinal products and fresh water via desalination. Supporting services include the absorption and recycling of human-generated waste products; much of the nitrogen

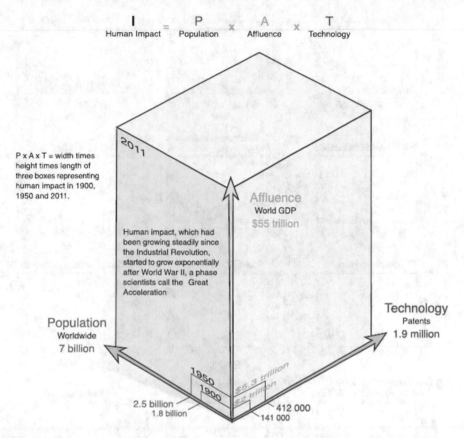

Fig. 13.2 I = PAT identity at the global scale from 1900 to the present. Note the difference in volume between the 1990–1950 period and the 1950–2011 period, which represents the Great Acceleration (Kolbert 2011)

and phosphorus waste from agricultural fertilizers and animal and human excrement ends, ultimately, in the coastal oceans, where they are metabolised. Regulating services include climate regulation via the uptake of atmospheric CO_2, but which also increases ocean acidity (Royal Society 2005), which in turn places stress on calcifying organisms such as corals and thus influences the provisioning service that coral reefs provide (Moberg/Folke 1999). More subtle is the role of the ocean circulation in establishing the global distribution patterns of heat and moisture and thus the patterns of water availability for human societies.

Returning to Fig. 13.1, some important changes have occurred in the characteristics of the human enterprise around the end of the twentieth century. Parts b, c, and d of the figure split the globally aggregated data for population, global GDP and the number of telephones into developed (OECD) and developing and transitional (non-OECD) countries for approximately the last 50 years. This disaggregation reveals two important features of the last decade. First, the post-2000 increase in growth rates of

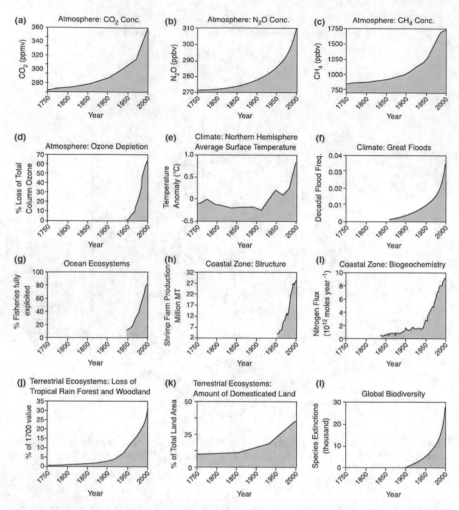

Fig. 13.3 Global-scale changes in the Earth System as a result of the dramatic increase in human activity: **a** atmospheric CO_2 concentration, **b** atmospheric N_2O concentration, **c** atmospheric CH_4 concentration, **d** percentage total column ozone loss over Antarctica, using the average annual total column ozone, 330, as a base, **e** northern hemisphere average surface temperature anomalies, **f** natural disasters after 1900 resulting in more than 10 people killed or more than 100 people affected, **g** percentage of global fisheries either fully exploited overfished or collapsed, **h** annual shrimp production as a proxy for coastal zone alteration, **i** model-calculated partitioning of the human-induced nitrogen perturbation fluxes in the global coastal margin for the period since 1850, **j** loss of tropical rainforest and woodland, as estimated for tropical Africa, Latin America and South and Southeast Asia, **k** amount of land converted to pasture and cropland, and **l** mathematically calculated rate of extinction (Steffen et al. 2004, and references therein)

some non-OECD economies (e.g., China and India) is evident, but the OECD countries still accounted for about 75% of the world's economic activity. On the other hand, the non-OECD countries continue to dominate the trend in population growth. Comparing these two trends demonstrates that consumption in the OECD countries, rather than population growth in the rest of the world, has been the more important driver of change during the Great Acceleration, including the most recent decade, as shown also in Fig. 13.2.

The second feature is the encouraging 'leapfrogging' of the non-OECD countries in some aspects of their development pathway, compared to the earlier development pathway of the OECD countries. For example, the increase in telephones over the past decade (Fig. 13.1d) has been dominated by the sharp rise in the number of phones in the developing world, with most of these being mobile phones rather than landlines (Canning and Farahani 2007; Canning 1998; ITU 2010). Much more challenging, however, is for the non-OECD countries to leapfrog the OECD fossil fuel intensive energy development pathways and thus decouple greenhouse gas emissions from strong economic growth.

Equity issues are also apparent in the changing pattern of CO_2 emissions (Raupach et al. 2007). While the wealthy countries of the OECD dominated emissions for much of the twentieth century, the share of emissions from developing countries rose rapidly to 40% of the annual total by 2004. By 2008 China had become the world's largest emitter of CO_2, with India becoming the third largest. However, the world's wealthy countries account for 80% of the cumulative emissions of CO_2 since 1751; cumulative emissions are important for climate given the long lifetime of CO_2 in the atmosphere. The world's poorest countries, with a combined population of about 800 million, have contributed less than 1% of the cumulative emissions.

One other twenty-first century feature of the Anthropocene, a great paradox, involves life itself. Humanity has now come very close to synthesizing life with the construction in 2010 of a genome from its chemical constituents, which was then implanted successfully into a bacterium where it replaced the original DNA (Gibson et al. 2010). This costly, labor-intensive and time-consuming exercise is in stark contrast to the continuing decline in the Earth's existing biological diversity. In a recent study, 31 indicators of biodiversity change show no reduction in the rate of biodiversity decline from 1970 to 2010; furthermore, the rate of human response to the biodiversity decline has itself slowed over the past decade (Butchart et al. 2010). Understanding the trajectory of the human enterprise from our long past as hunter-gathers to the Great Acceleration and into the twenty-first century provides an essential context for the transformation from resource exploitation toward stewardship of the Earth System. This has evolved from its early state, hostile to human existence, to the one we know today. We now take from it the goods and services that underpin our lives, at a scale and rate that is eroding its capacity to support us (Fig. 13.4—photo: Irrigated agriculture).

Fig. 13.4 The human
domination of land systems
in the Anthropocene.
Irrigated landscape, USA
(photo: Azote)

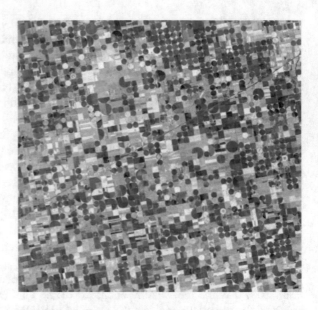

13.3 Understanding Planetary Dynamics: Earth as Our Life Support System

Humans have been in existence for only a very small fraction of the Earth's history. The planet's evolution has produced environments far different from that we know today—at least two episodes of near-complete freezing; much warmer periods than the present; atmospheric change from a chemically reducing to an oxidizing atmosphere; constant rearrangement of land and ocean; and a biology that has evolved from primitive beginnings into a succession of spectacular and diverse life forms. Figure 13.5 shows the temperature variation during the most recent 70 million years, with increasing higher temporal resolution from panels a to d.

Two features shown in Fig. 13.5 are particularly important for this analysis. First, the Quaternary as a whole thus clearly shows a systematic increase in long-term climate variability (panel b). Model simulations hint that the late Quaternary may represent a short, transient phase of climate instability toward a new stable state of a permanently glaciated, low-CO_2 world—a potential future now derailed by the injection of large amounts of greenhouse gases into the atmosphere (Crowley/Hyde 2008). Regardless, the late Quaternary clearly represents a time when the Earth System is unusually sensitive to being switched between strongly contrasting states by modest forcing agents or internal feedbacks.

Second, the Holocene, by comparison with the late Pleistocene, has shown remarkable climate stability (panel c; Petit et al. 1999), with relatively minor changes of a millennial and smaller-scale periodicity, such as the ca. 1 °C change from the Medieval Warm Period to the Little Ice Age, a mainly northern hemisphere phenomenon, but nothing remotely on the scale of a Dansgaard-Oeschger cycle

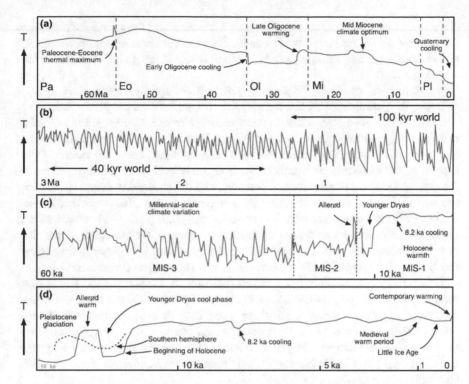

Fig. 13.5 Changes in global average surface temperature through Earth history, from ca. 70 million years ago to the present (adapted from Zalasiewicz and Williams 2009). **a** The most recent 70 million years, showing the long cooling trend to the present, coincident with decreasing atmospheric CO_2 concentrations; the Antarctic ice sheets formed about 34 million years ago and the northern hemisphere ice sheets about 2.5 million years ago. **b** The most recent 3 million years, encompassing the Quaternary period. The late Quaternary, the time during which Homo sapiens evolved, is characterized by ca. 100 000-year rhythmic oscillations between long, variable cold periods and much shorter warm intervals. The oscillations are triggered by subtle changes in the Earth's orbit but the temperature changes are driven by the waxing and waning of ice sheets and changes in greenhouse gas concentrations. **c** The most recent 60 000 years of Earth history, showing the transition from the most recent ice age into the much more stable Holocene about 12 000 years ago. The most recent ice age, which humans experienced, was characterized by repeated, rapid, severe, and abrupt changes in northern hemisphere climate (Dansgaard-Oeschger events), with changes in oceanic circulation, periodic major ice sheet collapses, 5–10 m scale sea-level changes, and regional changes in aridity/humidity. **d** The most recent 16 000 years of Earth history, showing the Holocene and the transition into it from the most recent ice age

(panel c). This stability has been maintained for longer than in the last three interglacials, and might naturally continue for at least 20 000 years, or perhaps even longer, based on orbital similarities with earlier 'long interglacials' (Berger/Loutre 2002).

The stable Holocene has proven to be a very accommodating global environment for the development of humanity; it has allowed agriculture, villages and larger settlements and more complex civilizations to develop and thrive. As we move into the Anthropocene, it is important to understand the envelope of natural variability that characterises the Holocene as a baseline to interpret the global changes that are now under way.

Some features of the Holocene state are well defined by palaeo-evidence. For example, the parameters that characterise the climate, such as temperature and CO_2 concentration, can be obtained with good accuracy from ice cores. Biome distribution before agriculture can be inferred from pollen records. Wetness or dryness of climate can be estimated by techniques such as speleothem (stalactite) records. Other aspects of the Holocene environment, such as the behaviour of the nitrogen cycle, the type and amount of atmospheric aerosols, and the changes in ocean circulation, are more difficult to discern. Together, the group of indicators that cover land, ocean, atmosphere and cryosphere and that consider physical, geological, chemical and biological processes define the environmental envelope of the Holocene. Thus, they characterise the only global environment that we are sure is 'safe operating space' for the complex, extensive civilization that *Homo sapiens* has constructed.

Biodiversity is a particularly important indicator for the state of the global environment. Although little is known about the relationship between biodiversity and the functioning of the Earth System, there is considerable evidence that more diverse ecosystems are more resilient to variability and change and underpin the provision of a large number of ecosystem services (MA 2005). Biodiversity may thus be as important as a stable climate in sustaining the environmental envelope of the Holocene.

Widespread biodiversity loss could affect the regulating services of the Earth System, given the importance of biological processes and feedbacks. Past biodiversity change can place the current mass extinction event into a longer term, Earth System context. Based on the fossil record, none of the five past mass extinction events are direct analogues for modern biodiversity loss, but their study can improve understanding of the role of biodiversity in Earth System dynamics (Erwin 2008). For instance, the partial loss of terrestrial herbivorous megafauna by hunting around the Pleistocene/Holocene boundary has been linked with regional temperature changes driven by consequent changes to vegetation and albedo (Doughty et al. 2010)

Such knowledge can inform questions about what types and how much biodiversity must be preserved in what regions to sustain the resilience of the Earth System or large parts of it. These questions will become more prominent as we move from a focus on exploiting Earth System goods and services to becoming active stewards of our own planetary life support system.

13.4   The Twenty-First Century Challenge: Toward Planetary Stewardship

The twenty-first century challenge is different from any other that humanity has faced. The planetary nature of the challenge is unique, and demands a global-scale solution that transcends national boundaries and cultural divides (Svedin 1998). The collision of the human enterprise with the rest of nature has occurred many times in the past at sub-global scales, leading to a new paradigm of integrated social-ecological systems (Folke et al. 2011). At the global scale, this paradigm challenges humanity to become active stewards of our own life support system (Kates et al. 2001; Young and Steffen 2009; Chapin et al. 2010). We are the first generation with the knowledge of how our activities influence the Earth System, and thus the first generation with the power and the responsibility to change our relationship with the planet.

The challenge for humanity is shown by a comparison of the *Human Development Index* (HDI), a measure of wellbeing, and the Ecological Footprint (Global Footprint Network 2011), an indicator of the human imprint on the global environment (Fig. 13.6). The world population-weighted average HDI rose to 0.68 in 2010 from 0.57 in 1990, continuing the upward trend from 1970, when it stood at 0.48 (UNDP 2010). The global ecological footprint has also been rising, leading to an overshoot of Earth's annual biocapacity in the mid-1970s, corresponding to 1.5 planets in 2007. The increase is mainly due to higher demand for CO_2 absorption generated primarily by fossil fuel energy usage (WWF 2010).

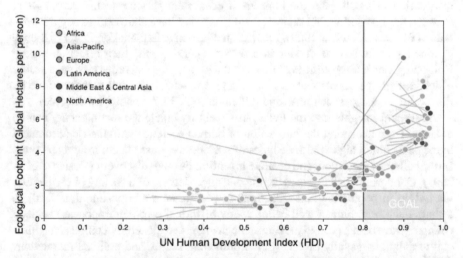

Fig. 13.6 National Human Development Index and Ecological Footprint trajectories, 1980–2007, compared with goal levels. (Global Footprint Network 2011) (see flash video at: <http://www.footprintnetwork.org/en/index.php/GFN/page/fighting_poverty_our_human_development_initiative/>)

The bottom right shaded area of Fig. 13.6 represents the 'sustainability quadrant', in which the HDI reaches an acceptably high value but the ecological footprint remains with the limits of one planet Earth (Global Footprint Network 2011). Currently, no country achieves these two levels simultaneously. However, a promising development, shown by the downwards-sloping trajectories of some countries, is that they have improved wellbeing while reducing both natural resource demand and pollution. At the aggregated global scale, however, the trends are clear. Population growth in combination with more intense resource use and growing pollution still sets the world as a whole on a pathway toward a growing total footprint (Global Footprint Network 2011).

In summary, human well-being has reached high levels in many countries while our planetary life support system is simultaneously being eroded. An analysis of this 'environmentalist's paradox', based on the assessment that 15 of 24 types of ecosystem services are in decline globally (MA 2005), concluded that provisioning services are currently more important than supporting and regulating services for human well-being, as measured by the rise in HDI over the 1970–2005 period (Raudsepp-Hearne et al. 2010). Thus, the benefits associated with food production (a provisioning service) currently outweigh the costs of declines in other services at the global scale.

Several questions have been raised about this conclusion. First, the HDI is too narrow, failing to incorporate cultural or psychological dimensions or security considerations and ignoring involuntary adaptation as a result of environmental deterioration and opportunity costs. Second, global aggregates mask the ways in which the distribution of wealth and the impacts of ecosystem service decline are skewed, between nations and within them, a factor that may have a strong bearing on wellbeing (Wilkinson/Pickett 2009). Finally, an alternative explanation is the existence of time lags between the decline in ecosystem services and their effect on human well-being, particularly time lags associated with geophysical processes, such as loss of ice in the large polar ice sheets and changes in ocean circulation that operate on timescales of decades, centuries and millennia (Fig. 13.7—photo: melting ice).

Additional insights into the twenty-first century challenge for humanity can be obtained from analysing the interaction of human societies with their environment in the past, which highlight three types of societal responses to environmental pressures: collapse, migration, and creative invention through discovery (Costanza et al. 2007). Collapse, which refers to the uncontrolled decline of a society or civilization via a drop in population and reductions in production and consumption, leads to a sharp decline in human well-being. Many historical (pre-Anthropocene) societal collapses have occurred, and the causal mechanisms are generally complex and difficult to untangle, usually involving environmental, social, and political interactions (Costanza et al. 2007, and references therein).

Some hypotheses regarding the causes of collapses in the past are particularly relevant to the Anthropocene. For example, increasing societal complexity in response to problems is an adaptive strategy at first, but as complexity increases, resilience is eroded and societies become more, rather than less, vulnerable to external shocks (Tainter 1998). Another hypothesis (Diamond 2005) proposes that societies collapse

Fig. 13.7 The planetary
boundary for climate change
is designed to avoid
significant loss of ice from
the large polar ice sheets.
Melting Greenland ice sheet
(photo: Bent Christensen,
Azote)

if core values become dysfunctional as the external world changes and they are
unable to recognise emerging problems. Such societies are locked into obsolete
values hindering, for example, the transition to new values supporting a reconnection
to the biosphere (Folke et al. 2011). A core value of post-World War II contemporary
society is ever-increasing material wealth generated by a growth-oriented economy
based on neoliberal economic principles and assumptions (McNeill 2000; Hibbard
et al. 2006), a value that has driven the Great Acceleration but that climate change
and other global changes are calling into question.

How likely are environmental pressures to trigger collapse in the contemporary
world? With a more interconnected world through trade, transportation and communication and with economic structures less reliant on local agricultural production,
vulnerability profiles of societies have been fundamentally altered. The nature of
human-environment interactions has also changed along several dimensions—scale,
speed, and complexity—which contribute to the new forms of vulnerability.

The increasingly *global scale* of environmental degradation in the Anthropocene
has led to some (partial) solutions (e.g., international treaties, international market

mechanisms), but the distribution of environmental and economic impacts are highly uneven. Whether the local impact is sufficient to cause local collapse, and whether local collapse can propagate rapidly throughout the globalised human enterprise, as in the global financial crisis, are important questions. On the other hand, a well-connected human enterprise could lead to increased knowledge and techniques for local adaptation, averting or containing local collapse before it can spread. Another qualitatively new problem is the 'democratic deficit' associated with international institutions, which are comprised of a collection of sovereign nation-states (Mason 2005; Bäckstrand et al. 2010). Human impacts on Earth System functioning cannot be resolved within individual jurisdictions alone; supranational cooperation is required.

Understanding the *speed* of environmental change is also important for distinguishing vulnerability to slow- onset versus quick-onset events. Many historical cases of collapse involved slow-onset or gradual change, where the rate of change was proportional to the pressure of the causal agent. Contemporary societies have a broad set of options to deal with such changes. While vulnerability to quick-onset events such as natural disasters has decreased in many respects (Parry et al. 2007), the frequency and intensity of extreme events are expected to increase with climate change (IPCC 2007; UNISDR 2009).

Concatenation of both slow- and quick-onset events, coupled with the increasing connectivity of the human enterprise, can lead to some unexpected global crises (Folke et al. 2011), such as the spikes in food prices (Fig. 13.8). The Earth System scale adds another twist to the concept of speed of change, as for the very large geophysical changes that have exceptionally long lag times but may then occur suddenly with potential devastating effects, as in the very abrupt warmings associated with the Pleistocene D/O events. Humanity, now largely in its post-agrarian phase of development, has no experience of dealing with such combinations of scale and speed of environmental change.

Finally, in addressing increasing *complexity,* Walker et al. (2009) argue that it is no longer useful to concentrate on environmental challenges and variables individually,

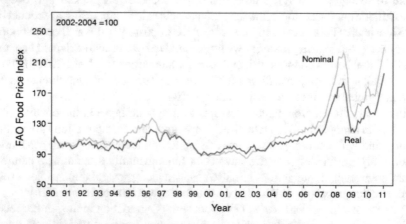

Fig. 13.8 FAO food price index, 1990–2010. *Source* FAO (2011)

but the challenge lies in the intertwining of multi-scale challenges across sectors (e.g., environment, demographics, pandemics, political unrest). An historical case occurred in fourteenth century Europe, when the Medieval Warm Period ended and was followed by colder and wetter growing seasons, a locust invasion, a millennial-scale flood and a pandemic (the Black Death) (Costanza et al. 2007). An oft-cited contemporary example is the food price crisis (Biggs et al. 2011). Climate change itself is an example of such a complex challenge. Multiple crises may coincide or trigger each other, and there is a need to move beyond narrow sectoral approaches toward more coherent and effective institutions that can deal with complex systems perspectives (UNISDR 2009; Walker et al. 2009).

The scale, speed and complexity of twenty-first century challenges suggest that responses based on marginal changes to the current trajectory of the human enterprise—'fiddling at the edges'—risk the collapse of large segments of the human population or of globalised contemporary society as whole. More transformational approaches may be required. Geo-engineering and reducing the human pressure on the Earth System at its source represent the end points of the spectrum in terms of philosophies, ethics, and strategies.

Geo-engineering—the deliberate manipulation or 'engineering' of an Earth System process—is sometimes argued to be an appropriate response to challenges posed by the Anthropocene, most often as a response to climate change. Manipulation of two different types of Earth System process are most often proposed: (i) those processes ultimately controlling the amount of heat entering the Earth's lower atmosphere (*solar radiation management*, SRM), and (ii) those affecting the amount of heat energy retained near the Earth's surface, that is, control of greenhouse gas concentration through manipulation of the global carbon cycle.

Both SRM and manipulation of the carbon cycle constitute a form of 'symptom treatment' rather than removal or reduction of the anthropogenic pressures leading to climate change. In particular, SRM targets only the temperature change by decreasing the heat input to the lower atmosphere through, for example, production of sulphate aerosols in the stratosphere (Crutzen 2006). This approach has no direct impact on atmospheric greenhouse gas concentrations, and other processes influenced by elevated concentrations of greenhouse gases, for example, ocean acidification (Royal Society 2005), would continue unchecked even if SRM managed to slow global temperature increases.

Approaches that manipulate the carbon cycle, such as carbon capture and storage, could slow the rate of increase of atmospheric greenhouse gas concentrations, or perhaps ultimately reduce the atmospheric concentration of CO_2. However, there are no proven mechanisms yet developed that would return the carbon removed from the atmosphere to a form as inert as the fossil fuels from which it was derived. Thus, although removed from the atmosphere, the carbon captured is stored biologically, in underground caverns or in the deep sea if the carbon capture is via chemical or mechanical means. Carbon stored in biological compartments is particularly vulnerable to return to the atmosphere with further climate change or with changes in human management.

In addition to CO_2, there are several other man-made greenhouse gases, which have contributed as much as 45% to the total man-made greenhouse effect. The life times of several of these gases (methane, ozone, HFCs) are short (<15 years) compared with the century to millennium time scales of CO_2 and hence actions to reduce their concentrations, possible with existing technologies, will lead to quick reduction in the total warming effect (Ramanathan/Xu 2010).

Nevertheless, it may become necessary to supplement efforts to reduce human emissions of greenhouse gases with geo-engineering to prevent severe anthropogenic climate change. If this strategy is required, then SRM mechanisms would probably be the more effective as the Earth System would respond more quickly to these than to manipulation of the carbon cycle (Richardson et al. 2011). However, in contrast to emissions reduction, the problem with geoengineering is 'not how to get countries to do it, (but) the fundamental question of who should decide whether and how geo-engineering should be attempted – a problem of governance' (Barrett 2008). Many potential forms of geo-engineering would be relatively inexpensive, could be carried out unilaterally and could potentially alter climate and living conditions in neighbouring countries. Thus, the potential geopolitical consequences of geo-engineering are enormous, and urgently require guiding principles for their application.

Those rows shaded in dark grey represent processes for which the proposed boundaries have already been transgressed (Rockström et al. 2009a, which also includes the individual references for the data presented in the table).

At the other end of the spectrum lie a number of alternative strategies to reduce or modify the human influence on the functioning of the Earth System at its source. The *Planetary Boundaries* (PB) approach (Rockström et al. 2009a, b) is a recent example that attempts to define a 'safe operating space' for humanity by analying the intrinsic dynamics of the Earth System and identifying points or levels relating to critical global-scale processes beyond which humanity should not go. The fundamental principle underlying the PB approach is that a Holocene-like state (Fig. 13.5, panel c; Petit et al. 1999) of the Earth System is the only one that we can be sure provides an accommodating environment for the development of humanity.

Nine planetary boundaries have been proposed (Table 13.1) which, if respected, would likely ensure that the Earth System remains in a Holocene-like state. Preliminary analyses (Rockström et al. 2009a, b) estimated quantitative boundaries for seven of the Earth System processes or elements—climate change, stratospheric ozone, ocean acidification, the nitrogen and phosphorus cycles, biodiversity loss, land-use change and freshwater use. For some of these it is a first attempt at quantifying boundaries of any kind, that is, quantifying the supply of some of the regulating and supporting Earth System services. There is insufficient knowledge to suggest quantitative boundaries for two of the processes—aerosol loading and chemical pollution. Rockström and colleagues estimate that three of the boundaries—those for climate change, the nitrogen cycle and biodiversity loss—have already been transgressed while we are approaching several others (Fig. 13.9).

Even if a scientific consensus around boundary definitions could be achieved, much more is required to achieve successful and effective global governance and stewardship (Richardson et al. 2011). Focusing on climate change, the outcomes of the COP15 meeting in Copenhagen in 2009 showed that (i) climate change has now been raised to an issue of high political priority internationally, and (ii) the road to achieving a legally binding international climate agreement, based on burden- or cost-sharing in the context of a global commons, is a long and complex one, with further steps beyond COP15 required to deliver such an agreement (Falkner et al. 2010; Richardson et al. 2011).

Recently, however, Ostrom (2010) has suggested that the traditional approach of collective action to climate change based on one international treaty may be misconceived. Addressing climate change through emission reductions can, for example also bring benefits at local and regional scales, such as improved air quality in metropolitan areas. This is particularly so for the emission of the short-term climate warming gases (ozone, methane, HFCs). This approach suggests that global governance and planetary stewardship could also be built in a multi-level, cumulative way by identifying

Table 13.1 The planetary boundaries

Earth-system process	Parameters	Proposed boundary	Current status	Pre-industrial value
Climate change	(i) Atmospheric carbon dioxide concentration (parts per million by volume)	350	387	280
	(ii) Change in radiative forcing (watts per meter sqaured)	1	1.5	0
Rate of biodiversity loss	Extinction rate (number of species per million species per year)	10	>100	0.1-1
Nitrogen cycle (part of a boundary with the phosphorous cycle)	Amount of N_2 removed form the atmosphere for human use (millions of tonnes per year)	35	121	0
Phosphorous cycle (part of a boundary with the Nitrogen cycle)	Quantity of P flowing into the oceans (millions of tonnes per year)	11	8.5-9.5	-1
Stratospheric ozone depletion	Concentration of ozone (Dobson unit)	276	283	290
Ocean acidification	Global mean saturation state of aragonite in surface sea water	2.75	2.90	3.44
Global freshwater use	Consumption of freshwater by humans (km^3 per year)	4,000	2,600	415
Change in land use	Percentage of global land cover converted to cropland	15	11.7	Low
Atmospheric aerosol loading	Overall particulate concentration in the atmosphere, on a regional basis	To be determined		
Chemical pollution	For example, amount emitted to, or concentration in, the global environment of persistent organic pollutants, plastics, endocrine disrupters, heavy metals and nuclear waste; or their effects on the functioning of ecosystems and the Earth System.	To be determined		

Boundaries for processes in dark grey have been crossed

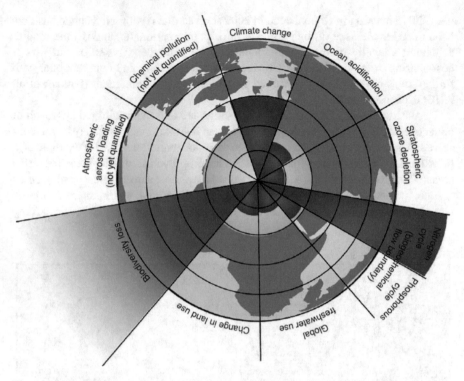

Fig. 13.9 The inner green shading represents the proposed safe operating space for nine plane-tary systems. The *red wedges* represent an estimate of the current position for each variable. The boundaries in three systems (rate of biodiversity loss, climate change and human interference with the nitrogen cycle) have already been exceeded (Rockström et al. 2009a)

where, when and for whom there are—or could be as a result of policy—incen-tives to act, independently of the international level (Liljenström and Svedin 2005). The resulting governance system would be 'polycentric', also allowing for more experimentation and learning.

Discussions on climate change, global change and global sustainability implic-itly assume that the current global environmental changes are perturbations of the stable Holocene state of the Earth System. The assumption is that effective gover-nance will turn the trajectory of the human enterprise toward long-term sustainability and the Earth System back toward a Holocene-like state. However, the concept of the Anthropocene, coupled with complex systems thinking, questions that assump-tion. The Anthropocene is a dynamic state of the Earth System, characterised by global environmental changes already significant enough to distinguish it from the Holocene, but with a momentum that continues to move it away from the Holocene at a geologically rapid rate.

13.5 The Anthropocene From a Complex Systems Perspective

Several pieces of evidence suggest that the Earth as a whole can be considered as a complex system, with the Holocene the most recent state of the system. This complex systems approach places the Anthropocene in a different perspective. Very long time-scales are associated with some features of the Anthropocene (Zalasiewicz et al. 2011). For example, millennial timescales are associated with significant changes in the large polar ice sheets and even longer timescales associated with recovery of mass extinctions of biological species. This suggests that the Anthropocene will not be a spike of a century or two's duration but may be evident for a geologically signif-icant period of time (Zalasiewicz et al. 2012). From a complex systems perspective (Scheffer 2009), this raises the possibility that the Anthropocene could become an alternative, more or less stable state of the Earth System.

The most striking feature of Earth System dynamics in the late Quaternary is the regular oscillation between two well-defined states: glacial phases and the shorter, intervening warm interglacials. This is characteristic behaviour of a complex system that has two stable states, or basins of attraction, between which it oscillates, that is, a limit cycle (Scheffer 2009). Figure 13.10 shows a 'stability landscape' in which a system (the ball) can move between two different states or basins of attraction (the valleys). A critical feature of a complex system is the existence of feedbacks that can move the system from an intermediate point toward either of the stable states. For the Earth System, the most important feedbacks that move the system toward either the

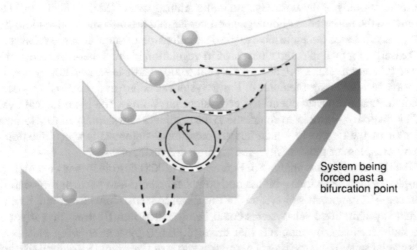

System being
forced past a
bifurcation point

Fig. 13.10 A stability landscape with two stable states. The valleys, or basins of attraction, in the landscape represent the stable states at several different conditions, while the hilltops represent unstable conditions as the system transitions from one state to another. If the size of the basin of attraction is small, resilience is small, and even a moderate perturbation may bring the system into the alternative basin of attraction. *Source* Scheffer (2009)

glacial or the interglacial state are (i) the release or uptake of greenhouse gases as the surface warms (net release) or cools (net uptake), and (ii) the change in reflectively as the ice sheets grow (increased reflectivity, cooling) or retreat (decreased reflectivity, warming). The interglacial state is much shorter than the glacial one, hinting that it is inherently less stable as regards duration, that is, it is a weaker basin of attraction or a shallower valley in Fig. 13.10.

Is the human perturbation to Earth System dynamics, which is pushing the system away from the glacial-interglacial limit cycle, strong and persistent enough to tip the system out of the Holocene stability domain and into an alternative, geologically long-lived, generally warmer state of the Earth System?

The degree of resilience, or the depth of the basin of attraction, of the Holocene state is at issue (cf. Fig. 13.10; Folke et al. 2010). The Holocene is already of significantly longer duration than the three previous interglacials, and, without human perturbation, may continue for many thousands of years. This suggests that the Holocene is inherently more stable than the three earlier interglacials. The tight regulation of global mean temperature through the Holocene and the strength of the land and ocean carbon sinks in absorbing over half of the human emissions of CO_2 (Le Quere et al. 2009; Raupach/Canadell 2010) is consistent with this suggestion, although the possible weakening of the ocean carbon sink over the last few decades (Le Quere et al. 2009) may challenge it.

Carbon cycle feedbacks highlight the role of the biosphere in contributing to the resilience of the Holocene. The role of biology in promoting homeostasis in the Earth System, that is, contributing to a strong basin of attraction around a state conducive to life, has been highlighted by James Lovelock (Lovelock 1979, 1988). An analysis of the functioning of the Earth System in the Anthropocene (Steffen et al. 2004) has confirmed the importance of biological processes, but has also shown that biological feedback processes can contribute to the destabilisation of states in large subsystems of the Earth, for example, the rapid shifts in vegetation in the Sahel-Sahara region of Africa (Claussen et al. 1999; deMenocal et al. 2000). Little is known, however, about the ways in which major features of Earth System structure and functioning—ocean circulation, atmospheric chemistry, ecosystem physiology, the hydrological cycle, and biodiversity—interact to contribute to the resilience and stability of the Holocene state, or of the degree to which human pressures—deforestation, acidification of ecosystems, loss of biodiversity—are eroding this resilience.

The tipping elements analysis of Lenton et al. (2008) describes many examples of complex system behaviour—bifurcation points, threshold-abrupt change, bi-stability domains—in important sub-systems of the Earth System. Considering the Earth as a single system, these sub-systems could be classified into (i) those that affect the two main feedback mechanisms that drive the Earth System between glacial and interglacial states (e.g., melting of permafrost; loss of Greenland ice sheet); (ii) those that may change the resilience of the Holocene state (e.g., loss of Amazon rainforest), and (iii) those that would affect humans but not the basin of attraction, or the overall stability, of the Holocene (e.g., the bistability of the Indian monsoon). Two of the tipping elements that already show signs of instability—the Greenland ice sheet and the Amazon rainforest—represent important sub-systems that, if tipped, would move the Earth System toward a warmer state, that is, away from the Holocene basin of attraction.

Evidence of complex system behaviour is common in the past record of Earth System dynamics. For example, the 100 000-year cycles between glacial and inter-glacial states, which are linked to rather weak orbital forcing with minor changes in solar insolation, may be an example of phase locking—the observed cycles are tuned to the time it takes for the large polar ice sheets to grow and decay (Scheffer 2009). Model studies suggest that the glacial-interglacial limit cycle is not particu-larly stable; the late Quaternary may be a geologically short phase of climate insta-bility that was headed toward a new stable state of a permanently glaciated, low-CO_2 world if we had not injected large amounts of greenhouse gases into the atmosphere (Crowley and Hyde 2008). Thus, the late Quaternary, including the Holocene, may represent a time when the Earth System is unusually sensitive to being switched between strongly contrasting states by modest forcing agents or internal feedbacks.

What are the implications of this complex systems perspective for the future of humanity? Will our attempts to achieve effective planetary stewardship slow and then halt the current trajectory further into the Anthropocene, eventually steering the Earth System back toward Holocene-like conditions and, in so doing, move contemporary civilization toward a new state of sustainability? Or is it already too late to return to a world of the Holocene that may be already lost? Is the Anthropocene, a one-way trip for humanity to an uncertain future in a new, much warmer—and very different—stable state of the Earth System? While these questions demand a greatly enhanced research effort, they reinforce the urgency for effective Earth System stewardship to maintain a global environment within which humanity can continue to develop in a humane and respectful fashion.

13.6   Conclusion – Key Messages

The challenges of the twenty-first century—resource constraints, financial insta-bility, inequalities within and between countries, environmental degradation—are a clear signal that "business-as-usual" cannot continue. We are passing into a new phase of human experience and entering a new world that will be qualitatively and quantitatively different from the one we have known.

The Anthropocene provides an independent measure of the scale and tempo of human-caused change—biodiversity loss, changes to the chemistry of atmosphere and ocean, urbanisation, globalisation—and places them in the deep time context of Earth history. The emerging Anthropocene world is warmer with a diminished ice cover, more sea and less land, changed precipitation patterns, a strongly modified and impoverished biosphere and human-dominated landscapes.

We are the first generation with widespread knowledge of how our activities influ-ence the Earth System, and thus the first generation with the power and the respon-sibility to change our relationship with the planet. Responsible stewardship entails emulating nature in terms of resource use and waste transformation and recycling, and the transformation of agricultural, energy and transport systems.

Effective planetary stewardship can be built around scientifically developed boundaries for critical Earth System processes that must be observed for the Earth System to remain within a Holocene-like state. An effective architecture of a governance system for planetary stewardship is likely to be polycentric and multi-level rather than centralised and hierarchical.

Effective planetary stewardship must be achieved quickly, as the momentum of the Anthropocene threatens to tip the complex Earth System out of the cyclic glacial-interglacial pattern during which *Homo sapiens* has evolved and developed. Without such stewardship, the Anthropocene threatens to become for humanity a one-way trip to an uncertain future in a new, but very different, state of the Earth System.

Acknowledgments The article is based on Steffen et al. (2011). The Anthropocene: from global change to planetary stewardship. Working Paper No. 2 prepared for the "3rd Nobel Laureate Symposium on Global Sustainability: Transforming the World in an Era of Global Change", in Stockholm, 16–19 May 2011, Stockholm Resilience Centre, the Royal Swedish Academy of Sciences, the Stockholm Environment Institute, the Beijer Institute of Ecological Economics and the Potsdam Institute for Climate Impact. We acknowledge support from Ebba och Sven Schwartz Stiftelse, Kjell and Marta Beijer Foundation, Formas, and Mistra through a core grant to the Stockholm Resilience Centre, a cross-faculty research centre at Stockholm University.

References

Arrhenius, S., 1896: "On the influence of carbonic acid in the air upon the temperature of the ground", in: *The London, Edinburgh and Dublin Philosophical Magazine and Journal of Science* (fifth series), 41: 237–275.

ASPO (Association of the Study of Peak Oil and Gas), 2010; at. www.peakoil.net.

Bäckstrand, K.; Khan,J.; Kronsell, A.; Lövbrand, E. (eds.), 2010: *Environmental politics and deliberative democracy: Examining the promise of new modes of governance* (Cheltenham: Edward Elgar).

Barnosky, A.D.; Matzke, N.; Tomiya, S.; Wogan, G.O.U.; Swartz, B.; Quental, T.B.; Marshall, C.; McGuire, J.L. et al., 2011: "Has the Earth's sixth mass extinction already arrived?", in: *Nature*, 471: 51–57.

Barrett, S., 2008: "The incredible economics of geoengineering", in: *Environmental & Resource Economics*, 39: 45–54.

Berger, A.; Loutre, M.F., 2002: "An exceptionally long interglacial ahead?", in: *Science*, 297: 1287–1288.

Biggs, D.; Biggs, R.; Dakos, V.; Scholes, R.; Schoon, M., 2011: "Are we entering an era of concatenated global crises?", in: *Ecology & Society*, 16,2: 27.

Butchart, S.H.M.; Walpole, M.; Collen, B.; van Strien, A.; Scharlemann, J.P.W.; Almond, R.E.A.; Baillie, J.E.M.; Bomhard, B. et al., 2010: "Global biodiversity: Indicators of recent declines", in: *Science*, 328: 1164–1168.

Canning, D., 1998: "A database of world stocks of infrastructure: 1950–1995", in: *The World Bank Economic Review*, 12: 529–548.

Canning, D.; Farahani, M., 2007: *A database of world stocks of infrastructure: Update 1950–2005*.

Castells, M., 2010: *The rise of the network society,* 2nd ed. (London: Blackwell).

Chapin III, F.S., S.R. Carpenter, G.P. Kofinas, C. Folke, N. Abel, W.C. Clark, P. Olsson, D.M. Stafford Smith, et al. 2010. Ecosystem stewardship: Sustainability strategies for a rapidly changing planet. *Trends in Ecology & Evolution* 25: 241–249.

Claussen, M.; Kubatzki, C.; Brovkin, V.; Ganopolski, A.; Hoelzmann, P.; Pachur, H.J., 1999: "Simulation of an abrupt change in Saharan vegetation at the end of the mid-Holocene", in: *Geophysical Research Letters*, 24: 2037–2040.

Cordell, D.; Drangert, J.-O.; White, S., 2009: "The story of phosphorus: Global food security and food for thought", in: *Global Environmental Change*, 19: 292–305.

Costanza, R.; Graumlich, L.; Steffen, W.; Crumley, C.; Dearing, J.; Hibbard, K.; Leemans, R.; Redman, C., et al. 2007: "Sustainability or collapse: What can we learn from integrating the history of humans and the rest of nature?", in: *Ambio*, 36: 522–527.

Crowley, T.J.; Hyde, W.T., 2008: "Transient nature of late Pleistocene climate variability", in: *Nature*, 456: 226–230.

Crutzen, P.J., 1995: "My life with O_3, NO_x and other YZO_xs", in: *Les Prix nobel* [The Nobel Prizes] (Stockholm: Almqvist & Wiksell International): 123–157.

Crutzen, P.J., 2002: "Geology of mankind: The Anthropocene", in: *Nature*, 415: 23.

Crutzen, P.J., 2006: "Albedo enhancement by stratospheric sulfur injections: A contribution to resolve a policy dilemma?", in: *Climatic Change*, 77: 211–219.

deMenocal, P.B.; Ortiz, J.; Guilderson, T.; Adkins, J.; Sarnthein, M.; Baker, L.; Yarusinki, M., 2000: "Abrupt onset and termination of the African Humid Period: Rapid climate response to gradual insolation forcing", in: *Quaternary Science Review*, 19: 347–361.

Diamond, J., 2005: *Collapse: How societies choose to fail or succeed* (New York: Viking).

Doughty, C.E.; Wolf, A.; Field, C.B., 2010: "Biophysical feedbacks between the Pleistocene megafauna extinction and climate: The first human-induced global warming?", in: *Geophysical Research Letters*, 37: L15703.

Ellis, E.C.; Klein Goldewijk, K.; Siebert, S.; Lightman, D.; Ramankutty, N., 2010: "Anthropogenic transformation of the biomes, 1700 to 2000", in: *Global Ecology and Biogeography*, 19: 589–606.

Erwin, D.H., 2008: "Macroevolution of ecosystem engineering, niche construction and diversity", in: *Trends in Ecology & Evolution*, 23: 304–310.

Falkner, R.; Stephan, H.; Vogler, J., 2010: "International climate policy after Copenhagen: Towards a 'building blocks' approach", in: *Global Policy*, 1: 252–262.

FAO, 2011: "FAO food price index".

Folke, C., S.R. Carpenter, B.H. Walker, M. Scheffer, F.S. Chapin III, and J. Rockstrom. 2010. Resilience thinking: Integrating resilience, adaptability and transformability. *Ecology and Society* 15: 20. http://www.ecologyandsociety.org/vol15/iss4/art20/.

Folke, C.; Jansson, A.; Rockström, J.; Olsson, P.; Carpenter, S.; Crepin, A-S.; Daily, G.; Ebbesson, J. et al., 2011: "Reconnecting to the Biosphere", in: *Ambio*; doi:https://doi.org/10.1007/s13280-011-0184-y.

Fraser, E.D.G.; Rimas, A., 2011: "The psychology of food riots", in: *Foreign Affairs*, 30 Jan 2011.

Galloway, J.N., and E.B. Cowling. 2002. Reactive nitrogen and the world: Two hundred years of change. *Ambio* 31: 64–71.

Gibson, D.G.; Glass, J.I.; Lartigue, C.; Noskov, V.N.; Chuang, R.-Y.; Algire, M.A.; Benders, G.A.; Montague, M.G. et al., 2010: "Creation of a bacterial cell controlled by a chemically synthesized gene", in: *Science Express*; doi:https://doi.org/10.1126/science.1190719.

Global Footprint Network, 2011: *Our human development initiative*; at: http://www.footprint network.org/en/index.php/GFN/page/fighting_poverty_our_human_development_initiative/ (22 February 2011).

Hibbard, K.A.; Crutzen, P.J.; Lambin, E.F.; Liverman, D.; Mantua, N.J.; McNeill, J.R.; Messerli, B.; Steffen, W., 2006: "Decadal interactions of humans and the environment", in: Costanza, R.; Graumlich, L.; Steffen, W. (eds.): *Integrated History and Future of People on Earth*. Dahlem workshop report 96 (Boston, MA: MIT Press): 341–375.

Intergovernmental Panel on Climate Change (IPCC), 2007: Solomon, S.; Qin, D.; Manning, M.; Chen, Z.; Marquis, M.; Averyt, K.; Tignor, M.M.B.; Miller Jr, H.L.; Chen Z., (eds.): *Climate*

Change 2007: The Physical Science Basis. Contribution of Working Group I to the Fourth Assess-
  ment Report of the Intergovernmental Panel on Climate Change (Cambridge, UK: Cambridge
  University Press).
IRRI, 2010: Paper on global food crisis—Section 1 (Los Banos: IRRI).
ITU, 2010: "ITU world telecommunication/ICT Indicators database"; at: http://www.itu.int/ITU-
  D/ICTEYE/Indicators/Indicators.aspx; http://www.itu.int/ITU-D/ict/statistics/index.html.
Kates, R.W.; Clark, W.C.; Corell, R.; Hall, J.M.; Jaeger, C.C.; Lowe, I.; McCarthy, J.J.; Schellnhuber,
  H.J et al., 2001: "Sustainability science", in: Science, 292: 641–642.
Kolbert, E., 2011: "Enter the Anthropocene: Age of man", in: National Geographic, 219: 60–77.
Le Quere, C.; Raupach, M.R.; Canadell, J.G.; Marland, G.; Bopp, L.; Ciais, P.; Conway, T.J.; Doney,
  S.C. et al., 2009: "Trends in the sources and sinks of carbon dioxide", in: Nature Geoscience 2:
  831–836.
Lenton, T.M.; Held, H.; Kriegler, E.; Hall, J.W.; Lucht, W.; Rahmstorf, S.; Schellnhuber, H.J., 2008:
  "Tipping elements in the Earth's climate system", in: Proceedings of the National Academic of
  Sciences, USA, 105: 1786–1793.
Liljenstrom, H.; Svedin, U. (eds.)., 2005: Micro meso, macroaddressing complex system couplings
  (New Jersey: World Scientific).
Lovelock, J.E., 1979: Gaia: A new look at life on Earth (Oxford, UK: Oxford University Press).
Lovelock, J.E., 1988: The ages of Gaia: A biography of our living earth (New York, NY: W.W.
  Norton & Co).
MA (Millennium Ecosystem Assessment), 2005: Ecosystems and human well-being: Synthesis
  (Washington, DC: Island Press).
Mason, M., 2005: The new accountability: Environmental responsibility across borders (London:
  Earthscan).
McNeill, J.R., 2000: Something new under the sun: An environmental history of the twentieth century
  world (London: W.W. Norton).
Moberg, F.; Folke, C., 1999: "Ecological services of coral reef ecosystems", in: Ecological
  Economics, 29: 215–233.
Ostrom, E., 2010: "Polycentric systems for coping with collective action and global environmental
  change", in: Global Environmental Change, 20: 550–557.
Parry, M.; Canziani, O.; Palutikof, J.; van der Linden, P.J.; Hanson, C.E. (eds.), 2007: Contribution
  of working group II to the fourth assessment report of the intergovernmental panel on climate
  change (Cambridge: Cambridge University Press).
Petit, J.R.; Jouzel, J.; Raynaud, D.; Barkov, N.I.; Barnola, J.-M.; Basile, I.; Bender, M.; Chappellaz,
  J. et al., 1999: "Climate and atmospheric history of the past 420,000 years from the Vostok ice
  core, Antarctica", in: Nature, 399: 429–436.
Ramanathan, V.; Feng, Y., 2008: "On avoiding dangerous anthropogenic interference with the
  climate system: Formidable challenges ahead", in: Proceedings of the National Academic of
  Sciences, USA, 105: 14245–14250.
Ramanathan, V.; Xu, Y., 2010: "The Copenhagen Accord for limiting global warming: Criteria,
  constraints, and available avenues", in: Proceedings of the National Academic of Sciences, USA,
  107: 8055–8062.
Raudsepp-Hearne, C.; Peterson, G.D.; Tengo, M.; Bennett, E.M.; Holland, T.; Benessaiah, K.;
  MacDonald, G.K.; Pfeifer, L., 2010: "Untangling the environmentalist's paradox: Why is human
  wellbeing increasing as ecosystem services degrade?", in: BioScience, 60: 576–589.
Raupach, M.R.; Canadell, J.G., 2010: "Carbon and the Anthropocene", in: Current Opinion in
  Environmental Sustainability, 2: 210–218.
Raupach, M.R.; Marland, G.; Ciais, P.; Le Quere, C.; Canadell, J.G.; Klepper, G.; Field, C.B., 2007:
  "Global and regional drivers of accelerating CO_2 emissions", in: Proceedings of the National
  Academic of Sciences, USA, 104: 10288–10293.
Richardson, K.; Steffen, W.; Liverman, D.; Barker, T.; Jotzo, F.; Kammen, D.; Leemans, R.; Lenton,
  T. et al., 2011: Climate change: Global risks, challenges and decisions (Cambridge: Cambridge
  University Press).

Rockström, J.; Steffen, W.; Noone, K.; Persson, A.; Chapin III, F.S.; Lambin, E.F.; Lenton, T.M.; Scheffer, M. et al., 2009a: "A safe operating space for humanity", in: *Nature* 461: 472–475.

Rockström, J.; Steffen, W.; Noone, K.; Persson, A.; Chapin III, F.S.; Lambin, E.F.; Lenton, T.M.; Scheffer, M. et al., 2009b: "Planetary boundaries: Exploring the safe operating space for humanity", in: *Ecology and Society*, 14: 32.

Royal Society, 2005: *Ocean acidification due to increasing atmospheric carbon dioxide, June 2005* (London: The Royal Society).

Ruddiman, W.F., 2003: The anthropogenic greenhouse gas era began thousands of years ago, in: *Climatic Change*, 61: 261–293.

Scheffer, M., 2009: *Critical transitions in nature and society* (Princeton: Princeton University Press).

Scheffer, M.; Carpenter, S.R.; Foley, J.A.; Folke, C.; Walker, B.H., 2001: "Catastrophic shifts in ecosystems", in: *Nature*, 413: 591–596.

Schellnhuber, H.J., 1999: "'Earth System' analysis and the second Copernican revolution", in: *Nature*, 402: C19–C23.

Sorrell, S.; Speirs, J.; Bentley, R.; Brandt, A.; Miller, R., 2009: *An assessment of the evidence for a near-term peak in global oil production* (London: UK Energy Research Centre).

Steffen, W.; Crutzen, P.J.; McNeill, J.R., 2007: "The Anthropocene: Are humans now overwhelming the great forces of Nature?", in: *Ambio*, 36: 614–621.

Steffen, W.; Sanderson, A.; Tyson, P.D.; Jager, J.; Matson, P.; Moore III, B.; Oldfield, F.; Richardson, K. et al., 2004: *Global change and the earth system: A planet under pressure*. The IGBP global change series (Berlin: Springer-Verlag).

Steffen, W.; Persson, A.: Deutsch, L.; Williams, M.; Zalasiewicz, J.; Folke, C.; Rockström, J.; Crumley, C.; Crutzen, P.; Gordon, L.; Molina, M.; Ramanathan, V.; Richardson, K.; Scheffer, M.; Schellnhuber, J.; Svedin, U., 2011: *The Anthropocene: From global change to planetary stewardship*. Working Paper No. 2. Prepared for the "3rd Nobel Laureate Symposium on Global Sustainability: Transforming the World in an Era of Global Change", in Stockholm, 16–19 May 2011. Stockholm Resilience Centre, the Royal Swedish Academy of Sciences, the Stockholm Environment Institute, the Beijer Institute of Ecological Economics and the Potsdam Institute for Climate Impact Research.

Svedin, U., 1998: "Implicit and explicit ethical norms in the environmental policy arena", in: *Ecological Economics*, 24: 299–309.

Sverdrup, H.U.; Ragnarsdottir, K.V., 2011: "Challenging the planetary boundaries II: Assessing the sustainable global population and phosphate supply, using a systems dynamics assessment model", in: *Applied Geochemistry*, 26: S307–S310.

Tainter, J.A., 1998: *The collapse of complex societies* (Cambridge: Cambridge University Press).

Taylor, M.S., 2009: "Innis lecture: Environmental crises: Past, present and future", in: *Canadian Journal of Economics*, 42: 1240–1275.

UNDP, 2010: *Human development report 2010—the real wealth of nations: pathways to human development*, 20th Anniversary Edition (New York: UNDP).

UNISDR (United Nations International Strategy for Disaster Reduction Secretariat), 2009: *Global assessment report on disaster risk reduction. Risk and poverty in a changing climate: Invest today for a safer tomorrow;* at: http://www.unisdr.org/we/inform/publications/9413.

U.S. Department of Agriculture Economic Research Service. USDA, 2010: *World Bank world development indicators* (Washington: USDA).

Walker, B.; Barrett, S.; Polasky, S.; Galaz, V.; Folke, C.; Engstrom, G.; Ackerman, F.; Arrow, K. et al., 2009: "Looming global-scale failures and missing institutions", in: *Science*, 325: 1345–1346.

WWF, 2010: *Living planet report 2010: Biodiversity, biocapacity and development* (Gland: WWF [Worldwide Fund for Nature]).

Wilkinson, R.; Pickett, K., 2009: *The spirit level—why more equal societies almost always do better* (London: Penguin).

Williams, M.; Zalasiewicz, J.; Haywood, A.; Ellis, M., 2011: "The Anthropocene: A new epoch of geological time?", in: *Philosophical Transactions of the Royal Society A*, 369: 835–1111 (special issue).

World Bank, 2010: "World Bank world development indicators".
Young, O.; Steffen, W., 2009: "The Earth System: Sustaining planetary life support systems", in: Chapin III, F.S.; Kofinas, G.P.; Folke, C. (eds.): *Principles of ecosystem stewardship: Resilience-based resource natural resource management in a changing world* (New York: Springer-Verlag): 295–315.
Zalasiewicz, J.; Crutzen, P.; Steffen, W., 2012: "Anthropocene", in: Gradstein, F.M. (ed.): *A geological time scale 2010* (Amsterdam: Elsevier).
Zalasiewicz, J.; Williams, M., 2009: "A geological history of climate change", in: Letcher, T.M. (ed.): *Climate change: Observed impacts on planet Earth* (Amsterdam: Elsevier B.V.): 127–142.
Zalasiewicz, J.; Williams, M.; Fortey, R.; Smith, A.; Barry, T.L.; Coe, A.L.; Brown, P.R.; Rawson, P.F. et al. 2011: "Stratigraphy of the Anthropocene", in: *Philosophical Transactions of the Royal Society* A, 369: 1036–1055.

Chapter 14
Atmospheric Chemistry and Climate in the Anthropocene (2012)

Klaus Lorenz, Paul J. Crutzen, Rattan Lal, and Klaus Töpfer

Abstract The effects of human activities are increasingly overwhelming the geologic, biological and chemical processes that drive changes in the abundance of trace and *greenhouse gases* (GHGs) in the atmosphere and Earth System behaviour. Thus, mankind has opened a new geological epoch or age – the Anthropocene. This development is driven by the strong population increase which may result in ten billion people by 2100. With this population increase, human demands for food and animal protein in particular, clean water, natural resources and nutrients such as fixed *nitrogen* (N) and *phosphorus* (P), land and energy will continue to increase strongly. It is also hypothesised that human enterprise is responsible for the sixth mass species extinction. Further, releases of gases such as *sulphur dioxide* (SO_2), *nitric oxide* (NO) and *chlorofluorocarbons* (CFC) into the atmosphere are several times higher than natural emissions. CFCs give rise to highly active radicals in the stratosphere which destroy *ozone* (O_3) by catalytic reactions. However, CFC emissions have been drastically reduced and stratospheric O_3 concentrations are increasing. In contrast, concentrations of GHGs such as *carbon dioxide* (CO_2), *methane* (CH_4) and *nitrous oxide* (N_2O) continue to increase and are now well above the preindustrial levels as increasing human demand for energy is met by burning fossil fuels. As a consequence, land and ocean temperatures are increasing, glaciers are melting and sea levels are rising. Strong reductions, in particular, of the anthropogenic CO_2 emissions are needed to minimise risks of future warming and its consequences as CO_2 is the major GHG with part of it having a long atmospheric residence time. However, CO_2 emissions continue to increase even during the periods of global economic crises and reduced industrial activities. Thus, geoengineering or climate engineering techniques are discussed to cool Earth indirectly by removing CO_2 from the atmosphere or directly by increasing backscattering of solar radiation into space. For example, injecting *sulphur* (S) into the stratosphere has been proposed as sulphate

---

This text was first published as: Lorenz, K.; Crutzen, P.J.; Lal, R.; Töpfer, K., 2012: "Atmospheric Chemistry and Climate in the Anthropocene", in: Lal, R.; Lorenz, K.; Hüttl, R.F.; Schneider, B.U.; von Braun, J. (eds.): *Recarbonization of the Biosphere: Ecosystems and the Global Carbon Cycle* (Dordrecht: Springer Science+Business Media B.V., April); https://doi.org/10.1007/978-94-007-4159-1_3. The permission was granted for Springer-Verlag by Ms. Alice Essenpreis, 28 May 2018.

---

particles reflect sunlight but many issues remain unresolved. Thus, a strong reduction in anthropogenic CO_2 emissions is needed to mitigate climate change.

Keywords Anthropocene · Atmospheric chemistry · Mitigating climate change · Greenhouse gases · Climate engineering · Energy production · Ozone layer · Black carbon · Chlorofluorocarbons · Methane · Nitrous oxide · Global biogeochemical cycles · Renewable energy · Nuclear energy · Carbon capture and storage · Solar radiation management

14.1 Introduction

The presence of life on Earth determines the chemical conditions of the atmosphere, the oceans, and Earth's crust (Schlesinger et al. 2011). In particular, Earths' radiative balance and atmospheric chemistry critically depend on atmospheric abundance of the trace gases *carbon dioxide* (CO_2), *ozone* (O_3), *methane* (CH_4), *nitrous oxide* (N_2O) and halogen-containing compounds (Seinfeld/Pandis 2006). Over the past two centuries, geologic, biological and chemical processes that drive changes in abundance of trace gases and Earth System behaviour have been overwhelmed by the rapidly increasing effects of human activities (Crutzen and Stoermer 2000; Steffen 2010). As the global population and economy are growing at an unprecedented rate, climate and atmospheric chemistry are being altered directly and indirectly (Crutzen 2002; Victor 2010). The central feature is the enormous increase in the use of fossil fuels since the onset of industrialisation (Steffen et al. 2007). Thus, the recent increases in atmospheric trace gases have occurred at an extraordinary pace, and the composition of the atmosphere is changing rapidly on the global scale (Seinfeld/Pandis 2006).

During the past three centuries the global population has increased tenfold to six billion, and quadrupled during the twentieth century (Potts 2009). With this exponential increase human demand for food is also increasing strongly. For example, from 1995 to 2005 cereal production increased from 1897 to 2200 million tons (20%), and meat production increased from 207 to 260 million tons (26%, FAO 2006a). Further, the number of cattle increased by a factor of four during the past century totalling now about 1.4 billion (FAO 2006b). There are about 20 billion farm animals worldwide. Global marine fish catch increased strongly from less than 20 million tons in the mid-twentieth century to about 80 million tons by 2000 (Watson/Pauly 2001). However, overfishing is one of the most serious conservation concerns in marine ecosystems but understanding which fish species are most at risk remains a challenge (Worm et al. 2009; Pinsky et al. 2011). The direct and indirect human pressures on the biosphere are strongly increasing along with the human demand for energy. For example, global energy production doubled over the past half-century while petroleum consumption increased 3.5-fold since 1960 (Steffen et al. 2007; Steffen 2010).

14.2   Changes in the Biosphere

The increasing demands of the growing population result in widespread land use and land use changes within the biosphere. The biosphere extends about 10 km above and 10 km below Earth's surface (Schlesinger et al. 2011). More than half of the terrestrial biosphere has been transformed from mostly wild land to anthropogenic land by human action (Ellis et al. 2010). In addition, the global population has become increasingly urban as urbanisation (i.e., the expansion of urban land uses, including commercial, industrial, and residential uses) increased more than tenfold in the past century (UNFPA 2007). About 0.5% of the global ice-free land area was urban land in 2002 but only 0.01% in 1700 (Ellis et al. 2010; Schneider et al. 2009). For the first time in history, more than 50% of the global population lives in urban centres, towns and settlements (UNFPA 2007).

The human alteration of Earth System behaviour includes also drastic changes in biodiversity and evolutionary processes. The evolution of new species typically takes at least hundreds of thousands of years. However, technology and human population growth also affect evolutionary trajectories and dramatically accelerates evolutionary change especially in commercially important, pest and disease organism (Palumbi 2001). This is apparent in antibiotic resistance, plant and insect resistance to pesticides, rapid changes in invasive species, life-history change in commercial fisheries, and pest adaptation to biological engineering products. Species extinction rates are also dramatically accelerated. Before humans existed, the average fossil species extinction rate was about 1.8 extinctions of species per million species-years (Barnosky et al. 2011). This is distinctively lower compared to the maximum observed rates during the last thousand years which are even above the late Pleistocene mega faunal diversity crash. Recent average species extinction rates are also too high compared to pre-anthropogenic averages. However, whether currently Earth's sixth mass extinction (i.e., loss of three-quarters of species in geological time interval) is under way is not known (Barnosky et al. 2011).

14.3   Human Alterations of Global Biogeochemical Cycles

Life has left its imprint on the chemistry of the planet resulting in characteristic biogeochemical cycling of elements driven mainly by microbial engines (Falkowski et al. 2008; Schlesinger et al. 2011). However, humans are increasingly altering global biogeochemical cycles. For example, prior to human intervention the Earth's *nitrogen* (N) cycle was almost entirely controlled by microbes (Falkowski et al. 2008). Atmospheric reactions and slow geological processes controlled Earth's earliest N cycle, and by about 2.7 billion years ago, a linked suite of microbial processes evolved to form the modern N cycle with robust natural feedbacks and controls (Canfield et al. 2010). However, the creation of reactive N (i.e., all N forms except non-reactive N_2) by humans for fertilizers is increasing every year, with drastic transformation of the global N cycle (Galloway et al. 2008). All biological systems need reactive N. As plant growth is usually constrained by soil N availability in most terrestrial ecosystems, increases in N fertilization and deposition may stimulate plant growth (LeBauer/Treseder 2008; Lu et al. 2011).

All the N used in food production is added to the environment, as is the N emitted to the atmosphere during fossil-fuel combustion (Gruber/Galloway 2008). In the 1990s, these two sources of anthropogenic N to the environment amounted to about 160 teragrams (Tg) N per year. Globally, this is more than that supplied by natural *biological N fixation* (BNF) on land (110 Tg N per year) or in the ocean (140 Tg N per year). Further, humans are likely to be responsible for doubling the turnover rates not only of the terrestrial N cycle but also of the N cycle of the entire Earth (Gruber/Galloway 2008). However, the fate of the human-enhanced N inputs to the land surface is little understood (Schlesinger 2009). For example, increases in terrestrial ecosystem N effluxes caused by N addition were much greater than those in plant and soil pools except soil nitrate (NO_3^-), suggesting a leaky terrestrial N system (Lu et al. 2011). Anthropogenic transformations of global N cycles interact also with C sequestration processes. However, the net effect of increase in N availability on the forest ecosystem *carbon* (C) balance, for example, is not well understood (Lorenz/Lal 2010).

Similar to N, *phosphorus* (P) is an essential macronutrient for plant growth, and considered important in determining the biodiversity and biomass of natural ecosystems (Cramer 2010). However, while P does not limit global primary production in the oceans, long-term degradation of soil and terrestrial ecosystems occurs over millennia in the absence of soil-resetting disturbance as P is lost to groundwater and by occlusion in strongly weathered soils (Walker/Syers 1976; Falkowski et al. 2000). The terrestrial ecosystem mass balance of P, in particular, is controlled by depletion, soil barriers and low P-parent material (Vitousek et al. 2010). Thus, as it is removed with harvest P is often a limiting nutrient in agriculture (Sanchez 2010). External P inputs may be required to sustain both primary productivity in terrestrial ecosystems and other biological processes (Vitousek et al. 2010). To overcome the P limitations, P inputs to the biosphere have increased about fourfold primarily due to mining of P compounds for fertilizer (Falkowski et al. 2000). Thus, human activities profoundly alter the global P cycle. Human release of P to the environment is causing widespread eutrophication of surface freshwaters (Carpenter/Bennett 2011). Biodiversity in many natural ecosystems such as Mediterranean terrestrial ecosystems is threatened by super-abundance of the formerly limiting resource P (Tilman et al. 2001). However, the global distribution of P is uneven, and soils of many regions remain P-deficient and those of others are P-saturated (MacDonald et al. 2011). This heterogeneity complicates major challenges for environmental management in the twenty-first century, i.e., the provision of food and high quality freshwater (Carpenter/Bennett 2011). Similar to N, whether the anthropogenic amplification of the P cycle enhances *net primary production* (NPP) of natural ecosystems and terrestrial C sequestration is not clearly understood (Lorenz/Lal 2010).

The enhanced global flows of P to the biosphere are the result of the growing consumption of inorganic P fertilizers derived from mining of non-renewable phosphate rock (Smil 2000). Fertilizers account for about 80% of global use of phosphate rock (Van Vuuren et al. 2010). To support agricultural production and meet growing demand for food there will be potentially not enough P in the future as global P demand may exceed global P supply (Cordell et al. 2009). However, when this 'peak P' is reached is difficult to predict as peaks in mineral resources are generally difficult to forecast (van Kauwenbergh 2010). Further, only four countries (i.e., Morocco and Western Sahara, China, Jordan, South Africa) control 80% of the world's global phosphate rock reserves that can be economically mined using current technologies (USGS 2010). Global implications are geopolitical 'shortages' in phosphate supply as have been observed in the past for crude oil and rare minerals. However, the non-renewable resource oil can be replaced by renewable energy sources but there is no replacement for non-renewable rock phosphates and no substitutes for P as plant nutrient in agriculture (Van Vuuren et al. 2010). Thus, sustainable P use and management is needed which includes maximising the efficient conversion of phosphate rock into fertilizer (Cordell et al. 2009). Further options include reducing the large flows of P to surface waters by recycling and recovering from municipal and other waste products and the efficient P use in agriculture including recycling of animal and human excreta (Cordell et al. 2009). Resolving agronomic P imbalances is particularly possible with more efficient use of P fertilizers (MacDonald et al. 2011). In summary, some of the biggest gains in the finite P resource can probably be made from the recovery and recycling of phosphates (Gilbert 2009).

14.4 Atmospheric Chemistry

The increasing anthropogenic pressures within the Earth System affect also atmospheric chemistry. The atmospheric concentrations of *greenhouse gases* (GHGs) and those of conventional air pollutant such as *sulphur dioxide* (SO_2), *nitrogen oxides* ($NO_x = NO + NO_2$), *carbon monoxide* (CO), primary carbonaceous particles of black carbon (BC), organic carbon (OC) and CH_4 are increasing (Cofala et al. 2007). For example, *sulphur* (S) is ubiquitous in the biosphere and often occurs in relatively high concentrations in fossil fuels such as coal and crude oil deposits (Smith et al. 2011). The widespread combustion of fossil fuels for energy production has greatly increased emissions of SO_2 into the atmosphere, with the anthropogenic component now substantially greater than natural emissions on a global basis. Global SO_2 emissions peaked in the early 1970s and decreased until 2000. However, in recent years emissions increased further due to increased emissions in China, India, the Middle-East, Brazil, and by international shipping, aviation and developing countries in general (RIVM 2005). About 120 Tg SO_2 have been emitted in 2005 by fossil fuel combustion and industrial processes but only about 4 Tg SO_2 in the same year by forest and grassland burning, and by agricultural waste burning (Smith et al. 2011). Emissions of SO_2 can result in sulfuric acid deposition that can be detrimental to ecosystems, harming aquatic animals and plants, and damaging to a wide range of terrestrial plant life. Further, SO_2 forms sulphate aerosols that have a significant effect on global and regional climate. Sulphate aerosols reflect sunlight into space and also act as condensation nuclei, which tend to make clouds more reflective and change their lifetimes, causing a net cooling (Smith et al. 2011).

Similar to the emissions of SO_2, global anthropogenic emissions of NO_x decreased in the early 1990s but increased afterwards to about 117 Tg NO_2 in 2005 (RIVM 2005). The most important anthropogenic NO_x sources are road transport (41% of anthropogenic NO_x emissions in 2000) followed by power plants, industry and non-road vehicles (21%, 16% and 13% respectively, Cofala et al. 2007). In addition to the 83 Tg NO_2 emitted in 2000, NO_x emissions from biomass burning, international shipping and aviation were responsible for 24.1, 10.0 and 2.5 Tg NO_2, respectively. Releases of NO_x may cause regional high surface (tropospheric) O_3 levels by the reaction of sunlight with air containing hydrocarbons and NO_x (Seinfeld/Pandis 2006). The O_3 in the lower atmosphere can have adverse effects on human health and plants.

The abundance of several important GHGs has substantially increased. For example, atmospheric CO_2 concentrations fluctuated between 180 and 300 *parts per million* (ppm) by volume over the glacial-interglacial cycles during the past 650,000 years (Jansen et al. 2007). However, about 100 years ago the levels of CO_2 began to increase markedly to about 390 ppm, and continue to rise at annual growth rates >2 ppm (Seinfeld/Pandis 2006; Tans 2011). Similar, atmospheric CH_4 concentrations fluctuated between 0.36 and 0.70 ppm during the last 420,000 years (Spahni et al. 2005). However, since the start of the industrial revolution CH_4 concentrations more than doubled to 1.78 ppm (Forster et al. 2007).

Among the GHGs are the almost inert *chlorofluorocarbon* (CFC) gases (Forster et al. 2007); their photochemical breakdown in the stratosphere including those of halons gives rise to highly reactive chlorine and bromine gases which destroy O_3 by catalytic reactions (Crutzen 2002). This anthropogenic O_3 destruction caused the Antarctic 'ozone hole' but CFCs are now regulated (Mäder et al. 2010). Gradually towards the end of the twenty-first century the amount of the CFCs will be diminished so much that N_2O will become the main source of NO_x, which in turn depletes O_3 by catalytic reactions (Ravishankara et al. 2009). Nitrogen oxides are known to catalytically destroy O_3 (Crutzen 1970). Reduction of stratospheric O_3 results in increased UV-B radiation from the sun which may lead to an enhanced risk of skin cancer.

14.5   Climate in the Anthropocene

With the dawn of settled agriculture with attendant deforestation, soil cultivation, spread of rice paddies and raising cattle, it is hypothesised that a trend of increase in atmospheric CO_2 concentration began 8000 years ago (Ruddiman 2003). Similarly, the rise in atmospheric CH_4 may have begun 5000 years ago but causes for the late Holocene rise in atmospheric CH_4 concentration are discussed controversially (Singarayer et al. 2011); However, since the beginning of the nineteenth century human activities have major and growing impacts on atmosphere and Earth. Thus, the current geological epoch has been appropriately named Anthropocene to emphasise the central role of mankind in geology and ecology within the current interglacial, the Holocene (the last 11,600 years; Crutzen and Stoermer 2000). Humans are now clearly affecting climate and can deliberately do so.

Climate forcings are imposed radiative perturbations of the Earth's energy balance, and can be of natural and anthropogenic origin (Newman et al. 2010). Primary forcings such as the orbital solar insulation changes are externally imposed on the climate systems. Secondary forcings are feedbacks within the climate system. Examples for secondary forcings are mineral dust, GHGs, land cover, sea ice, and continental ice and sea level which all impact the radiative balance of the atmosphere. The GHG concentrations for past millennia are well known from ice core analysis. The atmospheric concentrations of CO_2, CH_4 and N_2O have increased by 36%, 148% and 18%, respectively, since 1750 (IPCC 2007). Primarily, fossil fuel use and to a lower extent land use change cause the increase in CO_2 whereas agriculture is primarily responsible for the increase in CH_4 and N_2O. The CO_2 and CH_4 levels are much higher than at any time during the last 650,000 years. Further, for about 11,500 years before the industrial period, the concentration of N_2O varied only slightly but increased relatively rapidly toward the end of the twentieth century. The combined radiative forcing due to the increases in CO_2, CH_4 and N_2O is $+2.3$ W m^{-2} (IPCC 2007). Further, radiative forcing of $+0.35$ W m^{-2} is caused by tropospheric O_3 changes due to emissions of O_3-forming chemicals (NO_x, CO, and hydrocarbons). The direct radiative forcing due to changes in halocarbons is $+0.34$ W m^{-2}. Changes in surface albedo due to deposition of BC aerosols on snow cause a radiative forcing of $+0.1$ W m^{-2}. Changes in solar irradiance since 1750 are estimated to cause a radiative forcing of $+0.12$ W m^{-2}. In contrast, a cooling effect results from negative radiative forcing by anthropogenic contributions to aerosols (primarily SO_4^{2-}, organic C, BC, NO_3^- and dust), with a total direct radiative forcing of -0.5 W m^{-2} and an indirect cloud albedo forcing of -0.7 W m^{-2}. Further, changes in surface albedo due to land cover changes have also a cooling effect with a radiative forcing of -0.2 W m^{-2}. In summary, the global average net effect of human activities since 1750 has been one of warming, with a radiative forcing of $+1.6$ W m^{-2} (IPCC 2007).

14.6   The Evidence of Climate Change

Warming of the climate system is unequivocal (IPCC 2007). It is evident from observations of increases in global average air and ocean temperatures, widespread melting of snow and ice, and rising global average sea level. Specifically, global surface temperature has risen by 0.76 °C from 1850–1899 to 2001–2005. Since the late 1970s, the warming trend of the global surface temperature was 0.15–0.20 °C per decade (Hansen et al. 2010). Also increasing is the average temperature of the global ocean as shown by observations since 1961. Specifically, the ocean has been absorbing more than 80% of the heat added to the climate system. Thus, ocean water is expanding and this contributes to sea level rise (IPCC 2007).

The current warming is unusual when viewed from the millennial perspective provided by multiple lines of proxy evidence and the 160-year record of direct temperature measurements (Thompson 2010). Some of the strongest evidence that a large-scale, pervasive, and, in some cases, rapid change in Earth's climate system is underway is provided by the ongoing widespread melting of high-elevation glaciers and ice caps, particularly in low to middle latitudes. For example, observations of the twentieth and twenty-first centuries show that glaciers in the Andes, the Himalayas, and on Mount Kilimanjaro are shrinking. Ice cores retrieved from shrinking glaciers around the world confirm their continuous existence for periods ranging from hundreds of years to multiple millennia, suggesting that climatological conditions that dominate those regions today are different from those under which these ice fields originally accumulated and have been sustained (Thompson 2010). Further, summertime melting of Arctic sea-ice has accelerated far beyond the expectations of climate models (Allison et al. 2009).

The mean sea level remained nearly stable since the end of the last deglaciation about 3000 years ago (Lambeck et al. 2002). However, the sea level is now rising due to heat-induced ocean water expansion and due to widespread decreases in glaciers, ice caps and the ice sheets of Antarctica and Greenland (IPCC 2007). Specifically, since 1950 the sea level has risen by an average of 1.7 mm year^{-1}, and from 1993 to 2009 by 3.3 mm year^{-1} (Church/White 2006; Ablain et al. 2009). Contributions of ocean temperature change to the global mean sea level and those of glaciers are estimated to be both ~30% each for 1993–2009 (Cazenave/Llovel 2010; IPCC 2007). These sea level rises will almost certainly accelerate through the twenty-first century and beyond because of climate change (Nicholls/Cazenave 2010).

Further evidence of climate change is found in cold regions. Permafrost (perennially frozen ground) is a unique characteristic of polar regions and high mountains, and is fundamental to geomorphic processes and ecological development in tundra and boreal forests (Jorgenson et al. 2010). Thawing of permafrost affects surface hydrology and changes in soil drainage alter the degradation and accumulation of soil C and emissions of CO_2 and CH_4 (Christensen et al. 2004; Schuur et al. 2008; Turetsky et al. 2007). Moderate warming irreversibly thaws and decays permafrost C and initiate a positive permafrost C feedback on climate (Schaefer et al. 2011).

14.7 Mitigating Climate Change

14.7.1 Reductions in Anthropogenic Greenhouse Gas Emissions

Fossil fuel CO_2 emissions increased by 29% between 2000 and 2008 whereas emissions from land-use changes were nearly constant (Le Quéré et al. 2009). The fraction of CO_2 emissions that remained in the atmosphere has also increased to 45% in the past 50 years. Further, the contribution of CO_2 to the CO_2 equivalent concentration sum of all GHGs is projected to increase from 55% in 2005 to 75–85% by the end of this century (NRC 2011). However, efforts to mitigate climate change require in particular the stabilisation of atmospheric CO_2 concentrations as CO_2 is the single most important climate-relevant GHG in Earth's atmosphere (Lacis et al. 2010; Montzka et al. 2011). Thus, CO_2 emissions must be cut strongly, i.e., by more than 80% to stabilise atmospheric CO_2 concentrations for a century or so (House et al. 2008). Stabilisation of atmospheric CO_2 at any level requires in particular anthropogenic CO_2 emissions to go eventually to zero (Weaver 2011).

After CO_2, CH_4 is the second most important anthropogenic GHG in the atmosphere. Since CH_4 has a relatively short lifetime of about 9 years and it is very close to a steady state, reductions in its emissions would quickly benefit climate (Dlugokencky et al. 2011; Montzka et al. 2011). However, in contrast to CO_2 the sinks and, in particular, the sources for CH_4 are poorly quantified. The CH_4 emission rate for the 2000–2004 period has been estimated to be twice the preindustrial period, and 60–70% of the emissions during this period were of anthropogenic origin many of which related to agriculture (Denman et al. 2007; NRC 2011). Present-day CH_4 emissions are estimated to be composed of 64–76% of biogenic, 19–30% of fossil, and 4–6% of pyrogenic sources (Neef et al. 2010). After three decades with little change, atmospheric CH_4 concentrations are increasing since 2007 but the causes are debatable (Rigby et al. 2008; O'Connor et al. 2010; Heimann 2011). While Aydin et al. (2011) explained the slow-down of global atmospheric CH_4 growth during the past two decades by a decline in fossil-fuel emissions, Kai et al. (2011) explained it by a reduction in microbial CH_4 sources in the Northern Hemisphere (i.e., drying northern wetlands, decreasing emissions from rice agriculture in China; Heimann 2011). Further, the main reasons discussed for recent renewed growth are enhanced natural wetland emissions during 2007 and 2008 as a result of higher precipitation in the tropics and anomalously warmer temperatures in the Arctic (Bousquet et al. 2011). However, it is unclear why atmospheric CH_4 concentrations continued to increase in 2009 and 2010 (Montzka et al. 2011). Thus, the reduction in anthropogenic CH_4 emissions needed to stabilise atmospheric concentrations at the current level is unknown. Relative large emission reductions are possible for landfills, and the production of coal, oil and gas (van Vuuren et al. 2007). However, it will be difficult to bring anthropogenic CH_4 emissions to zero in the long term given the continuing need for agriculture to feed the world's population (NRC 2011). Further, significant increases in CH_4 emissions are likely in a future climate from wetland emissions, permafrost thaw, and destabilisation of marine hydrates (O'Connor et al. 2010).

N_2O is the fourth largest contributor to radiative forcing in the atmosphere, and second to CH_4 in radiative forcing among non-CO_2 GHGs (Denman et al. 2007). Sources of N_2O to the atmosphere from human activities are approximately equal to those from natural systems. Specifically, strong emissions occur in the Tropics with high temporal variability (Kort et al. 2011). Human activities that emit N_2O include transformation of fertilizer-N into N_2O and its subsequent emission from agricultural soils, biomass burning, raising cattle and some industrial activities but the release of N_2O is poorly understood (Manning et al. 2011). However, atmospheric concentrations of N_2O continue to rise linearly as most emissions are associated with feeding the world's growing population (Forster et al. 2007; Montzka et al. 2011).

The chlorofluorocarbons CFC-11 and CFC-12 are long-lived GHGs and were extensively used in the past as refrigeration agents and in other industrial processes (Forster et al. 2007). However, after their presence in the atmosphere was found to cause stratospheric O_3 depletion, an efficient reduction in global anthropogenic emissions was reached by the Montreal Protocol in 1987.

14.7.2 Reductions in Greenhouse Gas Emissions from Energy Production

The primary source of the increased atmospheric concentration of CO_2 since the preindustrial period is fossil fuel use for energy production (IPCC 2007). In 2004, CO_2 from fossil fuels contributed 56.6% to the total anthropogenic GHG emissions (Rogner et al. 2007). Further, emissions increased from 15,627 petagrams (Pg) CO_2 in 1973 to 29,236 Pg CO_2 in 2008 (IEA 2010). The recent growth in CO_2 emissions parallels a shift in the largest fuel emission source from oil to coal (Le Quéré et al. 2009). In addition to CO_2, fossil fuel use also contributes to the observed increase in atmospheric CH_4 concentration. Thus, energy savings must focus on a strong reduction in fossil fuel use as other energy sources were only responsible for 0.1–0.4% of total CO_2 emissions from energy use (IEA 2010). Improving efficiency in fossil fuel energy use will also contribute to reduced GHG emissions but behavioural interventions are required to improve energy efficiency (Allcott/Mullainathan 2010).

Renewable energy is an option for lowering GHG emissions from the energy system while still satisfying the global demand for energy services (Edenhofer et al. 2011). The most important renewable energy sources to mitigate climate change are bioenergy, direct solar energy, geothermal energy, hydropower, ocean and wind energy. Of the 12.9% of the total 492 *Exajoules* (EJ) of primary energy supply in 2008 provided by renewable energy, biomass contributed 10.2%, with the majority being traditional biomass used in cooking and heating applications in developing countries but with rapidly increasing use of modern biomass (i.e., all other biomass except traditional biomass). Hydropower represented 2.3% of renewable energy whereas other sources accounted for 0.4%. In particular, GHG emissions from renewable energy technologies for electricity generation are lower than those associated with fossil fuel options (Edenhofer et al. 2011). Although most current bioenergy systems result in GHG emission reductions, most biofuels produced through new processes (i.e., advanced or next generation biofuels) could provide higher GHG mitigation. However, land use changes and corresponding emissions and removals may also affect the GHG balance of biofuels. In total, renewable energy has a large potential to mitigate GHG emissions by 2050 with the largest contributions from modern biomass, wind and direct solar energy use (Edenhofer et al. 2011).

Nuclear energy has been viewed for a long time as playing an important role to meet increasing electricity demand while at the same time decreasing CO_2 emissions from energy production (Widder 2010). However, safety is a major concern and the secure disposal of spent nuclear fuel in a repository is unresolved. Further, the nuclear fuel cycle in many countries such as the United States is unsustainable. Thus, mere absence of GHG emissions may not be sufficient to assess nuclear power as mitigation strategy for climate change (Kopytko/Perkins 2011).

In particular, inland locations of nuclear power plants encounter great problems with interrupted operations while safety is of primary concern at coastal locations. The latter was recently emphasised by the March 11, 2011 disaster at the Fukushima Daiichi nuclear power facility located at the Pacific coast in Japan after a strong earthquake off-coast triggered a devastating tsunami. In the following days, massive explosions at the power plant scattered nuclear fuel around the site. The health effects of the release of large amounts of radioactive material especially with respect to contamination of food and water are uncertain (Butler 2011). Also, the clean-up process at and around Fukushima could last for many decades or even a century (Brumfiel 2011). As a consequence, Germany is looking at options to shut down all its nuclear power plants in the near future as risks for the human society appear to be unpredictable independent of any possible GHG savings by using nuclear power for electricity generation (Ethics Commission for a Safe Energy Supply 2011).

With the recent scepticism towards nuclear power, fossil fuels will account for even more than half of the projected increase in global energy consumption by 2035 with coal remaining the dominant fuel for electricity generation (IEA 2010). Thus, *carbon capture and storage* (CCS) technologies will be required to reduce increasing CO_2 emissions from coal-fired power plants. CCS refers to a range of technologies that aim to capture the CO_2 in fossil fuels either before or after combustion, and store it for very long time in underground formations such as depleted oil and gas reservoirs, deep saline formations and un-mineable coal seams (Global CSS Institute 2011). Currently, it is uncertain how much and when CCS can contribute to CO_2 sequestration as governments and industry are still in the early stages of implementing large scale international programmes to shorten the timeframe for the commercial deployment of CCS. These programmes are focused on the demonstration phase for developing and improving capture technologies in new industrial applications and proving the safe and secure long-term storage of CO_2. However, the demonstration phase is likely to last for over a decade and CO_2 emissions from existing and new coal-fired power plants will continue to strongly increase. For example, the world's first large-scale CCS project in the power sector, the Southern Company Integrated Gasification Combined Cycle project in the United States, is still under construction (Global CSS Institute 2011).

14.8 Climate Engineering

Global efforts to reduce GHG emissions have not yet been sufficiently successful to provide confidence that the reductions needed to avoid dangerous climate change will be achieved (The Royal Society 2009). For example, the CO_2 emission growth rate is increasing from increasing global economic activity and from increasing C intensity of the global economy while at the same time land and ocean based CO_2 sink activity appear to weaken (Canadell et al. 2007). Thus, additional actions involving climate engineering or geoengineering may become necessary to cool the Earth. Geoengineering can be defined as the deliberate large-scale intervention in the Earth's climate system in order to moderate global warming (The Royal Society 2009).

Geoengineering methods include *carbon dioxide removal* (CDR) techniques which remove CO_2 from the atmosphere and *solar radiation management* (SRM) techniques which reflect a small percentage of the sun's light and heat back into space. CDR techniques include (i) land use management to protect or enhance land C sinks, (ii) the use of biomass for C sequestration as well as a C neutral energy source, (iii) enhancement of natural weathering processes to remove CO_2 from the atmosphere, (iv) direct engineered capture of CO_2 from ambient air, and (v) the enhancement of oceanic uptake of CO_2, for example by fertilization of the oceans with naturally scarce nutrients, or by increasing upwelling processes (The Royal Society 2009). However, only permanent sequestration has the potential to decrease total cumulative CO_2 emissions over time and, thus, decrease the amount of climate change that occurs (Matthews 2010). By contrast, no permanent sequestration will generally only delay emissions as very slow leakage of C stored in ocean or geological reservoirs occurs. Thus, total cumulative emissions over time are not decreased and long-term temperature changes would be similar to a scenario without CDR.

SRM techniques include (i) increasing the surface reflectivity of the planet, by brightening human structures (e.g., by painting them white), planting of crops with a high reflectivity, or covering deserts with reflective material, (ii) enhancement of marine cloud reflectivity, and (iii) placing shields or deflectors in space to reduce the amount of solar energy reaching the Earth. However, little research has yet been done on most of the geoengineering methods and there have been no major directed programmes of research on the subject although CDR appears to be less risky than SRM. Otherwise, CDR may have a much slower effect on reducing global temperature, and SRM techniques may be ineffective in offsetting changes in rainfall patterns and storms (Irvine et al. 2010). In summary, there are major uncertainties regarding effectiveness, costs and environmental impacts of geoengineering methods

(The Royal Society 2009). The most appropriate response to human-induced climate change is to decrease the level of overall human intervention in the climate system by decreasing GHG emissions (Matthews 2010).

Among the SRM techniques discussed is mimicking the cooling effects of volcanic eruptions by injecting sulphate aerosols into the lower stratosphere (Crutzen 2006). Sulphate particles can act as cloud condensation nuclei and thereby influence the microphysical and optical properties of clouds, affecting regional precipitation patterns, and increasing cloud albedo. The deposition of 1–2 Tg S annually in the stratosphere would be sufficient to cool the climate similar to the cooling that occurred after release of 10 Tg S by the eruption of Mount Pinatubo in June 1991. However, the residence time of S in the stratosphere is only about 2 years and the required annual S inputs would be large and need to continue for more than a century. Further, unresolved issues are particle size, effects on long wave radiation, and possible O_3 loss through ClO/Cl reactions. Thus, the sulphate albedo scheme should only be used if climate changes drastically and no serious side effects occur. The reduction in anthropogenic GHG emissions must be the focus in climate change mitigation (Crutzen 2006).

14.9 Summary

Human population is drastically increasing with severe consequences for atmospheric chemistry and the Earth System behaviour. Ecosystem services such as the provision of food and water are increasingly exhausted by the ever increasing population. Atmospheric levels of trace gases are higher than during the pre-industrial period. Fortunately, emissions of *chlorofluorocarbons* (CFCs) are now regulated and *stratospheric ozone* (O_3) levels are recovering. However, increasing energy demand is met by burning fossil fuels and atmospheric concentrations of *greenhouse gases* (GHGs) are increasing. Thus, climate is warming and sea levels are rising as oceans are expanding and glaciers are melting. Without drastic reduction in CO_2 emissions, Earth is committed to long term warming with unpredictable consequences for human society. Deliberately reducing atmospheric CO_2 by geoengineering techniques has been proposed but actively reducing solar radiation reaching Earth's atmosphere may be a fast action method to cool Earth. However, environmental issues associated with geoengineering techniques such as the release of large amounts of S for more than a century in the stratosphere remain unresolved. Thus, efforts to mitigate climate change must focus on the reduction in anthropogenic CO_2 emissions.

References

Ablain, M.; Cazenave, A.; Valladeau, G. et al. 2009: "A new assessment of the error budget of global mean sea level rate estimated by satellite altimetry over 1993–2008", in: *Ocean Sci*, 5: 193–201.

Allcott, H.; Mullainathan, S., 2010: "Behavior and energy policy", in: *Science*, 327: 1204–1205.

Allison, I.; Bindoff, N.L.; Bindschadler, R.A. et al., 2009: *The Copenhagen diagnosis, 2009: updating the World on the latest climate science* (Sydney, Australia: The University of New South Wales, Climate Change Research Centre [CCRC]).

Aydin, M.; Verhulst, K.R.; Saltzman, E.S. et al., 2011: "Recent decreases in fossil-fuel emissions of ethane and methane derived from firn air", in: *Nature*, 476: 198–201.

Barnosky, A.D.; Matzke, N.; Tomiya, S. et al., 2011: "Has the Earth's sixth mass extinction already arrived?", in: *Nature*, 471: 51–57.

Bousquet, P.; Ringeval, B.; Pison, I. et al., 2011: "Source attribution of the changes in atmospheric methane for 2006–2008", in: *Atmos Chem Phys*, 11: 3689–3700.

Brumfiel, G., 2011: "Fukushima set for epic clean-up", in: *Nature*, 474: 135–136.

Butler, D., 2011: "Fukushima health risks scrutinized", in: *Nature*, 472: 13–14.

Canadell, J.G.; Le Quéré, C.; Raupach, M.R. et al., 2007: "Contributions to accelerating atmospheric CO_2 growth from economic activity, carbon intensity, and efficiency of natural sinks", in: *Proc Natl Acad Sci*, USA, 104: 18866–18870.

Canfield, D.E.; Glazer, A.N.; Falkowski, P.G., 2010: "The evolution and future of Earth's nitrogen cycle", in: *Science*, 323: 192–196.

Carpenter, S.R.; Bennett E.M., 2011: "Reconsideration of the planetary boundary for phosphorus", in: *Environ Res Lett*, 6: 014009.

Cazenave, A.; Llovel, W., 2010: "Contemporary sea level rise", in: *Annu Rev Mar Sci*, 2: 145–173.

Christensen, T.R.; Johansson, T.; Malmer, N. et al., 2004: "Thawing sub-arctic permafrost: effects on vegetation and methane emissions", in: *Geophys Res Lett*, 31: L04501.

Church, J.A.; White N.J., 2006: "A 20th century acceleration in global sea-level rise", in: *Geophys Res Lett*, 33: L01602.

Cofala, J.; Amann, M.; Klimont, Z. et al., 2007: "Scenarios of global anthropogenic emissions of air pollutants and methane until 2030", in: *Atmos Environ*, 41: 8486–8499.

Cordell, D.; Drangert, J.-O.; White, S., 2009: "The story of phosphorus: global food security and food for thought", in: *Glob Environ Chang*, 19: 292–305.

Cramer, M.D., 2010: "Phosphate as a limiting resource: introduction", in: *Plant Soil*, 334: 1–10.

Crutzen, P.J., 1970: "The influence of nitrogen *oxides* on the atmospheric ozone content", in: *QJR Meteorol Soc*, 96: 320–325.

Crutzen, P.J., 2002: "Geology of mankind", in: *Nature*, 415: 23.

Crutzen, P.J., 2006: "Albedo enhancement by stratospheric sulfur injections: a contribution to resolve a policy dilemma?", in: *Clim Chang*, 77: 211–219.

Crutzen, P.J.; Stoermer, E.F., 2000: "The 'anthropocene'", in: *IGBP Newsl*, 41: 17–18.

Denman, K.L.; Brasseur, G.; Chidthaisong, A. et al., 2007: "Couplings between changes in the climate system and biogeochemistry"; Chapter 7, in: Intergovernmental Panel on Climate Change (ed.): *Climate change 2007: the physical science basis* (Cambridge: Cambridge University Press).

Dlugokencky, E.J.; Nisbet, E.G.; Fisher, R. et al., 2011: "Global atmospheric methane: budget, changes and dangers", in: *Philos Trans R Soc A*, 369: 2058–2072.

Edenhofer, O.; Pichs-Madruga, R.; Sokona, Y. et al., 2011: "Summary for policymakers", in: Edenhofer, O.; Pichs-Madruga, R.; Sokona, Y. et al. (eds.): *IPCC special report on renewable energy sources and climate change mitigation* (Cambridge/New York: Cambridge University Press).

Ellis, E.C.; Goldewijk, K.K.; Siebert, S. et al., 2010: "Anthropogenic transformation of the biomes, 1700 to 2000", in: *Global Ecol Biogeogr*, 19: 589–606.

Ethics Commission for a Safe Energy Supply, 2011: "Germany's energy transition – a collective project for the future"; at: https://archiv.bundesregierung.de/resource/blob/656922/457334/784356871e5375b8bd74ba2a0bde22bf/2011-05-30-abschlussbericht-ethikkommission-en-data.pdf?download=1.

Falkowski, P.G.; Fenchel, T.; Delong, E.F., 2008: "The microbial engines that drive Earth's biogeochemical cycles", in: *Science*, 320: 1034–1039.

Falkowski, P.; Scholes, R.J.; Boyle, E. et al., 2000: "The global carbon cycle: a test of our knowledge of earth as a system", in: *Science*, 290: 291–296.

Food and Agriculture Organization of the United Nations (FAO), 2006a: *FAO statistical databases* (Rome: FAO); at: http://faostat.fao.org/default.aspx.

Food and Agriculture Organization of the United Nations (FAO), 2006b: *Livestock's long shadow: environmental issues and options* (Rome: Food and Agriculture Organization).

Forster, P.; Ramaswamy, V.; Artaxo, P. et al, 2007: "Changes in atmospheric constituents and in radiative forcing", in: Solomon, S.; Qin, D.; Manning, M. et al. (eds): *Climate change 2007: the physical science basis. Contribution of working group I to the fourth assessment report of the Intergovernmental Panel on Climate Change* (Cambridge/New York: Cambridge University Press).

Galloway, J.N.; Townsend, A.R.; Erisman, J.W., et al., 2008: "Transformation of the nitrogen cycle: recent trends, questions, and potential solutions", in: *Science*, 320: 889–892.

Gilbert, N., 2009: "The disappearing nutrient", in: *Nature*, 461: 716–718.

Global CCS Institute, 2011: *The global status of CCS: 2010* (Canberra, Australia).

Gruber, N.; Galloway, J.N., 2008: "An earth-system perspective of the global nitrogen cycle", in: *Nature*, 451: 293–296.

Hansen, J.; Ruedy, R.; Sato, M. et al., 2010: "Global surface temperature change", in: *Rev Geophys*, 48: RG4004; https://doi.org/10.1029/2010RG000345.

Heimann, M., 2011: "Enigma of the recent methane budget", in: *Nature*, 476: 157–158.

House, J.I.; Huntingford, C.; Knorr, W. et al., 2008: "What do recent advances in quantifying climate and carbon cycle uncertainties mean for climate policy?", in: *Environ Res Lett*, 3: 044002.

IEA, 2010: *Key world energy statistics* (Paris: Organisation for Economic Co-operation and Development/ International Energy Agency).

IPCC, 2007: "Summary for policymakers", in: Solomon, S.; Qin, D.; Manning, M.; Chen, Z.; Marquis, M.; Averyt, K.B.; Tignor, M.; Miller, H.L. (eds.): *Climate change 2007: the physical science basis. Contribution of working group I to the fourth assessment report of the intergovernmental panel on climate change* (Cambridge: Cambridge University Press).

Irvine, P.J.; Ridgwell, A.; Lunt, D.J., 2010: "Assessing the regional disparities in geoengineering impacts", in: *Geophys Res Lett*, 37: L18702; https://doi.org/10.1029/2010GL044447.

Jansen, E.; Overpeck, J.; Briffa, K.R. et al., 2007: "Palaeoclimate", in: Solomon, S.; Qin, D.; Manning, M. et al. (eds.): *Climate change 2007: the physical science basis. Contribution of working group I to the fourth assessment report of the Intergovernmental Panel on Climate Change* (Cambridge/New York: Cambridge University Press).

Jorgenson, M.T.; Romanovsky, V.; Harden, J. et al., 2010: "Resilience and vulnerability of permafrost to climate change", in: *Can J For Res*, 40: 1219–1236.

Kai, F.M.; Tyler, S.C.; Randerson, J.T. et al., 2011: "Reduced methane growth rate explained by decreased Northern Hemisphere microbial sources", in: Nature, 476:194–197.

Kopytko, N.; Perkins, J., 2011: "Climate change, nuclear power, and the adaptation-mitigation dilemma", in: *Energy Policy*, 39: 318–333.

Kort, E.A.; Patra, P.K.; Ishijima, K. et al., 2011: "Tropospheric distribution and variability of N_2O: evidence for strong tropical emissions", in: *Geophys Res Lett*, 38: L15806; https://doi.org/10.1029/2011GL047612.

Lacis, A.A.; Schmidt, G.A.; Rind, D. et al., 2010: "Atmospheric CO_2: principal control knob governing Earth's temperature", in: *Science*, 330: 356–359.

Lambeck, K.; Yokoyama, Y.; Purcell, T., 2002: "Into and out of the last glacial maximum: sea-level change during oxygen isotope stages 3 and 2", in: *Quaternary Sci Rev*, 21: 343–360.

LeBauer, D.S.; Treseder, K.K., 2008: "Nitrogen limitation of net primary productivity in terrestrial ecosystems is globally distributed", in: *Ecology*, 89: 371–379.

Le Quéré, C.; Raupach, M.R.; Canadell, J.G. et al., 2009: "Trends in the sources and sinks of carbon dioxide", in: *Nat Geosci*, 2: 831–836.

Lorenz, K.; Lal, R., 2010: *Carbon sequestration in forest ecosystems* (Dordrecht: Springer).

Lu, M.; Yang, Y.; Luo, Y. et al., 2011: "Responses of ecosystem nitrogen cycle to nitrogen addition: a meta-analysis", in: *New Phytol*, 189: 1040–1050.

MacDonald, G.K.; Bennett, E.M.; Potter, P.A. et al., 2011: "Agronomic phosphorus imbalances across the world's croplands", in: *Proc Natl Acad Sci*, USA, 108: 3086–3091.

Mäder, J.A.; Staehelin, J.; Peter, T. et al., 2010: "Evidence for the effectiveness of the Montreal Protocol to protect the ozone layer", in: *Atmos Chem Phys*, 10: 12161–12171.

Manning, A.C.; Nisbet, E.G.; Keeling, R.F. et al., 2011: "Greenhouse gases in the earth system: setting the agenda to 2030", in: *Philos Trans R Soc A*, 369: 1885–1890.

Matthews, H.D., 2010: "Can carbon cycle geoengineering be a useful complement to ambitious climate mitigation?", in: *Carbon Manage*, 1: 135–144.

Montzka, S.A.; Dlugokencky, E.J.; Butler, J.H., 2011: "Non-CO_2 greenhouse gases and climate change", in: *Nature*, 476: 43–50.

Neef, L.; van Weele, M.; van Velthoven, P., 2010: "Optimal estimation of the present day global methane budget", in: *Global Biogeochem Cycles*, 24: GB4024; https://doi.org/10.1029/2009GB003661.

Newman, L.; Kiefer, T.; Otto-Bliesner, B. et al., 2010: "The science and strategy of the past global changes (PAGES) project", in: *Curr Opin Environ Sustain*, 2: 193–201.

Nicholls, R.J.; Cazenave, A., 2010 "Sea-level rise and its impact on coastal zones", in: *Science*, 328: 1517–1520.

NRC, 2011: *Climate stabilisation targets: emissions, concentrations, and impacts over decades to millennia*. Committee on stabilisation targets for atmospheric greenhouse gas concentrations (Washington, DC: National Research Council, The National Academies Press).

O'Connor, F.M.; Boucher, O.; Gedney, N. et al., 2010: "Possible role of wetlands, permafrost, and methane hydrates in the methane cycle under future climate change: a review", in: *Rev Geophys*, 48: RG4005; https://doi.org/10.1029/2010RG000326.

Palumbi, S.R., 2001: "Humans as the world's greatest evolutionary force", in: *Science*, 293: 1786–1790.

Pinsky, M.L.; Jensen, O.P.; Ricard, D. et al., 2011: "Unexpected patterns of fisheries collapse in the world's oceans", in: *Proc Natl Acad Sci*, USA, 108: 8317–8322.

Potts, M., 2009: "What's next?", in: *Philos Trans R Soc B*, 364: 3115–3124.

Ravishankara, A.R.; Daniel, J.S.; Portmann, R.W., 2009: "Nitrous oxide (N_2O): the dominant ozone-depleting substance emitted in the 21st century", in: *Science*, 326: 123–125.

Rigby, M.; Prinn, R.G.; Fraser, P.J. et al., 2008: "Renewed growth of atmospheric methane", in: *Geophys Res Lett*, 35: L22805; https://doi.org/10.1029/2008GL036037.

RIVM, 2005: *Emission database for global atmospheric research-EDGAR* (Bilthoven, The Netherlands: RIVM).

Rogner, H.-H.; Zhou, D.; Bradley, R. et al., 2007: "Introduction", in: Metz, B.; Davidson, O.R.; Bosch, P.R.; Dave, R.; Meyer, L.A. (eds.): *Climate change 2007: mitigation. Contribution of working group III to the fourth assessment report of the Intergovernmental Panel on Climate Change* (Cambridge/New York: Cambridge University Press).

Ruddiman, W.F., 2003: "The anthropogenic greenhouse area began thousands of years ago", in: *Clim Chang*, 61: 261–293.

Sanchez, P.A., 2010: "Tripling crop yields in tropical Africa", in: *Nat Geosci*, 3: 299–300.

Schaefer, K.; Zhang, T.; Bruhwiler, L. et al., 2011: "Amount and timing of permafrost carbon release in response to climate warming", in: *Tellus*, 63B: 165–180.

Schlesinger, W.H., 2009: "On the fate of anthropogenic nitrogen", in: *Proc Natl Acad Sci*, USA, 106: 203–208.

Schlesinger, W.H.; Cole, J.J.; Finzi, A.C. et al., 2011: "Introduction to coupled biogeochemi-cal cycles", in: *Front Ecol Environ*, 9: 5–8.

Schneider, A.; Friedl, M.A.; Potere, D., 2009: "A new map of global urban extent from MODIS satellite data", in: *Environ Res Lett*, 4: 044003; https://doi.org/10.1088/1748-9326/4/4/044003.

Schuur, E.A.G.; Bockheim, J.; Canadell, J.G. et al., 2008: "Vulnerability of permafrost carbon to climate change: implications for the global carbon cycle", in: *Bioscience*, 58: 701–714.

Seinfeld, J.H.; Pandis, S.N., 2006: *Atmospheric chemistry and physics: from air pollution to climate change*, 2nd edn. (Hoboken: Wiley).

Singarayer, J.S.; Valdes, P.J.; Friedlingstein, P. et al., 2011: "Late Holocene methane rise caused by orbitally controlled increase in tropical sources", in: *Nature*, 470: 82–86.

Smil, V., 2000: "Phosphorus in the environment: natural flows and human interferences", in: *Annu Rev Energy Environ*, 25: 53–88.

Smith, S.J.; van Aardenne, J.; Klimont, Z. et al., 2011: "Anthropogenic sulfur dioxide emissions: 1850-2005", in: *Atmos Chem Phys*, 11: 1101–1116.

Spahni, R.; Chapellaz, J.; Stocker, T.F. et al., 2005: "Atmospheric methane and nitrous oxide of the late Pleistocene from Antarctic ice cores", in: *Science*, 310: 1317–1321.

Steffen, W., 2010: "Observed trends in earth system behavior", in: *WIREs Clim Chang*, 1: 428–449.

Steffen, W.; Crutzen, P.J.; McNeill, J.R., 2007: "The Anthropocene: are humans now overwhelming the great forces of nature?", in: *Ambio*, 36: 614–621.

Tans, P., 2011: "Trends in atmospheric carbon dioxide – global"; at: http://www.esrl.noaa.gov/gmd/ccgg/trends/.

The Royal Society, 2009: *Geoengineering the climate: science, governance and uncertainty* (London: The Royal Society, Science Policy).

Thompson, L.G., 2010: "Climate change: the evidence and our options", in: *Behav Anal*, 33: 153–170.

Tilman, D.; Fargione, J.; Wolff, B. et al., 2001: "Forecasting agriculturally driven global environmental change", in: *Science*, 292: 281–284.

Turetsky, M.R.; Wieder, R.K.; Vitt, D.H. et al., 2007: "The disappearance of relict permafrost in boreal North America: effects on peatland carbon storage and fluxes", in: *Glob Change Biol*, 13: 1922–1934.

United Nations Populations Fund (UNFPA), 2007: *State of world population 2007: unleashing the potential of urban growth* (New York: UNFPA). United Nations Population Fund, New York.

U.S. Geological Survey (USGS), 2010: *Mineral commodity summaries 2010* (Washington, DC: U.S. Geological Survey).

van Kauwenbergh, S.J., 2010: *World phosphate rock reserves and resources* (International Fertilizer Development Center, Muscle Shoals).

van Vuuren, D.P.; den Elzen, M.G.J.; Lucas, P.L. et al., 2007: "Stabilizing greenhouse gas concentrations at low levels: an assessment of reduction strategies and costs", in: *Clim Chang*, 81: 119–159.

van Vuuren, D.P.; Bouwman, A.F.; Beusen, A.H.W., 2010: "Phosphorus demand for the 1970–2100 period: a scenario analysis of resource depletion", in: *Glob Environ Chang*, 20: 428–439.

Victor, P., 2010: "Questioning economic growth", in: *Nature*, 468: 370–371.

Vitousek, P.M.; Porder, S.; Houlton, B.Z. et al., 2010: "Terrestrial phosphorus limitation: mechanisms, implications, and nitrogen-phosphorus interactions", in: *Ecol Appl*, 20: 5–15.

Walker, T.W.; Syers, J.K., 1976: "The fate of phosphorus during pedogenesis", in: *Geoderma*, 15: 1–19.

Watson, R.; Pauly, D., 2001: "Systematic distortions in world fisheries catch trends", in: *Nature*, 414: 534–536

Weaver, A.J., 2011: "Toward the second commitment period of the Kyoto Protocol", in: *Science*, 332: 395–396.

Widder, S., 2010: "Benefits and concerns of a closed nuclear fuel cycle", in: *J Renew Sustain Energy*, 2: 062801.

Worm, B.; Hilborn, R.; Baum, J.K. et al., 2009: "Rebuilding global fisheries", in: *Science*, 325: 578–585.

Chapter 15
Climate, Atmospheric Chemistry and Biogenic Processes in the Anthropocene (2012)

Paul J. Crutzen

15.1 Climate, Atmospheric Chemistry and Biogenic Processes in the Anthropocene

Figure 15.1 shows *polar stratospheric clouds* (PSC)—also called mother of pearl clouds. They are very beautiful and damaging at the same time. At very low temperatures (lower than −80 °C) in the stratosphere the ice particles forming PSC play a large role in ozone depletion. This is so because on the surface of the particles chlorine and bromine are converted into highly reactive catalytic forms, affecting the destruction of ozone.

A change in ozone concentration is something that was not foreseen, and there was no prediction from the scientific point of view in the beginning. Only later we recognised the dramatic role of human activity. We saw a change in ozone concentration in spring time at high altitudes (12–25 km) (see Figs. 15.2 and 15.3) in the Antarctic where we least expected it (Farman et al. 1985; Hofmann et al. 1987).

We were all flabbergasted and it took time to explain it scientifically. At the same time it clearly had a life-threatening dimension for humankind on earth. It was clear that something had to be done, and in the 1980s the CFC gases were banned from production. Nevertheless it will take several years to heal the ozone hole. The question now is: are we able to counter the effects of the other greenhouse gases released by human activities? Industrial and agricultural activities have grown dramatically especially since the second world war, the so-called 'great acceleration'. While the main activity has been in the developed world, the developing countries are following rapidly, especially in Asia. The dramatic increase in the quantity of

This text was first published as: Paul J. Crutzen, 2012: "Climate, Atmospheric Chemistry and Biogenic Processes in the Anthropocene", in: Kant. H.; Reinhardt, C. (eds.): *100 Jahre Kaiser-Wilhelm-/Max-Planck-Institut für Chemie (Otto-Hahn-Institut): Facetten seiner Geschichte*, Veröffentlichungen aus dem Archiv der Max-Planck-Gesellschaft, Vol. 22 (Berlin: MPG): 241–249. No permission was needed by this MPG text.

Fig. 15.1 Photograph of an Arctic polar stratospheric cloud (PSC) taken at Kiruna, Sweden, on 27 January 2000. *Credit* Ross Salawitch. NASA source

Fig. 15.2 Decrease of total ozone over the Halley Bay Station of the British Antarctic Survey

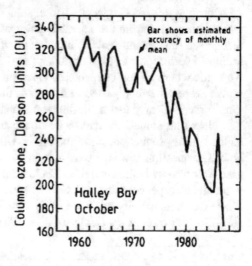

Fig. 15.3 Abrupt change of ozone concentration from maximum to the 'ozone hole' during just two months

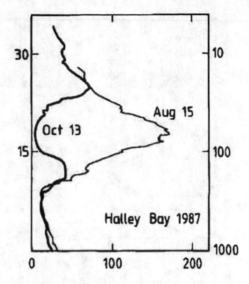

almost all human activities (see Fig. 15.4) since the 1950s has clear effects. During the past three centuries human population has increased tenfold to 7000 million. There are currently some 20 billion (20,000 million) of farm animals worldwide and urbanisation grew more than tenfold in the past century; almost half of the people live in cities and megacities. Industrial output increased 40 times during the past century; energy use 16 times and almost 50% of the land surface has been transformed by human action. Water use increased severalfold.

This has been described as the 'great acceleration' (see Fig. 15.4) and as the '1950s syndrome' (Pfister 2010) in other words, the rapid transition in the post-war period fueled by plenty of cheap energy. The demands of our lifestyle have some consequences unknown to many.

Why this happened since the 1950s is an unsolved question. It probably has to do with the stark increase of international financial corporations. While before 1940 much of international trade was still related to colonialism, in the second half of the twentieth century post-colonial and globalised commercial structures have been taking over. It seems that the causes of the observed changes have their origin in financial, economic and political doings. But changes are not just of a quantitative nature. There are qualitative alterations as well. Industry has introduced many thousands of newly synthesised compounds into the environment. Some of them are toxic, carcinogenic, mutagenic or reprotoxic. Even some which are not toxic at all can show negative effects, such as the almost inert *chlorofluorocarbons* (CFC) causing the ozone hole. With regard to biodiversity the impact of anthropogenic activities has been particularly rapid (Wilson 2002). It may well be that we now live in the age of the sixth mass extinction in the history of earth (Barnosky et al. 2001).

Also, anthropogenic measures cause evolutionary change in domestic animals. Antibiotics and pesticides speed up the normal evolutionary time scale and cost between 33 and 50 billion dollars a year in the USA alone (Palumbi 2001). Human

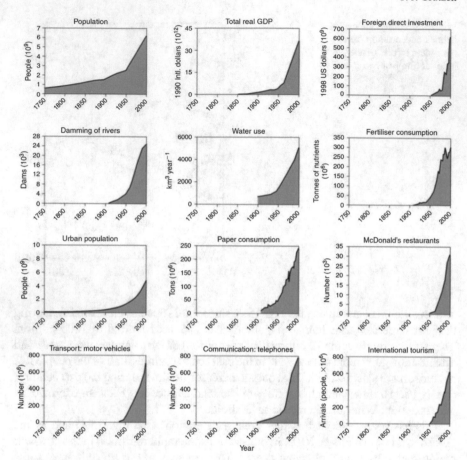

Fig. 15.4 The 'Great Acceleration'

intervention does not only change selection mechanisms and the biotic environment. It also has deep impact on geological features. Man-caused erosion by crop tillage and land uses for grazing and construction exceeds natural erosion by 15 times (Wilkinson 2005). The consequences of heavy use of raw materials can be felt in very unbalanced manners. For example, in some areas phosphates are badly needed, mainly for fertilizing. In other regions, however, the intensive application of chemical fertilizers has led to eutrophication. Negative and positive effects are not balanced well especially in the case of phosphates, a problem which has been neglected for a long time. If phosphorus is not recycled we are heading for catastrophe (Cordell et al. 2009) (Figs. 15.5 and 15.6).

Since the beginning of this accelerating impact of mankind's activities we have started a new geological epoch: the Anthropocene. Mankind is doing things that nature never did and influences actively the process of evolution. In addition prognoses and prediction play a much larger role in planning deliberate action, although

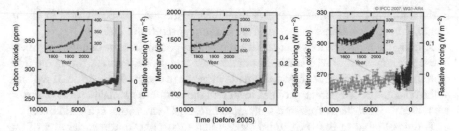

Fig. 15.5 Changes in Greenhouse Gases

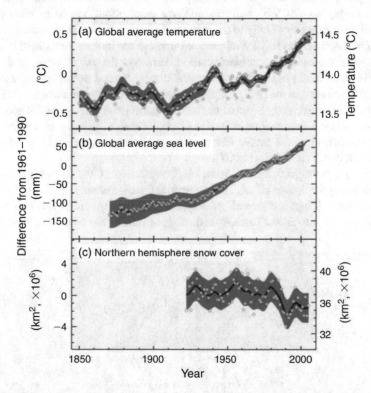

Fig. 15.6 Changes in temperature, sea level and northern hemisphere snow cover

prognosis dates back at least to the essay by Thomas Robert Malthus of 1798 on the principle of population, predicting the limits for the increasing of human population due to the bounded growth of resources (Malthus 1798). At the end of the nine-teenth century debates on the scarcity of nitrogen compounds for fertilizers were fierce and led to the invention of numerous technical processes for making use of the nitrogen available in the air (Stoltzenberg 1994, 2004). Technical solutions such as the Haber-Bosch-Process for the fixation of nitrogen have solved the problem only in the midterm perspective. Arguably they have made the situation worse in giving rise to almost unhindered increase of the modern agricultural industry.

Among the most obvious markers for the Anthropocene are the changes in greenhouse gas concentrations available from ice-core and modern data. Carbon dioxide, methane and nitrous oxide show a dramatic increase in the modern period. If we look at the changes of temperature, of sea level and the northern hemisphere snow cover, we recognise that the correlation with greenhouse gas emissions is a valid one.

A unique table of the IPCC third assessment report shows the variability of the anthropogenic impact (see Fig. 15.7). It states the global mean radiative forcing caused by atmospheric changes. Especially noteworthy here is that the level of scientific uncertainty is great. We simply do not know much about the consequences of our actions. This applies especially to the increased albedo effect. This backscattering of solar radiation from the surface of particles and clouds in the atmosphere is largely cooling, with the exception of black carbon. However the warming up of the earth climate is an evident phenomenon. Observations on air and ocean temperatures, on snow and ice cover and the rising global sea level are clear and unambiguous. Expectations of the rise of average global surface temperature are between 2.0 and 4.5 °C by the year 2100. Conservative estimates of sea level rise are between 19 and 58 cm in the same period, while Stefan Rahmstorf's prediction exceeds this by a factor of approximately 2.5 (Rahmstorf 2007).

Cleaning up the atmosphere from air pollution will warm the troposphere, enhancing the greenhouse effect. The opposite is the case when particles are added to the atmosphere. If that is done deliberately, the effect can go under geo-engineering. In a few cases stabilisation of atmospheric concentrations has occurred. For example

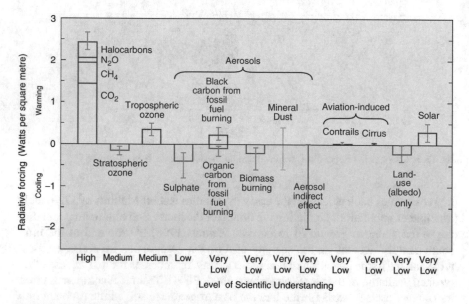

Fig. 15.7 Global mean radiative forcing of the climate system for the year 2000, relative to 1750

the banning of CFCs has met the required reduction. This is a non negligible part of the overall increase of greenhouse gases. However, for stabilizing concentrations of carbon dioxide and nitrous oxide at current levels, reductions in the emission of 60% in the case of carbon dioxide and 70–80% in the case of nitrous oxide have to be met. But the emissions still continue to increase. The conditions for the long term stabilisation of concentration of methane are not clear yet, which is especially due to the unknown effects of the thawing of permafrost.

There do exist a couple of feedback loops, both positive and negative. For example only a slight increase of acidity of the ocean water hinders carbonate-secreting organisms building their skeletons. As a consequence, the ocean water cannot contain as much carbon dioxide as before (Caldeira/Wickett 2003). Another example is that an increase of global temperature causes the melting of the permafrost which leads to additional releases of carbon dioxide and methane. This, of course, in turn leads to additional rises in temperature.

The scale of the problem is most visible when we study the situations in the Arctic. "Studies indicate that the Arctic oceans ice cover is about 40% thinner than 20–40 years ago" (Goss Levi 2000). Robert Corell, Chairman of the Arctic Climate Impact Assessment, stated in November 2004 that the pace of the dramatic climate change going on the Arctic is on the order of two to three times as much as in the rest of the world (Corell 2004) (Figs. 15.8 and 15.9).

The long term stability of the correlations of greenhouse gas forcing and the Vostok- Temperature clearly shows that we cannot escape a vicious cycle. Increase in greenhouse gas concentration is unequivocally related to the increase of temperature. Consequently there are two lessons to be learned: we have to reduce the emissions of greenhouse gases, and next to it we may choose to actively engage in counter-measures. A study by the Royal Society has investigated the efficacy of injecting very large amounts of sulfur into the stratosphere. We are talking about approximately 1–2

Fig. 15.8 Sudden thawing, here in the Noatak National Preserve, Alaska

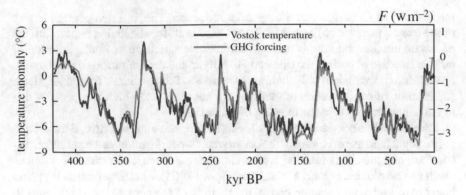

Fig. 15.9 Greenhouse gas forcing and Vostok temperature

Tg of sulphur per year. This very drastic action has to be followed up in the long term, should the cooling be affected. Due to the large uncertainties involved, we propose to study the resulting albedo scheme. But we think that it should only be applied if climate change will be dramatic. Too many questions are still unresolved. Among them are the particle effects on long wave radiation, ozone loss, cirrus effects, etc. Above all geo-engineering must not question our determination to reduce emissions of greenhouse gases.

We have shown that since the beginning of the industrial revolution in the mid eighteenth century human actions have influenced biological, geological and mete-orological features on earth. For the immediate future the anthropogenic impact will be felt regardless of any measures we will take. Thus, I have proposed to name a new geological epoch, the Anthropocene, taking into account the large and perma-nent impact of man. It is a curious incidence of history that at the very beginning of the Anthropocene stood two of its closely related decisive features: James Watts design of the steam engine in 1784 is the symbol of an unbound use of fossil fuels, and roughly at the same time the first scientific analyses indicated the start of the increase in concentration of carbon dioxide and methane in air. Thus, doing and measuring went hand in hand from the start. As measuring is the sine qua non for prognosis, we may state that two of the important characteristics of the Anthropocene have been with us for a long time. Science and industry seem to be mirror image twins also in this perspective. Scientists pondered on the impact of human interventions on the environment already in the nineteenth century. And in the first half of the twen-tieth century such diverse scholars as the Russian geologist Vladimir I. Vernadsky (1863–1945) and the catholic theologist and geologist Pierre Teilhard de Chardin (1881–1955) brought up the notion 'noosphere' encompassing the world of thought (Levit/Olsson 2006). We may share the implicit optimism that human brainpower will solve the challenging problems in front of us. However, the rapid quantitative expansion of mankind eats up much of our scientific and technical accomplishments. Especially frightening is the fact that all of what has happened so far has been caused by only a part of the world's population.

Alone by this, the earth will warm up by 2–4.5 °C in this century, according to the intergovernmental panel on climate change. Next to the risk involved in quantitative development, there are many qualitative uncertainties. For example if the ozone-destroying chlorine would chemically behave like the closely related element bromine we would not just see the ozone hole in Antarctic spring. We would experience it as a year-round and global effect. It was just luck and not our scientific intelligence, helping us out of a potential global catastrophe. This we know because the effects of halogens on the ozone layer have been studied intensively since 40 years. But there are many more scientific riddles waiting. We do not know the effects of the many toxic and carcinogenic substances that we release day by day into the environment. It is thus a mandatory task lying ahead for scientists and engineers to counsel society on the way towards a sustainable future. Because mankind is actively involved for the better or the worse, this will impart large-scale geoengineering projects. As these projects are both political and scientific in character we call for a new deal involving science and politics.

References

Barnosky, A.D. et al., 2001: "Has the Earth's sixth mass extinction already arrived?", in: *Nature*, 471: 51–57.

Caldeira, K.; Wickett, M.E., 2003: "Anthropogenic carbon and ocean pH", in: *Nature*, 425: 365.

Cordell, D. et al., 2009: "The story of Phosphorus: Global food security and food for thought", in: *Global Environmental Change*, 19: 292–305.

Corell, R.W., 2004: "Statement of the Arctic Climate Impact Assessment" (16 November).

Farman, J.C.; Gardiner, B.G.; Shanklin, J.D., 1985: "Large losses of total ozone in Antarctica reveal seasonal ClOx/NOx interaction", in: *Nature*, 315: 207–210.

Goss Levi, B., 2000: "The Decreasing Arctic Ice Cover", in: *Physics Today*, 53: 19.

Hofmann, D.J.; Harder, J.W.; Rolf, S.R.; Rosen, J.M., 1987: "Balloon-borne observations of the development and vertical structure of the Antarctic ozone hole in 1986", in: *Nature*, 326: 59–62.

Levit, G.S.; Olsson, Lennart, 2006: "'Evolution on rails'. Mechanisms and levels of orthogenesis", in: *Annals for the History and Philosophy of Biology*, 11: 97–136, 124–128.

Malthus, T.R., 1798: *An Essay on the Principle of Population* (London: J. Johnson in St. Paul's Churchyard).

Palumbi, S.R., 2001: "Humans as the world's greatest force", in: *Science*, 293: 1786–1790.

Pfister, C., 2010: "The '1950s Syndrome' and the transition from a slow-going to a rapid loss of global sustainability, in: Uekotter, F. (ed.): *Turning Points in Environmental History* (Pittsburgh: University of Pittsburgh Press): 90–117.

Rahmstorf, S., 2007: "A semi-empirical approach to projecting future sea-level rise", in: *Science*, 315: 368–370.

Stoltzenberg, Dietrich, 1994: *Fritz Haber. Chemiker, Nobelpreistrager, Deutscher, Jude* (Weinheim/New York/Basel/Cambridge: VCH Verlagsgesellschaft mbH): 133–186.

Stoltzenberg, Dietrich (English edition), 2004: *Fritz Haber: Chemist, Nobel Laureate, German, Jew: A Biography* (Philadelphia: Chemical Heritage Foundation).

Wilkinson, B.H., 2005: "Humans as geological agents", in: *Geology*, 33: 161–164.

Wilson, E.O., 2002: *The Future of Life* (New York: Alfred A. Knopf).

Chapter 16
The Palaeoanthropocene – The Beginnings of Anthropogenic Environmental Change (2013)

Stephen F. Foley, Detlef Gronenborn, Meinrat O. Andrea,
Joachim W. Kadereit, Jan Esper, Denis Scholz, Ulrich Pöschl,
Dorrit E. Jacob, Bernd R. Schöne, Rainer Schreg, Andreas Vött,
David Jordan, Jos Lelieveld, Christine G. Weller, Kurt W. Alt,
Sabine Gaudzinski-Windheuser, Kai-Christian Bruhn, Holger Tost,
Frank Sirocko, and Paul J. Crutzen

Abstract As efforts to recognise the Anthropocene as a new epoch of geological time are mounting, the controversial debate about the time of its beginning continues. Here, we suggest the term *Palaeoanthropocene* for the period between the first, barely recognizable, anthropogenic environmental changes and the industrial revolution when anthropogenically induced changes of climate, land use and biodiversity began to increase very rapidly. The concept of the Palaeoanthropocene recognises that humans are an integral part of the Earth system rather than merely an external forcing factor. The delineation of the beginning of the Palaeoanthropocene will require an increase in the understanding and precision of palaeoclimate indicators, the recognition of archaeological sites as environmental archives, and interlinking palaeoclimate, palaeoenvironmental changes and human development with changes in the distribution of Quaternary plant and animal species and socio-economic models of population subsistence and demise.

This text was first published as: Foley, S.F.; Gronenborn, D.; Andreae, M.O.; Kadereit, J.W.; Esper, J.; Scholz, D.; Pöschl, U.; Jacob, D.E.; Schöne, B.R.; Schreg, R.; Vött, A.; Jordan, D.; Lelieveld, J.; Weller, C.G.; Alt, K.W.; Gaudzinski-Windheuser, S.; Bruhn, K-C.; Tost, H.; Sirocko, F.; Crutzen, P.J., 2013: "The Palaeoanthropocene – The Beginnings of Anthropogenic Environmental Change", in: *Anthropocene*, 3: 83–88, open access at: https://www.blogs.uni-mainz.de/fb09climatology/files/2012/03/Foley_2013_Anthropocene.pdf. Corresponding author at: ARC Centre of Excellence for Core to Crust Fluid Systems and Department of Earth Sciences, Macquarie University, North Ryde, New South Wales 2109, Australia. Tel.: +61 298509452; *E-mail address:* stephen.foley@mq.edu.au (S.F. Foley). We are indebted to Anne Chin, Rong Fu, Xiaoping Yang, Jon Harbor and an anonymous reviewer for helpful comments on the manuscript. The concept of the Palaeoanthropocene grew during many discussions at the Geocycles Research Centre in Mainz.

16.1   The Anthropocene – Climate or Environment?

Eleven years after Crutzen (2002) suggested the term Anthropocene as a new epoch of geological time (Zalasiewicz et al. 2011a), the magnitude and timing of human-induced change on climate and environment have been widely debated, culminating in the establishment of this new Journal. Debate has centred around whether to use the industrial revolution as the start of the Anthropocene as suggested by Crutzen, or to include earlier anthropogenic effects on landscape, the environment (Ellis et al. 2013), and possibly climate (Ruddiman 2003, 2013), thus backdating it to the Neolithic revolution and possibly beyond Pleistocene megafauna extinctions around 50,000 years ago (Koch/Barnosky 2006). Here, we appeal for leaving the beginning of the Anthropocene at around 1780 AD; this time marks the beginning of immense rises in human population and carbon emissions as well as atmospheric CO_2 levels, the so-called 'great acceleration'. This also anchors the Anthropocene on the first measurements of atmospheric CO_2, confirming the maximum level of around 280 ppm recognised from ice cores to be typical for the centuries preceding the Anthropocene (Lüthi et al., 2008). The cause of the great acceleration was the *increase* in burning of fossil fuels: this did not begin in the 18th century, indeed coal was used 800 years earlier in China and already during Roman times in Britain (Hartwell 1962; Dearne/Branigan 1996), but the effects on atmospheric CO_2 are thought to have been less than 4 ppm until 1850 (Stocker et al. 2010). The Anthropocene marks the displacement of agriculture as the world's leading industry (Steffen et al. 2011).

However, the beginning of the Anthropocene is more controversial than its existence, and if we consider anthropogenic effects on the environment rather than on climate, there is abundant evidence for earlier events linked to human activities, including land use changes associated with the spread of agriculture, controlled fire, deforestation, changes in species distributions, and extinctions (Smith/Zeder 2013). The further one goes back in time, the more tenuous the links to human activities become, and the more uncertain it is that they caused any lasting effect.

The proposition of the Anthropocene as a geological epoch raises the question of what defines an epoch. To some extent this is a thought experiment applied to a time in the far future – the boundary needs to be recognizable in the geological record millions of years in the future, just as past boundaries are recognised. This requires changes of sufficient magnitude that can be accurately dated. It is interesting to note that the recent definition of the beginning of the Holocene with reference to ice cores (Walker et al. 2009) fails the criterion of being recognizable well into the future because of the geologically ephemeral nature of ice.

Some geological boundaries are characterised by distinct geochemical markers; for example, the iridium anomaly at the Cretaceous-Neogene boundary, which is thought to have been caused by a meteorite impact. The Anthropocene will leave numerous clear markers including synthetic organic compounds and radionuclides as well as sedimentological memories of sudden CO_2 release and ocean acidification (Zalasiewicz et al. 2011b). Many older geological boundaries were defined by disjunctures in the fossil record marked by first appearances or extinctions (Sedgwick 1852). However, the age of these has changed with improvements in radiometric age dating; for example, the beginning of the Cambrian has moved by 28 million years since 1980. There is abundant evidence that we are currently experiencing the Earth's sixth great mass extinction event (Barnosky et al. 2011), which will be another hallmark of the Anthropocene. The changes that mark the beginning of the Anthropocene are certainly changes of sufficient magnitude to justify a geological boundary (Steffen et al. 2011), whereas the gradual or small-scale changes in regional environments at earlier times were not.

16.2   The Palaeoanthropocene

The term *Palaeoanthropocene* is introduced here to mark the time interval before the industrial revolution during which anthropogenic effects on landscape and environment can be recognised but before the burning of fossil fuels produced a *huge crescendo* in anthropogenic effects. The beginning of the Palaeoanthropocene is difficult to define and will remain so: it is intended as a transitional period, which is not easily fixed in time. We emphasise that we do not intend it to compete for recognition as a geological epoch: it serves to delineate the time interval in which anthropogenic environmental change began to occur but in which changes were insufficient to leave a global record for millions of years. Although it covers a time period of interest to many scientific disciplines stretching from archaeology and anthropology to palaeobotany, palaeogeography, palaeoecology and palaeoclimate, its beginning is necessarily transitional on a global scale because it involves changes that are small in magnitude and regional in scale. The history of human interference with the environment can be represented on a logarithmic timescale (Fig. 16.1), resulting in three approximately equal areas. In the Anthropocene, major changes (orange) have been imposed on natural element cycles (black bar) that were typical of pre-human times. The Palaeoanthropocene includes the Holocene (beginning 11,700 years ago) and probably much of the Pleistocene (2.58 Ma), and may stretch from about the time of the first appearance of the genus *Homo* until the industrial revolution (Ruddiman). The arrows in Fig. 16.1 show the timescales normally considered by various scientific disciplines, emphasizing that only their integration can provide a complete picture. Anthropogenic influences on the environment taper out towards the beginning of the Palaeoanthropocene and get lost in the uncertainties of age determinations. The transition into the Anthropocene is much sharper, involving order of magnitude changes in a short time. The Palaeoanthropocene may seem to largely coincide with the Pleistocene and Quaternary, but these are defined stratigraphically without reference to the environmental effects of humans (Gibbard et al. 2010). Thus, the Palaeoanthropocene should not be anchored on any unit of the geological timescale, but instead be used to emphasise the as yet uncertain period in which humans measurably affected their environments.

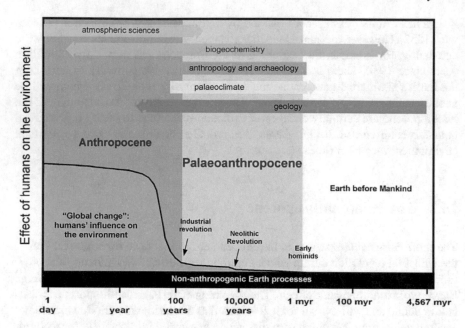

Fig. 16.1 Plotting the history of the Earth on a logarithmic scale gives three approximately equal sections for the Anthropocene, the Palaeoanthropocene and anthropogenically unaffected Earth processes. The Palaeoanthropocene is a period of small and regional effects that are more difficult to define and are currently hotly debated. It is also a time for which the research tools of several scientific disciplines overlap: the integration of results from all these disciplines will be essential to improve our understanding of processes in the Palaeoanthropocene

Human activities have always been interdependent with the functioning of natural processes. Climatic and environmental changes probably caused major migrations of humans throughout human prehistory (De Menocal 2001; Migowski et al. 2006), and conversely, the distribution of plants and animals has been strongly affected by human impacts on the environment (Parmesan 2006). It is important to view humans as an integral part of the Earth System in order to adequately understand inter-relationships and feedbacks between the Earth and humankind. The social perception of the environment and cultural behaviour are a crucial part of systemic interaction. In order to fully understand the transition to the Anthropocene, it is therefore essential to include human culture and its management of landscapes and material cycles into the Earth System concept.

There are several reasons for the diffuse beginning of the Palaeoanthropocene, particularly (1) limitations on the availability of environmental archives identifying events so far in the past; (2) the dampening of signals by the gradual saturation of reservoirs; and (3) the local to regional spatial scale at which these events occurred: populations grew gradually, and new technologies were introduced at different times from place to place.

Relatively little information has yet been extracted from natural archives in Palaeolithic and earlier times. For example, there may be a causal relationship between the arrival of humans and the extinction of Australian megafauna (Brook et al. 2007), but this is currently based on remarkably few localities that demonstrate the temporal coexistence of humans and now extinct species (Wroe/Field 2006; Field et al. 2013). Landscape burning may have been an important intermediary process (Bowman 1998). Humans and fire have always coexisted, but the deliberate use of fire may have caused the first appreciable anthropogenic effects on ecology. The habitual use of fire extends back further than 200,000 years (Karkanas et al. 2007) and possibly to almost 2 Ma (Bowman et al. 2009). However, exact dating is hampered by the currently high cost of precise ^{14}C dating, which restricts the number of age determinations, as well as the temporal restriction of ^{14}C to later periods. Further discoveries of fossils and archaeological remains will improve the temporal precision.

The dampening of signals have prevented thousands of years of wood burning and centuries of fossil fuel usage from being detectable as a significant increase in atmospheric carbon because other environmental carbon sinks had to be saturated before the surplus could be registered in the atmosphere. This is a recurring relationship between geochemical element sinks and atmospheric composition: the major rise of atmospheric oxygen in the early Proterozoic did not immediately follow the biogenic production of oxygen, but had to await the saturation of reduced geological formations before free oxygen could be released. Prior to this, banded iron formations and reduced paleosols dominated (Klein 2005; Rye/Holland 1998), to be replaced by oxygenated sediments (red beds) once the atmosphere became oxygenated. Geological processes are very slow, but the element reservoirs are enormous, allowing the potential to buffer anthropogenic increases in emissions. This may appear to render these increases harmless for a given period, but the exhaustion of buffers may lead to tipping points being reached with potentially grave consequences for humankind.

Scales in space and time form perhaps the most important distinction between the Palaeoanthropocene and the Anthropocene. Gas mixing rates in the atmosphere can be considered immediate on historical and geological time scales, and can therefore result in global changes. In contrast, the effects that humans have on their environment take place on a local scale, and these spread to regional events that will not immediately have global repercussions. Understanding the Palaeoanthropocene will require an *increased emphasis on more restricted temporal and spatial scales*. The concept of the Anthropocene has commonly been associated with *global* change, whereas Palaeoanthropocene studies must concentrate on regional issues. Regional studies may deal with human ecosystems as small as village ecosystems (Schreg 2013). Models of future climate change with regional resolution will also become more important, as local extremes are predicted in areas of high population density, such as the eastern Mediterranean (Lelieveld et al. 2012). For this reason, the beginning of the Palaeoanthropocene should not be assigned a global starting date, but instead is time-transgressive (Brown et al. 2013). It dissipates into a number of regional or local issues the further one moves back in time, varying with the history of each local environment and human society. When it comes to defining the beginning of anthropogenic effects on the environment, time appears to fray at the edges.

16.3   Studying the Palaeoanthropocene

The Palaeoanthropocene involves the interaction of humans with their environment, and so studying it is an interdisciplinary challenge encompassing climate, environmental and geological sciences as well as archaeology, anthropology and history, with improvements in one often prompted by viewpoints or methods imported from other sciences. Here, we briefly outline three areas where rapid progress can be expected.

16.3.1   Human Subsistence and Migration

The subsistence and migration of humans and their cultures is fundamental to understanding the interdependence between people, their environments and climatic conditions, and yet this is hampered by the scarcity of archaeological sites that can be dated precisely. Figure 16.2 illustrates the expansion of farming through Europe, but the reasons, particularly climatic or environmental factors, remain poorly understood.

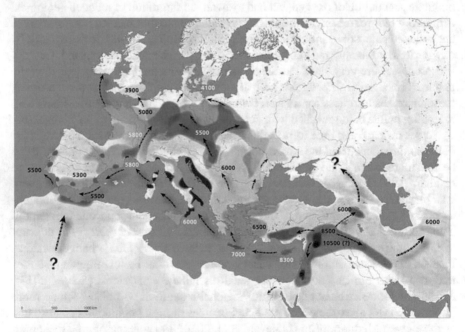

Fig. 16.2 Map showing the distribution of the first farming societies in western Eurasia together with dates for major human migration episodes (in years calibrated B.C.) beginning from the Fertile Crescent, which had a much more temperate climate at the time (after Gronenborn 2014). Greater temporal resolution and precision and further understanding of the reasons for migration will require input from many scientific disciplines to assess the relative importance of short-scale climate, environmental and species distributions

Prehistoric sites with human remains are known from the Palaeolithic, during which arctic species such as reindeer were amongst the main prey (Gaudzinski/Roebroeks 2000). The emergence of farming is related to the northward retreat of arctic conditions at the end of the last glacial period and thus to climate on a supra-regional scale. There are indications that early Holocene climate fluctuations may have paced the migration of farming populations (Weninger et al. 2009; Gronenborn 2014; Lemmen et al. 2011). However, the degree to which early farming populations caused measurable increases in greenhouse gases remains controversial (Kaplan et al. 2010; Ruddiman et al. 2011; Ruddiman 2013).

Food supplies have always played a central role in determining the migration and expansion of human populations in response to environmental and climate changes. Agricultural production of grains and the keeping of livestock gradually spread, leading to important societal changes and to new attitudes to the distribution of resources, stockpiling, territoriality and work distribution, resulting in the first major population increase in human history (Chamberlain 2006; Bocquet-Appel/Bar-Yosef 2008). Increasing population density led to new forms of interdependence between humans and nature such as crop failures and floods, which frequently ended in food shortages. Further technological innovations allowed further increases in population, which increased the risk of subsistence crises.

For a great proportion of their history, humans have been immediately dependent on their environment in terms of plants, animals and water supply. Changes in diet can be reconstructed using skeletal remains as a dietary archive and analysing radiogenic and stable isotopes, trace elements, and ancient DNA (Evans et al. 2006; Haak et al. 2008; Mannino et al. 2011). Radiogenic isotope systems are important in ascertaining the age, migration, geological substrate and diagenesis of bones and thus the relative importance of dietary and environmental factors. Ancient DNA analysis has recently allowed remarkable insights into movements, population mixing and evolution of the human genome (Haak et al. 2005; Burger et al. 2007; Bramanti et al. 2009; Haak et al. 2010; Brandt et al. 2013), providing a new temporal and spatial resolution for Palaeoanthropocene studies.

16.3.2   Regional Palaeoclimate

A main difference between the Palaeoanthropocene and the Anthropocene is the gradual switch from regional to global scale of anthropogenic influences. In Palaeolithic to Neolithic times, changes were related to fires, land use, and species extinctions, which are regional effects. In palaeoclimate research, the collection of longterm climate information has been emphasised because of the desire to model *global* changes in climate. Many of the archives are marine (e.g. Kennett/Ingram 1995), which may transmit a dampened signal in which extreme events are removed or minimised, particularly in the older time sections. Despite having more potential on short timescales, detailed continental records are commonly used only to derive average temperatures (Sukumar et al. 1993; Farrera et al. 1999). For Palaeoanthropocene climate studies, both regional and short time-scale information is needed to unravel the complex interplay of humans and their environment.

Ocean mixing processes are sluggish on anthropogenic time scales, resulting in dampened signals. Because it is the land on which people live, early land use changes will be recorded in continental archives first, promoting their importance over marine archives. Furthermore, continental archives preserve information on extreme events, permitting cross-referencing with archaeological records.

Periods of weeks to a year incorporate most of the hazards for human sustenance and survival, but are beyond the resolution of many palaeoclimate repositories. Although insignificant when the whole Quaternary is considered, this is the timescale of crop failures and subsistence crises (Büntgen et al. 2011). The integration of several proxies revealing the palaeoclimate of continental regions will increasingly permit annual to seasonal resolution, illuminating extreme natural events that may have been critical triggers for crises and migrations. We currently have only limited understanding of the spatial patterns of temperature, precipitation and drought variations in short-term extreme events and periods of rapid climate change throughout the Quaternary. The high temporal resolution that is becoming available from multiple continental palaeoclimate proxies will enable the closer study of time slices of single seasons to several years (Sirocko et al. 2013).

Speleothems can be dated with unprecedented precision over the last ~650,000 years by U-series methods (Scholz/Hoffmann 2011) representing a key archive for seamless climate reconstructions. The development of new proxies and archives, such as compound specific isotope ratios in lignin methoxyl groups in wood (Keppler et al. 2007), multi-proxy data derived from continental loess deposits and palaeosols (Sheldon/Tabor 2009), or Roman aqueduct sinter (Surmelihindi et al. 2013) will further strengthen multi-proxy approaches.

Biomineralisation needs to be considered in assessing past climate variability. Unexpected mismatches between temperature proxies illustrate that we know too little about the mechanisms by which climate and environmental information is recorded. Mineralising organisms exert specific physiological controls on the minerals they form so that the chemical behaviour of elements and isotopes used for climate reconstruction deviates from that expected in geochemical equilibrium. These 'vital effects' (Urey et al. 1951), occur in all living systems, describing an array of species-specific deviations from equilibrium compositions. Some bivalves begin the crystallisation process using amorphous calcium carbonate (Jacob et al. 2008, 2011), and amorphous precursor phases appear to be universally involved in biocarbonate and bioapatite formation. This affects the storage of temperature information, which may change during the lifetime of individual organisms (Schöne et al. 2011).

For all palaeoclimate reconstructions, the storage of data from individual proxies in central repositories will improve transparency and provide essential supplements to the publication of large data sets as figures.

16.3.3   Palaeoenvironmental Sciences

The clearing of forests to provide agricultural land may have already been widespread more than 3000 years ago (Kaplan et al. 2009), and may have had far-reaching impacts on palaeoecology and the evolution and distribution of plant and animal species. Much earlier, fire was used to control vegetation and may have affected species extinctions (Bowman 1998; Bowman et al. 2009).

We need to understand how Quaternary evolution would have progressed without the influence of humans. The Quaternary was a hotbed of evolution, and the spread of humans throughout Europe coincided with its re-colonisation by plants and animals after the end of the ice age (Comes/Kadereit 1998; Hewitt 1999). We also need to assess what the atmospheric composition would have been without human perturbation. This is possible for a number of trace gases such as CO_2 and CH_4 by analysing bubbles trapped in ice cores, but exceedingly difficult for other potent climate agents such as aerosol particles (Andreae 2007).

Modelling natural species distributions will further delineate changing ecological conditions, and may identify the beginnings of divergence of biodiversity from natural patterns. Models of niche evolution will integrate climate- and human-induced biological evolution with past environmental change, including dropping the assumption that the ecological requirements of species did not change in the relevant time span (Futuyma 2010). The projection of ecological niches into the past will be greatly refined by improved palaeoclimate chronologies. Approaches from ecological system theory help to provide new perceptions of sustainability and novel insights into the complex processes of societal collapse and triggering of migrations.

A further development assisting Palaeoanthropocene studies is the treatment of archaeological sites as environmental archives (Bridgland 2000; Tarasov et al. 2013). Integrated geomorphological, environmental and archaeological studies help to reveal the dimension, intensity and duration of how human societies exploited and changed natural environments and, conversely, how changing natural environments and landscapes provoked the adaptation of land use strategies. Examples are possible feedbacks between the climatically favoured expansion of savanna ecosystems beginning in the late Miocene, the acquisition of fire by early hominids and its influence on human evolution, and the eventual use of fire for landscape management in the late Pleistocene (Bowman et al. 2009).

The recognition of interactions between the regional and global scales is important since land use changes can have global effects (Foley et al. 2005). High-resolution regional data sets on vegetation, environment, climate and palaeoweather (integrating sedimentological and meteorological data; Pfahl et al. 2009) must be combined with models of land use and village ecosystem dynamics to achieve long-term perspectives on causality and complex system behaviour in human-environment systems (Dearing et al. 2010).

In summary, the term Palaeoanthropocene refers to the period from the beginning of human effects on the environment to the beginning of the Anthropocene, which should be reserved for the time after the great acceleration around 1780 AD. The

Palaeoanthropocene has a diffuse beginning that should not be anchored on geological boundaries, as it is linked to local events and annual to seasonal timescales that cannot be recognised globally. Progress in Palaeoanthropocene studies can be expected through greater precision in palaeoclimate reconstructions, particularly on continents, and it's coupling with studies of environmental archives, new fossil discoveries, species distributions and their integration into regional numerical models of climate and environment.

References

Andreae, M.O., 2007: "Aerosols before pollution", in: *Science*, 315: 50–51.

Barnosky, A.D.; Metzke, N.; Tomiya, S.; Wogan, G.O.U.; Swartz, B.; Quental, T.B.; Marshall, C.; Mcuire, J.L.; Lindsey, E.L.; Maguire, K.C.; Mersey, B.; Ferrer, E.A., 2011: "Has the Earth's sixth mass extinction event already arrived?", in: *Nature*, 471: 51–57.

Bocquet-Appel, J.P.; Bar-Yosef, O. (eds.), 2008: *The Neolithic Demographic Transition and Its Consequences* (Dordrecht, The Netherlands: Springer).

Bowman, D.M.J.S., 1998: Tansley Review No. 101: "The impact of aboriginal landscape burning on the Australian biota", in: *New Phytologist*, 140: 385–410.

Bowman, D.M.J.S.; Balch, J.K.; Artaxo, P.; Bond, W.J.; Carlson, J.M.; Cochrane, M.A.; D'Antonio, C.M.; DeFries, R.S.; Doyle, J.C.; Harrison, S.P.; Johnston, F.H.; Keeley, J.E.; Krawchuk, M.A.; Kull, C.A.; Marston, J.B.; Moritz, M.A.; Prentice, I.C.; Roos, C.I.; Scott, A.C.; Swetnam, T.W.; van der Werf, G.R.; Pyne, S.J., 2009: "Fire in the Earth system", in: *Science*, 324: 481–484.

Bramanti, B.; Thomas, M.G.; Haak, W.; Unterlaender, M.; Jores, P.; Tambets, K.; Antanaitis-Jacobs, I.; Haidle, M.N.; Jankauskas, R.; Kind, C.J.; Lueth, F.; Terberger, T.; Hiller, J.; Matsumura, S.; Forster, P.; Burger, J., 2009: "Genetic discontinuity between local hunter-gatherers and central Europe's first farmers", in: *Science*, 326: 137–140.

Brandt, G.; Haak, W.; Adler, C.J.; Roth, K.; Szecsenyi-Nagy, A.; Karimnia, S.; Moller- Rieker, S.; Meller, H.; Ganslmeier, R.; Friederich, S.; Dreselz, V.; Nicklisch, N.; Pickrell, J.K.; Sirocko, F.; Reich, D.; Cooper, A.; Alt, K.W., 2013: "The Genographic Consortium. Ancient DNA reveals key stages in the formation of central European mitochondrial genetic diversity", in: *Science*, 342: 257–261.

Bridgland, D.R., 2000: "River terrace systems in north-west Europe: an archive of environmental change, uplift and early human occupation", in: *Quaternary Science Reviews*, 19: 1293–1303.

Brook, B.; Bowman, D.J.S.; Bruney, D.; Flannery, T.; Gagan, M.; Gillespie, R.; Johnson, C.; Kershaw, A.P.; Magee, J.; Martin, P.; Miller, G.; Peiser, B.; Roberts, R.G., 2007: "Would the Australian megafauna have become extinct if humans had never colonized the continent? Comments on a review of the evidence for a human role in the extinction of Australian megafauna and an alternative explanation", in: *Quaternary Science Reviews*, 26: 560–564.

Brown, A.; Toms, P.; Carey, C.; Rhodes, E., 2013: "Geomorphology of the Anthropocene: time-transgressive discontinuities of human-induced alluviation", in: *Anthropocene*, 1: 3–13.

Büntgen, U.; Tegel, W.; Nicolussi, K.; McCormick, M.; Frank, D.; Trouet, V.; Kaplan, J.O.; Herzig, F.; Heussner, K.-U.; Wanner, H.; Luterbacher, J.; Esper, J., 2011: "2500 years of European climate variability and human susceptibility", in: *Science*, 331: 578–582.

Burger, J.; Kirchner, M.; Bramanti, B.; Haak, W.; Thomas, M.G., 2007: "Absence of the lactase-persistence associated allele in early Neolithic Europeans", in: *Proceedings of the National Academy of Sciences* 104: 3736–3741.

Chamberlain, A.T., 2006: *Demography in Archaeology*, Cambridge Manuals in Archaeology. (Cambridge: Cambridge University Press).

Comes, H.P.; Kadereit, J.W., 1998: "The effect of quaternary climatic changes on plant distribution and evolution", in: *Trends in Plant Science*, 3: 432–438.

Crutzen, P.J., 2002: "Geology of mankind", in: *Nature*, 415: 23.

De Menocal, P.B., 2001: "Cultural responses to climate change during the late Holocene", in: *Science*, 292: 667–673.

Dearing, J.A.; Braimoh, A.K.; Reenberg, A.; Turner, B.L.; van der Leeuw, S., 2010: "Complex land systems: the need for long time perspectives to assess their future", in: *Ecology and Society*, 15,4: 21.

Dearne, M.J.; Branigan, K., 1996: "The use of coal in Roman Britain", in: *The Antiquaries Journal*, 75: 71–105.

Ellis, E.C.; Kaplan, J.O.; Fuller, D.Q.; Vavrus, S.; Goldewijk, K.K.; Verburg, P.H., 2013: "Used planet: a global history", in: *Proceedings of the National Academy of Sciences*; at: https://doi.org/10.1073/pnas.1217241110 (Epub ahead of print).

Evans, J.A.; Chenery, C.A.; Fitzpatrick, A.P., 2006: "Bronze age childhood migration of individuals near Stonehenge, revealed by strontium and oxygen isotope tooth enamel analysis", in: *Archaeometry*, 48: 309–321.

Farrera, I.; Harrison, S.P.; Prentice, I.C.; Ramstein, G.; Guiot, J.; Bartlein, P.J.; Bonnefille, R.; Bush, M.; Cramer, W.; von Grefensetein, U.; Holmgren, K.; Hoogheimstra, H.; Hope, G.; Jolly, D.; Lauritzen, S.E.; Ono, Y.; Pinot, S.; Stute, M.; Yu, G., 1999: "Tropical climates at the last glacial maximum: a new synthesis of terrestrial palaeoclimate data: I. Vegetation, lake levels and geochemistry", in: *Climate Dynamics, 15: 823–856.*

Field, J.; Wroe, S.; Trueman, C.N.; Garvey, J.; Wyatt-Spratt, S., 2013: "Looking for the archaeological signature in Australian megafaunal extinctions", in: *Quaternary International*, 285: 76–88.

Foley, J.A.; DeFries, R.; Asner, G.P.; Barford, C.; Bonan, G.; Carpenter, S.R.; Chapin, F.S.; Coe, M.T.; Daily, G.C.; Gibbs, H.K.; Helkowski, J.H.; Holloway, T.; Howard, E.A.; Kucharik, C.J.; Monfreda, C.; Patz, J.A.; Prentice, I.C.; Ramankutty, N.; Snyder, P.K., 2005: "Global consequences of land use", in: *Science*, 309: 570–574.

Futuyma, D.J., 2010: "Evolutionary constraint and ecological consequences", in: *Evolution*, 64: 1865–1884.

Gaudzinski, S.; Rocbrocks, W., 2000: "Adults only: reindeer hunting at the Middle Palaeolithic site Salzgitter Lebenstedt, northern Germany", in: *Journal of Human Evolution, 38: 497–521.*

Gibbard, P.L.; Head, M.J.; Walker, M.J.C., 2010: "The Subcommission of Quaternary Stratigraphy, 2010. Formal ratification of the quaternary system/period and the Pleistocene series/epoch with a base at 2.58 Ma", in: *Journal of Quaternary Science*, 25: 96–102.

Gronenborn, D., 2010: "Climate, crises, and the neolithisation of Central Europe between IRD-events 6 and 4", in: Gronenborn, D.; Petrasch, J. (eds.): *The Spread of the Neolithic to Central Europe. RGZM – Tagungen*, 4,1/2: 61–80.

Gronenborn, D., 2014: "The persistence of hunting and gathering: Neolithic western temperate and central Europe", in: Cummings, V., et al. (eds.): *Oxford Handbook of the Archaeology and Anthropology of Hunter Gatherers* (Oxford: Oxford University Press, in press).

Haak, W.; Forster, P.; Bramanti, B.; Matsumura, S.; Brandt, G.; Tanzer, M.; Villems, R.; Renfrew, C.; Gronenborn, D.; Alt, K.W.; Burger, J., 2005: "Ancient DNA from the first European farmers in 7500-year-old Neolithic sites", in: *Science*, 310: 1016–1018.

Haak, W.; Brandt, G.; de Jong, H.N.; Meyer, C.; Ganslmeier, R.; Heyd, V.; Hawkesworth, C.; Pike, A.W.G.; Meller, H.; Alt, K.W., 2008: "Ancient DNA, strontium isotopes and osteological analyses shed light on social and kinship organization of the later Stone Age",in: *Proceedings of the National Academy of Sciences*, 105: 18226–18231.

Haak, W.; Balanovsky, O.; Sanchez, J.J.; Koshel, S.; Zaporozhchenko, V.; Adler, C.J.; Der Sarkissian, C.S.I.; Brandt, G.; Schwarz, C.; Nicklisch, N.; Dreseley, V.; Fritsch, B., Balanovska; E., Villems; R., Meller; H., Alt; K.W., Cooper, A., 2010: "The Genographic Consortium. Ancient DNA from European early neolithic farmers reveals their near eastern affinities", in: *PLoS Biology*, 8: 11; e1000536, https://doi.org/10.1371/journal.pbio.1000536.

Hartwell, R., 1962: "A revolution in the iron and coal industries during the Northern Sung", in: Journal of Asian Studies, 21: 153–162.

Hewitt, G.N., 1999: "Post-glacial re-colonization of European biota", in: Biological Journal of the Linnaean Society, 68: 87–112.

Jacob, D.E.; Soldati, A.L.; Wirth, R.; Huth, J.; Wehrmeister, U.; Hofmeister, W., 2008: "Nanostructure, composition and mechanisms of bivalve shell growth", in: Geochimica et Cosmochimica Acta, 72: 5401–5415.

Jacob, D.E.; Wirth, R.; Soldati, A.L.; Wehrmeister, U.; Schreiber, A., 2011: "Amorphous calcium carbonate in the shells of adult Unionoida", in: Journal of Structural Biology, 173: 241–249.

Kaplan, J.O.; Krumhardt, K.M.; Zimmermann, N., 2009: "The prehistoric and preindustrial deforestation of Europe", in: Quaternary Science Reviews, 28: 3016–3034.

Kaplan, J.O.; Krumhardt, K.M.; Ellis, E.C.; Ruddiman, W.F.; Lemmen, C.; Goldewijk, K.K., 2010: "Holocene carbon emissions as a result of anthropogenic land cover change", in: Holocene, 21: 775–791.

Karkanas, P.; Shahack-Groos, R.; Ayalon, A.; Bar-Matthews, M.; Barkai, R.; Frumkin, A.; Gopher, A.; Stiner, M.C., 2007: "Evidence for habitual use of fire at the end of the lower Paleolithic: site-formation processes at Qesem Cave, Israel", in: Journal of Human Evolution, 53: 197–212.

Kennett, J.P.; Ingram, B.L., 1995: "A 20,000-year record of ocean circulation and climate change from the Santa Barbara Basin", in: Nature, 377: 510–514.

Keppler, F.; Harper, D.B.; Kalin, R.M.; Meier-Augenstein, W.; Farmer, N.; Davis, S.; Schmidt, H.L.; Brown, D.M.; Hamilton, J.T.G., 2007: "Stable isotope ratios oflignin methoxyl groups as a paleoclimate proxy and constraint of the geographical origin of wood", in: New Phytologist, 176: 600–609.

Klein, C., 2005: "Some Precambrian banded iron formations (BIFs) from around the world: their age, geologic setting, mineralogy, metamorphism, geochemistry and origin", in: American Mineralogist, 90: 1473–1499.

Koch, P.L.; Barnosky, A.D., 2006: "Late quaternary extinctions: state of the debate. Annual Review of Ecology", in: Evolution and Systematics, 37: 215–250.

Lelieveld, J.; Hadjinocolaou, P.; Kostopoulou, E.; Chenoweth, J.; El Maayar, M.; Giannakopoulos, C.; Hannides, C.; Langa, M.A.; Tanarhte, M.; Tyrlis, E.; Xoplaki, E., 2012: "Climate change and impacts in the eastern Mediterranean and Middle East", in: Climatic Change, 114: 667–687.

Lemmen, C.; Gronenborn, D.; Wirtz, K., 2011: "A simulation ofthe Neolithic transition in Western Eurasia", in: Journal of Archaeological Science, 38: 3459–3470.

Lüthi, D.; le Floch, M.; Bereiter, B.; Bunier, T.; Barnola,J.-M.; Siegenthaler, U.; Raynaud, D.; Jouzel, J.; Fischer, H.; Kawamura, K.; Stocker, T.F., 2008: "High-resolution carbon dioxide concentration record 650,000-800,000 years before present", in: Nature, 453: 379–382.

Mannino, M.; Thomas, K.D.; Leng, M.J.; Di Salvo, R.; Richards, M.P., 2011: "Stuck to the shore? Investigating prehistoric hunter-gatherer subsistence, mobility and territoriality in a Mediterranean coastal landscape through isotope analyses on marine mollusc shell carbonates and human bone collagen", in: Quaternary International, 244: 88–104.

Migowski, C.; Stein, M.; Prasad, S.; Negendank, J.F.W.; Agnon, A., 2006: "Holocene climate variability and cultural evolution in the Near East from the Dead Sea sedimentary record", in: Quaternary Research, 66: 421–431.

Parmesan, C., 2006: "Ecological and evolutionary responses to recent climate change. Annual Review of Ecology", in: Evolution and Systematics, 37: 637–669.

Pfahl, S.; Sirocko, F.; Seelos, K.; Dietrich, S.; Walter, A.; Wernli, H., 2009: "A new windstorm proxy from lake sediments: a comparison of geological and meteorological data from western Germany for the period 1965–2001", in: Journal of Geophysical Research: Atmospheres, 114.

Ruddiman, W.F., 2003: "The anthropogenic greenhouse era began thousands of years ago", in: Climate Change, 61: 261–293.

Ruddiman, W.F., 2013: "The Anthropocene", in: Annual Reviews of Earth and Planetary Sciences, 41: 4–24.

Ruddiman, W.F.; Kutzbach, J.E.; Vavrus, S.J., 2011: "Can natural or anthropogenic explanations of late Holocene CO_2 and CH_4 increases be falsified?", in: *Holocene*, 21: 865–879.
Ruddiman, W.F., 2005: *Plows, Plagues and Petroleum* (Princeton, NJ, USA: Princeton University Press).
Rye, R.; Holland, H.D., 1998: "Paleosols and the rise of atmospheric oxygen: a critical review", in: *American Journal of Science*, 298: 621–672.
Scholz, D.; Hoffmann, D.L., 2011: "StalAge - an algorithm designed for construction of speleothem age models", in: *Quaternary Geochronology*, 6: 369–382.
Schöne, B.R.; Zhang, Z.J.; Radermacher, P.;Thebault, J.; Jacob, D.E.; Nunn, E.V.; Maurer, A.F., 2011: "Sr/Ca and Mg/Ca ratios of ontogenetically old, long-lived bivalve shells (Arctica islandica) and their function as paleotemperature proxies", in: *Palaeogeography, Palaeoclimatology, Palaeoecology*, 302: 52–64.
Schreg, R., 2013: "Ecological approaches in medieval rural archaeology", in: *European Journal of Archaeology*; at: https://doi.org/10.1179/1461957113Y.0000000045 (published 2017).
Sedgwick, A.M., 1852: "On the classification and nomenclature of the lower Palaeozoic rocks of England and Wales", in: *Quarterly Journal of the Geological Society*, 8: 136–168.
Sheldon, N.D.; Tabor, N.J., 2009: "Quantitative palaeoenvironmental and palaeoclimatic reconstruction using palaeosols", in: *Earth-Science Reviews*, 95: 1–52.
Sirocko, F.; Dietrich, S.; Veres, D.; Grootes, P.M.; Schaber-Mohr, K.; Seelos, K.; Nadeau, M.-J.; Kromer, B.; Rothacker, L.; Roehner, M.; Krbetschek, M.; Appleby, P.; Hambach, U.; Rolf, C.; Sudo, T.; Grim, S., 2013: "Multi-proxy dating of Holocene maar lakes and Pleistocenedry maar sediments in the Eifel, Germany", in: *Quaternary Science Reviews*, 62: 56–76.
Smith, B.D.; Zeder, M.A., 2013: "The onset of the Anthropocene", in: *Anthropocene*; at: https://doi.org/10.1016/j.ancene.2013.05.001 (Epub ahead of print).
Steffen, W.; Grinevald, J.; Crutzen, P.; McNeill, J., 2011: "The Anthropocene: conceptual and historical perspectives", in: *Philosophical Transactions of the Royal Society of London A*, 369: 842–867.
Stocker, B.; Strassman, K.; Joos, F., 2010: "Sensitivity of Holocene atmospheric CO_2 and the modern carbon budget to early human land use: analysis with a process-based model", in: *Biogeosciences Discussions*, 7: 921–952.
Sukumar, R.; Ramesh, R.; Pant, R.K.; Rajagopalan, G., 1993: "A delta C-13 record of late quaternary climate change from tropical peats in southern India", in: *Nature*, 364: 703–706.
Surmelihindi, G.; Passchier, C.W.; Spotl, C.; Kessener, P.; Betsmann, M.; Jacob, D.E.; Baykan, O.N., 2013: "Laminated carbonate deposits in Roman aqueducts: origin, processes and implications", in: *Sedimentology*, 60: 961–982.
Tarasov, P.E.; White, D.; Weber, A.W., 2013: "The Baikal-Hokkaido Archaeology project: environmental archives, proxies and reconstruction approaches", in: *Quaternary International*, 290: 1–2.
Urey, H.C.; Lowenstam, H.A.; Epstein, S.; McKinney, C.R., 1951: "Measurement of some palaeotemperatures and temperatures of the Upper Cretaceous of England, Denmark, and the southeastern United States", in: *Bulletin of the Geological Society of America*, 62: 399–416.
Walker, M.; Johnsen, S.; Rasmussen, S.O.; Popp, T.; Steffensen, J.-P.; Gibbard, P.; Hoek, W.; Lowe,J.; Andrews,J.; Bjorck, S.; Cwynar, L.C.; Hughen, K.; Kershaw, P.; Kromer, A.; Litt, T.; Lowe, D.J.; Nakagawa, T.; Newnham, R.; Schwander, J., 2009: "Formal definition and dating of the GSSP (Global Stratotype Section and Point) for the base of the Holocene using the Greenland NGRIP ice core, and selected auxiliary records", in: *Journal of Quaternary Science*, 24: 3–17.
Weninger, B.; Clare, L.; Rohling, E.J.; Bar-Yosef, O.; Bohner, U.; Budja, M.; Bundschuh, M.; Feurdean, A.; Gebel, H.-G.; Joris, O.; Linstadter, J.; Mayewski, P.; Muhlenbruch, T.; Reingruber, A.; Rollefson, G.; Schyle, D.; Thissen, L.; Todorova, H.; Zielhofer, C., 2009: " The impact of rapid climate change on prehistoric societies during the Holocene in the Eastern Mediterranean", in: *Documenta Praehistorica*, 36: 7–59.
Wroe, S.; Field, J., 2006: "A review of the evidence for a human role in the extinction of Australian megafauna and an alternative explanation", in: *Quaternary Science Reviews*, 25: 2692–2703.

Zalasiewicz, J.; Williams, M.; Haywood, A.; *Ellis*, M., 2011a: "The Anthropocene: a new epoch of geological time?", in: *Philosophical Transactions of the Royal* Society *of London A*, 369: 835–841.
Zalasiewicz, J.; Williams, M.; Fortey, R.; Smith, A.; Barry, T.L.; Coe, A.L.; Brown, P.R.; Rawson, P.F.; Gale, A.; Gibbard, P.; Gregory, F.J.; Hounslow, M.W.; Kerr, A.C.; Pearson, P.; Knox, R.; Powell, J.; Waters, C.; Marshall, J.; Oates, M.; Stone, P., 2011b: "Stratigraphy of the Anthropocene", in: *Philosophical Transactions of the Royal Society of London A*, 369: 1036–1055.

Chapter 17
Stratigraphic and Earth System Approaches to Defining the Anthropocene (2016)

Will Steffen, Reinhold Leinfelder, Jan Zalasiewicz, Colin N. Waters,
Mark Williams, Colin Summerhayes, Anthony D. Barnosky,
Alejandro Cearreta, Paul Crutzen, Matt Edgeworth, Erle C. Ellis,
Ian J. Fairchild, Agnieszka Galuszka, Jacques Grinevald, Alan Haywood,
Juliana Ivar do Sul, Catherine Jeandel, J. R. McNeill, Eric Odada,
Naomi Oreskes, Andrew Revkin, Daniel de B. Richter, James Syvitski,
Davor Vidas, Michael Wagreich, Scott L. Wing, Alexander P. Wolfe,
and H. J. Schellnhuber

Abstract Stratigraphy provides insights into the evolution and dynamics of the Earth System over its long history. With recent developments in Earth System science, changes in Earth System dynamics can now be observed directly and projected into the near future. An integration of the two approaches provides powerful insights into the nature and significance of contemporary changes to Earth. From both perspectives, the Earth has been pushed out of the Holocene Epoch by human activities, with the mid-20th century a strong candidate for the start date of the Anthropocene, the proposed new epoch in Earth history. Here we explore two contrasting scenarios for the future of the Anthropocene, recognizing that the Earth System has already undergone a substantial transition away from the Holocene state. A rapid shift of societies toward the UN Sustainable Development Goals could stabilise the Earth System in a state with more intense interglacial conditions than in the late Quaternary climate regime and with little further biospheric change. In contrast, a continuation of the present Anthropocene trajectory of growing human pressures will likely lead to biotic impoverishment and a much warmer climate with a significant loss of polar ice.

This text was first published as: Steffen, W.; Leinfelder, R.; Zalasiewicz, J.; Waters, C.N.; Williams, M.; Summerhayes, C.; Barnosky, A.D.; Cearreta, A.; Crutzen, P.; Edgeworth, M.; Ellis, E.C.; Fairchild, I.J.; Galuszka, A.; Grinevald, J.; Haywood, A.; do Sul, J.I.; Jeande, C.; McNeill, J.R.; Odada, E.; Oreskes, N.; Revkin, A.; de B. Richter, D.; Syvitski, J.; Vidas, D.; Wagreich, M.; Wing, S.L.; Wolfe, A.P.; Schellnhuber, H. J., 2016: "Stratigraphic and Earth System Approaches to Defining the Anthropocene", in: *Earth Future*, 4: 324–345. This is an open access article under the terms of the Creative Commons Attribution-NonCommercial-NoDerivs License, which permits use and distribution in any medium, provided the original work is properly cited, the use is non-commercial and no modifications or adaptations are made.

17.1   Introduction

The Anthropocene, the proposed new geological epoch in Earth history (Crutzen and Stoermer 2000; Crutzen 2002; Zalasiewicz et al. 2008), is challenging many areas of research in a variety of ways. The term and concept have been discussed within diverse disciplines in the natural sciences (e.g., Ellis et al. 2012; Gillings/Paulsen 2014; Capinha et al. 2015; Corlett 2015; Williams et al. 2015) and in the environmental humanities and social sciences (e.g., Chakrabarty 2009; Vidas 2011; Malm/Hornborg 2014; Fischer-Kowalski et al. 2014; Bai et al. 2015; Latour 2015; Vidas et al. 2015; Bonneuil/Fressoz 2016), with more interdisciplinary approaches also appearing (Braje 2015; Latour 2015; Maslin/Lewis 2015).

Although the proposal was initiated in the Earth System science community (Crutzen and Stoermer 2000; Steffen 2013; see Revkin 1992 for an earlier proposed 'Anthrocene'), recognition of the Anthropocene as an epoch following the Holocene necessitates that the proposal be grounded in the Geologic Time Scale, one of the cornerstones of geology. Subsequently, much work (Zalasiewicz et al. 2015; Waters et al. 2016, and references therein) has focused on testing whether the stratigraphic record of the Anthropocene is adequate for the formal definition of a new epoch following the protocols of the International Commission on Stratigraphy and its parent body, the International Union of Geological Sciences.

The result of this work has been a convergence of evidence and information obtained from Earth System science and from stratigraphy. Here we examine this convergence of approaches to define the Anthropocene, highlighting their changing relationship through time and the insights that each brings to examine the dynamics of the Earth System.

17.2   Historical Relationship Between Stratigraphy and Earth System Science

Earth System science is a highly interdisciplinary enterprise that aims to build a holistic understanding of our evolving planet (Lenton 2015). It arrived on the research landscape very recently (primarily since the 1980s), and some scholars have suggested that it represents an emerging paradigm (Malone/Roederer 1985; ICSU 1986; Grinevald 1987; Hamilton/Grinevald 2015). Earth System science builds on the long history of advances in the geosciences (Oldroyd 1996; Bard 2004; Galvez/Gaillardet 2012) and on more recent system-level thinking applied to the climate and the biosphere (Budyko 1986; Clark/Munn 1986; NASA 1988; Rambler et al. 1989).

An early pioneer of this holistic approach, following Alexander von Humboldt, was the Russian mineralogist and naturalist Vladimir I. Vernadsky, one of the founders of geochemistry and the creator of biogeochemistry (Vernadsky 1924, 1929, 1998). Vernadsky's research on biogeochemistry was central to the rise of Earth

System science, and to the scientific study of Earth's biosphere in general (Grinevald 1987; Polunin/Grinevald 1988; Smil 2002; Jorgensen 2010), including humanity as a new geological agent (Vernadsky 1924, 1945, 1998). Vernadsky's pioneering work largely languished for several decades, but global biogeochemical cycles reappeared prominently in the 1970s when James Lovelock, the father of the Gaia hypothesis, provided a complementary conceptual framework for the Earth as a system (Lovelock/Margulis 1974; Lovelock 1979, 1988). Contemporary Earth System science draws on a wide range of new tools and disciplinary expertise for directly observing and modelling the dynamics of the Earth System (cf. Sect. 17.4), emphasising the conceptual framework of complex-systems science, hence the emphasis on *System* in its name.

The Earth System is usually defined as a single, planetary-level complex system, with a multitude of interacting biotic and abiotic components, evolved over 4.54 billion years and which has existed in well-defined, planetary-level states with transitions between them (Schellnhuber 1998, 1999). A state is a distinct mode of operation persisting for tens of thousands to millions of years within some envelope of intrinsic variability. The Earth System is driven primarily by solar radiation and is influenced by other extrinsic factors, including changes in orbital parameters and occasional bolide strikes, as well as by its own internal dynamics in which the biosphere is a critical component.

Earth's mean temperature is determined primarily by its energy balance (Feulner 2012), including the key variables of solar insolation (increasing during Earth history), greenhouse gas forcing (broadly decreasing during Earth history) and albedo. The distribution of heat at the Earth's surface is modified by orbital variations and paleogeographic patterns driven by tectonics, which in turn can drive feedbacks that lead to whole-Earth changes in albedo or greenhouse gas forcing. Thus, over multi-million year timescales, Earth's climate shifts in response to gradual changes in continental configuration, the opening or closing of ocean gateways, and the plate tectonic or Wilson cycle, which, together, drive long-term changes to the carbon cycle and the biosphere. These long, slow changes modify the effects of solar forcing, not least by changing the balance between sources of CO_2 (from volcanic activity) and its sinks (starting with chemical weathering and progressing through sequestration in sediments), as documented for example by Berner (1999a, b, 2003), Franks et al. (2014) or Summerhayes (2015). Short-term abrupt changes are imposed by sudden aperiodic volcanic activity that may be as brief as a single volcanic eruption or as long as the life of a Large Igneous Province of the kind that gave rise to the Siberian Traps and the end-Permian extinction. These are aside from natural fluctuations of minor amplitude driven by orbital change or internal oscillations within the ocean-atmosphere system, such as El Niño events or the Pacific Decadal Oscillation.

Contemporary Earth System science has benefited greatly from evidence generated by the geosciences, particularly stratigraphy, the primary geoscience that has developed the 'book of records' of the Earth through time. The relationship between stratigraphy and Earth System science has been symbiotic and well defined: stratigraphy has been the generator of new knowledge about Earth history while Earth System science has interpreted that knowledge in a complex-systems framework that sometimes challenges geological interpretations of the stratigraphic record (e.g., Snowball Earth theory; Budyko 1969; Hoffman et al. 1998).

In terms of Earth history, this relationship was recently portrayed in Zalasiewicz et al. (2015):

> An effective geochronological and chronostratigraphical boundary often reflects a substantial change in the Earth system, so that the physical and chemical nature of the deposits, and their fossil contents, are recognizably different above and below the boundary. …To take (an) example, the boundary between the Ordovician and Silurian periods reflects a brief, intense glacial phase that triggered one of the 'Big Five' mass extinction events, and hence profoundly altered the biota (and fossil record) of the Earth.

Thus, the relationship is most useful to Earth System science when a stratigraphic boundary marks a substantial change in the planetary mode of operation. Such boundaries should mark a transition from one fundamental state of the Earth System to another, or, in other words, mark a regime shift (Scheffer/Carpenter 2003), although for the definition of many stratigraphic time boundaries this is not a prerequisite (see below). Some geological time units lower in the stratigraphic hierarchy maybe defined by, for example, some distinctive paleontological change that is not associated with any substantial Earth System change, as in the definition of the Aeronian Age of the Silurian Period discussed below. Zalasiewicz et al. (2015) went on to note:

> A stratigraphic time boundary, however arbitrary, needs as far as possible to be singular, globally synchronous and commonly understood.

Stratigraphy is valuable for Earth System science because it is also highly interdisciplinary, drawing information and insights from sedimentology, palaeontology, geochemistry, geochronology, archaeology, pedostratigraphy, palaeomagnetism, paleoclimatology, and other fields. The unifying thread that brings this wide array of relevant disciplines together is the stratigraphic handbook of the International Commission on Stratigraphy (Salvador 1994; Remane et al. 1996), which sets out the following definitions to guide stratigraphic research:

Lithostratigraphic unit. A body of rock established as a distinct entity based on its lithological characteristics. The boundaries of lithostratigraphic units may be effectively synchronous (as for instance with units comprising, or bounded by, volcanic ash layers) or they may be markedly time-transgressive (as in, for instance, a unit comprising a succession of beach deposits that follow a migrating coastline as sea-level changes.).

Chronostratigraphic classification. The organisation of rocks into units on the basis of their age or time of origin. The purpose of chronostratigraphic classification is to organise systematically the rocks forming the Earth's crust into named units (chronostratigraphic units) corresponding to intervals of geologic time (geochronological units) to serve as a basis for time-correlation and a reference system for recording events of geologic history.

Chronostratigraphic unit. A body of rocks that includes all rocks formed during a specific interval of geologic time, and only those rocks formed during that time span. Chronostratigraphic units are bounded by synchronous horizons. They are generally made up of stratified rocks, while the equivalent geochronological units (of Earth time) are inferred from them and may also be recognised within units of nonstratified rock such as polyphase metamorphic units (Zalasiewicz et al. 2013).

The following features of all chronostratigraphic unit definitions are important for the utility of chronostratigraphy for Earth System science.

1. A chronostratigraphic unit is typically represented by different types of sedimentary deposits that accumulate in environments ranging from land to deep sea, and which may be independently classified based on their physical characteristics into a hierarchy of lithostratigraphic or biostratigraphic units, the boundaries of which are commonly diachronous to various degrees (i.e., they cut across time planes). Such units are seldom entirely concordant with chronostratigraphic boundaries. Different kinds of time proxy evidence, such as guide fossils, geochemical patterns, and magnetic properties, may be used as approximations to time planes to help establish the boundaries of chronostratigraphic units.

2. A GSSP (*Global boundary Stratotype Section and Point*) or GSSA (*Global Standard Stratigraphic Age*; Gradstein et al. 2012) is used to define a synchronous horizon within strata around the globe, based on the boundary of a chronostratigraphic unit. In practice, there are always uncertainties in tracing this boundary worldwide, but the error bars narrow as dating precision improves. These boundaries help constrain the pattern in time and space of changes in the behaviour of the Earth System. For application to Earth System science, especially in identifying changes in the state of the system, having a globally synchronous boundary horizon is desirable, particularly for rapid or abrupt transitions. A central challenge, but also a remarkable advantage, to stratigraphers in the context of the Holocene-Anthropocene boundary is that the highly resolved timescale of human history (ca. 10,000 years) reveals diachroneity, sometimes on as fine a time scale as decades or even years, in the physical, chemical, and biological indicators of the transition. Such fine-scale diachroneity is ordinarily not detectable for older boundaries because time resolution is coarser.

3. In many cases a chronostratigraphic boundary and its associated lithostratigraphic (and/or biostratigraphic) unit(s) are broadly associated with a global shift in the state of the Earth System, commonly shown by marked changes in fossil assemblages and/or by changes in proxies for critical climate parameters. Although not all chronostratigraphic boundaries reflect a shift in the state of the Earth System, changes in the state of the Earth System should, in principle, result in a recognizable chronostratigraphic boundary. Examples of boundaries associated with an Earth System state shift are the transition from the Mesozoic to the Cenozoic (triggered largely by an asteroid impact that likely drove mass extinctions and reshaped the biosphere (Molina et al. 2006) and the onset of the Pleistocene ice ages (triggered by a coincidence of the Milankovitch orbital parameters with a paleogeography that attained requisite elevational and ocean-circulation patterns (Lunt et al. 2012)). The latter is an event that, while representing significant Earth System change reflected in new stratigraphic patterns

(Pillans/Naish 2004; Gibbard et al. 2005), is protracted and complex; hence, the base-Pleistocene boundary is placed with reference to the Gauss-Matuyama paleomagnetic boundary, not a major driver of Earth process *per se,* but a widely traceable horizon in strata within this key interval.

Over the last few centuries, geologists have assembled records of rocks and their various characteristics, for example their embedded fossils, and, more recently, their chemical, magnetic, and other properties. From this, they worked out time-based (i.e., chronostratigraphic) rock divisions based on clearly observable differences between a stratigraphic unit and the units above and below it, and used those to define geologic time (geochronological) units. They then correlated the chronostratigraphic (rock) units globally to refine and modify the Geologic Time Scale in tandem with improving knowledge of stratal successions. The heuristic rule for linking chronostratigraphy to Earth System dynamics is this: If the differences in attributes between units are large and evident across many areas of the Earth, or if at least the difference from the underlying strata to the overlying boundary layer is large, then the likelihood of a change in the state of the Earth System is high. Otherwise, only gradual or local changes might have taken place, but they happened to have created a detectable, near-synchronous horizon.

Simple heuristic rules have their limits. For example, some selected boundary-defining biostratigraphic events may not be associated with fundamental systemic changes, but nevertheless form good boundary-defining markers, as in the emergence of the distinctive triangulate monograptid graptolites used to recognise the beginning of the Aeronian Age of the Llandovery Epoch of the Silurian Period (Melchin et al. 2012). This evolutionary event appears not to correlate with wider changes in biota or Earth System functioning. On the other hand, chronostratigraphically useful changes that are individually trivial as regards Earth System dynamics (as with the signal used to define the Ordovician-Silurian boundary: Zalasiewicz/Williams 2014) may nevertheless prove to be useful for Earth System science by their association with a wider array of signals that reflect more fundamental change. While the Ordovician-Silurian boundary itself is based on a small change in paleo plankton composition that may not be important from an Earth System perspective, the boundary was preceded by changes driven by the onset and collapse of a particularly intense phase of a longer-lasting glaciation, in which the associated stratigraphic signals are regarded as having less precise power for correlation (Page et al. 2007; Hammarlund et al. 2012; Melchin et al. 2012). This large event likely represents a change in the state of the Earth System, even though it is not precisely coincident with a boundary in the Geologic Time Scale.

In summary, chronostratigraphy reveals the pattern of changes in Earth history, and leads to inferences about changes in the state of the Earth System. However, building a deeper understanding of the processes that drive the state changes requires theoretical as well as empirical investigations of the interacting components of the Earth System.

17.3 Unravelling Earth System Evolution From the Chronostratigraphic Record

17.3.1 Evolution of the Biosphere

The evolution of the biosphere can be divided into two fundamental stages. Between ~4 to 0.8 Ga (Ga = billion years ago), the biosphere comprised mostly of unicellular organisms occurring either individually or in colonies. This initial stage featured several important developments in biospheric functioning, such as the appearance of sulphur-reducing bacteria (Grassineau et al. 2006; Wacey et al. 2011; Bell et al. 2015) and the development of photosynthetic metabolic pathways (Grassineau et al. 2002; Payne et al. 2008; Allwood et al. 2009). From ~0.8 Ga molecular (genetic), fossil, trace fossil, and biomarker evidence supports the evolution of a biosphere with metazoans (animals). This led to the Cambrian adaptive radiation (or Cambrian explosion), in which skeletonised organisms become preserved in rock successions worldwide (Erwin et al. 2011). The rich fossil record of the past 600 million years provides additional evidence of major innovations in the Earth's biota and their inter-action with the abiotic components of the Earth System. For example, Neoproterozoic and Cambrian sedimentary strata provide the first evidence of motile bilateral organisms (e.g., Jensen 2003; Hou et al. 2004) as part of an evolutionary continuum that produced the complex trophic structures of the marine ecosystems of the Phanerozoic (Butterfield 2011).

The Ordovician to Devonian stratigraphic records show the rise of a complex terrestrial biosphere, first with nonvascular plants (Edwards et al. 1992; Wellman/Gray 2000; Wellman et al. 2003) and later with vascular plants that produced only spores (Hotton et al. 2001; Stein et al. 2007), followed by the rise of seed plants along with more complex seedless vascular plants and the growth of extensive forests (DiMichele et al. 1992; Stewart/Rothwell 1993; Greb et al. 2006). Regime shifts in the Earth's biosphere are reflected by mass extinction events (Barnosky et al. 2011) (arrows, Fig. 17.1), after which major alterations in the trajectory of evolution occurred, and in the relatively rapid transitions between the three 'evolutionary faunas' recognised by paleontologists as the Cambrian Fauna, the Paleozoic Fauna, and the Modern Fauna (Fig. 17.1). For more detail on the evolution of the biosphere, see Behrensmeyer et al. (1992), Stanley (1993), Nisbet and Fowler (2014) and Williams et al. (2016).

17.3.2 Evolution of the Climate System

The stratigraphic record, based on a wide variety of geological, paleontological, and geochemical proxies (Masson-Delmotte et al. 2013; Bradley 2015; Zalasiewicz/Williams 2016), also provides the evidence needed to infer changes in the climate (Fig. 17.2). From the Archean to the present, homeostatic processes

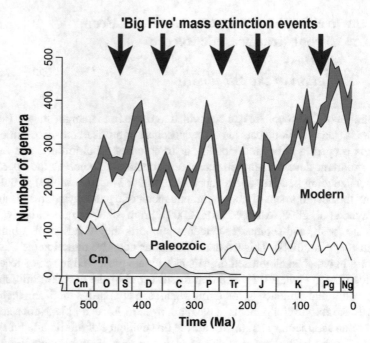

Fig. 17.1 Sampling-standardised Phanerozoic marine diversity curve (Alroy 2010), expressed as summed curves for constituent groups. Regime shifts in the Earth System are reflected in the transition from typical Cambrian (Cm) to Paleozoic to Modern marine faunas, and at mass extinction events (arrows). In this context 'Cambrian', 'Paleozoic', and 'Modern' do not refer to the respective time periods of the same name, but instead to evolutionary stages of the biota. Major alteration in the trajectory of evolution occurred at each of the mass extinctions, recognizable by the estimated loss of at least 75% of commonly fossilised marine species, after which previously uncommon clades became dominant (Barnosky et al. 2011) The dark gray area at top represents genera not assigned to one of the three evolutionary faunas. Ma = million years ago

have forced Earth's climate to remain within rather narrow temperature limits, unlike those of its neighbours Venus and Mars. That constraint has allowed the three phases of water—liquid, vapour, and solid—to coexist on the surface of the planet, providing a key precondition for the appearance and evolution of life.

The evolution of the climate system shows its highly systemic nature. This includes (i) the alternation between so-called greenhouse states (warm times when the poles were ice-free) and icehouse states (cold times with permanent polar and lower latitude sea ice and/or glacier ice), evident from late Archean times onward (Fig. 17.2); (ii) the evolution of the global carbon cycle that provides a critical link between the physical climate and the biosphere (Berner et al. 1983; Berner 1990, 1999a, b); and (iii) the Earth System's intrinsic negative feedback processes, coupled with lithosphere evolution (e.g., CO_2 release from within the Earth), that enable it to absorb and recover over the long term from marked temperature changes that cause severe glaciation (e.g., in the early and late Proterozoic—see Fairchild/Kennedy 2007).

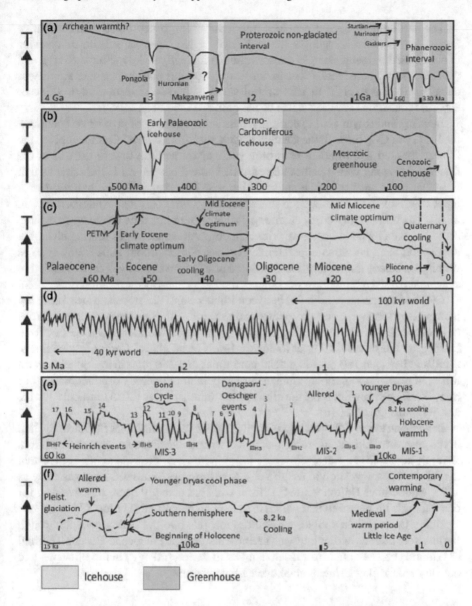

Fig. 17.2 Global climate variation at six different timescales (modified from Zalasiewicz and Williams, 2016 and references therein). On the left side of the figure, the letter 'T' denotes relative temperature, which can be taken as mean surface temperature for panels a, b, and f, while panels c-e are predicted on a reading of 'T' derived from the $\delta^{18}O$ of benthic marine foraminifera for different time frames of the Cenozoic, which for the intervals with permanent polar ice (within the Oligocene and younger) will record a combination of ice volume and ocean-floor temperature change. The hyperthermals of the Mesozoic (e.g., the Turonian) are not plotted. Ga = billion years ago; Ma = million years ago; ka = thousand years ago

The stratigraphic record provides important clues to key positive and negative feedback mechanisms, such as the influence of ice cover on albedo or changes in atmospheric greenhouse gas composition (principally CO_2, methane (CH_4), and water vapour). These feedbacks can, under appropriate conditions, either amplify or dampen external forcing, such as orbital variation and solar insolation, to drive or suppress transitions between states of the climate (Lunt et al. 2012).

Beerling and Royer (2011) compiled data from a variety of proxies to document the history of CO_2 through the Cenozoic. CO_2 rose from the end Cretaceous into the Eocene, peaked ~50 Ma (Ma = million years ago), then fell toward the end of the Eocene, following which values stayed rather low. This pattern is the same as that of global mean surface temperature, which rose to a peak ~50 Ma then fell to the point where the first Antarctic ice sheet formed ~34 Ma (see also Anagnostou et al. 2016). The ultimate driver for changing global CO_2 in this time frame was volcanic activity driven by plate tectonics, with changes in greenhouse gas forcing changing atmospheric and oceanic temperatures (e.g., see Kent/Muttoni 2008). However, as we discuss below, this pattern was disrupted briefly by a short-lived carbon injection event that caused a temporary warming at the Paleocene-Eocene boundary 56.0 Ma.

CO_2 and temperature both fell between Pliocene and Pleistocene times, probably in response to changing oceanic gateways (the rising of the Central American isthmus in Pliocene times, and the closing of the Indian Ocean-Mediterranean connection). During the Pleistocene, CO_2, at its lowest levels since glacial Carboniferous times, oscillated between 180 ppm in glacial cold times and 280 ppm in interglacial warm times, in response to periodic changes in temperature driven by orbital change. For more detail on the evolution of the climate, see Summerhayes (2015) and Zalasiewicz and Williams (2012, 2016).

In addition to providing essential knowledge on the evolution of the Earth System in the past, the stratigraphic record, coupled with mechanistic insights derived from Earth System science, can also provide insights into how the system might evolve in the future. The suggestion to use the past to inform the future was made as early as 1795, when James Hutton wrote "… from what has actually been, we have data for concluding with regard to that which is to happen hereafter" (Hutton 1795).

Building on Hutton's logic, we explore insights into the Anthropocene through the analyses of three earlier intervals in Earth history: the Paleocene-Eocene Thermal Maximum (PETM), the Mid-Piacenzian Warm Period (mPWP) in the Pliocene, and the Quaternary glacial-interglacial cycles.

17.3.3 Paleocene-Eocene Thermal Maximum

A major perturbation occurred at the epoch boundary between the Paleocene and the Eocene 56.0 Ma (Fig. 17.2c; note that here we adopt 56.0 Ma for the timing of the PETM, consistent with the Geological Time Scale (Gradstein et al. 2012)), which produced a sharp increase of 4–8 °C in global mean surface temperature within a few thousand years. The elevated temperature persisted for 0.1–0.2 million years and led to the extinction of 35–50% of the deep marine benthic foraminifera, and to continent-scale changes in the distributions of terrestrial plants and animals (Sluijs et al. 2007; McInerney/Wing 2011; Haywood et al. 2011; Winguth et al. 2012). The leading hypothesis to explain the PETM temperature spike is the geologically rapid (over a few thousand years) release of 3000–7000 Pg of carbon from methane hydrates in the sea floor, a release triggered by initial warming from other causes (Dickens et al. 1995; Dickens 2011; Bowen et al. 2015). As a result, oceans increased in acidity, the depth for calcium carbonate compensation became shallower (Zachos et al. 2008), and sea-level rose up to 15 m (Sluijs et al. 2008). The PETM has some parallels with the present anthropogenic increase of atmospheric CO_2 but the human impact is proceeding at a rate likely to be ten times higher (Cui et al. 2011; Haywood et al. 2011; Bowen et al. 2015; Zeebe et al. 2016). According to Zeebe et al. (2016), carbon release from anthropogenic sources reached ca. 10 PgC/year in 2014, which is an order of magnitude faster than the maximum sustained release of carbon in the PETM, which was <1.1 PgC/year. That makes the present anthropogenic release rate unprecedented in the past 66 million years, and puts the climate system in a 'no-analog' state that 'represents a fundamental challenge in constraining future climate projections'. In addition, Zeebe et al. (2016) point out that 'future ecosystem disruptions are likely to exceed the relatively limited extinctions observed at the PETM'.

As with the PETM, there is concern that an initial surface temperature rise caused by anthropogenic greenhouse gas emissions could trigger the release of significant amounts of carbon from methane hydrates, driving the temperature even higher (Lenton 2011; see White et al. 2013 for an alternative view).

17.3.4   Pliocene Epoch

A second interval of paleoclimate that informs some scenarios for the late 21st century climate is the *Mid-Piacenzian Warm Period*, mPWP (3.264–3.025 Ma; see Dowsett et al. 2013 for an overview) within the Pliocene Epoch, (5.33–2.58 Ma). Various proxies for warm (interglacial) intervals of the Pliocene suggest that atmospheric CO_2 concentration may have peaked around or slightly above 400 ppm (Pagani et al. 2005; Haywood et al. 2011 and references therein), similar to the current atmospheric concentration of CO_2, although Beerling and Royer (2011) provide some evidence for CO_2 concentrations having reached close to 450 ppm in the mPWP. During these warm intervals, global mean surface temperature was 2–3 °C higher than pre-industrial Holocene levels, and sea level is estimated to have been 10–20 m higher than today (Miller et al. 2012; Naish/Zwartz 2012). The warm intervals of the Pliocene, especially the mPWP, are viewed as important possible scenarios for late 21st century climate (Haywood et al. 2009 and references therein). In particular, contemporary warming may also lead to sea-level rises of 10 m or more, with a delay of several hundred years at least while the ocean warms to its full depth and ice caps equilibrate to raise temperatures (Clark et al. 2016).

17.3.5   The Quaternary Period: Complex-System Behaviour of the Climate

More recent stratigraphic records provide convincing evidence for the complex-system behaviour of Earth's climate. In particular, two Antarctic ice cores (Petit et al. 1999; EPICA 2004) display many striking features of Earth's climate that are characteristic of a single complex system (see Scheffer 2009 for more details on the complex-system behaviour of the Earth System). Records from these ice cores and from deep-sea cores provide evidence of:

i.   Two reasonably well-defined states of the system—ice ages (glacial states) and brief warm periods (interglacials);
ii.   Regular quasi-periodic transitions between the states (ca. 100,000-year modulations in the last 1.2 million years, ca. 40,000-year modulations earlier in the Quaternary; Fig. 17.2d), which is characteristic of phase locking of key internal system dynamics under relatively weak external forcing. In this case, the forcing was provided by minor astronomical modulation of incoming solar radiation patterns via variations in Earth's orbital eccentricity and precession, along with axial tilt;
iii. Tight coupling between temperature and greenhouse gas concentrations, typical of critical feedback processes within a system that lead to tipping points when feedbacks switch from negative (self-limiting) to positive (self-reinforcing) (Parrenin et al. 2013; and
iv.   Limit-cycle behaviour that defines clear upper and lower limits for the fluctuations in temperature, CO_2, and CH_4.

Despite the abrupt climate oscillations of the Quaternary (17.2d), the biosphere showed no marked long-term change through this time. In fact, there was little elevation in extinction rates until the megafaunal extinctions of the latest Pleistocene and early Holocene (Koch/Barnosky 2006; Barnosky et al. 2011). These extinctions appear to have resulted from interactions due to the coincidence of end-Pleistocene climate change with the trans-continental migration of rapidly increasing numbers of *Homo sapiens* into ecosystems that had never encountered them before (Brook/Barnosky 2012).

17.3.6   Biosphere-Climate Interaction—The Earth System

The climate and the biosphere are two highly intertwined, aggregate components of the whole-Earth System—a single complex system—even though the evolution of those two components can be inferred somewhat independently from each other. The stratigraphic record provides the means by which a systematic integration of climate and biosphere evolution can be attempted—the evolution of the Earth as a system (Stanley 1993; Lenton et al. 2004; Lenton/Watson 2011; Stanley/Luciaz 2014; Lenton 2015). Complex-systems approaches have been applied by ecologists to

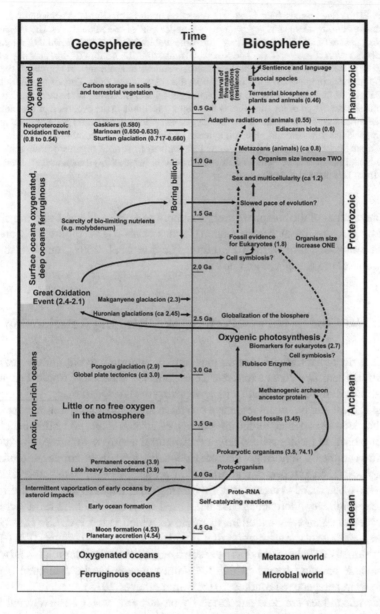

◄Fig. 17.3 Time line of geosphere-biosphere coevolution on Earth. Here the geosphere is defined as the atmosphere, hydrosphere, cryosphere, and upper part of the lithosphere. The biosphere is defined as the sum of all biota living at any one time and their interactions, including interactions and feedbacks with the geosphere. The time line runs from the bottom to top, starting with the accretion of planet Earth and ending at the present. Numbers indicate ages in billions of years ago (Ga). The major geological eons are indicated in the scale on the right. Left of the time line are major features of and changes in the state of the geosphere, including some perturbations from outside the system. Right of the time line is the major transitions in the evolution of the biosphere, plus some other significant appearances. The major transitions in evolution are given abbreviated descriptions. The arrows crossing the two spheres depict patterns of coevolution and the fact that they are a single system. Eusocial behaviour has evolved in several organism groups including arthropods and mammals, perhaps first in the Mesozoic, but possibly much earlier. Based on a concept from Lenton et al. (2004)

track coevolution of the biosphere and geosphere as a series of states and transitions, especially through the metazoan stage (Hughes et al. 2013). Figure 17.3 presents a visualisation of Earth System evolution (Lenton et al. 2004), emphasising the coevolution of the geosphere and biosphere.

17.4   The Contemporary Period—A Changing Relationship

Many traditional stratigraphic techniques remain important to our understanding of the Earth System. Ice cores extracted from polar ice sheets and tropical mountain glaciers have illuminated climatic changes during the late Quaternary. The Deep Sea Drilling Project and its successors have provided archives of long sediment columns from the deep ocean that underpin studies of stratigraphic and climatic change extending back to the Jurassic. Many other natural archives—marine, coastal, and lacustrine sediments, tree rings, charcoal deposits, long-lived corals, microfossils, paleontological, and archeological remains, ice cores, cave deposits, and historical records have been used to reconstruct environmental changes in the past. Multiproxy and interdisciplinary approaches have been used to reconstruct long-term records of environmental change, an excellent example of which is the PAGES (*Past Global Changes*) 2000-year reconstruction of global land surface temperature (PAGES 2 k Consortium 2013). Paleo-temperature reconstructions have been crucial in providing the Holocene baseline against which the Anthropocene can be evaluated from an Earth System perspective (Bradley 2015; Summerhayes 2015).

These techniques are now supported by an array of novel observational tools, particularly remote sensing technologies, which enable scientists to observe many important characteristics of the Earth System from space routinely and consistently. Scientists can now monitor, *inter alia,* the composition of the atmosphere; land-cover change; sea surface height, temperature, salinity, and biological productivity of the oceans; the temperature of the lower atmosphere; the standing biomass of forests; disturbances such as fire; and a wealth of human activities, including the night-time illumination of the planet and the rapid expansion of urban areas.

In situ measurements of Earth System processes have also increased dramatically in the last few decades. For example, the uptake of CO_2 by terrestrial ecosystems is directly measured by eddy covariance techniques. The climate is monitored in real time by a global network of stations that observe temperature, precipitation, solar radiation, wind speed, and other parameters. Through systems such as the Argo buoy network, oceanographers routinely measure the temperature, salinity, and chemical state of the ocean from the surface down to 2000 m. The flow of sediments, nitrogen, phosphorus, and other elements down river systems and into the coastal zone can be measured directly, as well as the subsidence of many of the world's large delta areas. Growing databases coupled with satellite observations show how humans have changed the terrestrial hydrological system by engineering the landscape (Syvitski/Kettner 2011), especially by building millions of small impoundments and thousands of large dams (ICOLD 2016).

More controversially, scientists also undertake manipulative experiments on critical processes of the Earth System. For example, large amounts of iron have been deposited onto nutrient-poor regions of the ocean to stimulate CO_2 uptake by phytoplankton (Boyd et al. 2007); *in situ* terrestrial ecosystems have been treated with excess CO_2 over many years (e.g., Ainsworth/Long 2005); soils have been warmed to observe changes in microbial activity (e.g., Knorr et al. 2005); and the species richness of grassland ecosystems has been altered to explore the effect on ecosystem functioning (e.g., Tilman et al. 2006).

The dynamics of the Earth System can also be simulated using a spectrum of computational modelling approaches. These include simulating climate system dynamics using *General Circulation Models* (GCMs); this approach forms the basis for the IPCC (Intergovernmental Panel on Climate Change) projections of possible future changes of the climate system (Flato et al. 2013). GCMs are now being tested by their application to modelling past climate change, with some success (e.g., Valdes 2011; Lunt et al. 2012), and are increasingly incorporating more detailed dynamics of the biosphere. Coupling of GCMs with (mostly economic) models of human systems creates *Integrated Assessment Models* (IAMs). Other approaches include *Earth system Models of Intermediate Complexity* (EMICs) and simple conceptual or other models of reduced complexity (Claussen et al. 2002). EMICs and conceptual models are useful for exploring nonlinear systems behaviour (i.e., tipping points and abrupt shifts) and for simulating Earth System dynamics over very long timeframes. Projections on geological timescales are underpinned by analyses of changes in insolation received by Earth (Berger/Loutre 2002; Laskar et al. 2010).

Earth System science has benefited from the formation of large international research networks, such as the *World Climate Research Programme* (WCRP), the *International Geosphere-Biosphere Programme* (IGBP: Seitzinger et al. 2015), the *International Human Dimensions Programme on Global Environment Change* (IHDP), *Diversitas*, a global programme on biodiversity change, the International Polar Year (2007–2009), and the global Earth's Critical Zone Network. Several of these have recently evolved into Future Earth www.futureearth.org, a single, interdisciplinary research programme on the Earth System, fully incorporating the human dimensions of the system.

It is no accident that the proposal for the Anthropocene Epoch (Crutzen and Stoermer 2000; Crutzen 2002) arose out of the Earth System science community, in particular out of the synthesis project of the IGBP. That project assembled a wealth of observations on recent changes to the Earth System and set them against the paleoenvironmental record of the Holocene, concluding that the Earth System was now operating in a 'no analogue state' (Steffen et al. 2004). For the first time, a major shift in the state of the Earth System was proposed on the basis of direct observations of changes in the Earth System, without specific reference to evidence in the stratigraphic record.

Simultaneously, stratigraphy was experiencing a revolutionary increase in the types of materials and proxies that could be observed in the records of the very recent past, leading to essentially real-time stratigraphy. These include both the many artifacts of human activities over millennia in archeological strata (Edgeworth et al. 2015; Zalasiewicz et al. 2015; Williams et al. 2016), and the rapidly developing stratigraphic record emerging from the technosphere (Haff 2014) that will form archeological strata of the future. The latter include unique markers such as radionuclides, new forms of metals (e.g., aluminum), spheroidal carbonaceous particles from the combustion of fossil fuels, concrete, and plastics (Zalasiewicz et al. 2016) and synthetic fibres (Waters et al. 2016). There has also been an increasing number of studies on high-resolution bio- and chemo-stratigraphic records of the last few centuries and decades (e.g., Wolfe et al. 2013). This high resolution, data-rich condition has, in part, triggered a growing array of options for defining the Anthropocene and its start date (e.g., Crutzen 2002; Ruddiman 2013; Lewis and Maslin 2015; Zalasiewicz et al. 2015; Waters et al. 2016).

In summary, the relationship between stratigraphy and Earth System science is now much closer and more effective than it was just a few decades ago. Earth System science has a wealth of contemporary data to assess changes in the Earth System and to test predictions arising from theoretical grounds. It is this wealth of direct Earth System data that has led to the proposal for the Anthropocene Epoch. The challenge is to turn this rapidly expanding body of data in stratigraphy and Earth System science into a productive partnership that can define a significant change to the state of the planet consistent with both the Geologic Time Scale and Earth System science.

17.5   Defining the Anthropocene by Integrating Stratigraphic and Earth System Approaches

17.5.1   Stratigraphic Anthropocene

The stratigraphic approach to defining the Anthropocene is clear (Waters et al. 2016):

> Have humans changed the Earth system to such an extent that recent and currently forming
> geological deposits include a signature that is distinct from those of the Holocene and earlier

epochs, which will remain in the geological record? If so, when did this stratigraphic signal (not necessarily the first detectable anthropogenic change) become recognizable worldwide?

A new time interval in Earth history can be defined only when globally synchronous stratigraphic signals related to the structure and functioning of the Earth System are clearly outside the Holocene norm, a new time interval in Earth history can be defined. There is an overwhelming amount of stratigraphic evidence that the Earth System is indeed now structurally and functionally outside the Holocene norm. This evidence includes novel materials such as elemental aluminum, concrete, plastics, and geochemicals; carbonaceous particles from fossil fuel combustion; widespread human-driven changes to sediment deposits; artificial radionuclides; marked rises in greenhouse gas concentrations in ice cores; and trans-global alteration of biological species assemblages (Waters et al. 2016 and references therein).

Determination of a start date for the stratigraphic Anthropocene requires an examination of how the magnitude and rate of contemporary Earth System change, driven largely by human impact, may be best represented by optimal selection of a stratigraphic marker or markers to allow tracing of a synchronous boundary globally. Human environmental impacts began almost as soon as *Homo sapiens* appeared on the Earth. A rich array of stratigraphically relevant materials record these impacts, starting with the megafaunal extinctions of the latest Pleistocene, continuing through early agricultural activities that changed landscapes and emitted CO_2 and CH_4 to the atmosphere (Ellis et al. 2012; Edgeworth et al. 2015; Ruddiman et al. 2015), and increasing significantly with the advent of the late 1700s industrial revolution (Steffen et al. 2007). Globally recognizable, geosynchronous change clearly began in the mid-20th century at the beginning of the Great Acceleration (Hibbard et al. 2006; Steffen et al. 2015a; McNeill/Engelke 2016), which marks a step change in human activity.

There are precedents for utilizing not only the type but also the degree of change in the stratigraphic record to determine chronostratigraphic boundaries. For example, in the late 1820s, the Italian geologist Giambattista Brocchi used percentages of living molluscan forms in fossil assemblages to subdivide the strata of the Apennines. British geologist Charles Lyell followed Brocchi, extending his work across Europe. As noted in Summerhayes (2015):

> By 1828, following Brocchi, he (Lyell) had used the percentages of modern molluscs in each epoch, and the relations of strata to one another, to subdivide the Tertiary Period into several geological Epochs…. In the *Principles of Geology* (1830-33) (Lyell) named the four periods of the Tertiary as Eocene ('dawn of the recent', with 3.5% modern species), Miocene (with 17% modern species), Early Pliocene (with 35–50% modern species) and Late Pliocene (with 90-95% modern species).

Choosing the boundary between the Holocene and the Anthropocene at the mid-20th century is consistent with Lyell's approach in defining subdivisions within the Tertiary based on percentage or degree of change rather than simply on presence or absence of change.

Moreover, the observed differences between strata often indicate enhanced *rates* of change across the boundary. This is most clearly illustrated in the mass extinction events that coincide with some geologic boundaries (Fig. 17.1), when extinction rates rise to at least tens of times above background rates (Barnosky et al. 2011). Contemporary rates of change in both the biosphere and the climate are particularly striking. At present, extinction rates are at least tens (and possibly hundreds) of times above background rates (Miller et al. 1999; Barnosky et al. 2012; Pimm et al. 2014; Ceballos et al. 2015). The rate of species translocations around the globe, resulting in homogenisation of the world's biota and in new ecosystems, has risen sharply above the norm prior to extensive intercontinental shipping and air travel (McNeeley 2001; Williams et al. 2015). Climate-triggered species movement, causing marked shifting of biogeographic ranges, rivals or exceeds the changes evident at both the beginning and end of the Pleistocene, and in the near future such changes may be an order-of-magnitude faster than any at the last glacial-interglacial transition (Diffenbaugh/Field 2013).

The carbon cycle, a critical link between the biosphere and the climate, is now changing at rates 200 times above long-term background levels (Berner 2003; De Paolo et al. 2008; Archer et al. 2009). Atmospheric CO_2 concentration has risen over the past two decades about 100 times faster than the most rapid rate during the last glacial termination (Wolff 2011), and about 10 times faster than the maximum rate of carbon out-gassing during the PETM about 56.0 Ma (Zeebe et al. 2016). In terms of climate, the rate of increase in global average temperature since 1970 is about 170 times the Holocene baseline rate over the past 7000 years, and in the opposite direction (Marcott et al. 2013; NOAA 2016). These accelerated rates are evident in stratigraphic signals, suggesting that human forcings since the mid-20th century are triggering as big a change to the Earth System as the transitions from the Pliocene into the Pleistocene, and then into the Holocene (Barnosky et al. 2012), though not (yet) as big as those which coincided with the ends of the Permian, Cretaceous, and Eocene (Summerhayes 2015).

17.5.2 Earth System Anthropocene

A simple ball-and-cup depiction of complex-system dynamics, which captures the concepts of an envelope of natural variability, a basin of attraction, and a regime shift (Fig. 17.4), is useful in conceptualizing the Earth System approach to defining the Anthropocene.

Determining the start date for the Anthropocene from an Earth System science perspective requires a consideration of both the Holocene envelope of natural variability and the Holocene basin of attraction. The former represents the limit of natural variability of the Earth System (e.g., climatic and intrinsic biosphere variability that occurs in the absence of major human perturbations), shown in Fig. 17.4 as the horizontal broken green line. Perturbations of the Earth System, such as those driven by more intensive human activity of agriculture and then the industrial revolution, can, up to a point, push the Earth beyond the limits of natural variability while remaining within the Holocene basin of attraction, that is, within a state of the Earth System that

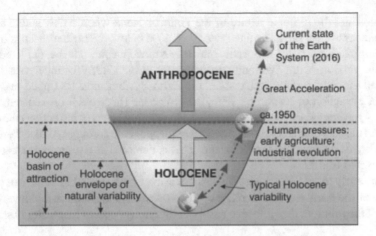

Fig. 17.4 A ball-and-cup depiction of the Earth System definition of the Anthropocene, showing the Holocene envelope of natural variability and basin of attraction. The basin of attraction is more difficult to define than the envelope of variability and so its position is represented here with a higher degree of uncertainty

is still recognizable structurally and functionally as being the Holocene and within which negative feedbacks are still dominant (Schellnhuber 2009).

In Fig. 17.4, these emerging human pressures are positioned between the Holocene envelope of variability and the top of the Holocene basin of attraction, indicating a transition period of growing human activity that moves the Earth System toward the Anthropocene, but not yet into it. However, the beginning of the Great Acceleration marks a sharp step change in the nature, magnitude, and rate of human pressures on the Earth System, driving impacts that push the system beyond the Holocene basin of attraction (Steffen et al. 2015a). We base this judgment on (i) the fact that the current atmospheric concentration of CO_2 of 400 ppm is far higher than at any other time during the last 800,000 years at least (PAGES 2016) and (ii) the rates of change of the climate system, described in detail above, which show that the system is in a strongly transient phase with significantly higher temperature and sea level virtually certain when equilibrium is finally re-established (e.g., Clark et al. 2016). That is, human forcing is now overwhelming the negative feedbacks that would keep the Earth System within the Holocene basin of attraction. As discussed above, the emerging stratigraphic evidence is consistent with this Earth System analysis and suggests that a mid-20th century start date for the Anthropocene is optimal (Zalasiewicz et al. 2015; Waters et al. 2016, and references therein).

In summary, the stratigraphic definition of the Anthropocene is virtually identical to the Earth System definition. In Fig. 17.4 the stratigraphic Holocene Epoch (and Series) is represented by the area below the broken red line; the stratigraphic Anthropocene Epoch (and Series) by the area above the broken red line. The Earth System Anthropocene in Fig. 17.4 is shown not as a stable state but as a trajectory away from the Holocene; the ultimate nature of the Anthropocene when a new stable state is achieved cannot yet be determined; see Sect. 17.6 below for further discussion.

The transition of the Holocene into the Anthropocene can be depicted by a dynamic version of the ball-and-cup metaphor (Fig. 17.5). To undergo a regime shift and move to another stable state, the Earth System must be tipped out of the basin of attraction of its current state, the Holocene (horizontal broken red line in Fig. 17.4). Alternatively (or concurrently), the existing basin of attraction (the cup in Fig. 17.4) is substantially reconfigured by anthropogenic forcings so that there is no possibility of returning to the Holocene. This is depicted as a progressive flattening of the cup in Fig. 17.5.

In the early to mid-Holocene, the basin of attraction is deep, but as human perturbations of the Earth System increase by development of agriculture and later by the industrial revolution, the Holocene basin of attraction becomes increasingly shallow.

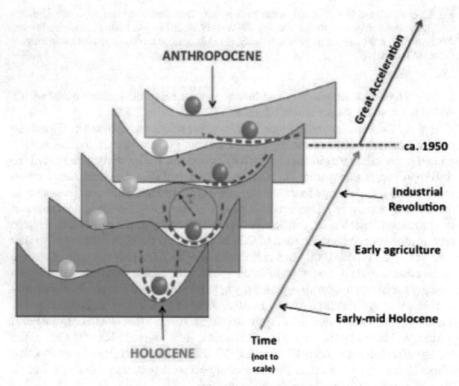

Fig. 17.5 A ball-and-cup depiction of a regime shift. The cup on the right represents a stable basin of attraction (the Holocene) and the orange ball, the state of the Earth System. The cup on the left and the pink ball represent a potential state (the Anthropocene) of the Earth System. Under gradual anthropogenic forcing, the cup becomes shallower and finally disappears (a threshold, ca. 1950), causing the ball to roll to the left (the regime shift) into the trajectory of the Anthropocene toward a potential future basin of attraction. The symbol τ represents the response time of the system to small perturbations. Adapted from Lenton et al. (2008)

The step change in the structure and functioning of the Earth System with the beginning of the Great Acceleration in the mid-20th century, clearly evident in the stratigraphic record (Zalasiewicz et al. 2015; Waters et al. 2016) and in direct observations of Earth System change (Steffen et al. 2015a), is represented by the final flattening and disappearance of the Holocene cup. This represents the crossing of a threshold into the trajectory of the Anthropocene toward a potential future basin of attraction.

The irrevocable nature of the regime shift away from the Holocene is clear. The Earth's biosphere may be approaching a third fundamental stage of evolution (Williams et al. 2015; the first two, as noted above, being a microbial stage from ~4 to 0.8 Ga and thereafter a metazoan stage), and the climate is in an interval of rapid, and possibly, irreversible change. With the amount of CO_2 currently in the atmosphere, the planet will continue to warm, driving a long-term rise in sea level even if emissions of CO_2 ceased immediately (Masson-Delmotte et al. 2013; Clark et al. 2016). Past rises in sea level have taken considerably longer to reach equilibrium than the rise in surface air temperature. For example, warming due to orbital influences ended around 11,700 ka, but sea level continued to rise, by an additional 45 m, for a further 5000 years as ice sheets continued to melt (Clark et al. 2016).

It is clear from both chronostratigraphic and Earth System perspectives that the Earth has entered the Anthropocene, and the mid-20th century is the most convincing start date (Waters et al. 2016). Moreover, the Earth System is still in a phase of rapid change and the outcome is not yet clear; there is no sign that the system is anywhere near a stable or quasi-stable state. In the next section, we explore two possible trajectories and states of the Earth System in the Anthropocene.

17.6   The Future Trajectory of the Anthropocene

The ability of Earth System science to project changes into the future offers some interesting insights into the trajectory of the Anthropocene. Clearly, this trajectory is influenced strongly by human agency in addition to natural processes and feedbacks inherent in the Earth System, and so cannot be predicted with any confidence. Furthermore, it is not clear whether a scenario characterised by a transition from one well-defined state of the Earth System, the Holocene, to another well-defined state is plausible, given that the geological climate record shows a broad range of dynamics, such as transitions, aberrations, perturbations, singular events, and a great deal of variability overall. For example, following cessation of CO_2 emissions at the PETM, 56.0 Ma, the system reverted eventually to its former baseline over a period of around 100,000 years. Nevertheless, two contrasting state-and-transition scenarios, focusing on the climate and the biosphere, may provide insights into the spectrum of potential futures (Box 17.1; Fig. 17.6).

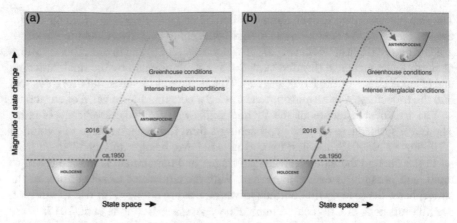

Fig. 17.6 Two of the many possible scenarios for the Anthropocene, relative to the Holocene

Box 17.1: Two Contrasting Trajectories for the Anthropocene

Figure 17.6 uses the ball-and-cup metaphor of Fig. 17.4, with the vertical axis representing the magnitude of change in the Earth System as estimated by both climate and biosphere indicators and the horizontal axis representing the state space of the system. The position of the Anthropocene state indicates relative difference from the Holocene state. The putative states of the Anthropocene (the cups) represent distinct modes of operation of the Earth System that emerge after the system's strongly transient phase ends and more stable conditions develop. We assume that these more stable conditions would persist for many millennia within some envelope of intrinsic variability.

The scenario in Fig. 17.6a is broadly consistent with the Sustainable Development Goals (UN 2015) and the 2015 Paris climate targets (2015 UNFCCC, n.d.), and is based on rapid and deep reductions in greenhouse gas emissions and a radical turnaround in human exploitation of the biosphere. In this scenario, we assume that the climate is significantly warmer than that of the Holocene, but remains in more intense interglacial conditions with most of the Antarctic ice sheet intact; here the intensity of an interglacial is defined by a range of indicators representing different aspects of the Earth System (e.g., proxies for insolation, astronomical parameters, maximum CO_2 and CH_4 concentrations, global average surface temperature anomaly; see (PAGES 2016). In this putative state of the Earth System, biodiversity does not decline much with respect to current conditions.

Figure 17.6b is an unmitigated Anthropocene scenario that assumes an ongoing increase in human pressures on the climate and the biosphere. Several tipping points in the Earth System are crossed, producing a possibly irreversible transition out of the late Quaternary regime of glacial-interglacial cycles and toward the Earth's sixth great extinction event. This leads to a climate that is much warmer than anything resembling an interglacial state (PAGES 2016), with smaller polar ice sheets, a much higher sea level and a vastly changed biosphere. With an ongoing source of CO_2, this state of the Earth System could persist for millions of years, as similarly warm states have in the past (Fig. 17.2).

Realising the Sustainable Development scenario in Box 17.1 (Fig. 17.6a) requires a significant and rapid shift in the nature and magnitude of human perturbations to the climate and the biosphere (Rockström et al. 2009; Griggs et al. 2013; Steffen et al. 2015b); it aims to keep the planet in a state in which human societies can continue to develop and thrive much as they did throughout most of the Holocene. This scenario is, in essence, an Earth System interpretation of the policy goal to avoid 'dangerous interference with the climate system' (e.g., WBGU 2011). This underpins the policy target of limiting mean global surface temperature rise to less than 2 °C above the pre-industrial level (UNFCCC 2010), and the more aspirational 1.5 °C target specified at the recent COP21 meeting in Paris (UNFCCC 2015; Schellnhuber et al. 2016).

Currently stated national commitments for greenhouse gas emission reductions fall considerably short of what is required to have a reasonable probability of meeting the 2 °C target, let alone the 1.5 °C target (Meinshausen et al. 2009; IPCC 2013; Climate Interactive and MIT 2015). However, if global society is able to achieve the required deep decarbonisation, the temperature rise would likely peak late this century and decline very slowly over many millennia (Solomon et al. 2009). In that scenario, the climate would be beyond the orbital control of the late Quaternary, in the sense that warming would be more intense than that of any of the interglacial intervals of the late Quaternary (Fig. 17.6a). The IPCC has concluded that "It is virtually certain that orbital forcing will be unable to trigger widespread glaciation during the next 1000 years. Paleoclimate records indicate that, for orbital configurations close to the present one, glacial inceptions only occurred for atmospheric CO_2 concentrations significantly lower than pre-industrial levels. Climate models simulate no glacial inception during the next 50,000 years if CO_2 concentrations remain above 300 ppm" (Masson-Delmotte et al. 2013).

Consistent with the IPCC assessment, the recent model results of Ganopolski et al. (2016) suggest that anthropogenic atmospheric CO_2 from fossil fuel combustion may have already shifted the Earth System enough to postpone the next glacial inception for at least another 100,000 years. Furthermore, Clark et al. (2016) suggest that even with rapid decarbonisation, a significant, long-term rise of sea level is still very likely. Last, if fossil fuel emissions continue on a business-as-usual trajectory for several more decades, deep-ocean acidification is likely irreversible on a millennial scale. Even 'negative emissions'—massive implementation of anthropogenic CO_2 removal from the atmosphere—would not restore the marine environment for millennia (Mathesius et al. 2015).

Regarding the biosphere, the Earth may be approaching a third fundamental stage of evolution because of a wide range of human pressures (Williams et al. 2015).

The contemporary biosphere differs significantly from previous stages of evolution due to many anthropogenic modifications and perturbations. These include global homogenisation of flora and fauna; human appropriation of 25–40% of net primary production (likely to increase along with population growth); extensive use of fossil fuels to break through photosynthetic energy barriers; human-directed evolution of other species; and increasing interaction of the biosphere with technological systems (Haff 2014; Williams et al. 2015).

The degree of stabilisation of biospheric change equivalent to that needed to stabilise the climate system would require ecosystem restoration and careful stewardship, a rapid reduction in the extinction rate, innovative approaches to agricultural production, full recycling of nutrients such as nitrogen and phosphorus and other materials, the spread of 'living (green) infrastructure' in urban areas, and so on. This scenario requires a fundamental change in the nature of the anthroposphere, so that its dynamics become more synergistic with those of the biosphere (Williams et al. 2015). Yet even this dramatic shift could not undo the past alteration of the biosphere relative to the Holocene, an alteration that already represents a regime shift in the Earth System.

Figure 17.6b illustrates an alternative possible future for the Anthropocene, an extension of its current trajectory that could be called an unmitigated Anthropocene trajectory. Here greenhouse gas emissions continue to rise or are maintained at high levels for several decades at least, driving the global average surface temperature rise toward 2 °C by mid-21st century. Warming continues to weaken the terrestrial and marine carbon sinks that sequester a significant fraction of anthropogenic emissions (Raupach et al. 2014), further amplifying the warming. The climate system begins to cross tipping points (Lenton et al. 2008), triggering, for example, the loss of much of the Amazon rainforest (Hirota et al. 2011), a completely or nearly ice-free Arctic Ocean in summer (Kirtman et al. 2013), and a rapid increase in outgassing of CH_4 and CO_2 from thawing permafrost (Ciais et al. 2013). These positive feedbacks would accelerate the warming so that even deep cuts in greenhouse gas emissions could no longer bend the warming trajectory. This scenario would likely push the Earth System into a much longer, much warmer, persistent state (Ganopolski et al. 2016). An ultimate sea-level rise of tens of meters would become unavoidable (Dutton et al. 2015; Clark et al. 2016).

Given continuance of a supply of CO_2 or some equivalent, this new state of the Earth System could persist for millions of years. As Lenton and Williams (2013) note, the glacial-interglacial cycles that typify the late Quaternary climate could be a rare condition of potential instability in the Earth System, where positive feedbacks dominate during the transitions between the glacial and interglacial states and relatively small external forcing triggers the passage from one state to another. More common in Earth history are long periods of very slow change (e.g., the greenhouse phases of Fig. 17.2b), where negative feedbacks dominate and the Earth System is generally insensitive to perturbations. While variations in the pattern of solar radiation are still reflected in climate variability during these periods, they are less prominent than during the pronounced Quaternary-style climate oscillations (Gale et al. 1999; Naish et al. 2001, 2009).

The equivalent trajectory for the biosphere, if sustained over millions of years, might represent a third stage of evolution of life on Earth, characterised by the many changes described above (Williams et al. 2015). If continuation of these trends were also to increase the likelihood of a sixth mass extinction event within the Phanerozoic Eon (Barnosky et al. 2011), then that extinction would mark a major biostratigraphical boundary in the geological record, perhaps comparable to that separating Mesozoic and Cenozoic biotas.

17.7 Conclusion

Over the past several decades, the relationship between stratigraphy and Earth System science has changed significantly. Diverse tools now allow scientists to observe, record, test, and model Earth System processes directly and to synthesise them into the overarching concept of the Earth as a single complex system. Stratigraphy has benefitted from an increase of methods, data, and conceptual and explanatory frameworks; Earth System science has consequently benefitted from new types of stratigraphical inputs. The type of materials and proxies that can be examined in the sedimentary record has expanded greatly to include a plethora of materials of exclusively human origin. Much greater temporal and spatial resolution of various layers in the stratal record can now be achieved, especially in the recent past. The concept of the Anthropocene has provided a trigger for the Earth System science and stratigraphic communities to integrate their knowledge, tools, skills, and rapidly growing masses of data in unprecedented ways.

From both Earth System science and stratigraphic perspectives, the Earth has been pushed out of the pre-industrial Holocene norm by human activities. This has led scientists to ask: How will the Anthropocene evolve? Even with a rapid and decisive shift of contemporary human societies toward sustainable development, the Anthropocene will remain a distinctly different epoch from the Holocene.

The current trajectory of human societies would lead to an Anthropocene that is a much warmer and biotically different state of the Earth System, one that is no longer governed by the late Quaternary regime of glacial-interglacial cycles, and with far fewer species. Earth in a much warmer greenhouse state would be nothing new. However, it would be novel for *Homo sapiens,* which evolved only 200,000 years ago. Under this scenario, the Earth System would be markedly different from the one humans now know, and from the state that supported the development of human civilization. Which trajectory the Anthropocene follows depends on the decisions and actions of global society today, and over the next few decades.

Acknowledgments This paper is a contribution of the Anthropocene Working Group. The complex systems interpretation of Earth System dynamics has drawn heavily on the work and insights of Tim Lenton and Marten Scheffer. We are grateful to Greg Heath for assistance with Figs. 17.4 and 17.6. We thank the anonymous reviewers for their insightful comments that have contributed to improve this paper. All data for this paper are properly cited and referred to in the reference list.

References

Ainsworth, E.A.; Long, S.P., 2005: "What have we learned from 15years of free-air CO_2 enrichment (FACE)? A meta-analytic review of the responses of photosynthesis, canopy properties and plant production to rising CO_2", in: *New Phytol.*, 165, 2: 351–372; https://doi.org/10.1111/j.1469-8137.2004.01224.x.

Allwood, A.C.; Grotzinger, J.P.; Knoll, A.H.; Burch, I.W.; Anderson, M.S.; Coleman, M.L.; Kanik, I., 2009: "Controls on development and diversity of early Archean stromatolites", in: *Proc. Natl. Acad. Sci. USA*, 106: 9548–9555; https://doi.org/10.1073/pnas.0903323106.

Alroy, J., 2010: "The shifting balance of diversity among major marine animal groups", in: *Science*, 329: 1191–1194; https://doi.org/10.1126/science.1189910.

Anagnostou, E.; John, E.H.; Edgar, K.M.; Foster, G.L.; Ridgwell, A.; Inglis, G.N.; Pancost, R.D.; Lunt, D.J.; Pearson, P.N., 2016: "Changing atmospheric CO_2 concentration was the primary driver of early Cenozoic climate", in: *Nature*, 533: 380–384; https://doi.org/10.1038/nature17423.

Archer, D. et al., 2009: "Atmospheric lifetime of fossil fuel carbon dioxide", in: *Ann. Rev. EarthPlanet. Sci.*, 37: 117–134; https://doi.org/10.1146/annurev.earth.031208.100206.

Bai, X., et al., 2015: "Plausible and desirable futures in the Anthropocene: a new research agenda", in: *Global Environ. Change*, 39: 351–362; https://doi.org/10.1016/j.gloenvcha.2015.09.017.

Bard, E., 2004: "Greenhouse effect and ice ages: historical perspective", in: *C. R. Geosci.*, 336; 603–638; https://doi.org/10.1016/j.crte.2004.02.005.

Barnosky, A. D., et al., 2011: "Has the Earth's sixth mass extinction already arrived?", in: *Nature*, 471: 51–57; https://doi.org/10.1038/nature09678.

Barnosky, A. D., et al., 2012: "Approaching a state-shift in the biosphere, in: *Nature*, 486: 52–56; https://doi.org/10.1038/nature11018.

Beerling, D.J.; Royer, D.L., 2011: "Convergent Cenozoic CO_2 history, in: *Nat. Geosci.*, 4: 418–420; https://doi.org/10.1038/ngeo1186.

Behrensmeyer, A.K.; Damuth, J.D.; DiMichele, W.A.; Potts, R.; Sues, H.-D.; Wing, S.L., 1992: *Terrestrial Ecosystems through Time. Evolutionary Palaeocology of Terrestrial Plants and Animals* (Chicago, Ill. – London, U.K.: Univ. of Chicago Press).

Bell, E.A.; Boehnke, P.; Harrison, T.M.; Mao, W.L., 2015: "Potentially biogenic carbon preserved in a 4.1 billion-year-old zircon", in: *Proc. Natl. Acad. Sci. USA*, 112, 47: 14518–14521; at: https://doi.org/10.1073/pnas.1517557112.

Berger, A.; Loutre, M.F., 2002: "An exceptionally long interglacial ahead?", in: *Science*, 297: 1287–1288; https://doi.org/10.1126/science.1076120.

Berner, R.A., 1990: "Atmospheric carbon dioxide levels over Phanerozoic time", in: *Science*, 249: 1382–1386; https://doi.org/10.1126/science.249.4975.1382.

Berner, R.A., 1999a: "A new look at the long-term carbon cycle", in: *GSA Today*, 11, 9: 1–6.

Berner, R.A., 1999b: "Atmospheric oxygen over Phanerozoic time", in: *Proc. Natl. Acad. Sci. USA*, 96, 20: 10955–10957; https://doi.org/10.1073/pnas.96.20.10955.

Berner, R.A., 2003: "The long-term carbon cycle, fossil fuels and atmospheric composition", in: *Nature*, 426: 323–326; https://doi.org/10.1038/nature02131.

Berner, R.A.; Lasaga, A.C.; Garrels, R.M., 1983: "The carbonate-silicate geochemical cycle and its effect on atmospheric carbon dioxide over the past 100 million years", in: *Am. J. Sci.*, 283: 641–683; https://doi.org/10.2475/ajs.283.7.641.

Bonneuil, C.; Fressoz, J.-B., 2016: *The Shock of the Anthropocene: The Earth, History and Us* (London, U.K., Verso).

Bowen, G.J.; Maibauer, B.J.; Kraus, M.J.; Rohl, U.; Westerhold, T.; Steimke, A.; Gingerich, P.D.; Wing, S.L.; Clyde, W.C., 2015: "Two massive, rapid releases of carbon during the onset of the Palaeocene-Eocene thermal maximum", in: *Nat. Geosci.*, 8: 44–47; https://doi.org/10.1038/NGE O2316.

Boyd, P.W. et al., 2007: "Mesoscale iron enrichment experiments 1993–2005: synthesis and future directions", in: *Science*, 315: 612–617; https://doi.org/10.1126/science.1131669.

Bradley, R.S., 2015: *Paleoclimatology: Reconstructing Climates of the Quaternary,* 3rd ed., (Amsterdam: Elsevier).

Braje, T.J., 2015: "Earth Systems, human agency, and the Anthropocene: Planet Earth in the human age", in: *J. Archaeol. Res.,* 23, 3: 369–396; https://doi.org/10.1007/s10814-015-9087-y.

Brook, B.W.; *Barnosky*, A.D., 2012: "Quaternary extinctions and their linkto climate change", in: Hannah, L. 8ed.): *Saving a Million Species: Extinction Risk from Climate Change* (Washington, D.C.: Island Press): 179–198.

Budyko, M.I., 1969: "The effect of solar radiation variations on the climate of the Earth", in: *Tellus,* 21, 5: 611–619; https://doi.org/10.1111/j.2153-3490.1969.tb00466.x.

Budyko, M.I., 1986: *"The Evolution of the Biosphere* (Dordrecht: Reidel).

Butterfield, N.J., 2011: "Animals and the invention of the Phanerozoic Earth System", in: *Trends Ecol. Evol.,* 26: 81–87; https://doi.org/10.1016/j.tree.2010.11.012.

Capinha, C.; Essl, F.; Seebens, H.; Moser, D.; Pereira, H.M., 2015: "The dispersal of alien species redefines biogeography in the Anthropocene, in: *Science,* 348, 6240: 1248–1251; https://doi.org/10.1126/science.aaa8913.

Ceballos, G.; Ehrlich, P.R.; *Barnosky*, A.D.; Garcia, A.; Pringle, R.M.; Palmer, T.M., 2015: "Accelerated modern human-induced species losses: entering the sixth mass extinction", in: *Sci. Adv.,* 1, 5: e1400253; https://doi.org/10.1126/sciadv.1400253.

Chakrabarty, D., 2009: "The Climate of history: four theses", in: *Cri tInq,* 35, 2: 197–222; https://doi.org/10.1086/596640.

Ciais, P. et al., 2013: "Carbon and other biogeochemical cycles", in: Stocker, T.F. et al. (eds.), *Climate Change 2013: The Physical Science Basis, Contribution of Working Group I to the Fifth Assessment Report of the Intergovernmental Panel on Climate Change* (Cambridge – New York: Cambridge Univ. Press): 465–570; https://doi.org/10.1017/CBO9781107415324.015.

Clark, W.C.; Munn, R.E. (eds.), 1986: *Sustainable Development of the Biosphere* (Laxenburg – Cambridge: IIASA – Cambridge Univ. Press).

Clark, P.U. et al., 2016: "Consequences of twenty-first-century policy for multi-millennial climate and sea-level change", in: *Nat. Clim. Change,* 6: 360–369; https://doi.org/10.1038/nclimate2923.

Claussen, M. et al., 2002: "Earth System models of intermediate complexity: closing the gap in the spectrum of climate system models", in: *Clim. Dynam.,* 18, 7: 579–586; https://doi.org/10.1007/s00382-001-0200-1.

Climate Interactive – MIT, 2015: "Climate Scoreboard", at: https://www.climateinteractive.org/tools/scoreboard/.

Corlett, R.T., 2015: "The Anthropocene concept in ecology and conservation", in: *Trends Ecol. Evol.,* 30, 1: 36–41; https://doi.org/10.1016/j.tree.2014.10.007.

Crutzen, P.J., 2002: "Geology of mankind—the Anthropocene", in: *Nature,* 415: 23; https://doi.org/10.1038/415023a.

Crutzen, P.J.; Stoermer, E.F., 2000: "The Anthropocene", in: *Global Change Newslett.,* 41: 17–18.

Cui, Y.; Kump, L.R.; Ridgwell, A.J.; Charles, A.J.; Junium, C.K.; Diefendorf, A.F.; Freeman, K.H.; Urban, N.M.; Harding, I.C., 2011: "Slow release of fossil carbon during the Palaeocene-Eocene thermal maximum", in: *Nat. Geosci.,* 4: 481–485; https://doi.org/10.0138/ngeo1179.

DePaolo, D.J.; Cerling, T.E.; Hemming, S.R.; Knoll, A.H.; Richter, F.M.; Royden, L.H.; Rudnick, R.L.; Stixrude, L.; Trefil, J.S., 2008: *Origin and Evolution of Earth: Research Questions for a Changing Planet* (Washington, D.C.: The National Academies Press).

Dickens, G.R., 2011: "Methane release from gas hydrate systems during the Paleocene-Eocene thermal maximum and other past hyperthermal events: setting appropriate parameters for discussion", in: *Clim. Past Discuss.,* 7, 2: 1139–1174; https://doi.org/10.5194/cpd-7-1139-2011.

Dickens, G.R.; O'Neil, J.R.; Rea, D.K.; Owen, R.M., 1995: "Dissociation of oceanic methane hydrate as a cause of the carbon isotope excursion at the end of the Paleocene", in: *Paleoceanography,* 10, 6: 965–971; https://doi.org/10.1029/95PA02087.

Diffenbaugh, N.S.; Field, C.B., 2013: "Changes in ecologically critical terrestrial climate conditions", in: *Science,* 341: 486–492; https://doi.org/10.1126/science.1237123.

DiMichele, W.A.; Hook, R.W.; Beerbower, R.; Boy, J.A.; Gastaldo, R.A.; Hotton III, N.; Phillips, T.L.; Scheckler, S.E.; Shear, W.A.; Sues, H.-D.; Behrensmeyer, A.K.; Damuth, J.D.; DiMichele, W.A.; Potts, R.; Sues, H.-D.; Wing, S.L. (eds.), 1992: "Paleozoic terrestrial ecosystems", in: *Terrestrial Ecosystems through Time* (Chicago, Ill.: Univ. Chicago Press): 205–325.

Dowsett, H.J.; Robinson, M.M.; Stoll, D.K.; Foley, K.M.; Johnson, A.L.A.; Williams, M.; Riesselman, C.R., 2013: "The PRISM (Pliocene palaeoclimate) reconstruction: time for a paradigm shift", in: *Phil. Trans. Roy Soc. Lond. A,* 371: 20120524; https://doi.org/10.1098/rsta.2012.0524.

Dutton, A.; Carlson, A.E.; Long, A.J.; Milne, G.A.; Clark, P.U.; DeConto, R.; Horton, B.P.; Rahmstorf, S.; Raymo, M.E., 2015: "Sea-level rise due to polar ice-sheet mass loss during past warm periods", in: *Science,* 349, 6244: 153; https://doi.org/10.1126/science.aaa4019.

Edgeworth, M.; deB Richter, D.; Waters, C.N.; Haff, P.; Neal, C.; Price, S.J., 2015: "Diachronous beginnings of the Anthropocene: the lower bounding surface of anthropogenic deposits", in: *Anthropocene Rev.,* 2, 1: 1–26; https://doi.org/10.1177/2053019614565394.

Edwards, D.; Davies, K.L.; Axe, L., 1992: "A vascular conducting strand in the early land plant *Cooksonia*", in: *Nature,* 357: 683–685; https://doi.org/10.1038/357683a0.

Ellis, E.C.; Antill, E.C.; Kreft, H., 2012: "All is not loss: plant biodiversity in the Anthropocene, in: *PLoSOne,* 7, 1: e30535, https://doi.org/10.1371/journal.pone.0030535.

EPICA (European Project for Ice Coring in Antarctica) Community Members, 2004: "Eight glacial cycles from an Antarctic ice core", in: *Nature,* 429: 623–628; https://doi.org/10.1038/nature 02599.

Erwin, D.H.; Laflamme, M.; Tweedt, S.M.; Sperling, E.A.; Pisani, D.; Peterson, K.J., 2011: "The Cambrian conundrum: early divergence and later ecological success in the early history of animals", in: *Science,* 334: 1091–1097; https://doi.org/10.1126/science.1206375.

Fairchild, I.J.; Kennedy, M.J., 2007: "Neoproterozoic glaciation in the Earth System", in: *J. Geol. Soc. Lond.,* 164: 895–921; https://doi.org/10.1144/0016-76492006-191.

Feulner, G., 2012: "The faint young sun problem", in: *Rev. Geophys.,* 50: RG2006; https://doi.org/ 10.1029/2011RG000375.

Fischer-Kowalski, M.; Krausmann, F.; Pallua, I., 2014: "A sociometabolic reading of the Anthropocene: modes of subsistence, population size and human impact on Earth", in: *Anthropocene Rev.,* 7, 1: 8–33; https://doi.org/10.1177/2053019613518033.

Flato, G. et al., 2013: "Evaluation of Climate Models", in: Stocker, T.F. et al. (eds.): *Climate Change 2013: The Physical Science Basis,* Contribution of Working Group I to the Fifth Assessment Report of the Intergovernmental Panel on Climate Change (Cambridge – New York: Cambridge Univ. Press).

Franks, P.J.; Royer, D.L.; Beerling, D.J.; Van deWater, P.K.; Cantrill, D.J.; Barbour, M.M.; Berry, J.A., 2014: "New constraints on atmospheric CO_2 concentration for the Phanerozoic, in: *Geophys. Res. Lett.,* 47: 4685–4694; https://doi.org/10.1002/2014GL060457.

Gale, A.S.; Young, J.R.; Shackleton, N.J.; Crowhurst, S.J.; Wray, D.S., 1999: "Orbital tuning of the Cenomanian marly chalk successions: towards a Milankovitch time-scale for the late Cretaceous", in: *Phil. Trans. Roy. Soc. Lond. A,* 357:1815–1829; https://doi.org/10.1098/rsta.1999.0402.

Galvez, M.E.; Gaillardet, J., 2012: "Historical constraints on the origin of the carbon cycle concept", in: *C. R. Geosci.,* 344, 11–12: 549–567; https://doi.org/10.1016/j.crte.2012.10.006.

Ganopolski, A.; Winkelmann, R.; Schellnhuber, H.J., 2016: "Critical insolation-CO_2 relation for diagnosing past and future glacial inception, in: *Nature,* 529: 200–203; https://doi.org/10.1038/ nature16494.

Gibbard, P.L. et al., 2005: "What status for the Quaternary?", in: *Boreas,* 34, 1–6; https://doi.org/ 10.1080/03009480510012854.

Gillings, M. R.; Paulsen, I.T., 2014: "Microbiology of the Anthropocene", in: *Anthropocene,* 5: 1–8; https://doi.org/10.1016/j.ancene.2014.06.004.

Gradstein, F.; Ogg, G.; Schmitz, M. (eds.), 2012: *The Geological Time Scale 2072,* vol. 7 (Amsterdam: Elsevier): 77.

Grassineau, N.V.; Nisbet, E.G.; Fowler, C.M.R.; Bickle, M.J.; Lowry, D.; Chapman, H.J.; Mattey, D.P.; Abell, P.; Yong, J.; Martin, A., 2002: "Stable isotopes in the Archaean Belingwe belt,

Zimbabwe: evidence for a diverse microbial mat ecology", in: *Spec Publ Geol Soc, Lond,* 79: 309–328; https://doi.org/10.1144/GSL.SP.2002.199.01.15.

Grassineau, N.; Abell, P.; Appel, P.W.U.; Lowry, D.; Nisbet, E., 2006: "Early life signatures in sulfur and carbon isotopes from Isua, Barberton, Wabigoon (Steep Rock), and Belingwe Greenstone Belts (3.8 to 2.7 Ga)", in: Kesler, S E.; Ohmoto, H. (eds.): *Evolution of Early Earth's Atmosphere, Hydrosphere, and Biosphere—Constraints from Ore Deposits,* vol. 798 (Boulder, Colo.: Geological Society of America): 33–52.

Greb, S.F.; DiMichele, W.A.; Gastaldo, R.A., 2006: "Evolution and importance of wetlands in earth history", in: *Geol. Soc. Am. Special Papers,* 399: 1–40.

Griggs, D.; Stafford Smith, M.; Gaffney, O.; Rockström, J.; Ohman, M.C.; Shyamsundar, P.; Steffen, W.; Glaser, G.; Kanie, N.; Noble, I., 2013: "Sustainable development goals for people and planet", in: *Nature,* 495: 305–307; https://doi.org/10.1038/495305a.

Grinevald, J., 1987: "On a holistic concept for deep and global ecology: the biosphere", in: *Fundamenta Scientiae,* 8,2: 197–226.

Haff, P.K., 2014: "Humans and technology in the Anthropocene. Six rules", in: *Anthropocene Rev.,* 7:, 126–136; https://doi.org/10.1177/2053019614530575.

Hamilton, C.; Grinevald, J., 2015: "Was the Anthropocene anticipated?", in: *Anthropocene Rev.,* 2:59–72; https://doi.org/10.1177/2053019614567155.

Hammarlund, E.U.; Dahl, T.W.; Harper, D.A.T.; Bond, D.P.G.; Nielsen, A.T.; Bjerrum, C.J.; Schovsbo, N.H.; Schönlaub, H.P.; Zalasiewicz, J.A.; Canfield, D.E., 2012: "A sulfidic driver for the end-Ordovician mass extinction", in: *Earth Planet. Sci. Lett,* 331–332: 128–139; https://doi.org/10.1016/j.epsl.2012.02.024.

Haywood, A. M.; Dowsett, H.J.; Valdes, P.J., 2009: "The Pliocene. A vision of Earth in the late twenty-first century?", in: *Phil. .Trans. Roy. Soc. Lond. A*: 367: 3–204, edited thematic set.

Haywood, A.M.; Ridgwell, A.; Lunt, D.L.; Hill, D.J.; Pound, M.J.; Dowsett, H.J.; Dolan, A.M.; Francis, J.E.; Williams, M., 2011: "Are there pre-Quaternary geological analogues for a future greenhouse gas-induced global warming?", in: *Phil. Trans. Roy. Soc. Lond. A,* 369: 933–956; https://doi.org/10.1098/rsta.2010.0317.

Hibbard, K.A.; Crutzen, P.J.; Lambin, E.F.; Liverman, D.M.; Mantua, N.J.; McNeill, J.R.; Messerli, B.; Steffen, W., 2006: "Decadal interactions of humans and the environment", in: Costanza, R.; Graumlich, L.; Steffen, W. (eds.): *Integrated History and Future of People on Earth,* Dahlem Workshop Report 96 (Cambridge, Mass.: The MIT Press): 341–375.

Hirota, M.; Holmgren, N.M.; Van Nes, E.H.; Scheffer, M., 2011: "Global resilience of tropical forest and savanna to critical transitions", in: *Science,* 334: 232–235; https://doi.org/10.1126/science.1210657.

Hoffman, P.F.; Kaufman, A.J.; Halverson, G.P.; Schrag, D.P., 1998: "A neoproterozoic snowball earth", in: *Science,* 287: 1342–1346; https://doi.org/10.1126/science.281.5381.1342.

Hotton, C.L.; Hueber, F.M.; Griffing, D.H.; Bridge, J.S., 2001: "Early terrestrial plant environments: an example from the Emsian of Gaspe, Canada", in: Gensel, P.G.; Edwards, D. (eds.): *Plants Invade the Land: Evolutionary and Environmental Perspectives* (New York: Columbia Univ. Press): 179–212.

Hou, X.-G.; Aldridge, R.; Bergstrom, J.; David, J.S.; Siveter, D.J.; Feng, X.-H., 2004: *"The Cambrian Fossils of Chengjiang, China: The Flowering of Early Animal Life* (Oxford: Wiley Blackwell).

Hughes, T.P.; Carpenter, S.; Rockström, J.; Scheffer, M.; Walker, B., 2013: "Multiscale regime shifts and planetary boundaries", in: *Trends Ecol. Evol.,* 28: 389–395; https://doi.org/10.1016/j.tree.2013.05.019.

Hutton, J., 1795: *Theory of the Earth with Proofs and Illustrations* (In four parts): vol. I, 620 pp., vol. II, 567 pp., vol. III (Edinburgh: Geological Society); vol. 7899 (London: Geological Society).

ICOLD (International Commission of Large Dams Registry), 2016: at: http://www.icold-cigb.org/GB/World_register/general_synthesis.asp..

ICSU, 1986: *"The International Geosphere Biosphere Programme: A Study of Global Change,* Final report of the Ad Hoc Planning Group, Prepared for the 21st General Assembly, Berne, 14–19 September 1986 (Paris: International Council of Scientific Unions).

IPCC, 2013: "Summary for Policymakers", in: Stocker et al., T.F. (eds.): *Climate Change 2013: The Physical Science Basis, Contribution of Working Group I to the Fifth Assessment Report of the Intergovernmental Panel on Climate Change* (Cambridge – New York: Cambridge Univ. Press).

Jensen, S., 2003: "The Proterozoic and earliest Cambrian trace fossil record: patterns, problems and perspectives", in: *Integr. Comp. Biol.,* 43: 219–228; https://doi.org/10.1093/icb/43.1.219.

Jørgensen, S.E., (ed.), 2010: *"Global Ecology: A Derivative of Encyclopedia of Ecology,* (Amsterdam – Boston, Mass: Elsevier and Academic Press).

Kent, D.V.; Muttoni, G., 2008: "Equatorial convergence of India and early Cenozoic climate trends", in: *Proc. Natl. Acad. Sci. USA,* 705, 42: 16065–16070; https://doi.org/10.1073/pnas.0805382105.

Kirtman B., et al., 2013: "Near-term climate change: projections and predictability, in: Stocker, T.F. et al. (eds.): *Climate Change 2073: The Physical Science Basis, Contribution of Working Group I to the Fifth Assessment Report of the Intergovernmental Panel on Climate Change* (Cambridge – New York: Cambridge Univ. Press): 465–570; https://doi.org/10.1017/CBO978 1107415324.015.

Knorr, W.; Prentice, I.C.; House, J.I.; Holland, E.A., 2005: "Long-term sensitivity of soil carbon turnover to warming", in: *Nature,* 433: 7023: 298–301; https://doi.org/10.1038/nature03226.

Koch, P.L.; Barnosky, A.D., 2006: "Late quaternary extinctions: state of the debate, in: *Ann. Rev. Ecol. Evol. System.,* 37: 215–250; https://doi.org/10.1146/annurev.ecolsys.34.011802.132415.

Laskar, J.; Fienga, A.; Gastineau, M.; Manche; H., 2010: "A new orbital solution for the long-term motion of the Earth", in: *Astron. Astrophys.,* 532: A89; https://doi.org/10.1051/0004-6361/201 116836.

Latour, B., 2015: *Face à Gaïa: Huit Conférences sur le Nouveau Régime Climatique* (Paris: La Découverte).

Lenton, T.M., 2011: "Tipping elements: jokers in the pack", in: Richardson, K.; Steffen, W.; Liverman, D. (eds): *Climate Change: Global Risks, Challenges and Decisions* (Cambridge: Cambridge Univ. Press): 163–201.

Lenton, T.M., 2015: *Earth System Science. A Very Short Introduction* (Oxford: Oxford Univ. Press).

Lenton, T M.; Watson, A.J., 2011: *Revolutions That Made the Earth* (Oxford: Oxford Univ. Press).

Lenton, T.M.; Williams, H.T.P., 2013: "On the origin of planetary-scale tipping points", in: *Trends Ecol. Evol.,* 28: 380–382; https://doi.org/10.1016/j.tree.2013.06.001.

Lenton, T.M. et al., 2004: "Long-term geosphere-biosphere coevolution and astrobiology", in: Schellnhuber, H.J.; Crutzen, P.J.; Clark, W.C.; Claussen, M.; Held, H. (eds.): *Earth System Analysis for Sustainability* (Cambridge, Mass.: The MIT Press): 110–139.

Lenton, T.M.; Held, H.; Kiegler, E.; Hall, J.W.; Lucht, W.; Rahmstorf, S.; Schellnhuber, H.J., 2008: "Tipping elements in the Earth's climate system", in: *Proc. Natl. Acad. Sci. USA,* 705: 1786–1793; https://doi.org/10.1073/pnas.0705414105.

Lewis, S.L.; Maslin, M.A., 2015: "Defining the Anthropocene", in: *Nature,* 579: 171–180; https://doi.org/10.1038/nature14258; pmid: 25762280.

Lovelock, J.E., 1979: *GAIA: A New Look at Life on Earth* (Oxford: Oxford Univ. Press; new edition, 1995).

Lovelock, J. E., 1988: *The Ages of Gaia: A Biography of Our Living Earth* (New York: W.W. Norton & Co; new edition, 1995).

Lovelock, J.; Margulis, L., 1974: "Atmospheric homeostasis by and for the biosphere: the Gaia hypothesis", in: *Tellus,* 26, 1–2: 2–10; https://doi.org/10.1111/j.2153-3490.1974.tb01946.x.

Lunt, D. J. et al., 2012: "A model-data comparison for a multi-model ensemble of early Eocene atmosphere-ocean simulations: EoMIP", in: *Clim. Past,* 8: 1717–1736; https://doi.org/10.5194/cp-8-1717-2012.

Malm, A.; Hornborg, A., 2014: "The geology of mankind? A Critique of the Anthropocene narrative", in: *Anthropocene Rev.,* 7, 1: 62–69; https://doi.org/10.1177/2053019613516291.

Malone, T.F.; Roederer, J.G. (eds.), 1985: *Global Change. The Proceedings of a Symposium sponsored by the International Council of Scientific Unions (ICSU) during its 20th General Assembly in Ottawa, Canada on September 25, 1984* (Cambridge: ICSU Press – Cambridge Univ. Press).

Marcott, S.A.; Shakun, J.D.; Clark, P.U.; Mix, A., 2013: "A reconstruction of regional and global temperature for the past 11,300 years", in: *Science,* 339, 6124: 1198–1201; https://doi.org/10.1126/science.1228026.

Maslin, M.A.; Lewis, S.L., 2015: "Anthropocene: Earth System, geological, philosophical and political paradigm shifts", in: *Anthropocene Rev.,* 2, 2: 108–116; https://doi.org/10.1177/2053019615588791.

Masson-Delmotte, V. et al., 2013: "Information from paleoclimate archives", in: Stocker, T.F. et al. (eds.): *Climate Change 2073: The Physical Science Basis, Contribution of Working Group I to the Fifth Assessment Report of the Intergovernmental Panel on Climate Change* (Cambridge – New York: Cambridge Univ. Press): 383–464.

Mathesius, S.; Hofmann, M.; Caldeira, K.; Schellnhuber, H.J., 2015: "Long-term response of oceans to CO_2 removal from the atmosphere", in: *Nat. Clim. Change,* 5, 12: 1107–1113; https://doi.org/10.1038/nclimate2729.

McInerney, F.A.; Wing, S.L., 2011: "The Paleocene-Eocene thermal maximum—a perturbation of carbon cycle, climate, and biosphere with implications for the future", in: *Ann. Rev. Earth Planet. Sci.,* 39: 489–516; https://doi.org/10.1146/annurev-earth-040610-133431.

McNeill, J.R.; Engelke, P., 2016: *The Great Acceleration* (Cambridge Mass.: Harvard Univ. Press).

Meinshausen, M.; Meinshausen, N.; Hare, W.; Raper, S.C.B.; Frieler, K.; Knutti, R.; Frame, D.J.; Allen, M.R., 2009: "Greenhouse gas emission targets for limiting global warming to 2°C", in: *Nature,* 458: 1158–1162; https://doi.org/10.1038/nature08017.

Melchin, M.J.; Sadler, P.M.; Cramer, B.D., 2012: "The Silurian period", in: Gradstein, F.; Ogg, G.; Schmitz, M. (eds.): *The Geological Time Scale 2072* (Amsterdam: Elsevier): 526–558.

Miller, G. H.; Magee, J.W.; Johnson, B.J.; Fogel, M.L.; Spooner, N.A.; McCulloch, M.T.; Ayliffe, L.K., 1999: "Pleistocene extinction of *Genyornis newtoni:* human impact on Australian megafauna", in: *Science,* 283: 205–208; https://doi.org/10.1126/science.283.5399.205.

Miller, K.G.; Wright, J.D.; Browning, J.V.; Kulpecz, A.; Kominz, M.; Naish, T.R.; Cramer, B.S.; Rosenthal, Y.; Peltier, W.R.; Sosdian, S., 2012: "High tide of the warm Pliocene: implications of global sea level for Antarctic deglaciation", in: *Geology,* 40: 407–410; https://doi.org/10.1130/G32869.1.

Molina, E.; Alegret, L.; Arenillas, I.; Arz, J.A.; Gallala, N.; Hardenbol, J.; von Salis, K.; Steurbaut, E.; Vandenberghe, N.; Zaghbib-Turki, D., 2006: "The Global Boundary Stratotype Section for the base of the Danian Stage (Paleocene, Paleogene, 'Tertiary', Cenozoic) at El Kef, Tunisia — original definition and revision", in: *Episodes,* 29,4: 263–273.

Naish, T.; Zwartz, D., 2012: "Palaeoclimate: looking back to the future", in: *Nat. Clim. Change,* 2: 317–318; https://doi.org/10.1038/nclimate1504.

Naish, T.R. et al., 2001: "Orbitally induced oscillations in the East Antarctic ice sheet at the Oligocene/Miocene boundary", in: *Nature,* 473: 719–723; https://doi.org/10.1038/35099534.

Naish, T.R., et al., 2009: "Oliquity-paced Pliocene West Antarctic Ice Sheet oscillations", *Nature,* 458: 322–329; https://doi.org/10.1038/nature07867.

NASA Earth System Sciences Committee, 1988: "*Earth System Science: A Closer View* (Washington, D. C.: NASA Advisory Council).

Nisbet, E.G.; Fowler, C.M.R., 2014: "The early history of life", in: Holland, D.; Turekian, K.K. (eds.): *Treatise on Geochemistry* vol. 70, 2nd ed. (Oxford: Elsevier): 1–42.

NOAA, 2016: "State of the Climate: Global Analysis for Annual 2015, National Centers for Environmental Information"; at: http://www.ncdc.noaa.gov/sotc/global/201513.

Oldroyd, D., 1996: *Thinking about the Earth: A History of Ideas in Geology* (London, U.K.: Athlone).

Pagani, M.; Zachos, J.C.; Freeman, K.H.; Tipple, B.; Bohaty, S., 2005: "Marked decline in atmospheric carbon dioxide concentrations during the Paleogene", in: *Science,* 309: 600–603; https://doi.org/10.1126/science.1110063.

Page, A.; Zalasiewicz, J.A.; Williams, M.; Popov, L.E., 2007: "Were transgressive black shales a negative feedback modulating glacioeustasy in the Early Palaeozoic Icehouse?" in: Williams, M.; Haywood, A.M.; Gregory, F.J.; Schmidt, D.N. (eds.): *Deep-Time Perspectives on Climate Change: Marrying the Signal from Computer Models and Biological Proxies,* Special Publications (London, U.K.: The Geological Society – The Micropalaeontological Society): 123–156.

PAGES (Past Interglacials Working Group of PAGES), 2016: "Interglacials of the last 800,000 years", in: *Rev. Geophys.,* 54: 162–219; https://doi.org/10.1002/2015RG000482.

PAGES 2 K Consortium, 2013: "Continental-scale temperature variability during the past two millennia", in: *Nat. Geosci.,* 6: 339–346; <https://doi.org/10.1038/ngeo1797<.

Parrenin, F.; Masson-Delmotte, V.; Kohler, P.; Raynaud, D.; Paillard, D.; Schwander, J.; Barbante, C.; Landais, A.; Wegner, A.; Jouzel, J., 2013: "Synchronous change of atmospheric CO_2 and Antarctic temperature during the last deglacial warming", in: *Science,* 339, 6123: 1060–1063; https://doi.org/10.1126/science.1226368.

Payne, J.L. et al., 2008: "Two-phase increase in the maximum size of life over 3.5 billion years reflects biological innovation and environmental opportunity", in: *Proc. Natl. Acad. Sci. USA,* 106: 24–27; https://doi.org/10.1073/pnas.0806314106.

Petit, J.R. et al., 1999: "Climate and atmospheric history of the past 420,000 years from the Vostok ice core, Antarctica", in: *Nature,* 399: 429–436; https://doi.org/10.1038/20859.

Pillans, B.; Naish, T., 2004: "Defining the quaternary", in: *Quat. Sci. Rev.,* 23: 2271–2282; https://doi.org/10.1016/j.quascirev.2004.07.006.

Pimm, S.L.; Jenkins, C.N.; Abell, R.; Brooks, T.M.; Gittleman, J.L.; Joppa, L.N.; Raven, R.H.; Roberts, C.M.; Sexton, J.O., 2014: The biodiversity of species and their rates of extinction, distribution, and protection", in: *Science,* 344, 6187: 987; https://doi.org/10.1126/science.124 6752.

Polunin, N.; Grinevald, J., 1988: "Vernadsky and biospheral ecology", in: *Environ. Conservation,* 75, 2: 117–123; https://doi.org/10.1017/S0376892900028915.

Rambler, M.B.; Margulis, L.; Fester, R. (eds.), 1989: "*Global Ecology: Toward a Science of the Biosphere* (Boston, Mass.: Academic Press).

Raupach, M.R.; Gloor, M.; Sarmiento, J.L.; Canadell, J.G.; Frölicher, T.L.; Gasser, T.; Houghton, R.A.; Le Quere, C.; Trudinger, C.M. 2014: "The declining uptake rate of atmospheric CO_2 by land and ocean sinks", in: *Biogeosciences,* 11: 3453–3475; <https://doi.org/10.5194/bg-11-3453-2014.

Remane, J. et al., 1996: "Revised guidelines for the establishment of global chronostratigraphic standards by the International Commission on Stratigraphy (ICS), in: *Episodes,* 19,3: 77–81.

Revkin, A., 1992: *Global Warming: Understanding the Forecast* (New York: Abbeville Press).

Rockström, J. et al., 2009: "A safe operating space for humanity", in: *Nature,* 461: 472–475; https://doi.org/10.1038/461472a.

Ruddiman, W.F., 2013: "The Anthropocene", in: *Annu. Rev. Earth Planet. Sci.,* 41: 45–68; https://doi.org/10.1146/annurev-earth-050212-123944.

Ruddiman, W.F. et al., 2015: "Late Holocene climate: natural or anthropogenic?", in: *Rev. Geophys.,* 54: 93–118; https://doi.org/10.1002/2015RG000503.

Salvador, A. (ed.), 1994: "*International Stratigraphic Guide—A Guide to Stratigraphic Classification, Terminology and Procedure,* 2nd ed. (Boulder, Colo: International Union of Geological Sciences and the Geological Society of America).

Scheffer, M., 2009: *Critical Transitions in Nature and Society* (Princeton, N. J.: Princeton Univ. Press).

Scheffer, M.; Carpenter, S., 2003: "Catastrophic regime shifts in ecosystems: linking the theory to observation", in: *Trends Ecol. Evol.,* 18: 656; https://doi.org/10.1016/j.tree.2003.09.002.

Schellnhuber, H.J. (1998: "Discourse: Earth System analysis: the scope of the challenge, in *Earth System Analysis,* edited by H.J. Schellnhuber and V. Wetzel, pp. 3 -195, Springer-Verlag, Berlin, Heidelberg and New York.

Schellnhuber, H.J., 1999: "'Earth System' analysis and the second Copernican revolution", *Nature,* 402: C19–C23; https://doi.org/10.1038/35011515.

Schellnhuber, H.J. 2009: "Tipping elements in the Earth System", in: *Proc. Natl. Acad. Sci. USA,* 106, 49: 20561–20563; https://doi.org/10.1073/pnas.0911106106.

Schellnhuber, H.J.; Rahmstorf, S.; Winkelmann, R., 2016: "Why the right climate target was agreed in Paris", in: *Nat. Clim. Change,* 6, 653; https://doi.org/10.1038/nclimate3013.

Seitzinger, S.P., et al., 2015: "International Geosphere-Biosphere Programme and Earth System science: three decades of co-evolution", in: *Anthropocene,* 12: 3–16; https://doi.org/10.1016/j.ancene.2016.01.001.

Sluijs, A.; Bowen, G.J.; Brinkhuis, H.; Lourens, L.J.; Thomas, E., 2007: "The Palaeocene-Eocene Thermal Maximum super greenhouse: biotic and geochemical signatures, age models and mechanisms of global change", in: Williams, M.; Haywood, A.M.; Gregory, F.J.; Schmidt, D.N. (eds.): *Deep Time Perspectives on Climate Change: Marrying the Signal From Computer Models and Biological Proxies,* Special Publications (London, U. K.: The Geological Society, The Micropalaeontological Society): 323–347.

Sluijs, A. et al., 2008: "Eustatic variations during the Paleocene-Eocene greenhouse world", in: *Paleoceanography,* 23: PA4216; https://doi.org/10.1029/2008PA001615.

Smil, V., 2002: *The Earth's Biosphere: Evolution, Dynamics, and Change* (Cambridge, Mass.: The MIT Press).

Solomon, S.; Plattner, G.-K.; Knutti, R.; Friedlingstein, P., 2009: "Irreversible climate change due to carbon dioxide emissions", in: *Proc. Natl. Acad. Sci. USA,* 106: 1704–1709; https://doi.org/10.1073/pnas.0812721106.

Stanley, S.M., 1993: *Exploring Earth and Life through Time* (New York: W.H. Freeman).

Stanley, S.M.; Luciaz, J.A., 2014: *Earth System History,* 4th ed. (New York: Macmillan).

Steffen, W., 2013: "Commentary: Paul J. Crutzen and Eugene F. Stoermer: 'The Anthropocene' (2000)", in: Robin, L.; Sorlin, S.; Warde, P. (eds.): *The Future of Nature,* (New Haven, Conn. – London: Yale Univ. Press): 486–490.

Steffen, W. et al., 2004: *Global Change and the Earth System: A Planet under Pressure,* The IGBP Book Series (Berlin – Heidelberg – New York: Springer-Verlag).

Steffen, W.; Crutzen, P.J.; McNeill, J.R., 2007: "The Anthropocene: are humans now overwhelming the great forces of Nature?", in: *Ambio,* 36: 614–621; https://doi.org/10.1579/0044-7447(2007)36[614:TAAHNO]2.0.CO;2.

Steffen, W.; Broadgate, W.; Deutsch, L.; Gaffney, O.; Ludwig, C., 2015a: "The trajectory of the Anthropocene: The Great Acceleration", in: *Anthropocene Rev.,* 2, 1: 81–98; https://doi.org/10.1177/2053019614564785.

Steffen, W. et al., 2015b: "Planetary boundaries: guiding human development on a changing planet", in: *Science,* 347, 6223: 736; https://doi.org/10.1126/science.1259855.

Stein, W.E.; Mannolini, F.; Hernick, L.V.; Landing, E.; Berry, C.M., 2007: "Giant cladoxylopsid trees resolve the enigma of the Earth's earliest forest stumps at Gilboa", in: *Nature,* 446, 7138: 904–907; https://doi.org/10.1038/nature05705.

Stewart, W.N.; Rothwell. G.W., 1993: *Paleobotany and the Evolution of Plants,* 2nd ed., (Cambridge: Cambridge Univ. Press).

Summerhayes, C.P.; 2015: *Earth's Climate Evolution* (Oxford: Wiley/Blackwell).

Syvitski, J. P.M.; Kettner, A.J., 2011: "Sediment flux and the Anthropocene", in: *Phil. Trans. Roy. Soc. Lond. A,* 369: 957–997.

Tilman, D.; Reich, P.B.; Knops, J.M., 2006: "Biodiversity and ecosystem stability in a decade-long grassland experiment", in: *Nature,* 441, 7093: 629–632; <https://doi.org/10.1038/nature04742.

UN (United Nations General Assembly), 2015: "Transforming our world: the 2030 Agenda for Sustainable Development. Resolution adopted by the General Assembly on 25 September 2015, A/RES/70/1, 21 October 2015"; at: <http://www.un.org/ga/search/view_doc.asp?symbol=A/RES/70/1&Lang=E.

UNFCCC (United Nations Framework Convention on Climate Change), 2010: "The Cancun Agreements"; at: http://cancun.unfccc.int/cancun-agreements/significanceof-the-key-agreements-reached-at-cancun/.

UNFCCC (United Nations Framework Convention on Climate Change), 2015: "Conference of the Parties: Durban Platform for Enhanced Action (decision 1/CP.17) Adoption of a protocol, another legal instrument, or an agreed outcome with legal force under the Convention applicable to all Parties"; at: http://www.cop21.gouv.fr/wpcontent/uploads/2015/12/l09r01.pdf.

Valdes, P.J., 2011: "Built for stability", in: *Nat. Geosci.,* 4: 414–416; https://doi.org/10.1038/nge o1200.

Vernadsky, V.I., 1924: *La Géochimie* (Paris: Librairie Felix Alcan, Nouvelle Collection scientifique).

Vernadsky, V.I., 1929: *La Biosphère* (Paris: Librairie Felix Alcan, Nouvelle Collection scientifique), second revised and expanded edition.

Vernadsky, V.I., 1945: "The Biosphere and the Noösphere", in: *Am. Sci.,* 33,1: 1–12.

Vernadsky, V.I., 1998: *"The Biosphere,* foreword by Lynn Margulis et al., introduction by Jacques Grinevald, translated by David Langmuir, revised and annotated by Mark A. S. McMenamin; A Peter Nevraumont Book (New York: Copernicus/Springer-Verlag).

Vidas, D., 2011: "The Anthropocene and the international law of the sea", in: *Phil. Trans. Roy. Soc. Lond. A,* 369: 909–925; https://doi.org/10.1098/rsta.2010.0326.

Vidas, D.; Fauchald, O.K.; Jensen, O.; Tvedt, M.W., 2015: "International law for the Anthropocene? Shifting perspectives in regulation of the oceans, environment and genetic resources", in: *Anthropocene,* 9: 1–13; https://doi.org/10.1016/j.ancene.2015.06.003.

Wacey, D.; Kilburn, M.R.; Saunders, M.; Cliff, J.; Brasier, M.D., 2011: "Microfossils of sulphur-metabolizing cells in 3.4-billion-year-old rocks of Western Australia", in: *Nat. Geosci.,* 4: 698–702; https://doi.org/10.1038/ngeo1238.

Waters, C.N. et al., 2016: "The Anthropocene is functionally and stratigraphically distinct from the Holocene", in: *Science,* 351, 6269: 137; https://doi.org/10.1126/science.aad2622.

WBGU (Schellnhuber, H.J.; Messner, D.; Leggewie, C.; Leinfelder, R.; Nakicenovic, N.; Rahmstorf, S.; Schlacke, S.; Schmid, J.; Schubert, R. [eds.]), 2011: *World in Transition—A Social Contract for Sustainability* (Berlin: German Advisory Council on Global Change (WBGU); at: http://www.wbgu.de/en/flagship-reports/fr-2011-a-social-contract/.

Wellman, C.; Gray, J., 2000: "The microfossil record of early land plants", in: *Phil. Trans. Roy. Soc. Lond. B,* 355: 707–732; https://doi.org/10.1098/rstb.2000.0612.

Wellman, C.; Osterloff, P.L.; Mohiuddin, U., 2003: "Fragments of the earliest land plants", in: *Nature,* 425: 282–285; https://doi.org/10.1038/nature01884.

White, J.W.C. et al., 2013: *Abrupt Impacts of Climate Change, Anticipating Surprises* (Washington, D. C.: National Academies Press).

Williams, M.; Zalasiewicz, J.; Haff, P.K.; Schwägerl, C.; *Barnosky,* A.D.; *Ellis,* E.C., 2015: "The Anthropocene biosphere", in: *Anthropocene Rev.,* 2, 3: 196–219; https://doi.org/10.1177/205301 9615591020.

Williams, M. et al., 2016: "The Anthropocene: a conspicuous stratigraphical signal of anthropogenic changes in production and consumption across the biosphere", in: *Earth's Future,* 4: 34–53; https://doi.org/10.1002/2015EF000339.

Winguth, A.M.; Thomas, E.; Winguth, C., 2012: "Global decline in ocean ventilation, oxygenation, and productivity during the Paleocene-Eocene thermal maximum: implications for the benthic extinction", in: *Geology,* 40, 3: 263–266; https://doi.org/10.1130/G32529.1.

Wolfe, A.P. et al., 2013: "Stratigraphic expressions of the Holocene-Anthropocene transition revealed in sediments from remote lakes", in: *Earth Sci. Rev.,* 116: 17–34; https://doi.org/10.1016/j.earscirev.2012.11.001.

Wolff, E.W., 2011: "Greenhouse gases in the Earth system: a palaeoclimate perspective", in: *Phil. Trans. Roy. Soc. Lond. A,* 369, 2133–2147; https://doi.org/10.1098/rsta.2010.0225;pmid: 21502180.

Zachos, J.C.; Dickens, G.R.; Zeebe, R.E., 2008: "An early Cenozoic perspective on greenhouse warming and carbon-cycle dynamics", in: *Nature,* 451: 279–283; https://doi.org/10.1038/nature 06588.

Zalasiewicz, J.; Williams, M., 2012: *The Goldilocks Planet—The Four Billion Year Story of Earth's Climate* (Oxford: Oxford Univ. Press).

Zalasiewicz, J.; Williams. M., 2014: "The Anthropocene: a comparison with the Ordovician-Silurian boundary", in: *Rendiconti Lincei—ScienzeFisichee Naturali,* 25, 1: 5–12; <https://doi.org/10.1007/s12210-013-0265-x.

Zalasiewicz, J.; Williams, M., 2016: "Climate change through Earth's history", in: Letcher, T.M. (ed.): *Climate Change: Observed Impacts on Planet Earth* (Amsterdam: Elsevier): 3–17.

Zalasiewicz, J. et al., 2008: "Are we now living in the Anthropocene?", in: *GSA Today,* 18: 4–8; https://doi.org/10.1130/GSAT01802A.1.

Zalasiewicz, J.; Cita, M.B.; Hilgen, F.; Pratt, B.R.; Strasser, A.T.J.; Weissert, H., 2013: "Chronostratigraphy and geochronology: a proposed realignment", in: *GSA Today,* 23, 3: 4–8, https://doi.org/10.1130/GSATG160A.1.

Zalasiewicz, J., et al., 2015: "When did the Anthropocene begin? A mid-twentieth century boundary level is stratigraphically optimal", in: *Quaternary Int.,* 383: 196–203; https://doi.org/10.1016/j.quaint.2014.11.045.

Zalasiewicz, J. et al., 2016: "The geological cycle of plastics and their use as a stratigraphic indicator of the Anthropocene", in: *Anthropocene,* 13: 4–17; https://doi.org/10.1016/j.ancene.2016.01.002.

Zeebe, R.E.; Ridgwell, A.; Zachos, J.C., 2016: "Anthropogenic carbon release rate unprecedented during the past 66 million years", in: *Nat. Geosci.,* 9: 325–329; https://doi.org/10.1038/ngeo2681.

Chapter 18
Was Breaking the Taboo on Research on Climate Engineering via Albedo Modification a Moral Hazard, or a Moral Imperative? (2016/2017)

M. Lawrence and Paul J. Crutzen

Abstract The topic of increasing the reflectivity of the Earth as a measure to counteract global warming has been the subject of high-level discussions and preliminary research since several decades, though prior to the early 2000s there was only very limited research on the topic. This changed in the mid-2000s, particularly following the publication of a special section of *Climatic Change* with a lead paper by Crutzen (2006), which posited the idea of stratospheric aerosol injections as a possible solution to a policy dilemma. The discussions around the publication of Crutzen (2006) demonstrated how contentious the topic was at that time. The special section of *Climatic Change* contributed to breaking the 'taboo' on albedo modification research that was perceived at that time, and scientific publications on the topic have since proliferated, including the development of several large national and international projects, and the publication of several assessment reports over the last decade. Here we reflect on the background and main conclusions of the publications in 2006, the developments since then, and on some of the main developments over the next decade that we anticipate for research and dialogue in support of decision-making and policy development processes.

This text was first published as: Lawrence, M.G.; Crutzen, P.J., 2016/2017: "Was Breaking the Taboo on Research on Climate Engineering via Albedo Modification a Moral Hazard, or a Moral Imperative?", in: *Earth's Future,* 5: 136–143; https://doi.org/10.1002/2016EF000463. This is an open access article under the terms of the Creative Commons Attribution-NonCommercial-NoDerivs License, which permits use and distribution in any medium, provided the original work is properly cited, the use is non-commercial and no modifications or adaptations are made. The preparation of this article was completed by ML at the Institute for Advanced Sustainability Studies, Potsdam, Germany, which is funded by the Federal Ministry of Education and Research, Germany, the Ministry of Science, Research and Culture, Brandenburg, Germany and by PC at the Max Planck Institute for Chemistry, Mainz, Germany. The authors would like to thank many colleagues in the climate engineering research community who have entered into countless discussions with the authors on this topic surrounding the development of the 2006 papers and the years since then. They are also grateful for the very thoughtful referee comments, which helped to focus the manuscript and also to bring in a few new aspects. The authors declare no conflict of interest, except that MGL works with Miranda Boettcher and Stefan Schafer who also serve as the publication's editors. Therefore, an independent third party was chosen by the journal to lead the decision-making editorial process of this paper.

18.1 Introduction

Anthropogenic climate change poses vast and rapidly growing societal challenges, as described by the IPCC in their most recent assessment reports (IPCC 2013, 2014a, b), While efforts to address climate change have sensibly primarily focused on mitigation and adaptation, over the last decades increasing attention has being given to various other possibilities, including

1. actively removing carbon dioxide from the atmosphere, and
2. cooling the Earth by either reflecting more sunlight back to space by various means, or by modifying cirrus cloud properties to allow more terrestrial infrared radiation to escape to space.

These two types of proposed techniques, though notably different, are often grouped together under terms like 'climate engineering', 'climate intervention', or 'geoengineering'. These techniques are rarely proposed as standalone replacements for mitigation, rather as supplemental possibilities to reduce the overall severity and impacts of climate change.

Here we focus on the second type of intervention noted above, which is commonly called either 'albedo modification' (where the planetary albedo is defined as the fractional amount of sunlight reflected back to space) or 'solar radiation management', or simply 'radiation management'. We employ the former term, which is also used in the two most recent major assessments. A wide range of techniques has been proposed for increasing the planetary albedo, ranging from painting surfaces white to placing mirrors in orbit between the Earth and the sun. The technique which has been most discussed is the possibility of injecting aerosol particles or their precursors into the stratosphere. This specific idea dates back over 40 years (Budyko 1974, 1977), and for the more general issue of planetary albedo modification, high-level discussions and preliminary research date back to at least the early 1960s (Keith 2000; Fleming 2010). However, prior to the early 2000s, research on this and other techniques for modifying the planetary albedo was very limited (see Fig. 18.1), despite efforts to counteract this by raising awareness in the broader academic community of the issue and its complex facets. For instance, a special section of *Climatic Change* was published in 1996, featuring six papers by prominent researchers covering the scientific, economic, political, and ethical dimensions of the issue (Bodansky 1996; Dickinson 1996; Jamieson 1996; Marland 1996; Schelling 1996; Schneider 1996).

Figure 18.1 shows how this changed in the 2000s, with a very rapid increase in the number of publications during the course of the decade. In the middle of this, and perceived as one of the key factors leading to the more open discourse was the publication of a second special section in *Climatic Change* in 2006, 10 years after the first. This special section again included six papers (Bengtsson 2006; Cicerone 2006; Crutzen 2006; Kiehl 2006; Lawrence 2006; MacCracken 2006) (noteworthy is that this time all were natural scientists), in the form of editorial essays covering roughly the same wide range of issues as in the first special section. Nearly all the authors concluded that it is sensible to conduct research on this topic, though all

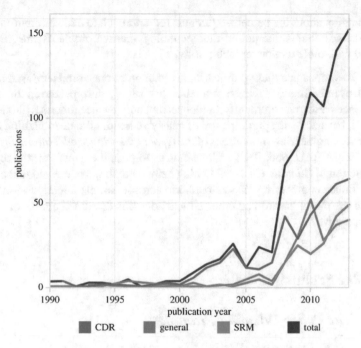

Fig. 18.1 Trends in scientific publications on climate engineering (number of publications per year, indexed in *Web of Science*). *Source* Oldham et al. (2014)

expressed considerable caution, for various reasons. Several authors also noted a sense of 'taboo' toward research which had been apparent in the broader climate and atmospheric science communities; for example, Cicerone (2006) noted: "I am aware that various individuals have opposed the publication of Crutzen's paper, even after peer review and revisions, for various and sincere reasons that are not wholly scientific", and Lawrence (2006) noted that "... serious scientific research into geoengineering possibilities, such as discussed in the publications by Crutzen (2006) and Cicerone (2006), is not at all condoned by the overall climate and atmospheric chemistry research communities".

This sense of taboo was based on a range of arguments against research on albedo modification that have been raised by the broader scientific community, including:

1. the so-called moral hazard issue, that is, the possibility that research on climate engineering could be perceived as an implicit legitimisation, and thus reduce the motivation for mitigating anthropogenic emissions;
2. the concern that reducing temperatures by albedo modification could distract from other impacts of a fossil-fuel-based economy and the resulting CO_2 emissions, such as ocean acidification;
3. the 'slippery slope' concern that research into understanding the potential effectiveness could cascade toward the development and deployment of the techniques under investigation; and

4. contention about the perceived 'techno-fix' approach to address environmental challenges, that is, the notion that technology-caused problems can simply be fixed with more and better technology.

In the 10 years since these publications, although climate engineering remains a very controversial issue, the sense of a taboo has largely disappeared in the broader Earth sciences research community, evidenced not only by the numerous publications (Fig. 18.1), but also by the participation of nearly all major climate modelling centres in the climate engineering simulations of the *Geoengineering Model Intercomparison Project* (GeoMIP, see Sect. 18.3). This leads us to pose the question: was the 2006 special section in *Climatic Change*—which helped break the perceived taboo on the topic of climate engineering via stratospheric aerosol particle injections—at risk of leading to the moral hazard and other ethical concerns discussed above, or was it a moral imperative?

18.2   The Framing in 2006

18.2.1   The Policy Dilemma Framing

Although the essay by Crutzen (2006) was quite influential, the primary line of argumentation that was employed has largely been forgotten, and has hardly been used since then. At that time, which was during the run-up to the fourth assessment report of the IPCC (2007), with the rapidly growing awareness and acceptance of the seriousness of global warming and its anticipated impacts, policy makers were hesitant to do anything that would exacerbate global warming. At the same time, there was a growing realisation of the significant role that lower tropospheric reflecting aerosol particles played in cooling the climate. Reflecting particles are composed of sulphate, nitrate, organic carbon, and other compounds; only black carbon, a primary component of soot, is strongly absorbing, and has a significant warming effect on the climate. The reflecting particles not only directly reflect sunlight, but they also affect clouds, generally causing them to also be more reflective of sunlight. Overall, the reflecting particles cool the climate on average by a similar magnitude to the warming by carbon dioxide (nearly 1.5 W m^{-2}), and thus they partially 'mask' global warming. Reducing the emissions of reflecting particles and their precursor substances (such as sulfur dioxide, from which sulphate particles form) would thus partially 'de-mask' the cooling effect, exacerbating global warming. However, at the same time, these particles contribute substantially to negative impacts on human health and ecosystems (as well as reducing visibility, damaging monuments and other structures, etc.).

This contrast between beneficial and harmful effects was the main premise for Crutzen (2006): "This creates a dilemma for environmental policy makers, because the required emission reductions of SO_2, and also anthropogenic organics (except black carbon), as dictated by health and ecological considerations, add to global warming and associated negative consequences, such as sea level rise, caused by the greenhouse gases." Looking back, it is now known that the dilemma was actually considerably worse than was realised at that time, since many of the health impacts of air pollution were not yet known then. Crutzen (2006) relied on studies available at that time that indicated air pollution contributed toward approximately 500,000 premature deaths annually. Studies since then have shown that that number was likely underestimating the impacts by about a factor of 10, with current estimates being that air pollution contributes to about 6–7 million premature deaths annually (*United Nations Environmental Programme* [UNEP] 2011; Lelieveld et al. 2015). Air pollution is now recognised as the single largest environmental cause of premature death (Lim et al. 2012). But even with the factor of 10 underestimate of the health impacts in 2006, the situation was sufficient for Crutzen (2006) to propose that "... although by far not the best solution, the usefulness of artificially enhancing earth's albedo and thereby cooling climate by adding sunlight reflecting aerosol in the stratosphere (Budyko 1977; *National Academy of Sciences* (NAS), 1992) might again be explored and debated as a way to defuse the Catch-22 situation just presented and additionally counteract the climate forcing of growing CO_2 emissions".

This main aspect of Crutzen (2006)—considering the contrast between the health impacts and the 'masking' of global warming, and thus the concomitant policy dilemma—was hardly addressed in the accompanying editorials published in the *Climatic Change* special section. It has also hardly been used since then as a line of argumentation in the discourse around albedo modification (we are not aware of it being brought into any substantial discussions in at least the past 5 years). The impacts of air pollutants on climate and on health and ecosystems and other sectors are still getting increasing international attention, for example, in the context of the so-called *short-lived climate-forcing pollutants*, or SLCPs (Ramanathan/Feng 2009; Shindell et al. 2012; Schmale et al. 2014), which has led to the development of a rapidly growing international consortium to address these issues, the *Climate and Clean Air Coalition* http://www.ccacoalition.org. In this context, the attention has turned primarily toward the opportunities for rapidly reducing climate impacts by reducing the emissions of the short-lived warming pollutants. The role of the cooling pollutants is acknowledged, but has generally left the focus of the discussions. This is especially due to results showing that the warming pollutants are apparently playing a significant role in climate change (Ramanathan/Feng 2009), with a recent major review (Bond et al. 2013) indicating the warming effect is perhaps much larger than previously acknowledged in many studies. Warming short-lived pollutants thus present a much more appealing 'solution' to the policy dilemma discussed by Crutzen (2006) than consideration of albedo modification: if the warming pollutants can be reduced at the same rate or faster than the cooling pollutants, then the impacts on health will be substantially reduced, and the net impact on the climate would remain neutral or cooling, rather than warming. Thus, in the years since the 2006 special section, the discussions of climate-cooling and climate-warming pollutants have ended up usually being kept very distinct from discussions of albedo modification.

18.2.2 The Research Needs and Hazards Framing

However, while the article by Crutzen (2006) did not end up contributing directly to solving the particular policy dilemma that was specifically addressed, it did contribute substantially to opening up the other part of the discourse that it was focused on, which was already pointed out by Dickinson (1996), 10 years before in the previous special section: "As global greenhouse warming continues to intensify, it is likely that demands to employ technologies of climate engineering will become increasingly insistent." In this context, Crutzen (2006) posited: "Given the grossly disappointing international political response to the required greenhouse gas emissions ... research on the feasibility and environmental consequences of climate engineering of the kind presented in this paper, which might need to be deployed in future, should not be tabooed."

It is interesting to ask would a manuscript of the nature of Crutzen (2006) nevertheless have been written, even if the policy dilemma discussed in the previous section had not existed? Given the recognition already by Dickinson (1996), and by Crutzen (2006) and many others of the impending situation of increasing political relevance of alternate methods for reducing climate change, beyond mitigation (and removing carbon dioxide), it seems probable that such a manuscript would have eventually been written. This is more relevant than ever with the unanimous agreement in the Paris Agreement to try to keep global temperature increase well below 2 °C. In light of that, the motivation to write such a paper would still exist if a lasting, broad taboo on the topic of albedo modification were evident within the research community, since breaking that taboo would help ensure that political discussions and decisions could be based on good scientific knowledge, rather than an unnecessarily high degree of uncertainty and speculation.

Whether or not to break the perceived taboo at that time—and how to go about this responsibly—was the primary focus of the five commentaries that were published along with Crutzen (2006). These give a cross-section of some of the key perspectives that were shaping the discourse at that time. We thus summarise here some of the main points in each of these commentaries, in the order of publication.

Cicerone (2006) was primarily concerned with supporting research, and in the process making sure to adequately address the 'moral hazard' issue by imposing a moratorium on implementation, so that research could be conducted in a framework that would be intended to decouple it from concerns about premature implementation or reducing motivation for mitigation:

> Here, I write in support of his (Crutzen's) call for research on geoengineering and propose a framework for future progress in which supporting and opposing viewpoints can be heard and incorporated. I also propose that research on geoengineering be considered separately from actual implementation, and I suggest a path in that direction.

Kiehl (2006) was also in support of research, but with a substantial concern about whether or not we could adequately judge the quality of model simulations used as a basis for that research:

For this reason, I support Crutzen's argument that more detailed and comprehensive modelling studies be carried out with regards to experiments. But my concern is that all models have their limitations (e.g. note the inability of models to predict the appearance of the Antarctic ozone hole before it was observed). When will we know a model is 'good enough' to go out and perform a real experiment?

Bengtsson (2006) was the most sceptical of all about prospects of future research, not so much from the perspective of objecting to research on the topic due to the moral hazard or other concerns about the impacts of research, but rather because of a strong notion that albedo modification will likely prove to be ineffective, and even if it did work, would present major societal challenges, in particular maintaining it over the long timescales that carbon dioxide would remain in the atmosphere:

So in conclusion, I do consider it more feasible to succeed in solving the world's energy problem, which is the main cause to the present concern about climate change, than to successfully manage a geo-engineering experiment on this scale and magnitude, which even if it works is unable to solve all problems with the very high concentration of greenhouse gases in the atmosphere.

MacCracken (2006), in contrast, was relatively strongly in support of research, with a primary focus more on alleviating regional than global climate change effects (e.g., reducing polar ice sheet melting):

... continued and increasing emissions of greenhouse gases merit very serious control efforts. In addition, however, as Crutzen suggests, research would be a prudent option, both on the globally uniform approach to geoengineering proposed by Crutzen and, in my view, on dealing with specific changes, some of which have already begun, in a way that might benefit the world community of nations.

Finally, Lawrence (2006) focused on the responsibility of the scientific community in the face of a potentially urgent future political need for knowledge:

... we may eventually reach the state of extreme climate change where the overall international sentiment is in favour of applying geoengineering. If we do not conduct careful research now, we will not be prepared to advise politicians on how to best approach large-scale geoengineering applications—including providing sound information on the various risks involved, and on which ideas should not be pursued further.

Going beyond this impression of a fundamental moral imperative is the concern about potential 'covert', 'clandestine', or 'unilateral' applications of climate engineering, which makes research even more imperative. Nevertheless, concern about the moral hazard and other risks of climate engineering research were noted to be significant, and in order to address these it was concluded that "... the pertinent question is not 'To speak or not to speak?', but rather 'How to speak?' about geoengineering", with a focus on care in using accurate scientific terminology, and explicit discussion of values and norms behind assessments and recommendations.

18.3 Resulting Developments

Since shortly after 2006, research and scientific publications on the topic of albedo modification have proliferated (see Fig. 18.1). The 2006 *Climatic Change* special section was soon followed by the first major assessment report, conducted by the Royal Society (Shepherd et al. 2009), as well as by a plethora of further assessment reports from various national and international perspectives and focused on various aspects of the topic (e.g., Blackstock et al. 2009; Gordon 2010; UK House of Commons Science and Technology Committee 2010; Ginzky et al. 2011; Rickels et al. 2011; Bodle et al. 2013; Caviezel/Revermann 2014; McNutt et al. 2015a, b; Schafer et al. 2015; Klepper et al. 2016). The topic has also been addressed by all three Working Groups in the IPCC's Fifth Assessment Report (IPCC 2013, 2014a, b).

Numerous national and international projects have also developed. The majority of this work has been model based. Experimental work directly on albedo modification techniques and their effects and side effects has been very limited thus far, although considerable work has gone into conceiving possible future field experiments (Keith et al. 2014). Among the modelling studies, we would particularly like to highlight GeoMIP (Kravitz et al. 2011), and a smaller predecessor, *Implications and Risks of Novel Options to Limit Climate Change* (IMPLICC) (Schmidt et al. 2012), which was funded by the European Commission. The GeoMIP simulations build on CMIP, the Coupled Model Intercomparison Project, which provides the community simulations that are primarily employed in the IPCC Assessment Reports. The first phase of GeoMIP corresponded to the CMIP5 simulations for the Fifth Assessment Report, with nearly all of the CMIP5 models having completed the GeoMIP simulations by now. A second phase of GeoMIP is now being conducted corresponding to CMIP6 for the Sixth IPCC Assessment Report. IMPLICC and GeoMIP have contributed substantially to the understanding of how to set up useful CE simulations that are comparable across a wide array of models, helping to improve the robustness of the model results, and giving an initial sense of uncertainty in these results. The GeoMIP project shows the degree to which research on albedo modification has become 'normalised' among the scientific research community, in contrast to having been perceived as a largely tabooed (or at least broadly avoided) topic prior to 2006, as discussed above. Interestingly, the GeoMIP project also, at least partly, addresses all of the primary concerns brought out by the articles published in the 2006 *Climatic Change* special section, corresponding to the main points noted above:

1. It provides for a substantial research platform and scientific exchange forum which is kept distinct from the research and development toward actual implementation.
2. It supports the call by Kiehl (2006) for 'detailed and comprehensive modelling studies', while providing a forum for comparatively assessing model results as a valuable (though in itself inadequate) step toward assessing the model limitations and when models will be 'good enough'.

3.   It shows, based on all simulations thus far, and in contrast to some initial expectations, that if the technology for aerosol injection or other forms of increasing the planetary albedo were to actually work, then albedo modification would indeed be effective in reducing the global mean surface temperatures, and although there would be significant regional variability in the climate response, it appears that some techniques including stratospheric aerosol injection would be able to reduce the risk due to climate change in most regions of the world simultaneously. In addition, the GeoMIP protocol includes simulations to examine the effects of a potential, unintended abrupt termination of albedo modification, which was a major source of concern expressed by Bengtsson (2006).
4.   It provides a forum in the second phase for encouraging the development community simulations of new focal topics, including those focused especially on alleviating regional climate change effects.
5.   It encourages 'careful research now' … 'in order to' … "be prepared to advise politicians on how to best approach large-scale geoengineering applications— including providing sound information on the various risks involved, and on which ideas should not be pursued further", as recommended by Lawrence (2006).

On the whole, the GeoMIP simulations and numerous other separate studies conducted by the scientific modelling community in the last 10 years have given an initial impression which has been summarised by Irvine et al. (2014), who concluded that "Although the evidence from model studies about the impacts of SRM geoengineering is, at present, limited, the initial evidence broadly indicates that SRM deployed to cool the climate could potentially reduce many of the physical impacts of climate change as well as the risk of crossing tipping points", and furthermore that the GeoMIP results make it clear that "SRM is no panacea; it would introduce new risks and would shift the overall burden of risks, which might pose substantial political problems …," and that "… to minimise the risks posed by climate change, mitigation will need to be pursued vigorously".

18.4   Future Developments

The last decade of intensifying investigation of albedo modification by the Earth system sciences community has thus provided very valuable information—but also challenging in many ways. First, the numerous studies conducted on the scientific and engineering aspects have not yet found a major fundamental scientific or engineering reason why stratospheric aerosol injections would definitely not work to cool the climate (albeit regionally unevenly, but—according to the model studies thus far— likely feasible in a way that could reduce the overall impacts and risks due to climate change nearly everywhere). Second, no side effects that would be a real 'showstopper' have been uncovered yet. This is quite in contrast to the situation for some other proposed forms of climate engineering, especially ocean iron fertilization, for which the effectiveness and potential has been brought more and more into question through field experiments, and numerous potentially very detrimental side effects have been uncovered (Lawrence et al. 2008). Of course, still uncovering major side effects of albedo modification techniques cannot be ruled out, as was already pointed out by Crutzen (2006); "The chances of unexpected climate effects should not be underrated, as clearly shown by the sudden and unpredicted development of the antarctic ozone hole." There is still much to be learned from model studies, and the second phase of GeoMIP and many other independent studies are expected to bring us further important insights. However, to us, going beyond further model studies, these results of the last decade point toward two main future developments:

1. First, a broad, well-informed sociopolitical dialogue is needed to determine whether humanity as a whole is likely to actually someday provide broad support for the pursuit of full-fledged climate engineering—and if so, in what forms, for what purposes, and for how long—or if it will be a topic like human cloning or genetic engineering, which is open for general discussion, but under a broad societal taboo for any form of experimental research or steps toward realisation.
2. Second, depending partly on how the societal dialogue develops, and in support of a better information basis for what will potentially be a protracted inter-national political debate, it is likely that the scientific community will pursue field experimentation to help clarify many uncertainties; if so, then an adequate public funding and governance framework (in soft and hard forms) urgently needs to be developed (Morgan et al. 2013; Parson/Keith 2013; Schafer et al. 2013; Victor et al. 2013).

Here we posed the question: was breaking the perceived taboo on the topic of stratospheric aerosol particle injections at risk of leading to the moral hazard and other ethical concerns, or was it a moral imperative? The Paris Agreement has provided evidence that the overall international political intention to reduce greenhouse gas emissions has not diminished, and that while there may be an overoptimism about the possibilities of removing large amounts of carbon dioxide from the atmosphere via various means, there is no evidence of hope being built on employing albedo modification to achieve the Paris Agreement goals. Thus we generally conclude that the moral hazard risk has thus far largely been avoided on an international political scale. There is also no clear evidence that we are aware of that the other ethical concerns (e.g., the 'slippery slope' and 'techno-fix' arguments) are manifesting in the scientific literature or in climate policy. However, given the balance of results of model studies over the last decade, as discussed in the previous sections, and the challenging directions that this implies both for future research and also for sociopolitical aspects, especially public perception and the development of good governance principles, we have to conclude that the overall verdict is still out. The responsibility still resides with the scientific community to conduct research and engage in the broader dialogue in a responsible way, so that whatever the outcome, historians will hopefully look back and conclude that it was indeed of value—and in that sense a moral imperative—to begin carefully investigating this topic at this point in our history. Perhaps this will already be clear by the time of the next special section like this one, which, following those in 1996 and 2006, should be due in 2026.

References

Bengtsson, L. 2006: "Geo-engineering to confine climate change: Is it at all feasible?", in: *Clim. Change,* 77,3–4: 229–234.
Blackstock, J.J.; Battisti, D.; Caldeira, K.; Eardley, D.E.; Katz, J.I.; Keith, D.W.; Aristides, A.N.P.; Schrag, D.P.; Socolow, R.H.; Koonin, S.E., 2009: *Climate Engineering Responses to Climate Emergencies* (Santa Barbara, Calif.: Novim).
Bodansky, D., 1996: "May we engineer the climate?", in: *Clim. Change*, 33: 309–321; https://doi.org/10.1007/BF00142579.
Bodle, R.; Oberthur, S.; Donat, L.; Homann, G.; Sina, S.; Tedsen E., 2013: *Options and Proposals for the International Governance of Geoengineering* (Berlin: Ecologic Institute).
Bond, T.C. et al., 2013: "Bounding the role of black carbon in the climate system: a scientific assessment", in: *J. Geophys. Res.,* 118,11: 5380–5552; https://doi.org/10.1002/jgrd.50171.
Budyko, M.I., 1974: *Climate and Life* (New York: Academic Press).
Budyko, M. I., 1977: *Climatic Changes* (Washington, D.C.: American Geophysical Union).
Caviezel, C.; Revermann, C., 2014: *Climate Engineering: Kann und Soll Man Die Erderwärmung Technisch Eindämmen?* (Berlin: Büro fur Technikfolgenabschätzung beim Deutschen Bundestag).
Cicerone, R.J., 2006: "Geoengineering: Encouraging research and overseeing implementation", in: *Clim. Change*, 77,3–4: 221–226; https://doi.org/10.1007/s10584-006-9102-x.
Crutzen, P J., 2006: "Albedo enhancement by stratospheric sulfur injections: A contribution to resolve a policy dilemma?", in: *Clim. Change,* 77, 3–4: 211–219; https://doi.org/10.1007/s10584-006-9101-y.
Dickinson, R.E., 1996: "Climate engineering: A review of aerosol approaches to changing the global energy balance", in: *Clim. Change,* 33: 279–290; https://doi.org/10.1007/BF00142576.
Fleming, J.R., 2010: *Fixing the Sky: The Checkered History of Weather and Climate Control* (New York: Columbia Univ. Press).
Ginzky, H.; Herrmann, F.; Kartschall, K.; Leujak, W.; Lipsius, K.; Mäder, C.; Schwermer, S.; Straube, G., 2011: *Geoengineering: Effective climate protection or megalomania?* (Dessau; Umweltbundesamt).
Gordon, B., 2010: *Engineering the Climate: Research Needs and Strategies for International Coordination* (Washington D.C.: U.S. House of Representatives Science and Technology Committee).
IPCC, 2007: "The physical science basis", in: Solomon, S.; Qin, D.; Manning, M.; Chen, Z.; Marquis, M.; Averyt, K.B.; Tignor, M.; Miller, H.L. (ed.): *Contribution of Working Group I to the Fourth Assessment Report of the Intergovernmental Panel on Climate Change,* (Cambridge, U.K.: Cambridge Univ. Press).
IPCC, 2013: "Summary for policymakers", in: Stocker, T.F.; Qin, D.; Plattner, G.K. et al. (eds.); *Climate Change 2013: The Physical Science Basis. Contribution of Working Group I to the Fifth Assessment Report of the Intergovernmental Panel on Climate Change* (Cambridge, U.K.: Cambridge Univ. Press): 3–29.
IPCC, 2014a: "Summary for policymakers", in: Field, C.B.; Barros, V.R.; Dokken, D.J. et al. (eds.) *Climate Change 2014: Impacts, Adaptation, and Vulnerability. Contribution of Working Group II to the Fifth Assessment Report of the Intergovernmental Panel on Climate Change* (Cambridge, U.K.: Cambridge Univ. Press): 1–32.

IPCC, 2014b: *Climate change 2014: Mitigation of climate change, in Contribution of Working Group III to the Fifth Assessment Report of the Intergovernmental Panel on Climate Change* (Cambridge, U.K.: Cambridge Univ. Press).

Irvine, P.; Schäfer, S.; Lawrence, M.G., 2014: "Solar radiation management could be a game changer", in: *Nature,* 4: 842; https://doi.org/10.1038/nclimate2360.

Jamieson, D., 1996: "Ethics and intentional climate change", in: *Clim. Change,* 33: 323–336; https://doi.org/10.1007/BF00142580.

Keith, D.W., 2000: "Geoengineering the climate: History and prospect", in: *Annu. Rev. Energy Environ.,* 25: 245–284; https://doi.org/10.1146/annurev.energy.25.1.245.

Keith, D.W.; Duren, R.; MacMartin, D.G., 2014: "Field experiments on solar geoengineering: Report of a workshop exploring a representative research portfolio", in: *Philos. Trans. R. Soc. A,* 372: 14; https://doi.org/10.1098/rsta.2014.0175.

Kiehl, J.T., 2006: "Geoengineering climate change: Treating the symptom over the cause?", in: *Clim. Change,* 77, 3–4: 227–228; https://doi.org/10.1007/s10584-006-9132-4.

Klepper, G.; Dovern, J.; Rickels, W.; Barben, D.; Goeschel, T.; Harnisch, S.; Heyen, D.; Janich, N.; Maas, A.; Matzner, N.; Scheffran, J.; Uther, S., 2016: "Herausforderungen climate engineering - Bewertung neuer Optionen für den Klimaschutz", in: *Kieler Beiträge zur Wirtschaftspolitik Nr. 8* (Kiel, Germany: Institut für Weltwirtschaft (IfW); at: http://hdl.handle.net/10419/142241.

Kravitz, B.; Robock, A.; Boucher, O.; Schmidt, H.; Taylor, K.E.; Stenchikov, G.; Schulz, M., 2011: "The geoengineering model intercomparison project (GeoMIP)", in: *Atmos. Sci. Lett.,* 12, 2: 162–167; https://doi.org/10.1002/asl.316.

Lawrence, M.G.; Rayfuse, R.; Gjerde, K., 2008: "Climate change mitigation by geoengineering, potential side effects, and the need for an extended legal framework: The case of ocean iron fertilization", in: *ECF Background Paper 1*: 25–29 (European Climate Forum e.V., Potsdam, Germany; at: http://www.globalclimateforum.org/fileadmin/ecf-documents/publications/ecf-background-papers/ecf-background-paper-01_2008_global-investments-for-climate-and-energy-security.pdf.

Lawrence, M.G., 2006: "The geoengineering dilemma: To speak or not to speak", in: *Clim. Change,* 77, 3–4: 245–248; https://doi.org/10.1007/s10584-006-9131-5.

Lelieveld, J.; Evans, J.S.; Fnais, M.; Giannadaki, D.; Pozzer, A., 2015: "The contribution of outdoor air pollution sources to premature mortality on a global scale", in: *Nature,* 525: 367–371; https://doi.org/10.1038/nature15371.

Lim, S., et al., 2012: "A comparative risk assessment of burden of disease and injury attributable to 67 risk factors and risk factor clusters in 21 regions, 1990-2010: a systematic analysis for the Global Burden of Disease Study 2010", in: *Lancet,* 380: 2224–2260; https://doi.org/10.1016/S0140-6736(12)61766-8.

MacCracken, M.C., 2006: "Geoengineering: Worthy of cautious evaluation?", in: *Clim. Change,* 77, 3–4: 235–243; https://doi.org/10.1007/s10584-006-9130-6.

Marland, G., 1996: "Could we/should we engineer the Earth's climate?", in: *Clim. Change,* 33: 275–278; https://doi.org/10.1007/BF00142575.

McNutt, M.K., et al., 2015a: *Climate Intervention: Carbon Dioxide Removal and and Reliable Sequestration* (Washington, D.C.: National Research Council of the National Academies).

McNutt, M.K., et al., 2015b: *Climate Intervention: Reflecting Sunlight to Cool Earth* (Washington, D.C.: National Research Council of the National Academies).

Morgan, M.G.; Victor, D.G.; Apt, J.; Steinbruner, J.; Ricke, K.L., 2013: *The Truth about Geoengineering,* (Washington, D.C.: Foreign Affairs, Council on Foreign Relations).

National Academy of Sciences (NAS), 1992: *Policy Implications of Greenhouse Warming: Mitigation, Adaptation, and the Science Base, Panel on Policy Implications of Greenhouse Warming* (Washington, D.C.: National Academy Press, Committee on Science, Engineering, and Public Policy).

Oldham, P.; Szerszynksi, B.; Stilgoe, J.; Brown, C.; Eacott, B.; Yuille, A., 2014: "Mapping the landscape of climate engineering", in: *Philos. Trans. R. Soc. A,* 372: 20; https://doi.org/10.1098/rsta.2014.0065.

Parson, E.A.; Keith, D.W., 2013: "End the deadlock on governance of geoengineering research", in: *Science,* 339: 1278–1279; https://doi.org/10.1126/science.1232527.

Ramanathan, V.; Feng, Y., 2009: "Air pollution, greenhouse gases and climate change: Global and regional perspectives", in: *Atmos. Environ.,* 43: 37–50; https://doi.org/10.1016/j.atmosenv.2008.09.063.

Rickels, W. et al., 2011: *Large-Scale Intentional Interventions into the Climate System? Assessing the Climate Engineering Debate. Scoping Report Conducted on Behalf of the German Federal Ministry of Education and Research* (Kiel, Germany: Kiel Earth Inst.).

Schäfer, S.; Irvine, P.J.; Hubert, A.-M.; Reichwein, D.; Low, S.; Stelzer, H.; Maas, A.; Lawrence, M.G., 2013: "Field tests of solar climate engineering", in: *Nat. Clim. Change,* 3:, 766; https://doi.org/10.1038/nclimate1987.

Schäfer, S.; Lawrence, M.; Stelzer, H.; Born, W.; Low, S. (eds.), 2015: The European transdisciplinary assessment of climate engineering (EuTRACE): *Removing greenhouse gases from the atmosphere and reflecting sunlight away from earth* (European Union, Funded by the European Union's Seventh Framework Programme under Grant Agreement 306993).

Schelling, T.C., 1996: "The economic diplomacy of geoengineering", in: *Clim. Change,* 33: 303–307.

Schmale, J.; Shindell, D.; Schneidemesser, E. von; Chabay, I.; Lawrence, M.G., 2014: "Air pollution: Clean up our skies", in: *Nature,* 515: 335–337; https://doi.org/10.1038/515335a.

Schmidt, H., et al., 2012: "Solar irradiance reduction to counteract radiative forcing from a quadrupling of CO_2: Climate responses simulated by four earth system models", in: *Earth Syst. Dyn.,* 3: 63–78; https://doi.org/10.5194/esd-3-63-2012.

Schneider, S.H., 1996: "Geoengineering: Could - Or should - We do it?", in: *Clim. Change,* 33: 291–302.

Shepherd, J., et al., 2009: *Geoengineering the Climate: Science, Governance and Uncertainty,* (London, U.K.: Royal Society).

Shindell, D., et al., 2012: "Simultaneously mitigating near-term climate change and improving human health and food security", in: *Science,* 335: 183–189; at: https://doi.org/10.1126/science.1210026.

UK House of Commons Science and Technology Committee, 2010: *The Regulation of Geoengineering,* (London, U.K.: UK House of Commons).

United Nations Environmental Programme (UNEP) and World Meteorological Organization (WMO), 2011: *Integrated Assessment of Black Carbon and Tropospheric Ozone* (Nairobi: UNEP, WMO, UNON Publishing Services Section).

Victor, D.G.; Morgan, M.G.; Apt, J.; Steinbruner, J., 2013: "The Truth About Geoengineering: Science Fiction and Science Fact", in: *Foreign Affairs* (27 March 2013); at: https://www.foreignaffairs.com/articles/global-commons/2013-03-27/truth-about-geoengineering.

Chapter 19
Declaration of the Health of People, Health of Planet and Our Responsibility-Climate Change, Air Pollution and Health Workshop (2017)

Pontifical Academy of Sciences

19.1 Statement of the Problem

With unchecked climate change and air pollution, the very fabric of life on Earth, including that of humans, is at grave risk. We propose scalable solutions to avoid such catastrophic changes. There is less than a decade to put these solutions in place to preserve our quality of life for generations to come. The time to act is now.

We human beings are creating a new and dangerous phase of Earth's history that has been termed the Anthropocene. The term refers to the immense effects of human activity on all aspects of the Earth's physical systems and on life on the planet. We are dangerously warming the planet, leaving behind the climate in which civilization developed. With accelerating climate change, we put ourselves at grave risk of massive crop failures, new and re-emerging infectious diseases, heat extremes, droughts, mega-storms, floods and sharply rising sea levels. The economic activities that contribute to global warming are also wreaking other profound damages, including air and water pollution, deforestation, and massive land degradation, causing a rate of species extinction unprecedented for the past 65 million years, and a dire threat to human health through increases in heart disease, stroke, pulmonary disease, mental health, infections and cancer. Climate change threatens to exacerbate the current unprecedented flow of displacement of people and add to human misery by stoking violence and conflict.

This text: "Declaration of the Health of People, Health of Planet and Our Responsibility-Climate Change, Air Pollution and Health Workshop" is generally accessible at: http://www.pas.va/content/accademia/en/events/2017/health/declaration.html. It is republished with the permission of the Pontifical Academy of Sciences. This declaration is based on the data and concepts presented at the workshop: "Health of People, Health of Planet and Our Responsibility-Climate Change, Air Pollution and Health". The workshop was organised by the Pontifical Academy of Sciences, Casina Pio IV, Vatican City, 2–4 November 2017, Casina Pio I.

The poorest of the planet, who are still relying on 19th century technologies to meet basic needs such as cooking and heating, are bearing a heavy brunt of the damages caused by the economic activities of the rich. The rich too are bearing heavy costs of increased flooding, mega-storms, heat extremes, droughts and major forest fires. Climate change and air pollution strike down the rich and poor alike.

19.2   Principal Findings

- Burning of fossil fuels and solid biomass release hazardous chemicals to the air.
- Climate change caused by fossil fuels and other human activities poses an existential threat to Homo sapiens and contributes to mass extinction of species. In addition, air pollution caused by the same activities is a major cause of premature death globally.
  Supporting data are summarised in the attached background section. Climate change and air pollution are closely interlinked because emissions of air pollutants and climate-altering greenhouse gases and other pollutants arise largely from humanity's use of fossil fuels and biomass fuels, with additional contributions from agriculture and land-use change. This interlinkage multiplies the costs arising from our current dangerous trajectory, yet it can also amplify the benefits of a rapid transition to sustainable energy and land use. An integrated plan to drastically reduce climate change and air pollution is essential.
- Regions that have reduced air pollution have achieved marked improvements in human health as a result.
  We have already emitted enough pollutants to warm the climate to dangerous levels (warming by 1.5 °C or more). The warming as well as the droughts caused by climate change, combined with the unsustainable use of aquifers and surface water, pose grave threats to availability of fresh water and food security. By moving rapidly to a zero-carbon energy system – replacing coal, oil and gas with wind, solar, geothermal and other zero-carbon energy sources, drastically reducing emissions of all other climate altering pollutants and by adopting sustainable land use practices, humanity can prevent catastrophic climate change, while cutting the huge disease burden caused by air pollution and climate change.
- We advocate a mitigation approach that factors in the low probability-high impact warming projections such as the one in twenty chances of a 6 °C warming by 2100.

19.3   Proposed Solutions

We declare that governments and other stakeholders should urgently undertake the scalable and practical solutions listed below:

1. Health must be central to policies that stabilise climate change below dangerous levels, drive zero-carbon as well as zero-air pollution and prevent ecosystem disruptions.
2. All nations should implement with urgency the global commitments made in Agenda 2030 (including the Sustainable Development Goals) and the Paris Climate Agreement.
3. Decarbonise the energy system as early as possible and no later than mid-century, shifting from coal, oil and gas to wind, solar, geothermal and other zero-carbon energy sources;
4. The rich not only expeditiously shift to safe energy and land use practices, but also provide financing to the poor for the costs of adapting to climate change;
5. Rapidly reduce hazardous air pollutants, including the short-lived climate pollutants methane, ozone, black carbon, and hydro fluorocarbons;
6. End deforestation and degradation and restore degraded lands to protect biodiversity, reduce carbon emissions and to absorb atmospheric carbon into natural sinks;
7. In order to accelerate decarbonisation there should be effective carbon pricing informed by estimates of the social cost of carbon, including the health effects of air pollution;
8. Promote research and development of technologies to remove carbon dioxide directly from the atmosphere for deployment if necessary;
9. Forge collaboration between health and climate sciences to create a powerful alliance for sustainability;
10. Promote behavioural changes beneficial for human health and protective of the environment such as increased consumption of plant-based diets;
11. Educate and empower the young to become the leaders of sustainable development;
12. Promote an alliance with society that brings together scientists, policy makers, healthcare providers, faith/spiritual leaders, communities and foundations to foster the societal transformation necessary to achieve our goals in the spirit of Pope Francis's encyclical *Laudato Si'*.

To implement these 12 solutions, we call on health professionals to: engage, educate and advocate for climate mitigation and undertake preventive public health actions vis-à-vis air pollution and climate change; inform the public of the high health risks of air pollution and climate change. The health sector should assume its obligation in shaping a healthy future. We call for a substantial improvement in energy efficiency; and electrification of the global vehicle fleet and all other downstream uses of fossil fuels. Ensure clean energy benefits also protect society's most vulnerable communities. There are numerous living laboratories including tens of cities, many universities, Chile, California and Sweden, who have embarked on a pathway to cut both air pollution and climate change. These thriving models have already created 8 million jobs in a low carbon economy, enhanced the wellbeing of their citizens and shown that such measures can both sustain economic growth and deliver tangible health benefits for their citizens.

19.4   End of Declaration

What follows is a summary of the data and concepts on air pollution and climate change as described at the workshop; the last IPCC report published in 2013; and the new data that were published since 2013, including several reports by the LANCET commissions and WHO.

The full declaration with author names can be found *here*.

Signatories

1.   Monsignor Marcelo Sánchez Sorondo (PAS Chancellor)
2.   Joachim von Braun (PAS President & UOB)
3.   Margaret Archer (PASS President)
4.   Veerabhadran Ramanathan (PAS & UCSD)
5.   Partha Dasgupta (PASS & CU)
6.   Peter Raven (PAS & President Emeritus, Missouri Botanical Garden)
7.   Jeffrey Sachs (UN SDSN)
8.   Edmund G. Brown Jr. (Governor of California)
9.   Kevin de León (President of the California State Senate)
10.   Alberto Rodriguez Saá (Gobernador de la Provincia de San Luis, República Argentina)
11.   Scott Peters, Congressman (Member of the U.S. House of Representatives from California's 52nd district)
12.   Sir Andy Haines (London School of Hygiene and Tropical Medicine)
13.   Jos Lelieveld (Max Planck Institute for Chemistry, Germany)
14.   The Rev. Mitchell C. Hescox (President/CEO, The Evangelical Environmental Network)
15.   Leith Anderson (President, National Association of Evangelicals)
16.   Bishop Alastair Redfern (Church of England)
17.   Giuseppe card. Betori (Archbishop of Florence, Italy)
18.   Don Matteo Galloni (President, Comunità Amore e Libertà)
19.   Br Guy Consolmagno SJ (PAS & Specola Vaticana)
20.   Msgr. Cesare Pasini (PAS & Prefect, Apostolic Vatican Library)
21.   Msgr. Roland Minnerath (PASS & Archbishop of Dijon)
22.   Werner Arber (PAS, Nobel laureate in Physiology or Medicine)
23.   Yuan T. Lee (PAS, Nobel laureate in Chemistry)
24.   Paul J. Crutzen (PAS, Nobel laureate in Chemistry)
25.   Gerhard Ertl (PAS, Nobel laureate in Chemistry)
26.   Klaus von Klitzing (PAS, Nobel laureate in Physics, Director, Max-Planck-Institut FKF)
27.   Aaron Ciechanover (PAS & Nobel laureate in Chemistry, TICC, The Rappaport Faculty of Medicine and Research Institute, Technion-Israel Institute of Technology)
28.   Mario Molina (PAS & Nobel laureate in Chemistry)

29. William Phillips (PAS & Nobel laureate in Physics, Joint Quantum Institute, National Institute of Standards and Technology and University of Maryland)
30. Stephen Hawking (PAS)
31. John (Hans Joachim) Schellnhuber (PAS, Potsdam Institute for Climate Impact Research, Germany)
32. Ignacio Rodríguez Iturbe (PAS & Distinguished University Professor and TEES Distinguished Research Professor, Texas A&M University)
33. Francis L. Delmonico (PAS)
34. Ingo Potrykus (PAS)
35. Antonio Battro (PAS)
36. Michael Sela (PAS)
37. Helen M. Blau (PAS & Donald E. and Delia B. Baxter Foundation Professor, Director, Baxter Laboratory for Stem Cell Biology, Stanford University School of Medicine)
38. Takashi Gojobori (PAS & National Institute of Genetics, Japan)
39. Lord Martin Rees (PAS)
40. Albert Eschenmoser (PAS)
41. Sir Salvador Moncada (PAS & FRS, FMedSci Cancer Domain Director, School of Medical Sciences, Manchester Cancer Research Centre)
42. Yves Coppens (PAS)
43. Govind Swarup (PAS)
44. Suzanne Cory (PAS)
45. Yuri Manin (PAS)
46. Rafael Vicuña (PAS)
47. Luis Caffarelli (PAS)
48. Chintamani N.R. Rao (PAS)
49. Roald Sagdeev (PAS)
50. Wael Al-Delaimy (UCSD Institute for Public Health)
51. Fonna Forman (UCSD Center on Global Justice)
52. Erminia M Guarneri (President, Academy of Integrative Health and Medicine, Treasurer Miraglo Foundation)
53. Howard Frumkin (University of Washington School of Public Health)
54. Ulrich Pöschl (Max Planck Institute for Chemistry)
55. Daniel M. Kammen (Professor of Energy, UC Berkeley)
56. Nithya Ramanathan (Nexleaf Analytics)
57. Marcelo M. Suárez-Orozco (UCLA Wasserman Dean & Distinguished Professor of Education)
58. Bess H. Marcus (Dean, Brown University School of Public Health)
59. Jonathan M. Samet (Dean, Colorado School of Public Health)
60. Glen G. Scorgie (Professor of Theology and Ethics, Bethel Seminary San Diego)
61. Conrado Estol (Director, Heart and Brain Medicine -MECyC, Buenos Aires, Argentina)
62. Edward Maibach (George Mason University)

63. Lise Van Susteren (Advisory Board; Center for Health and the Global Environment; Harvard T.H. Chan School of Public Health)
64. Jeremy Farrar (Director, Wellcome Trust)
65. Rauni Prittinen King (Miraglo Foundation, San Diego, California)
66. Manuel Frávega (Organismo Provincial para el Desarrollo Sostenible de la Provincia de Buenos Aires, Argentina)
67. Qiyong Liu (Chief Scientist for Health and Climate Change in China)
68. Maria Neira (Director, Department of Public Health, Environmental and Social Determinants of Health, WHO)
69. Leslie Parker (REIL)
70. Emilio Chuvieco (Professor of Geography, Satellite Earth Observation, University of Alcalá, Spain)
71. Antonella Litta (International Society of Doctors for the Environment – Isde)
72. Justin Farrell (Yale University, School of Forestry and Environmental Studies)
73. Philip J. Landrigan, MD, M.Sc., FAAP (President, Collegium Ramazzini)
74. Mark Miller MD, MPH (President, International Society for Children's Health and the Environment – ISCHE)
75. Pauline Mendola (President, American College of Epidemiology)
76. Jack Ende (President, American College of Physicians)
77. Jeanne A. Conry (past President, American College of Obstetrics and Gynecology)
78. Jesper Hallas (President, International Society for Pharmacoepidemiology)
79. Manolis Kogevinas, MD, Ph.D. (President, International Society for Environmental Epidemiology – ISEE)
80. Debra Saliba, MD, MPH, AGSF (President, American Geriatrics Society)
81. Peter Yellowlees MBBS, MD (Professor of Psychiatry, and Vice Chair for Faculty Development & President, American Telemedicine Association & Department of Psychiatry, UC Davis)
82. Mark G. Lawrence (Managing Scientific Director, Institute for Advanced Sustainability Studies – IASS)
83. Bradly Jacobs MD MPH (CEO and Executive Medical Director, BlueWave Medicine)
84. Virgilio Viana (Director General, Sustainable Amazon Foundation – FAS)
85. Bill McKibbon (co-founder 350.org)
86. Kathleen Rogers (President, Earth Day Network)
87. Georges C. Benjamin, MD (Executive Director, American Public Health Association)
88. Saul Levin, M.D. (Chief Executive Officer; Medical Director, American Psychiatric Association)
89. Robert Perkowitz (President, EcoAmerica)
90. Gene Baur (President and Co-Founder, The Farm Sanctuary)
91. Rabbi Arthur Waskow (Founder, Shalom Center)
92. Bruce Friedrich (Director, The Good Food Institute)
93. Sahar Alsahlani (CEO, Religions for Peace)
94. Constance Hanson (Director, Christians Caring for Creation)

95. Pierre-Yves Fux (Ambassador of Switzerland to the Holy See)
96. Vittorio Hösle (PASS & Notre Dame)
97. Stefano Zamagni (PASS)
98. Lubomír Mlčoch (PASS)
99. Vittorio Possenti (PASS)
100. Rocco Buttiglione (PASS)
101. John McEldowney (PASS)
102. Pierpaolo Donati (PASS)
103. Janne Haaland Matlary (PASS & University of Oslo)
104. Juan José Llach (PASS)
105. José T. Raga (PASS)
106. Louis Sabourin (PASS)
107. Gualserio Zamperini (Consul General in Tunis, Italy)
108. Dario Nardella (Mayor of Florence, Italy)
109. Saifallah Lasram (Mayor of Tunis, Tunisia)
110. Leoluca Orlando (Mayor of Palermo, Italy)
111. Janez Fajfar (Mayor of Bled, Slovenia)
112. Nemanja Pajić (Mayor of Šabac, Serbia)
113. Vladimir Jokić (Mayor of Kotor, Montenegro)
114. Driss El Azami El Idrissi (Mayor of Fès, Morocco)
115. Michelle Sol (Mayor of Nuevo Cuscatlán, El Salvador)
116. Renato Accorinti (Mayor of Messina, Italy)
117. Luca Menesini (Mayor of Capannori, Italy)
118. Gheorge Falca (Mayor of Arad, Romania)
119. Giga Nikoleishvili (Deputy Mayor of Tiblisi, Georgia)
120. Bojan Režun (Deputy Mayor of Idrija, Slovenia)
121. Petar Đakonović (Mayor's Advisor, Kotor, Montenegro)
122. Laura Bocancios (Mayor's Chief of Staff, Arad, Romania)
123. Eleftherios Papagiannakis (Deputy Mayor, Athens, Greece)
124. Anna Paola Concia (Florence City Councillor, Italy)
125. Fadhel Moussa (Member of the Constitutional Assembly, Tunis, Tunisia)
126. Gregor Prezelj (Director of Municipal Administration of Idrija, Slovenia)
127. Bojanab Mladenović (Mayor's Cabinet Delegate, Šabac, Serbia)
128. Mika Annaken (Turku City Manager of Internal Affairs, Finland)
129. Suzanna Ruta-Clarisse (Head of the Spatial Development Department, Nijmegen, Netherlands)
130. Lorraine Spiteri (Malta Confederation of Women's Organisations, MCWO, Birgu Local Council)
131. Zrinka Raguz (Mayor's Cabinet Delegate, Dubrovnik, Croatia)
132. Katsunobu Kubo (Vice Chairperson, Kyoto City Assembly, Japan)
133. Jovan Ristić (Delegate, Kotor, Montenegro)
134. Emanuele Finardi (Milan, Italy).

Acknowledgements We especially thank the global leaders who spoke at the workshop: Honorable Jerry Brown, Governor of California, Honorable Governor Alberto Rodríguez Saá, the Governor of San Luis, Argentina, Honorable Dr. Marcelo Mena, Minister of Environment of Chile, Honorable Kevin de León, President Pro Tempore of California Senate, and Honorable Scott Peters of the US House of Representatives. We also thank the contributions of the faith leaders: Rev Leith Anderson, President of the National Association for Evangelicals, USA; Rev Alastair Redfern, Bishop of Derby, UK; Rev Mitch Hescox, CEO of Evangelical Environmental Network, USA. We thank Dr. Jeremy Farrar, CEO of the Wellcome Trust for his contributions as a speaker and for thoughtful edits of the document. We acknowledge the major contributions to the drafting of the declaration by Drs.: Maria Neira (WHO), Andy Haines (London School of Hygiene and Tropical Medicine) and Jos Lelieveld (Max Planck Inst for Chemistry, Mainz). For a list of speakers and panelists at the symposium, please see the agenda of the meeting attached at the end of this document. We are thankful to the sponsors of the workshop: Maria Neira of WHO; Drs. Bess Marcus and Michael Pratt of Institute of Public Health at the University of California at San Diego; Drs. Erminia Guarneri and Rauni King of the Miraglo Foundation.

Chapter 20
Transition to a Safe Anthropocene (2017), Foreword to Well Under 2 °C: Fast Action Policies to Protect People and the Planet from Extreme Climate Change

Paul J. Crutzen

We are clearly living in the Anthropocene–but then, what exactly is the Anthropocene? Even after thinking about that for many years, it still is not really clear to me. We have not yet found a clean, quasi-mathematical definition, which is also reflected by the extensive discussions about what starting point to assign. The idea of the Anthropocene–the age of Humans–all started from the simple idea that humanity has moved out of the Holocene and taken over from nature in shaping the face of the planet. But then it quickly becomes more abstract than one might think at first. And it becomes more complicated and extensive than only the scientific discourse: the Anthropocene concept is being used socially and culturally, for example in Art, and even in the context of what it means for religions. A fun question to consider is: how would the world look if we had never brought about the Anthropocene? What if our global society had grown in a way that was built from the beginning on renewable energy, on circular economies and on environmentally and societally low-impact consumption? We rarely think about such an alternate present, but maybe that can give us insight into how we can go about pursuing the noble goal of Sustainable Development–which in many ways is just as abstract as the Anthropocene, maybe even more so.

Climate change, one of the main indicators of the Anthropocene, is also an abstract concept for many – something "out there", in the future predicted by complicated climate models. But sadly, it is becoming less abstract with every passing year, given the mounting evidence ranging from temperature records to melting glaciers

This Foreword to the report: *Well under 2 °C: Fast action policies to protect people and the planet from extreme climate* was first published in 2017 by the Institute for Governance and Sustainable Development (IGSD) and the Committee to Prevent Extreme Climate Change (CPECC); at: https://www.ccacoalition.org/en/resources/well-under-2-degrees-celsius-fast-action-policies-protect-people-and-planet-extreme.

and ice sheets to sea level rise. And within the Anthropocene, climate change is intricately linked to many of the other grand challenges that we face. For example, rising temperatures and shifting precipitation patterns can affect agriculture, animal husbandry, and fisheries, severely threatening food security. Climate change is also a challenging justice issue, since the poor and future generations are mostly the ones who will be worst affected. Furthermore, climate change can lead to immigration and conflicts or wars over borders and resources like water, which in turn can hinder international efforts towards disarmament and peace. Even education is affected by climate change, since in farming regions where the changing climate is leading to water scarcity, we are already seeing children not going to school because they have to spend hours a day helping the families carry water from far away. And in turn, issues like hunger, poverty, lack of education and cultural and national conflicts make it challenging to put the attention needed into transitioning to technologies and lifestyles that cause less emissions of CO_2 and other climate-forcing gases and particles. A viscious cycle can develop within these connections.

The enormity of these challenges is reflected in the first impression I got in looking through this report: it just feels as though it's too much, far beyond the simple, elegant, mathematics-based solution that I would like as a scientist. Probably many people feel this way when they first look at a report like this. But don't give up too easily! After I looked through it again, and another time, then found it actually got very easy to go through and get an overview, not only of the problems, but also of the pathways to solving them. The report is very nicely structured, with its four building blocks, three levers, and 10 scalable solutions. It creates an appetite for reading it, and at the same time, provides valuable food for thought. And one thought I had while reading the report is that in many ways, it's like reading about the Anthropocene, due to how it all fits together and its grand scope–even though it's really about climate change, and not much is explicitly said about the Anthropocene. But it doesn't really need to be mentioned much explicitly, since it's woven in the fabric of the report: the authors did not have on blinders, looking only at climate change and its physical, chemical, and biological basis, but kept their eyes on the broader humanistic and societal aspects that are so central to the Anthropocene.

Nevertheless, despite the best effort of the authors to make an understandable and convincing case, I'm very concerned that humanity collectively will not be wise enough to follow the straightforward solutions to the extent laid out here. Of course, it's not black and white: doing some is better than doing none. And in the course of this, it will be very important to continue to lay out this scientific basis, so that we know better and better how to apply the full constructive talents of human beings (which, sadly, are harder to apply than the destructive talents). In order to do this well, we're going to have to learn to make better use of that great gift we are given: the human brain. Supplied with the right conditions, a healthy human brain can think much better than under challenging conditions and massive stress. We still need to learn new and better ways to think, to apply our minds–especially to be able to really get our minds around such massive issue as climate change in the larger context of the Anthropocene. This may require taking a serious step back, and becoming more reflective about how our own thoughts work. If we can learn to do this, then not only will we be able to forecast a safe Anthropocene, but perhaps even more importantly: a beautiful Anthropocene.

Paul Crutzen

Nobel Laureate, Chemistry 1995

Part II
Reflections and Review of Global Debate on the Anthropocene

Chapter 21
The Anthropocene – Reflections (January 2018)

Abraham Horowitz

To the general public, Geological Time Scale does not appear to be a major concept of interest. With the exception of geologists, in particular those concerned with the order and relative position of strata and their relationship to the geological time scale, this may even apply to the scientific community at large. However, this statement does not apply to the specific case of the Anthropocene – The Age of Men.

At the beginning of this century, Paul Crutzen and Eugen Stoermer (Crutzen and Stoermer (2000), Crutzen (2002)) made a case for adding a new epoch to the geological time table to mark the great acceleration of human activities that induce global changes in the Earth's systems. They coined the name Anthropocene. As might have been anticipated, within the relevant scientific community, this proposal was received with reservations and objections. The ensuing debate between proponents and opponents thus became quite passionate and at times even bitter. This outcome brought more and more scientists and other academics from various disciplines to join the debate. Finally, extensive mass media coverage of the Anthropocene exposed large segments of the general public to its meaning and to what it stands for and even to the scholarly differences of opinion about its necessity and starting point.

The diversity and scope of human thought areas and human activities in which the Anthropocene has been used as a sort of buzzword is illustrated by the following representative sample of article titles, conference presentations blogs etc.: "*Architecture in the Anthropocene*", "*Art in the Anthropocene*", "The ethical guide to the Anthropocene", "*Philosophy of the Anthropocene*", "Marxism and the Anthropocene", "*Nature interactions in the Anthropocene*", "Science, Socialism and the Anthropocene", "Human Politics and Governance in the Anthropocene", "*The Anthropocene and Environmental Humanities* …." and "Music in the Anthropocene". Clearly, throughout the history of stratigraphy, nothing like the reaction to the proposal of introducing the Anthropocene has ever taken place before.

This is an unpublished and incomplete text.

S. Benner et al. (eds.), *Paul J. Crutzen and the Anthropocene: A New Epoch in Earth's History*, The Anthropocene: Politik–Economics–Society–Science 1, https://doi.org/10.1007/978-3-030-82202-6_21

The immediate question that arises therefore is what makes the Anthropocene so unique? The answer to this question requires a closer look at the geological time scale and the characteristic features of the Anthropocene. Excluding the Holocene, the present epoch, all previously named periods included in the official geological time scale that covers the 4.6 billion years of the Earth's existence started and ended in very remote past and lasted for millions of years. With the exception of Earth scientists, one would not expect the great majority of the modern academic community and general public to really care or get excited about what exactly happened at such a distant time. According to a recent recommendation of the *Anthropocene Working Group* (AWG), the Anthropocene's beginning should be set in the middle of the previous century. Formal approval of this recommendation will put an end to the Holocene epoch that began some 11,700 years ago, rendering it by far the shortest epoch ever. Furthermore, in this case, the Anthropocene's past amounts to approximately 70 years. In other words, in the terms of a normalized geological time scale, if the Holocene is looked upon as a two-day-old newborn, then the Anthropocene has just been conceived. Had this comparison been made with the lengths of other epochs in the Cenozoic era to which the Holocene belongs, the starting points of these two epochs would merge and become completely indistinguishable[1].

This analogy might appear to be simply a (futile) trivial numbers game. However, it allows us to highlight one of the basic differences between the Anthropocene and the other periods in the geological time scale: **the Anthropocene has no past. It is about the present and the future.**

This statement thus leads to the next point that needs clarification: to whom do the present and future relate? As implied in the original proposal, the Anthropocene is about the interconnected future of mankind and this planet. The concept is not concerned with the fate of mankind 3.7 billion years from now, when our planet will be swallowed by the decaying sun. Nor does it deal with apocalyptic calamities such as a collision with a comet or an asteroid, a supervolcano eruption or other remote doomsday scenarios. The Anthropocene is about the very near future of mankind and the planet. How near?

The promoters of the Anthropocene concept do not claim the gift of prophecy or some kind of divine connection. They simply extrapolate the recent trends in human activities and their effects on our planet and its ability to sustain human life. In this respect, they have clearly established the central role that mankind plays now and will play in the immediate critical future–the next hundred years or so[2]. Thus, there is nothing unexpected in the widespread public attention that the Anthropocene has attracted. After all, isn't it what is expected from people who may face the dire consequences of the processes that characterize the Anthropocene? Also, should they not be concerned about the future of their children and grandchildren?

Thus, for the general public, the term Anthropocene more than anything else is a warning and a call for immediate action. As such, it can be seen as a symbolic

---

[1] All other epochs in the Cenozoic era to which the Holocene belongs were 1.8–22.9 million years long.

[2] Stephen Hawking has been reported to make this estimate.

declaration and for many, including scholars from various disciplines, it is much more important than just its stratigraphic significance. In other words, the idea of humanity's responsibility to act in order to avert self-inflicted major catastrophes, including those that could lead to the end of mankind's existence on this planet, is at the core of the Anthropocene concept. In this respect, the Anthropocene can be looked upon as "The Age of Mankind's Responsibility".[3]

At this stage, the concept becomes rather complicated and not as simple as it first appeared. What does the word mankind stand for? Is it humanity in the sense of human beings only or do we also include in this term being human? In other words, is our goal to protect and preserve only the basic human anatomical characteristics that have not changed over the past 200,000 years, or should we include characteristic human attributes such as the intellectual, cultural and behavioural heritage of present day civilization that has developed over the past 6,000 years or so?

As there is no Anthropocene without *Anthropos*, it is obvious that physical extinction of the human race will put an end to this era. When considering non-physical human attributes, the situation is more difficult because there are different opinions as to what being human means and requires.[4] Furthermore, the very essence of the modern approach *is not to keep a lid on future intellectual developments* and, at the same time, to observe certain basic human principles as moral guidelines. *Individual and collective responsibility belongs to this category of basic principles.*

Schematically, the self-inflicted threats to mankind's existence can be broadly divided into two categories. *First*, indirect threats are human actions that affect living conditions to such an extent that life of *Homo sapiens* is no longer possible. *Second*, direct threats are human actions that alter human mental faculties and behavioural patterns to the extent that a new post-human species is created. Global scale environmental dangers that the Anthropocene is concerned with belong to the first category and the potential consequences of the implementations of *artificial intelligence* (AI) and *genetic engineering* (GE) belong to the second.

Different as they are, both types of threats have a common cause: extremely rapid scientific and technological progress and its potential consequences. In the sense used here, a threat is based on the probability that something undesired might occur. Its occurrence is not definite. Thus, in the case of man-induced environmental global catastrophe, the probability that it will completely wipe out mankind is quite low compared to humankind's replacement by an AI-controlled species. Borrowing from the terminology introduced by Bostrom (2002), most environmental threats belong to the category of "global endurable risk"[5], while misuse of AI represents an "existential risk". *In spite of all the different classifications of the dangers that face humanity, the prevailing belief is that a point of no return has not yet been reached. Currently, therefore, mankind is in a position that still allows it to take preventive steps that will*

---

[3] Or "The Age of Responsibility", similar to the "Age of Reason".

[4] As an illustration of the difficulties including the logical and linguistic ones consider the following question – When a human behaves inhumanly does he still remain a human being?

[5] This approach excludes runaway scenarios like large global temperature increase. (see there)

reverse the observed trends and avoid the threats posed by uncontrolled scientific and technological progress.

The last statement made here implies that advances in science and especially in technology do not necessarily bring progress. The central position that progress has in any discussion of mankind's future in the Anthropocene thus warrants a critical examination of what it means and stands for. In the *Oxford English Dictionary* progress is defined as

a.   Forward or onward movement towards a destination.
b.   Development towards an improved or more advanced condition

When speaking about mankind and global problems, can we really say that we know what our destination as a species is? Religions seemingly have the answer, philosophy attempts to give answers. The truth is that we don't know. Nevertheless, mankind can choose its destination and pursue it. We have no reason to believe that evolution has ended with the creation of the human race. Still, as is true for other species, it would appear that, at least for the present, self-preservation is a general goal behind which mankind can unite.

Based on this premise, any scientific and technological activity that might endanger the existence of mankind or its well-being needs to be controlled on a global scale and requires worldwide cooperation. However, because of differing political and economical interests, ideological and cultural differences as well as many other obstacles, even the achievement of declarative non-abiding statements and agreements requires considerable efforts and compromises. The following rather extreme example illustrates the type of attitudes that must be bridged:

> "Scientific and technological progress is the foundation of social progress. Under capitalism, however, scientific and technological progress is accomplished chiefly in the interests of the ruling class, is used for militaristic, misanthropic purposes, and frequently is accompanied by regression of spiritual values and destruction of human individuality. Under socialism, scientific and technological progress is accomplished in the interests of all the people, and the successful development of science and technology promotes the solution of the body of economic and social tasks of communist construction and the creation of the material and spiritual prerequisites for comprehensive and harmonious development of the individual".
> *The Great Soviet Encyclopaedia* (1979)

No wonder that, when it comes to operative decisions and actual implementation of what has been agreed upon, the gaps between the various differences and conflicting attitudes emerge and are extremely difficult to bridge. For instance, the Paris Agreement of 2015 regarding climate change was concluded by 196 countries, including the United States. It came into effect at the end of 2016 and set a goal of reducing carbon emissions. US president Donald Trump did not take long to reveal his real priorities. Soon after, he made it very clear that he is not going to pursue climate change policies that put the US economy at risk and overturned the moratorium on coal mining in the US.

A key part of the second definition of progress is the use of the word improved.[6] In technology evaluation, in addition to the assessment of measurable performance, this definition allows other considerations to be applied. *Improved means better and thus it opens the door to human values and ethics as vital elements in the appraisal of potential advantages and disadvantages of a technology. Many over-enthusiastic proponents of uncontrolled pursuit of new technologies tend to disregard these aspects.*

Philosophers and social scientists as well as religious leaders are the ones concerned with ethical problems of human society. Faced with the question: "What good is technological progress without moral progress?" they would certainly agree with the clear, concise and straight to the point answer given by the Queen of Jordan: "It is nothing more than the illusion of progress".[7]

To put it in more general terms, scientific and in particular technological achievements do not automatically qualify as bringing progress. Depending on their usage, they can serve both good and bad causes.[8] Therefore, in order to belong to this category, these achievements must be carefully scrutinized, applying criteria that enable the evaluation of all their potential application modes and, as in medicine, the possible "side effects".

In theory and on paper, these appear to be rational and achievable requirements. Not so in reality. First, as psychology and history teach us, in many instances rationality does not appear to play a dominant role in a human individual nor in collective decisions and actions. Second, as past experience shows, some potential applications of specific scientific knowledge and new specific technology are not always possible to envisage in advance. This is also true when their direct and indirect effects on human life, the "side effects", are considered. Third, there is a lack of transparency as a result of deliberately imposed secrecy necessary to protect various interests. These interests include protecting intellectual property rights as well as economic and/or military advantages. In some cases of scientific research, information is also not accessible because of the priority rule[9] that rewards those who are first to make a discovery.

Quite often, opponents of controls and regulations appear to be champions of freedom. They use the argument embodied in President Kennedy's statement: "The best road to progress is freedom's road"[10] as justification for their stand and to

---

[6] Quite interestingly, the term "advanced" in this dictionary is defined as "Far on or ahead in development or progress": a circular definition that is not especially illuminating.

[7] Queen Raia al Abdullah, *The Huffington Post*, 2016.

[8] The use of technologies for both civil and military serves as a good example for what has been said here. The export of such dual-use products and technologies is restricted by many countries. For example, the EU regulation lists ten diverse categories including nuclear materials, facilities and equipment, materials processing, sensors and lasers, computers, electronics and others (REGULATION (EU) No 388/2012 OF THE EUROPEAN PARLIAMENT).

[9] "The Role of the Priority Rule in Science" Michael Strevens, Journal of Philosophy 100: 55–79, 2003.

[10] Speech by Senator John F. Kennedy, Civic Center, Charleston, WV – (Advance Release Text), September 19, 1960.

disguise their real interests. Variations of this "technique" that uses the word freedom to justify opposition to controls and restrictions are found in areas other than scientific and technological R&D. The widespread use of this kind of argumentation and the position taken here in favour of the need for additional controls in the Anthropocene warrant further examination of what freedom is and what its boundaries are.

Freedom as a general term has many connotations and its meaning and limitations have been extensively discussed by philosophers and social scientists. In the present context of controls in the Anthropocene, Wiener's (1954) definition may be most appropriate: *"You can do whatever you want, as long as it does not hurt any other person"*.[11,12] The key element in this definition is the provision "as long as" that sets a limit to personal freedom. Freedom is not anarchy and therefore in any human society, be it a small group or all of humanity, the members and sub-entities must abide by the rules that reflect the norms and needs, whatever they may be. In large communities, states as well as the whole human society, abiding rules are institutionalized by laws and regulations that set the boundaries of freedom. *Thus, unlimited freedom does not exist in any society.*[13] *That is to say, restriction of freedom is an inherent part of every social system.*

Opposition to control often takes a more practical ground-to-earth form. It rides the anti-bureaucracy wave (spirit) that today is quite popular. To be enacted and enforced, controls need bureaucracy. The argument against controls and regulations in this case rests on the generally accepted premise that bureaucracy is responsible for their enforcement and thus more control means more bureaucracy. The red tape associated with bureaucracy, its sometimes blind adherence to rules and its tendency to have a life and interests of its own, as well as other shortcomings, are well known.[14] As in other areas, the effect of these faults in the case of scientific research and technological development includes increased costs, loss of time and resultant loss or delay in the availability of products and technologies that advance and benefit society and mankind. There is some truth to this argument; however, as noted above, technology does not always bring progress and, as will be shown below, in many cases deceleration of technological developments is justified in spite of its cost.

Last but not least, totalitarian regimes are characterized by a desire to maximize controls. Democracies do not deny the need for controls but aspire to keep them at the necessary minimum that will preserve creativity and initiative. This principle is a guideline, the application of which is the subject of debate and controversy and which must be examined on a case-by-case basis. *Still, with all the differences, one fundamental truth about which there is common agreement is that, in case of emergencies,*

---

[11] Wiener, N. (1950/1954). The Human Use of Human Beings: Cybernetics and Society. Houghton Mifflin, 1950. (Second Edition Revised, Doubleday Anchor, 1954.).

[12] Essentially, this is a rewording of Kant's concept of freedom, according to which all persons have an innate right to as much freedom as can be reconciled with the freedom of everyone else. Immanuel Kant, *Grounding for the Metaphysics of Morals* 39–42 (James W. Ellington trans., 1981).

[13] This privilege can be enjoyed by isolated individuals only: – Daniel Defoe's *Robinson Crusoe* in literature and some despots that flourished in the previous century, such as Hitler, Stalin, Mao and Pol Pot, to name a few.

[14] Parkinson.

excessive emergency powers must be granted and applied. Accordingly, priorities change and some freedoms might have to be curtailed and controls might have to be enhanced or relaxed, depending on the nature and cause of the emergency.[15] *What the Anthropocene concept attempts to convey is the sense of emergency in which mankind now finds itself.*[16]

The time scale for processes extends much longer than a single event. Therefore, urgent intervention may take years to start. A case in point is global climate change. The threat of global temperature increase due to the release of greenhouse gases was recognized by Arrhenius more than 120 years ago. Where do we stand now? Assuming that, in the best case, the Paris Agreement will be followed to the letter by all signatory states, the global average temperature will level off at about 1.5 °C above its pre-industrial value. Even if this were to happen, it will take centuries to reverse some of the effects of global warming, such as glacier melting. The immense cost and human suffering during this period indicate that in order to reduce these and other effects, much more drastic steps than those that are planned for the immediate future need to be agreed upon and enacted[17].

References

"Human Politics and Governance in the Anthropocene"; at: https://ecpr.eu/Events/SectionDetails. aspx?SectionID=254&EventID=14.

"Science, Socialism and the Anthropocene", at: https://climateandcapitalism.com/2016/08/08/soc ialism-and-the-anthropocene/.

Bostrom, Nick, 2002: "Existential Risks: Analyzing Human Extinction Scenarios and Related Hazards", in: *Journal of Evolution and Technology*, 9,1: 1–31.

Castree, Noel, 2014: "The Anthropocene and the Environmental Humanities: Extending the Conversation", in: *Environmental Humanities*, 5,1 (November): 233–260. https://doi.org/10.1215/220 11919-3615496

Crutzen, P.J., 2002: "Geology of Mankind: The Anthropocene", in: *Nature*, 415: 23.

Crutzen, Paul J.; Stoermer, Eugene F., 2000: "The 'Anthropocene'", in: *Global Change Newsletter*, 41: 17–18 [see in this vol. Chap. 2 below].

---

[15] Dictators in Rome plague quarantine.

[16] It should be noted that the time scale for processes extends much longer than single events. Therefore urgent intervention may take years to start. A case in point is global climate change. The threat of global temperature increase due to release of greenhouse gases was recognized by Arrhenius more than 120 years ago. Where do we stand now? Even if the Paris Agreement will be abided to the letter and by all signatory states, the global temperature will level off at about 1.5 °C higher temperature than its pre-industrial value and most aspects of climate change will persist for many centuries, even if emissions are stopped.

[17] Only some days after sending the present text as draft to Paul Crutzen and Ulrich Pöschl Abraham Horowitz deceased in January 2018. He commented that he "did not imagine how extensively the term and the concept of Anthropocene, and what is associated with it, have attracted the academic and public interest and discourse" and that he felt "overrun by the avalanche of relevant facts, debates and issues (on the Anthropocene) that he thought that many of them should be addressed."

Davis, Heather; Turpin, Etienne (Eds.), 2014: *Art in the Anthropocene: Encounters Among Aesthetics, Politics, Environments and Epistemologies* (Ann Arbor: Michigan Publishing University of Michigan Library, Open Humanities Press).

Defoe, Daniel, 1719: *Robinson Crusoe* (London: W. Taylor).

Glaser, Marion; Krause, Gesche; Ratter, Beate M. W.; Welp, Martin (eds.), 2012: *Human-Nature Interactions in the Anthropocene: Potentials of Social-Ecological Systems Analysis* (Routledge Studies in Environment, Culture, and Society, Band 1).

Kant, Immanuel, 1981: *Grounding for the Metaphysics of Morals – On a Supposed Right to Lie because of Philanthropic Concerns* (Cambridte – Indianapolis, in: Hackett Publishing Company, Inc.).

Kennedy, John F., 1960: Speech by Senator, Civic Center, Charleston, WV, 19 September 1960; at: https://www.jfklibrary.org/archives/other-resources/john-f-kennedy-speeches/charleston-wv-19600919.

Luther Adams, John, 2015: "Music in the Anthropocene"; at: https://www.news.ucsb.edu/2015/015 471/music-anthropocene.

Stevenson, Angus, 2010: *Oxford Dictionary of English* (Oxford: Oxford University Press).

Raffnsøe, Sverre, 2016: *Philosophy of the Anthropocene – The Human Turn* (London: Palgrave Macmillan).

Royle, Camilla, 2016: "Marxism and the Anthropocene", at: http://isj.org.uk/marxism-and-the-anthropocene/.

Siegle, Lucy, 2016: "The ethical guide to the Anthropocene", in: *The Guardian*; at: https://www.theguardian.com/environment/2016/dec/11/the-ethical-guide-to-the-anthropocene.

Strevens, Michael, 2003: "The Role of the Priority Rule in Science", in: *Journal of Philosophy*, 100: 55–79, 2003

The Great Soviet Encyclopedia [3rd ed.], 1979: *The Great Soviet Encyclopedia* (Moscow); http://russian-world-citizens.blogspot.com/2016/09/the-great-soviet-encyclopedia-in.html

Turpin, Etienne (Ed.), 2013: *Architecture in the Anthropocene, Encounters Among Design, Deep Time, Science and Philosophy* (Ann Arbor: Michigan Publishing University of Michigan Library, Open Humanities Press).

Wiener, Norbert, 1950/1954: *The Human Use of Human Beings: Cybernetics and Society* (Boston: Houghton Mifflin, 1950; New York: Doubleday Anchor, 2nd ed. revised, 1954).

Chapter 22
The Anthropocene Concept in the Natural and Social Sciences, the Humanities and Law – A Bibliometric Analysis and a Qualitative Interpretation (2000–2020)

Hans Günter Brauch

22.1 Introduction

The rapid global spread of the Anthropocene concept across disciplines, languages, cultures and religions has been extraordinary and is unique in scientific history for a basic concept.[1] By 31 December 2020–barely 2 decades after Paul J. Crutzen[2] coined

PD Hans Guenter Brauch is a German political scientist and international relations specialist who taught as an Adj. Prof. at the Otto-Suhr-Institute of Political Science of the Free University of Berlin (2000–2012). Since 1987 he has been chairman of *Peace Research and European Security Studies* (AFES-PRESS), an international non-profit scientific society http://www.afes-press.de/, and in May 2020 he also became chairman of the board of the *Hans Günter Brauch Foundation on Peace and Ecology in the Anthropocene* (HGBS), a non-profit foundation under private law. Since 2000 he has established 5 multidisciplinary English language book series with Springer Nature in Heidelberg, of which he is the sole editor.

[1] The author is grateful for the critical comments and helpful suggestions of Prof. Will Steffen, Australian National University, Canberra; Prof. Jan Zalasiewicz and Hon. Prof. Dr. Colin Waters, University of Leicester, the 2 successive chairmen of the Anthropocene Working Group, Prof. Dr. Jürgen Scheffran (University of Hamburg), and Dr. Gregor Lax, a co-editor of this volume.

[2] Paul J. Crutzen's major publications on the Anthropocene since 2000 include: Crutzen (2002a, 2002b, 2002c, 2006, 2007, 2010, 2012); Crutzen/Ramanathan (2004): Crutzen/Steffen (2003); Crutzen/Schwägerl (2011); Crutzen and Stoermer (2000); Ajai et al. (2011); Steffen et al. (2004a): 313–340; Steffen et al. (2007): 614–621; Zalasiewicz et al. (2015): 196–203; Zalasiewicz et al. (2010): 2228–2231; Zalasiewicz et al. (2010): 6008. See also Chaps. 2–20 in this book.

© The Author(s), under exclusive license to Springer Nature Switzerland AG 2021   289
S. Benner et al. (eds.), *Paul J. Crutzen and the Anthropocene: A New Epoch in Earth's History*, The Anthropocene: Politik–Economics–Society–Science 1,
https://doi.org/10.1007/978-3-030-82202-6_22

the Anthropocene concept on 23 February 2000 at an *International Geosphere-Biosphere Programme* (IGBP) scientific committee meeting in Cuernavaca, Mexico–nearly 15,000 texts listed in the *World Catalogue*[3], 1000 books (*Amazon*)[4] and more than 5000 peer-reviewed scientific texts in both the *Web of Science*[5] and *Scopus*[6] have addressed and discussed the Anthropocene concept.

This chapter will address the following research questions:

- How has the scientific literature on the Anthropocene developed within the global change and the geological community and in the natural sciences over just 2 decades?
- How did the Anthropocene concept spread from a slowly emerging debate in the natural sciences during the first decade to the social sciences, the humanities, law, the applied sciences (including medicine) and engineering?
- How did this spread occur during the 4 sequences of 5 years and were there scientific networks and communities and individual key authors who were instrumental in overcoming disciplinary boundaries?
- What role did 2 multidisciplinary scientific networks, the *International Geosphere-Biosphere Programme* (IGBP) and the *Anthropocene Working Group* (AWG) play in spreading the Anthropocene concept globally and across disciplinary boundaries?
- Who were the key scientific authors who were instrumental in initiating the scientific debate and whose work was highly cited in the peer-reviewed scientific literature?

When seeking to answer these questions, neither the approach of concept history (Koselleck 2002, Müller and Schmieder 2016) nor the different historical (Landwehr 2009) and sociological approaches of discourse analysis (Keller 2008, 2011, 2012) is suitable for assessing a debate that has spread across most disciplinary boundaries. The approach of evaluating the role of an "epistemic community" (Haas 1990, 1992) in analysing and explaining the rapid spread of this concept and initiating many rich scientific and intellectual discourses within and across disciplines does not appear promising either.

Therefore this author uses a bibliometric approach based on the available bibliographical and bibliometric data to structure and explain the emergence of this new concept in the natural and social sciences, the humanities and law, including literature, the performing arts and the applied sciences in medicine and engineering.

---

[3] On 17 November 2020 the *World Catalogue* counted the following occurrences of the term 'Anthropocene' in titles: all formats 13,363 documents, therof 10,479 articles and 3030 chapters, 1867 books, of which 1079 were printed books, 714 ebooks, 229 theses/dissertations, 3 manuscripts, 2 large prints and one continually updated source, 771 downloadable archival materials, 320 articles in encyclopedias, 119 computer files and 35 pieces of music.

[4] On 17 November 2020 Amazon.com counted 1000 items on the Anthropocene.

[5] On 17 November 2020 the *Web of Science Core Collection* of Clarivate counted a total of 5252 titles.

[6] On 17 November 2020 *Scopus* of Elsevier counted a total of 5128 documents.

This analysis relies on 3 databases which offer bibliographic and bibliometric data and are accessible through private companies with different scientific and commercial interests. The *World Catalogue* (a global group of participating libraries which have contributed their catalogued bibliographical resources) is freely available to everybody, while the specialised indexing databases of *Web of Science* (Clarivate Analytics) and *Scopus* (Elsevier) can be consulted by individuals or institutions who pay for access through subscriptions (limited previews of their data can also be obtained by creating a free account).

This chapter offers first a brief review of the emergence of the Anthropocene concept (22.1.1), focusing on when the Anthropocene Epoch began (22.1.2) and the vote of the Anthropocene Working Group of May 2019 on the next steps of decision-making within the geological scientific community (22.1.3).

In the second part of this study, different approaches to analysing the term or concept of Anthropocene–concept history, discourse analysis, epistemic community–are discussed from a social science perspective. None can explain the rapid spread of the Anthropocene concept across scientific disciplines within just 2 decades (2000–2020).

Based on a bibliometric analysis of the indexed peer-reviewed literature, this chapter argues that the *Anthropocene Working Group* (AWG), as an international multidisciplinary scientific network, and key experts were instrumental in spreading this concept globally within the scientific realm. The Anthropocene has also been taken up by societal institutions and used in debates on global environmental change, sustainability transition and a global green deal which aims to achieve climate neutral production and consumption by 2050. This bibliometric analysis of the spread of the Anthropocene concept and literature from 2000 until mid February 2021 (22.3) is based on publications on the Anthropocene assessed by *Web of Science* and *Scopus* (22.3.1). These are listed by year (22.3.2), by research areas and disciplines (22.3.3), by countries and languages (22.3.4), by type, journals and book series (22.3.5), by authors and editors (22.3.6), and by organisations and universities (22.3.7).

In the fourth part, a quantitative thematic analysis of Anthropocene literature is offered. It examines the literature across 4 phases–2001–2005, 2006–2010, 2011–2015 and 2016–2020 – (22.4.1), and in the natural sciences (22.4.2), the social sciences (22.4.3), the humanities (22.4.4), national and international law (22.4.5), linguistics and related literatures (22.4.6), the performing arts (22.4.7), medicine and health (22.4.8), and technology and engineering (22.4.9).

The concluding part (22.5) presents the results of this bibliometric analysis of the use of the Anthropocene as a scientific concept and its global spread across scientific disciplines. As many authors in the social sciences and humanities have conceptualised the Anthropocene according to their specific research context, a genealogy of the Anthropocene concept was not possible. A network analysis focusing on the *International Geosphere-Biosphere Programme* (IGBP 1987–2014) and the *Anthropocene Working Group* (AWG 2009 onwards) will be applied to interpret the results of the quantitative bibliometric analysis.

22.1.1 Crutzen's Coining of the Anthropocene Concept in 2000

© Photo of Paul J. Crutzen (2013) provided by MPIC

During a meeting of the IGBP in Cuernavaca (Morelos, Mexico), probably on 23 February 2000, Paul Crutzen disagreed with a colleague who referred to the Holocene when discussing recent factors in global change dominated by human actions, and exclaimed, "We are not in the Holocene any more… We are in the… the Anthropocene!" He recalled this debate when speaking to a BBC Journalist (Falcon-Lang 2011), and this exchange was later confirmed by Will Steffen (2013), who had witnessed it.

> I was at a conference where someone said something about the Holocene. I suddenly thought this was wrong. The world has changed too much. No, we are in the Anthropocene. I just made up the word on the spur of the moment. Everyone was shocked (cited by Lewis and Maslin 2018: 21).

Helmuth Tritschler (2015), who curated the first exhibition at the German Technological Museum in Munich on the Anthropocene, wrote:

> Crutzen has repeated this story of a sudden flash of insight, a 'eureka' moment of our day, on multiple occasions (Schwägerl 2015), and Will Steffen has confirmed it, thus codifying an attractive founding myth about the origins of the term (Steffen 2013: 486). Crutzen tells an additional story as part of this founding myth: after colleagues suggested that he should claim ownership of the term, he discovered that Stoermer had come up with the term independently. The joint authorship of the article in the *IGBP Newsletter* [Crutzen and Stoermer 2000] that initiated the entire current debate was a reflection of this co-creation. … To him, the new term offers a powerful tool to validate the ongoing process of anthropogenic climate change, which has impacted the Earth in such an enduring manner that it needs a new geological era to depict it properly (Tritschler 2016).

The Australian Earth System scientist Will Steffen,[7] who had worked closely with Paul Crutzen when Steffen was the executive director (1998–2004) of the *International Geosphere-Biosphere Programme* (IGBP), a global research programme

---

[7] Steffen was born and educated in the United States but has lived and worked in Canberra, Australia, since then. He is an Australian citizen.

studying changes to the Earth System as a whole, was present at the meeting in Cuernavaca and has since referred to this episode in many lectures.[8]

22.1.2   When did the Anthropocene Epoch of Earth's History Start?

The starting date of the Anthropocene has been contested in scientific debates since the year 2000. At least 5 different times have been identified as the start of the Anthropocene: (a) The *Industrial Revolution* (1750–1850); (b) the *Neolithic or Agricultural Revolution* (12,000 BP); (c) the Impact of the *Columbian Exchange* (1492–1600); (d) the beginning of the *Nuclear Era* and the nuclear fallout from nuclear testing (1945) that coincided with the end of World War II and the gradual evolution of a new political, economic, security and ideological order; and (e) the period of the *Great Acceleration* (since 1950).

Crutzen and Stoermer (2000) initially suggested that the Anthropocene began with the Industrial Revolution, and especially with Watt's invention of the steam engine in 1782. The British palaeobiologist Jan Zalasiewicz (2008, 2008a), who chaired the *Anthropocene Working Group* (AWG) from its inception in 2009 until 2020, shared Crutzen's initial proposal, arguing that "a point can be made for [the Anthropocene's] consideration as a formal epoch in that, since the start of the Industrial Revolution, Earth has endured changes sufficient to leave a global stratigraphic signature distinct from that of the Holocene". A decade after the first publication of the term (Crutzen and Stoermer 2000), Steffen et al. (2011) argued that "the term Anthropocene … suggests that the Earth has left its natural geological epoch … called the Holocene. Human activities have become so pervasive and profound that they rival the great forces of Nature and are pushing the Earth into planetary *terra incognita*" (Angus 2016). Later, influenced by the Dahlem workshop (Steffen et al. 2004), the results of the *Millennium Ecosystem Assessment* (MEA 2005), and projects of the IGBP, Steffen, Crutzen and McNeill (2005) agreed that "the twentieth century can be characterised by global change processes of a magnitude which has never occurred in human history" (Steffen et al. 2015).

A minority of natural scientists, especially William F. Ruddiman (2003, 2005, 2016), argued that humans have impacted on the climate since the Neolithic or

---

[8] In lectures, oral interventions, and discussions Will Steffen has referred to Crutzen's intervention during a discussion at an IGBP meeting in Cuernavaca which he attended on 23 February 2000; see the hyperlinks to 5 lectures (2010–2019) by Will Steffen in Attachment VI, part III. The *IGBP Newsletter* with Crutzen and Stoermer (May 2000: 17–18) is at: http://www.igbp.net/download/18.316f1832132347017758000140l/1376383088452/NL41. pdf. A synthesis of the IGBP Series by Steffen et al. (2004: 81–142) can be downloaded free at: http://www.igbp.net/download/18.56b5e28e137d8d8c09380001694/1376383141875/SpringerI GBPSynthesisSteffenetal2004_web.pdf. This book has a chapter on "The Anthropocene Era: How Humans are Changing the Earth System".

Agricultural Revolution (Lewis and Maslin 2015, 2018: 141–142; Renn 2020: 348–354).

> Ruddiman claims that the Anthropocene has had a significant human impact on greenhouse gas emissions, which began not in the industrial era, but rather 8000 years ago, as ancient farmers cleared forests to grow crops. Ruddiman's work has, in turn, been challenged with data from an earlier interglaciation. … Although 8000 years ago the planet sustained a few million people, it was still fundamentally pristine. This claim is the basis for an assertion that an early date for the proposed Anthropocene term does account for a substantial human footprint on Earth.[9]

Since the sixteenth century the successive scientific revolutions (Renn 2020: 118–142) in astronomy triggered by Copernicus, Kepler and Galileo, and the new discoveries by Newton, Bacon, Leibniz et al. triggered turning points in science history which fundamentally changed the world-view and challenged the orthodoxy of the Catholic Church.

Lewis and Maslin (2016: 9f.) organised their book on the *Human Planet* around 4 themes relevant to the Anthropocene: "The environmental changes caused by human activity have increased to a point that today human actions constitute a new force of nature." They identified 4 major transitions related to (a) *energy use and its environmental impacts* on the Earth System; (b) the gradual start of colonialism since Columbus reached the Americas in 1492, which resulted in a '*Columbian Exchange*' of species (animals and plants) and the emergence of a global capitalist economy; (c) the availability and *use of coal* as a result of mining; and (d) the *Great Acceleration* after the end of World War II, partly based on the easy access to cheap fossil fuel energy sources. Lewis and Maslin (2016: 13) argued:

> that the start of the Anthropocene should be dated to the *Orbis Spike*, a trough in carbon dioxide levels associated with the arrival of Europeans in the Americas. Reaching a minimum around 1610, global carbon dioxide levels were depressed below 285 parts per million, largely as a result of sequestration due to forest regrowth in the Americas. This was likely caused by indigenous peoples abandoning farmland following a sharp population decline due to initial contact with European diseases – around 50 million people or 90% of the indigenous population may have succumbed.[10] … Maslin and Lewis … also go on to say that associating the Anthropocene with European arrival in the Americas makes sense given that the continent's colonisation was instrumental in the development of global trade networks and the capitalist economy, which played a significant role in initiating the Industrial Revolution and the Great Acceleration.[11]

In its discussions, the AWG later provided its reasons for not accepting the earliest 2 turning points in human history (Zalasiewicz et al. 2015) and instead opting for

---

[9] 'Anthropocene', in: *Wikipedia*; at: https://en.wikipedia.org/wiki/Anthropocene (31 May 2020).

[10] One reviewer wrote me that "this has subsequently proven to be not true. Carbonyl sulphide measurements from the Law Dome ice core show that the drop in CO_2 levels was not due to forest regrowth but rather to cooling-induced terrestrial uptake. The reference is Rubino et al. (2016) 'Low atmospheric CO_2 levels during the Little Ice Age due to cooling-induced terrestrial uptake', in: *Nature Geoscience*; 10.1038/ NGEO2769."

[11] "The Anthropocene", in: *Wikipedia*; at: https://en.wikipedia.org/wiki/Anthropocene (31 May 2020).

2 more recent turning points that were instrumental in triggering a more intrusive global environmental and anthropogenic climate change: the start of the *Nuclear Era* or *Atomic Age* (since 1945) that caused a fundamental change in military strategy and politics during the Cold War and created for the first time the ability to destroy a significant part of humankind during a major nuclear war (Zalasiewicz et al. 2015); and the *Great Acceleration*[12] (since 1950) that was made possible by the US-planned new world order and its multilateral global leadership role.

The period from 1940 to 1950 was a major turning point in international politics. It has been closely analysed in international political, military and security history, and also in economic, social and human history (Notter 1949; Brauch 1976, 1977).

After the first nuclear fission by Otto Hahn and Lise Meitner in Berlin, Albert Einstein and Leo Szilard persuaded President F.D. Roosevelt to launch the top secret "Manhattan District" Project (1942–1945), which "produced the atomic bomb" that "created not only a new technology, but also a new kind of socio-economic complex of technological, political, economic, and knowledge structures that seems impossible to abolish" (Renn 2020: 373). For the first time humankind had the military tools to destroy the human species in a major nuclear war. This new 'super weapon' was first tested secretly on 16 July 1945 in Almogordo in New Mexico and employed on 6 August 1945 against Hiroshima and on 9 August 2020 against Nagasaki, thus forcing Japan to accept unconditional defeat in World War II.

Between 1940 and 1950 the US role in world politics had fundamentally changed from an inward-centred, isolationist country during the 1920s and 1930s to a globally orientated political, military, economic and intellectual 'superpower' that had gradually adopted an active world leadership role during World War II and was instrumental in crafting, developing and enforcing its new role as the leader of the so-called 'free world' and the antipode of the Soviet Union in the post-World War II world. In this decade a fundamental structural change occurred in American foreign, defence, armament, arms control and disarmament policy during World War II and at the beginning of the Cold War (Brauch 1976, 1977: 341). But in 1945 there was no scientific and political awareness of a *global environmental change* or anthropogenic *climate change*. Policy-makers were similarly unaware of the 'environment' and 'development' as policy issues to be addressed by the United Nations Charter (1945). The "social construction of reality" (Berger/Luckman 1966, 1967), of science (Hacking 1999) and of scientific knowledge (Mendelsohn 1977) on *Global Environmental Change* (GEC) and *Climate Change* (CC) and the *Anthropocene* concept had not yet been proposed and could thus have no impact on the framing and development of the Post-World War II order. But by 1999 the Anthropocene had definitely 'occurred', using all definitions of the term.

---

[12] Jan Zalasiewicz and Colin Waters, the chairmen if the AWG, clarified this in their comment of 17 February 2021: "We linked the beginning of the Anthropocene with the Great Acceleration because the processes of the latter drove the main part of the Earth System change that took it out of a Holocene trajectory; the development of nuclear bombs were not chosen per se, as representing the Nuclear Age, but because they led to a globally detectable, very near-synchronous signal in strata that could be used to mark the base of the Anthropocene as a geological unit."

These ecological problems had not yet been socially constructed and had thus not existed in the minds of the decision-makers. Therefore, little ecological reflection in academia and within society on these issues was then possible.[13]

Although as far back as 1896 Svante Arrhenius (Rodhe et al. 1997) had claimed a link between the burning of fossil fuels and the concentration of greenhouse gases, especially *carbon dioxide* (CO_2), in the atmosphere, the scientific debate on climate change did not take off until the 1970s, and the argument of a transition in Earth's history to the Anthropocene Epoch was not suggested until 2000.

Later, Jan Zalasiewicz (2015, 2018) and his colleagues[14] linked the Anthropocene with the start of the nuclear age when they suggested:

> The entry into the nuclear age (signalled by the detonation of the Trinity A-bomb in New Mexico on July 16, 1945) could be used to mark the beginning of the Anthropocene because of the stratigraphic evidence from artificial radionuclides but also from the sediment of global industrialisation such as fly ash, concrete and plastic (Renn 2020: 360).

Lewis and Maslin (2018: 257–258) supported this argument when they argued:

> The most ubiquitous signature of humans in geological deposits is probably the fallout from nuclear bomb tests. This could provide a very clear 'golden spike' for the Anthropocene, as the fallout was global. The clearest of these signals, and among the best understood scientifically, is the radioactive isotope of carbon, carbon-14, which reaches a maximum just after the 1963 Partial Test Ban Treaty outlawed above-ground weapons testing (Lewis and Maslin 2015: 171–180).

Thus, in 1945 two turning points – in *geological time*, with the transition from the Holocene to the Anthropocene, and in *political time* of *longue durée* (Braudel 1969, 1972) – coincided with the end of World War II, which resulted in a fundamental change in the international diplomatic, peace and security order (Brauch 1976). For the first time in Earth's history, the manifold human economic and military activities directly interfered with the Earth System.

Several scientists inspired by an IGBP project argued that the transition to the Anthropocene was triggered by the 'Great Acceleration' in Earth System and socio-economic trends[15] that began around 1950 and intensified after 1990 (Brauch 2021). Despite these global contextual changes since the 1950s and the intensified acceleration since 1990, only a few industrialised countries implemented the quantitative reductions in GHGs and CO_2 emissions into the atmosphere in accordance with the Kyoto Protocol (1997–2012). In 1990 and 1991 a window of opportunity opened with the co-existence of global, regional and national challenges in the realms of policies on peace, international security, international political economy and the global environment. Global environmental governance emerged in late 1988. However, momentum stalled somewhat in 1998, when the US Senate refused to ratify the Kyoto Protocol (1997).

---

[13] Jürgen Scheffran (2019, 2020) pointed to several early comments by natural scientists on the relationship between nuclear weapons and geoengineering.

[14] Zalasiewicz et al. (2015); Waters et al. (2018): 379–422.

[15] See: "Great Acceleration", at: http://www.igbp.net/globalchange/greatacceleration.4.1b8ae2051 2db692f2a680001630.html (27 January 2021).

Steffen et al. (2007: 616–620) initially distinguished between 2 stages of the Anthropocene: (a) the *industrial era* (1800–1945), and (b) the *Great Acceleration* (1946–2015). They identified 3 philosophical responses – (a) *business as usual*, (b) *mitigation*, and (c) *geoengineering* – any of which "can raise serious ethical questions and intense debate", given "the possibility for unintended and unanticipated side-effects that could have severe consequences. The cure could be worse than the disease." They concluded that "The Great Acceleration is reaching criticality", but pointed to "evidence for radically different directions built around innovative, knowledge-based solutions".

According to *Future Earth*, there has been a "synchronous acceleration of trends from the 1950s to the present day … with little sign of abatement".[16] The emergence of the Anthropocene reflects the impacts of the US-dominated world order since 1945, entailing the nuclear era, a world trade system based on 'open doors', cheap fossil energy, and the 'American way of life', or a Western lifestyle based on abundance.

Since 1945 humankind has for the first time threatened its own survival through the fallout from atmospheric nuclear tests and a dominant economic system that has resulted in a massive increase in the burning of cheap fossil energy sources. However, there is a major difference between the two existential threats. While the first threat is influenced by rival superpowers or geopolitical camps, the second is the result of our own individual and joint economic behaviour through the individual burning of fossil energy resources.

The *nuclear era* and the *Anthropocene* are two sides of the same coin that gradually emerged between 1940 and 1950. However, they are addressed by different disciplines and research programmes: the first has been primarily tackled in the social sciences, political science and international relations via security and peace research, while the second has been primarily developed in the natural sciences via environmental research studies focusing on *global environmental change*.

22.1.3   The Vote of the Anthropocene Working Group of 2019

The professional organisations of geologists (the *International Union of Geological Scientists*, the *International Commission on Stratigraphy* (ICS) and the *Subcommission on Quaternary Stratigraphy*) that are entitled to determine the start and end of epochs of Earth's history set up the *Anthropocene Working Group* (AWG) in 2009. In April 2016 the AWG voted to recommend "the Anthropocene as the new geological age" to the *International Union of Geological Sciences* (IUGS) at its congress in Cape Town in August 2016 and to "proceed towards a formal proposal to define the Anthropocene Epoch in the Geological Time Scale".

---

[16] See the figures on the 24 IGBP categories of the 'Great Acceleration' on the website of Future Earth; at: https://futureearth.org/2015/01/16/the-great-acceleration/ (1 December 2019); Steffen et. al. (2015); Steffen et al. (2004).

On 21 May 2019, 29 of the 34 members of the AWG voted that an official proposal on the Anthropocene as a new geological epoch with a starting point in the mid-twentieth century should be made to the ICS (Box 22.1).

Box 22.1: "Working Group on the 'Anthropocene': Results of binding vote by AWG, released 21st May 2019", in: *AWG Newsletter No. 10*: 1–2. *Source* http://quaternary.stratigraphy.org/working-groups/anthropocene/ (19 December 2019).

Following guidance from the Subcommission on Quaternary Stratigraphy and the International Commission on Stratigraphy, the AWG have completed a binding vote to affirm some of the key questions that were voted on and agreed at the IGC Cape Town meeting in 2016. The details are as follows:

No. of potential voting members: 34. No. required to be quorate (60%): 21. No. of votes received: 33 (97% of voting membership).

Q1. Should the Anthropocene be treated as a formal chrono-stratigraphic unit defined by a GSSP?

29 voted in favour (88% of votes cast); 4 voted against; no abstentions.

Q2. Should the primary guide for the base of the Anthropocene be one of the stratigraphic signals around the mid-twentieth century of the Common Era?

29 voted in favour (88% of votes cast); 4 voted against; no abstentions.

Both votes exceed the 60% supermajority of cast votes required to be agreed by the Anthropocene Working Group as the official stance of the group and will guide their subsequent analysis.

The AWG offers this definition and status of the Anthropocene on its website (Box 22.2).

Box 22.2: What is the Anthropocene? – current definition and status. *Source* http://quaternary.stratigraphy.org/working-groups/anthropocene/ (19 December 2019).

- The '*Anthropocene*' is a term widely used since its coining by Paul Crutzen and Eugene Stoermer in 2000 to denote the present geological time interval, in which many conditions and processes on Earth are profoundly altered by human impact. This impact has intensified significantly since the onset of industrialization, taking us out of the Earth System state typical of the Holocene Epoch that post-dates the last glaciation.
- The '*Anthropocene*' has developed a range of meanings among vastly different scholarly communities. Here we examine the Anthropocene as a geological time (chronostratigraphic) unit and potential addition to the Geological Time Scale, consistent with Crutzen and Stoermer's original proposal. The *Anthropocene Working Group* (AWG) is charged with this

task as a component body of the *Subcommission on Quaternary Stratigraphy* (SQS) which is itself a constituent body of the *International Commission on Stratigraphy* (ICS).

- Phenomena associated with the *Anthropocene* include: an order-of-magnitude increase in erosion and sediment transport associated with urbanization and agriculture; marked and abrupt anthropogenic perturbations of the cycles of elements such as carbon, nitrogen, phosphorus and various metals together with new chemical compounds; environmental changes generated by these perturbations, including global warming, sea-level rise, ocean acidification and spreading oceanic 'dead zones'; rapid changes in the biosphere both on land and in the sea, as a result of habitat loss, predation, explosion of domestic animal populations and species invasions; and the proliferation and global dispersion of many new 'minerals' and 'rocks' including concrete, fly ash and plastics, and the myriad 'technofossils' produced from these and other materials.
- Many of these changes will persist for millennia or longer, and are altering the trajectory of the Earth System, some with permanent effect. They are being reflected in a distinctive body of geological strata now accumulating, with potential to be preserved into the far future.
- The *Anthropocene* is not currently a formally defined geological unit within the Geological Time Scale; officially we still live within the Meghalayan Age of the Holocene Epoch. A proposal to formalise the *Anthropocene* is being developed by the AWG.

Based on preliminary recommendations made by the AWG in 2016, this proposal is being developed on the following basis:

1. It is being considered at series/epoch level (and so its base/beginning would terminate the Holocene Series/Epoch as well as Meghalayan Stage/Age);
2. It would be defined by the standard means for a unit of the Geological Time Scale, via a Global boundary Stratotype Section and Point (GSSP), colloquially known as a 'golden spike';
3. Its beginning would be optimally placed in the mid-20th century, coinciding with the array of geological proxy signals preserved within recently accumulated strata and resulting from the 'Great Acceleration' of population growth, industrialization and globalization;
4. The sharpest and most globally synchronous of these signals, that may form a primary marker, is made by the artificial radionuclides spread worldwide by the thermonuclear bomb tests from the early 1950s.

Analyses of potential 'golden spike' locations are underway. The resultant proposal, when made, would need supermajority (60%) agreement by the AWG and its parent bodies (successively the SQS and ICS) and ratification by the

Executive Committee of the *International Union of Geological Sciences*. The
success of any such proposal is not guaranteed.

- Broadly, to be accepted as a formal geological time term the **Anthropocene**
  needs to be (a) scientifically justified, i.e. the 'geological signal' currently
  being produced in strata now forming must be significantly large, clear
  and distinctive; sufficient evidence has now been gathered to demonstrate
  this phenomenon (b) useful as a formal term to the scientific community.
  In terms of (b), the currently informal term '*Anthropocene*' has already
  proven highly useful to the global change and Earth System science research
  communities and thus will continue to be used. Its value as a formal
  geological time term to other communities continues to be discussed.
- The **Anthropocene** has emerged as a popular scientific term used by
  scientists, the scientifically engaged public and the media to designate the
  period of Earth's history during which humans have a decisive influence
  on the state, dynamics and future of the Earth System. It is widely agreed
  that the Earth is currently in such a state. The term has also been used
  in a non-chronostratigraphic context to be an informal term to denote a
  broader interpretation of anthropogenic impact on the planet that is markedly
  diachronous, reaching back many millennia. In geology, such an interpretation is already encompassed by lithostratigraphy, in which the character
  of stratified rocks is based solely on their physical features and not by age.
  Such an interpretation represents a concept sharply distinct from the Anthropocene as a chronostratigraphic unit, though it can be complementary with
  it.

22.2   Analysing the Anthropocene Concept

Earlier concepts preceding the Anthropocene in the 19th and 20th centuries and alternative concepts to the Anthropocene are noted below (22.2.1), and 3 methodological
approaches that were considered unpromising for an interpretation of the rapid spread
of the Anthropocene concept are discussed (22.2.2). Because of their unsuitability, a
bibliometric analysis (22.2.3) was adopted to answer the specific research questions
posed in this chapter.

Meeting of the Anthropocene Working Group on 7–8 September 2018 at the Max Planck Institute for Chemistry (MPIC) in Mainz. *Source* Used with the permission of © MPIC

Meeting of the Anthropocene Working Group on 7–8 September 2018 at the Max Planck Institute for Chemistry (MPIC) in Mainz. *Source* Used with the permission of © MPIC

22.2.1 Predecessors and Alternatives to the Anthropocene Concept

Based on Greek origins, the Anthropocene concept is a combination of *anthropo-* from *anthropos* (ancient Greek: ἄνθρωπος), meaning 'human', and *-cene* from *kainos* (ancient Greek: καινός), meaning 'new' or 'recent'.[17] Various earlier analogue concepts and alternatives to the Anthropocene concept were suggested in the literature that will be reviewed first.

In their first text on the Anthropocene, Crutzen and Stoermer (2000) reviewed several earlier concepts that were suggested in the 19th and 20th centuries. As early as 1873, the Italian geologist *Antonio Stoppani* (1873) pointed to the increasing power and effect of humanity on the Earth's systems and referred to an *anthropozoic era*.[18] Even earlier, in 1864, G.P. Marsh (1864, 1965) addressed the growing influence of humankind on nature in his book *Man and Nature*; and in 1926, the Russian geologist, V.I. Vernadsky (1929, 1945, 1986, 1998) recognised the increasing power of mankind as part of the biosphere when referring to "the direction in which the processes of evolution must proceed, namely towards increasing consciousness and thought, and forms having greater and greater influence on their surroundings."

Later, in 1924, the Jesuit priest Pierre Teilhard de Chardin and E. Le Roy used the term *noosphere* (Levit 2001; cited by Renn 2020: 5, 380), signifying the world of thought, for the growing role of humankind's brain-power and technological talents in shaping its own future and environment (Crutzen and Stoermer 2000) through a tenfold increase in human population during the past 3 centuries and a tenfold increase in urbanisation during the past century, while "30–50% of the land surface has been transformed by human action". Vitousek et al. (1997) stated that "human activity has increased the species extinction rate by thousand to ten thousand fold in the tropical rain forests" (Wilson 1992, 2002). From a biological perspective, H. Markl (1986, cited by Ehlers 2008) referred to an *Anthropozoikum* due to the dominance of human beings as an essential element of the present time. In the year 2000 Crutzen and Stoermer (2000) argued in the *IGBP Newsletter*:

> Considering these and many other major and still growing impacts of human activities on Earth and atmosphere, and at all, including global, scales, it seems to us more than appropriate to emphasise the central role of mankind in geology and ecology by proposing to use the term *anthropocene* for the current geological epoch. The impacts of current human activities will continue over long periods. ... Without major catastrophes like an enormous volcanic eruption, an unexpected epidemic, a large-scale nuclear war, an asteroid impact, a new ice

---

[17] On the Greek origins and meanings of the term Anthropos, access: Frederik Arends' (1943–2020) oral contribution on "Conceptual Origins: 'Anthropos' and 'Politik' as seen by the Greeks"; at: http://www.afes-press-books.de/html/Brainstorming_Mosbach.html [use the password: 'anthropocene2017' to listen]. An obituary of Dr. Jacob Frederik Martinus Arends, who passed away on 3 June 2020, is at: http://www.afes-press-books.de/html/PDFs/2020/Obituary_Jacob_Frederik_Martinus_Arends.pdf. See also: http://www.afes-pressbooks.de/html/PDFs/2020/Obituary_Jacob_Frederik_Martinus_Arends.pdf.

[18] 'Anthropocene', in: *Wikipedia*; at: https://en.wikipedia.org/wiki/Anthropocene (7 June 2020); Steffen et al. (2011): 842–67; Stoppani (1873).

age, or continued plundering of Earth's resources by partially still primitive technology (the last four dangers can, however, be prevented in a real functioning noosphere) mankind will remain a major geological force for many millennia, maybe millions of years, to come.

Two decades later, the German historian Jürgen Renn (2020: ix) offered this definition of the Anthropocene "as the new geological epoch of humankind, defined by the profound and lasting impact of human activities on the Earth system. The Anthropocene is thus the ultimate context for a history of knowledge and the natural vanishing point for an investigation of cultural evolution from a global perspective." Renn (2020: 5–6) noted that the "the idea of humankind as a planetary force" had already been posited by Comte de Buffon (1778), who stated 12 years prior to the French Revolution that "the entire face of the Earth bears the imprint of human power".

Simon Dalby (2016) identified 3 different perspectives for the framing of the Anthropocene: "the good, the bad and the ugly".

> While ecomodernists argue that current circumstances present opportunities and possibilities for a thriving future for humanity, a 'good Anthropocene', critics suggest that the future will be bad for at least most of humanity as we accelerate the sixth extinction event on the planet. The geopolitics of all this, which may be very ugly in coming decades, requires much further elucidation of the common Anthropocene tropes currently in circulation. … How the Anthropocene is interpreted, and who gets to invoke which framing of the new human age, matters greatly both for the planet and for particular parts of humanity. All of which is now a key theme in the discussions of political ecology that requires careful evaluation of how geology has recently become so important in both global politics and discussions of humanity's future, and how scholars from various disciplines might now usefully contribute to the discussion.

The notion of a 'good Anthropocene' has been extensively used by 'cornucopian' (Gledisch 2003) approaches and authors, among them the so-called *ecomodernists* (e.g. The Breakthrough Institute),[19] who argue that liberal capitalism can cope with these challenges with technological fixes (*geoengineering*) and technological innovation without any change to Western lifestyles.

The 'bad Anthropocene' is associated with the negative consequences of 'dangerous' or 'catastrophic' climate change (Schellnhuber et al. 2006) and with 'hothouse Earth' (Steffen et al. 2018). In his *Anthropocene Geopolitics* Dalby (2020) also referred to the ugly geopolitical consequences of global environmental change issues on security, conflicts and wars (Scheffran et al. 2012).

For Arias-Maldonado (2019: 137–138), the Anthropocene concept is "in itself as ambitious as it is ambiguous, and the more so in relation to normative concerns". He has proposed "a reframing of the good Anthropocene, so that it is separated from the ecomodernist vision and presented not only as sustainable but as the result of a collective reflection about the way in which socio-natural relations should be organised".

Several alternatives to the geological concept of the Anthropocene have been suggested, primarily in the social sciences, including the *Anthrocene* (Revkin 1992),

---

[19] See at: https://thebreakthrough.org/ (25 November 2020).

the *Homogenocene* (Samways 1999), the *Capitalocene*, the *Chthulucene* (Haraway 2016a, 2016b), the *Econocene* (Norgaard 2013), the *Technocene* (Hornborg 2015), the *Misanthropocene* (Patel 2013), the *Manthropocene* (Raworth 2014; di Chiro 2017), the *Necrocene* (McBrien 2016), the *Novacene* (Lovelock/Appleyard 2019), the *Myxocene* (Pauly 2010), the *Plasticene* (Stager 2011), the *Plantationocene* (Davis et al. 2019), and others (Moore 2016: 6).[20]

The AWG members Zalasiewicz et al. (2019a: 15) concluded that as a geological unit

> attempts to 'design' a name that might better symbolise its essence … would have little significance. … There is considerable congruence between the meaning of the Anthropocene as originally devised and used in the Earth System Science community and the Anthropocene as considered geologically, as a chronostratigraphic unit (Steffen et al. 2016; Zalasiewiecz et al. 2017a). This, together with the way the name has become quickly established in the literature, suggests that the term Anthropocene should be retained with this meaning – with appropriate qualifications as needed when it is necessary to distinguish it from other meanings and interpretations of the word.

Among the concepts that have been suggested as alternatives to the Anthropocene, the *Capitalocene* (Moore 2016; Altvater 2016; Hartley 2016; Parenti 2016) has been discussed most by both Marxists and critical political economists, while some ecologists prefer the *Novacene*, coined by the founder of the *Gaia hypothesis*. In discussing nature, history and the crisis of capitalism, Jason W. Moore (2016a: 3) noted that "the Anthropocene concept entwines human history and natural history – even if the 'how' and the 'why' remain unclear and hotly debated". Moore and his collaborators argue "for reconstructions that point to a new way of thinking humanity-in-nature, and nature-in humanity" (Moore (2016a: 5). For Moore, the "Capitalocene does not stand for capitalism as an economic and social system". For him this term "signifies capitalism as a way of organising nature – as a multispecies, situated, capitalist world-ecology" (Moore 2016a: 6).

For Dryzek and Pickering (2019: 13–14, 59–60), the *Capitalocene* "ties itself too closely … to [an] aspect of human social organisation". Its proponents "blame certain groups (e.g. capitalists) for global environmental change and its associated injustices". Kelly (2019: 68–69, 74) stated that the adherents of the Capitalocene addressed the "conjoined histories of capitalism and ecological exploitation". Moore and Patel (2018) suggested that the early winners of modern capitalism should repay "a form of ecological reparation for past wrongs". In the view of Hickmann et al. (2019: 4), the proponents of the *Capitalocene* "highlight the causal role of capitalism for irreversible environmental impacts and the emergence of the new epoch". Lewis and Maslin (2016: 444 n.21) agree with Moore

> that the development of the Anthropocene is tied to the modern world and various forms of capitalism. Calling this the Capitalocene is, in our view, not correct because the Anthropocene will last so far into the future – perhaps millions of years – that it may well encompass other future modes of living that are not based on the hallmarks of a capitalist social system.

---

[20] I am grateful to Colin Waters for bringing to my attention: Zalasiewicz et al. (2021) (in press) citing Halle/Millon (2020).

From a critical Marxist perspective, Ian Angus (2016: 232) stated that "capitalism is a 600-year old social and economic system, while the Anthropocene is a 60-year old Earth System epoch ... Insisting on a different word ... can only cause confusion, and direct attention away from far more important issues."

The science historian Renn (2020: 382) argued that "the Anthropocene is now a 'Capitalocene', in the sense that the economic, social and political structures of financialised capitalism drive and fundamentally shape current anthropogenic environmental changes, forcing nature into a world external to the logic of capital, where it serves as a provider of resources and a dumping ground for waste and emissions." However, Renn is opposed to relabelling the Anthropocene the Capitalocene because this would tear down "an important bridge between the natural sciences and the social sciences and humanities".

From a legal perspective, Sally Wheeler (2017: 293) prefers the 'Anthropocene' to a 'Capitalocene', as the latter term "does risk disguising the reality of the Anthropocene. This reality is that the effects of the Anthropocene will be felt by all, albeit unevenly. Thus, 'Anthropos', emphasising the totality of the crisis, should be the underpinning idea rather than what might be seen as a focus on a limited capital-accruing or benefiting class." Of the many suggestions to replace 'Anthropocene', only the 'Capitalocene' has been selectively discussed in the literature, in which a majority argued that the 'Capitalocene' should not be used instead of the 'Anthropocene', which has been increasingly used across most scientific disciplines.

In *The Anthropocene as a Geological Time Unit*, Zalasiewicz et al. (2019c: 1), key members of the AWG, present evidence for defining the Anthropocene as a geological epoch and "demonstrate the extent to which the term has practical utility in the field of geology, in the field of natural science generally, and to the wider academic community".

22.2.2 Three Less Promising Methodological Frameworks

The Anthropocene concept was introduced in the natural sciences and by the Anthropocene Working Group as a *new epoch of Earth's history*, while this author (Brauch 2021) discussed the Anthropocene as a *context*, *turning point* and *challenge* for political and human history since 1945 and for policies aiming at a "European Green Deal"[21] by 2050 or a "sustainability transition" (Brauch et al. 2016) until the year 2100.

---

[21] See: European Council, 12–13 December 2019; at: https://www.consilium.europa.eu/en/policies/climate-change/; see European Commission: "2050 long-term strategy"; at: https://ec.europa.eu/clima/policies/strategies/2050_en (20 January 2020); Communication From The Commission To The European Parliament, The European Council, The Council, The European Economic And Social Committee, The Committee Of The Regions And The European Investment Bank: *A Clean Planet for All: A European strategic long-term vision for a prosperous, modern, competitive and climate neutral economy*. COM/2018/773 final.

3 methodological frameworks were considered for the analysis and interpretation of the emergence and spread of the Anthropocene concept:

- *Concept* or *conceptual history*: Compared with the "concept history" approach (Koselleck 2006; Müller and Schmieder 2016) for concepts like 'peace' and 'security' (Brauch et al. 2008, 2009, 2011) that developed over millennia or centuries, the history of the 'Anthropocene concept' is limited to 2 decades, which is why this approach is not promising for the Anthropocene
- Qualitative approaches of *discourse analysis*, including SCAD (Keller et al. 2018), were not considered advisable, given the complexity of the emerging debates in the social sciences, although they may offer attractive frameworks for a qualitative reconstruction of the discussions within a single discipline (e.g. political science), a specialisation (international relations) or a research programme (security, peace, environment and development studies).
- Interpreting the spread of the Anthropocene debate through the role of the Anthropocene Working Group in the framework of an "*Epistemic Community*", a concept that was developed by Peter Haas (1990, 1992) in international relations, was not appropriate as the AWG lacked a political agenda. The analysis of the AWG as a 'scientific network of experts' or as a 'community of knowledge' may be an attractive approach for the future.

To cope with these multiple methodological challenges, a quantitative *bibliometric* approach (22.2.3) seems preferable.

The history of concepts (*Begriffsgeschichte*) was instrumental in a major German editorial project on key historical concepts (Brunner et al. 1972–1997). Koselleck (1979, 1989, 1994, 1996, 2000, 2002, 2006) addressed the complex interlinks between the temporal features of events, structures and concepts in human (societal) history, and also the dualism between experience and concepts. Political and scientific concepts are used within a complex context (Koselleck 2006). These concepts have a temporal and systematic structure; they embody and reflect the time when they were used and are thus historical documents of the persistent change in the history of short events (*histoire des événements*) and long structures (Braudel's 1949, 1969, 1972 *histoire de la longue durée*). Concepts are influenced by manifold perceptions and interpretations of events that only rarely change the basic structures of societal and political features.

Müller and Schmieder (2016: 581–582) offered a multidisciplinary overview of the multiple discourses on and approaches to concept history and historical semantics in philosophy, history (as history of ideas), the social sciences, communication science, the history of science and knowledge, and cultural history and studies, in which they briefly discussed Crutzen's proposal for a transition from the Holocene to the Anthropocene Epoch of Earth's history. Müller and Schmidt (2016: 582) interpreted Crutzen's conceptual proposal as the replacement of social frameworks with technological designs of a 'sustainable society' (low carbon society, or solar era) and as a shift from the authoritative interpretation in favour of disciplines from the natural sciences. They claimed that Crutzen's conceptual use indicates an indifference to concepts used in the social sciences that reflect specific modes of production whose unlimited expansion has caused the 'ecological problem'. They argue that Crutzen's definition of a new epoch is based on a world-view that favours technical problem solutions, as indicated by his early flirtation with *geo-engineering*.

There has been a rapid spread in the scientific and popular debate on the Anthropocene concept. More than a thousand books, reports and other publications, primarily in English, are available from 12 national Amazon stores, and at the time of writing there are about 5000 peer-reviewed journal articles, book chapters and contributions to proceedings (as indexed by the *Web of Science* and *Scopus* since 2005).

A major role in the rapid spread of the Anthropocene concept – especially between 2016 and 2020 (Table 22.1) – has been played by the *Anthropocene Working Group* (AWG), which has existed since 2009, and by its members, who have been active as authors and editors in setting the scientific agenda. Given the volume of literature on the Anthropocene that has been published across most scientific disciplines since 2000, conceptual mapping using qualitative discourse analysis for a *sociology of knowledge approach to discourse* (SKAD) was considered undoable.[22]

22.2.3 A Bibliometric or Scientometric Analysis

Paul Otlet first introduced the term *bibliométrie* in 1934, defining it as "the measurement of all aspects related to the publication and reading of books and documents".[23] Alan Pritchard (1969) defined *bibliometrics* as "the application of mathematics and statistical methods to books and other media of communication". According to the University of Leeds (UK), "bibliometrics analyses the impact of research outputs using quantitative measures. Bibliometrics complements qualitative indicators of research impact such as peer review, funding received, and the number of patents and awards granted. Together they assess the quality and impact of research." This web definition lists the following types and common bibliometric measures: (a) *Citation counts* (Google Scholar, *Scopus* and *Web of Science*), (b) *H-index*, (c) *Journal Impact Factor*.[24]

Scientometrics, the field of study "which concerns itself with measuring and analysing scientific literature",[25] was invented by Eugene Garfield, the founder of the Institute for Scientific Information, who "created the Science Citation Index"

---

[22] See: Keller et al. (2018) and the detailed review by Simon Smith, "Discursively Constructed Realities: Exploring and Extending the Sociology of Knowledge Approach to Discourse (SKAD)", *Lectures: Les Notes Critiques*; at: http://journals.openedition.org/lectures/34700 (1 Nov. 2020).

[23] Otlet, Paul, 1934: *Traité De Documentation: Le Livre Sur Le Livre, Théorie Et Pratique* (Mons: Editiones Mundaneum); Rousseau, Ronald, 2014: "Library Science: Forgotten Founder of Bibliometrics", *Nature*, 510 (7504): 218.

[24] University of Leeds: "Measuring Research Impact: What is Bibliometrics?"; at: https://library.leeds.ac.uk/info/1406/researcher_support/17/measuring_research_impact (1 November 2020).

[25] Wikipedia: 'Scientometrics', at: https://en.wikipedia.org/wiki/Scientometrics (1 November 2020).

and in 1978 the academic journal, *Scientometrics*, which deals with the scientific and empirical study of science and its outcomes. The author of this chapter used the tools and information provided by both the *Web of Science* and *Scopus* to search the 2 databases for the peer-reviewed English language journal literature on the Anthropocene, with the aim of achieving a quantitative mapping of the global debate based on their different selection criteria.

22.3 Bibliometric Analysis of the Anthropocene Literature (May 2000–January 2021)

Independently of the debate among natural scientists and members of the AWG, the Anthropocene as a new scientific concept was taken up, discussed and associated with a much wider understanding than the initial one used by geologists, global environment change specialists and Earth science scholars referred to above.

For a purely quantitative analysis of the 'term', 'keyword' and 'word in the source title' of the Anthropocene concept for the time period from February 2000 to 31 January 2021, the following sources will be systematically analysed and compared:

- The books listed by *Amazon* in the US, UK, Canada, Australia, India, Germany, France, Spain, Mexico, Brazil, China and Japan (based on the websites of 12 national stores);
- *The World Catalogue* maintained by a global association of libraries;
- *The Web of Science* of Clarivate Analytics;
- *Scopus* maintained by Elsevier, the largest publisher of scientific journals.

As these companies have used different research strategies for terms (e.g. keywords, words in titles etc.) the quantitative results cannot be compared, but they indicate the rapid spread of the primarily academic interest in the term and the scientific concept of the Anthropocene. The quantitative analysis is based on a review of the websites of these companies that sell books (*Amazon.com*), publish (*Elsevier*) and assess scientific literature (*Web of Science* of Clarivate Analytics and *Scopus* maintained by Elsevier). The analysis was carried out by the author by visiting the websites of these companies between 28 September 2020 and 1 February 2021.

22.3.1 Publications on the Anthropocene Offered by Amazon

Amazon is the largest global internet bookstore which offers a comprehensive list of available books, e-books (Kindle) and other information media (videos, CDs, DVDs), including both academic and popular publications primarily in English but also in other languages. However, there is no assessment made by Amazon. In 12 national Amazon stores approximately a thousand academic and popular titles were listed under books on the Anthropocene, most of them in English (584–891), followed by French (38–127), German (4–94), Spanish (1–103), Portuguese (9–203), and Italian (1–11), but only 1–3 titles in Chinese by 30 September 2020. While most of these books were originally written in English, many English texts have been translated into French, Spanish, German, Portuguese, Italian and Chinese.

22.3.2   Anthropocene Literature in the World Catalogue

The *World Catalogue* (World Cat) is the largest book catalogue globally "that itemizes the collections of 17,900 libraries in 123 countries and territories that participate in the OCLC global cooperative … [that] was founded in 1967 … to develop the union catalog technology that would later evolve into WorldCat; the first catalog records were added in 1971. … The subscribing member libraries collectively maintain WorldCat's database, the world's largest bibliographic database. … As of July 2020, WorldCat contained almost 500 million bibliographic records in 483 languages, representing over 3 billion physical and digital library assets, and the WorldCat persons dataset (mined from WorldCat) included over 100 million people."[26]

The World Catalogue is operated by OCLC "under principles and systems of shared governance across several groups: Global and Regional Councils, the Board of Trustees, and the OCLC Executive Management Team".[27] Furthermore,

> The OCLC cooperative is not owned by anyone. A non-profit, library cooperative, OCLC is a membership organization whose public purposes are to further access to the world's information and reduce the rate of rise of library costs. Librarians guide the OCLC cooperative and help shape its services and direction. In contrast to a for-profit corporation, there are no shareholders. The cooperative is governed by a 15-member Board of Trustees, more than half of whom are librarians. A 48-delegate Global Council, all of whom are elected by member libraries, meet at least once a year in person to articulate the interests and concerns of OCLC member libraries. Global Council delegates elect six Board members and ratify amendments to the OCLC Code of Regulations and Articles of Incorporation.[28]

On 8 October 2020 the WorldCat listed for the years 2000–2020 a total of 30,274 media, 23,741 articles, 5896 book chapters, 4178 books (of which 2255 were printed books), 2110 e-books, 461 university publications, 331 lexicon articles, 36 videos, 35 e-videos and 7 Blu-rays. On 9 October 2020 a separate 'keyword search' on the Anthropocene referred to 30,211 items, while a mere 'title search' on the Anthropocene listed only 13,476 entries into the WorldCat for the years 2000–2021.

A second search on 30 January 2021 listed a total of 32,840 Anthropocene results. There were 25,234 articles, 6193 chapters, 2814 downloadable articles, 4166 books, of which 2354 were print books, 2125 e-books, 504 theses and dissertations, 2681

---

[26] See 'WorldCat', at: https://en.wikipedia.org/wiki/WorldCat (9 October 2020).

[27] See at: https://www.oclc.org/en/about/leadership.html (1 February 2021). OCLC's headquarter is at: OCLC, Inc., 6565 Kilgour Place, Dublin, Ohio 43017-3395, USA. There are no specific selection criteria for library content.

[28] See at: https://www.oclc.org/en/about/finance.html (1 February 2021).

archival materials, 368 encyclopedia articles, 280 computer files and 76 videos. According to this second search, 1977 items were in English, 225 in French, 39 in German, 16 in Chinese, 10 in Spanish, 7 in Japanese, 3 in Korean, 2 in Polish and 2 in Danish. Among the topics or disciplines the *Worldcat* listed these numbers: Sociology (344), Geography and Earth Sciences (330), Language and Linguistics (234), Art & Architecture (225), Business & Economics (169), Philosophy & Religion (167), Biological Sciences (147), Engineering & Technology (110), Political Science (95), History & Auxiliary (87), Law (87), Anthropology (46), Education (40), Agriculture (37) and Physical Sciences (31).

22.3.3   Indexing Companies (Web of Science and Scopus)

Two major scientific indexing companies regularly survey and assess the peer-reviewed scientific literature–primarily published in journals that were processed since 1955 by Dr. Garfield who founded the *Institute for Scientific Information* (ISI) and since 2004 by *Scopus*.

22.3.3.1   Anthropocene Titles Listed in the *Web of Science*

In 1955, Dr. Eugene Garfield introduced citation indexing and searching as a tool of scientific research, which a few decades later resulted in the *Web of Science* website "which provides subscription-based access to multiple databases that provide comprehensive citation data for many different academic disciplines". This task was originally produced by the *Institute for Scientific Information* (ISI) and is currently maintained by *Clarivate Analytics*, which took over the "Intellectual Property and Science Business of *Thomson Reuters*" in 2009.[29] *Clarivate Analytics* offers "solutions that drive the entire lifecycle of innovation: scientific and academic research, patent intelligence and compliance standards, pharmaceutical and biotech intelligence, and trademark, domain and brand protection".

Its content "is determined by an evaluation and selection process based on the following criteria: impact, influence, timeliness, peer review, and geographic representation". *Clarivate Analytics* relies on the subscriptions of the academic libraries of universities and major research institutes. Its "*Web of Science* Core Collection contains over 21,100 peer-reviewed, high-quality scholarly journals published worldwide … in over 250 sciences, social sciences, and arts & humanities disciplines".[30]

---

[29] Wikipedia: "*Web of Science*".

[30] See *Web of Science* Core Collection at: https://clarivate.com/webofsciencegroup/solutions/web-of-science-core-collection/ (29 September 2020). See more on "The Institute for Scientific Information", at: https://clarivate.com/webofsciencegroup/solutions/isi-institute-for-scientific-information/ (29 September 2020).

The *Web of Science Core Collection* provides citation indexes of journal literature on the natural and social sciences and the humanities. It includes the *Science Citation Index* and the *Social Sciences Citation Index* (SSCI) (1900-present); the *Arts & Humanities Citation Index* (A&HCI) (1975-present); the *Conference Proceedings Citation Index – Science* (CPCI-S) and the *Conference Proceedings Citation Index – Social Science & Humaniti*es (CPCI-SSH) (1990-present); the *Book Citation Index – Science* (BKCI-S) and the *Book Citation Index – Social Sciences & Humanities* (BKCI-SSH) (2005-present); the *Emerging Sources Citation Index* (ESCI) (2015-present); and 2 indices on Chemistry: *Current Chemical Reactions* (1985-present) and the *Index Chemicus* (1993-present).

In the autumn of 2020 its book citation index included about 60,000 books, and the intention is "to expand the index coverage with an additional 10,000 new titles each year". Those included in 2014 were from the social and behavioural sciences (38%), the arts and humanities (21%), physics and chemistry (12%), engineering, computing and technology (13%), life sciences (6%), clinical medicine (5%), and agriculture and biology (5%). On 1 February 2021, "The Book Citation Index includes over 116,000 editorially selected books, with 10,000 new books added each year. Containing more than 53.2 million cited references, coverage dates back from 2005 to present. The Book Citation Index is multidisciplinary, covering disciplines across the sciences, social sciences, and arts & humanities."[31]

According to its mission statement, the *Web of Science* Core Collection is "curated by an expert team of in-house editors, is a trusted, high-quality, definitive database for journals, books and conference proceedings. We are guided by the legacy of Dr. Eugene Garfield, inventor of the world's first citation index: we remain true to his principles of objectivity, selectivity and collection dynamics, but also adapt and respond to technological advances and changes in the publishing landscape."[32] It aims to be publisher-neutral and conducts in-house curation underpinned by the category knowledge of its editors. The journal selection is based on 24 quality and 4 impact criteria. Only journals, books and conference proceedings are included.

Basic searches are by topic (searches title, abstract, author, keywords and keywords plus). On 29 September 2020 the basic search for 'Anthropocene' offered a total of 5080 documents in the *Web of Science* Core Collection. By 18 November 2020 this number had increased to 5265 documents.

On 30 January 2021 a third search for the term 'Anthropocene' in the *Web of Science* Core Collection using the 'title' category resulted in 5549 items. When the search was limited to the years 2001 to 2020 the results were 5531 items, 4038 journal articles, 614 book chapters, 506 editorial materials, 412 reviews, and 305 book reviews. Of the total results, 5275 were in English, 67 in Spanish, 36 in German, 36 in Portuguese, 34 in Italian, 31 in French, 15 in Russian, 11 in Polish, 5 in Turkish and 4 in Chinese. The 10 most recorded countries for publications issued between

---

[31] See at: https://clarivate.com/webofsciencegroup/solutions/webofscience-bkci/ (1 February 2021).

[32] See at: https://clarivate.com/webofsciencegroup/solutions/web-of-science-core-collection/ (1 February 2021).

2001 and 2020 were: the USA (2026), the UK (937), Australia (681), Canada (519), Germany (462), Sweden (262), France (260), The Netherlands (220), the People's Republic of China (214) and Spain (212).

22.3.3.2   Anthropocene in Titles and Keywords Listed in *Scopus*

Scopus was launched in November 2004 by Elsevier, which owns and maintains *Scopus* as the largest global scientific publisher, especially of peer-reviewed scientific journals. According to its own definition, "with over 25,100 titles from more than 5000 international publishers, *Scopus* delivers the most comprehensive overview of the world's research output in the fields of science, technology, medicine, social science, and arts and humanities". *Scopus* claims to be "the largest indexer of global research content" in journals, books and conference papers.[33] Its *Content Selection & Advisory Board* (CSAB) continuously reviews suggestions as to which types of content may be listed. It states: "Only serial titles may be suggested to the content selection and advisory board for inclusion in *Scopus*. Serials include journals, book series or conference series." For journals to qualify for selection, they must:

- Consist of peer-reviewed content and have a publicly available description of the peer review process.
- Be published on a regular basis and have an *International Standard Serial Number* (ISSN) as registered with the *ISSN International Centre.*
- Have content that is relevant for and readable by an international audience, meaning: have references in Roman script and have English language abstracts and titles.
- Have a publicly available publication ethics and publication malpractice statement.

Of *Scopus* total serial publications, 60.5% are active journals, 2.1% are active book series, 0.7% are trade journals, and the remaining 36.6% are inactive journals, book series and trade journals. *Scopus* has more than 76.8 million core records, of which 51.3 million records are post-1995 (67.5%) and 25.3 million records pre-1996 (32.5%) dating back to 1788. Approximately 3 million new records are added each year (5500/day).[34]

Scopus listed 4945 documents on the Anthropocene between 2001 and 28 September 2020 for the publication years 2001–2020, which slightly differed from the *Core Web of Science* total of 5080 documents. A second search on 30 January 2021 for the publication years 2001–2020 resulted in 5531 documents on the *Web of Science* and 5311 document on *Scopus*. Of thedocuments listed on Scopus during the 20-year period, 5100 were in English, 50 in French, 46 in Spanish, 42 in German,

---

[33] See on *Scopus*; at: https://www.elsevier.com/solutions/scopus/how-scopus-works/content/content-policy-and-selection (29 September 2020).

[34] These data rely on *Scopus: Content Coverage Guide*; see at: https://www.elsevier.com/?a=69451 (2 Oct. 2020)

42 in Portuguese, 18 in Italian, 11 in Chinese, 7 in Russian and 5 in Croatian. The 10 most represented countries for the authors of these texts were the USA (1904), the UK (976), Australia (643), Canada (462), Germany (452), France (274), Sweden (256), The Netherlands (202), Spain (198), and China (180).

22.3.3.3 Anthropocene Texts Listed by Year in '*Web of Science*' and '*Scopus*' (2001–2020)

The annual additions of titles on the Anthropocene are compared below with the *World Catalogue*, which lists all documents that have been reported in national catalogues. This occasionally results in multiple entries or repetitions, which is why the actual figures are lower than those that have been listed. Although the indexing methodology and the types, scope of sources and disciplines slightly differ on the '*Web of Science*' and '*Scopus*', duplicate listings are unlikely.

Based on a search on 1 February 2021 for the years 2000 to 31 December 2020, during the first decade (2000–2010) the *World Catalogue* listed 802 Anthropocene documents by keyword and 351 documents by title, the *Web of Science* listed 115 titles on the Anthropocene, and *Scopus* lised 137 documents. During the second decade (2011–2020) the *World Catalogue* listed 31,802 Anthropocene texts by keyword and 14,639 by title, the *Web of Science* listed 4409 by topic, and *Scopus* listed 5172 documents by themes (Table 22.1).

Table 22.1 Anthropocene texts by year in the World Catalogue, the *Web of Science*, and *Scopus* (2000–2020)

Year	World Catalogue (1 February 2021)		Web of Science	Sco-pus	Year	World Catalogue (1 February 2021)		Web of Science	Scopus
			(1 Feb. 2021)					1 Feb. 2021)	
	keyword	title	topic	theme		keyword	title	topic	theme
2000	2		0						
2001	5	3	1	1	2011	401	196	55	63
2002	20	14	4	3	2012	467	201	64	88
2003	15	9	3	3	2013	883	439	106	158
2004	30	18	2	3	2014	1,645	945	264	326
2005	61	34	5	5	2015	2,692	1,438	434	442
Subtotal [2001-2005]	[133]	[78]	[15]	[15]	Subtotal [2011-2015]	[6,088]	[3,219]	[923]	[1,077]
2006	85	48	8	15	2016	3,526	1,747	555	482
2007	95	46	13	15	2017	4,915	2,375	811	776
2008	133	48	20	22	2018	5,712	2,550	830	733
2009	124	31	20	27	2019	6,568	2,500	1,170	986
2010	232	100	39	45	2020	4,993	2,248	1,127	1,118
Subtotal [2006-2010] [2001-2010]	[669] [802]	[273] [351]	[100] [115]	[124] [139]	Total [2016-2020] [2011-2020]	[25,714] [31,802]	[11,420] [14,639]	[4,493] [5,416]	[4,095] [5,172]
Total [2001-2020]	[32,604]	[14,990]	[5,426]	[5,309]	Total (2001-2020)	[32,604]	[14,990]	[5,426]	5,309

Sources Websites of the WorldCat, *Web of Science* and *Scopus* (1 February 2021)

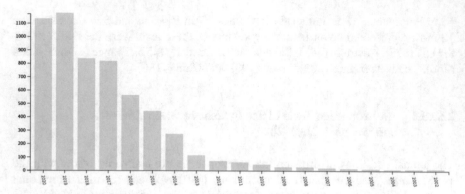

Fig. 22.1 Anthropocene Texts Listed in the *Web of Science* Core Collection (2001–2020): 5530 on 1 February 2021. *Source Web of Science*

According to the World Catalogue, there was a slow annual increase in the number of Anthropocene documents that were listed between 2001 and 2015, resulting in 6890 documents by keyword and 3570 documents by title. The *Web of Science* listed by topic 1038 documents on the Anthropocene for the first 15 years and *Scopus* listed by theme 1214 such documents. The highest increase in publications on the Anthropocene occurred between 2016 and 31 December 2020 in the *World Catalogue*, with 25,714 documents by keyword and 11,420 by title. In the fourth 5-year period the number of Anthropocene documents on the *Web of Sci*ence increased by 3,486 documents by topic and on *Scopus* by 4095 documents by theme (Table 22.1). The highest number of new titles on the Anthropocene occurred in 2018 on the *World Catalogue* (2550 documents), in 2019 on the *Web if Science* (1170 documents), and in 2020 on *Scopus* (1118 documents). Table 22.1 summarises the documents counted on the Anthropocene concept between February 2000 and 31 December 2020 on the *World Catalogue*, based on keyword and title searches, and on the 2 databases of the selected peer-reviewed literature on the *Web of Science* (Clarivate Analytics, Fig. 22.1) and *Scopus* (Elsevier, Fig. 22.2).

The data are split into 4 time periods. In 2000–2005 the concept was hardly noted, with *World Cat* listing by title only 78; *Web of Science* 15; and *Scopus* 15. During the second phase (2006–2010) the focus of the scientific community on the Anthropocene slightly increased – *World Cat* by title: 100; *Web of Science*: 39; and *Scopus*: 45 – especially after the *Anthropocene Working Group* (AWG) was established in 2009. During the third phase (2011–2015) the number of publications on the Anthropocene significantly increased, especially between 2013 and 2015: *World Cat* by title: 1438; *Web of Science*: 434; and *Scopus* 442. The highest increase in publications on the Anthropocene was between 2016 and 31 December 2020: *World Cat* by title: 11,420; *Web of Science*: 3486; and *Scopus* 4095. Figures 22.1 and 22.2 illustrate the annual increase in documents on the Anthropocene between February 2000 and 31 December 2020.

Documents by year

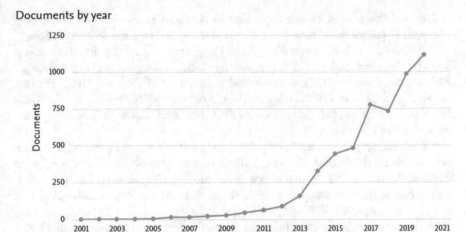

Fig. 22.2 Anthropocene Texts Listed on *Scopus* (2001–2000): 5311 on 1 February 2021. *Source Scopus* website, at: https://www.scopus.com (1 February 2021)

22.3.4   Anthropocene Texts Listed by Research Areas and Disciplines on the Web of Science and Scopus (2001–2020)

An analysis of the references to the Anthropocene in the titles for all entries between 2000 and 2020 resulted, according to the *World Cat* based on title searches, in 14,990 items. These were automatically categorised by the following disciplines or topics: Sociology (306), Geography and Earth Sciences (253), Language, Linguistics and Literature (122), Arts & Architecture (107), Philosophy and Religion (102), Biological Sciences (101), Business & Economics (97), Political Science (71), Engineering & Technology (59), History & Auxiliary Sciences (50), Law (47), Anthropology (28), Education (28), Agriculture (27), Physical Sciences (26), Music (17), Medicine (15), Library Science, Generalities and Reference (14), Performing Arts (9), Psychology (6), Government Documents (5), Health Profession and Public Health (3), Physical Education and Recreation (2), Chemistry (1), Mathematics (1), Medicine by Discipline (1), and Health Facilities, Nursing and History (1).[35]

Both the *Web of Science* and *Scopus* have used different categories, definitions and names for their titles for the time period from 2001 to 2020, which makes direct comparison difficult (Table 22.4). While the data for Environmental Sciences and Ecology on the *Web of Science* (1,702) are relatively similar to the titles indexed under Environmental Science on *Scopus* (1,787), the titles listed under Social Sciences on the *Web of Science* (285) significantly differ from those listed under Social Sciences

---

[35] See at: https://www.worldcat.org/search?q=ti%3AAnthropocene&fq=yr%3A2000..2021+%3E+&dblist=638&fc=s0:_50&qt=show_more_s0%3A&cookie (9 October 2020).

on *Scopus* (2086). Also, titles categorised as Arts and Humanities on the *Web of Science* (198) differ from those listed by *Scopus* for the same disciplines (1236).

The *Web of Science* lists as disciplines Geography (332), Physical Geography (303), Government and Law (257), Philosophy (186), Anthropology (177), Sociology (126), Cultural Studies (131), International Relations (105), History (103), History, Philosophy of Science (103), Religion (90), Archaeology (66), Area Studies (52), Development Studies (45), Urban Studies (36), and Social Issues (34). None of these research areas are listed on *Scopus*.

In contrast, *Scopus* lists a lot or research areas in the Natural Sciences (Biochemistry, Genetics, Molecular Biology, 190; Physics, Astronomy, 52), the Applied Sciences (Medicine 152; Immunology, Microbiology 57; Nursing, 13; Veterinary, 7; Pharmacology, Toxicology, 6; Health Professions, 5) and Engineering (223) that are not listed under the *Web of Science* (Table 22.5). Given the difference in categories, comparing the indexed titles on *Web of Science* and *Scopus* by discipline and research areas offers a more balanced snapshot (based on a search on 9 October 2020).

22.3.5 Anthropocene Texts Listed by Countries and Language on the Web of Science and Scopus (2001–2020)

The indexing of the titles in terms of the countries of their authors confirms a large degree of similarity. In both assessments the USA is top, with 1883 entries on the *Web of Science* and 1801 texts on *Scopus*. This is followed by the United Kingdom with 927 texts on *Scopus* and a total of 1039 titles *Web of Science* (England: 879, Scotland: 109, Wales: 51). In third place is Australia with 621 titles on *Web of Science* and 609 on *Scopus*, followed by Canada in fourth place with 478 titles on the *Web of Science* and 438 texts on *Scopus*. These 4 Anglophone countries account for 4021 titles on the *Web of Science* and 3775 on *Scopus* (based on a search on 30 September 2020).

Among the first 50 places, other English-speaking countries of the former British Empire amount to 401 (South Africa: 148; New Zealand: 81; India: 68; Ireland: 40; Singapore 22, Kenya: 23; Malaysia: 19) in the records of the *Web of Science* and a total of 406 texts (South Africa: 138; New Zealand: 94; India: 71; Ireland: 40; Singapore 25; Malaysia: 20; Kenya: 18) on *Scopus*. The selected countries of the Anglophone world account for 4422 titles (86.91%) on the *Web of Science* and 4,181 texts (84.55%) on *Scopus*. These data confirm the bibliometric trends of global academic publishing that have also been used by UNESCO and many other international organisations.

Germany is in 5th place on the *Web of Science* with 417 titles, and in the same position on *Scopus* with 414 texts. France takes 7th place on the *Web of Science* with 240 titles, and 6th place on *Scopus* with 250 texts. While countries from the

Francophone world in the Global South are missing, some authors from Belgium and Switzerland are included. China, as one of the two most populous countries, ranks in tenth place on both the *Web of Science* (191 titles) and *Scopus* (180 texts). If Taiwan, as the second Chinese-speaking country, is added, the number of Chinese Anthropocene titles reaches 240 titles on the *Web of Science* and 208 texts on *Scopus*. The Hispanic world is represented on the *Web of Science* by 7 countries (Spain: 192; Mexico: 58; Chile: 55; Argentina: 31; Ecuador: 14; Panama: 12; Peru: 19) which together produced 372 titles, and on *Scopus* by 362 texts produced by: Spain: 180; Mexico: 48; Chile: 41; Argentina: 36; Columbia: 24; Panama: 11, Peru: 11; and Puerto Rico (a Spanish speaking territory of the USA): 10. The Scandinavian countries and the Netherlands, representing 6 countries (Sweden: 265; Netherlands: 198; Norway: 123, Denmark: 119; Finland: 82, Iceland: 15), together reached 802 titles on the *Web of Science* and 728 texts on *Scopus* (Sweden: 233, Netherlands: 178; Norway: 126; Denmark: 105; Finland: 71; Iceland: 15). Japan (74) and South Korea (30) are listed as producing 104 titles on the *Web of Science* and 101 texts (Japan: 72; South Korea: 29) on *Scopus*. For Russia the *Web of Science* listed only 31 Anthropocene texts, and on *Scopus* even fewer – only 24. In contrast, for Poland the *Web of Science* indexed 65 Anthropocene texts, and *Scopus* 70. The data of both US-based companies were produced according to criteria which they made publicly available. The differences in the data may reflect the 2 organisations' different approaches to assessing information. In neither case do they aim to provide a globally representative mapping of the literature on the Anthropocene (Table 22.5).

The media referred to in the *WorldCat* were published in the following major languages: English (23,444), not determined (865), French (649), German (227), Spanish (163), Japanese (66), Italian (41), Portuguese (34), Chinese (22), Polish (21), Swedish (15), Korean (10), Croatian (8), Russian (8), Danish (7), Slovenian (6), Afrikaans (5), Czech (4), Dutch (4), Catalan (3), several languages (3), Slovakian (3), Ndonga (2), Turkish (2), Ukrainian (2), Afar (1), Galician (1), Greek (1), Malay (1), Nauruan (1), Norwegian (1), Romanian (1), Serbian (1), and Tamil (1).[36]

The selection bias was even more obvious in the languages represented on the *Web of Science* and *Scopus*. During the period surveyed, the vast majority of Anthropocene titles were in English on both the *Web of Science* (4856 texts out of 5080, representing 95.59%) and *Scopus* (4777 texts out of 4945, representing 96.6%). Spanish was in second place on both the *Web of Science* (61 titles) and *Scopus* (40 titles). Italian took the third place on the *Web of Science* (32 titles) while France was in this position on *Scopus* (39). On both the *Web of Science* (31) and *Scopus* (38) Portuguese was in fourth place, while German followed in fifth place with 30 and 27 Anthropocene titles. The *Web of Science* noted only 10 texts in Russian, 4 in Chinese and 3 in Korean, while *Scopus* listed 9 texts in Chinese, 7 in Russian and 4 in Japanese. The *Web of Science* did not index any text on the Anthropocene in Japanese, but *Scopus* listed 4. Several Slavonic languages seemed to be overrepresented on the *Web of Science* (Polish: 11; Czech: 3; Croatian: 2; Slovenian: 1) and *Scopus* (Polish:

---

[36] See at: https://www.worldcat.org/search?q=kw%3AAnthropocene&fq=yr%3A2000..2020+% 3E+&dblist=638&fc=s0:_25&qt=show_more_s0%3A&cookie (9 October 2020).

5, Croatian: 4; Bosnian: 3; Slovak: 3; Moldavian: 2; Moldovan: 2, Romanian: 2; Slovenian: 2). (Table 22.6). Do these selections represent systematic criteria or are they more indicative of the language competence of the experts who were doing the selection and indexing?

22.3.6 Anthropocene Texts Listed by Type, Journals and Book Series on the Web of Science and Scopus (2001–2021)

The types of texts on the Anthropocene differed between the 2 indexing companies. While on both the *Web of Science* (3955 or 3690) and *Scopus* (2955, plus 674 reviews, 158 editorials, 125 notes, 40 letters and 36 short services) peer-reviewed journal articles prevailed, on both the number of indexed books, book chapters, and conference proceedings was strictly limited. The *Web of Science* included a total of 748 books (44 books, 547 book chapters) while *Scopus* listed 135 books and 649 book chapters (see Table 22.7).

The names of the first 50 peer-reviewed journals included on the *Web of Science* and *Scopus* show many similarities but also some differences (Table 22.8). 36 of the first 50 journals listed on the *Web of Science* and *Scopus* were identical (in yellow), 12 of the remaining journals were only listed on the *Web of Science* (in green) and 14 were only included on *Scopus* (in blue).

Only the *Web of Science* listed 139 book series. Table 22.6 includes the first 50 major book series containing references to the Anthropocene, in order of the number of book series represented, led by *Springer Nature* (17 series), *Taylor and Francis* (12 series), *Annual Reviews* (4 series), *Wiley* (2 series) and *Elsevier* (2 Series). The remaining 13 series are presented by 4 *University Presses* (MIT Press, Virginia, California and Ohio University Presses) and 2 scientific societies (Geological Society, London and IAHS). The publishers of the remaining 6 book series are: Bloomsbury, Il Sileno Ed., W.W. Norton & Co., Atlantis Press, Emerald Publishing and Brill, while the publisher of the series on Humanity and the Sea could not be identified.

22.3.7 Anthropocene Texts Listed by Authors and Editors on the Web of Science and Scopus (2001–2021)

Among the authors with between 58 and 10 entries on *WorldCat* were: *Jan Zalasiewicz* (58), David Chandler (33), Andrew Y. Glikson (30), *Paul J. Crutzen* (25), Simon Dalby (24), Hans Günter Brauch (23), Noel Castree (23), Lourdes Arizpe Schlosser (21), Manuel Arias-Maldonado (19), Vincent Blok (19), Frank Biermann (18), James Bowen (15), *Erle C. Ellis* (15), Úrsula Oswald Spring (15), Nigel Clark (13), *Will Steffen* (13), Williston Byron (13), Daniele Revel (12), Anne Chin (11),

Choi Moon (11), S.A. Elias (11), Jamie Lorimer (11), Claire Colebrook (10), and Janet Mcintyre Mills (10). [Members of AWG are in *italics*.][37]

A total of 37 of the first 50 authors (Table 22.10) were listed on both the *Web of Science* and *Scopus*, while 13 authors (Anonymous, A.Y. Glikson, D.D. Richter, O.R. Young, N. Clark, A.S. Goudie, H.A. Viles, R.E. Kim, J.J. Schmidt, D. Chakrabarty, E.A. Hadly, R.J. Hobbs, R. Ison) were only included on the *Web of Science* and 12 (T.J. Braje, A. Chin, J.M. Erlandson, I.J. Fairchild, U. Pöschl, M. Arias-Maldonado, M. Glaser, C. Hamilton, F. Huettmann, A.E. Lugo, A.V. Norström and P. Tarolli) were only noted on *Scopus*.

The *Web of Science* team has also indexed publications by scientific editors of the books included in their analysis. They listed the first 100 editors (by record count), the first 50 of whom are included in Table 22.11.

22.3.8 Anthropocene Texts Listed by Organisations and Universities on the Web of Science and Scopus (2001–2021)

Only the *Web of Science* and *Scopus* have included a list of universities and research organisations with whom the authors of scholarly publications on the Anthropocene have been associated (Table 22.12). 38 of the first 50 universities were included in both indexes.

The *Web of Science* lists in addition the following 12 academic institutions: 3. Centre National de la Recherche Scientifique ([CNRS] France), 7. State University System of Florida (USA), 17. Helmholtz Association (Germany), 18. Max Planck Society (Germany), 20. Arizona State University, Tempe (USA), 23. National Environment Research Council ([NERC] UK), 36. Pennsylvania PCSHE (USA), 44. University of North Carolina, 47. IRD, France, 48. Stellenbosch University (South Africa), 49. US Department of Agriculture, and 50. University of Waterloo (Canada).

Scopus has added 18 other institutions that were not on the *Web of Science*: 23. University of Melbourne (Australia), 28. James Cook University (Australia), 22. University of Arizona (USA), 30. University of Cornell (USA), 34. University of Helsinki (Finland), 35. University of Sydney (Australia), 37. University of Vienna (Austria), 38. University of Western Australia, 39. Lund University (Sweden), 42. University of Victoria (Australia), 43. U.S. Geological Survey, 44. Griffith University (Australia), 45. McGill University (Canada), 46. University of Florida (USA), 47. University of Tasmania (Australia), 48. Macquarie University (Australia), 49. University of Cape Town (South Africa), and 50. University of Copenhagen (Denmark).

The 3 US-based information-related companies have different goals with regard to the emergence of the scientific and popular work on the Anthropocene concept during

---

[37] See at: https://www.worldcat.org/search?q=kw%3AAnthropocene&fq=yr%3A2000..2020+%3E+&dblist=638&fc=s0:_25&qt=show_more_s0%3A&cookie (9 October 2020).

the past 2 decades. *Amazon's* interest has been to sell popular and scholarly books and publications in English and other languages to its global customers and thus offers a broad overview of the available information on the Anthropocene. However, the broadest overview of publications that have been released since spring 2000 is located in the *World Catalogue*. The *Web of Science* and *Scopus* serve the academic market of universities and research institutes following their distinct selection with a major focus on literature in peer-reviewed journals.

22.3.9   Websites on the Anthropocene – List of Hyperlinks

On 10 October 2020 Google presented a total of 4,670,000 results on the word 'Anthropocene', many of which are related to the bibliographic references analysed by the 3 US-based information processing companies. A second search on 3 February 2021 resulted in 6,260,000 references. Above (Sect. 22.3.5) a selection of websites with hyperlinks are offered. In Section 6 hyperlinks to videos on the Anthropocene are provided. Most of these websites are in English, but there are also many in French, German, Spanish, and Portuguese. Some websites offer short videos, slides, figures and graphics. Most websites are educational, while a few aim to outline the state of scientific knowledge on the Anthropocene.

22.4   Bibliometric Analyses of the Anthropocene in the Literature Listed Above by Scientific Discipline

There has been no consistent scientific discourse on the Anthropocene, but several parallel and overlapping discourses have evolved within and across scientific disciplines.[38] For many authors the Anthropocene has become a major turning point in Earth's history or a temporal context for their specific field and theme. Based on the bibliometric and scientometric methods used by the *Web of Science* and *Scopus* for their listings of the peer-reviewed scientific literature on the Anthropocene, this section will be limited to a structural quantitative review of the scientific literature on the Anthropocene in different disciplines.

This statistical disciplinary analysis may be a bibliographic first step towards a discourse analysis of the Anthropocene in this author's own discipline of political science, his research subdiscipline in international relations and his specific research programmes in peace, security and environmental studies (Brauch 2022).[39] This part

---

[38] Gildersleeve/Kleinhesselink (2017); Ruuska et al. (2019).

[39] See discourse analyses on the Anthropocene in social and political science: Ingram et al. (2015); Di Chiro (2016); Fagan (2016); Howard-Williams (2017); Demos (2017); Baldwin et al. (2019); Rubin (2020); Lövbrand et al. (2020).

offers quantitative and thematic analyses of the Anthropocene literature within and between disciplines:

- 4 phases of the gradual increase in the Anthropocene concept (22.4.1);
- the Anthropocene concept in the *Natural Sciences* (Geology, Global Environmental Change, Meteorology, Biology, Atmospheric Physics, and Chemistry; 22.4.2);
- in the *Social Sciences* (Anthropology, Human Geography, Political Science, Sociology, Psychology and Economics; 22.4.3);
- in the *Humanities* (Archaeology, History, Philosophy, Theology, Linguistics and Literature; 22.4.4);
- in *National and International Law* (22.4.5);
- in *Linguistics* and selected *Literatures* (22.4.6);
- in the *Performing Arts* (Painting, Sculpture, Photography, Films, Theatre; 22.4.7);
- in *Medicine* and the *Health Professions* (22.4.8);
- in *Technology* and *Engineering* (22.4.9).

22.4.1 Four Phases of the Gradual Increase in the Anthropocene

22.4.1.1 The Early Anthropocene Debate (2000–2005)

During the first 5 years (2000–2005) of the debate on the Anthropocene the *World Cat* listed under 'title' only 78 entries on the Anthropocene, while both the *Web of Science* and *Scopus* listed 15 documents. Among these early documents many texts were written by Crutzen (2002a, 2002b, 2002c) or co-authored by Crutzen with his colleagues, who were physicists and Earth System scientists (Will Steffen [2003, 2003a, 2004], Hans Joachim Schellnhuber [2004], the chemist V. Ramanathan [2004]). In this first 5-year period 3 single-authored texts (Chaps. 3, 4, 5) and 4 co-authored texts (Chaps. 2, 6, 7, 8) by Paul J. Crutzen were included.

During this first period the Anthropocene concept was discussed by meteorologists and Earth scientists (Martin Claussen et al. 2005), geographers Eckart Ehlers and Thomas Kraft (2005), hydrologists (Michel Meybeck; Charles Vörösmarty), and soil specialists (Blum/Eswaran 2004), and an environmental and Earth science journalist (Randy Showstack 2004).

These early book publications publicised the Anthropocene concept especially in the Natural Sciences (Steffen et al. 2002; Steffen et al. 2004; Schellnhuber et al. 2004; Ehlers/Krafft 2005; Oldfield 2005; Crossland et al. 2005). Only one French book in the Social Sciences picked up this discussion (Sinaï 2005).

Among the early peer-reviewed journals that published articles on the Anthropocene were:

- *Nature* (Crutzen 2002) and an anonymous editorial comment (2003);
- *Journal de Physique IV (Proceedings)*, 12,10 (2002): 1–5;

- *Journal of Environmental Science and Health Part A – Toxic/Hazardous Substances & Environmental Engineering* 37,4 (2006): 423–424;
- *Climatic Change,* 61,3 (2003): 251-257; 69,2-3 (2005): 409–417;
- *Philosophical Transactions of the Royal Society B – Biological Sciences,* 358,1440 (2003): 1935–1955;
- *New Perspectives Quarterly,* 22,2 (2005): 14–16;
- *Beijing Review,* 48,1 (6 January): 48;
- *American Journal of Science,* 305,9 (2005): 875–918;
- *Proceedings of the National Academy of Sciences of the USA,* 102,12 (2005): 4397–4402.

In the early phase (2000–2005) the scientific contributions on the Anthropocene were limited to the natural sciences. In the *Web of Science* 3 out of 15 peer-reviewed texts were authored by Crutzen and one was co-authored with Will Steffen. In *Scopus* 2 out of 15 peer-reviewed texts were authored by Crutzen and one was co-authored with Will Steffen. Both listed the following journals not noted in the *World Cat*:

- *Comptes Rendus Geoscience,* 337,1–2 (2005): 107–123;
- *Hydrological Processes,* 19,1 (2005): 331–338;
- *Geochimica et Cosmochimica Acta,* 68,19 (2004): 3807–3826;
- *Water Science and Technology,* 49,7 (2004): 73–83;
- *Aquatic Sciences,* 64,4 (2002): 376–393;
- *Scientia Marina,* 65,2 (2001): 85–105;
- *Journal of Infrastructure Systems,* 10,3 (2004): 79–86;
- *Water Science and Technology,* 49,7 (2004): 73–83.

22.4.1.2 Geologists Join the Anthropocene Debate (2006–2010)

During the second 5 years (2006–2010), the *World Catalogue* listed 273 Anthropocene publications under titles, the *Web of Science* included 100 documents, and *Scopus* listed 122. One single-authored text (Chap. 10) and one co-authored text by Crutzen from this second phase are included in this anthology. Several additional co-authored texts by Crutzen (Steffen/McNeill 2007; Steffen et al. 2010) could not be included because the publishers did not grant a free licence. In 2009 members of the *Anthropocene Working Group* (AWG) started to contribute texts on the Anthropocene from geological and Earth science perspectives (Zalasiewicz et al. 2008, 2008a), and gradually texts on the Anthropocene were published in the social sciences (see 22.4.3) and the humanities (22.4.4), including literature (poetry, novels, 22.4.6), but no specific discourses emerged within and across scientific disciplines.

Between 2006 and 2010 the *Web of Science* included a total of 65 articles, 13 papers in proceedings, 11 book chapters, 11 reviews and 10 editorial items, while *Scopus* assessed 61 articles, 20 book chapters, 11 conference papers, 11 reviews, 6 editorials, 6 notes, 3 books, 3 short surveys, 2 letters and 1 conference review. The *Web of Science* listed one author with 6 publications (B. Allenby), one with 5 texts (*W. Steffen*), one with 4 texts (P.J. Crutzen), 3 with 3 texts (R. Agniotri, S. Kurian, and Aqvi Swa), and 9 authors with 2 texts (S.C. Doney, A.J. Kettner, Y. Saito, D.S. Schimel, F. Schlutz, *J-P-M Syvitski, M. Williams,* A.P. Wolfe, *J. Zalasiewicz).*

Scopus included 2 authors with 5 texts each (B. Allenby, *W. Steffen),* one with 4 texts (B. Gomez) and 8 authors with 3 publications (R. Agnihotri, *P.J. Crutzen,* E. Ehlers, A.J. Kettner, S. Kurian, S.W.A. Naqvi, J.P.M. *Syvitsky,* and M. Zalewski).

A few books discussed the Anthropocene during the second 5-year period (Ehlers/Krafft 2006; Mackenzie 2006; Slaymaker 2006; Gautier 2008; Sammonds 2007; Grinevald 2008; Zalasiewicz et al. 2010a, 2010b; Vidas 2010). Increasingly, some volumes addressed aspects of relevance to the social and policy sciences (Bindé et al., 2007; Pearce, 2007; Sachs 2007, 2008; Smith 2008; Oswald Spring et al. 2009). According to *World Cat,* the peer-reviewed journals in the natural sciences that published articles on the Anthropocene included:

- *Carbon Balance and Management,* 1,1 (2006) [*Will Steffen*];
- *Global Biogeochemical Cycles,* 20,1 (2006): n.p. [A.J. Andersson et al.];
- *Repères Transdisciplinaires,* 2007 [*Grinevald*];
- *International Journal of Ecohydrology & Hydrobiology,* 7,2 (2007): 99–100 [Zalewski];
- *Global Environmental Change – Guildford,* 17,2, (2007): 149–151 [Kotchen, Young];
- *Geophysical Research Letters,* 34,18 (September 2007) [Mahowald, N.M.];
- *Ambio,* 36,8, (2007): 614–621 [*Steffen, Crutzen, McNeill*];
- *Journal of Experimental Botany,* 59,7: 1489–1502 [Osmond, Neales, Stange];
- *Annual Review of Environment and Resources,* 32,1 (2007): 31–66 [Doney];
- *Journal of Paleolimnology,* 37,4 (2007): 591–602 [Verburg];
- *GSA Today,* 18,2: 4–8 [*Zalasiewicz; Gibbard; Waters*];
- *Revista Internacional de Sostenibilidad, Tecnología y Humanismo,* 2008,3: 145–157;
- *Rangelands,* 30,5 (2008): 31–35 [deBuys];
- *Chemical & Engineering News,* 86,5 (2008): 3 [Baum Rudy];
- *Current science,* 95,1 (2008): 18–19 [Rajendran];
- *Mineralogical Magazine,* 72,1 (2008): 487 [Plant/Mckinlay/Voulvoulis];
- *Eos, Transactions American Geophysical Union,* 89,37 (9 September 2008): 343 [Safford];
- *Eos, Transactions American Geophysical Union,* 90,49 (8 December 2009): 473 [*Ellis/Haff*]: "Earth Science in the Anthropocene: New Epoch, New Paradigm, New Responsibilities";
- *Journal Open University Geological Society,* 30,2 (2009): 31–34 [*Williams/ Zalasiewicz*]: "**Enter the Anthropocene: An Epoch of Time Characterised by Humans**";
- *Advances in Geoecology,* 40 (2009): 1–16 [Eswaran; Reich; Vearasilp].

In addition to the above publications, the *Web of Science* included the following journals with at least 3 texts on the Anthropocene:

- *Global Biogeochemical Cycles* (4);

- *Journal of Paleolimnology* (4);
- *Biogeosciences* (3);
- *Current Opinion in Environmental Sustainability* (3);
- *Environmental Science Technology* (3);
- *International Legal Regime of Areas Beyond National Jurisdiction Current and Future Developments* (3);
- *Nova et Vetera Iuris Gentium Series: A Modern International Law* (3);

Scopus added the following journals with at least 3 texts on the Anthropocene:

- *Earth System Science in the Anthropocene* (8);
- *Journal of Paleolimnology* (4);
- *IEEE International Symposium on Electronics and the Environment* (3);
- *Intellectual Property Law Library* (3);

Journals that published texts on the Anthropocene in the social sciences (2006–2010) were:

- *American Behavioral Scientist*, 52,1 (2008): 107–140 [Allenby/Brad];
- *Gaia*, 17,3 (2008): 326–328 [Ratter/Welp], and 17,1 (2008): 77–80 [Glaser] in German;
- *SSRN Electronic Journal* (2009) [Rayfuse];
- *Global Environmental Change*, 19 (2009): 7–13 [Lövbrand/Stripple/Wiman], "Earth System Governmentality: Reflections on Science in the Anthropocene";
- *Millennium* (2010) [Hamilton/Scott], "Foucault's End of History: The Temporality of Governmentality and its End in the Anthropocene";
- *Erde: Zeitschrift der Gesellschaft für Erdkunde zu Berlin*, 141,4 (2010): 361–383 [Ehlers], "The Anthropocene – a New Chance for Geography?";
- *Political Studies* (2010) [Eckersley], "Geopolitan Democracy in the Anthropocene".

Among the first journals in the humanities that discussed the Anthropocene were:

- *The Virginia Quarterly Review*, 83,2: 266–267 [Nurkse (Poetry)];
- *Australian Humanities Review* (Online, 2009) [Rigby], "Writing in the Anthropocene: Idle Chatter or Ecoprophetic Witness?".

There was no clear pattern of disciplinary interest in the Anthropocene during the first 10 years and it is difficult to observe any emerging discourses, except in the Earth sciences, global environmental change and geology. However, beyond the natural sciences, there was great curiosity about the Anthropocene in the social sciences and the humanities as well as in the applied sciences.

22.4.1.3   The Anthropocene Debate Spreads Within the Natural Sciences (2011–2015)

During the third time period (2011–2015) the Anthropocene debate started to spread globally across many disciplines. The *World Catalogue* listed 6088 Anthropocene texts under keywords and 3219 publications under titles. The *Web of Science* included 434 references and *Scopus* 442.

During this period the *Web of Science* listed the following authors of 6 or more texts: *J. Zalasiewicz* (18), *M. Williams* (15), *W. Steffen* (12), *A.D. Barnosky* (11), *E.C. Ellis* (9), *C. Waters* (9), F. Biermann (8), S. Dalby (7), *A. Cearreta* (6), Rockström (6), *J.P.M. Syvitski* (6), Thomas, C.D. (6), and O.R. Young (6).

On *Scopus* between 2011 and 2015 the 13 most documented authors on the Anthropocene were: *J. Zalasiewicz* (21), *M. Williams* (11), S. Dalby (11), *A.D. Barnosky* (10), *W. Steffen* (10), *C. Waters* (10), *D. Vidas* (9), N. Castree (8), *E.C. Ellis* (8), *J.P.M. Syvitski* (8), *A. Cearreta* (6), J.M. Erlandson (6), and C.D. Thomas (6).

On both databases during the third phase the leading members of the *Anthropocene Working Group* (AWG) became the intellectual leaders of this knowledge network, especially the long-time chair, the palaeobiologist *Jan Zalasiewicz* (UK). According to the *Web of Science*, between 2009 and 2020 key AWG co-authors wrote a total of 105 publications, with their involvement being as follows: *J. Zalasiewicz* (43), *W. Steffen* (38), *M. Williams* (36), *E.C Ellis* (27), J.C. Svenning (26), *C.N. Waters* (26), L.J. Kotzé (24), *A.D. Barnovsky* (22), *A. Carreta* (22).

According to the *World Catalogue* for the years 2009 to 2020 the most counted authors with the 'Anthropocene' in the title between 2009 and 2020 were: D. Chandler (29), *J.A. Zalasiewicz* (39), A.Y. Glikson (27), *P.J.* Crutzen (25), M. Arias Maldonado (21), L. Arizpe Schlosse (21), *E.C. Ellis* (21), F. Biermann (20), S. Dalby (19), V. Blok (18), N. Castree (17), J. Bowen (15), H.G. Brauch (15), S.A. Elias (13), J. Lorimer (13), B. Williston (13), A. Barry (12), J. Baichwal (12), N. Clark (12), A. Federau (12), L. Gillson (12), and *W. Steffen* (12).

22.4.1.4   The Anthropocene Debate Proliferated Across Disciplinary Boundaries (2016–2020)

During the fourth period (2016–2020) under titles the *World Catalogue* listed 11,420 Anthropocene publications. On 3 February 2021 the *Web of Science* included a total of 4,493 references and *Scopus* 4095 texts (Table 22.1). During this phase the *Web of Science* included 3,319 articles on the Anthropocene, 488 book chapters, 384 editorial materials, 324 reviews, 286 book reviews, 145 early access texts, 118 proceedings papers and 43 books.

Among the authors with more than 13 texts on the *Web of Science* during this period (2016–2020) were: J.C. Svenning (24); L.J. Kotzé (23), *W. Steffen* (23), *J. Zalasiewicz* (23), *M. Williams* (19), *E.C. Ellis* (18), C.N. Collins (17), *A. Cearretaa* (16), F. Biermann (13), D. Chandler (13), C. Folke (13), *R. Leinfelder* (13), J. Rockström (13), and *M. Wagreich* (13).

On 3 February 2021 the *Scopus* list of the most prolific authors during the fourth period included: *J. Zalasiewicz* (25), J.C. Svenning (22), *M. Williams* (21), *E.C. Ellis* (19), *W. Steffen* (19), *A. Cearreta* (17), *C.N. Waters* (15), D. Chandler (12), S. Dalby (12), C. Folke (12), L.J. Kotzé (12), S.J. Cooke (11), *R. Leinfelder* (11), *J.R. McNeill* (11), *M. Wagreich* (11), *A.D. Barnovsky* (10), F. Biermann (10), J. Rockström (10) and *J. Syvitski* (10).

The *Web of Science* listed 43 books on the Anthropocene which were published between 2016 and 2020.*Scopus* listed 108 books for this period.[40] The selection of these books on the Anthropocene by the *Web of Science* (43) and *Scopus* (108) for the period from 2016 and 2020 did not follow any systematic assessment nor does it reflect any comprehensive transdisciplinary discourse on the new epoch of Earth's history. Instead they tend to focus on the temporal concept of the Anthropocene in order to provide a recontextualisation of the themes addressed, or on an observed turning point within the discipline analysed or the theme addressed by the authors.

22.4.2   The Anthropocene Concept in the Natural Sciences

Of the thousand or so publications listed under the Anthropocene on *Amazon,* about half of the titles (537) were listed under *Science and Mathematics,* 415 under *Nature and Ecology,* 371 under *Earth Sciences,* 363 under *Environment and Nature,* 170 under *Biological Sciences,* 46 under *History and Philosophy of Science* and 43 under *Evolution.*

On 30 September 2020 the *World Catalogue* categorised more than 13,386 publications searched by 'title' under *Biological Sciences* (94), *Physical Sciences* (26), *Chemistry* (1) and *Mathematics* (1). In a second search on 3 February 2021 the *World Catalogue* listed 15,192 documents of all formats on the Anthropocene, which it categorised by these topics in the *Natural Sciences*: Geography & Earth Sciences

---

[40] See these books: Vaz (2020); Eriksen/Ballard (2020); Torrescano-Valle et al. (2019); Leane/Mc Gee (2019); McLean (2019); Diogo et al. (2019); De Lucia (2019); Usher (2019); Scobie (2019); Arias-Maldonado/Trachtenberg (2019a), (2019b); Horn/Bergthaller (2019); Barry/Keane (2019); Moore (2019); Washington (2019); Southgate (2019); Fletcher (2019); Jaria-Manzano/Borràs (2019); Goodenough/Waite (2019); Giblett (2019); Sultana/Loftus (2019); Ezban (2019); Chan (2018); Cosens/Gunderson (2018); Stephenson (2018); Macilenti (2018); Chandler (2018); Tobias/Morrison (2018); Fay (2018); Hickmann et al. (2018); Costa (2018); Bowen/Gleeson (2018); Kaczmarczyk (2018); Grandcolas/Maurel (2018); Birch et al. (2018); Wood (2018a), (2018b); Neyrat/Neyrat (2018); Grusin (2018); González-Ruibal (2018); Erdelen/Richardson (2018); Nyman (2018); Taylor/Pacini-Ketchabaw (2018); Cheetham (2018); McEwan (2018); Marouby (2018); Yilamu (2017); Philippopoulos-Mihalopoulos/Brooks (2017); Higgins (2017); Kelly et al. (2017); Figueroa (2017); Ison (2017); MacGregor (2017); Westra et al. (2017); Buck/Langan (2017); Cromsigt et al. (2017); Dellasala (2017); Denny (2017); Hardt (2017); Tyszczuk (2017); Cronin (2017); Stevens et al. (2017); Blok (2017); Malone et al. (2017); Fishel (2017); Pennycook (2017); Mason/Lang (2017); Munns (2017); Viola/Franchini (2017); Baghel et al. (2017); Keim (2017); Meyer (2016); Everard (2016); Alaimo (2016); Harvey et al. (2016); Palidda (2016); Haraway (2016a), (2016b); Connor (2016): Heesterman/Heesterman (2016); Noye (2016); Adamson/Davis (2016); Issberner/Léna (2016).

(250), Biological Sciences (98), Physical Sciences (25), Chemistry (1), and Mathematics (1); in the *Social Sciences*: Sociology (309), Business & Economics (99), Political Science (73), Law (49), Anthropology (30), Psychology (5); in the *Humanities*: Art & Architecture (109), History & Auxiliary; (46); Language, Linguistics and Literature (129), Philosophy & Religion (102), Education (26), Music (18), Library Science (16), and Performing Arts (8); and in the *Applied Sciences*: Engineering & Technology, (57), Agriculture (29), Medicine (14), Health Profession (3), Physical Education, (2), Medicine by Discipline (1), and Health Facilities (1).

In a first search on 30 September 2020, the *Web of Science* with its then 5,080 entries (3 February 2021: 5569) grouped 1,702 titles under *Environmental Sciences and Ecology*, 608 under *Geology*, 303 under *Physical Geography*, 121 under *Meteorology and Atmospheric Science*, 142 under *Biodiversity Conservation*, 110 under *Marine and Freshwater Resources*, 103 under *History and Philosophy of Science*, 87 under *Evolutionary Biology*, 60 under *Chemistry*, 60 under *Plant Sciences*, 42 under *Oceanography*, 37 under *Zoology*, 30 under *Biochemistry and Molecular Biology*, 30 under *Genetics and Heredity* and 22 under *Geochemistry and Geophysics*.

In a second search on 3 February 2021, the *Web of Science* with its then 5,569 documents grouped them under Environmental Sciences (946), Environmental Studies (923), Geo-sciences: Multidisciplinary (576), Ecology (529), Geography (371), Geography: Physical (312) Multidisciplinary Sciences (253), Green Sustainable Science Technology (226), Literature (220), Philosophy (207), Humanities: Multidisciplinary (202), Political Science (186), Social Sciences: Interdisciplinary (174), Educational Research (173), Biodiversity Conservation (161), Cultural Studies (139), Sociology (139), Biology (133), Meteorology Atmospheric Sciences (127, History (118), History: Philosophy of Science (112), Water Resources (112), Law (111), International Relations (109), Marine Freshwater Biology (103), Evolutionary Biology (99), Religion (95), Literary Theory Criticism (94), Ethics (91), Economics (83), Art (74), Engineering: Environmental (71), Regional Urban Planning (71), Geology (70), Archaeology (69), Plant Sciences (65), Architecture (60), Area Studies (58), Management (54), Public Environmental Occupational Health (52), Chemistry: Multidisciplinary (48), Development Studies (48), Oceanography (47), Communication (46), Forestry (45), Hospitality Leisure Sport Tourism (45), Literary Reviews (40), Social Issues (40), Film Radio Television (39), Urban Studies (38), Zoology (38), Soil Science (34), Biochemistry: Molecular Biology (33), Education Scientific Disciplines (31), Genetics Heredity (31), Limnology (31), Paleontology (29), Asian Studies (27), Agriculture Multidisciplinary (26), Theatre (25), Language & Linguistics (24), Geochemistry: Geophysics (23), Public Administration (21), Women's Studies (21), Cell Biology (20), Criminology: Penology (20), Fisheries (20), Literature: American (20), Music (19), Medicine: General Internal (17), Entomology (16), Literature: Slavic (15), Social Sciences: Biomedical (14), Computer Science Interdisciplinary Applications (14), Business (13), Information Science/Library Science (13), Remote Sensing (13), Energy Fuels (12) Engineering: Multidisciplinary (12), Literature: German, Dutch, Scandinavian (12), Agronomy (11), Linguistics (11), Microbiology (11), Transportation (11), Poetry (10), Veterinary Sciences (10), Agricultural Economics Policy (9), Engineering: Civil (9), Medical Ethics (9), Chemistry:

Physical (8), Computer Science Theory Methods (8), History of Social Sciences (8), Infectious Diseases (8), Psychology: Multidisciplinary (8), Engineering: Electrical Electronic (7), Engineering: Industrial (7), Ethnic Studies (7), and Astronomy: Astrophysics (6).

Finally, *Scopus* applied a different categorisation to its titles, using during the first search on the Anthropocene on 30 September 2020: *Environmental Science* (1,787 titles), *Earth and Planetary Science* (1,021), *Agriculture and Biological Science* (936), *Biochemistry, Generics, Molecular Biology* (190), *Chemistry* (79), *Immunology and Microbiology* (57), *Physics and Astronomy* (52), *Mathematics* (29) and Neuroscience (16 on 4 March 2021).

In a second search on 3 February 2021 *Scopus* listed 5,389 documents on the Anthropocene, which it categorised by the following subject areas: *Social Sciences* (2278), *Environmental Science* (1923), *Arts and Humanities* (1380), *Earth and Planetary Sciences* (1099), *Agricultural and Biological Sciences* (1017), *Engineering* (254), *Business, Management and Accounting* (241), *Economics, Econometrics and Finance* (241), *Biochemistry, Genetics and Molecular Biology* (200), *Medicine* (174), *Multidisciplinary* (173), *Energy* (118), *Computer Science* (98), *Chemistry* (85), *Immunology and Microbiology* (61), *Psychology* (53), *Physics and Astronomy* (51), *Mathematics* (31), *Decision Sciences* (28), *Materials Science* (19), *Chemical Engineering* (18), *Nursing* (17), *Neuroscience* (16), *Pharmacology, Toxicology and Pharmaceutics* (8), *Veterinary* (8) and *Health Professions* (6).

As the categories used by *Amazon*, the *World Catalogue*, the *Web of Science* and *Scopus* differed, it is not possible to select a sample of publications (books and journals) for each discipline as a precondition for reconstructing an ongoing discourse within a scientific discipline or across scientific disciplines. The following bibliometric analysis includes only the peer-reviewed texts on the *Web of Science* and *Scopus* for which abstracts were provided.

22.4.2.1 The Anthropocene Concept in Geology and in Earth Systems Science on Global Environmental Change Issues

On 21 October 2020 the *Web of Science* listed 559 texts on *multidisciplinary geosciences*, 261 on *physical geography* and 63 texts on *geology*, or together a total of 617 texts. 10 of the 13 most counted authors in these disciplines were members of the AWG in 2019: *J. Zalasiewicz* (20), *M. Williams* (19), *A.D. Barnosky* (18), *C. Waters* (18), *W. Steffen* (17), *A. Cearreta* (14), *M. Wagreich* (12), *M. Edgeworth* (10), *E.C Ellis* (10), and *R. Leinfelder* (10), who jointly wrote 48 texts. Between 2012 and 2020 this small group constituted the core scientific knowledge network that was instrumental in discussions about the formal acceptance of the term as a new epoch in Earth's history by the geological community.

In a text on "The Anthropocene" in *Geologic Time Scale*, Zalasiewicz et al. (2012: 1033–1040) commented that the

Anthropocene is a currently informal term to signify a contemporary time interval in which surface geological processes are dominated by human activities, now being studied by an ICS working group as regards potential formalization within the Geological Time Scale. Its developing stratigraphic signature includes components that are lithostratigraphic, biostratigraphic, and chemostratigraphic; and these vary from being approximately synchronous to strongly diachronous. Formalization will depend upon both scientific justification and utility to working scientists, and upon the choice of an effective boundary, whether by GSSP or GSSA.

Two years later, *Waters* et al. (2014) discussed whether there was "A stratigraphical basis for the Anthropocene?" in a Special Publication of the Geological Society, London (Vol. 395: 1–21):

Recognition of intimate feedback mechanisms linking changes across the atmosphere, biosphere, geosphere and hydrosphere demonstrates the pervasive nature of humankind's influence, perhaps to the point that we have fashioned a new geological epoch, the Anthropocene. To what extent will these changes be evident as long-lasting signatures in the geological record? To establish the Anthropocene as a formal chronostratigraphical unit it is necessary to consider a spectrum of indicators of anthropogenically-induced environmental change and determine how these show as stratigraphic signals that can be used to characterise an Anthropocene unit and to recognise its base. It is important to consider these signals against a context of Holocene and earlier stratigraphic patterns. Here we review the parameters used by stratigraphers to identify chronostratigraphical units and how these could apply to the definition of the Anthropocene. The onset of the range of signatures is diachronous, though many show maximum signatures which post-date 1945, leading to the suggestion that this date may be a suitable age for the start of the Anthropocene.

The same Special Publication of the Geological Society, London includes 7 additional texts co-authored by: *Zalasiewicz, Williams* and *Waters* ("Can an Anthropocene Series be Defined and Recognized?": 39–53); Ford, Price, Cooper et al. ("An Assessment of Lithostratigraphy for Anthropogenic Deposits": 55–89); *Edgeworth* ("The Relationship between Archaeological Stratigraphy and Artificial Ground and its Significance in the Anthropocene": 91–108); *Zalasiewicz, Kryza* and *Williams* ("The Mineral Signature of the Anthropocene in its Deep-Time Context": 109–117); *Williams, Zalasiewicz, Waters* et al. ("Is the Fossil Record of Complex Animal Behaviour a Stratigraphical Analogue for the Anthropocene?": 143–148); *Barnosky* ("Palaeontological Evidence for Defining the Anthropocene": 149–165); and *Galuszka*, Migaszewski and *Zalasiewicz* ("Assessing the Anthropocene with Geochemical Methods": 221–238).

Additional members of the *Anthropocene Working Group* (AWG) supplied case material providing evidence from sediments that a change in Earth's history has been taking place since the mid-20th century. Their findings appeared in 3 texts that were presented at the *1st International Congress on Stratigraphy* (STRATI) in Lisbon on 1–7 July 2013 and subsequently published in *Springer Geology* in 2014. These were by *Wagreich* ("Do Old Mining Waste Deposits from Austria Define an 'Old' Anthropocene?"); *Waters, Zalasiewicz, Williams* et al. ("Evidence for a Stratigraphic Basis for the Anthropocene"); and *Zalasiewicz, Waters* and *Williams* ("Potential Formalization of the Anthropocene: A Progress Report"). 3 geological texts were published by members of the AWG in *Quaternary International*: (1) Garcia-Artola, *Cearreta*,

Leorri: "Relative Sea-Level Changes in the Basque Coast (Northern Spain, Bay of Biscay) during the Holocene and Anthropocene: The Urdaibai Estuary Case" (364 [2015]: 172–180); (2) Irabien, Garcia-Artola, *Cearreta* et al.: "Chemostratigraphic and Lithostratigraphic Signatures of the Anthropocene in Estuarine Areas from the Eastern Cantabrian Coast (N. Spain)" (364 [2015]: 196–205); and (3) *Zalasiewicz, Waters, Williams* et al.: "When Did the Anthropocene Begin? A Mid-Twentieth-Century Boundary Level is Stratigraphically Optimal" (383 [2015]: 196–203). Two texts were published by an AWG member in *Cuaternario y Geomorfologia*: (1) *Cearreta*: "The Anthropocene Era and the Necessary Steps for its Possible Formalization after the 35th International Geological Congress" (30,3–4 [2016]: 5–8); and (2) Silva, Bardaji, Roquero, *Cearreta* et al.: "The Quaternary Period: The Geological History of Prehistory" (31,3–4 [2017]: 113–154). One text was published in the *Newsletters on Stratigraphy* by *Zalasiewicz, Waters*, Wolfe et al.: "Making the Case for a Formal Anthropocene Epoch: An Analysis of Ongoing Critiques" (50,2 [2017]: 205–226). Another text was published by *Zalasiewicz, Waters, Williams* et al. on: "The Stratigraphical Signature of the Anthropocene in England and its Wider Context", in: *Proceedings of the Geologists Association* (129,3 [2018]: 482–491). One geological text by Terrington, Silva, *Waters* et al., "Quantifying Anthropogenic Modification of the Shallow Geosphere in Central London, UK", appeared in: *Geomorphology* (319 [2018]: 15–34). AWG member *Ellis* published "Physical Geography in the Anthropocene" in: *Progress in Physical Geography – Earth and Environment* (41,5 [2017]: 525–532). *Waters, Zalasiewicz, Summerhayes* et al. published "Global Boundary Stratotype Section and Point (GSSP) for the Anthropocene Series: Where and How To Look for Potential Candidates", in: *Earth-Science Reviews* (178 [2018]: 379–429).

The next 18 texts were published in the *Anthropocene Review*[41] primarily by one or several AWG members (in *italics*) together with experts from other disciplines on issues of global environmental change.[42] Since 2014 these texts have helped to broaden the theme of the Anthropocene to the Earth Sciences in general and to Global Environmental Change issues. AWG members also published their results in

[41] *The Anthropocene Review* is published by SAGE as "a trans-disciplinary journal issued 3 times per year, bringing together peer-reviewed articles on all aspects of research pertaining to the Anthropocene, from Earth and environmental sciences, social sciences, material sciences, and humanities". In 2020 the Editor was: Robert Costanza, Crawford School of Public Policy, ANU, Australia and the North American Editor was AWG member Anthony D. Barnosky, Stanford University, USA.

[42] See the texts here: Oldfield et al. (2014); Zalasiewicz et al. (2017); Oldfield/Steffen (2014); Barnosky/Hadly (2014); Barnosky et al. (2014a), (2014b), (2014c); Edgeworth et al. (2015); Steffen et al. (2015); Zalasiewicz et al. (2015a); Williams et al. (2015); Zalasiewicz et al. (2015b); Donges et al. (2017); Gaffney/Steffen (2017); Fox et al. (2017); Gaffney (2017); Wagreich/Draganits (2018); Cooper et al. (2018).

the Journal *Anthropocene*.[43] Finally, key members of the AWG additionally published their results in the journal *Earth's Future*.[44]

The core group of active AWG members has placed the Anthropocene on the scientific map of Geology, Earth Systems Science and Physical Geography, and AWG members have also co-authored many texts with leading authors from related research communities primarily in the Natural Sciences but also in the Social Sciences, thus forging an extended scientific network that has contributed to the rapid spread of texts on the Anthropocene, especially since 2017.

Scopus assessed 1031 texts in *Earth and Planetary Sciences*, 83 in *Chemistry* and 51 in Physics and *Astronomy*, which amounted to 1140 documented results. In 2019 13 of the 15 most counted authors in these disciplines were AWG members: *J. Zalasiewicz*, (32), *M. Williams* (31), *W. Steffen* (22), *A. Cearreta* (20), *C.N. Waters* (20), *A.D. Barnosky* (18), *J.P.M. Syvitski* (13), *E.C. Ellis* (12), *M. Wagreich* (12), *R. Leinfelder* (11), *M. Edgeworth* (10), *C. Summerhayes* (10), and *D. Vidas* (10), who produced 86 documents. Only those that were excluded by the *Web of Science* are listed in the footnote.[45]

---

[43] The journal *Anthropocene* is published by Elsevier and its Editor-in-Chief is Anne Chin, University of Colorado Denver, Denver, Colorado, US. It is described as "an interdisciplinary peer-reviewed journal answering questions about the nature, scale and extent of interactions between *people* and *Earth* processes and systems. The scope of the journal includes the significance of human activities in altering Earth's *landscapes, oceans, the atmosphere, cryosphere, and ecosystems* over a range of time and space scales – from global phenomena over geologic eras to single isolated events – including the linkages, couplings, and feedbacks among physical, chemical, and biological components of Earth systems. The journal also addresses how such alterations can have profound effects on, and implications for, human society. As the scale and pace of human interactions with Earth systems have intensified in recent decades, understanding human-induced alterations in the past and present is critical to our ability to anticipate, mitigate, and adapt to changes in the future. The journal aims to provide a venue to focus research findings, discussions, and debates toward advancing predictive understanding of human interactions with Earth systems – one of the grand challenges of our time"; at: https://www.-journals.elsevier.com/anthropocene (22 October 2020). Zalasiewicz et al. (2016); Zalasiewicz et al. (2017).

[44] The journal *Earth's Future* is published by Wiley as a "transdisciplinary, Gold Open Access journal examining the state of the planet and its inhabitants, sustainable and resilient societies, the science of the Anthropocene, and predictions of our common future through research articles, reviews and commentaries"; at: https://agupubs.onlinelibrary.wiley.com/journal/23284277?tabAct ivePane= (22 October 2020). See the texts: Williams et al. 2016; Steffen et al. 2016.

[45] Steffen (2003); Steffen (2006); Syvitski/Saito (2007); Kettner et al. (2008); Ellis/Haff (2009); Steffen (2010); Zalasiewicz et al. (2010a), (2010b); Zalasiewicz et al. (2011b); Steffen et al. (2011a); Vidas (2011); Syvitski/Kettner (2011); Zalasiewicz et al. (2011a); Zalasiewicz et al. (2012); Ellis et al. (2013); Syvitski et al. (2013); Rogers et al. (2013); Overeem et al. (2013); Zalasiewicz/Williams (2014); Waters et al. (2014a); Barnosky et al. (2014c); Zalasiewicz et al. (2014b); Barnosky (2014); Waters et al. (2014c); Barnosky et al. (2014a); Williams et al. (2014); Zalasiewicz et al. (2015b); Williams et al. (2015b); Vidas et al. (2015); Seitzinger et al. (2015); Cearreta (2015); Zalasiewicz/Williams (2016); Williams et al. (2016); Cearreta et al. (2017); Gaffney/Steffen (2017); Cearreta et al. (2018); Zalasiewicz et al. (2018b); Waters et al. (2018a); Zalasiewicz et al. (2018a); Williams et al. (2018); Summerhayes/Zalasiewicz (2018); Irabien et al. (2018); Terrington et al. (2018); Altuna et al. (2019); Wright et al. (2019a), (2019b); Tarolli et al. (2019); Ibáñez et al. (2019);

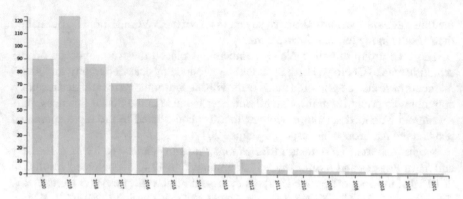

Fig. 22.3 Number of titles on the Anthropocene in biology that were recorded on the *Web of Science* from 2001 to 2020

Since its establishment in 2009 the members of the *Anthropocene Working Group* (AWG) have been a very active scientific network in the Natural Sciences, especially Geology, Earth Sciences and subjects relating to global environmental and climate change. As a professional organisation within Geology, the AWG performed multiple functions: its members published new knowledge, advanced the debates in the respective communities, and spread the scientific results to other disciplines within and beyond the Natural Sciences to the Social Sciences, Humanities and Law.

22.4.2.2 The Anthropocene Concept in Biology and in Biosciences

The *Web of Science* also assessed titles on the Anthropocene in Biology, which on 5 October 2020 were categorised under *Biodiversity Conservation* (147), *Biology* (128), *Marine and Freshwater Biology* (87), *Plant Sciences* (61), *Forestry* (43), *Zoology* (36), *Biochemistry and Molecular Biology* (31). *Cell Biology* (19), *Microbiology* (10), and *Biotechnology: Applied Microbiology* (7) which together referred to 493 texts. On 23 October 2020, 342 of these entries were articles, 79 were reviews, 60 were editorial material, 43 were book chapters and 18 were papers from conference proceedings. Research on the Anthropocene took off in Biology with 17 texts in 2014 and reached its highest level in 2019 with 123 titles (Fig. 22.3).

The most recorded 25 authors in Biology contributed 72 texts. Among the top 100 authors was only one AWG member (*E.C. Ellis*), who contributed 4 texts.[46]

Scopus categorised Anthropocene titles in Biology differently, using the categories *Agriculture and Biological Sciences* (936), *Biochemistry, Genetics, Molecular Biology* (190), and *Immunology and Microbiology* (57), which together produced 1027 document results. Among the top 100 authors listed, there were only 4 AWG

Edgeworth et al. (2019); Zalasiewicz et al. (2019b); Sames et al. (2020); Sáez-Muñoz et al. (2020); Randin et al. (2020); Himson et al. (2020); Bianchi et al. (2020); Irabien et al. (2020).

[46] Seddon et al. 2014; Martin et al. 2014b; Campagnaro et al. 2019; Ellis 2019.

members: *E.C. Ellis* (9). *J. Zalasiewicz* (3), *M. Williams* (3) and *A.D. Barnovsky* (1) with a total of 11 document results.[47]

Not surprisingly, the scientists in the AWG played only a minor role as authors and co-authors in Biology, and their peer-reviewed publications on the Anthropocene started a few years later in 2014. The highly cited hot paper on the Anthropocene in Biology, with 180 citations (June 2019 – October 2020), was authored by Reid, Andrea J.; Carlson, Andrew K.; Creed, Irena F.; et al., 2019: "Emerging Threats and Persistent Conservation Challenges for Freshwater Biodiversity", in: *Biological Reviews* (94,3: 849–873), which addressed a totally different theme:

> In the 12 years since Dudgeon et al. (2006) reviewed major pressures on freshwater ecosystems, the biodiversity crisis in the world's lakes, reservoirs, rivers, streams and wetlands has deepened. While lakes, reservoirs and rivers cover only 2.3% of the Earth's surface, these ecosystems host at least 9.5% of the Earth's described animal species. Furthermore, using the World Wide Fund for Nature's Living Planet Index, freshwater population declines (83% between 1970 and 2014) continue to outpace contemporaneous declines in marine or terrestrial systems. The *Anthropocene* has brought multiple new and varied threats that disproportionately impact freshwater systems. We document 12 emerging threats to freshwater biodiversity that are either entirely new since 2006 or have since intensified: (i) changing climates; (ii) e-commerce and invasions; (iii) infectious diseases; (iv) harmful algal blooms; (v) expanding hydropower; (vi) emerging contaminants; (vii) engineered nanomaterials; (viii) microplastic pollution; (ix) light and noise; (x) freshwater salinisation; (xi) declining calcium; and (xii) cumulative stressors. Effects are evidenced for amphibians, fishes, invertebrates, microbes, plants, turtles and waterbirds, with potential for ecosystem-level changes through bottom-up and top-down processes. In our highly uncertain future, the net effects of these threats raise serious concerns for freshwater ecosystems. However, we also highlight opportunities for conservation gains as a result of novel management tools (e.g. environmental flows, environmental DNA) and specific conservation-oriented actions (e.g. dam removal, habitat protection policies, managed relocation of species) that have been met with varying levels of success. Moving forward, we advocate hybrid approaches that manage fresh waters as crucial ecosystems for human life support as well as essential hotspots of biodiversity and ecological function. Efforts to reverse global trends in freshwater degradation now depend on bridging an immense gap between the aspirations of conservation biologists and the accelerating rate of species endangerment.

In this text the Anthropocene was interpreted as a major 'challenge' or 'threat' for Earth's ecosystems. In Biology the Anthropocene stands for a 'turning point' or for a 'new context' which nature and its ecosystems are facing. In the Biosciences and Biology J.C. Svenning (Denmark, 25–24), M. Dornelas (UK, 0:7) and F. Huettmann (USA, 0:7) were the key authors of texts on the Anthropocene.

---

[47] Ellis et al. (2012); Zalasiewicz/Williams (2014); Seddon et al. (2014); Martin (2014b); Ellis (2015); Bennett et al. (2016); Zalasiewicz et al. (2017i); Cantrell et al. (2017); Campagnaro et al. (2019); Ellis (2019); Himson et al. (2020).

22.4.3 The Anthropocene: An Emerging Concept in the Social Sciences: Common and Distinct Features

Out of approximately 1000 publications connected with the Anthropocene on *Amazon* in late September 2020 about half (512) of the titles were listed under *Politics and Social Sciences* and 259 were subcategorised under *Social Sciences*, 185 under *Politics & Government*, 140 under *Philosophy*, 74 under *Anthropology* and 73 under Sociology. *Amazon* also listed 139 titles under *Business and Money*. On 5 February 2021 this categorisation did not exist any longer.

The *World Catalogue* categorised more than 15,251 publications on the Anthropocene on 5 February 2021, searched by title under *Sociology* (310), *Geography & Earth Sciences* (250), *Business & Economics* (98), *Political Science* (74), *Anthropology* (30), *Education* (26), and *Psychology* (5).

The *Web of Science,* with 5,574 Anthropocene entries on 5 February 2021, listed 371 titles under *Geography*, 207 under *Philosophy*, 190 under *Anthropology*, 186 under Political Science, 174 under *Social Sciences: Interdisciplinary,* 173 under *Education and Educational Research*, 139 under *Sociology*, 139 under *Cultural Studies*, 111 under Law, 109 under *International Relations*, 83 under Economics, 58 under *Area Studies*, 48 under *Development Studies*, 46 under *Communication*, 40 under *Social Issues*, 31 under *Education Scientific Disciplines,* 27 under *Asian Studies,* 21 under Women's Studies, 21 under *Public Administration,* 13 under *Business* and 8 under *Multidisciplinary Psychology*.

On 5 February 2021, *Scopus* again used different categories for its total of 5,398 Anthropocene results. Under Social Sciences there were 2,282 Anthropocene titles, subcategorised under headings such as *Business, Management and Accounting* (241 titles), *Economics, Econometrics and Finance* (210 titles), *Psychology* (53 titles), and *Decision Sciences* (28 titles).

The peer-reviewed Social Sciences texts from selected disciplines listed on the *Web of Science* and *Scopus* on 5 February 2021 are analysed below: *Human Geography* (22.4.3.1), *Anthropology* (22.4.3.2), *Political Science* (22.4.3.3), *Sociology* (22.4.3.4), *Economics* (22.4.3.5), *Psychology* (22.4.3.6), and *Education* (22.4.3.7).

22.4.3.1 The Anthropocene Concept in Human Geography

On 5 February 2021 the *Web of Science* listed the following titles on the Anthropocene under *Geography* (371), Physical Geography (312) and *Regional Urban Planning* (71). A total of 430 documents included 331 articles, 46 book chapters, 39 editorial materials and 26 book reviews. Among the top 100 most counted authors were S. Dalby (8), J.J. Schmidt (7) and K. Yusoff (7), and AWG members: *W. Steffen* (4), *J. Syvitski* (3) and *J. McNeill* (2), who published 7 of the 430 texts in Geography, 11 of which were highly cited papers.[48] The 20 most counted authors in

---

[48] Gibson-Graham et al. 2019a, 2019b; Norman/Steffen 2018; Brondizio/*Syvitski* 2016; Brondizio et al. 2016; Verburg et al. 2016; Bai et al. 2016.

Geography contributed 81 texts, while the 10 most counted authors (co)-authored 56 texts. Among them were: S. Dalby (8), J.J. Schmitt (7), K. Yusoff (7), N. Castree (6), J. Lorimer (6), J.K. Gibson-Graham (5), S. Wakefield (5), K. Barry (4), D. Houston (4), and R.A. Slaughter (4). *Scopus* did not use (Human) Geography as a category for the texts it assessed on the Anthropocene.

22.4.3.2 The Anthropocene Concept in Anthropology

On 7 February 2020 the *Web of Science* listed the following titles on the Anthropocene under Anthropology (190), Cultural Studies (139), and Ethnic Studies (7), which resulted in 314 documents of the selected texts. The 10 most highly counted authors out of a total of 100 were: N. Bubandt (Denmark, 6), K. Yusoff (UK, 5), T.J. Braje (USA, 4), Arizpe Schlosser (Mexico, 4), N. Clark, (UK, 3), M. Denny (Germany/USA, 3), A. Fuentes (USA, 4), C. Howe (USA, 3), C. Isenhour (USA, 3) and M. Lock (Canada, 3). The most highly cited Anthropology text on the Anthropocene, with 60 citations, was by Nigel Clark and Kathryn Yusoff (2017): "Geosocial Formations and the Anthropocene", in: *Theory Culture & Society* (34,2–3: 3–23). Among the top 100 most counted authors were 4 AWG members who co-authored 2 texts.[49]

Scopus did not use Anthropology as a category for the texts it assessed on the Anthropocene. However, a direct search on 5 February 2021 referred to 1,018 counted results. Among the 7 most highly counted authors were 3 AWG members: *E.C. Ellis* (USA, 11), *M. Williams* (UK, 7) and *J. Zalasiewicz* (UK, 7), and 4 other colleagues: T.J. Braje (USA, 11), J.M. Erlandson (USA, 10), A. Fuentes (USA, 7), T.C. Rick (7) who produced a total of 45 texts together.[50]

In Anthropology the top key authors of texts on the Anthropocene on the *Web of Science* were: N. Bubandt (Denmark, 6), K. Yusoff (UK, 5), and L. Arizpe Schlosser (Mexico, 4), and on *Scopus* T.J. Braje (USA, 11), J.M. Erlandson (USA, 10), A. Fuentes, (7) and T.C. Rick (7).

---

[49] Zalasiewicz et al. 2017b; Zalasiewicz/Waters 2016.

[50] Fuentes (2012); Braje/Erlandson (2013a), (2013b); Erlandson (2013); Ellis et al. (2013a); Rick et al. (2013); Ellis et al. (2013b); Rick et al. (2014); Malone et al. (2014); Hofman et al. (2015); Williams et al. (2015); Ellis (2015); Braje (2015); Williams et al. (2016); Braje (2016a), (2016b); Hofman et al. (2016); Zalasiewicz et al. (2017f); Fuentes (2017); Williams et al. (2017); Fuentes/Baynes-Rock (2017); Braje et al. (2017); Fox et al. (2017); Ellis et al. (2018); Hofman/Rick (2018); Bauer/Ellis (2018); Fitzpatrick/Erlandson (2018); Rick et al. (2018); Bennett et al. (2018); Tarolli et al. (2019); Fuentes (2019); Edgeworth et al. (2019); Zalasiewicz et al. (2019b); Fuentes (2020); Rick et al. (2020); Marcantonio/Fuentes (2020); Rick/Sandweiss (2020); Braje/Lauer (2020); Gasparin et al. (2020); Rick et al. (2020).

22.4.3.3   The Anthropocene Concept in Political Science

On 5 February 2021 the *Web of Science* listed the following Political Science titles on the Anthropocene under *Political Science* (186), *Social Sciences Interdisciplinary* (174), *International Relations* (109), *Area Studies* (58), *Development Studies* (48), *Social Issues* (40), *Urban Studies* (38), *Asian Studies* (27), *Women's Studies* (21), and *Public Administration* (21), which together resulted in 617 titles consisting of articles (443), book chapters (141), book reviews (60), editorial material (56), and proceedings papers (36).

There was no AWG member among the top 100 Political Science authors (by record count). The top 17 of the 100 listed authors who contributed Political Science titles on the Anthropocene were: F. Biermann (9), S. Dalby (8), O.R. Young (8), M. Arias-Maldonado (5), J.B. Foster (4), R.W. Luke (4), Ú. Oswald Spring (4), I. Angus (3), H.G. Brauch (3), W. Gabardi (3), J. Gupta (3), T. Hickmann (3), R. Karlsson (3), L. Partzsch (3), P. Pattberg (3), J.C. Pereira (3), and S. Weiland (3), who between them produced 53 documents.[51] On 4 March 2021 the *Web of Science* counted 347 titles in the Social Sciences, 179 of which were listed under *Political Science*, 58 under *International Relations*, 332 under *Environment Studies*, 210 under *Environment Politics*, 6 under *Peace Studies*, 53 under *Security Studies* and 17 under *Gender Studies*.

Scopus categorised its titles differently. On 5 February 2021 it counted 2,282 Social Science documents on the Anthropocene, 1,690 titles of which were listed under *Political Science*, 307 under *International Relations*, 1,261 under *Environment Studies*, 105 under *Peace Studies*, 924 under *Development Studies*, 335 under *Security Studies*, and 233 under *Gender Studies*. Due to the totally different categorisation, the data on Political Science are not comparable and may not be representative of the Political Science publications on the Anthropocene.

The Anthropocene: Impact on International Relations

On 6 February 2021 the *Web of Science* counted a total of 110 *International Relations* titles on the Anthropocene, 88 of which were articles, 29 book chapters, 12 book

---

[51] Dalby (2011); Young (2013); Arias-Maldonado (2013); Dalby (2013a), (2013b); Brown/Schmidt (2014); Biermann (2014); Biermann/Young (2014a), (2014b): vii; (2014c): ix-xii; (2014d): 203–213; Dalby (2014a), (2014b): 1–9; Angus (2015): 1–11; Luke (2015b): 3–14; Luke (2015a): 139–162; Arias-Maldonado (2015): 83–102; Foster (2016): 393–421; Brauch/Oswald Spring (2017): 3–66; Dalby (2017b): 67–88; Oswald Spring (2017): 161–185; Brauch (2017b): 187–236; Oswald Spring (2017): 675–704; Oswald Spring et al. (2017): 887–927; Biermann/Young (2016): xi–xii; Biermann (2016a): 405–420; Kim/van Asselt (2016): 473–495; Biermann (2016b): 178–178; Gupta/Vegelin (2016): 433–448; Gabardi (2017); Pereira (2017): e018; Foster (2017): 439–458; Pereira (2017): e018; Gabardi (2017c): 11–29; Gabardi (2017b): 175–185; Karlsson (2018): 77–84; Symons/Karlsson (2018): 685–704; Dalby (2018a): 718–742; Megoran/Dalby (2018): 251–276; Luke (2018): 9–24; Hickmann et al. (2019a): 1–12; Hickmann et al. (2019b): 237–251; Arias–Maldonado (2019b): 790–791; Angus (2019): 3–30; Arias-Maldonado (2020b): 1024–1041; Dalby (2020a): 144–160; Foster (2020): 1–17; Angus (2020): 50–54; Gupta/Schmeier (2020b): 731–747.

reviews, 7 early access and 6 proceedings papers. The top 10 Anthropocene authors categorised under *International Relations* on the *Web of Science* were: D. Chandler (6), S. Dalby (5), S. Kalantzakos (4), R.E. Kim (4), L. Arizpe Schlosser (4), Ú. Oswald Spring (4), H.G. Brauch, (3), S. Hamilton (3), J.C. Pereira (3), and D. Simangan (3), who contributed 37 of the 110 International Relations titles listed on the *Web of Science*.

On the same day *Scopus* listed 504 *International Relations* titles on the Anthropocene. The 16 most counted International Relations authors on *Scopus* were: D. Chandler (10), S. Dalby (4), *E.C. Ellis* (4), S. Hamilton (4), R. Karlsson, (4), F. Biermann (3), A. Burke (3), V. Galaz (3), C. Gibson (3), D Houston (3), L.J. Kotzé (3), M. Logan 83), P. Pattberg (3), C. Shearing (3), D. Simangan (3), and A. Taylor (3), who published a total of 58 texts.[52] Only one International Relations document on *Scopus* was written by an AWG member (*E.C. Ellis*).

Dahlia Simangan (2020), one of the authors who wrote on International Relations scholars focusing on the Anthropocene and on their specific theoretical discourses argued that

> several disciplines outside the natural sciences, including International Relations (IR), have engaged with the Anthropocene discourse in order to theorize their relevance and translate their practical value in this new phase of the Earth's history. Some IR scholars have called for a post-humanist IR, planet politics, a cosmopolitan view, and ecological security, among other approaches, to recalibrate the theoretical foundations of the discipline, making it more attuned to the realities of the Anthropocene. Existing discussions, however, tend to universalize human experience and gravitate towards western ontologies and epistemologies of living in the Anthropocene. Within this burgeoning scholarship, how is the IR discipline engaging with the Anthropocene discourse? Although the Anthropocene has become a new theoretical landscape for the conceptual broadening of conventional IR subjects, this review reveals the need for sustained discussion that highlights the differentiated human experiences in the Anthropocene. The existing IR publications on the Anthropocene locate the non-spatial narratives of vulnerability and historical injustice, the non-modernist understanding of nature, the agency of the vulnerable, and the amplification of security issues in the Anthropocene. It is in amplifying these narratives that the IR discipline can broaden and diversify the discourse on the Anthropocene and, therefore, affirm its relevance in this new geological age.

---

[52] Bauer/Ellis (2018); Biermann (2014); Biermann et al. (2016); Biermann/Lövbrand (2019a); Burke (2019); Burke/Fishel (2019); Burke et al. (2016); Chandler (2017), (2018), (2019a), (2019b), (2019c), (2019d), (2019e); Chandler et al. (2018); Chandler/Pugh (2020); Chandler/Reid (2018); Cutter-Mackenzie-Knowles et al. (2019); Dalby (2013c), (2015a); Ellis (2017); Ellis et al. (2019); Froestad/Shearing (2017); Galaz (2014); Galaz et al. (2016); Gibson (2019a), (2019b); Hamilton (2017), (2018a), (2018b), (2019); Hickmann et al. (2018): 1–12, (2020); Holley et al. (2020), (2018); Houston et al. (2018); Karlsson (2017), (2018); Keys et al. (2019); Klocker et al. (2018); Kotzé (2015), (2019a); Kotzé/Muzangaza (2018); Logan/Khatun (2019); Malone et al. (2020); McGregor/Houston (2018); Megoran/Dalby (2018): 251–276; Melo Zurita et al. (2018); Pattberg/Widerberg (2015); Simangan (2019), (2020a), (2020b); Symons/Karlsson (2015), (2018); Tarolli et al. (2019); Taylor (2017), (2020); Taylor/Pacini-Ketchabaw (2018).

The Anthropocene: A Theme in Environmental Politics

On 7 February 2021 the *Web of Science* counted a total of 5578 documents on the Anthropocene, including 926 titles under Environment Studies. Within this category 30 titles were listed under *Environmental Policy*, 26 of which were listed as articles, 3 as proceedings papers, 1 as a book chapter, 1 as early access, and 1 as a book review. Of the 48 documents listed under *Environmental Politics*, 40 were listed as articles, 4 as book chapters, 3 as editorial materials, 2 as book reviews, and 2 as early access texts. The 3 most counted Environment Politics authors on the *Web of Science* were: V. Galvaz (2), E. Lövbrand (2), and A. Machin (2), who contributed 6 of the 48 titles listed on the *Web of Science*.[53] Under the 30 titles on Environment Policy the 4 most counted authors were: F. Biermann (3), M. Arias-Maldonado (2), X.M. Bai (2), and M. Glaser, who authored 5 of the 30 titles.[54]

On the same day, out of a total of 5411 Anthropocene documents on *Scopus,* 1931 titles were listed under *Environment Science*, and thereof 512 under *Environment Politics* and 830 under *Environment Policy*. The 10 most counted Anthropocene authors on Environment Politics on *Scopus* were: F. Biermann (9), C. Folke (7), M. Arias-Maldonado (5), T. Biggs (5), V. Galaz (5), A.V. Norström (5), X. Bai (4), C. Crumley (4), *E.C. Ellis* (4), and M. Glaser (4), who contributed 36 Environmental Politics results on the Anthropocene.[55] On the *Web of Science* there was no AWG member among the 10 most counted authors in these categoris, and on *Scopus* there was only one (*Ellis*).

The Anthropocene: A Theme in Development Studies, Policy and Politics

On 7 February 2021 the *Web of Science* counted 48 *Development Studies* titles on the Anthropocene, 25 of which were articles, 14 proceedings papers, 7 book chapters, 4 book reviews, and 3 reviews. The 8 most counted authors in this category on the *Web of Science* were: J. Gupta (2); T. Hickmann (2), R.E. Kim (2), J.W. Moore, (2),

---

[53] Galaz (2012): Art. No. 24; Uhrqvist/Loevbrand (2014): 339–356; Loevbrand et al. (2015): 211–218; Galaz et al. (2016): 189–221; Machin (2019a): 1–6; Machin (2019b): 347–365.

[54] Arias-Maldonado (2016), (2020c); Bai et al. (2016); Biermann (2020); Biermann et al. (2016).

[55] Biermann et al. (2010): 202–208; Steffen et al. (2011b): 739–761; Biermann (2012): 6–17; Galaz (2012); Palsson et al. (2013):3–13; Arias-Maldonado (2013): 428–446; Barthel et al. (2013): 1142–1152; Galaz (2014): 1–189; Norström et al. (2014): 8; Seddon et al. (2014): 256–267; Glaser/Glaeser (2014): 2039–2052; Crumley et al. (2015): 1721–1723; Arias-Maldonado (2016): 1–17; Folke et al. (2016): 41; Biermann et al. (2016): 341–350; Bai et al. (2016): 351–362; Brondizio et al. (2016): 318–327; Bennett et al. (2016): 441–448; Österblom et al. (2017): 54–61; Gordon et al. (2017): 100–201; Leach et al. (2018): e13; Pereira et al. (2018): 19; Gerhardinger et al. (2018): 395; Reyers et al. (2018): 267–289; Ellis/Mehrabi (2019): 22–30; Keys et al. (2019): 667–673; Bluemling et al. (2020): 365–373; Arias–Maldonado (2020c): 2538; Raudsepp-Hearne et al. (2020): 605–661; Hanusch/Biermann (2020), 19–41; Jiménez-Aceituno et al. (2020): 729–743; Arias-Maldonado (2020a): 97–112; Arias-Maldonado (2020b): 1024–1041.

L. Partzsch (2), P. Pattberg (2), A. Underdal (2), and S. Weiland (2), who wrote 8 of the titles listed on the *Web of Science*.[56]

On the same day *Scopus* documented 2348 *Development Studies* titles on the Anthropocene, 1516 under *Development Policy* and 1171 under *Development Politics*. On 7 February 2021 the 9 most counted Anthropocene authors listed under *Development Politics* on *Scopus* were: S. Dalby (15), F. Biermann (15), C. Folke (10), *E.C. Ellis* (8), J. Gupta (8), N. Castree (7), V. Galaz (7), P. Pattberg (7), and *W. Steffen* (7), who published a total of 71 texts.[57]

The Anthropocene: A Theme in Peace Studies

On 7 February 2021 the *Web of Science* counted 5 titles on the Anthropocene under Peace Studies, 4 of which were listed as articles, 3 as book chapters and 1 as a proceedings paper. Only 4 authors contributed on this theme: H.G. Brauch (3), Ú. Oswald Spring (2), S. Dalby (1), and N. Megoran (1), who contributed a total of 5 texts.[58]

On the same day *Scopus* documented 189 titles on the Anthropocene under *Peace Studies*. The 3 most counted authors on *Scopus* were: S. Dalby (6), R. Karlsson (4), and D. Chandler (3), who published a total of 13 texts.[59]

---

[56] Hickmann et al. (2019a): 1–12; Hickmann et al. (2019a): 237–251; Moore (2018): 237–279; Moore (2017): 594–630; Young et al. (2017): 53–74; Underdal/Kim (2017): 241–258; Gupta et al. (2019): 25–50; Gupta et al. (2015): 541–559.

[57] Biermann et al. (2010): 202–208; Steffen et al. (2011): 739–761; Biermann et al. (2012): 1306–1309; Biermann (2012): 6–17; Dalby (2013d): 184–192; Dalby (2013b): 38–47; Gupta et al. (2013): 573–580; Johnson et al. (2014): 439–456; Biermann (2014b): 57–61; Biermann (2014a): 1–268; Dalby (2014a): 3–16; Seddon et al. (2014): 256–267; Dalby (2014b): 1–9; Pattberg/Widerberg (2015): 684–705; Dalby (2015b): 190–201; Gupta et al. (2015): 541–559; Ellis (2015): 287–331; Dalby (2016a): 33–51; Dalby (2016b): 29–48; Burke et al. (2016): 499–523; Folke et al. (2016): 41; Pattberg/Zelli (2016a): 1–254; Pattberg/Zelli (2016b): 1–12; Gupta/Vegelin (2016): 433–448; Biermann et al. (2016): 341–350; Bai et al. (2016): 351–362; Brondizio et al. (2016): 318–327; Bennett et al. (2016): 441–448; Dalby (2017a): 233–252; Zalasiewicz et al. (2017i): 205–226; Österblom et al. (2017): 54–61; Pouw/Gupta (2017): 104–108; Gordon et al. (2017): 100–201; Hickmann et al. (2018a): 1–260; Hickmann et al. (2018b): 237–251; Gupta (2018): 259–274; Dalby (2018c): 173–188; Hickmann et al. (2018a): 1–12; Waters et al. (2018b): 379–429; Bauer/Ellis (2018): 209–227; Megoran/Dalby (2018): 251–276; Biermann (2018): 467–478; Steffen et al. (2018): 8252–8259; Reyers et al. (2018): 267–289; Dalby (2018b): 307–317; Sterner et al. (2019): 14–21; Campagnaro et al. (2019): 117–124; *Ellis*/Mehrabi (2019): 22–30; Ellis et al. (2019): 86–94; Dalby (2019): 3287–3300; Gupta/Schmeier (2020b): 731–747; Bluemling et al. (2020): 365–373; Dalby (2020a): 144–160; Hanusch/Biermann (2020): 19; Castree (2015a), (2015b), (2015c), (2015d), (2017a), (2017b), (2019); Galaz (2012), (2014), (2019); Galaz et al. (2016); Keys et al. (2019); Polasky et al. (2020); Reyers et al. (2018).

[58] Oswald Spring, 2016: 3–66; Brauch 2016b: 187–236; Brauch 2019: 175–234; Oswald Spring, 2016: 675–704; Megoran/Dalby, 2018: 251–276.

[59] Dalby (2013d): 184–192; Dalby (2013b): 38–47; Dalby (2014b): 1–9; Dalby (2015b): 190–201; Karlsson (2016): 23–32; Karlsson (2017): 550–562; Karlsson (2018): 77–84; Chandler (2018): 1–244; Megoran/Dalby (2018): 251–276; Symons/Karlsson (2018): 685–704; Chandler (2019b): 26–44; Dalby (2020a): 144–160; Chandler (2020a): 195–214.

The Anthropocene: A Theme in Security Studies

The *Web of Science* counted a total of 5,578 documents on the Anthropocene on 7 February 2021, 50 of which were listed under *Security Studies*; 36 documents in this category were articles, 7 book chapters, 5 editorial material, 5 proceedings papers and 5 reviews. The 7 most counted authors in this category on the *Web of Science* were: H.G. Brauch (3), C. Harrington (2), C.D. Shearing (2), N.P. Simpson (2), C.J Vörosmarty (2), and S. Wakefield (2), who contributed 12 titles.[60]

Scopus documented on 7 February 2021 a total of 658 titles on the Anthropocene in Security Studies. The six most counted authors in *Scopus* were: Chandler, D. (10), Dalby, S. (9), Rockström, J. (8) and *Steffen*, W. (7), Biermann, F. (6) and Folke, C. (6) who contributed 37 document results.[61]

The Anthropocene: A Theme in Gender Studies

Out of the 5,578 Anthropocene documents on the *Web of Science* on 7 February 2021, 17 titles were categorised under *Gender Studies*; 14 of these were listed as articles, 2 as editorial material, 2 as proceedings papers and 1 as a book chapter. The first 6 of the 31 authors in this category on the *Web of Science* each had one text listed. These authors were J. Antero, A. Arlander, C. Asberg, G. Berthelot, H.G. Brauch, and M.J. Bunch, who contributed to a total of 5 papers.[62]

Out of a total of 5411 Anthropocene documents listed on *Scopus* on 7 February 2021, 395 titles were categorised as *Gender Studies*. The 11 most counted authors in this category on *Scopus* were: L. Head (4), J. Atchison (3), S. Bergmann (3), S. Ergene (3), A. Fuentes (3), A. Grear (3), S.W. Schwartz (3), L. Stevens (3), P. Tait (3), A. Taylor (3), and E. de Freitas (3), who published a total of 30 Gender Studies

---

[60] Voeroesmarty et al. (2013): 539–550; Voeroesmarty (2015): 94–109; Brauch/Oswald Spring (2016): 3–66; Brauch (2016b): 187–236; Oswald Spring (2016): 675–704; Harrington (2016): 478–498; Lecavalier/Harrington (2017): 529–546; Wakefield (2018): e12377; Brauch (2019): 175–234; Simpson et al. (2019a): 100–196; Simpson et al. (2019b): 257–276; Wakefield (2020): UNSP 102–148.

[61] Steffen et al. (2011a): 842–867; Steffen et al. (2011b): 739–761; Dalby (2013d): 184–192; Dalby (2013b): 38–47; Dalby (2014b): 1–9; Rockström et al. (2014): 1249–1261; Dalby (2015b): 190–201; Burke et al. (2016): 499–523; Verburg et al. (2016): 328–340; Bai et al. (2016): 351–362; Brondizio et al. (2016): 318–327; Dalby (2017a): 233–252; Rockström et al. (2017): 4–17; Chandler (2017): 113–130; Gordon et al. (2017): 100–201; Chandler et al. (2018): 190–208; Chandler (2018): 1–244; Megoran/Dalby (2018): 251–276; Steffen et al. (2018): 8252–8259; Dalby (2018b): 307–317; Chandler (2019e): 381–387; Chandler (2019a): 209–229; Falkenmark et al. (2019): 100009; Chandler (2019b): 26–44; Willett et al. (2019): 447–492; Chandler (2019c): 695–706; Chandler (2019d): 301–312: Chandler/Pugh (2020): 65–72; Dalby (2020a): 144–160; Gleeson et al. (2020): e2019WR024957; Chandler (2020): 195–214; Biermann (2014a), (2014b); Biermann et al. (2016); Biermann/Lövbrand (2019a): 1–22; Reyers et al. (2018): 267–289.

[62] Arlander 2020; Marck et al. 2019: 591–599; Asberg 2017: 185–204; Bunch 2016: 614–632; Brauch/Oswald Spring 2016: 3–66.

texts on the Anthropocene. There was no overlap between the most highly counted authors on the *Web of Science* and *Scopus*.[63]

22.4.3.4   The Anthropocene Concept in Sociology

On 7 February 2021 the *Web of Science* categorised a total of 55 titles on the Anthropocene under *Sociology*, 49 of which were articles, 5 reviews, 4 book chapters, 4 early access and 1 editorial matter. The 11 most counted authors in this category on the *Web of Science* were: R. Barbanti (2), A. Calanchi (2), P. Debaeke (2), A. Farina (2), M. Hamelin (2), P. Heikkurinen (2), A. Henri (2), E. Lichtfouse (2), M. Navarrete (2), T. Ruuska (2), and A.M. Stoner (2), who contributed 8 titles.[64]

Out of 5106 Anthropocene documents on *Scopus* on 7 February 2021, 608 titles were listed under *Sociology*. The 11 most counted authors in this category on *Scopus* were: D. Chandler (7), F. Biermann (6), J. Atchison (5), A. Hornborg (4), X. Bai (4), S. Dalby (4), V. Galaz (4), L. Head (4), A.J. Hoffman (4), *W. Steffen* (4), and B. Szerszynski (4), who produced a total of 45 documents. There was again no overlap between the highly counted authors on the 2 databases.[65]

---

[63] Fuentes (2012): 101–117; Malone et al. (2014): 8–29; Head (2014): 313–320; Head et al. (2015): 311–318; Head/Atchison (2015a): 169–182; Taylor/Pacini–Ketchabaw (2015): 507–529; Grear (2015): 225–249: Head (2016): 1–181; Stevens et al. (2017): 1–271; Stevens et al. (2017): 1–22; Calás et al. (2017): 197–228; Schwartz (2017): 180–197; Taylor (2017): 1448–1461; Ergene et al. (2018): 222–245; Tait (2018): 149–164; Grear (2018): 297–316; Schwartz (2019a): 264–284; Schwartz (2019b): 73–93; Stevens (2019): 89–97; Atchison (2019): 735–748; Grear (2020): Article in Press; de Freitas/Truman (2020): Article in Press; Ergene et al. (2020): Article in Press; Fuentes (2020): Article in Press; Taylor (2020): 340–358; de Freitas/Weaver (2020): 195–202; de Freitas (2020): 203–212; Bergmann (2015); Bergmann/Vähäkangas (2020b).

[64] Lichtfouse et al. (2010): 1–10; Lichtfouse et al. (2011): 3–14; Heikkurinen et al. (2016); Calanchi (2017a): 205–216; Calanchi/Barbanti (2017b): 67–79; Stoner/Melathopoulos (2018): 105–132; Heikkurinen et al. (2020): 5159; Stoner (2020): Article Number: 0896920520958099.

[65] Dalby (2013c): 561–567; Palsson et al. (2013): 3–13; Dalby (2013b): 38–47; Malm/Hornborg (2014): 62–69; Galaz (2014): 1–189; Hoffman/Jennings (2015): 8–31; Head et al. (2015): 311–318; Steffen et al. (2015): 81–98; Head/Atchison (2015a): 169–182; Szerszynski (2015): 177–183; Head/Atchison (2015b): 225–234; Head (2016): 1–181; Galaz et al. (2016): 189–221; Bai et al. (2016): 351–362; Biermann et al. (2016): 341–350; Brondizio et al. (2016): 318–327; Hornborg (2016): 61–76; Szerszynski (2017a): 111–131; Dalby (2017a): 233–252; Hornborg (2017a): 95–110; Chandler (2017): 113–130; Hornborg (2017b): 61–77; Szerszynski (2017b): 92–102; Zalasiewicz et al. (2017a): 55–60; Hoffman/Jennings (2018): Article in Press; Leach et al. (2018): e13; Chandler 2018: 1–244; Megoran/Dalby 2018: 251–276; Chandler/Reid 2018: 251–268; Atchison (2018): 25–46; Reyers et al. (2018): 267–289; Jennings/Hoffman (2019): Article in Press; Chandler (2019a): 209–229, (2019b): 26–44; Keys et al. (2019): 667–673; Atchison (2019): 735–748; Ergene et al. (2020): Article in Press; Bluemling et al. (2020): 365–373; Chandler/Pugh (2020): 65–72; Hanusch/Biermann (2020): 19–41; Chandler (2020a): 195–214; Hoffman/Jennings (2020), (2021); Barret et al. (2020); Biermann (2020); Schill et al. (2019).

22.4.3.5   The Anthropocene Concept in Economics

On 7 February 2021 the *Web of Science* listed 83 of its 5578 Anthropocene titles under *Economics*, 54 under *Management*, and 13 under *Business*, which resulted in a total in 142 Anthropocene documents on Economics. The top 28 of the first 100 authors were A.J. Hoffman (5), P.D. Jennings (3), R.A. Slaughter (3), J.M. Alcaraz (2), G. Austin (2), J. Cameron (2), R. Edgeman (2), S. Ergene (2), J. Freund (2), Gasparin (2), J.K. Gibson-Graham (2), L. Giessen (2), W. Green (2), J. Gupta (2), J. Haqq-Misra (2), S. Healy (2), P. Heikkurinen (2), R. Ison (2), W. Kupers (2), D.C. Lane (2), *J. McNeill* (2), K. Nicolopoulou (2), D. Nyberg (2), L. Rickards (2), B. Singer (2), S. Waddock (2), M. Williams (2), and C. Wright (2), who produced a total of 39 documents.[66]

On the same day *Scopus* listed 241 titles on the Anthropocene under *Business, Management* and *Accounting* and 213 under *Economics, Econometrics and Finance*, which together amounted to 328 separate documents. The top 5 most highly counted Anthropocene authors in Economics were: P. Heikkurinen (6), P. Pattberg (6), M. Gren (5), A.J. Hoffman (5), and Huijbens (4), who created 23 texts.[67]

22.4.3.6   The Anthropocene Concept in Psychology

On 7 February 2021 the *Web of Science* listed a total of 174 Anthropocene titles under *Social Sciences Interdisciplinary* and 8 titles under *Psychology Multidisciplinary*, which produced 182 results and 21 titles on the Anthropocene directly in

---

[66] Slaughter (2012): 119–126; Hoffman/Ehrenfeld (2015): 228–246; Hoffman/Jennings (2015): 8–31; Freund (2015): 1529–1551; Alcaraz et al. (2016): 313–332; Edgeman et al. (2016): 858–868; Ison/Shelley (2016): 589–594; Ison (2016): 595–613; Lane (2016): 633–650; Austin (2017): 1–22; (2017): 95–118; Singer/Giessen (2017): 69–79; Kearnes/Rickards (2017): 48–58; Singer/Giessen (2017): 178; Hoffman/Jennings (2018a): 1–93; Ergene et al. (2018): 222–245; Slaughter (2018): 1–19; Wright et al. (2018): 455–471; Gibson–Graham et al. (2019b): 1–21; Gibson–Graham et al. (2019a): 27–29; Haqq-Misra (2019): 1–3; Mullan/Haqq-Misra (2019): 4–17; Lane (2019): 233–243; Heikkurinen et al. (2019b): 106,369; Heikkurinen et al. (2019): Art. No.: UNSP 1086026619881145; De Cock et al. (2019): Art. No.: UNSP 1350508419883377; Waddock (2020): 189–201; Edgeman (2020): 469–482; Ergene et al. (2020): Art. No.: 0170840620937892; Slaughter (2020): Art. No.: 102496; Waddock/Kuenkel (2020): 342–358; Ergene et al. (2020): Gasparin et al. (2020a); Gasparin et al. (2020b); Gupta/Schmeier (2020b); Gupta/Vegelin (2016): 433–448; Hoffman/Ehrenfeld (2015); Hoffman/Jennings (2015), (2018); (2021).

[67] Gren/Huijbens (2014): 6–22; Hoffman/Jennings (2015): 8–31; Gren/Huijbens (2016b): 1–204; Gren/Huijbens (2016a): 189–199; Huijbens/Gren (2016): 1–13; Gren (2016): 171–188; Pattberg/Zelli (2016a): 1–254; Zelli/Pattberg (2016): 231–242; Pattberg/Zelli (2016b): 1–12; Heikkurinen et al. (2016): 705–714; Heikkurinen (2017a): 1–193; Heikkurinen (2017b): 1–4; Heikkurinen (2017c): 7–15; Hoffman/Ehrenfeld (2017): 228–246; Hoffman/Jennings (2018a): Article in Press; Hickmann et al. (2018a): 1–260; Hickmann et al. (2018b): 237–251; Hickmann et al. (2018c): 1–12; Heikkurinen et al. (2019): Article in Press; Jennings/Hoffman (2019): Article in Press; Heikkurinen et al. (2019): 106369; Ergene et al. (2020): Article in Press.

Psychology.[68] On the same day *Scopus* documented 353 titles on the Anthropocene in *Psychology.* The top 9 authors were: C. Folke (5), M. Glaser (4), M. Adams (3), J.T. Bruskotter (3), M. Di Paola (3), K. Moser (3), M. Scheffer (3), W. Steffen (3), and G. Waitt (3), who produced 25 documents.[69]

22.4.3.7   The Anthropocene Concept in Education

On 8 February 2021, 204 of the *Web of Science's* 5,578 Anthropocene documents were listed under *Education,* 173 under *Education and Educational Research*, and 31 under *Education and Scientific Disciplines*, across 189 titles altogether. Discussions on the Anthropocene within these scientific disciplines took off in 2014 (8), 2017 (10), 2018 (2), 2019 (6) and 2020 (2), but none of the authors in 2019 was an AWG member. A direct search on the Anthropocene in *Education* produced 204 titles. The 7 most counted authors of these were: R. Irvin (5), D. Rousell (5), R. Affifi (4), J. Gilbert (4), P.G. Mahaffy (4), F. Nxumalo (4) and J.J. Wallin (4), who together produced 24 texts.[70] On 8 February 2021 a direct search on *Scopus* on the Anthropocene in *Education* produced, out of a total of 5,416 results, 839 publications. The top 4 authors were: N. Castree (6), S.J. Cooke (6), J. Gupta (6) and A. Taylor (6), who produced a total of 24 documents.[71]

---

[68] Doyle (2008): 3–7; Walton/Shaw (2015): 1–3; Kashima (2016): 4–20; Pyhala et al. (2016): 25; Shaw/Bonnett (2016): 565–579; Hudspeth (2017): 407–418; Groes (2017): 973–993; Berzonsky/Moser (2017): 15–23; De Almeida/da Silva Carvalho (2018): 299–326; Adams (2018): e12375; Brick (2019): 485–487; Bernard (2019): 193–201; Merrill/Giamarelos (2019): Article 7; Ingram et al. (2015): 492–503; Schill et al. (2019): 1075–1082; Sundar et al. (2020): 1238–1250; Pasquero/Poletto (2020): 120–140, Article UNSP 1478077120922941; Barrios-O'Neill (2020); Hoffman/Jennings (2021); Adam-Hernandez/Harteisen (2020); Ramsay (2020); Slesinger (2020).

[69] Steffen et al. (2011b): 739–761; Glaser et al. (2012): 193–222; Di Paola/Garasic (2013): 59–81; Di Paola (2015): 183–207; Bai et al. (2016): 351–362; Brondizio et al. (2016): 318–327; Waitt/Harada (2016): 1079–1100; Di Paola (2017): 119–126; Glaser et al. (2018): 34; Aswan et al. (2018): 192–202; Carpenter et al. (2019): 23; Duffy et al. (2019): 378–384; Schill et al. (2019): 1075–1082; Barrett et al. (2020): 6300–6307; Waitt et al. (2020): 2131–2146; Adams (2017), (2019), (2020); Batavia et al. (2020); Carpenter et al. (2019): 23; Folke et al. (2016): 41; George et al. (2016); Manfredo et al. (2020); Moser (2018d), (2019a), (2020).

[70] Mahaffy (2014): 463–465; Mahaffy et al. (2014): 2488–2494; Gilbert (2016): 187–201; Nxumalo/Pacini–Ketchabaw (2017): 1414–1426; Mahaffy et al. (2017): 1027–1035; Hebrides et al. (2018a): 1–22; Hebrides et al. (2018b): 23–50; Hebrides et al. (2018c): 51–62; Nelson et al. (2018): 4–14; Nxumalo (2018): 148–159; Nxumalo/Ross (2019): 502–524; Irwin (2020a): 361–375; Irwin (2020b): 834–834; Mahaffy et al. (2019:): 2730–2741; Affifi (2020): 1435–1452; Hebrides et al. (2018a): 1–22, (2018b), (2018c): 51–62, (2018b): 23–50; Lasczik et al. (2020): 146–168; Rousell (2016), (2020); Rousell/Cutter-Mackenzie-Knowles (2019): 84–96; Rousell/Cutter-Mackenzie/Foster (2017): 654–669; Wallin (2015), (2016), (2017), (2020).

[71] Castree (2014): 464-476; Castree (2015): 244–254; Castree (2015b): 66–75; Taylor/Pacini-Ketchabaw (2015): 507–529; Castree (2015c): 301–316; Taylor (2017): 61–75; Castree (2017a): 52–74; Taylor (2017a): 1448–1461; Castree (2017b): 160–182; Taylor/Pacini-Ketchabaw (2018): 1–135; Hammerschlag et al. (2019): 369–383. Cooke et al. (2019): 845–856; Jeanson et al. (2020): 99–108; Holder et al. (2020): 137–151; Kraftl et al. (2020): 333–339; Taylor (2020): 340–358; Cooke et al. (2020): 108589; Biermann et al. (2012): 1306–1307; Cuadrado–Quesada et al. (2018):

22.4.4   The Anthropocene: A Concept in the Humanities

Out of about 1,000 Anthropocene publications listed on *Amazon* in September 2020, 82 were categorised as *History* and 29 as *Religion and Spirituality*. On 9 February 2021 the *World Catalogue* listed the following categories among more than 15,306 Anthropocene publications searched by title: *Language, Linguistics and Literature* (132), *Art & Architecture* (106), *Philosophy & Religion* (101), *History & Auxiliary* (47), *Library Science* (17), *Performing Arts* (8), and *Physical Education* (2).

On 9 February 2021 the *Web of Science* listed 208 its 5,580 Anthropocene entries as *Philosophy*, 202 as *Humanities Multidisciplinary*, 119 as *History*, 112 as *History and Philosophy of Science*, 95 as *Religion*, and 69 as *Archaeology*. *Scopus* again used a different categorisation within its total of 5,418 Anthropocene results, listing 1,385 titles under *Arts and Humanities*.

22.4.4.1   The Anthropocene Concept in Archaeology

On 9 February 2021 the *Web of Science* counted a total of 69 titles in *Archaeology*, 53 of which were articles, 14 editorial material, 10 book chapters and 1 a book. The top 5 authors by record count were: T.J. Braje (5), A.Y. Glikson (5), H.J. Schellnhuber (4), S. Lawrence (3), and C. Whitmore (3), who contributed a total of 16 titles.[72] A direct search on *Scopus* on the Anthropocene in Archaeology on 9 February 2021 produced a total of 642 results in which the 3 most prolific authors were: *J. Zalasiewicz* (16), *E.C. Ellis* (15), and *M. Williams* (15), who produced a total of 28 document results.While AWG members played hardly any role in this category on the *Web of Science*, the *Scopus* listings were dominated by authors who were active in the AWG in 2019.

22.4.4.2   The Anthropocene Concept in History

On 9 February 2021 the *Web of Science* listed 119 of its 5,580 Anthropocene entries under *History* and 112 under *History and Philosophy of Science* in 228 different documents. Its 5 authors with the highest record count were: D. Salottolo (5), W. Wright (5), F.A. Jonsson (4), J. Renn (4), and Z.B. Simon (4), who together produced 22 documents.[73]

475–489; Gupta (2015): 311–315, (2018): 259–274; Gupta et al. (2015): 541–559; Gupta/Schmeier (2020b): 731–747; Taylor (2017): 1448–1461.

[72] Burstrom et al. (2011): 40–88; Article: PII 938514980; Glikson (2014): 1–174; Schellnhuber (2014a): vii–viii; Schellnhuber (2014b): ix–x; Schellnhuber (2014c): 71–74; Schellnhuber (2014d): 75–90; Braje (2015): 369–396; Braje (2016b): 504–512; Braje (2016a): 517–518; Witmore (2018): 26–46; Campbell et al. (2018): 65–84; Witmore (2019): 136–153; Rick et al. (2020); Hil et al. (2020); Lawrence et al. (2016): 1348–1362; (2017): 49–65.

[73] Wright (2017): 668-695, (2018a): 110–122, (2018b): 110–122; (2018c): 110–122; (2018d); Carofalo et al. (2019): 21–31; Salottolo (2019a): 50–75, (2019b): 323–328, (2019c): 200–250, (2020):

On 9 February 2021 a direct search on *Scopus* for the Anthropocene in *History* produced a total of 3,311 document results. The 6 most counted authors in this category were: *J. Zalasiewicz* (37), *M. Williams* (32), *W. Steffen*, (31), *E.C. Ellis* (21), J.C. Svenning (21), and *C.N. Waters* (21), who together produced 93 documents. The only historian among the 10 most counted authors was *J.R. McNeill* (12).[74]

There was a fundamental difference between the Anthropocene in History sources counted on the *Web of Science* and those referred to on *Scopus*. While authors representing the AWG were missing in the records of the *Web of Science*, they were dominant in *Scopus*.

22.4.4.3   The Anthropocene Concept in Philosophy

On 9 February 2021 the *Web of Science* listed 208 of its 5,590 Anthropocene entries under *Philosophy* and 112 under *History and Philosophy of Science,* and counted 318 different documents across these categories. The 9 authors with the highest record count were: C. Colebrook (6), S. Raffnsoe (5), D. Salottolo (5), S. Baranzoni (4), M. Dipaola (4), T.W. Luke, (4), J. Renn (4), P. Vignola (4), and W. Wright (4), who accounted for 38 records on the Anthropocene in Philosophy, while the 3 most counted authors produced the 15 texts listed in this footnote.[75]

On 9 February 2021, in its 5,418 entries on the Anthropocene, *Scopus* listed 1,329 texts under *Philosophy.* The top 10 authors were: D. Chandler (11), M. Arias-Maldonado, (7), K. Yusoff (7), F. Biermann (6), V. Blok (6), C. Colebrook (6), C. Hamilton (6), P. Heikkurinen (6), K. Moser (6), and Z.B. Simon (6), who produced 68 documents.[76] Between them, the 2 most counted authors, D. Chandler and M.

350–385; Renn (2020a): 1–561, (2020b): 3–22, (2020c): ix; (2020d), Renn (2020d): 377–407; Jonsson (2012): 679–696, (2013): 462–472, (2019): 83–94; Otter et al. (2018): 568–596; Simon (2015): 819–834, (2018): 105–125, (2019): 171–184; Tamm/Simon (2020): 285–309.

[74] Steffen et al. (2007): 614–621; McNeill (2015): 51–82; Zalasiewicz et al. (2015b): 117–127; Waters et al. (2015): 46–57; Waters et al. (2016): aad2622; Zalasiewicz et al. (2016b): 4–17; Williams et al. (2016): 34–53; Steffen et al. (2016): 324–345; McNeill (2016): 117–128; Zalasiewicz et al. (2017i): 205–226; Zalasiewicz et al. (2017j): 9–22; Zalasiewicz et al. (2017a): 55–60; Waters et al. (2018b): 379–429; McNeill (2019): 200–210; Zalasiewicz et al. (2019b): 319–333.

[75] Colebrook (2012): 185; Colebrook/Weinstein (2015): 167–178; Raffnsoe (2016a): 2–7; Raffnsoe (2016b): 8–19; Raffnsoe (2016c): 20–27; Raffnsoe (2016d): 29–34; (2016e): CP1–75; Colebrook (2016b): 440–454; Carofalo et al. (2019): 21–31; Salottolo (2019a): 50–75; Salottolo (2019b): 323–328; Salottolo (2019c): 200–250; Colebrook (2019): 40; Salottolo (2020): 350–385; Colebrook (2020a): 327–348, (2020b): 135–141.

[76] Arias-Maldonado (2013): 428–446, (2016b): 31–46, (2019a): 137–150, (2019b): 50–66, (2020a): 97–112; Arias- Maldonado/Trachtenberg (2019); Bai et al. (2016): 351–362; Biermann (2020); Blok (2017a), (2017b): 127–149, (2018), (2020): 583–591; Brondizio et al. (2016) 318–327; Chandler (2017): 113–130, (2018), (2019a): 209–229, (2019b): 26–44, (2019c): 695–706, (2019d): 301–312, (2019d): 381–387, (2020): 195–214, (2020b): 26–32; Chandler/Pugh 2020: 65–72; Chandler/Reid (2018): 251–268; Clark/Yusoff (2017): 3–23; Colebrook (2012): 185–209, (2017): 115–136, (2019a): 175–195, Colebrook (2020a): 327–348; Hamilton (2015): 32–43, (2016): 93–106, (2020): 110–119; Hamilton et al. (2015); Heikkurinen et al. (2016): 705–714; Heikkurinen et al. (2020): 5159; Heikkurinen et al. (2019b):,106369; Hoły–Łuczaj/Blok (2019): 325–346; Moser (2018a): 32–47, (2018b): 95–110,: (2018c): 129–147, (2019a): 113–131, (2019b): 63–78, (2020); Ruuska et al. (2020): 2617; Simon (2017): 239–245, (2018b): 105–125, (2020a): 184–199, (2020b):

Arias-Maldonado, published a total of 19 texts.There was no AWG member among the top 100 authors listed on the *Web of Science* under *Philosophy,* but there were 2 Philosophy texts by AWG member *E.C. Ellis* on *Scopus.*

22.4.4.4   The Anthropocene Concept in Theology and Religion

On 10 February 2021 the *Web of Science* listed 95 Anthropocene texts on *Religion* among its total of 5559 Anthropocene entries. The top 7 authors, producing a total of 18 texts, were: L.H. Sideris (5), E.M. Conradie (3), W.A. Baumann (2), F. Clingerman (2), W. Jenkins (2), E.L. Simmons (2), and K. Stone (2).[77]

On 10 February 2021, 393 of the 5,418 Anthropocene entries on *Scopus* were listed under *Religion.* Its 8 most highly counted authors in this category were: D. Chandler (7), L.H. Sideris (5), S. Bergmann (4), N. Bubandt (3), F. Clingerman (3), F. Huettmann (3), L.J. Kotzé (3), and J.J. Schmidt (3), who produced a total of 31 documents.[78]

22.4.5   The Anthropocene in National and International Law

In September 2020, out of approximately 1,000 Anthropocene publications, *Amazon* categorised 34 titles under *Law,* 23 of which were listed under *Natural Resources Law.* On 10 February 2021, out of more than 15,409 publications searched by title, the *World Catalogue* listed 49 texts under *Law.*

The *Web of Science* listed 114 texts on the *Anthropocene in Law* among its 5,590 Anthropocene entries on 10 February 2021. The 7 authors with the highest text count in this category were L.J. Kotzé (20), R.E. Kim (7), A. Grear (4), D. Matthews (4), T. Stephens (4), K. Bosselmann (3), and *D. Vidas* (3), who between them produced

---

377–389; Tamm/Simon (2020): 285–309; Yusoff (2013): 779–795, (2016): 3–2, (2017): 105–127, (2018): 255–276; (2020); Zwier/Blok (2017): 222–242, (2019): 621–646.

[77] Bauman (2015): 742–754, (2020): Article 16; Clingerman (2014): 6–21, (2016): 225–237; Conradie (2019), (2020); Jenkins (2018): 441–462, (2020): Article 215; Sideris (2015): 136–153, (2017): 399–419, (2019): 455–478, 2020: Article 293; Sideris/Whalen–Bridge (2019): 409–413; Simmons (2014): 271–273, (2015): 742–754, (2018a): 99–106; 2; Stone (2018): 297–320, (2019): 236–247.

[78] Sideris (2013): 147–162; Schmidt (2013): 105–112; Kotzé (2014): 252–275; Clingerman (2014): 6–21; Swanson et al. (2015): 149–166; Sideris (2015): 136–153; Clingerman (2016): 225–237; Haraway et al. (2016): 535–564; Schmidt et al. (2016): 188–200; Sideris (2016): 399–419; Chandler (2017): 113–130; Schmidt (2017): 373–380; Clingerman/O'Brien (2017): e480; Chandler (2018): 1–244; Kotzé/French (2018): 811–838; Sideris/Whalen–Bridge (2019): 409–413; Kotzé (2019): 50–74; Chandler (2019a): 209–229; Chandler (2019b): 26–44; Chandler (2019c): 695–706; Tsing et al. (2019): S186–S197; Chandler (2019d): 301–312; Huettmann/Regmi (2020): 3–23; Huettmann/Regmi (2020a): 877–886; Huettmann (2020): 497–520; Chandler (2020a): 195–214; Sideris (2020): 1–16; Bergmann (2015a): 32–44, (2015b): 389–404, (2020): 60–180; Bergmann/Vähäkangas (2020b): 1–14.

42 different texts.[79] On the same day the *Web of Science* listed 39 texts that dealt with *Environmental Law*, 27 texts that specifically focused on *International Law*, 5 on *National Law*, and 3 on *National Environmental Law*.

On 10 February 2021 in its 5,418 entries on the Anthropocene *Scopus* listed a total of 1,220 document results on *Law*. The top 6 authors of law texts focusing on the Anthropocene were: *M. Williams* (19), *J. Zalasiewicz* (19), *W. Steffen* (18), *D. Vidas* (16), *E.C. Ellis* (15), and L.J. Kotzé (14), who produced 55 document results.[80] Only *Vidas* and Kotzé are lawyers. On the same day *Scopus* listed 1,220 texts which dealt with law, of which 1,092 texts focused on *Environmental Law*, 932 texts focused specifically on *International Law*, 724 documents on *National Law*, and 687 on *National Environmental Law*.

22.4.6   The Anthropocene in Linguistics and in Literature

In September 2020 *Amazon* listed 182 Anthropocene titles under *Literature and Fiction*, 90 of which were categorised as *Literary Criticism*. Out of more than 15,408 Anthropocene publications searched by title, the *World Catalogue* categorised 132 texts under *Language Linguistics and Literature*.

On 11 February 2021 the *Web of Science* categorised 221 of its 5,598 Anthropocene entries as *Literature*, 94 as *Literature Theory Criticism*, 40 as *Literary Reviews*, 24 as *Language Linguistics*, 20 as *American Literature*, 15 as *Slavic Literature*, 12 as *German, Dutch and Scandinavian Literature*, 11 as *Linguistics*, and 10

---

[79] Vidas/Schei (2011): 3–15; Kim/Bosselmann (2013): Kim/Mackey (2014): 5–24; Kim (2015): 202–205; Kim/Bosselmann (2015): 194–208; Grear (2015): 225–249; Kotzé (2016a): 241–271; Stephens (2016): 153–176; Kotzé (2016b): 1–282; Kotzé (2016c): 1–20; Kotzé (2016d): ii; Kotzé (2016e): 21–42; Kotzé (2016f): 176–200; Kotzé (2016g): 201–245; Kotzé (2016h): 246–247; Kim/van Asselt (2016): 473–495; Kim (2016): 15–26; Freestone et al. (2019): 5–35; Vidas (2017): 101–123; Kotzé (2017a):   vii–xv; Stephens (2017): 31–54; Kotzé (2017a): 189–218; Bosselmann (2017): 241–265; Grear (2017): 77–95; Kotzé/Calzadilla (2017): 401–433; Kotzé (2018a): 13–33; Stephens (2018): 31–45; Kotzé (2018b): 41–65; Kotzé/Muzangaza (2018):   278–292; Kotzé/French (2018): 811–838; Grear (2019): 297–315; Lubbe/Kotzé (2019):   76–99; Matthews (2019): 665–691; Kotzé (2019f): 367–382; Kotzé (2019b): 437–458; Matthews (2019): Article UNSP 1743872119871830; Kotzé (2019c): 213–236; Birrell/Matthews (2020): 275–292; Grear (2020): 351–366; Birrell/Matthews (2020a): 233–238; Kim/Kotze (2020).

[80] Steffen et al. (2007): 614–621; Vidas (2010); Vidas (2011): 1–608; Zalasiewicz et al. (2010b): 2228–2231; Vidas (2011): 909–925; Zalasiewicz et al. (2011b): 835–841; Steffen et al. (2011): 842–867; Vidas/Schei (2011): 3–15; Zalasiewicz/Williams (2011): 19–35; Vidas/Schei (2011): 3–15; Zalasiewicz/Williams (2011): 19–35; Zalasiewicz et al. (2012): 1033–1040; Kotzé (2014): 252–275; Vidas (2014): 70–84; Zalasiewicz et al. (2014c): 39–53; Martin et al. (2014): 745–755; Williams et al. (2014): 57–63; Zalasiewicz et al. (2015b): 117–127; Zalasiewicz et al. (2015c): 196–220; Steffen et al. (2015): 81–98; Kotzé (2015): 171–191; Vidas et al. (2015): 1–13; Ellis (2015): 287–331; Kotzé (2016a): 241–271; Waters et al. (2016): aad2622; Zalasiewicz et al. (2016b): 4–17; Williams et al. (2016): 34–53; Kotzé/French (2018): 811–838; Kotzé (2019a): 50–74; Lubbe/Kotzé (2019):   76–99; Kotzé (2019b): 437–458; Kotzé (2019c): 213–236; Sterner et al. (2019): 14–21; Campagnaro et al. (2019): 117–124; Tarolli et al. (2019): 95–128; Kotzé (2019d): 62–85; Zalasiewicz et al. (2019): 319–333; Ellis/Mehrabi (2019): 22–30; Ellis et al. (2019): 86–94; Kotzé (2019e): 6796; Kotzé (2020): 75–104; Ellis et al. (2020): 129; Gasparin et al. (2020): 385–405. 2.

as *Poetry*. This combined sample resulted in 377 indexed texts. The 12 most counted texts were written by: S.C. Estok (6), L. Keller (5), T. Bristow (3), M. Caracciolo (3), D. Chakrabarty (3), T. Clark (3), J.P. Deleage (3), H. Jennings (3), H.I. Sullivan (3), J.O. Taylor (3), P. Vermeulen (3), and D. Woods (3), who, based on this sample, produced 41 texts.[81]

Finally, with its different categorisation, on 11 February 2021 *Scopus* listed 1,089 of its 5,418 Anthropocene titles under *Literature*. The 9 most counted authors were: K. Moser (7), D. Chandler (6), S.C. Estok (6), H.I. Sullivan (6), J. Gupta (5), P. Vermeulen (5), C. Colebrook (4), *E.C. Ellis* (4) and C. Folke (4), who were responsible for a total of 47 of the results in this category.[82]

22.4.7   The Anthropocene in the Performing Arts

In September 2020, out of approximately 1,000 Anthropocene publications, *Amazon* listed 83 under *Arts and Photography,* 26 of which were subcategorised under *Art History and Criticism*, 16 under *Photography and Video*, and 9 under *Individual Artists*. Out of 15,408 Anthropocene publications searched by title, the *World Catalogue* listed 106 texts under *Art & Architecture*, 8 under *Performing Arts*, and 18 under *Music*.

On 11 February 2021, the *Web of Science* listed 77 of its 5,598 Anthropocene entries under *Art*, 60 under *Architecture*, 40 under *Film, Radio and Television*, 19

---

[81] Deleage (2013): 23–23; Chakrabarty (2014b): 168–199; Taylor (2014): 73–82; Woods (2014): 133–142; Chakrabarty (2014a): 245–250; Bristow (2015a): 1–139; Bristow (2015b): 1–+; Vermeulen (2015a): 1–182; Bristow (2015c): 107–123; Vermeulen (2015b): 68–81; Keller (2016): 47–63; Sullivan (2016): 285–304; Chakrabarty (2016): 377–397; (2017): 1–284; Keller (2017b): 31–60; Keller (2017c): 208–238; Keller (2017d): 239–244; Vermeulen (2017b): 867–885; Schaumann/Sullivan (2017): 7–21; Sullivan (2017b): 25–44; Estok/Chou (2017): 3–11; Estok (2017a): 33–50; Deleage (2017): 22–23; Estok (2018a): 1–197; Taylor (2018b): 108–133; Estok (2018): 37–52; Woods (2018): 502–504; Taylor (2018a): 573–577; Caracciolo/Lambert (2019): 45–63; Woods (2019): 6–29; Jennings (2019a), 131–152; Jennings (2019b): 16–33; Jennings (2019c): 191–210; Caracciolo et al. (2019): 221–240, Art. No.: UNSP 0963947019865450; Caracciolo (2019): 270–289; Estok (2020b): 27–39; Estok (2020a): 1–7; Clark (2012): 5–6, 2013: 5; (2019): 17.

[82] Schaumann/Sullivan (2011): 105–109; Colebrook (2012): 185–209; Ellis et al. (2013b): 7978–7985; Gupta et al. (2013): 573–580; Vermeulen (2015b): 68–81; Vermeulen (2015a): 1–182; Gupta et al. (2015): 541–559; Williams et al. (2015a): 196–219; Ellis (2015): 287–331; Colebrook (2016a): 307–315, Sullivan (2016c): 47–59; Sullivan (2016b): 285–304; McCarthy et al. (2016): 1–357; Gupta/Vegelin (2016); 433–448; Ellis et al. (2016): 192–193; Pouw/Gupta (2017): 104–108; Estok (2017a): 33–50; Estok/Chou (2017): 3–11; Chandler (2017): 113–130; Estok (2017b): 3–16; Vermeulen (2017a): 181–200; Vermeulen (2017b): 867–885; Colebrook (2017): 115–136; Sullivan (2017a): 115–124; Gupta (2018): 259–274; Moser (2018a): 32–47; Moser (2018b): 95–110; Chandler (2018): 1–244; Moser (2018c): 129–147; Chandler/Reid (2018): 251–268; Estok (2018b): 37–52; Craps et al. (2018): 498–515; Moser (2018d): 405–421; Moser (2019a): 113–131; Chandler (2019a): 209–229; Sullivan (2019): 152–167; Ellis/Mehrabi (2019): 22–30; Moser (2019a): 63–78; Chandler/Pugh (2019): 65–72; Randin et al. (2020): 111626; Chandler (2020a): 195–214; Estok (2020a): Estok (2020b): 27–39; Gordon et al. (2017); Reyers et al. (2018); Schill et al. (2019); Barrett et al. (2020).

under *Music*, and 25 under *Theatre*. On the same day *Scopus* did not specifically list the *Arts* among its 5,418 Anthropocene document results, but it counted 1,385 texts for *Arts and Humanities*. Direct searches for *Art* produced 1,187 document results, *Architecture* 209 texts, *Film, Radio and Television* 2 titles, *Music* 93 texts, *Theater* 119, and *Theatre* 8.

22.4.8   The Anthropocene in Medicine and Health

Amazon did not categorise any of its approximately 1,000 Anthropocene publications as *Medicine and Health* in September 2020. On 11 February 2021, out of 15,408 Anthropocene publications searched by title, the *World Catalogue* listed 15 texts under Medicine, 3 under *Health Profession* and 1 each under *Medicine by Discipline* and *Health Facilities*.

On 11 February 2021 the *Web of Science* listed 5,598 Anthropocene documents, 17 of which were categorised as *Medicine General Internal*, 10 as *Veterinary Sciences,* and 9 as *Medical Ethics*. On the same day, out of 5,418 document results on the Anthropocene, *Scopus* listed 1,209 entries under *Health*, 311 texts under *Medicine*, 18 under *Nursing*, and 20 under *Pharmacology and Toxicology*.

22.4.9   The Anthropocene in Technology and Engineering

Out of about 1000 publications listed under the Anthropocene, Amazon categorised 101 titles under *Engineering* and *Transportation*. On 11 February 2021 the *World Catalogue* listed 57 of its 15,408 Anthropocene publications searched by title under *Engineering and Technology* and 29 under *Agriculture*.

On 11 February 2021 the *Web of Science* listed 227 of its 5,598 Anthropocene entries under *Green Sustainable Science Technology*, 113 under *Water Resources*, 71 under *Engineering Environmental,* 26 under *Agriculture Multidisciplinary,* 14 under *Computer Science: Interdisciplinary Application,* 12 under *Engineering Multidisciplinary,* 9 under *Engineering Civil,* 7 under *Engineering Electrical Electronic*, and 8 under *Computer Science Theory: Methods*. On the same day *Scopus* listed 256 of its 5,418 Anthropocene results under *Engineering*, 119 under *Energy*, 98 under *Computer Science,* 19 under *Materials Science*, and 18 under *Chemical Engineering*.

22.5   Qualitative Interpretation of the Results of the Bibliometric Analysis: the Emergence and Spread of the Anthropocene Concept over Two Decades

This chapter has aimed to provide a bibliometric mapping of the Anthropocene concept in the Natural and Social Sciences and the Humanities since Paul J. Crutzen coined the concept on 23 February 2000 during an IGBP meeting in Cuernavaca in Mexico. The analysis was exclusively quantitative, based on samples taken between September 2020 to mid-February 2021 from 4 databases that are owned and managed by 3 US companies and an Association of participating libraries: *Amazon*, the *World Catalogue*, and the 2 assessments of peer-reviewed literature by the *Web of Science* (Clarivate Analytics) and *Scopus* (Elsevier), which pursue different aims:

(a).   to document books, book chapters, journal articles and conference papers that are bibliographically hosted by participating libraries around the globe (*World Catalogue*);
(b).   to sell books, primarily in English language through 12 national online stores (*Amazon*);
(c).   to assess peer-reviewed literature and their citations (*Web of Science* and *Scopus*).

With very few exceptions, both the *Web of Science* and *Scopus* almost exclusively assessed texts in English peer-reviewed academic journals and excluded most books (even in English), reports, and grey academic and societal literature. Consequently, these databases do not provide an accurate picture of the global output of publications on the Anthropocene.

Although on 18 February 2021 the *Web of Science* (5,611) and *Scopus* (5,426) assessed nearly the same number of publications, the selected key authors by discipline significantly differed. Nevertheless, the databases are unique in documenting the quality literature primarily in English language journals, published in countries of the Global North. Authors whose work has been published in other languages or by publishers in the Global South are underrepresented.

The Anthropocene concept that was coined by the Nobel Laureate and atmospheric chemist Paul J. Crutzen and disseminated in publications by himself and a few close colleagues during the first 5 years of the millennium (2000–2005) was noted in the *World Catalogue* in 78 entries (including duplicates), while the *Web of Science* and *Scopus* listed only 15 titles each that were not identical.

- How did this modest recognition of the Anthropocene concept during the first 5-year period (2000–2005) increase during the second 5-year period (2006–2010), and were scientific communities, networks and key authors instrumental in overcoming disciplinary boundaries and increasing the number of publications on the Anthropocene?

During the second 5-year period (2006–2010) the *World Catalogue* counted an additional 273 texts, while the *Web of Science* assessed 100 and *Scopus* 124 titles. During the first decade (2000–2010) a total of 351 publications on the Anthropocene were counted on the *WorldCat*, while the *Web of Science* indexed 115 and *Scopus* 139 texts.

The take-off occurred during the third assessment period (2011–2015) with an increase in titles on the Anthropocene on the *World Cat* by 3219, on the *Web of Science* by 923 and on *Scopus* by 1,077 documents.

During the fourth phase (2016–2020) the rise of the Anthropocene texts further increased exponentially by 11,420 on the *World Cat,* by 4,493 on the *Web of Science,* and by 4,095 on *Scopus* (Table 22.1). Thus, the exponential increase in the peer-reviewed English-language Anthropocene literature across scientific disciplinary and political country boundaries, especially during the fourth 5-year period (2016–2020), is unique for a scientific concept in the Natural Sciences.

The geologists Jan Zalasiewicz and Colin Waters, successive chairs of the *Anthropocene Working Group* (AWG), and the Earth System scientist, Will Steffen, interpreted this unique experience in their Obituary of Paul J. Crutzen for the *Scientific American*:

> It may seem extraordinary that one person's life, and, as a consequence, so many other peoples' lives, can be so radically reshaped by a moment's irritation. But, with Paul Crutzen, ... the extraordinary had come to be habitual. ... His sudden realization that humanity had very recently stumbled into a new geological epoch of its own making, the Anthropocene, created reverberations that continue to shake not only the world of science but that of all of scholarship, now spilling into political and economic discourse worldwide.

> The beginning of his foray into redefining Earth's geological history is already legendary. At a meeting in 2000 of the Scientific Committee of the International Geosphere-Biosphere Program (IGBP) in Cuernevaca, Mexico, ... Crutzen was listening, with increasing exasperation, to evidence of how global environmental parameters were dramatically changing in recent decades, in what was repeatedly referred to as the late Holocene Epoch .... His exasperation spilled over into an interjection that we were no longer in the Holocene but in ... (pausing to try to think of the appropriate word) ... the Anthropocene.

> The on-the-spot improvisation caught the attention of the audience, crystallizing the growing realization that the Earth system had recently begun to change at a much more dramatic rate and scale than through many previous millennia of slowly growing human occupation of the planet. Crutzen, characteristically, developed his idea both energetically and generously.

> From this beginning, the Anthropocene rapidly evolved. The IGBP/Earth system science community quickly adopted it as a central framing concept for much of their work, using the term as a de facto geological epoch succeeding the Holocene, with little understanding of the lengthy and elaborate protocols needed to formally change any part of the geologic time scale. A few years later, geologists ... becoming aware of the expanding use of the term, began formally analyzing the term to see whether it really could satisfy all the geological protocols. The process continues, with the formal outcome uncertain ... but it is already clear that Crutzen's intuition was correct. The Anthropocene is real (Zalasiewicz/Waters/Steffen, 2021).[83]

---

[83] Jan Zalasiewicz, Colin Waters, Will Steffen: "Remembering the Extraordinary Scientist Paul Crutzen (1933–2021)", at: Obituary, 5 February 2021, © 2021 Scientific American, Springer Nature America, Inc; see at: http://www.afes-press-books.de/html/Crutzen_obituary_global.htm.

This bibliometric analysis has documented how the scientific literature on the Anthropocene initially started in the global change community in the framework of the *International Geosphere and Biosphere Programme* (IGBP) network and was later – primarily since 2009 – taken up and promoted by the geological community and in the natural sciences within 2 decades. But the bibliometric assessment can only partly explain what has contributed to this rapid growth during the second decade and especially during the fourth 5-year period (2016–2020). This requires a qualitative interpretation and explanation of a conceptual creation process that is still underway.

Key milestones in the development of the Anthropocene debate during the first decade (2000–2010) were:

- Crutzen's intervention during a scientfic debate of the Scientific Committee of the IGBP in Cuernavaca, probably on 23 February 2000;
- The *Amsterdam Conference* in 10–13 July 2001 organised by the 4 scientific programmes in the emerging *global environmental change* debate;
- The *Dahlem Workshop* 25–30 May 2003 in Berlin with the Earth System Science Community on *Earth System Analysis for Sustainability*;
- The early journal and book publications by a group of scholars in the global change and Earth System Science communities: Paul J. Crutzen (Chemistry), Will Steffen (IGBP, Earth System Science), and Eckart Ehlers (IHDP, Geography).
- The establishment in 2009 of the *Anthropocene Working Group* as a multidisciplinary group of scientists, primarily in Geology as roving members and other Natural Sciences, with a few representatives from the Social Sciences, Humanities and Law as advisory members.

A purely bibliometric analysis is insufficient to answer the remaining qualitative questions:

- What role did 3 multidisciplinary scientific networks – the *International Geosphere and Biosphere Programme* (IGBP), the *Earth Systems Science Community* and the *Anthropocene Working Group* (AWG) – play in spreading the Anthropocene concept globally and across disciplinary boundaries?

My thesis is that the Anthropocene concept emerged in the framework of 3 successive and partly overlapping multidisciplinary scientific networks:

- The *International Geophysical and Biological Programme* (IGBP),[84] one of the 4 big scientific programmes besides the *World Climate Research Programme*

---

[84] On the *International Geophysical and Biological Programme* (IGBP), see: Noon/Nobre/Seitzinger 2001; see website at: http://www.igbp.net/about.4.6285fa5a12be4b4 03968000417.html.

(WCRP),[85] the *International Human Dimension Programme* (IHDP)[86] and *Diversitas*,[87] which were complemented by the *Earth Systems Science Partnership* (ESSP)[88] after the Amsterdam Conference (2001) and were all replaced by *Future Earth*[89] between 2013 and 2015.

- The *Earth System Science* (ESS) or the *Earth System Analysis* (ESA) *Community*,[90] which met during the Amsterdam Conference in 2001 and at the Dahlem Workshop on *Earth System Analysis for Sustainability* (May 2003). Several large research institutes had a significant impact on this community, among them the *Potsdam Institute on Climate Change Impact Research* (PIK), which was founded in 1991, and the *Stockholm Resilence Centre*, which was formed in 2007 by Stockholm University, the Beijer Institute of Ecological Economics.[91]
- The *Anthropocene Working Group* (AWG), which was set up as a multidisciplinarily composed study group in 2009 and tasked with assessing evidence that the geological epoch of the Anthropocene is succeeding the Holocene.[92]

22.5.1   International Geosphere-Biosphere Programme (IGBP): The Anthropocene's Birthplace

The Stockholm Environment Conference (1972) was instrumental in the establishment of the *United Nations Environment Programme* (UNEP) in Nairobi. It also triggered scientific interest in the possibility that humans are impacting on the climate of the Earth. In 1979 a group of scientists led by the Swedish meteorologist Bert Bolin set up the *World Climate Research Programme* (WCRP), and in 1987 a team of researchers led by Bert Bolin, James McCarthy, *Paul* Crutzen, H. Oeschger and others, successfully argued for an international research programme to investigate global change, which resulted in *the International Geosphere-Biosphere Programme*

---

[85] On the *World Climate Research Programme* (WCRP), see: https://www.wcrp-climate.org/; Church/Asrar/Busalacci/Arndt 2011.

[86] On the *International Human Dimension Programme* (IHDP), see Falkenhayn/Rechkemmer/Young 2011; see at: https://en.wikipedia.org/wiki/International_Human_D imensions_Programme; Ehlers 2016; Arizpe 2016.

[87] On *Diversitas*, see at: https://www.diversitas-international.org/; Walther/Larigauderie/Loreau 2011.

[88] *Earth Systems Science Partnership* (ESSP); see at: https://en.wikipedia.org/wiki/Earth_System_ Science_ Partnership>; Leemans/Rice/Henderson-Sellers/Noone 2011.

[89] *Future Earth;* see at: https://futureearth.org/.

[90] On *Earth Systems Science* (ESS), see at: https://en.wikipedia.org/wiki/Earth_system_science; Steffen et al. 2020; on *Earth System Analysis* (ESA), see: Schellnhuber/Wenzel 1998; Schellnhuber et al. 2004.

[91] On the *Potsdam Institute on Climate Change Impact Research* (PIK), see at: https://www.pik-pot sdam.de /en/home>. On the *Stockholm Resilience Centre* see at: https://www.stockholmresilience. org/ (13 February 2021).

[92] All Newsletters of the AWG and an overview of its publications are accessible at: http://quater nary.stratigraphy.org/working-groups/anthropocene/> (13 February 2021).

(IGBP). In 1991 *Diversitas* was established by *the United Nations Educational, Scientific and Cultural Organization* (UNESCO), the *Scientific Committee on Problems of the Environment* (SCOPE) and the *International Union of Biological Science* (IUBS).[93] In 1996 the *International Human Dimensions Programme* (IHDP) was set up as the fourth global change science programme.

During a meeting of the IGBP's Scientific Committee in late February 2000 in Cuernevaca (Mexico) at which Will Steffen, the IGBP's Executive Director (1998–2004) was present, the IGBPs Vice-Chair, Nobel laureate Paul J. Crutzen, coined the Anthropocene concept.

On 10–13 July 2001 IGBP, IHDP, the WCRP and Diversitas held a major conference in Amsterdam that produced the historic *Amsterdam Declaration on Earth System Science* addressing the Challenges of a Changing Earth.[94] The conference was attended by 1400 scientists from 105 countries, and resulted in a volume of Proceedings in the IGBP Series co-edited by Will Steffen, Jill Jäger, David J. Carson and Clare Bradshaw under the title *Challenges of a Changing Earth*. It included Crutzen's text on "Atmospheric Chemistry in the 'Anthropocene'".

An anthology in the IGBP Series co-edited by W. Steffen, R.A. Sanderson, P.D. Tyson, J. Jäger, P.A. Matson, B. Moore III, F. Oldfield, K. Richardson, H-J Schellnhuber, B.L. Turner, and R.J. Wasson (2004), called *Global Change and the Earth System: A Planet under Pressure*, included a chapter on "The Anthropocene Era: How Humans are Changing the Earth System" (81–141).[95] A year later the IGBP series included a volume on the *Coastal Fluxes in the Anthropocene* that was co-edited by C.J. Crossland, H.H. Kremer, H.J Lindeboom, J.I. Marshall Crossland, and M.D.A Le Tissier.

The *International Human Dimension Programme* (IHDP) and its former chairman (1996–1999) and leading member Eckart Ehlers et al. published the first books on the Anthropocene in English and in German that developed the Anthropocene concept within Geography during the first (2005) and second 5-year periods (2006, 2008). Ehlers (2016: 387) later argued that after 2000 "the propagation of the Anthropocene (Crutzen and Stoermer 2000) and of a geology of mankind (Crutzen 2002) … had a deep impact on IHDP's sustainability discussions and its discourses on how individuals and societies cope or are forced to cope with climate change, risks and disasters".

Thus, the 4 Science programmes (IGBP, IHDP, WCRP, Diversitas), the *Earth System Science Partnership* (ESSP; since 2001) and *Future Earth* (2012–2015) have

---

[93] See "Diversitas", at: https://en.wikipedia.org/wiki/Diversitas (18 February 2021).

[94] See the Amsterdam Declaration of 13 July 2001 that was signed by the chairs of IGBP (Berrien Moore III), IHDP (Arild Underdal), WCRP (Peter Lemke) and Diversitas (Michel Loreau), at: http://www.igbp.net/about/history/2001amsterdamdeclarationonearthsystemscience.4.1b8ae20512db692f2a680001312.html.

[95] The book can be downloaded without any charge at: http://www.igbp.net/download/18.56b5e28e137d8d8c09380001694/1376383141875/SpringerIGBPSynthesisSteffenetal2004_web.pdf (13 February 2021).

taken up, discussed, further developed and promoted this debate within the global environmental change research community, e.g. in *Anthropocene Magazine*.[96]

22.5.2   Earth System Science in the Anthropocene

Earth System Science (ESS) has been defined as "the application of systems science to the Earth" that "considers interactions and 'feedbacks', through material and energy fluxes, between the Earth's sub-systems' cycles, processes and 'spheres'." It aims at a holistic interpretation of nature that was inspired in the nineteenth century by Alexander von Humboldt and in the twentieth century by Vladimir Vernadsky (1863–1945), and since the mid-1960s by James Lovelock, who pointed to a feedback mechanism within the Earth system that was later called the 'Gaia hypothesis'.[97]

Earth System Analysis (ESA) was suggested by Schellnhuber and Wenzel (1998) of the *Potsdam Institute of Climate Impact Research* (PIK) as a scientific approach for "Integrating Science for Sustainability". According to PIK's research department, headed by Stefan Rahmsdorf and Wolfgang Lucht, ESS or ESA "focuses on the understanding and modelling of the physical and biogeochemical processes that govern the Earth System and its responses to human interference". PIK's research has been guided by "four major themes": tipping points, planetary boundaries, earth trajectories and extreme events. PIK's *FutureLab Earth Resilience in the Anthropocene* (ERA-lab) "drives this research forward with the goal of outlining the properties of, and pathways towards, a stabilized Earth System in the Anthropocene. It is dedicated to assessing the risk of cascading interactions of tipping elements in the Earth System and providing a framework for conceptualising, monitoring and modelling Earth System resilience."[98]

The Amsterdam Conference of July 2001 had an impact on both the Dahlem workshop of physicists, chemists, and meteorologists in May 2003 and the work of the *German National Committee on Global Change Research* (NKGCF) that later became the *German Committee Future Earth* (DKN) that coordinated research, while the *German Advisory Council on Global Change* (WBGU) has been advising the federal government since 1992.[99]

---

[96] See at: https://futureearth.org/publications/anthropocene-magazine/ (13 February 2021).

[97] See "Earth Systems Science", at: https://en.wikipedia.org/wiki/Earth_system_science (13 February 2021).

[98] See PIK, at: https://www.pik-potsdam.de/en/institute/departments/earth-system-analysis/research (12 February 2021).

[99] See DKN at: https://www.dkn-future-earth.org/about_us/us/index.php.en (13 February 2021). In 1992 the *German Advisory Council on Global Change* (WBGU) was created as an independent, scientific advisory body to the German Federal Government. Its Flagship Report 2021 addressed: *Rethinking Land in the Anthropocene: From Separation to Integration*. See "WBGU" at: https://www.wbgu.de/en/; the 2021 report is at: https://www.wbgu.de/en/publications/publication/landshift (18 February 2021).

The intensive discussion of the ESS or ESA in the context of the Anthropocene may have begun in May 2003 at the Dahlem Workshop on *Earth System Analysis for Sustainability* (Schellnhuber/Crutzen/Clark/Claussen/Held 2004), where the emerging theoretical debate on Earth System Science and analysis was increasingly framed within the context of the Anthropocene (Crutzen/Ramanathan 2004: 265–202). It was followed by the "Group Report on Earth Systems Dynamics in the Anthropocene" (Steffen 2004: 313–340).

The German geographers Eckart Ehlers and Thomas Kraft (2005, 2006; see also Ehlers 2008) also conceptually framed the Earth Systems Science within the new context of the 'Anthropocene'. In their introductory chapter they concluded that "the Anthropocene demands an Earth System Science, which understands Science in a framework that encompasses all aspects of Global Change that may be well suited to provide essential answers for identifying suitable pathways towards sustainability" (Ehlers/Kraft 2005a: 11).

Fifteen years later, Steffen, Richardson, Rockström, Schellnhuber, Dube, Dutreuil, Lenton and Lubchenco (2020) discussed

> the emergence and evolution of ESS, outlining the importance of these developments in advancing our understanding of global change. Inspired by early work on biosphere-geosphere interactions and by novel perspectives such as the Gaia hypothesis, ESS emerged in the 1980s following demands for a new 'science of the Earth'. The International Geosphere-Biosphere Programme soon followed, leading to an unprecedented level of international commitment and disciplinary integration. ESS has produced new concepts and frameworks central to the global- change discourse, including the Anthropocene, tipping elements and planetary boundaries. Moving forward, the grand challenge for ESS is to achieve a deep integration of biophysical processes and human dynamics to build a truly unified understanding of the Earth System.[100]

In the discussion of Earth System Science the Anthropocene has become

> the most influential concept … to describe the new geological epoch in which humans are the primary determinants of biospheric and climatic change …The Anthropocene has become an exceptionally powerful unifying concept that places climate change, biodiversity loss, pollution and other environmental issues, as well as social issues such as high consumption, growing inequalities and urbanization, within the same framework. Importantly, the Anthropocene is building the foundation for a deeper integration of the natural sciences, social sciences and humanities, and contributing to the development of sustainability science through research on the origins of the Anthropocene and its potential future trajectories (Steffen/Richardson/ Rockström, et al., 2020: 59).

Within 2 decades, the Anthropocene has become a conceptual framework and *context* for 2 rapidly developing research areas on global change and Earth System Science. The third scientific debate within the geological community understood the Anthropocene as a *turning point* in Earth's history, from the Holocene to the Anthropocene, which coincided with a major structural change in human and political history in the mid-20th century between 16 July 1945 (the first test of a nuclear bomb) and 1950, the start of the Great Acceleration that coincided with the major change

---

[100] See: Steffen, Richardson, Rockström, et al., 2020: "The Emergence and Evolution of Earth System Science", in: *Nature Reviews Earth & Environment*, 1 (January): 54 –63.

in post-war US global military, economic and political strategy (Brauch 2021) that was fundamental in the setting up of the post World War II international political, economic and security order during the Cold War.

22.5.3   The Anthropocene Working Group Members as a Scientific Knowledge Community

By 2010 the Anthropocene concept was discussed within the 2 overlapping communities focusing on global change and Earth Systems Science research, still primarily from the standpoint of the Natural Sciences.

How did the Anthropocene concept spread during the third 5-year period (2011–2015) from a slowly emerging debate in the Natural Sciences to Geology and the Social Sciences, the Humanities, Law, the Applied Sciences (including Medicine) and Engineering?

In 2008, a group of British geologists led by Jan Zalasiewicz and Mark Williams of the University of Leicester and Alan Smith of the University of Cambridge et al. (2008) raised the provoking question "Are we now living in the Anthropocene" in the *Geological Survey of America* (GSA), and suggested that:

> The term Anthropocene, proposed and increasingly employed to denote the current interval of anthropogenic global environmental change, may be discussed on stratigraphic grounds. … A case can be made for its consideration as a formal epoch in that, since the start of the Industrial Revolution, Earth has endured changes sufficient to leave a global stratigraphic signature distinct from that of the Holocene or of previous Pleistocene inter-glacial phases, encompassing novel biotic, sedimentary, and geochemical change. These changes … are sufficiently distinct and robustly established for suggestions of a Holocene–Anthropocene boundary in the recent historical past to be geologically reasonable. The boundary may be defined either via Global Stratigraphic Section and Point ("golden spike") locations or by adopting a numerical date. Formal adoption of this term in the near future will largely depend on its utility, particularly to Earth scientists working on late Holocene successions. This datum, from the perspective of the far future, will most probably approximate a distinctive stratigraphic boundary (Zalasiewicz/ Williams/Smith et al., 2008: 4).

Taking up Crutzen's proposal of 2002 (Chap. 3 in this volume) these primarily British geologists concluded:

> Sufficient evidence has emerged of stratigraphically significant change (both elapsed and imminent) for recognition of the Anthropocene – currently a vivid yet informal metaphor of global environmental change – as a new geological epoch to be considered for formalization by international discussion. The base of the Anthropocene may be defined by a GSSP in sediments or ice cores or simply by a numerical date (Zalasiewicz/Williams/Smith et al., 2008: 7).

Inspired by Crutzen's proposal, in December 2009, in Newsletter 1 of the *Anthropocene Working Group* (AWG),[101] its chairman Jan Zalasiewicz argued:

---

[101] See at: http://quaternary.stratigraphy.org/wp-content/uploads/2018/08/Anthropocene-Working-Group-Newsletter-No1-2009.pdf (13 February 2021).

that we are now living in a new geological interval of time – the Anthropocene – that is dominated by human activities. Since then, the term (currently informal) has been widely used by a range of Earth and environmental scientists, and the concept has been further examined (e.g. Steffen et al. 2008). Recent assessment by the Stratigraphy Commission of the Geological Society of London (Zalasiewicz et al. 2008) suggested a case for formally incorporating the term into the Geological Time Scale. The rationale behind this suggestion was recognition of the wide-ranging effects of anthropogenic influence on stratigraphically significant parameters: global atmospheric composition and temperature, cryosphere stability, ocean chemistry (e.g. acidification, anoxia) and sea level, biodiversity, landscape processes and near-surface sedimentation. The resulting anthropogenically driven changes – it was suggested – will almost certainly be reflected in a distinctive geological record.

Zalasiewicz pointed to the task of the AWG:

to critically consider the case for a formal Anthropocene, and to make recommendations to our parent body (the Subcommission on Quaternary Stratigraphy (SQS) of the International Commission on Stratigraphy (ICS), through them to the ICS itself, and then on to its parent body, the International Union of Geological Sciences (IUGS). A further body to be (at least) consulted and kept informed is the International Union of Quaternary Studies (INQUA), a body equal in rank to the IUGS.

Zalasiewicz outlined the AWG's work programme in its first newsletter:

We have to critically compare the current degree and rate of environmental change, caused by anthropogenic processes, with the environmental perturbations of the geological past, to assess the scale of current global change in geological terms. …

In the longer term… we need, in sum, to establish whether meaningful comparisons of this type can be made, and whether the scale and rate of past and current global change – biological, physical, chemical – can be established, and preferably quantified, sufficiently to justify (or deny) the formal use of the term 'Anthropocene'. We need also to consider the question of utility: whether more precise definition and formalization of the term 'Anthropocene' would serve a useful purpose for Earth scientists. If a formal Anthropocene was justified, consideration should be made of its *hierarchical level* (Age/Epoch/Period). If it is to be an Age, then it would be a component unit of the Holocene; if on the other hand it is to be an Epoch or Period we would be suggesting that the Holocene as a geological unit had terminated – a decision with considerable scientific and likely political implications. … Decisions on this question by us (and, up the line, by other ICS/IUGS bodies) need perhaps be guided by a combination of geological comparison (estimated scale of change) and utility to working Earth scientists – i.e. essentially within the scientific domain. We should be aware of wider implications, though (arguably) not influenced by these.

As a formal body within the geological community, since its formation the AWG's membership (2009, 2019)[102] has been both multidisciplinary (Geology, Chemistry, Geography, Earth and Environment Studies, Oceanography, Space Research, Biology, History, Archaeology) and international (the UK, the Netherlands, the USA, Brazil, Kenya, South Africa, Australia, Croatia, Norway, China, Spain, Poland, Switzerland, Canada, Germany, France, Austria). In its tenth Newsletter it distinguished between voting and advisory members.[103]

---

[102] The first 9 Newsletters for the years 2009 to 2019 are at: http://quaternary.stratigraphy.org/working-groups/anthropocene/.

[103] The tenth Newsletter will be published in 2021. A draft version was made available to the author.

The debate on the Anthropocene steadily expanded after the AWG members started to publish co-authored texts on the Anthropocene. Since 2009, the AWG may be defined as a multidisciplinary *Scientific Knowledge Community* (SKC) that differs from the early scientific networks (IGBP, IHDP, ESSP) and from "epistemic communities" (Haas 1990, 1992). While the AWG pursued a specific task in the geological community, it also contributed to the spread of this new term across scientific boundaries in the second decade, when AWG members took the lead in introducing the Anthropocene in the *Natural Sciences* and later co-published with authors in the *Social Sciences*, the *Humanities* and *Law* (Table 22.2). 8 AWG members were among the most prolific authors on the Anthropocene on both the *Web of Science* and *Scopus*, and 24 of the 35 AWG members of 2019 were among the top 100 Anthropocene authors on the *Web of Science* and *Scopus*.

Explanation of the colours used in the following Table.

Natural Sciences	Social Sciences	Humanities	Law	Language, Linguistics, Literature	Applied Sciences
Earth Sciences, Global Environmental Change Geology Biology & Biosciences	Human Geography Anthropology Political Science Sociology Economics Psychology	Archaeology History Philosophy Theology/Religion Performing Arts Music	National Law International Law Environmental Law	Literature & Fiction Literary Criticism Linguistics Poetry National Literatures	Medicine Health Engineering Technology

Interest in the fields of biology and biological diversity in the Anthropocene has increased since 2012. The Anthropocene concept rapidly spread across disciplinary, national and language boundaries between 2011 and 2015 and increased exponentially between 2016 and 2020 by more than tripling the number of entries in the *World Catalogue*, the *Web of Science* and *Scopus*. While the keywords or disciplines in which the authors were noted differed significantly between the *Web of Science* and *Scopus*, the overall quantitative results were similar. However, *Scopus* recognised AWG members more beyond the Natural Sciences.

As the originator of the Anthropocene concept, *Paul J.* Crutzen had 16 texts listed on the *Web of Science* on 13 February 2021, 3 of which were highly cited (1,241, 761, 596), and 17 on *Scopus*, 7 of which had more than 200 citations (1,405, 868, 662, 454, 278, 234, 232).

Jan Zalasiewicz,[104] a palaeobiologist in the Department of Geology at Leicester University (UK), was the chairman of the AWG (2009-2020) and since 2020 has chaired the *Subcommission on Quaternary Stratigraphy* (SQS). *Jan Zalasiewicz* had 37 [on 13 February 2021: 43] documents listed on the *Web of Science* and 46 [47] on *Scopus*, and was the most prolific AWG author. His Anthropocene texts were listed under Geology, Earth Sciences, and Global Environmental Change, and also noted in Biology, Archaeology, History, and Law. 5 of his Anthropocene papers on the

---

[104] See the publications of *Jan Zalasiewicz* at: https://www2.le.ac.uk/departments/geology/people/zalasiewicz-ja and in the references at the end of this chapter. In November 2020 he was Emeritus Professor of Palaeobiology at Leicester University (UK); Email: jaz1@le.ac.uk.

Table 22.2 AWG members as lead authors on the Anthropocene in different scientific disciplines, according to the *Web of Science* and *Scopus*, based on data from 30 September 2020 and 13 February 2021

No.	Rank of AWG members in counted documents by Source (Web of Science and Scopus)				AWG members as Lead Authors in scientific disciplines by source			
	Leading AWG Members	Web [W]	Scopus [S]	W+S	Disciplines	Web	Scopus	W+S
	Crutzen, Paul J.	**[16]**	**[17]**	**[33]**				
1	Zalasiewiscz, Jan (UK)	37 [43]	46 [47]	83 [90]	NS: Geology, Earth Science, GEC	20	32	52
					NS: Biology		3	3
					H: Archaeology	0	6	6
					H: History		36	36
					L: Law (Nat./Intern.)	0	18	18
2	Steffen, Will (USA, AU)	39 [41]	35 [35]	74 [76]	NS: Geology, Earth Science, GEC	17	22	39
					SS: Human Geography	4		
					SS: Security Studies	0	7	
					SS: Sociology	0	4	4
					SS: Psychology	0	3	3
					H: History	0	31	31
					L: Law (Nat./Intern.)	0	18	18
3	Willams, Mark (UK)	32 [36]	40 [41]	72 [77]	NS: Geology, Earth Science, GEC	19	31	50
					NS: Biology	0	3	3
					H: Archaeology	0	15	15
					H: History	0	31	31
					L: Law (Nat./Intern.)	0	18	18
4	Ellis, Erle (USA)	27 [27]	28 [28]	55 [55]	NS: Geology, Earth Science, GEC	10	12	22
					NS: Biology	4	9	13
					SS: Political Science	4	?	
					SS: Intern. Relations	4	?	
					SS: Envir. Politics	4	?	
					H: Archaeology	0	15	15
					H: History	0	21	
					L: Law (Nat./Intern.)	0	18	18
					L: Law (Nat./Intern.)	0	15	15
					LL: Linguistics, Literatures			
5	Syvitski, Jaia (USA)	22 [22]	26 [27]	48 [49]	NS: Geology, Earth Science, GEC	?	13	
					SS: Human Geography	3		
6	Barnosky, Anthony (USA)	22 [22]	20 [22]	42 [44]	NS: Geology, Earth Science, GEC	18	18	36
					NS: Biology		1	1
					H: History	0	17	17
7	Cearreta, Alejandro (E)	21 [22]	23 [24]	43 [46]	NS: Geology, Earth Science, GEC	14	20	34
					H: History	0	18	18
8	Vidas, Davor (Norway)	14 [15]	18 [18]	32 [33]	NS: Geology, Earth Science, GEC	3	10	13
					L: Law (Nat./Intern.)	3	16	19
9	McNeill, John R. (USA)	13 [13]	15 [15]	28 [28]	SS: Human Geography	2	2	4
					H: History	2	15	17
10	Waters, Colin (UK)	21 [26]	25 [25]	46 [51]	NS: Geology, Earth Science, GEC	18	20	38
					H: History	0	21	21

Source Compiled by the author using data from the *Web of Science* and *Scopus*

Web of Science were highly cited (683, 596, 286, 227, 157), and 4 of his 46 papers indexed by *Scopus* had more than 200 citations (710, 662, 330, 268). Zalasiewicz's first peer-reviewed text on the Anthropocene was dated 2010, according to both the *Web of Science* and *Scopus*.

Will Steffen,[105] by training an inorganic chemist who specialises in Earth System Science and Global Environmental Change and was formerly a professor at ANU (Canberra, Australia) and Executive Director of IGBP in Stockholm working closely with the Resilience Centre in Stockholm (Sweden) and the Potsdam Institute on Climate Impact Studies (Germany), had 39 [41] documents listed on the *Web of Science* and 35 on *Scopus*. His texts were primarily on his own areas of expertise (Earth System Science and Global Environmental Change) but also on the Social Sciences (Human Geography, Security Studies, Sociology and Psychology) as well as History and Law. 12 of Will Steffen's 39 Anthropocene papers on the *Web of Science* were highly cited (761, 596, 102, 708, 227, 683, 157, 84, 117, 107, 427, 37), and 8 of his 35 [35] documents on *Scopus* had more than 200 citations (1,405, 868, 789, 710, 479, 454, 268, 232). *Steffen's* authored and co-authored publications had the highest number of citations of the 10 leading AWG authors on both the *Web of Science* and *Scopus*.

Mark Williams,[106] a geologist of Leicester University (UK), achieved 32 [36] document counts on the *Web of Science* and 40 [41] on *Scopus*. His texts were categorised under Geology, Earth Sciences and Global Environmental Change, and also noted in Biology, Archaeology, History and Law. 5 of *Mark Williams's* 32 Anthropocene papers on the *Web of Science* were highly cited (286, 596, 227, 683, 157), and 3 of his 40 papers indexed by *Scopus* had more than 200 citations (710, 662, 268).

Erle C. Ellis,[107] an American environmental scientist at the University of Maryland, Baltimore County, had a total of 27 [27] texts on the *Web of Science* and 28 [28] on *Scopus*. His texts were noted in most disciplines beyond the disciplinary boundaries of Environmental Studies – in Biology, Political Science, International Relations, Environmental Politics, Archaeology, History, Law, and Linguistics and Literature. 4 of *Erle C. Ellis's* 27 Anthropocene papers on the *Web of Science* were highly cited (683, 343, 214, 204), and 3 of his 28 papers indexed by *Scopus* had more than 200 citations (710, 371, 204).

---

[105] See the biography of Prof. Will Steffen at: https://climate.anu.edu.au/about/people/academics/professor-will-steffen. Selected publications are in the references at the end of this chapter. In November 2020 he was Emeritus Professor at the Fenner School of Environment & Society in Canberra (Australia).

[106] In November 2020 Prof. *Mark Williams* was Professor of Palaeobiology at the University of Leicester, where he was working on the Anthropocene, the biosphere, evolution, and life. His selected publications on the Anthropocene are in the references at the end of this chapter.

[107] Prof. Dr. Erle C. Ellis is a Professor at the Department of Geography and Environmental Systems, University of Maryland, Baltimore County, USA, see at: https://ges.umbc.edu/ellis/. His selected publications are in the references at the end of this chapter.

Jaia Syvitski,[108] an American oceanographer and geologist at Colorado University, had 22 texts indexed on the *Web of Science* and 26 [27] on *Scopus*, and a few of his texts were also assessed under Human Geography. 2 of *Jaia Syvitski's* 12 Anthropocene papers on the *Web of Science* were highly cited (348, 267), and 3 of his 20 papers indexed by *Scopus* had more than 200 citations (710, 479, 268).

Anthony Barnosky, an American ecologist, geologist and biologist at Stanford University, had 22 publications on the Anthropocene listed on the *Web of Science* and 20 [22] on *Scopus*. His texts were also indexed under Biology and History. 4 of *Anthony Barnosky's* 22 Anthropocene papers on the *Web of Science* were highly cited (683, 427, 227, 157), and 4 of his 20 papers indexed by *Scopus* had more than 200 citations (710, 376, 271, 268).

Alejandro Cearreta, a natural scientist at the University of the Basque Country in Spain, had 21 [22] Anthropocene documents counted on the *Web of Science* and 23 [24] on *Scopus*. On both databases these documents were categorised under Geology, Earth Science, Global Environmental Change and History. 3 of *Alejandro Cearreta's* 21 Anthropocene papers on the *Web of Science* were highly cited (683, 227, 157). Of his 23 papers that were indexed by *Scopus* two texts had more than 200 citations (710, 268).

Davor Vidas,[109] Director, Marine Affairs and Law of the Sea Programme at the Fridtjof Nansen Institute in Norway, has 3 single-authored texts on the Anthropocene listed on the *Web of Science* and 16 texts written with several co-authors on *Scopus*. 3 of *Davor Vidas's* 14 [15] Anthropocene papers on the *Web of Science* were highly cited (683, 227), and 2 of his 18 papers indexed by *Scopus* had more than 200 citations (710, 268).

John R. McNeill,[110] an American environment historian at Georgetown University in Washington, D.C., had 2 History texts indexed on the *Web of Science* and 15 on *Scopus*, and 2 texts were also indexed under Human Geography on both the *Web of Science* and *Scopus*. 2 of *John R. McNeill's* 13 [13] Anthropocene papers on the *Web of Science* were highly cited (683, 227), and 2 of his 15 [15] papers indexed by *Scopus* had more than 200 citations (1,405, 710).

Colin Waters,[111] a geologist at Leicester University (UK), had 21 [26] texts on the Anthropocene listed on the *Web of Science* and 26 [26] on *Scopus*. In addition, 25 [25] other documents were also counted on *Scopus*. 3 of *Colin Waters's* 21 Anthropocene papers listed on the *Web of Science* on 30 September 2020 were highly cited (1,241,

---

[108] Information about Prof. *Jaia Syvitski* is at: https://instaar.colorado.edu/people/jaia-syvitski/ and https://en.wikipedia.org/wiki/James_Syvitski His selected references are at the end of this chapter.

[109] Information about Research Prof. *Davor Vidas* is at: https://www.fni.no/search/category294. html?q=Davor+VidascategoryID=294. His selected publications on the Anthropocene are in the references at the end of this chapter.

[110] More information about Prof. *John R. McNeill* is at: https://gufaculty360.georgetown.edu/s/con tact/00336000014Ri2hAAC/john-mc-neill. His selected publications on the Anthropocene are in the references at the end of this chapter.

[111] The publications of Hon. Prof. *Colin Waters* are listed in the references at the end of this chapter. He has succeeded Prof. Zalasaieweicz as AWG chairman.

683), and 2 of his 25 papers indexed by *Scopus* had more than 200 citations (710, 268).

Paul J. Crutzen, a Dutch citizen working and living in Germany, and the 10 members of the AWG who were listed in Table 22.2 by one or both indexes were noted across their disciplinary boundaries. 5 of them were Americans, 3 were British and one each was working in Spain and Norway. All publications that were indexed were in English.

Since its establishment in 2009 the AWG and its 35 members (2019) have become a *Scientific Knowledge Community* (SKC) of distinguished primarily natural scientists who have created new peer-reviewed scientific knowledge that has been increasingly noted and whose publications are counted by both the *Web of Science* and *Scopus*. Many of these authored and primarily co-authored texts by AWG members in collaboration with additional scholars have been produced since 2009/2010, first by natural scientists of multiple disciplines and since 2013/2014 by social scientists and authors from the Humanities, Law and the Applied Sciences. Within a decade (2009–2019) the AWG also reached a decision that was supported by nearly 90% of its members and has resulted in a proposal to make an initial decision by 2022 within the geological scientific community on whether the Anthropocene is a new epoch of Earth's history that terminates and succeeds the Holocene epoch.

22.5.4   Leading Scientific Developers and Promoters of the Anthropocene Concept in the Natural Sciences: Paul J. Crutzen, Will Steffen and Jan Zalasiewicz (2000–2020)

While Paul Crutzen and the ten most prolific AWG members writing on the Anthropocene were widely recognised, discussed and cited beyond their own scientific discplines since the second five year period (2006–2010), other key authors in the Social Sciences, Humanities, in Law and in the applied sciences intiated and contributed since the third five year period (2011–2015) to specific discourses within their respective disciplines that were also noted and discussed both within and beyond their disciplines that were widely spreading during the fourth five year period (2016–2020).

After Paul J. Crutzen's death on 28 January 2021, on 13 February 2021 the *Web of Science* reported 16 of his peer-reviewed texts on the Anthropocene for 2002 (3), 2003 (1), 2007 (1), 2010 (3), 2011 (2), 2012 (1), 2013 (1), 2015 (1), 2016 (2), and 2017 (1). The citation report noted that Crutzen's 16 reported Anthropocene texts had 50 citations by 2010, about 440 in 2015 and between 500 and 550 in the years 2017 to 2020. The total citations of Crutzen's 16 Anthropocene texts on the *Web of Science* had reached 3,945 citations (3925 without self-citations) by 13 February 2021 (Fig. 22.4).

Scopus listed 17 of Crutzen's peer-reviewed texts on the Anthropocene for 2002 (2), 2003 (1), 2006 (1), 2007 (1), 2010 (3), 2011 (2), 2012 (1), 2013 (3), 2015 (1), 2016

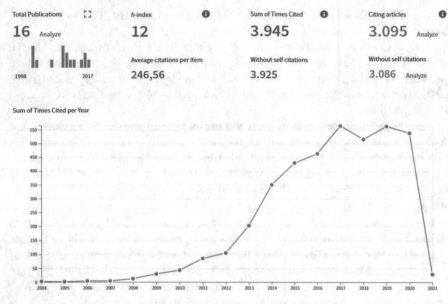

Fig. 22.4 Citation of Paul J. Crutzen's 16 reported documents on the Anthropocene on the *Web of Science* (2004–2020). *Web of Science* database (13 February 2021)

(1), and 2017 (1). On 13 February 2021 the *Scopus* citation report listed a total 4,822 citations (without any self-citations) of Crutzen's 17 texts on the Anthropocene. The number of citations reached 14 in 2008, 49 in 2010, 531 in 2015 and 650 in 2020.

On 13 February 2021 the *Web of Science* reported a total of 41 peer-reviewed texts on the Anthropocene by Will Steffen for the years 2003 to 2020: 2003 (1), 2007 (1), 2008 (1), 2010 (3), 2011 (3), 2012 (1), 2013 (1), 2014 (4), 2015 (3), 2016 (8), 2017 (7), 2018 (3), 2019 (4), and 2020 (1). Between 2004 and 2021 Steffen achieved a total of 7,864 citations (7,737 without self-citations), starting with 2 citations in 2004 and reaching 35 citations in 2010, 547 in 2015 and 1,648 in 2020 (Fig. 22.5).

On 13 February 2021 *Scopus* listed 35 texts by Will Steffen on the Anthropocene, which had reached a total of 7,522 citations (without any self-citations), starting with 2 citations in 2004 and reaching 44 citations in 2010, 554 citations in 2015, and 1,486 citations in 2020.

On 13 February 2021 the *Web of Science* reported a total of 43 peer-reviewed texts by Jan Zalasiewicz on the Anthropocene for the years 2010 to 2020, which had reached 3,479 citations, or 3,298 without self-citations (Fig. 22.5), starting with 2 in 2010 and reaching 261 in 2015 and 631 in 2020. On 13 February 2021 *Scopus* listed 47 texts by Jan Zalasiewicz on the Anthropocene for the years 2010 to 2021, which had reached 3,422 citations without self-citations, starting with 22 citations in 2010 and reaching 248 in 2015 and 625 in 2020 (Fig. 22.6).

The 3 lead authors of the AWG with texts on the Anthropocene concept were Paul J. Crutzen, who proposed the concept, Will Steffen, a leading scientist within

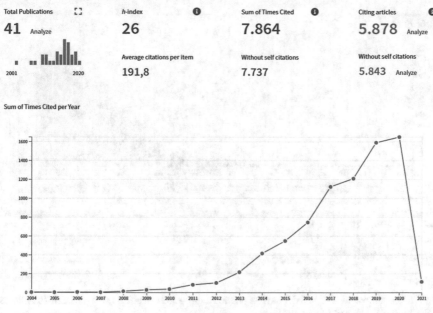

Fig. 22.5 Citation of Will Steffen's 41 reported documents on the Anthropocene on the *Web of Science* (2004–2021). *Web of Science* database (13 February 2021)

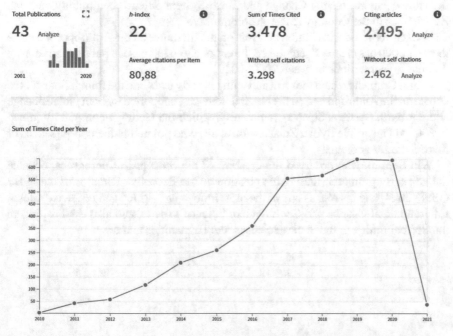

Fig. 22.6 Citation of Jan Zalasiewicz's 43 reported documents on the Anthropocene on the *Web of Science* (2010–2021). *Web of Science* database (13 February 2021)

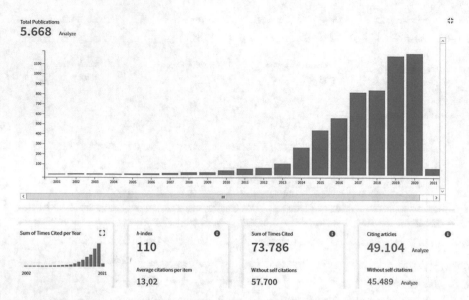

Fig. 22.7 Citations of all 5,668 documents on the Anthropocene from 2000 to 4 March 2021 listed in the *Web of Science* (4 March 2021)

the Global Environmental Change and Earth System Science community, and Jan Zalasiewicz in the Geology research community.

The bibliometric and citation analysis on the literature on the Anthropocene since the year 2000 has clearly documented the exponential increase in publications during the four 5-year periods between 2000–2020.

Paul J. Crutzen created the term and with his early texts laid the foundations for the debate on the Anthropocene in the Natural Sciences, supported by both global change and the Earth Systems Science community, which the Geology community joined in 2008 via an article by Jan Zalasiewicz et al., who pointed to the consequences of this debate for geologists.

Will Steffen, who observed the creation of the concept in Cuernavaca, became the key early promoter, linking his position as the Executive Director of the IGBP (1998–2004) and his role as one of the most innovative, skilled and respected Global Environmental Change and Earth System Science experts, and also establishing an intellectual bridge to the Social Sciences, the Humanities and Law.

P.J. Crutzen and J. Zalasiewicz. © C.Waters

Figure 22.7 documents the annual citations of all 5,668 documents on the Anthropocene from 2000 to 4 March 2021 listed in the *Web of Science* (4 March 2021). That amounted to 57,700 documents without self citations.

Jan Zalasiewicz's strategy of setting up the Anthropocene Working Group as a broad multidisciplinary scientific network housed within the Geology community, with both Paul J. Crutzen and Will Steffen as founder members complemented by the environmental historian John McNeill and the international lawyer Davor Vidas, further broadened the institutional base that promoted and spread the Anthropocene debate to the Social Sciences, the Humanities, Law and the Applied Sciences during the second decade, especially during the fourth 5-year period (2015–2020).

22.5.5 Key Scientific Authors in Biology, the Social Sciences, the Humanities, the Performing Arts, Law and Applied Sciences (Medicine, Health, Engineering and Technology)

The last research question concerns the key scientific authors who were instrumental in initiating the scientific debate on the Anthropocene in Biology, the Social Sciences, the Humanities, the Performing Arts, Law and the Applied Sciences (Medicine, Health, Engineering, and Technology) and whose work was highly cited in the peer-reviewed scientific literature beyond the Natural Sciences?

Given the quantity of Anthropocene references on *Amazon* and the *World Catalogue* and in the peer-reviewed literature indexed and assessed by the *Web of Science* and *Scopus*, it was not possible to reconstruct the genealogy of the spread of the Anthropocene concept during the first two decades (February 2000–2020) or to offer a qualitative review and assessment of the multiple scientific debates and discourses that have emerged in many other scientific disciplines. Nevertheless, based on the bibliometric analysis of the Anthropocene literature produced during the first two decades (2000–2020), it is possible to identify the lead authors (besides AWG members) whose work was highly cited in the Social Sciences, the Humanities and Law. The criteria for inclusion in Table 22.3 were:

(a)   the publications had to have a significant impact in the 2 citation analyses of the *Web of Science* and in *Scopus* (Table 22.3).
(b)   the Anthropocene publications had to be by the lead authors of at least 2 disciplines whose publications were included in the previous section (4).

In contrast to the Natural Sciences, there has been no specific 'scientific knowledge community' similar to the *Anthropocene Working Group* (AWG) in the Social Sciences, the Humanities, Law, or the Applied Sciences (Medicine, Health, Engineering, Technology, Development etc.).

A major contribution to the rapid spread of publications on the Anthropocene was made by *scientific lead authors*. The high number of citations which their peer-reviewed publications (journal articles, book chapters and books) have achieved has helped to spread the scientific debates and discourses on the new Anthropocene Epoch:

1.   *Johan Rockström* is an environment and sustainability expert from Sweden who has been the founding director of the *Resilience Centre* (RC) in Stockholm and who in 2018 became Co-Director of the *Potsdam Institute for Climate Impact Research* (PIK) in Germany. He is also Professor of Earth System Science at the University of Potsdam. By 14 November 2020, 20 of his Anthropocene texts indexed by the *Web of Science* (2011–2020) had achieved 3,617 citations, and his 15 publications on *Scopus* had achieved 2,966 citations. Besides his own expertise, his publications were also noted as the most highly counted publications on the Anthropocene in Security Studies.[112]

2.   *Carl Folke*, an ecological economist from Sweden, holds the positions of Science Director and Chair of the Board of the *Stockholm Resilience Centre* and Director of the *Beijer Institute of Ecological Economics* at the Royal Swedish Academy of Sciences.[113] Besides his own discipline of Ecological Economics, his work was also noted as making the top contributions to Environment and Development Politics and Psychology. On 12 November 2020 the *Web of Science* counted 1,860 citations of his 16 Anthropocene publications, while his 11 Anthropocene texts on *Scopus* achieved 1,974 citations.

3.   *Hans Joachim Schellnhuber* was the founding Director of the *Potsdam Institute for Climate Impact Research* (PIK) in Germany (1992–2018). Since his retirement he has been a Distinguished Visiting Professor at Tsinghua University (China) and a member of the Pontifical Academy of Sciences, the German

---

[112] On Johan Rockström, see: https://www.pik-potsdam.de/members/johanro and https://en.wikipedia.org/wiki/Johan_Rockstr%C3%B6m (14 November 2020).

[113] On Carl Folke, see: https://www.stockholmresilience.org/contact-us/staff/2008-01-15-folke.html and https://en.wikipedia.org/wiki/Carl_Folke (14 November 2020).

Table 22.3 Top Lead Authors of Publications on the Anthropocene (excluding members of the *Anthropocene Working Group*). *Source* Compiled by the author based on information from *Web of Science* and *Scopus* databases on 12–14 November 2020

Name	Web of Science	Author discipline	Scopus	Other disciplines	Country	Lead Author (in at least 2 disciplines)
Svenning, J.C.	25	Biosciences, Biodiversity	24	History	Denmark	2 disciplines (851 cit., WoS, 2014–2020; 889 cit., Scopus, 2016–2020)
Kotze, L.J.	23	Law	14	International Relations, Philosophy	S. Africa	3 disciplines (58 cit., WoS, 2015–2020; 183 cit., Scopus, 2016–2020)
Biermann, F.	20	Political Science	12	Environmental Politics, Dev. Politics, Sociology	Germany, Netherlands	4 disciplines (700 cit., WoS, 2010–2020; 887 cit., Scopus, 2012–2020)
Rockström, J.	20	Environment	15	Security Studies	Sweden	2 disciplines (3,617 cit., WoS, 2011–2021; 2,966 cit., Scopus, 2010–2020)
Dalby, S.	17	Geography	23	Political Science, Internat. Relations, Dev. Politics, Peace Studies, Security Studies, Sociology, History	Canada, Ireland, UK	8 disciplines (373 cit., WoS, 2013–2020; 463 cit., Scopus, 2013–2020)
Folke, C.	16	Ecological Economics	11	Env. Politics, Dev. Politics, Psychology	Sweden	4 disciplines (1,860 cit., WoS, 2011–2020; 1,974 cit., Scopus, 2011–2020)
Glikson, A.Y.	15	Geosciences		Archaeology	Australia	2 disciplines (10 cit., WoS)

(continued)

Table 22.3 (continued)

Name	Web of Science	Author discipline	Scopus	Other disciplines	Country	Lead Author (in at least 2 disciplines)
Chandler, D.	14	Social Sciences	12	Political Science, Internat. Relations, Security Studies, Sociology, Philosophy, Religion, Linguistics & Literature	UK/USA	7 disciplines (89 cit., WoS, 2017–2020; 48 cit., Scopus, 2017–2020)
Yusoff, K.	12	Environment		Philosophy	UK	2 disciplines (333 cit., WoS, 2013–2020)
Biggs, R.	11	Ctr.Complex Systems Transition	9	Env. Politics	S Africa Sweden	2 disciplines (510 cit., WoS, 2016–2020)
Cooke. S.J.	11	Environment		Education	Canada	2 disciplines (267 cit., WoS, 2018–2020)
Kim, R.E.	10	Sustainable Development		Political Science, Internat. Relations, Dev. Politics, Law	Netherlands	4 disciplines (116 cit., WoS, 2013–2020)
Braje, T.J.	9	Anthropology	11	Archaeology	USA	2 disciplines (121 cit., WoS, 2015–2020; 275 cit., Scopus, 2013–4-March 2021)
Castree, N.	9	Geography		Education	UK	2 disciplines (135 cit., WoS, 2015–2020)
Chakrabarty, D.	9	History	9 [4-3-2021]	Linguistics & Literature	USA	2 disciplines (78 cit., WoS, 2017–2020; 110 cit., Scopus, 2014–2020)

(continued)

Table 22.3 (continued)

Name	Web of Science	Author discipline	Scopus	Other disciplines	Country	Lead Author (in at least 2 disciplines)
Gupta, A.	9	Hydrology		Political Science, Dev. Politics, Ling., & Literature	Netherlands	4 disciplines (557 cit., WoS, 2010–2020
Schellnhuber, H.J.	9	Physics, Earth Science	6	Archaeology	Germany	2 disciplines (1,111 cit., WoS, 2011–2020; 1,155 cit., Scopus, 2012–2020)
Glaser, M.	8	Sociology	9 [4-3- 2021]	Psychology	Germany	2 disciplines (277 cit., WoS, 2011–2020; 359 cit., Scopus, 2013–2021)
Hamilton, C.		Earth Science	9	Philosophy	Australia	2 disciplines (377 cit., Scopus, 2016–2020)
Huettmann, F.		Philosophy	9	Biosciences	USA	2 disciplines (106 cit., Scopus, 2016–2020)
Pattberg, P.		Political Science	9	International Relations, Dev. Politics, Economics	Germany	4 disciplines (315 cit., Scopus, 2016–2020)
Gibson-Graham	7	Anthropology	8	Economics	Australia	2 disciplines (293 cit., WoS, 2011–2020; 306 cit., Scopus, 2010–2020)
Heikkurinen, P.		Earth Systems Science	8	Sociology, Economics, Philosophy,	Finland	4 disciplines (50 cit., Scopus, 2016–2020)
Colebrook, C.	7	Earth Systems Science	7	Philosophy, Linguistics & Literature	USA	3 disciplines (47 cit., WoS, 2013–2020; 11 cit., Scopus, 2016–2020))

(continued)

Table 22.3 (continued)

Name	Web of Science	Author discipline	Scopus	Other disciplines	Country	Lead Author (in at least 2 disciplines)
Galaz, V.		Sociology	7	Internat. Relations, Env. Politics,	Sweden	3 disciplines (270 cit., Scopus, 2016–2020)
Grear, A.	7	Earth Systems Science	7	Gender Studies, Law	UK	3 disciplines (55 cit., WoS, 2016–2020; 43 cit., Scopus, 2016–2020)
Arias-Maldonado M.	7	Political Science		Environmental Politics, Philosophy,	Spain	3 disciplines (42 cit., WoS, 2013–2020; 61 cit., Scopus, 2013–2020)
Rick, T.	7	Biosciences	7	Anthropology	USA	2 disciplines (110 cit., WoS, 2015–2020; 116 cit., Scopus, 2014–2020)
Erlandson, J.	7	Biosciences	7	Anthropology	USA	2 disciplines (348 cit., WoS, 2015–2020)
Oswald Spring, U	4	Political Science		Internat. Relations, Peace, Security & Gender Studies	Mexico	5 disciplines (10 cit., WoS, 2019).
Renn, J.	4	History		Philosophy	Germany	2 disciplines (5 cit., WoS, 2020).
Arizpe Schlosser, L.	4	Anthropology		Political Science, Internat. Relations	Mexico	3 disciplines (1 cit., WoS, 2019)
Brauch, H.G.	3	Political Science		Internat. Relations, Peace Studies, Security Studies, Gender Studies	Germany	5 disciplines (5 cit., WoS, 2019).

National Academy Leopoldina, and the US National Academy of Sciences.[114] His 9 publications on the Anthropocene on the *Web of Science* achieved 1,111 citations, and his 6 Anthropocene texts indexed on *Scopus* received 1,154 citations.

4. *Jens-Christian Svenning* is Centre Director and Professor in the Department of Biology at Aarhus University in Denmark.[115] The *Web of Science* counted 25 of his texts on the Anthropocene, which had 851 citations, while *Scopus* included 24 Anthropocene texts with 889 citations. His texts were also highly noted under History.

5. *Frank Biermann* is a research professor of Global Sustainability Governance at the Copernicus Institute of Sustainable Development at Utrecht University, The Netherlands. Biermann pioneered the 'Earth System' governance paradigm in Global Change research in 2005 and was the founder and first chair (2008–2018) of the Earth System Governance Project.[116] His 20 texts on the Anthropocene on the *Web of Science* achieved 700 citations and his 12 texts noted on *Scopus* had 887 citations. As well as in his own discipline of Political Science, his publications were among the lead texts noted in Environment, Development Politics and Sociology.

6. *Joyeeta Gupta* is a professor of Human Geography at Amsterdam University in the Netherlands.[117] Her 9 publications on the Anthropocene that were included on the *Web of Science* achieved 557 citations and her 10 texts indexed on *Scopus* had 505 citations. Besides Human Geography, her Anthropocene texts were most highly counted in the categories of Political Science, Development Politics. and Linguistics and Literature.

7. *Reinette Biggs* of the Centre for Complex Systems Transition at Stellenbosch University in South Africa[118] had 11 Anthropocene publications on the *Web of Science* which achieved 510 citations (2016–2020), and 9 texts on *Scopus* with 412 citations. Besides her own discipline, her Anthropocene texts were most highly counted in Environment Politics.

8. *Jon M. Erlandson* of the University of Oregon and its Museum for Natural and Cultural History has 7 Anthropocene texts included on the *Web of Science* which obtained 348 citations (2015–2020), and 10 documents on *Scopus* with 252 citations.[119] In addition to his own discipline of Biosciences, his work was highly noted in Archaeology.

---

[114] On Hans Joachim Schellnhuber, see: https://www.pik-potsdam.de/members/john/homepage and https://en.wikipedia.org/wiki/Hans_Joachim_Schellnhuber (14 November 2020).

[115] On J.C. Svenning, see: https://pure.au.dk/portal/en/persons/jc-svenning(33c3c4e2-57ab-478c-889d-06c594e57b8b).html (14 November 2020).

[116] On Frank Biermann, see: https://www.uu.nl/medewerkers/FHBBiermann (14 November 2020).

[117] On Joyeeta Gupta, see: https://www.uva.nl/content/nieuws/hoogleraarsbenoemingen/2012/11/gupta-joyeeta.html (14 November 2020).

[118] On R. Biggs, see: https://www0.sun.ac.za/cst/person/reinette-oonsie-biggs/.

[119] On J.M. Erlandson, see: https://mnch.uoregon.edu/leadership/jon-m-erlandson (14 November 2020).

9. *Clive Hamilton* is a public intellectual and Professor of Public Ethics at the Charles Sturt University in Australia.[120] He had 9 texts on the Anthropocene included on *Scopus* which achieved 365 citations between 2015 and 2020.[121]

10. *Kathryn Yusoff* is Professor of Inhuman Geography at the Queen Mary University of London. She had 12 Anthropocene texts included on the *Web of Science* which achieved 333 citations (2013–2020) and 11 publications on *Scopus* with 419 citations (2014–2020). Her Anthropocene publications were also noted in philosophy.

11. *Philipp Pattberg* (Political Science, Germany, The Netherlands) is Professor of Transnational Environmental Governance at the Free University of Amsterdam, where he serves as Director of the *Institute for Environmental Studies* (IVM) in the Department of Environmental Policy Analysis and Director of the *Amsterdam Sustainability Institute* (ASI). He had 9 Anthropocene texts listed on *Scopus*, which noted his work in International Relations, Development Politics and Economics.[122]

12. *J. Katherine Gibson-Graham* is a Geographer at the Institute for Culture and Society at Western Sydney University (Australia)[123] whose work on the Anthropocene was also noted in Economics. She had 7 texts with 293 citations included on the *Web of Science* and 8 documents with 306 citations indexed on *Scopus*.

The next list provides details of the scientific lead authors whose publications were counted most in 3 or more scientific disciplines and research programmes in the disciplines reviewed above.

1. *Simon* Dalby[124] is a Professor of Geography and Environmental Studies at Wilfrid Laurier University in Canada and also a citizen of the UK and Ireland. He had 23 Anthropocene publications with 463 citations included on *Scopus* and 17 texts with 373 citations on the *Web of Science*. Apart from his discipline of Human Geography, they were noted in Political Science, International Relations, Development Politics, Peace and Security Studies, Sociology, and History (2013–2020).

2. *David Chandler*[125] is a Professor of International Relations who has worked at the universities of Warwick, Westminster (UK) and Monash (Australia) and at the Oak Ridge National Laboratory (USA). He had 14 Anthropocene publications with 89 citations (2017–2020) on the *Web of Science* and 12 with 48 citations on *Scopus* that were noted in his own field of International Relations

---

[120] On C. Hamilton, see: https://en.wikipedia.org/wiki/Clive_Hamilton (15 November 2020).

[121] On K. Yusoff, see: https://www.qmul.ac.uk/geog/staff/yusoffk.html (15 November 2020).

[122] On P. Pattberg, see: https://research.vu.nl/en/persons/ph-pattberg (15 November 2020).

[123] On J. K. Gibson-Graham, see: https://www.westernsydney.edu.au/ics/people/researchers/kather ine_gibson (15 November 2020).

[124] On S. Dalby, see: https://www.balsillieschool.ca/simon-dalby/ and https://en.wikipedia.org/wiki/Simon_Dalby.

[125] On D. Chandler, see: https://www.westminster.ac.uk/about-us/our-people/directory/chandler-david and http://www.davidchandler.org/biography/ (15 November 2020).

and also in 6 disciplines and research programmes: Political Science, Security Studies, Sociology, Philosophy, Religion and Theology, and Linguistics and Literature. He is Editor of the open access online journal: *Anthropocenes – Human, Inhuman, Posthuman.*

3. *Rakhyun E. Kim*[126] is an Assistant Professor with dual expertise in International Law and International Relations at the Copernicus Institute of Sustainable Development at Utrecht University in the Netherlands. His work was noted in the fields of Political Science, International Relations, Development Politics, and Law. His 10 Anthropocene texts on the *Web of Science* achieved 116 citations.

4. *Pasi* Heikkurinen[127] is a Lecturer in the Department of Economics and Management at the *Helsinki Institute of Sustainability Science* (HELSUS) in Finland. He had 8 texts on the Anthropocene included on *Scopus* with 30 citations that were noted in Sociology, Economics and Philosophy.

5. *Victor Galaz,* the Deputy Director and an Associate Professor at the Stockholm Resilience Centre,[128] had 7 Anthropocene texts on *Scopus* that achieved 219 citations (2012–2020). His work on the Anthropocene was noted in Sociology, International Relations and Environment Politics.

6. *Louis J. Kotzé* is a Research Professor in the Faculty of Law, North-West University (South Africa)[129] who had 23 texts with 58 citations included on the *Web of Science* and 14 texts with 183 citations indexed by *Scopus*. His work was noted in International Relations and Philosophy as well as his own discipline of Law.

7. *Claire Colebrook*[130] is the Edwin Erle Sparks Professor of English, Philosophy, and Women's, Gender and Sexuality Studies at Pennsylvania State University (USA), where she specialises in contemporary literature, visual culture, and theory and cultural studies. She had 7 texts with 47 citations (2013–2020) on the *Web of Science* and 7 texts with 11 citations on *Scopus*.

8. *Manuel Arias-Maldonado*[131] is a Spanish Political Scientist at the University of Malaga whose work was noted in Environment Policy and Philosophy. His 9 Anthropocene texts on the *Web of Science* had 42 citations (2013–2020) and his 9 documents on *Scopus* had 61 citations.

---

[126] On R.E. Kim, see: https://www.uu.nl/medewerkers/RKim (15 November 2020).

[127] On P. Heikkurinen, see: https://www.helsinki.fi/en/people/people-finder/pasi-heikkurinen-912 2307 and https://researchportal.helsinki.fi/en/persons/pasi-heikkurinen (15 November 2020).

[128] On V. Galaz, see: https://www.stockholmresilience.org/contact-us/staff/2008-01-10-galaz.html (15 November 2020).

[129] On L.J. Kotzé, see: https://nwu.academia.edu/LouisKotze and http://law.nwu.ac.za/sites/ law.nwu.ac.za/ files/files/Law/eng/CurriculumVitaeKotzeJune2020.pdf (15 November 2020).

[130] On C. Colebrook, see: http://english.la.psu.edu/faculty-staff/cmc30 and https://philpeople.org/ profiles/claire-colebrook/publications?order=viewings(15 November 2020).

[131] On M. Arias-Maldonado, see: https://uma.academia.edu/ManuelAriasMaldonado and https:// www.carsoncenter.uni-muenchen.de/fellows/sof/members/alumni_scholars/manuel_maldonado/ index.html (15 November 2020).

9. *Anna Grear*[132] is a Professor of Law in the School of Law and Politics at Cardiff University, UK. She had 7 Anthropocene texts with 55 citations (2016–2020) on the *Web of Science* and 7 texts with 43 citations (2016–2020) on *Scopus*.

10. *Marion Glaser* is a working group leader on Social Ecological Systems Analysis at the Leibniz Centre for Marine Tropical Researcch in Bremen (Germany).[133] She has 8 Anthropocene publications with 277 citations listed on the *Web of Science* and 9 texts with 248 citations included on *Scopus*. Her work was also noted in Marine Research and Psychology.

11. *Steven J. Cooke* is a Professor in the Department of Biology at Carleton University in Ottawa (Canada).[134] He has 11 Anthropocene texts with 267 citations on the *Web of Science* and 11 texts with 243 citations on *Scopus*. In addition to his own discipline of Biology, his work was noted in the fields of Environment and Education.

12. *Todd J. Braje* is a Professor of Anthropological Archaeology specialising in long-term human-environmental interactions at San Diego University (USA).[135] On the Anthropocene he has 9 texts with 121 citations (2015–2020) included on the *Web of Science* and 11 documents with 227 citations on *Scopus*.

While, as part of a *Scientific Knowledge Community* (SKC), the 10 most counted AWG members (Table 22.2) have been intensively noted, the 24 *Top Lead Authors* (TLA) of publications on the Anthropocene (Table 22.3) who came primarily from the Social Sciences and a very few from the Humanities and one from Law, have primarily been noted and cited by authors from the Social Sciences and the Humanities (Philosophy [8], Archaeology [3], History [2], Religion and Theology [1], Literature [3]), and only one from Biology and Law.

Thus, during the first two phases (2000–2010) the spread of the Anthropocene occurred almost exclusively within the Natural Sciences. During the third phase (2011–2016) the publications on the Anthropocene in the Social Sciences, the Humanities and Law took root and intensified during the fourth phase (2016–2020).

The number of citations of the Top Lead Authors from the Natural Sciences was also much higher than the citations of authors from the Social Sciences, the Humanities and Law, according to the high number of peer-reviewed publications on the Anthropocene in the *Web of Science* Core Collection (5,248 texts on 15 November 2020 and 5,603 on 14 February 2021) and on *Scopus* (5,124 documents on 15 November 2020 and 5,421 on 14 February 2021).

With his comment "We are now in the Anthropocene", in late February 2020 Paul J. Crutzen triggered a global debate across most scientific boundaries, cultures and religions. While the Anthropocene has been discussed as a new epoch in Earth's

---

[132] On A. Grear, see: https://www.cardiff.ac.uk/people/view/478859-grear-anna (15 November 2020).

[133] On M. Glaser, see: https://www.leibniz-zmt.de/de/mitarbeiter/marion-glaser.html (15 November 2020).

[134] On S.J. Cooke, see: https://carleton.ca/biology/people/steven-j-cooke/ (15 November 2020).

[135] On T.J. Braje, see: https://anthropology.sdsu.edu/people/braje and https://anthropology.sdsu.edu/_resources/docs/faculty/CV_Braje.pdf. (15 November 2020).

history that started between 1945 (start of the nuclear era) and 1950 when the process of Great Acceleration set in, the Anthropocene as the "good, bad and ugly", as Simon Dalby phrased it in November 2015, has acquired different meanings in the Social Sciences:

> Whether all this will turn out well, a 'good Anthropocene' (as initially suggested by *Erle Ellis*, 2011a) or disastrously for humanity, a 'bad Anthropocene' as so many invocations of imminent catastrophe suggest, has been a matter of considerable and sometimes acrimonious discussion recently … The Anthropocene is now more than a proposed new geological epoch that marks the transformation of the Earth System wrought by humanity; it has become a contentious term and a lightening rod for political and philosophical arguments about what needs to be done, the future of humanity, the potential of technology and the prospects for civilization. It is so in part because the future is not determined and, while the Anthropocene has already turned out to be bad for many species that have been made extinct as a result of human action, whether it will turn out to be bad for humanity in the future depends in part on what is decided by the rich and powerful parts of our species in coming decades. Various futures are possible for humanity, but given the geological-scale transformations that have triggered the Earth System sciences in the first place, not all options are available. Only some are possible, but which ones depend on current political choices (Dalby 2015: 34).

22.5.6   Methodological Shortcomings of the Bibliometric Sources

The bibliometric data on the Anthropocene provided by the *Web of Science* (Analytica) and *Scopus* (Elsevier) on which this quantitative analysis and qualitative interpretation are based have major shortcomings, which is why they cannot be regarded as offering a truly representative assessment of global scholarship on the Anthropocene:

1. All 4 sources (databases) on which this analysis is based – *Amazon*, *World Cat*, *Web of Science* and *Scopus* – use English primarily or nearly exlusively for all documents covered. The texts noted in other languages are not included in the analysis.
2. *Web of Science* (Analytica) and *Scopus* (Elsevier) focus their analysis of texts primarily on authors from English language countries and, to a lesser extent, on authors from non-Anglophone countries in the Global North who were working in or associated with academic institutions in English-speaking regions and countries.
3. Research and publications in English by authors from outside the Global North and in other languages are less likely to be included and cited (even in the publications of the IPCC).
4. These peer-reviewed publication data are often used for the ranking of universities, and Anglophone countries clearly dominate.
5. Scholars from the Global South who are associated with academic institutions in the Global South and publish in peer-reviewed journals and book series that are sometimes published in these regions are often less likely to be included.

6. Thus, publications on the Anthropocene from many parts of the world that do not meet the selection criteria of either the *Web of Science* (Analytica) or *Scopus* (Elsevier) are absent from this and many other bibliometric analyses and are therefore not globally recognised.
7. Enhancing the representativeness of global research output – not only with regard to the literature on the Anthropocene – remains a major challenge for the future.

Despite these and other shortcomings, this multidisciplinary quantative assessment based on bibliometric data was able to achieve the goal of documenting the rapid increase in publications on the Anthropocene in just two decades and how this concept has been taken up far beyond the Natural Sciences represented by the unique *Scientific Knowledge Community* (SKC) of the *Anthropocene Working Group* (AWG), which has been instrumental in spreading the Anthropocene concept and its messages beyond Geology, the Earth Sciences and Global Environmental Change issues.

Since the the third time period (2011–2015) a few distinguished *Scientific Lead Authors* (SLA), among them highly distinguished and recognised scholars in the Natural and Social Sciences, Humanities, Law, the Performing Arts, Medicine and Health, Engineering, and Technology Development, have initiated and supported the rapid spread of a scientific concept of a relatively small scientific community to a concept that, just two decades since its inception, is used by scholars in nearly all scientific disciplines, cultures and religions.

This has generated multiple qualitative debates and more narrow discourses on the Anthropocene that have developed their own specific definitions and meanings that differ from the understanding of the Anthropocene as a stratigraphic epoch in Earth's history. Thus, the Anthropocene has become a global, multidisciplinary catchword that has accelerated human-nature relations on the scientific and increasingly on the societal and political agenda.

22.5.7 Paul J. Crutzen's Scientific and Human Legacy

While I was working on this concluding chapter of this anthology containing selected English texts by Paul J. Crutzen on the Anthropocene, Crutzen passed away on 28 January 2021. The condolences[136] offered by his colleagues and friends and the obituaries[137] prepared by the institutions with which he was associated, the global press and many scientific journals pay testament to his manifold achievements.

They have stressed his pioneering research on the depletion of the ozone layer, for which he was honoured in 1995 with the Nobel Prize in Chemistry, along with Mario

---

[136] For a list of condolences, see: https://www.mpic.de/4677594/trauer-um-paul-crutzen#Condolences.

[137] For a collection of obituaries, see: http://afes-press-books.de/html/SpringerBriefs_PSP_Crutzen.htm.

Molina (Mexico) and Sherwood Rowland (USA); his joint work with John Birks on the Nuclear Winter (1982), which had an impact on the nuclear strategic debate in the USA and USSR during the 1980s; and his proposal of the Anthropocene on 23 February 2020, which triggered a rethinking of the relationship and feedbacks of the activities of humankind on nature and on the Earth's history that addresses the key *challenges* humankind is facing now and in future decades, centuries and millennia.

While, for geologists, the Anthropocene is a *turning point* as a new epoch in Earth's history, in the Natural and Social Sciences, the Humanities, Law and the Applied Sciences it has become a new *context* for rethinking and a *challenge* or *threat* for human livelihoods and survival in the future.

Addressing the multiple challenges and threats of human actions to nature through anthropogenic global environmental change and climate change, biodiversity loss, soil erosion, desertification, and water pollution, scarcity and poverty must not result in *catastrophism* but in forward-looking preventive action aimed at sustainability transition and sustainable peace (Brauch/Oswald Spring/Grin/Scheffran 2016), possibly from the perspective of a "Peace Ecology in the Anthropocene" (Brauch 2021). This requires both rethinking and action to respond to human-induced catastrophes and disasters. In their obituary of Crutzen in the *Scientific American*, Jan Zalasiewicz, Colin Waters, and Will Steffen wrote:

> Over the last century, the Earth has acquired a striking and indelible geological record of human-driven perturbation, and been sharply set on a new trajectory, towards a warmer, more biologically impoverished and polluted state – one that will be more difficult for humanity to thrive in. This sobering realization quickly spread from the sciences to the humanities, provoking reimagination of their disciplines to incorporate the Earth no longer as a passive and stable backcloth for the human adventure, but as an active, hyperresponsive and dangerous actor.
>
> Crutzen was indispensable to this scientific revolution (for such it is). … He was, simply, the right person (of immense and deserved authority) making a conceptual leap at the right time (when sufficient evidence had built up) and in the right company (as a central figure in the highly active and international community studying contemporary global change). Moreover, this community had come to consider our planet holistically, as an integrated Earth system. His concerns over the growing realities of global warming led to a controversial foray into theoretical geoengineering, proposing injection of sulfur gases into the atmosphere to reduce insolation.[138]

22.5.8   Outlook: From a Multidisciplinary Bibliometric Analysis Towards a Review of Discourses in Single Disciplines

From a Peace Ecology perspective, the starting point of the Anthropocene Epoch (1945–1950) is a new *context*, a fundamental *turning point* and a potentially

---

[138] Zalasiewicz et al. (2021): "Remembering the Extraordinary Scientist Paul Crutzen (1933–2021)", © 2021 Scientific American, Springer Nature America, Inc.; see: https://www.scientificamerican.com/article/remembering-the-extraordinary-scientist-paul-crutzen-1933-2021/.

dangerous social *challenge* for political and human history which humankind may face during the twenty-first century, when effects of the anthropogenic interventions of humankind on the Earth System are increasingly observed, addressed and debated.

This author plans to try to reconstruct the qualitative evolution of the Anthropocene discourse in his own discipline of *Political Science,* his subdiscipline of *International Relations,* his preferred research programmes of Peace, Security and Environmental Studies (Brauch 2022), and his two proposed linkage projects of *Political Geoecology* (Brauch 2016a, 2016b) in the Natural and Social Sciences and *Peace Ecology* (Brauch 2021) in the Social Sciences. The manifold challenges that are associated with the Anthropocene require a response from the political actors, processes, fields, institutions and norms, which may necessitate a precautionary, forward-looking and preventive ecological peace policy.

22.6 Appendix

The following tables and figures are daily snapshots to illustrate the data in the databases of the *Web of Science* and *Scopus* (22.9 and 22.13).

Table 22.4 Anthropocene texts by research areas and scientific disciplines in *the Web of Science* and in *Scopus* (2001–2021)

No.	Discipline	Web of Science	No.	Discipline	Scopus	No.	Discipline	Web	No.	Discipline	Scopus
1	Env. Sciences, Ecology	1,702	1	Social Sciences	2,086	26	Evolutionary Biology	87	26	Health professions	5
2	Geology	608	2	Environmental Science	1787	27	Public Administration	80			
3	Science, Technoloy et al.	449	3	Arts, Humanities	1236	28	Agriculture	77			
4	Literature	353	4	Earth & Planet. Sc.	1021	29	Archaeology	66			
5	Geography	332	5	Agricult. & Biol.Sc.	936	30	Art	66			
6	Physical Geography	303	6	Business, Management, Accounting	230	31	Chemistry	60			
7	Social Sciences, et al.	285	7	Engineering	223	32	Plant Sciences	60			
8	Government, law	257	8	Economics, Econometrics, Finance	200	33	Architecture	52			
9	Arts, Humanities et al.	198	9	Biochemistry, Genetics, Mol. Biol.	190	34	Area Studies	52			
10	Philosophy	186	10	Multdisciplinary	162	35	Public Env. Occupat. Health	47			
11	Anthropology	177	11	Medicine	158	36	Development Studies	45			
12	Education, Educational Res.	161	12	Energy	106	37	Forestry	42			
13	Life Sciences, Biomedicine	127	13	Computer Science	82	38	Oceanography	42			
14	Sociology	126	14	Chemistry	79	39	Communication	41			
15	Meteorology, Atmosph. Scienc.	121	15	Immunology, Microbiology	57	40	Zoology	37			
16	Biodiversity, Conservation	142	16	Physics, Astronomy	52	41	Urban Studies	36			
17	Cultural Studies	131	17	Psychology	48	42	Social Issues	34			
18	Business, Economics	130	18	Mathematics	29	43	Biochemistry Molecular Biology	30			
19	Marine, Freshwater Biology	110	19	Decision Sciences	27	44	Genetics Heredity	30			
20	Water Resources	107	20	Material Sciences	19	45	Paleontology	28			
21	International Relations	105	21	Chem. Engineering	18	46	Film, Radio, Televison	27			
22	History	103	22	Neuroscience	15	47	Linguistics	27			
23	History, Philosophy of Sc.	103	23	Nursing	13	48	Computer Science	25			
24	Engineering	98	24	Veterinary	7	49	Asian Studies	23			
25	Religion	90	25	Pharmacology, Toxicology	6	50	Geochemisty, Geophysics	22			

Sources Websites of *Web of Science* and *Scopus* (30 September 2020). Explanation: Green in *Web of Science* and blue in *Scopus*.

Table 22.5 Anthropocene texts by countries and languages in the *Web of Science* and in *Scopus* (2001–2021)

No.	Countries	Web	No.	Countries	Web	No.	Countries	Scopus	No.	Countries	Scopus
1	USA	1883	26	Portugal	58	1	USA	1801	26	Chile	41
2	England	879	27	Chile	55	2	United Kingdom	927	27	Ireland	40
3	Australia	621	28	Wales	51	3	Australia	609	28	Taiwan	39
4	Canada	478	29	Ireland	40	4	Canada	438	29	Argentina	36
5	Germany	417	30	Taiwan	39	5	Germany	414	30		
6	Sweden	265	31	Argentina	31	6	France	250	31	Czech Republic	29
7	France	240	32	Russia	31	7	Sweden	233	32	Singapore	25
8	Netherlands	198	33	South Korea	30	8	Spain	180	33	Columbia	24
9	Spain	192	34	Czech Republic	25	9	Netherlands	178	34	Russian Federat.	24
10	Peopl. R. China	191	35	Kenya	23	10	China	169	35	Israel	21
11	Italy	162	36	Singapore	22	11	Italy	152	36	Malaysia	20
12	South Africa	148	37	Israel	21	12	Brazil	149	37	Kenya	18
13	Brazil	129	38	Malaysia	19	13	South Africa	138	38	Turkey	18
14	Switzerland	125	39	Turkey	17	14	Norway	126	39	Indonesia	16
15	Norway	123	40	Greece	15	15	Switzerland	122	40	Croatia	15
16	Denmark	119	41	Iceland	15	16	Denmark	105	41	Iceland	15
17	Scotland	109	42	Indonesia	15	17	New Zealand	94	42	Greece	14
18	Austria	83	43	Ecuador	14	18	Belgium	72	43	Hungary	12
19	Finland	82	44	Panama	12	19	Japan	72	44	Philippines	12
20	New Zealand	81	45	Romania	12	20	Finland	71	45	Panama	11
21	Japan	74	46	Thailand	11	21	India	71	46	Peru	11
22	Belgium	73	47	Hungary	10	22	Poland	70	47	Puerto Rico	10
23	Poland	65	48	Peru	10	23	Austria	65	48	Slovakia	10
24	India	68	49	Philippines	10	24	Portugal	52	49	Saudi Arabia	9
25	Mexico	58	50	Slovakia	10	25	Mexico	48	50	Thailand	9
No.	Countries	Web	No.	Countries	Web	No.	Countries	Scopus	No.	Countries	Scopus

Sources Websites of *Web of Science* and *Scopus* (30 September 2020).

Table 22.6 Anthropocene texts by languages in the *World Cat*, the *Web of Science* and in *Scopus* (2001–2021)

No.	Languages	World Cat	No.	Languages	Web of Science	No.	Languages	Scopus
1	English	23,444	1	English	4,856	1	English	4,777
2	not determined	865	2	Spanish	61	2	Spanish	40
3	French	649	3	Italian	32	3	French	39
4	German	227	4	Portuguese	31	4	Portuguese	38
5	Spanish	167	5	German	30	5	German	27
6	Japanese	66	6	French	27	6	Italian	18
7	Italian	41	7	Polish	11	7	Chinese	9
8	Portuguese	34	8	Russian	10	8	Russian	7
9	Chinese	22	9	Turkish	5	9	Polish	5
10	Polish	21	10	Chinese	4	10	Croatian	4
11	Swedish	15	11	Czech	3	11	Japanese	4
12	Korean	10	12	Korean	3	12	Afrikaans	3
13	Croatian	8	13	Croatian	2	13	Bosnian	3
14	Russian	8	14	Dutch	2	14	Slovakian	3
15	Danish	7	15	Catalan	1	15	Moldavian	2
16	Slovenian	6	16	Danish	1	16	Moldovan	2
17	Afrikaans	5	17	Slovenian	1	17	Romanian	2
18	Czech	4	18			18	Slovenian	2
19	Danish	4	19			19	Malay	1
20	Catalan	3	20			20	Turkish	1
21	Slovakian	3	Total	English(95.6%)	5080		English(96.6%)	4945
22	Ndonga	2						
23	Turkish	2						
24	Ukrainian	2						
25	Afar	1						
26	Galician	1						
27	Modern Greek	1						
28	Malay	1						
29	Nauru	1						
30	Norwegian	1						
31	Romanian	1						
32	Serbian	1						
33	Tamil	1						
Total	English 23,444 (77.44%)	30,274						

Sources Websites of *Web of Science* and *Scopus* (30 September 2020)

Table 22.7 Anthropocene texts by type in the *Web of Science* and in *Scopus* (2001–2021). *Sources* Websites of *Web of Science* and *Scopus* (30 September 2020)

No.	Type (Total 5,080)	Web of Sc.	No.	Types (4,945)	Scopus	No.	Types (Total 4,945)	Web of Sc.
1	Article	3,690	1	Article	2,955	1	Journal	3,955
2	Book chapter	547	2	Review	674	2	Book	748
3	Editorial Material	476	3	Book Chapter	649	3	Book Series	139
4	Review	381	4	Editorial	158	4	Conference Proceeding	88
5	Book Review	282	5	Conference Paper	141	5	Trade Journal	15
6	Proceedings paper	169	6	Book	135			
7	Early Access	116	7	Note	125			
8	Book	44	8	Letter	40			
9	Letter	36	9	Short Survey	36			
10	Meeting Abstract	16	10	Erratum	18			
11	Correction	15	11	Conference Review	9			
12	News item	8	12	Data paper	1			
13	Poetry	5	13	Undefined	4			
14	Data paper	2						
15	Art Exhibit Review	1						
16	Biographical Item	1						
17	Fiction Creative prose	1						
18	Film Review	1						
19	Music Performance Review	1						
20	Reprint							
21	Theater Review	1						

Table 22.8 Anthropocene texts by journals in the *Web of Science* and in *Scopus* (2001–2021)

No.	Journals	Web of Sc.	No.	Journals	Web of Sc.	No.	Journals	Scopus	No.	Journals	Scopus
1	Anthropocene Review	108	26	Geolog. Society Special Publicat.	18	1	Anthropocene Review	106	26	Religions	18
2	Anthropocene	57	27	Biological Conservation	17	2	Anthropocene	98	27	Biological Conservations	17
3	Sustainability	46	28	Earths Future	17	3	Encyclopedia of the Anthropocene	86	28	Geological Society Special Publicat.	17
4	Proceedings of the National Academy of US	40	29	Elementa Science of the Anthropocene	17	4	Sustainability Switzerland	37	29	Angelaki Journal of Theor.et.ical Humanities	16
5	Ecology & Society	37	30	Env. Law & Governance for the Anthr.	17	5	Proceedings of NAS of US	36	30	Environm ent & Planning D Society Space	16
6	Holocene	36	31	Envitonment Values	17	6	Quatenary Sc. Rev.	36	31	European Journal of Social Theory	16
7	Quatenary Sc. Rev.	36	32	Frontiers in Ecology & the Environ.	17	7	Holocene	34	32	Global Ecology and Biogeography	16
8	Environm. Humanities	35	33	Geographical Research	17	8	Science of the Total Environment	34	33	Proceedings of the 59th Meeting of the International Society for the Systems Sciences	16
9	Science of the Total Environment	35	34	Plos One	17	9	Philosophical Transactions of the Royal Soc. B	30	34	Progress in Phys. Geography	16
10	Philosophical Transactions of the Royal Society B	29	35	Routledge Env. Humanities	17	10	Ecology & Society	27	35	Anthropocene Debate and Political Science	15
11	Science	29	36	Stratigraphical Basis for the Anthropocene	17	11	Nature	27	36	Conservation for the Anthropocene Ocean	15
12	Geoforum	28	37	Angelaki Journal of Theor.et. Humanities	16	12	Trends in Ecology & Evolution	27	37	Frontiers in Ecology and the Environment	15
13	Scientific Reports	28	38	Earth Surface Proc. & Landforms	16	13	Science	26	38	Global environmental Change	15
14	Current Opinion of Env. Sustainability	26	39	Europ. Journ. of Social Theory	17	14	Scientific Reports	26	39	Green Letters	15
15	Nature	26	40	Global Ecology & Biogeography	17	15	Elementa	23	40	Plos On	15
16	Theory-Culture-Society	26	41	Proceedins of the Royal Society B Biol. Sc.	16	16	Current Opinion of Env. Sustainab.	22	41	Ecology and Evolution	14
17	Trends in Ecology Evolution	25	42	Routledge Research on Global Environm. Government	16	17	Geomorphology	22	42	Global and Planetary Change	14
18	Geomorphology	23	43	Env. History	15	18	Global Change Biology	22	43	Revista Virtual de Quimica	14
19	Global Change Biology	23	44	Global & Planetary Change	15	19	Theory, Culture, Society	22	44	Telos	14
20	Annals of the American Assoc. of Geographers	22	45	Glob.Env.Change Human & Pol Dim.	15	20	Annals of the American Assoc. of Geographers	21	45	Dialogues in Human Geography	13
21	Biosciences	21	46	New Scientist	15	21	Geoforum	21	46	Earth Surface Proc. & Landforms	13
22	Religion	19	47	Progress in Phys. Geography –Earth & Env.	15	22	Biosciences	19	47	Environmental Politics and Governance in the Anthropocene	13
23	Current Biology	18	48	Ambio	14	23	Curr. Biology	18	48	Geographical Research	13
24	Environm.ent & Planning D Society Space	18	49	Anthropocene Deb. & Political Science	14	24	Env., Science, Technol.	18	49	Rendiconti Lincei	13
25	Env., Science, Technology	18	50	Aust. Journal of Env. Educ.	14	25	Journ. of Contemporary Archaeology	18	50	Zygon	13

Sources Websites of the *Web of Science* and *Scopus* (30 September 2020).

Table 22.9 The first fifty Anthropocene texts by peer-reviewed book series in the *Web of Science* (2001–2021)

No.	Book Series Title	Publisher	No. indexed	No.	Book Series Title	Publisher	No. indexed
1	Geological Society Special Publication	Geological Society, London	18	26	IAHS	IAHS Press (UK)	4
2	Routledge Environmental Humanities	Taylor & Francis	17	27	Routledge Companions	Taylor & Francis	4
3	Routledge Research in Global Environmental Governance	Taylor & Francis	16	28	Routledge International Handbooks	Taylor & Francis	4
4	Annual Review of Environment and Resources	Annual Reviews	13	29	Routledge Studies in Environmental Policy	Taylor & Francis	4
5	International Library of Environmental Agriculture and Food Ethics	Springer Nature	13	30	Archaeological Orientations	Taylor & Francis	3
6	Palgrave Pivot	Springer Nature	13	31	Critical Food Studies	Taylor & Francis	3
7	Palgrave Studies in Educational Futures	Springer Nature	13	32	Ecological Studies Analyses and Synthesis	Springer Nature	3
8	Transnational Law and Governance	Taylor & Francis	12	33	Environmental Cultures	Bloomsbury	3
9	Annual Review of Anthropology	Annual Reviews	11	34	Geographies of the Anthropocene	Il Sileno Ed.	3
10	Contemporary Geographies of Leisure Tourism and Mobility	Taylor & Francis	11	35	Geophysical Monograph Series	Wiley-Blackwell Publish.	3
11	Earth System Governance	MIT Press	10	36	Humanity and the Sea	unknown	3
12	Modern Approaches in Solid Earth Sciences	Springer Nature	10	37	New Directions in Rhetoric and Materiality	Ohio State Univ. Press	3
13	Hexagon Series on Human and Environmental Security and Peace	Springer Nature	9	38	Nova et Vetera Iuris Gentium Series. A Modern International Law	Springer Nature	3
14	Literatures Cultures and the Environment	Springer Nature	8	39	Procedia Engineering	Elsevier BV	3
15	Anthropocene: Politik, Economics, Science	Springer Nature	7	40	Springer Transactions in Civil and Environmental Engineering	Springer Nature	3
16	Routledge Literature Companions	Taylor & Francis	7	41	State of the World	W.W. Norton & Co.	3
17	Routledge Studies in Sustainability	Taylor & Francis	7	42	Advances in Asian Human Environmental Research	Springer Nature	2
18	Annual Review of Ecology Evolution and Systematics	Annual Reviews	6	43	Advances in Intelligent Systems and Computing	Springer Nature	2
19	Geophysical Monograph Series	Wiley-Blackwell Publish.	6	44	Advances in Social Science Education and Humanities Research	Atlantis Press	2
20	Under the Sign of Nature	Univ. of Virginia Press	6	45	Annual Review of Political Science	Annual Reviews	2
21	Critical Environments, Nature, Science, Politics	University of California Press	5	46	Coastal Research Library	Springer Nature	2
22	Ecology and Ethics	Springer Nature	5	47	Current Perspectives in Social Theory	Emerald Publishing	2
23	Routledge Advances in Art and Visual Studies	Taylor & Francis	5	48	Developments in Soil Science	Elsevier Ltd.	2
24	Springer Geology	Springer Nature	5	49	Educational Futures	Brill Publ.	2
25	Springer Briefs in Earth Sciences	Springer Nature		50	Geocriticism and Spatial Literary Studies	Springer Nature	2

Sources Websites of *Web of Science* (5 October 2020)

Table 22.10 Anthropocene texts by authors in the *Web of Science* and in *Scopus* (2001–2021)

No. AWG	Authors	Web of Sc.	No. AWG	Authors	Web of Sc.	No. AWG	Authors	Scopus	No AWG	Authors	Scopus
1	Steffen, W.	39	26	Yusoff, K.	12	1	Zalasiewicz, J.	46	26	Folke, C.	11
2	Zalasiewicz, J.	37	27	Biggs, R.	11	2	Williams, M.	40	27	Meybeck, M.	11
3	Williams, M.	31	28	Clark, N.	11	3	Steffen, W.	35	28	Jusoff, K.	11
4	Ellis, E.C.	27	29	Cooke, S.J.	11	4	Ellis, EC	28	29	Allenby, B.	10
5	Svenning, J.C.	25	30	Galetti, M.	11	5	Waters, CN	25	30	Chin, A.	10
6	Kotzé, L.J.	23	31	Galuszka, A.	11	6	Svenning, JC	24	31	Crutzen, P.J.	10
7	Barnosky, A.D	22	32	Goudie, A.S.	11	7	Cearreta, A	23	32	Erlandson, J.M.	10
8	Waters, C.N.	21	33	Viles, H.A.	11	8	Dalby, S.	23	33	Fairchild, I. J.	10
9	Biermann, F.	20	34	Allenby, B.	10	9	Barnosky, AD	20	34	Grinevald, J.	10
10	Cearreta, A.	20	35	Grinevald, J.	10	10	Vidas, D	18	35	Pöschl, U.	10
11	Rockström, J.	19	36	Kim, R.E.	10	11	Syvitski, JPM	16	36	Syvitski, J.P.M.	10
12	Dalby, S.	17	37	Meybeck, M.	10	12	McNeill, JR	15	37	Thomas, C.D.	10
13	Folke, C.	16	38	Schmidt, J.J.	10	13	Rockström, J.	15	38	Waters, C.N.	10
14	Anonymous	15	39	Summerhayes,C	10	14	Kotze, L.J.	14	39	Arias-Maldonado, M.	9
15	Glikson, A.Y.	15	40	Syvitski, J	10	15	Wagreich, M.	14	40	Biggs, R.	9
16	Chandler, D.	14	41	Thomas, C.D.	10	16	Edgeworth, M.	13	41	Galetti, M.	9
17	Edgeworth, M.	14	42	Castree, N.	9	17	Leinfelder, R.	13	42	Glaser, M.	9
18	Vidas, D.	14	43	Chakrabarty, D.	9	18	Wolfe, A.P.	13	43	Gupta, J.	9
19	Wagreich, M.	14	44	Crutzen, P.J.	9	19	Biermann, F.	12	44	Hamilton, C.	9
20	McNeill, J.R.	13	45	Gupta, J.	9	20	Chandler, D.	12	45	Head, M. J.	9
21	Richter, D.D.	13	46	Hadly, E.A.	9	21	Galuszka, A.	12	46	Huettmann, F.	9
22	Wolfe, A.P.	13	47	Head, M. J.	9	22	Summerhayes, C.	12	47	Jeandel, C.	9
23	Leinfelder, R.	12	48	Hobbs, R. J.	9	23	Braje, T.J.	11	48	Lugo, A.E.	9
24	Syvitski, J.P.M.	12	49	Ison, R.	9	24	Castree, N.	11	49	Norström, A.V.	9
25	Young, O.R.	12	50	Jeandel, C.	9	25	Cooke, S.J.	11	50	Tarolli, P.	9

Sources Websites of *Web of Science* and *Scopus* (30 September 2020).

Table 22.11 Fifty major editors of books in the Anthropocene in the *Web of Science*

No.	Editor	University	Country (residence, work)	No.indexed	No.	Editor	University	Country (residence, work)	No. indexed
1	Kotzé, L.J.	North-West Univ. Lincoln Univ.	S. Africa UK	18	26	Schaumann, C.	Emory Univ.	USA	8
2	Ellis, M.	British Geological Survey, AWG	USA	17	27	Scheffran, J.	Hamburg Univ: AFES-PRESS	Germany	8
3	Snelling, A.M.	British Geological Survey	UK	17	28	Christensen, J.	Aarhus Univ. Hospital	Denmark	7
4	Waters, C.N.	British Geological Survey, Univ. Leicester, AWG	UK	17	29	Corcoran, P.B.	Bristol University	UK	7
5	Williams, M.	Univ.Leicester	UK	17	30	Dalbotten, D.	University of Minnesota	USA	7
6	Zalasiewicz, J.A.	Leicester Univ., AWG	UK	17	31	Hamilton, P.	Science Museum of Minnesota, Saint Paul	USA	7
7	Hickmann, T.	Potsdam Univ.	Germany	14	32	Heise, U.K.	UCLA	USA	7
8	Partzsch, L.	Univ. of Freiburg	Germany	14	33	Niemann, M.	Univ. of Wisconsin-Madison,	USA	7
9	Pattberg, P.	Free University of Amsterdam	Germany	14	34	Roerig, G.	U.of Minnesota	USA	7
10	Weiland, S.	Lille Catholic Univ.	Germany	14	35	Tomich, T.P.	Editor, AR	USA	7
11	Brauch, H.G.	Free Univ., Berlin AFES-PRESS	Germany	12	36	Wals, A.E.J.	Editor, Wage-ningen Academic press	Netherlands	7
12	Gadgill, A.	Editor		12	37	Austin, G.	Univ. of Cambridge	UK	6

(continued)

Table 22.11 (continued)

No.	Editor	University	Country (residence, work)	No.indexed	No.	Editor	University	Country (residence, work)	No. indexed
13	Heikkurinen, P.	Univ. Leeds	UK	12	38	Bixler, R.P.	Ed, Univ. Press Colorado	USA	6
14	Oswald Spring, U.	UNAM, CRIM AFES-PRESS	Mexico	12	39	Futuyma, D.J.	Ed., Annual Review of Ecology, Evolution, and Systematics		6
15	Brenneis, D.	Editor (?)		11	40	Leane, E.	Univ. of Tasmania	Australia	6
16	Gren, M.	Linnaeus Univ,	Sweden	10	41	Mc Gee, J.	Univ. of Tasmania	Australia	6
17	Huijbens, E.H.	Univ. Akureyri	Iceland	10	42	Miller, C.	Ed, Univ. Press Colorado	USA	6
18	Jagodzinski, J.	Univ. Alberta	Canada	10	43	Sample, V.A.	Ed, Univ. Press Colorado	USA	6
19	Grin, J.	Amsterdam Univ., AFES-PRESS	The Netherlands	9	44	Weakland, J.P.	Ed, Univ. Press Colorado	USA	6
20	Levin, P.S.	Univ. Washington	USA	9	45	Adger, W.N.	Univ. of Exeter	UK	5
21	Poe, M.R.	Univ. Washington	USA	9	46	Arnesto, J.J.	Dordrecht, Springer		5
22	Strier, K.B.	Editor		9	47	Callicott, J.B.	Univ. of North Texas	USA	5
23	Sullivan, H.I.	Hamilton College	USA	9	48	Finney, S.	Ed, Springer Geology	??	5
24	Bovenkerk, B.	Univ.Wageningen	Netherlands	8	49	Gremaud, A.S.N.	Ed., Routledge	UK	5
25	Keulartz. J.	Radboud Univ Nijmegen	Netherlands	8	50	Hanson, S.E.	Ed, Delta in the Anthropocene	UK	5

Source Website of *Web of Science*

Table 22.12 Anthropocene texts by universities and organisations in the *Web of Science* and in *Scopus* (2001–2021). *Sources* Websites of *Web of Science* and *Scopus* (30 September 2020)

No.	Authors	Web of S.	No.	Authors	Web of Sc.	No.	Authors	Scopus	No.	Authors	Scopus
1	Univ. of California	235	26	Univ. of Texas	53	1	Australian Nat. Univ	92	26	Univ. of California Santa Barbara	43
2	Univ. of London	171	27	Univ. of Washington Seattle	52	2	Univ. of Oxford	92	27	Univ. of Maryland	43
3	CNRS (France)	113	28	Wageningen Univ.	52	3	Stockholm Univ.	83	28	James Cook Univ.	42
4	Univ. of Oxford	94	29	INRAE	50	4	Univ. of Cambridge	72	29	Univ. of Arizona	42
5	Aust. Nat. Univ	92	30	Lancaster Univ.	50	5	UC London	61	30	Univ. of Cornell	41
6	Stockholm Univ.	85	31	Univ. of British Columbia	50	6	Aarhus Univ.	60	31	Univ. of Utrecht	41
7	State Univ. Syst. of Florida	80	32	Univ. of California Berkeley	50	7	Chinese Academy of Science	60	32	Harvard Univ.	40
8	Univ. of Cambridge	74	33	Univ. of Exeter	50	8	Stanford University	59	33	CSIC	39
9	University System of Maryland	68	34	University of Leicester	50	9	Arizona State Univ.	58	34	Univ. of Helsinki	39
10	UC London	64	35	Durham Univ.	49	10	Duke	54	35	Univ. Sydney	39
11	Aarhus Univ.	63	36	Pennsylvania PCSHE	49	11	Univ. of Toronto	54	36	Univ. Texas Austin	38
12	Smithsonian Inst.	63	37	Univ. of Minnesota-Twin Cities	49	12	Univ. of Washington	53	37	Univ. Vienna	38
13	Chinese Academy of Science	62	38	Harvard Univ	47	13	Univ. Colorado	50	38	Univ. of Western Australia	38
14	CSIC	60	39	Univ. of Colorado Boulder	47	14	Univ. Exeter	50	39	Lund Univ.	37
15	Stanford University	60	40	NERC Brit. Geological Survey	46	15	Univ. Leicester	50	40	Smithsonian Inst.	37
16	Arizona State Univ.	58	41	Univ. of Alberta	45	16	Univ. Minnesota	50	41	Univ. Calif Davis	37
17	Helmholtz Assoc.	58	42	Univ. of. Califor. Santa Barbara	45	17	University of British Columbia	49	42	Univ. of Victoria	37
18	Max Planck Society	58	43	Univ. of Leeds	45	18	Univ. Calf. Berkeley	49	43	US Geolog. Survey	37
19	Univ. of Toronto	58	44	Univ.of N.Carol.	45	19	Univ. of Durham	49	44	Griffith Univ.	36
20	Arizona State Univ. Tempe	57	45	University of Queensland	45	20	Univ. Lancaster	48	45	Mc Gill Univ.	36
21	Univ. of Colorado System	57	46	Utrecht Univ.	44	21	Univ. of Alberta	45	46	Univ. of Florida	36
22	Duke Univ.	54	47	IRD, France	43	22	Univ. Leeds	45	47	Univ. of Tasmania	36
23	NERC Nat. Env. Research Council	54	48	Stellenbosch Un.	43	23	Univ. of Melbourne	45	48	Macquarie Univ.	35
24	Univ of Washington	54	49	USDA	43	24	Univ. Queensland	45	49	Univ.of Cape Town	35
25	Univ. of Minnesota System	53	50	University of Waterloo	43	25	British Geological Survey	44	50	University of Copenhagen	35

Table 22.13 Members of the Working Group on the Anthropocene (AWG), *Subcommission on Quaternary Stratigraphy* in 2009 and in 2019

No.	Name (2019)	Member in 2009 [2021]	Function	Discipline	University or Institute Affiliation	Country	Author Web	Author Scopus
1	An Zhisheng (Advisory)	2009		Geography	Chinese Academy of Science	China	2	
2	Anthony Barnosky (Voting)	—		Ecology, Geology Biology	Stanford University	USA	22	20
3	Alejandro Cearreta (Voting)	—		?	Univ. del País Vasco	Spain	20	23
4	Paul Crutzen (†) (Honorary, deceased)	2009	Nobel Laureate	Chemistry	MPIC	Netherlands Germany	9	10
5	Andy Cundy (Voting)	2020	School of Ocean and Earth Science University of Southampton	Environmental Radioactivity	*Nat. Oceanography Centre, Southampton*	UK		
6	Matt Edgeworth (Advisory)	—			Leicester Univ.;	UK	14	13
7	Erle Ellis (Advisory)	2009 —		Geography	Univ. of Maryland (2009) UMBC	USA Canada	27	28
	Mike Ellis (left)	2009		Geology	British Geological Survey	UK		?
8	Ian Fairchild (Voting)	—	School of Geography, Earth & Environmental Sciences		Univ. of Birmingham	UK	8	10
9	Agnieszka Gałuszka (Voting)	—	Institute of Chemistry	Geology	Jan Kochanowski University	Poland	11	12

(continued)

Table 22.13 (continued)

No.	Name (2019)	Member in 2009 [2021]	Function	Discipline	University or Institute Affiliation	Country	Author	
							Web	Scopus
10	Philip Gibbard (Voting)	2009	Secretary-General, ICS,	Geography	Cambridge Univ.	UK	7	6
11	Jacques Grinevald (Advisory)	—			Graduate Institute, Geneva	Switzerland	10	10
12	Peter Haff (Advisory)	—	Nicholas School of the Environment		Duke Univ.	USA	7	7
13	Irka Hajdas (Voting)	—	Laboratory of Ion Beam Physics		ETH Zuerich	Switzerland		
	Alan Haywood (left)	2009		Earth & Env.	Univ. of Leeds	UK		
14	Han Yongming (Voting)	2020	The Institute of the Earth Environment,		Chinese Academy of Sciences	China		
15	Martin J. Head (Voting)	—		Geologist	Brock Univ.	Canada	9	9
16	Juliana Assunção Ivar do Sul (Advisory)	—			iowarnemuende	Germany		5
17	Catherine Jeandel (Advisory)	—			egos.obs-mip.fr	France	9	9
	Andrew Kerr (left)	2009			Univ. of Cardiff	UK		

(continued)

Table 22.13 (continued)

No.	Name (2019)	Member in 2009 [2021]	Function	Discipline	University or Institute Affiliation	Country	Author Web	Scopus
18	Reinhold Leinfelder (Voting)	—			Free Univ. Berlin	Germany	12	13
19	Francine McCarthy (Voting)	2020	Depart. Biological Sciences Environmental Sustainability Research Centre		Brock Univ.	Canada		
20	John McNeill (Advisory)	—	History Department	EnvironmentalHistory	Georgetown Univ.	USA	13	15
	Carlos Nobre (left)	2009		Space research	INPE	Brazil	0	0
21	Eric Odada (Advisory)	2009	Geology Department	Geology	Univ. of Nairobi	Kenya		
22	Naomi Oreskes (Advisory)	—	Departm.of the History of Science	History	Harvard Univ.	USA	7	7
23	Clément Poirier (Advisory)	—			Normandie Université	France	8	8
	Simon Price (left)	2009			British Geological Survey	UK		
24	Dan Richter (Advisory)	—			Duke Univ.	USA	13	7

(continued)

Table 22.13 (continued)

No.	Name (2019)	Member in 2009 [2021]	Function	Discipline	University or Institute Affiliation	Country	Author Web	Author Scopus
25	Neil Rose (Voting)	—	Environmental Change Research Centre	Department of Geography	Univ. College London	UK		
26	Yoshiki Saito (Voting)	2020			shimane-u.ac.	Japan		
	Mary Scholes (left)	2009			Witwatersrand	South Africa		
27	Bill Shotyk (Advisory)	—	Department of Renewable Resources		Univ. Alberta	Canada		
28	Will Steffen (Advisory)	2009		Earth System Science	Australian National Univ. (ANU)	Australia	39	35
29	Colin Summerhayes (Voting)	—			Cambridge U.	UK	10	12
30	Jaia Syvitski (Voting)	—	Institute of Arctic & Alpine Research		Colorado Univ.	USA	12	16
31	Simon Turner	2020	AWG Secretary		Univ.Col. London	UK		
32	Davor Vidas (Advisory)	2009	Research Professor, Senior Fellow	Marine Affairs & Law of Sea Programme	Fritjov Nansen Institute	Norway	14	18
33	Michael Wagreich (Voting)	—	Department of Geology,		Vienna Univ.	Austria	14	14

(continued)

Table 22.13 (continued)

No.	Name (2019)	Member in 2009 [2021]	Function	Discipline	University or Institute Affiliation	Country	Author	
							Web	Scopus
	Mike Walker (left)	2009		Geology	Univ. of Wales	UK		
34	Colin Waters (Voting)	2011	AWG Chair		Leicester Univ.	UK	21	25
35	Mark Williams (Voting)	2009	School of Geography	Geology	Leicester Univ.	UK	31	40
36	Scott Wing	—	Dept. of Paleobiology,	Museum of Na tural History	Smithsonian Institution	USA		
37	Jan Zalasiewicz (Voting)	2009	Chair, SQS	Geology	Leicester Univ.	UK	37	46
38	Jens Zinke (Voting)	2020			Leicester Univ.	UK		

Source http://quaternary.stratigraphy.org/working-groups/anthropocene/ (7 October 2020 and 18 February 2021).

Meeting of the Anthropocene Working Group at the Max Planck Institute for Chemistry in Mainz on 16 March 2017 hosted by Paul Crutzen with AWG members from University of Leicester (Jan Zalasieweicz, Colin Waters and Mark Williams), from Norway (Davor Vidas), from MPIC (Ulrich Pöschl, Jos Lelieveld, Susanne Benner, Bettina Weber and Astrid Kaltenbach, from the MPIWG (Jürgen Renn), of Haus der Kulturen der Welt (Bernhard Scherer and Christof Rosol) and from IASS (Mark Lawrence and Franz Mauelshagen). *Source AWG Newsletter* No. 7, December 2017, p. 6. © C. Waters.

Multidiciplinary Bibliography on the Anthropocene Literature (2000–2020)

The name of the members of the Anthropocene Working Group listed in 2020 are in *italics*.

References

Adam-Hernandez, Alistair; Harteisen, Ulrich, 2020: "A Proposed Framework for Rural Resilience - How can peripheral village communities in Europe shape change?", in: *Ager-Revista De Estudios Sobre Despoblacion Y Desarrollo Rural*, 28: 7–42.

Adams, M., 2018: "Towards a critical psychology of human-animal relations", in: *Social and Personality Psychology Compass*, 12, 4:e12375.

Adamson, J.; Davis, M., 2016: *Humanities for the Environment: Integrating Knowledge, Forging New Constellations of Practice* (London: Routledge).

Ajai, L.B.; Breashears, D.; *Crutzen*, P.J.; Fuzzi, S.; Haeberli, W.; Immerzeel, W.W.; Kaser, G.; Kennel, C.; Kulkarni, A.; Pachauri, R.; Painter, T.H.; Rabassa, J.; Ramanathan, V.; Robock, A.; Rubbia, C.; Russell, L.; Sánchez Sorondo, M.; Schellnhuber, H.J.; Sorooshian, S.; Stocker, T.F.; Thompson, L.G.; Toon, O.B.; Zaelke, D.; Mittelstraß, J., 2011: *Fate of Mountain Glaciers in the Anthropocene. A Report by the Working Group Commissioned by the Pontifical Academy of Sciences*.

Alaimo, S., 2016: *Exposed: Environmental politics and pleasures in posthuman times* (Minneapolis, MN: Minnesota University Press).

Alcaraz, J.M.; Sugars, K.; Nicolopoulou, K.; et al., 2016: "Cosmopolitanism or globalization: the Anthropocene turn", in: *Society and Business Review*, 11, 3: 313–332.

Altuna, N.E.B., *Cearreta, A.*, Irabien, M.J., (...), Soualili, K., Hilario, A., 2019: "Environmental evolution of the Deba estuary (Basque Coast Geopark) during the Holocene and Anthropocene [Evolución ambiental del estuario del Deba (Geoparque de la Costa Vasca) durante el Holoceno y el Antropoceno]", in: *Geogaceta*, 66: 63–66.

Altvater, E., 2016: "The Capitlocene, or Geoengineering against Capitalism's Planetary Boundaries", in: Moore, Jason M. (Ed.): *Anthropocene or Capitalocene? Nature, History, and the Crisis of Capitalism* (Oakland, CA: PM Press): 138–153.

Angus, I., 2015: 1–11: "When Did the Anthropocene Begin ... and Why Does It Matter?", in: *Monthly Review-An Independent Socialist Magazine*, 67, 4: 1–11.

Angus, I., 2016: *Facing the Anthropocene: Fossil Capitalism and the Crisis of the Earth System* (New York: Monthly Review Press).

Angus, I., 2019: 3–30: "Superbugs in the Anthropocene - A Profit-Driven Plague", in: *Monthly Review-An Independent Socialist Magazine*, 71,2: 3–30.

Angus, I., 2020: 50–54: "Facing the Anthropocene An Update", in: *Monthly Review-An Independent Socialist Magazine*, 72,6: 50–54.

Arias-Maldonado, M., 2013: "Rethinking Sustainability in the Anthropocene", in: *Environmental Politics*, 22,3: 428–446.

Arias-Maldonado, M., 2015: "Spelling the end of nature? Making sense of the anthropocene", in: *Telos*, 172: 83–102.

Arias-Maldonado, M., 2016a: "The anthropocenic turn: Theorizing sustainability in a postnatural age", in: *Sustainability* (Switzerland), 8,1: 1–17.

Arias-Maldonado, M., 2016b: "Nature and the Anthropocene: The sense of an ending?", in: *Environmental Politics and Governance in the Anthropocene: Institutions and legitimacy in a complex world*: 31–46.

Arias-Maldonado, M., 2019a: "Towards a good anthropocene?", in: *Rethinking the Environment for the Anthropocene: Political Theory and Socionatural Relations in the New Geological Epoch*: 137–150.

Arias-Maldonado, M., 2019b: "Urgency in the Anthropocene", in: *Environmental Politics*, 28,4: 790–791.

Arias-Maldonado, M., 2019c: "The 'Anthropocene' in philosophy: The neo-material turn and the question of nature", in: Biermann, F., Lövbrand, E. (Eds.): *Anthropocene Encounters: New Directions in Green Political Thinking* (Cambridge: Cambridge University Press): 50–66.

Arias-Maldonado, M., 2020a: "Bedrock or social construction? What Anthropocene science means for political theory", in: *Anthropocene Review*, 7,2: 97–112.

Arias-Maldonado, M., 2020b: "Blooming landscapes? The paradox of utopian thinking in the Anthropocene", in: *Environmental Politics*, 29,6: 1024–1041.

Arias-Maldonado, M., 2020c: "Sustainability in the anthropocene: Between extinction and populism", in: *Sustainability* (Switzerland), 12,6: 2538.

Arias-Maldonado, M., Trachtenberg, Z., 2019a: "Introduction", in: Arias-Maldonado, M.; Trachtenberg, Z. (Ed.): *Rethinking the Environment for the Anthropocene: Political Theory and Socionatural Relations in the New Geological Epoch* (London: Routledge): 1–16.

Arias-Maldonado, M.; Trachtenberg, Z. (Ed.), 2019b: *Rethinking the Environment for the Anthropocene: Political Theory and Socionatural Relations in the New Geological Epoch* (Abingdon: Routledge).

Arizpe, L.; Price, M.F.; Worcester, R., 2016: "The First Decade of Initiatives for the Research on the Human Dimensions of Global (Environmental) Change (1986–1995)", in: Brauch, H.G.; Oswald Spring, Ú.; Grin, J.; Scheffran; J. (Eds.): *Handbook on Sustainability Transition and Sustainable Peace* (Cham: Springer International Publishing Switzerland): 349–358.

Arizpe Schlosser, L., 2019a: *Culture, International Transactions and the Anthropocene* (Cham: Springer International Publishing).

Arizpe Schlosser, L., 2019b: "Culture, International Transactions and the Anthropocene – Introduction", in: *Culture, International Transactions and the Anthropocene* (Cham: Springer International Publishing): 1–26.

Arizpe Schlosser, L., 2019c: "Culture and the Anthropocene", in: *Culture, International Transactions and the Anthropocene* (Cham: Springer International Publishing): 267–292.

Arizpe Schlosser, L., 2019d: "Culture, International Transactions and the Anthropocene Conclusions", in: *Culture, International Transactions And The Anthropocene* (Cham: Springer International Publishing): 293–314.

Arlander, Annette, 2020: "Performing with Plants in the Ob-scene Anthropocene ", in: *Nordic Theatre Studies*, 32,1: 121–142.

Asberg, C., 2017: "Feminist Posthumanities in the Anthropocene: Forays into the Postnatural", in: *Journal of Posthuman Studies-Philosophy Technology Media*, 1,2: 185–204.

Aswani, S., Basurto, X., Ferse, S., (...), Vaccaro, I., Christie, P., 2018: "Marine resource management and conservation in the Anthropocene", in: *Environmental Conservation*, 45,2: 192–202.

Atchison, J., 2018: "Thriving in the anthropocene: Understanding human-weed relations and invasive plant management using theories of practice", in: Maller, Cecily, Strengers, Yolande (Eds.): *Social Practices and Dynamic Non-Humans: Nature, Materials and Technologies* (Basingstoke: Palgrave Macmillan.): 25–46.

Atchison, J., 2019: "Between disgust and indifference: Affective and emotional relations with carp (Cyprinus carpio) in Australia", in: *Transactions of the Institute of British Geographers*, 44,4: 735–748.

Austin, G., 2017a: "Africa and the Anthropocene", Annual Pierre du Bois Conference on Economic Development in the Anthropocene - Perspectives on Asia and Africa, Grad. Inst. Int. & Dev. Studies, Geneva, Switzerland, 26–27 Sep. 2014, in: *Economic Development and Environmental History in the Anthropocene: Perspectives on Asia and Africa* (London: Bloomsbury): 95–118.

Austin, G., 2017b: "Introduction", Annual Pierre du Bois Conference on Economic Development in the Anthropocene - Perspectives on Asia and Africa Location: Grad Inst Int & Dev Studies, Geneva, Switzerland, 26–27 Sep. 2014, in: *Economic Development and Environmental History in the Anthropocene: Perspectives on Asia And Africa* (London: Bloomsbury): 1–22.

Azizi, D.; Biermann, F.; Kim, R.E., 2019: "Policy Integration for Sustainable Development through Multilateral Environmental Agreements. An Empirical Analysis, 2007–2016", in: *Global Governance*, 25,3: 445–475.

Baghel, R.; Stepan, L.; Hill, J.K.W., 2017: *Water, Knowledge and the Environment in Asia: Epistemologies, Practices and Locales* (London: Taylor and Francis).

Bai, X.; van der Leeuw, S.; O'Brien, K.; (...); *Steffen*, W.; *Syvitski*, J., 2016: "Plausible and desirable futures in the Anthropocene: A new research agenda", in: *Global Environmental Change-Human and Policy Dimensions*, 39: 351–362.

Baldwin, A.; Fröhlich, C.; Rothe, D., 2019: "From climate migration to anthropocene mobilities: shifting the debate", in: *Mobilities*, 14,3.

Barnosky, A.D., 2014: "Palaeontological evidence for defining the Anthropocene", in: *Geological Society Special Publication*, 395,1: 149–165.

Barnosky, A.D.; Brown, J.H.; Daily, G.C.; (...); Stenseth, N.C.; Wake, M.H., 2014a: "Introducing the scientific consensus on maintaining humanity's life support systems in the 21st century: Information for policy makers", in: *Anthropocene Review*, 1,1: 78–109.

Barnosky, A.D.; Hadly, E.A. Dirzo, R.., et al., 2014b: "Translating science for decision makers to help navigate the Anthropocene ", in: *The Anthropocene Review,* 1,2: 160–170.

Barnosky, A.D.; Hadly, E.A., 2014: "Problem solving in the Anthropocene", in: *Anthropocene Review*, 1,1: 76–77.

Barnosky, A.D.; Holmes, M.; Kirchholtes, R....; Villavicencio, N.A.; Wogan, G.O.U., 2014c: "Prelude to the Anthropocene: Two new North American Land Mammal Ages (NALMAs)", in: *Anthropocene Review*, 1,3: 225–242.

Barrett, S.; Dasgupta, A.; Dasgupta, P.; (...); Walker, B., Wilen, J., 2020: "Social dimensions of fertility behavior and consumption patterns in the Anthropocene", in: *Proceedings of the National Academy of Sciences of the United States of America*, 117,12: 6300–6307.

Barrios-O'Neill, Daniel, 2020: "Focus and social contagion of environmental organization advocacy on Twitter", in: *Conservation Biology*, 0,0: 1–8.

Barry, K.; Keane, J., 2019: *Creative Measures of the Anthropocene: Art, Mobilities, and Participatory Geographies* (Basingstoke: Palgrave Macmillan).

Barthel, S.; Crumley, C.; Svedin, U. , 2013: "Bio-cultural refugia - Safeguarding diversity of practices for food security and biodiversity", in: *Global Environmental Change*, 23,5: 1142–1152.

Bauer, A.M., *Ellis*, E.C., 2018: "The anthropocene divide: Obscuring understanding of social-environmental change", in: *Current Anthropology*, 59,2: 209–227.

Bauman, Whitney A., 2015: "Climate Weirding and Queering Nature: Getting Beyond the Anthropocene", in: *Religions*, 6,2: 742–754.

Bauman, Whitney A., 2020: "Returning Faith to Knowledge: Earthlings after the Anthropocene", in: *Religions*; 11,4: Article 169.

Behie, A.M.; Teichroeb, J.A.; Malone, N. (Eds.), 2019: *Primate Research and Conservation in the Anthropocene*, Book Series: Cambridge Studies in Biological and Evolutionary Anthropology (Cambridge: Cambridge Univ. Press).

Bennett, C.E.; Thomas, R.; *Williams*, M.; (...); Burton, E.J.; Marume, U., 2018: "The broiler chicken as a signal of a human reconfigured biosphere", in: *Royal Society Open Science*, 5,12: 180325.

Bennett, E.M.; Solan, M.; Biggs, R.; (...); Vervoort, J.M.; Xu, J., 2016: "Bright spots: seeds of a good Anthropocene, in: *Frontiers in Ecology and the Environment*, 14,8: 441–448.

Berger, Peter L.; Luckmann, Thomas, 1966, 1967: *The Social Construction of Reality: A Treatise in the Sociology of Knowledge* (Garden City, NY: Anchor).

Bergmann, S., 2015a: "The legacy of Trinitarian cosmology in the Anthropocene", in: *Studia Theologica - Nordic Journal of Theology* , 69,1: 32–44.

Bergmann, S., 2015b: "Sustainable development, climate change and religion", in: *The Routledge Handbook of Religions and Global Development* (London: Routledge): 389–404.

Bergmann, S., 2020: "Theology in the anthropocene - and beyond?", in: *Contextual Theology: Skills and Practices of Liberating Faith* (London: Routledge): 60–180.

Bergmann, S., Vähäkangas, M., 2020a: *Contextual Theology: Skills and Practices of Liberating Faith* (London: Routledge).

Bergmann, S., Vähäkangas, M., 2020b: "Doing situated theology: Introductory remarks about the history, method, and diversity of contextual theology", in: *Contextual Theology: Skills and Practices of Liberating Faith* (London: Routledge): 1–14.

Bernard, P., 2019: "Health psychology at the age of Anthropocene", in: *Health Psychology and Behavioral Medicine*, 7,1: 193–201.

Berzonsky, C.L.; Moser, S.C., 2017: "Becoming homo sapiens sapiens: Mapping the psycho-cultural transformation in the anthropocene", in: *Anthropocene*, 20: 15–23.

Bianchi, T.S.; Arndt, S.; Austin, W.E.N.; (…); Smith, R.W.; *Syvitski*, J., 2020: "Fjords as Aquatic Critical Zones (ACZs)", in: *Earth-Science Reviews*, 203,10314.

Biermann, F., 2012: "Greening the united nations charter: World politics in the anthropocene", in: *Environment*, 54,3: 6–17.

Biermann, F, 2014a: *Earth System Governance: World Politics in the Anthropocene* (Cambridge, Mass.: MIT Press).

Biermann, F., 2014b: "The anthropocene: A governance perspective", in: *Anthropocene Review*, 1,1:. 57–61.

Biermann, F., 2016a: "Epilogue Politics for a New Earth: Governing in the 'Anthropocene'", in: Nicholson, S.; Jinnah, S. (Eds.): *New Earth Politics: Essays From The Anthropocene* (Cambridge, MA: The MIT Press): 405–420.

Biermann, F., 2016b: "Response to John S. Dryzek's review of Earth System Governance: World Politics in the Anthropocene", in: *Perspectives on Politics*, 14,1: 178.

Biermann, F., 2018: "Global governance in the 'anthropocene'", in: Brown, C.; Eckersley, R. (Eds.): *The Oxford Handbook of International Political Theory* (New York-Oxford: Oxford University Press): 467–478.

Biermann, Frank, 2020: "The future of 'environmental' policy in the Anthropocene: time for a paradigm shift", in: *Environmental Politics* (Nov.).

Biermann, F.; Abbott, K.: Andresen: S.; (…); Brock, A.; Zondervan, R., 2012: "Navigating the anthropocene: Improving earth system governance", in: *Science*, 335,6074: 1306–1307.

Biermann, F.; Betsill, M.M.; Vieira, S.C.; (…); Yanda, P.Z.; Zondervan, R., 2010: "Navigating the anthropocene: The Earth System Governance Project Strategy Paper", in: *Current Opinion in Environmental Sustainability*, 2,3: 202–208.

Biermann, F.; Bai, X.; Bondre, N.; (…); Seitzinger, S.; Seto, K.C., 2016: "Down to Earth: Contextualizing the Anthropocene", in: *Global Environmental Change*, 39: 341–350.

Biermann, F., Lövbrand, E., 2019a: "Encountering the "Anthropocene": Setting the scene", in: Biermann, F., Lövbrand, E. (Eds.): *Anthropocene Encounters: New Directions in Green Political Thinking* (Cambridge – New York: Cambridge University Press): 1–22.

Biermann, F., Lövbrand, E. (Eds.), 2019b: *Anthropocene Encounters: New Directions in Green Political Thinking* (Cambridge – New York: Cambridge University Press).

Biermann, F.; Young, O.R., 2014a: "Earth System Governance World Politics in the Anthropocene Introduction", in: Nicholson, S.; Jinnah, S. (Eds.): *Earth System Governance: World Politics in the Anthropocene* (Cambridge, Mass.: MIT Press): 1–14.

Biermann, F.; Young, O.R., 2014b: "Earth System Governance World Politics in the Anthropocene Foreword", in: Nicholson, S.; Jinnah, S. (Eds.): *Earth System Governance: World Politics in the Anthropocene* (Cambridge, Mass.: MIT Press): vii+.

Biermann, F.; Young, O.R., 2014c: "Earth System Governance World Politics in the Anthropocene Foreword", in: Nicholson, S.; Jinnah, S. (Eds.): *Earth System Governance: World Politics in the Anthropocene* (Cambridge, Mass.: MIT Press): ix–xii.

Biermann, F.; Young, O.R., 2014d: "Earth System Governance World Politics in the Anthropocene Foreword", in: Nicholson, S.; Jinnah, S. (Eds.): *Earth System Governance: World Politics in the Anthropocene* (Cambridge, Mass.: MIT Press): 203–213.

Biermann, F.; Young, O.R., 2016: "New Earth Politics Essays from the Anthropocene Series Foreword", in: Nicholson, S.; Jinnah, S. (Eds.): *Earth System Governance: World Politics in the Anthropocene* (Cambridge, Mass.: MIT Press): XI–XII.

Bindé, Jérôme et al., 2007: *Making peace with the earth: what future for the human species and the planet?* (New York: Berghahn).

Birch, B.C.; Lapsley, J.E.; Moe-Lobeda, C.D.; Rasmussen, L.L., 2018: *Bible and Ethics in the Christian Life: A New Conversation* (Minneapolis, MN: Fortress Press).

Birrell, K.; Matthews, D., 2020a: "Laws for the Anthropocene: Orientations, Encounters, Imaginaries", in: *Law and Critique*, 31: 233–238.

Birrell, Kathleen; Matthews, Daniel, 2020b: "Restorying Laws for the Anthropocene: Rights, Obligations and an Ethics of Encounter", in: *Law and Critique*, 31,3: 275–292.

Blok, V., 2017a: *Ernst Jünger's Philosophy of Technology: Heidegger and the Poetics of the Anthropocene* (London: Taylor and Francis).

Blok, V., 2017b: "Earthing technology: Towards an eco-centric concept of biomimetc technologies in the Anthropocene", in: *Techne: Research in Philosophy and Technology*, 21,2–3: 127–149.

Blok, V., 2018: "Technocratic Management Versus Ethical Leadership - Redefining Responsible Professionalism in the Agri-Food Sector in the Anthropocene, in: *Journal of Agricultural and Environmental Ethics*, 31: 583–591.

Blok, V., 2020: "Technocratic Management Versus Ethical Leadership Redefining Responsible Professionalism in the Agri-Food Sector in the Anthropocene", in: *Journal of Agricultural and Environmental Ethics*, 31,2–3: 583–591.

Bluemling, B.; Kim, R.E.; Biermann, F., 2020: "Seeding the clouds to reach the sky: Will China's weather modification practices support the legitimization of climate engineering?", in: *Ambio*, 49,1: 365–373.

Bosselmann, K., 2017: "The Imperative of Ecological Integrity: Conceptualising a Fundamental Legal Norm for a New 'World System' in the Anthropocene". in: Kotzé, L.J. (Ed.): *Environmental Law and Governance For The Anthropocene* (Oxford: Hart): 241–265.

Bowen, W.M., Gleeson, R.E., 2018: *The Evolution of Human Settlements: From Pleistocene Origins to Anthropocene Prospects* (Basingstoke: Palgrave Macmillan).

Braje, T.J., 2015: "Earth Systems, Human Agency, and the Anthropocene: Planet Earth in the Human Age", in: *Journal of Archaeological Research*, 23,4: 369–396.

Braje, T.J., 2016a: "A mid-twentieth-century Anthropocene makes the Holocene more important than ever", in: *Antiquity*, 90,350: 517–518.

Braje, T.J., 2016b: "Evaluating the Anthropocene: Is there something useful about a geological epoch of humans?", in: *Antiquity*, 90,350: 504–512.

Braje, T.J.; Erlandson, J.M., 2013a: "Human acceleration of animal and plant extinctions: A late pleistocene, holocene, and anthropocene continuum", in: *Anthropocene* , 4: 14–23.

Braje, T.J.; Erlandson, J.M., 2013b: "Looking forward, looking back: Humans, anthropogenic change, and the Anthropocene", in: *Anthropocene*, 4: 116–121.

Braje, T.J.; Lauer, M., 2020: "A meaningful anthropocene?: Golden spikes, transitions, boundary objects, and anthropogenic seascapes", *Sustainability* (Switzerland): 12,16: 6459.

Braje, T.J.; Leppard, T.P.; Fitzpatrick, S.M.; Erlandson, J.M., 2017: "Archaeology, historical ecology and anthropogenic island ecosystems", in: *Environmental Conservation*, 44,3: 286–297.

Brauch, H.G., 1976: *Struktureller Wandel and Rüstungspolitik der USA (1940–1950). Zur Weltführungsrolle and ihren innenpolitischen Bedingungen* (Inaugural Dissertation zur Erlangung des Doktorgrades der Philosophie der Philosophischen-Historischen Fakultät der Ruprecht-Karl Universität zu Heidelberg, 30 June).

Brauch, H.G., 1977: *Struktureller Wandel and Rüstungspolitik der USA (1940–1950). Zur Weltführungsrolle and ihren innenpolitischen Bedingungen* (Ann Arbor - London: University Microfilms).

Brauch, H.G., 2008: "Introduction: Globalization and Environmental Challenges: Reconceptualizing security in the 21st Century", in: Brauch, H.G.; Oswald Spring, Ú.; Mesjasz, C.; Grin, J.; Dunay, P.; Behera, N.C.; Chourou, B.; Kameri-Mbote, P.; Liotta, P.H. (Eds.): *Globalization and*

Environmental Challenges: Reconceptualizing Security in the 21st Century (Berlin - Heidelberg: Springer-Verlag): 27–43.

Brauch, H.G., 2009: "Securitizing Global Environmental Change", in: Brauch, H.G.; Oswald Spring, Ú.; Grin, J.; Mesjasz, C.; Kameri-Mbote, P.; Behera, N.C.; Chourou, B.; Krummenacher, H. (Eds.), 2009: *Facing Global Environmental Change: Environmental, Human, Energy, Food, Health and Water Security Concepts* (Berlin—Heidelberg—New York: Springer-Verlag): 65–102.

Brauch, H.G., 2016a: "Historical Times and Turning Points in a Turbulent Century: 1914, 1945, 1989 and 2014?", in: Brauch, H.G.; Oswald Spring, Ú.; Bennett, J.; Serrano Oswald, S.E. (Eds.): *Addressing Global Environmental Challenges from a Peace Ecology Perspective* (Cham: Springer International Publishing Switzerland): 11–54.

Brauch, H.G., 2016b: "Sustainable Peace in the Anthropocene: Towards Political Geoecology and Peace Ecology", in: Brauch, H.G.; Oswald Spring, Ú.; Grin, J.; Scheffran; J. (Eds.), 2016: *Handbook on Sustainability Transition and Sustainable Peace* (Cham: Springer International Publishing Switzerland): 187–236.

Brauch, H.G., 2019: "Sustainable Peace Through Sustainability Transition as Transformative Science: A Peace Ecology Perspective in the Anthropocene", Brauch, H.G.; Oswald Spring, U.; (Eds.): *Climate Change, Disasters, Sustainability Transition and Peace in the Anthropocene* (Cham: Springer International Publishing): 175-234.

Brauch, H.G., 2021: "Peace Ecology in the Anthropocene", in: Oswald Spring, Ú.; Brauch, H.G. (Eds.): *Decolonising Conflicts, Security, Peace, Gender, Environment and Development in the Anthropocene* (Cham: Springer Nature Switzerland): 51-186.

Brauch, H.G., 2022: *Anthropocene Politik: Policy, Politics and Polity in International Relations, Development, Environment, Peace and Security Studies* (Cham: Springer International Publishing).

Brauch, H.G. (Ed.), 2023: *Addressing, Analysing and Responding to the Anthropocene: Turning Point, Context, Challenges, Tasks, Opportunities and Obligations for Policies, Politics and Polity* (Cham: Springer Nature Switzerland) [Working Title].

Brauch, H.G.; Dalby, S.; Oswald Spring, U., 2011: "Political Geoecology for the Anthropocene", in: Brauch, H.G.; Oswald Spring, Ú.; Mesjasz, C.; Grin, J.; Kameri-Mbote, P.; Chourou, B.; Dunay, P.; Birkmann, J. (Eds.): *Coping with Global Environmental Change, Disasters and Security – Threats, Challenges, Vulnerabilities and Risks* (Berlin – Heidelberg – New York: Springer-Verlag, 2011): 1453–1486.

Brauch, H.G.; Oswald Spring, Ú., 2017: "Sustainability Transition and Sustainable Peace: Scientific and Policy Context, Scientific Concepts and Dimensions", in: Brauch, H.G.; Oswald Spring, Ú.; Grin, J.; Scheffran; J. (Eds.): *Handbook on Sustainability Transition and Sustainable Peace* (Cham—Heidelberg—New York—Dordrecht—London: Springer): 3–66.

Brauch, H.G.; Oswald Spring, Ú.; Grin, J.; Scheffran, J. (Eds.), 2016: *Handbook of Sustainability Transition and Sustainable Peace* (Cham: Springer International Publishing Switzerland).

Braudel, F., 1966: *La Méditerranée et le monde méditerranéen a l'époque de Philippe II* (Paris: Armand Colin).

Braudel, Fernand, 1969: "Histoire et science sociales. La longue durée", in: *Écrits Sur l'Histoire* (Paris: Flammarion): 41–84.

Braudel, F., 1972: *The Mediterranean and the Mediterranean World in the Age of Philip II*, 2 volumes (New York: Harper & Row).

Brick, C., 2019: "A modest proposal for restoration ecology", in: *Restoration Ecology*, 27,3: 485–487.

Bristow, T., 2015a: "Anthropocene Lyric: An Affective Geography of Poetry, Person, Place", in: *Anthropocene Lyric: An Affective Geography of Poetry, Person, Place*, Palgrave Pivot (Basingstoke: Palgrave): 1–139.

Bristow, Tom, 2015b: "The Anthropocene Lyric", in: *Anthropocene Lyric: An Affective Geography of Poetry, Person, Place*, Palgrave Pivot (Basingstoke: Palgrave):1–+.

Bristow, Tom, 2015c: "The Anthropocene Lyric", in: *Anthropocene Lyric: An Affective Geography of Poetry, Person, Place*, Palgrave Pivot (Basingstoke: Palgrave): 107–123.

Brondizio, E.S.; O'Brien, K.; Bai, X.; (...); Palma-Oliveira, J.; Chen, C.-T.A., 2016: "Reconceptualizing the Anthropocene: A call for collaboration", in: *Global Environmental Change*, 39: 318–327.

Brondizio, Eduardo S.; *Syvitski*, James, 2016: "Editorial: The Anthropocene", in: *Global Environmental Change-Human and Policy Dimensions*, 39: 316–317.

Brown, P.G.; Schmidt, J.J., 2014: "Living in the Anthropocene: Business as Usual, or Compassionate Retreat?", in: *State of the World 2014: Governing For Sustainability* (Washington, D.C: Island Press): 63–71.

Brunner, Otto; Conze, Werner; Koselleck, Reinhart (Eds.), 1972–1997: *Geschichtliche Grundbegriffe* (Stuttgart: Klett-Cotta).

Buck, B.H.; Langan, R., 2017: *Aquaculture Perspective of Multi-Use Sites in the Open Ocean: The Untapped Potential for Marine Resources in the Anthropocene* (Cham: Springer International Publishing).

Bunch, M.J., 2016: "Ecosystem Approaches to Health and Well-Being: Navigating Complexity, Promoting Health in Social-Ecological Systems", in: *Systems Research and Behavioral Science*, 33,5 (Sep.–Oct.): 614–632.

Burke, A., 2019: "Blue screen biosphere: The absent presence of biodiversity in international law", in: *International Political Sociology*, 13,3: 33–351.

Burke, A., Fishel, S., 2019: "Power, world politics, and thing-systems in the Anthropocene", in: Biermann, F., Lövbrand, E. (Eds.): *Anthropocene Encounters: New Directions in Green Political Thinking* (Cambridge – New York: Cambridge University Press), 87–108.

Burke, A.; Fishel, S.; Mitchell, A.; Dalby, S.; Levine, D.J., 2016: "Planet Politics: A Manifesto from the End of IR", in: *Millennium: Journal of International Studies*, 44,3: 499.

Calanchi, A.; Farina, A.; Barbanti, R., 2017a: "An Eco-Critical Cultural Approach to Mars Colonization" in: *Forum for World Literature Studies*, 9,2: 205–216.

Calanchi, A.; Farina, A.; Barbanti, R., 2017b: "An Eco-critical Cultural Approach to Mars Colonization", in: *Interdisciplinary Studies of Literature*, 1,2: 67–79.

Calás, M.B.; Smircich, L.; Ergene, S., 2017: "Postfeminism as new materialisms: A future unlike the present?", in: *Postfeminism and Organization*, 197–228.

Campagnaro, T.; Sitzia, T.; Bridgewater, P.; Evans, D.: *Ellis*, E.C., 2019: "Half Earth or Whole Earth: What Can Natura 2000 Teach Us?", in: *BioScience*, 69,2: 117–124.

Campbell, B.; Braje, T.J.; Whitaker, S.G., 2018: "Trans-Holocene Human Impacts on California Mussels (Mytilus Californianus) Historical ecological management implications from the Northern Channel Islands", in: *Multispecies Archaeology, Archaeological Orientations*: 65–84.

Cantrell, B.; Martin, L.J.; *Ellis*, E.C., 2017: "Designing Autonomy: Opportunities for New Wildness in the Anthropocene", in: *Trends in Ecology and Evolution*, 32,3: 156–166.

Caracciolo, M., 2019: "Form, Science, and Narrative in the Anthropocene", in: *Narrative*, 27,3: 270–289.

Caracciolo, M.; Ionescu, A.; Fransoo, R., 2019: "Metaphorical patterns in Anthropocene fiction", in: *Language and Literature*, 28,3: 221–240: Art. No.: UNSP 0963947019865450.

Caracciolo, M.; Lambert, S., 2019: "Narrative Bodies and Nonhuman Transformations", in: *SubStance*, 48,3: 45–63.

Carofalo, V.; Salottolo, D.; Saito, K., 2019: "Anthropocene and Ecosocialism: A Perspective", in: *S&F-Scienzaefilosofia It*, 21: 21–31.

Carpenter, S.R.; Folke, C.; Scheffer, M.; Westley, F.R., 2019: "Dancing on the volcano: Social exploration in times of discontent", in: *Ecology and Society*, 24,1: 23.

Castree, N., 2014 : "The Anthropocene and Geography III: Future Directions", in: *Geography Compass*, 8,7: 464–476.

Castree, N., 2015a: "Geographers and the Discourse of an Earth Transformed: Influencing the Intellectual Weather or Changing the Intellectual Climate?", in: *Geographical Research*, 53,3: 244–254.

Castree, N., 2015b: "The Anthropocene: A primer for geographers", in: *Geography*, 100: 66–75.

Castree, N., 2015c: "Changing the Anthropo(s)cene: Geographers, global environmental change and the politics of knowledge", in: *Dialogues in Human Geography*, 5,3: 301–316.

Castree, N., 2015d: "Coproducing global change research and geography: The means and ends of engagement", in: *Dialogues in Human Geography*, 5,3: 343–348.

Castree, N., 2017a: "Unfree Radicals: Geoscientists, the Anthropocene, and Left Politics", in: *Antipode*, 49: 52–74.

Castree, N., 2017b: "Speaking for the 'people disciplines': Global change science and its human dimensions", in: *Anthropocene Review*, 4,3: 160–182.

Castree, N., 2019: "The "Anthropocene" in global change science: Expertise, the earth, and the future of humanity", in: Biermann, F., Lövbrand, E. (Eds.): *Anthropocene Encounters: New Directions in Green Political Thinking* (Cambridge – NewYork: Cambridge University Press): 25–49.

Cearreta, A., 2015: The Anthropocene and the steps necessary for its possible formalization after the 35th International Geological Congress (2016) | [El Antropoceno y los pasos necesarios para su posible formalización tras el 35º Congreso Geológico Internacional (2016)], in: *Cuaternario y Geomorfologia*, 30,3–4: 5–8.

Cearreta, A.; Irabien, M.J.; Arozamena, J.G., 2018: "Recent anthropogenic transformation of the Pasaia bay (Guipuzcoa, N. Spain): Multiproxy analysis of its sedimentary record | [Transformación antrópica reciente de la bahía de Pasaia (Guipúzcoa): Análisis multidisciplinar de su registro sedimentario]", in: *Geogaceta*, 64: 107–110.

Cearreta, A.; Irabien, M.J.; Arozamena, J.G.; Kortabitarte, I.; González-Lanchas, A., 2017: "The Anthropocene geological record in the Abra of Bilbao: evidences of its natural and human history | [El registro geológico antropoceno en el Abra de Bilbao: evidencias de su historia natural y humana], in: *Geogaceta*, 61: 11–14.

Chakrabarty, D., 2014a: "Baucom's Critique: A Brief Response", in: *Cambridge Journal of Postcolonial Literary Inquiry*, 1,2: 245–250.

Chakrabarty, Dipesh, 2014b: "The climate of history: four theses", in: *Teksty Drugie*, 5: 168–199.

Chakrabarty, Dipesh, 2016: "Humanities in the Anthropocene: The Crisis of an Enduring Kantian Fable", in: *New Literary History*, 47, 2–3: 377–397.

Chan, J.K.H., 2018: *Urban Ethics in the Anthropocene: The Moral Dimensions of Six Emerging Conditions in Contemporary Urbanism* (London: Palgrave Macmillan).

Chandler, D., 2017: "Securing the Anthropocene? International policy experiments in digital hacktivism: A case study of Jakarta", in: *Security Dialogue*, 48,2: 113–130.

Chandler, D., 2018: *Ontopolitics in the Anthropocene: An Introduction to Mapping, Sensing and Hacking* (London: Taylor and Francis).

Chandler, D., 2019a: "Building community resilience in the anthropocene: A study of international policy experiments with digital technology in Jakarta", in: *Resilience in Social, Cultural and Political Spheres*: 209–229.

Chandler, D., 2019b: "The Transvaluation of Critique in the Anthropocene", in: *Global Society*, 33,1: 26–44.

Chandler, D., 2019c: "The death of hope? Affirmation in the Anthropocene", in: *Globalizations*, 16,5: 695–706.

Chandler, D., 2019d: "Rethinking the ambiguities of abstraction in the Anthropocene", in: *Distinktion*, 20,3: 301–312.

Chandler, D., 2019e: "Forum 2: the migrant climate: resilience, adaptation and the ontopolitics of mobility in the Anthropocene", in: *Mobilities*, 14,3: 381–387.

Chandler, D., 2020a: "Security through societal resilience: Contemporary challenges in the Anthropocene", in: *Contemporary Security Policy*, 41,2: 195–214.

Chandler, D., 2020b: "Biopolitics 2.0: Reclaiming the power of life in the anthropocene The Microbial State: Global Thriving and the Body Politic Stefanie R Fishel Minneapolis, University of Minnesota Press, 2017-- Postcolonial Biology: Psyche and Flesh after Empire Deepika Bahri Minneapolis, University of Minnesota Press, 2017", in: *Contemporary Political Theory*, 19,1: 14–20.

Chandler, D., 2020c: "The coronavirus: Biopolitics and the rise of 'anthropocene authoritarianism'", in: *Russia in Global Affairs*, 18,2: 26–32.

Chandler, D.; Cudworth, E.: Hobden, S., 2018; "Anthropocene, capitalocene and liberal cosmopolitan IR: A response to Burke et al.'s 'Planet Politics'", in: *Millennium: Journal of International Studies*, 46,2: 190–208.

Chandler, D.; Pugh, J., 2019: "Islands of relationality and resilience: The shifting stakes of the Anthropocene", in: *Area*, 52,1: 65–72.

Chandler, D.; Reid, J., 2018: 'Being in Being': Contesting the Ontopolitics of Indigeneity , in: *European Legacy*, 23,3: 251–268.

Cheetham, M.A., 2018: *Landscape into Eco Art: Articulations of Nature Since the '60s* (Philadelphia, PA: Penn State University Press).

Church, J.A.; Asrar, G.R.; Busalacci, A.J.; Arndt, C.E., 2011: "Climate Information for Coping with Environmental Change: Contributions of the *World Climate Research Programme* (WCRP)", in: Brauch, H.G.; Oswald Spring, Ú; Mesjasz, C., et al. (Eds.): *Coping with Global Environmental Change, Disasters and Security – Threats, Challenges, Vulnerabilities and Risks* (Berlin – Heidelberg – New York: Springer-Verlag): 1257–1270.

Clark, Timothy, 2012: "Deconstruction in the Anthropocene", in: *Oxford Literary Review*, 34,2: 5–6.

Clark, Timothy, 2013: "What on World is the Earth?: The Anthropocene and Fictions of the World", in: *Oxford Literary Review*, 35,1: 5.

Clark, Timothy, 2019: "The 'Anthropocene'? Nature and Complexity", in: *Value Of Ecocriticism*: 17.

Clark, N., Yusoff, K., 2017: "Geosocial Formations and the Anthropocene", in: *Theory, Culture and Society*, 34,2–3: 3–23.

Clary Lemon, J., 2019: *Planting The Anthropocene: Rhetorics of Nature culture* (Logan: Utah State University Press).

Clingerman, F., 2014: "Geoengineering, theology, and the meaning of being human", in: *Zygon*, 49,1: 6–21.

Clingerman, F., 2016: "Place and the Hermeneutics of the Anthropocene", in: *Worldviews: Environment, Culture, Religion*, 20,3: 225–237.

Clingerman, F.; O'Brien, K.J., 2017: "Is climate change a new kind of problem? The role of theology and imagination in climate ethics", in: *Wiley Interdisciplinary Reviews: Climate* Change, 8,5: e480.

Clouse, C., 2021: *Climate-adaptive design in high mountain villages: Ladakh in transition*, (London: Taylor and Francis).

Colebrook, C., 2012: "Not Symbiosis, Not Now: Why Anthropogenic Change Is Not Really Human", in: *Oxford Literary Review*, 34,2: 185–209.

Colebrook, C., 2016a: "Losing the self? Subjectivity in the digital age", in: Groes, S. (Ed.): *Memory in the Twenty-First Century: New Critical Perspectives from the Arts, Humanities, and Sciences* (London: Palgrave Macmillan): 307–315.

Colebrook, C., 2016b: "'A Grandiose Time of Coexistence': Stratigraphy of the Anthropocene", in: *Deleuze Studies*, 10,4: 440–454.

Colebrook, C., 2017: "The twilight of the Anthropocene: Sustaining literature", in: *Literature and sustainability: Concept, text and culture* (.): 115–136.

Colebrook, C., 2019a: "Slavery and the Trumpocene: It's Not the End of the World", in: *Oxford Literary Review*, 41,1: 40.

Colebrook, C., 2019b: "A Cut In Relationality: Art at the end of the world", in: *Angelaki - Journal of the Theoretical Humanities*, 24,3: 175–195.

Colebrook, C., 2020a: "Extinction, Deterritorialisation and End Times: Peak Deleuze", in: *Deleuze and Guattari Studies*, 14,3: 327–348.

Colebrook, C., 2020b: "Fire, Flood and Pestilence as the Condition for the Possibility of the Human", in: *Derrida Today*, 13,2: 135–141.

Colebrook, C., 2020c: "Fast Violence, Revolutionary Violence: Black Lives Matter and the 2020 Pandemic", in: *Journal of Bioethical Inquiry*, 17,4: 495–499. Colebrook, C.; Weinstein, J., 2015: "Anthropocene Feminisms: Rethinking the Unthinkable Introduction", in: *Philosophia - A Journal Of Continental Feminism*, 5–2: 167–178.

Connor, L.H., 2016: *Climate Change and Anthropos: Planet, People and Places* (London: Routledge).

Conradie, Ernst M., 2019: "Is Christian humanism what is needed in the Age of the Anthropocene?", in: *Stellenbosch Theological Journal*, 5,3.

Conradie, Ernst M., 2020: "The Four Tasks Of Christian Ecotheology: Revisiting The Current Debate", in: *Scriptura - International Journal of Bible Religion and Theology in Southern Africa*, 119,1.

Cooke, S.J.; Bergman, J.N.; Nyboer, E.A.; (…); Van de Riet, K.; Vermaire, J.C., 2020: "Overcoming the concrete conquest of aquatic ecosystems", in: *Biological Conservation*, 247: 108589.

Cooke, S.J.; Twardek, W.M.; Reid, A.J.; (…); Hyder, K.; Danylchuk, A.J., 2019: "Searching for responsible and sustainable recreational fisheries in the Anthropocene", in: *Journal of Fish Biology*, 94,6: 845–856.

Cooper, Brown, Price et al., 2018: "Humans are the most significant global geomorphological driving force of the 21st century", in: *Anthropocene Review*, 5,3: 222–229.

Cosens, B.; Gunderson, L.H., 2018: *Practical Panarchy for Adaptive Water Governance: Linking Law to Social-Ecological Resilience* (Cham: Springer International Publishing).

Craps, S.; Crownshaw, R.; Wenzel, J.; (…); Colebrook, C.; Nardizzi, V., 2018: "Memory studies and the Anthropocene: A roundtable", in: *Memory Studies*, 11,4: 498–515.

Cromsigt, J.P.G.M.; Archibald, S.; Owen-Smith, N., 2017: *Conserving Africa's Mega-Diversity in the Anthropocene: The Hluhluwe-iMfolozi Park Story* (Cambridge: Cambridge University Press).

Cronin, M., 2017: *Eco-Translation: Translation and Ecology in the Age of the Anthropocene* (London: Taylor and Francis).

Crosslan, C.J.; Kremer, H.H.; Lindeboom, H.J.; Marshall Crossland, J.I.; Tissier, M,D.A. (Eds.), 2005: *Coastal Fluxes in the Anthropocene: The Land-Ocean Interactions in the Coastal Zone Project of the International Geosphere-Biosphere Programme* (Berlin Heidelberg: Springer-Verlag).

Crumley, C.; Laparidou, S.; Ramsey, M.; Rosen, A.M., 2015: "A view from the past to the future: Concluding remarks on the 'The Anthropocene in the Longue Durée'", in: *Holocene*, 25,10: 1721–1723.

Crutzen, P. J., 2002a: "Geology of mankind – The Anthropocene", in: *Nature*, 415: 23.

Crutzen, P. J., 2002b: "The "anthropocene"", in: ERCA, Vol. 5. Boutron, C. (Ed.): *From the Impacts of Human Activities on our Climate and Environment to the Mysteries of Titan* (Paris: EDP Sciences): 1–5.

Crutzen, P. J., 2002c: "The effects of industrial and agricultural practices on atmospheric chemistry and climate during the Anthropocene", in: *J. Environ. Sci. Health*, 37: 423–424.

Crutzen, P. J., 2002d, 2003: Atmospheric Chemistry in the "Anthropocene", in: *Steffen*, W.; Jäger, J.; Carson, D.J.; Bradshaw, C. (Eds.): *Challenges of a Changing Earth. Proceedings of the Global Change Open Science Conference*, Amsterdam, The Netherlands, 10–13 July 2001 (Berlin-Heidelberg: Springer Verlag, 2002): 45–48.

Crutzen, P. J., 2006: The "Anthropocene", in: Ehlers, E.; Krafft, T. (Eds.): *Earth System Science in the Anthropocene – Emerging Issues and Problems* (Heidelberg: Springer): 13–18.

Crutzen, P.J., 2007: "Atmospheric chemistry and climate in the Anthropocene", in: Bindé, J. (Ed.), *Making Peace with the Earth – What Future for the Human Species and the Planet?* (Oxford: Berghahn Books, UNESCO Publishing):113–120.

Crutzen, P. J., 2010: "Anthropocene man", in: *Nature*, 467: S10.

Crutzen, P.J., 2012: "Climate, Atmospheric Chemistry and Biogenic Processes in the Anthropocene", in: Kant, H.; Reinhardt, C. (Eds): *100 Jahre Kaiser-Wilhelm-/Max-Planck-Institut für Chemie (Otto-Hahn-Institut, Archiv der Max-Planck-Gesellschaft*, Vol. 22.

Crutzen, P.J.; Ramanathan, V., 2004: "Atmospheric Chemistry and Climate in the Anthropocene. Where are we Heading?", in: Schellnhuber, H.J; *Crutzen*, P.J.; Clark, W.C.; Claussen, M.; Held, H. (Eds.): *Earth System Analysis for Sustainablility. Dahlem Workshop Report* (Cambridge, CA: MIT Press): 265–292.

Crutzen, P.J.; Schwägerl, C., 2011: "Living in the Anthropocene: Toward a New Global Ethos", in: *Yale Environment*, 360.

Crutzen, P.J., *Steffen*, W., 2003: "How long have we been in the Anthropocene era? An Editorial Comment", in: *Climatic Change* , 61,3: 251–257.

Crutzen, P.J.; Stoermer, E.F., 2000: "The Anthropocene", in: *IGBP Newsletter*, 41: 17–18.

Cutter-Mackenzie-Knowles, A., Lasczik, A., Logan, M., Wilks, J., Turner, A., 2019: "Touchstones for deterritorializing the socioecological learner", in: *Touchstones for Deterritorializing Socioecological Learning: The Anthropocene, Posthumanism and Common Worlds as Creative Milieux* (Basingstoke: Palgrave):1–26.

Cutter Mackenzie Knowles, A.; Lasczik, A.; Wilks, J.; et al. (Eds.), 2020: *Touchstones for Deterritorializing Socioecological Learning: The Anthropocene, Posthumanism and Common Worlds as Creative Milieux* (Basingstoke: Palgrave).

Dalby, S., 2011: "Geographies of the International System: Globalization, Empire and the Anthropocene", in: *International Studies: Interdisciplinary Approaches*: 125–148.

Dalby, S., 2013a: "Geopolitics in the Anthropocene: A reply to Clark, Kahn and Lehman", in: *Political Geography* , 37: 56–57.

Dalby, S., 2013b: "The geopolitics of climate change", in: *Political Geography*, 37: 38–47.

Dalby, S., 2013c: "Peace in the Anthropocene", in: *Peace Review*, 25,4: 561–567.

Dalby, S., 2013d: "Biopolitics and climate security in the Anthropocene", in: *Geoforum*, 49: 184–192.

Dalby, S., 2014a: "Environmental Geopolitics in the Twenty-first Century", in: *Alternatives,* 39,1: 3–16.

Dalby, S., 2014b: "Rethinking Geopolitics: Climate Security in the Anthropocene", in: *Global Policy,* 5,1: 1–9.

Dalby, S., 2015a: "Anthropocene Discourse: Geopolitics after Environment", draft presentation.

Dalby, S., 2015b: "Geoengineering: The next era of geopolitics?", in: *Geography Compass*, 9,4: 190–201.

Dalby, S., 2016a: "Framing the Anthropocene: The Good, the Bad and the Ugly", in: *Anthropocene Review*, 3,1: 33–51.

Dalby, S., 2016b: "Climate security in the Anthropocene: 'Scaling up' the human niche", in: *Reimagining Climate Change* : 29–48.

Dalby, S., 2017a: "Anthropocene Formations: Environmental Security, Geopolitics and Disaster", in: *Theory, Culture and Society*, 34,2–3: 233–252.

Dalby, S., 2017b: "Contextual Changes in Earth History: From the Holocene to the Anthropocene - Implications for Sustainable Development and for Strategies of Sustainable Transition Sustainability Transition and Sustainable Peace: Scientific and Policy Context, Scientific Concepts and Dimensions", in: Brauch, H.G.; Oswald Spring, Ú.; Grin, J.; Scheffran; J. (Eds.): *Handbook on Sustainability Transition And Sustainable Peace* (Cham: Springer International Publishing): 67–88.

Dalby, S., 2018a: "Firepower: Geopolitical Cultures in the Anthropocene", in: *Geopolitics,* 23,3: 718–742.

Dalby, S., 2018b: "Climate change, Gaia and the Anthropocene", in: *Handbook on the Geographies of Globalization* : 307–317.

Dalby, S., 2018c: "The anthropocene thesis", in: *The Oxford Handbook of Global Studies*: 173–188.

Dalby, S., 2019: "Anthropocene Discourse: Geopolitics after environment", in: *Handbook of the Changing World Language Map*, 1: 3287–3300.

Dalby, S., 2020a: "Bordering sustainability in the Anthropocene", in: *Territory, Politics, Governance*, 8,2: 144–160.

Dalby, S., 2020b: *Anthropocene Geopolitics: Globalization, Security, Sustainability Politics and Public Policy* (Ottawa: University of Ottawa Press).

Davis, J.; Moulton, A.A.; Van Sant, L.; Williams, B., 2019: "Anthropocene, Capitalocene, … Plantationocene?: A Manifesto for Ecological Justice in an Age of Global Crises", in: *Geography Compass*, 13,5 (May): e12438.

De Almeida, R.; da Silva Carvalho, P.G., 2018: "Healthy People Living on a Healthy Planet-The Role of Education of Consciousness for Integration as an Instrument of Health Promotion", in: Azeiteiro, U.M.; Akerman, M.; Leal Filho, W.; Setti, A.F.F.; Brandli, L.L. (Eds.): *Lifelong Learning And Education In Healthy And Sustainable Cities* (Cham: Springer Publisher): 299–326.

De Cock, C.; Nyberg, D.; Wright, C., 2019: "Disrupting climate change futures: Conceptual tools for lost histories", in: *Organization*, Art. No.: UNSP 1350508419883377.

de Freitas, E., 2020: "Science Studies and the Metamorphic Multiple Earth: Bruno Latour's Risky Diplomacy", in: *Cultural Studies - Critical Methodologies*, 20,3: 203–212.

de Freitas, E.; Truman, S.E., 2020: "New Empiricisms in the Anthropocene: Thinking With Speculative Fiction About Science and Social Inquiry", in: *Qualitative Inquiry*, Article in Press.

de Freitas, E.; Weaver, J.A., 2020: "Rethinking Social Inquiry in the Wake of Science Studies: Transdisciplinary Pursuits in Times of Climate Change, Information Flows, and Fading Democracies", in: *Cultural Studies - Critical Methodologies*, 20,3: 195–202.

De Lucia, V., 2019: *The 'Ecosystem Approach' in International Environmental Law: Genealogy and Biopolitics* (London: Taylor and Francis).

De Souza, M.A.T.; Costa, D.M., 2018: *Historical Archaeology and Environment* (Cham: Springer International Publishing).

Deleage, Jean-Paul, 2011: "Journey in the Anthropocene: This new era where we are the heroes", in: *Quinzaine Litteraire*, 1033: 23–23.

Deleage, Jean-Paul, 2013: "The Event Anthropocene: Earth, History and We", in: *Quinzaine Litteraire*, 1093: 23–23.

Deleage, Jean-Paul, 2017: "Geopolitics of a Disordered Planet - The Shock of the Anthropocene", in: *Nouvelle Quinzaine Litteraire*, 1183: 3257: 22–23.

Dellasala, D.A.; Goldstein; M.I. (Eds.), 2017: *Encyclopedia of the Anthropocene* (Oxford [England] ; Waltham MA: Elsevier).

Demos, T. J., 2017: *Against the Anthropocene Visual Culture and Environment Today* (Cambridge, Mass., MIT Press).

Denny, M., 2017: *Making the Most of the Anthropocene: Facing the Future* (Baltimore: Johns Hopkins University Press).

Di Chiro, G., 2016: **"Environmental Justice and the Anthropocene Meme"**, **in:** Gabrielson, Teena; Hall, Cheryl; Meyer, John M.; Schlosberg, David (Eds.): *The Oxford Handbook of Environmental Political Theory*, Oxford Handbooks Online; at: https://www.oxfordhandbooks.com/view/10.1093/oxfordhb/9780199685271.001.0001/oxfordhb-9780199685271-e-18 (2 November 2020).

Di Chiro, G., 2017: "Welcome to the White (M)Anthropocene? A Feminist-Environmental Critique", in: MacGregor, S. (Ed.): *Routledge Handbook of Gender and Environment* (London: Routledge): 487–505.

Di Paola, M., 2015: "Virtues for the anthropocene", in: *Environmental Values*, 24,2: 183–207.

Di Paola, M., 2017: "Virtue", in : *Encyclopedia of the Anthropocene*, 1–5: 119–126.

Di Paola, M., Garasic, M.D., 2013: "The dark side of sustainability: Avoiding and shortening lives in the anthropocene", in: *Rivista di Studi sulla Sostenibilita*, 2: 59–81.

Diogo, M.P.; Simões, A.; Rodrigues, A.D.; Scarso, D., 2019: *Gardens and human agency in the anthropocene* (London: Taylor and Francis).

DiPaola, M., 2017: *Ethics and Politics of the Built Environment: Gardens of the Anthropocene* (Cham: Springer International Publishing).

Donges, S.F.; Lucht, W..; Mueller-Hansen, F.., et al., 2017: "The technosphere in Earth System analysis: A coevolutionary perspective", in: *The Anthropocene Review* 4,1: 23–33.

Doyle, K.O., 2008 : "Thinking differently" about the new media – Introduction", in: *American Behavioral Scientist*, 52,1: 3–7.

Dryzek, John S.; Pickering, Jonathan, 2019: *The Politics of the Anthropocene* (Oxford: Oxford University Press).

Duffy, M., Gallagher, M., Waitt, G., 2019: "Emotional and affective geographies of sustainable community leadership: A visceral approach", in: *Geoforum*, 106: 378–384.

Dufresne, T., 2019: *Democracy of Suffering: Life on the Edge of Catastrophe, Philosophy in the Anthropocene* (Montreal: McGill-Queens Univ. Press).

Edgeman, R., 2020: "Urgent evolution: excellence and wicked Anthropocene Age challenges", in: *Total Quality Management & Business Excellence*, 31,5–6: 469–482.

Edgeman, Rick; Neely, Andy; Eskildsen, Jacob, 2016: "Paths to sustainable enterprise excellence", in: *Journal of Modelling in Management*, 11,4: 858–868.

Edgeworth, M., *Ellis*, E.C., *Gibbard*, P., Neal, C., *Ellis*, M., 2019: "The chronostratigraphic method is unsuitable for determining the start of the Anthropocene", in: *Progress in Physical Geography*, 43,3: 334–344.

Edgeworth, M.; *Richter*, D.; *Waters*, C., et al., 2015: "Diachronous beginnings of the Anthropocene: The lower bounding surface of anthropogenic deposits", in: *The Anthropocene Review*, 2,1: 33–58.

Ehlers, Eckart, 2008: *Das Anthropozän: Die Erde im Zeitalter des Menschen* (Darmstadt: Wissenschaftliche Buchgemeinschaft).

Ehlers, E., 2016: "From HDP to IHDP: "Evolution of the International Humans Dimensions of Global Environmental Change Programme", in: Brauch, H.G.; Oswald Spring, Ú.; Grin, J.; Scheffran; J. (Eds.): *Handbook on Sustainability Transition and Sustainable Peace* (Cham: Springer International Publishing Switzerland): 359–376.

Ehlers, Eckart; Krafft, Thomas, 2005, 2006: *Earth system science in the anthropocene: Emerging issues and problems* (Berlin-Heidelberg: Springer).

Ellis, E.C., 2015: "Ecology in an anthropogenic biosphere", in: *Ecological Monographs*, 85,3: 287–331.

Ellis, E.C., 2017: "Physical geography in the Anthropocene", in: *Progress in Physical Geography*, 41,5: 525–532.

Ellis, E.C., 2019: "Evolution: Biodiversity in the Anthropocene", in: *Current Biology*, 29,17: R831–R833.

Ellis, E.C.; Antill, E.C.; Kreft, H., 2012: "All is not loss: Plant biodiversity in the anthropocene", in: *PLoS ONE*, 7,1: e30535.

Ellis, E.C.; Beusen, A.H.W.; Goldewijk, K.K., 2020: "Anthropogenic biomes: 10,000 BCE to 2015 CE", in: *Land*, 9,5: 129.

Ellis, E.C.; Fuller, D.Q.; Kaplan, J.O.; Lutters, W.G., 2013: "Dating the Anthropocene: Towards an empirical global history of human transformation of the terrestrial biosphere", in: *Elementa*, 1,000018.

Ellis, E.C.; Haff, P.K., 2009 : "Earth science in the anthropocene: New Epoch, new Paradigm, new responsibilities", in: *Eos*, 90,49: 473;

Ellis, E.C.; Kaplan, J.O.; Fuller, D.Q.; (...), Goldewijk, K.K.; Verburg, P.H., 2013: "Used planet: A global history", in: *Proceedings of the National Academy of Sciences of the United States of America*, 110,20: 7978–7985.

Ellis, E.C.; Magliocca, N.R.; Stevens, C.J.; Fuller, D.Q., 2018: "Evolving the Anthropocene: linking multi-level selection with long-term social–ecological change", in: *Sustainability Science*, 13,1: 119–128.

Ellis, E.C.; Maslin, M.; Boivin, N.; Bauer, A., 2016: "Involve social scientists in defining the Anthropocene", in: *Nature*, 540,7632: 192–193.

Ellis, E.C.; Mehrabi, Z., 2019: "Half Earth: promises, pitfalls, and prospects of dedicating Half of Earth's land to conservation", in: *Current Opinion in Environmental Sustainability*, 38: 22–30.

Ellis, E.C.; Pascual, U.; Mertz, O., 2019: "Ecosystem services and nature's contribution to people: negotiating diverse values and trade-offs in land systems", in: *Current Opinion in Environmental Sustainability*; 38: 86–94.

Erdelen, W.R.; Richardson, J.G., 2018: *Managing Complexity: Earth Systems and Strategies for the Future* (Abingdon: Taylor and Francis).

Ergene, S.; Banerjee, S.B.; Hoffman, A.J., 2020: "(Un)Sustainability and Organization Studies: Towards a Radical Engagement", in: *Organization Studies*, Art. No.: 0170840620937892.

Ergene, S.; Calás, M.B.; Smircich, L., 2018: "Ecologies of Sustainable Concerns: Organization Theorizing for the Anthropocene", in: *Gender, Work and Organization*, 25,3: 222–245.

Eriksen, C.; Ballard, S., 2020, *Alliances in the anthropocene: Fire, plants, and people* (Basingstoke: Palgrave).

Erlandson, J.M., 2013: "Shell middens and other anthropogenic soils as global stratigraphic signatures of the Anthropocene", in: *Anthropocene*, 4: 24–32.

Erlandson, J.M., Braje, T.J., 2013: "Archeology and the anthropocene", in: *Anthropocene*, 4: 1–7.

Estok, S.C., 2017a: "Anthropocene, What Anthropocene? The City and the Epoch in a Fine Balance and the Dog", in: *Concentric-Literary and Cultural Studies*, 43,1: 33–50.

Estok, S.C., 2017b: "Pollution, sci-fi, and the sublime", in: *Tamkang Review*, 47,2: 3–16;

Estok, S.C., 2018a: *Ecophobia Hypothesis* (Abingdon: Routledge).

Estok, S.C., 2018b: "Hollow Ecology and Anthropocene Scales of Measurement", in: *Mosaic-An Interdisciplinary Critical Journal*, 51,3: 37–52.

Estok, S.C., 2020a, "Introduction to special cluster on 'The body and the Anthropocene'", in: *Neohelicon*, 47,1: 1–7.

Estok, S.C., 2020b: "Corporeality, hyper-consciousness, and the Anthropocene ecoGothic: slime and ecophobia", in: *Neohelicon*, 47,1: 27–39.

Estok, S.C., Chou, S.S., 2017: "Foreword: The city and the Anthropocene", in: *Concentric: Literary and Cultural Studies*, 43,1: 3–11.

Everard, M., 2016: *The Ecosystems Revolution* (London: Palgrave).

Ezban, M., 2019: *Aquaculture Landscapes: Fish Farms and the Public Realm* (Abingdon: Taylor and Francis).

Fagan, Madeleine, 2016: "Security in the anthropocene: Environment, ecology, escape", in: *European Journal of International Relations*, 1–23.

Falcon-Lang, H., 2011: "Anthropocene: How humans created a new geological age"; at: *BBC website*, 11 May.

Falkenhayn, L. von; Rechkemmer, A.; Young, O. T., 2011: "The International Human Dimension Programme on Global Environmental Change – Taking Stock and Moving Forward" in: Brauch, H.G.; Oswald Spring, U; Mcsjasz, C., et al. (Eds): *Coping with Global Environmental Change, Disasters and Security – Threats, Challenges, Vulnerabilities and Risks* (Berlin – Heidelberg – New York: Springer-Verlag): 1221–1234.

Falkenmark, M.; Wang-Erlandsson, L.; Rockström, J., 2019: "Understanding of water resilience in the Anthropocene", in: *Journal of Hydrology*, X, 2:100009.

Fay, J., 2018: *Inhospitable World: Cinema in the Time of the Anthropocene* (Oxford: Oxford University Press).

Figueroa, A., 2017: *Economics of the Anthropocene Age* (Cham: Springer International Publishing).

Fishel, S.R., 2017: *The Microbial State: Global Thriving and the Body Politic* (Minneapolis, MN: University of Minnesota Press).

Fitzpatrick, S.M.; Erlandson, J.M., 2018: "Island Archaeology, Model Systems, the Anthropocene, and How the Past Informs the Future", in: *Journal of Island and Coastal Archaeology*, 13,2: 279–295.

Fletcher, A.L., 2019: *De-Extinction and the Genomics Revolution: Life on Demand* (Basingstoke: Palgrave).

Folke, C.; Biggs, R.; Norström, A.V.; Reyers, B.; Rockström, J., 2016: "Social-ecological resilience and biosphere-based sustainability science", in: *Ecology and Society*, 21,3: 41.

Foster, J.B., 2016: "Marxism in the Anthropocene: Dialectical Rifts on the Left", in: *International Critical Thought*, 6,3: 393–421.

Foster, J.B., 2017: "The Earth-System Crisis and Ecological Civilization: A Marxian View", in: *International Critical Thought*, 7,4: 439–458.

Foster, J.B., 2020: "Engels's Dialectics of Nature in the Anthropocene", in: *Monthly Review-An Independent Socialist Magazine*, 72,6: 1–17.

Foster, J.B.; Holleman, H.; Clark, B., 2019: "Imperialism in the Anthropocene", in: *Monthly Review-An Independent Socialist Magazine*, 71, 3: 70–88.

Fox, T., Pope, M., *Ellis*, E.C., 2017: "Engineering the Anthropocene: Scalable social networks and resilience building in human evolutionary timescales", in: *Anthropocene Review*, 4,3: 199–215.

Freestone, D.; *Vidas*, D.; Camprubi, A.T., 2019: "Sea Level Rise and Impacts on Maritime Zones and Limits The Work of the ILA Committee on International Law and Sea Level Rise", in: *Korean Journal of International and Comparative Law*, 5,1: 5–35.

Freund, J., 2015: "Rev. Billy vs. the Market: a sane man in a world of omnipotent fantasies", in: *Journal of Marketing Management*, 31,13–14: 1529–1551.

Froestad, J., Shearing, C., 2017: "Energy and the Anthropocene: security challenges and solutions", in: *Crime, Law and Social Change*, 68,5: 515–528.

Fuentes, A., 2012: "Ethnoprimatology and the anthropology of the human-primate interface", in: *Annual Review of Anthropology*, 41: 101–117.

Fuentes, A., 2017: "Human niche, human behaviour, human nature", in: *Interface Focus*, 7,5: 20160136.

Fuentes, A., 2019: "Holobionts, Multispecies Ecologies, and the Biopolitics of Care: Emerging Landscapes of Praxis in a Medical Anthropology of the Anthropocene", in: *Medical Anthropology Quarterly*, 33,1: 156–162.

Fuentes, A., 2020: "Biological anthropology's critical engagement with genomics, evolution, race/racism, and ourselves: Opportunities and challenges to making a difference in the academy and the world", in: *American Journal of Physical Anthropology*, Article in Press;

Fuentes, A.; Baynes-Rock, M., 2017: "Anthropogenic landscapes, human action and the process of co-construction with other species: Making anthromes in the Anthropocene", in: *Land*, 6,1: 15.

Gabardi, W., 2017a: *Next Social Contract: Animals, the Anthropocene, and Biopolitics* (Philadelphia, PA: Temple Univ. Press).

Gabardi; W., 2017b: "Animals, The Anthropocene, And Biopolitics Conclusion", in: Gabardi, W: *Next Social Contract: Animals, the Anthropocene, and Biopolitics* (Philadelphia, PA: Temple Univ. Press): 175–185.

Gabardi; W., 2017c: "The Anthropocene Hypothesis", in: Gabardi, W: *Next Social Contract: Animals, the Anthropocene, and Biopolitics* (Philadelphia, PA: Temple Univ. Press): 11–29.

Gaffney, O., *Steffen*, W., 2017a: "The Anthropocene equation", 4,1: 53–61.

Gaffney, O.; *Steffen*, W., 2017b: "Mathematics and the Anthropocene equation: Comment on Gaffney, O. and Steffen, W (2017): The Anthropocene equation. *The Anthropocene Review* 4:53-61 Response", 4,3: 264–265.

Gaffney, O.; *Steffen*, W., 2017c: "Response to Heijung et al.", in: *Anthropocene Review*, 4,3: 264–265.

Galaz, V., 2012: "Geo-engineering, Governance, and Social-Ecological Systems: Critical Issues and Joint Research Needs", in: *Ecology and Society*, 17,1: Art. No. 24.

Galaz, V., 2014: *Global Environmental Governance, Technology and Politics: The Anthropocene Gap*: 1–189.

Galaz, V., 2019: "Time and politics in the Anthropocene: Too fast, too slow?", in: Biermann, F., Lövbrand, E. (Eds.): *Anthropocene Encounters: New Directions in Green Political Thinking* (Cambridge – New York: Cambridge University Press): 109–127.

Galaz, V.; Österblom, H.; Bodin, Ö.; Crona, B., 2016: "Global networks and global change-induced tipping points", in: *International Environmental Agreements: Politics, Law and Economics*, 16,2: 189–221.

Gasparin, M.; Brown, S.D.; Green, W.; (...); Williams, M.; Zalasiewicz, J., 2020: "The business school in the anthropocene: Parasite logic and pataphysical reasoning for a working earth", in: *Academy of Management Learning and Education*, 19,3: 385–405.

Gasparin, Marta; Green, William; Schinckus, Christophe, 2020: "Slow design-driven innovation: A response to our future in the Anthropocene epoch", In: *Creativity and Innovation Management*, 29,4: 551–565.

Gautier, C., et al., 2008: *Facing climate change together* (Cambridge: Cambridge UP).

Gerhardinger, L.C.; Gorris, P.; Gonçalves, L.R.; (...); Zondervan, R.; Glavovic, B.C., 2018: "Healing Brazil's Blue Amazon: The role of knowledge networks in nurturing cross-scale transformations at the frontlines of ocean sustainability", in: *Frontiers in Marine Science*, 4 (January): 395.

Giblett, R., 2019: *Psychoanalytic Ecology: The Talking Cure for Environmental Illness and* Health (London: Taylor and Francis).

Gibson, C., 2019a: "A sound track to ecological crisis: Tracing guitars all the way back to the tree", in: *Popular Music*, 38,2: 183-203.

Gibson, C., 2019b: "Critical tourism studies: new directions for volatile times", in: *Tourism Geographies*.

Gibson-Graham, J. K.; Cameron, J.; Healy, S.; et al., 2019a: "Economic Geography and Ethical Action in the Anthropocene: A Rejoinder", in: *Economic Geography*, 95,1: 27–29.

Gibson-Graham, J. K.; Cameron, J.; Healy, S.; et al., 2019b: "Roepke Lecture in Economic Geography-Economic Geography, Manufacturing, and Ethical Action in the Anthropocene", in: *Economic Geography*, 95,1: 1–21;

Gilbert, J., 2016: "Transforming Science Education for the Anthropocene-Is It Possible?", in: *Research In Science Education*, 46, 2: 187–201.

Gildersleeve R.E., Kleinhesselink K., 2017: "A Mangled Educational Policy Discourse Analysis for the Anthropocene". In: Lester J., Lochmiller C., Gabriel R. (eds): *Discursive Perspectives on Education Policy and Implementation*. (Cham: Palgrave Macmillan). https://doi.org/10.1007/978-3-319-58984-8_6

Glaser, M.; Glaeser, B., 2014: "Towards a framework for cross-scale and multi-level analysis of coastal and marine social-ecological systems dynamics", in: *Regional Environmental Change*, 14,6: 2039–2052.

Glaser, M.; Krause, G.; Glaeser, B., 2012: "Towards global sustainability analysis in the anthropocene", in: *Human-Nature Interactions in the Anthropocene: Potentials of Social-Ecological Systems Analysis*, 9780203123195: 193–222.

Glaser, M.; Plass-Johnson, J.G.; Ferse, S.C.A.; (...); Teichberg, M.; Reuter, H., 2018: "Breaking resilience for a sustainable future: Thoughts for the anthropocene", in: *Frontiers in Marine Science*, 5(Feb): 34.

Gleditsch, Nils Petter, 2003: "Environmental Conflict: Neomalthusians vs. Cornucopians ", in: Brauch, H.G.; Liotta, P.H; Marquina, A.; Rogers, P.; Selim, M. El-Sayed (Eds.): *Security and Environment in the Mediterranean. Conceptualising Security and Environmental Conflicts* (Berlin-Heidelberg: Springer 2003): 477–485.

Gleeson, T.; Wang-Erlandsson, L.; Porkka, M.; (...); Falkenmark, M.; Famiglietti, J.S., 2020: "Illuminating water cycle modifications and Earth system resilience in the Anthropocene", in: *Water Resources Research*, 56,4: e2019WR024957.

Glikson, A.Y., 2014: "Evolution of the Atmosphere, Fire and the Anthropocene Climate Event Horizon", in: *Evolution of the Atmosphere, Fire and the Anthropocene Climate Event Horizon*, SpringerBriefs in Earth Sciences (Dordrecht: Springer Netherlands).

Glikson, A.Y., 2017: *Plutocene: Blueprints for a Post-Anthropocene Greenhouse Earth* (Cham: Springer International Publishing).

González-Ruibal, A., 2018: *An Archaeology of the Contemporary Era* (London: Taylor and Francis).

Goodenough, A.; Waite, S., 2019: *Wellbeing from Woodland: A Critical Exploration of Links between Trees and Human Health* (Basingstoke: Palgrave).

Gordon, L.J.; Bignet, V.; Crona, B.; (...), Rockström, J.; Queiroz, C., 2017: "Rewiring food systems to enhance human health and biosphere stewardship", in: *Environmental Research Letters*, 12,10: 100201.

Grandcolas, P.; Maurel, M.-C., 2018: *Biodiversity and Evolution* (Amsterdam: Elsevier).

Grear, A., 2015: "Deconstructing Anthropos: A Critical Legal Reflection on Anthropocentric' Law and Anthropocene 'Humanity'", in: *Law and Critique*, 26,3: 225–249.

Grear, A., 2017: "'Anthropocene, Capitalocene, Chthulucene': Re-encountering Environmental Law and its 'Subject' with Haraway and New Materialism", in: *Environmental Law and Governance for the Anthropocene*: 77–95.

Grear, A., 2018: "Anthropocene 'time'? - A reflection on temporalities in the 'New Age of the Human'", in: *Routledge Handbook of Law and Theory* (London: Routledge): 297–316.

Grear, A., 2019: "'Anthropocene "Time"?' - A reflection on temporalities in the 'New Age of the Human'", in: *Routledge Handbook of Law and Theory*, Book Series: Routledge Handbooks: 297–315.

Grear, A., 2020: "Legal Imaginaries and the Anthropocene: 'Of' and 'For'", in: *Law and Critique*, Article in Press.

Gren, M., 2016: "Mapping the anthropocene and tourism", in: Gren, M.; Huijbens, E.H., (Eds.): *Tourism and the Anthropocene* (London: Routledge): 171–188.

Gren, M.; Huijbens, E.H., 2014: "Tourism and the Anthropocene", in: *Scandinavian Journal of Hospitality and Tourism*, 14,1: 6–22.

Gren, M.; Huijbens, E.H., 2016a: "The anthropocene and tourism destinations", in: *Tourism and the Anthropocene* (London: Routledge): 189–199.

Gren, M.; Huijbens, E.H., 2016b: *Tourism and the Anthropocene* (London: Routledge).

Griffiths, M., 2017: *New Poetics of Climate Change: Modernist Aesthetics for a Warming World* (London: Bloomsbury Publ. Inc.).

Grinevald, Jacques, 2008: *La biosphère de l'anthropocène: climat et pétrole, la double menace; repères transdisciplinaires (1824–2007)* (Genève: Georg).

Groes, S., 2017: "'I love Alaska': posthuman subjectivity and memory on the final frontier of our ecological crisis", in: *Textual Practice*, 31,5: 973-993.

Grusin, R., 2018: *After Extinction* (Minneapolis, MN: University of Minnesota Press).

Gupta, J., 2015: "Growth, the Environment, and Development in the Anthropocene", in: *Current History*, 114,775: 311–315.

Gupta, J., 2018: "Sharing our water: Inclusive development and glocal water justice in the anthropocene", in: *Water Justice*: 259–274.

Gupta, J.; Chu, E., 2018 : "Inclusive Development and Climate Change: The Geopolitics of Fossil Fuel Risks in Developing Countries", in: *African and Asian Studies*, 17, 1–2: 90–114;

Gupta, J.; Hordijk, M.; Vegelin, C., 2019: "An Inclusive Development Perspective on Development Studies in the Anthropocene", in: *Building Development Studies For The New Millennium*, EADI Global Development Series (Cham: Springer International Publishing): 25–50.

Gupta, J.; Pahl-Wostl, C.; Zondervan, R., 2013: "'Glocal' water governance: A multi-level challenge in the anthropocene", in: *Current Opinion in Environmental Sustainability*, 5,6: 573–580.

Gupta, J.; Pouw, N.R.M.; Ros-Tonen, M.A.F., 2015: "Towards an Elaborated Theory of Inclusive Development", in: *European Journal of Development Research*, 27,4: 541–559.

Gupta, J.; Schmeier, S., 2020a: "Future proofing the principle of no significant harm", in: *International Environmental Agreements: Politics, Law and Economics*, 20: 731–747.

Gupta, Joyeeta; Schmeier, Susanne, 2020b: *International Environmental Agreements-Politics Law and Economics*, 20,4: 731–747.

Gupta, J.; Vegelin, C., 2016: "Sustainable development goals and inclusive development", in: *International Environmental Agreements: Politics, Law and Economics*, 16,3: 433–448.

Haas, Peter M., 1990: *Saving the Mediterranean: The Politics of International Environmental Cooperation* (New York: Columbia University Press).

Haas, Peter M., 1992: "Introduction: epistemic communities and international policy coordination", in: *International Organization,* 46,1 (Winter): 1–35.

Hacking, Ian, 1999: *The Social Construction of What?* (Cambridge, MA: Harvard University Press).

Hallé, C.; Milon, A-S., 2020: "The infinity of the Anthropocene: A (Hi)story with a thousand names", in: Latour, B.; Weibel, P. (Eds.): *Critical Zones: The Science and Politics of Landing on Earth.* (Cambridge, MA: MIT Press): 44–49.

Hamilton, C., 2015: "Human destiny in the Anthropocene", in: Hamilton, C.; Gemenne, F.; Bonneuil, C. (Eds.): *The Anthropocene and the Global Environmental Crisis: Rethinking Modernity in a New Epoch* (London: Routledge): 32–43.

Hamilton, C., 2016: "The anthropocene as rupture", in: *Anthropocene Review*, 3,2: 93–106.

Hamilton, S., 2017: "Securing ourselves from ourselves? The paradox of "entanglement" in the Anthropocene", in: *Crime, Law and Social Change*, 68,5: 579–595.

Hamilton, S., 2018a: "Foucault's end of history: The temporality of governmentality and its end in the anthropocene", in: *Millennium: Journal of International Studies*, 46,3; 371–395.

Hamilton, S., 2018b: The measure of all things? The Anthropocene as a global biopolitics of carbon", in: *European Journal of International Relations*, 24,1: 33–57.

Hamilton, S., 2019: "I am uncertain, but We are not: A new subjectivity of the Anthropocene", in: *Review of International Studies*, 45,4: 607–626.

Hamilton, C., 2020: "Towards a Fifth Ontology for the Anthropocene", in: *Angelaki - Journal of the Theoretical Humanities*, 25,4: 110–119.

Hamilton, C., Bonneuil, C., Gemenne, F., 2015: *The Anthropocene and the Global Environmental Crisis: Rethinking Modernity in a New Epoch* (London: Routledge).

Hammerschlag, N.; Schmitz, O.J.; Flecker, A.S.; (...); Skubel, R.; Cooke, S.J., 2019: "Ecosystem Function and Services of Aquatic Predators in the Anthropocene", in: *Trends in Ecology and Evolution*, 34,4: 369–383.

Hanusch, F.; Biermann, F., 2020: "Deep-time organizations: Learning institutional longevity from history", in: *Anthropocene Review*, 7,1: 19–41.

Haqq-Misra, Jacob, 2019: "Introduction: Detectability of future Earth", in: *Futures*, 106: 1–3.

Haraway, D.; Ishikawa, N.; Gilbert, S.F.; (...), Tsing, A.L.; Bubandt, N., 2016: "Anthropologists Are Talking – About the Anthropocene", in: *Ethnos*, 81,3: 535–564.

Haraway, Donna J., 2016a: "Staying with Trouble: Anthropocene, Capitalocene, Chthulucene", in: Moore, Jason M. (Ed.): *Anthropocene or Capitalocene? Nature, History, and the Crisis of Capitalism* (Oakland, CA: PM Press).

Haraway, Donna J., 2016b: *Manifestly Haraway* (Minneapolis, MN: University of Minnesota Press).

Hardt, J.N., 2017: *Environmental Security in the Anthropocene: Assessing Theory and Practice* (Abingdon: Routledge).

Harrington, C., 2016: "The Ends of the World: International Relations and the Anthropocene", in: *Millennium*, 44,3: 478–498.

Hartley, D., 2016: "Anthropocene, Capitalocene, and the Problem of Culture", in: Moore, Jason M. (Ed.): *Anthropocene or Capitalocene? Nature, History, and the Crisis of Capitalism* (Oakland, CA: PM Press): 154–165.

Harvey, P., Jensen, C.B., Morita, A., 2016: *Infrastructures and Social Complexity: A Companion* (Abingdon: Routledge).

Harvey, M., 2019: *Utopia in the Anthropocene: A Change Plan for a Sustainable and Equitable World* (London: Routledge).

Head, L., 2014: "The Anthropoceneans", in: *Geographical Research*, 53,3: 313–320.

Head, L., 2016: *Hope and Grief in the Anthropocene: Re-Conceptualising Human-Nature Relations* (Abingdon: Routledge): 1–181.

Head, L.; Atchison, J., 2015a: "Entangled invasive lives: Indigenous Invasive Plant Management in northern Australia", in: *Geografiska Annaler, Series B: Human Geography*, 97,2: 169–182.

Head, L.; Atchison, J., 2015b: "Governing invasive plants: Policy and practice in managing the Gamba grass (Andropogon gayanus) - Bushfire nexus in northern Australia", in: *Land Use Policy*, 47: 225–234;

Head, L.; Larson, B.M.H.; Hobbs, R.: (...): Kull, C.; Rangan, H., 2015: "Living with invasive plants in the anthropocene: The importance of understanding practice and experience", in: *Conservation and Society*, 13,3: 311–318.

Hebrides, I.; Affifi, R.; Blenkinsop, S.; et al., 2018a: "Why Wild Pedagogies?", Conference: Wild Pedagogies Sailing Colloquium Location: Oban, Scotland, May 2017, in: *Wild Pedagogies: Touchstones For Re-Negotiating Education And The Environment In The Anthropocene,* Book Series: Palgrave Studies in Educational Futures (London: Palgrave Macmillan): 1–22.

Hebrides, I.; Affifi, R.; Blenkinsop, S.; et al., 2018b: "On Wilderness", Conference: Wild Pedagogies Sailing Colloquium Location: Oban, Scotland, May 2017, in: *Wild Pedagogies: Touchstones For*

Re-Negotiating Education And The Environment In The Anthropocene, Book Series: Palgrave Studies in Educational Futures (London: Palgrave Macmillan): 23–50.

Hebrides, I.; Affifi, R.; Blenkinsop, S.; et al., 2018c: "On the Anthropocene", Conference: Wild Pedagogies Sailing Colloquium Location: Oban, Scotland, May 2017, in: *Wild Pedagogies: Touchstones For Re-Negotiating Education And The Environment In The Anthropocene,* Book Series: Palgrave Studies in Educational Futures (London: Palgrave Macmillan): 51–62.

Hedin, G.; Gremaud, A.S.N., 2018: *Artistic Visions of the Anthropocene North: Climate Change and Nature in Art* (London: Routledge).

Heesterman, A.R.G., Heesterman, W.H., 2016: *Transformation of Collective Intelligences: Perspective of Transhumanism* (Hoboken, N.J.: Wiley).

Heikkurinen, P. (Ed.), 2017a: *Sustainability and Peaceful Coexistence for the Anthropocene,* (London: Routledge).

Heikkurinen, P., 2017b: "Introduction", in: *Sustainability and Peaceful Coexistence for the Anthropocene* (London: Routledge): 1–4.

Heikkurinen, P., 2017c "On the emergence of peaceful coexistence", in: *Sustainability and Peaceful Coexistence for the Anthropocene* (London: Routledge): 7–15.

Heikkurinen, P.; Clegg, S.; Pinnington, A.H.; et al., 2019a: "Managing the Anthropocene: Relational Agency and Power to Respect Planetary Boundaries", in: *Organization & Environment,* Art. No.: UNSP 1086026619881145.

Heikkurinen, P.; Ruuska, T.; Wilen, K.; et al., 2019b: "The Anthropocene exit: Reconciling discursive tensions on the new geological epoch", in: *Ecological Economics,* 64: Art. No.: 106369.

Heikkurinen, P.; Ruuska, T.; Wilén, K.; Ulvila, M., 2019c: "The Anthropocene exit: Reconciling discursive tensions on the new geological epoch", in: *Ecological Economics,* 164: 106369.

Heikkurinen, P.; Rinkinen, J.; Järvensivu, T.; Wilén, K.; Ruuska, T., 2016: "Organising in the Anthropocene: An ontological outline for ecocentric theorising", in: *Journal of Cleaner Production,* 113: 705–714.

Heikkurinen, P.; Ruuska, T.; Valtonen, A.; et al., 2020: "Time and Mobility after the Anthropocene", in: *Sustainability,* 12,12: 5159.

Hickmann, T.; Partzsch, L.: Pattberg, P.; Weiland, S., 2018a: "Introduction: A political science perspective on the anthropocene", in: Hickmann, T.; Partsch, L.; Pattberg, P.; Weiland, S. (Eds.): *The Anthropocene Debate and Political Science* (London - New York: Routledge): 1–12.

Hickmann, T.; Partzsch, L.; Pattberg, P.; Weiland, S., 2018b: "Conclusion: Towards a 'deep debate' on the anthropocene", in: Hickmann, T.; Partsch, L.; Pattberg, P.; Weiland, S. (Eds.): *The Anthropocene Debate and Political Science* (London - New York: Routledge): 237–251.

Hickmann, T.; Partsch, L.; Pattberg, P.; Weiland, S. (Eds.), 2019a: *The Anthropocene Debate and Political Science* (London – New York: Routledge).

Hickmann, T.; Partzsch, L.; Pattberg, P.; et al., 2019b: "Conclusion Towards a 'deep debate' on the Anthropocene", in: *Anthropocene Debate And Political Science:* 237–251.

Higgins, D., 2017: *British Romanticism, Climate Change, and the Anthropocene: Writing Tambora* (London: Palgrave Macmillan).

Hil, Greg; Lawrence, Susan; Smith, Diana. 2020: "Remade ground: Modelling historical elevation change across Melbourne's Hoddle Grid", in: *Australian Archaeology* (early access).

Himson, S.J.; Kinsey, N.P.; Aldridge, D.C.; *Williams,* M.; *Zalasiewicz,* J., 2020: "Invasive mollusc faunas of the River Thames exemplify biostratigraphical characterization of the Anthropocene", in: *Lethaia,* 53,2: 267–279.

Hoffman, A.J.; Ehrenfeld, J.R., 2015: "The fourth wave Management science and practice in the age of the Anthropocene", in: *Corporate Stewardship: Achieving Sustainable Effectiveness:* 228–246.

Hoffman, A.J.; Jennings, P.D., 2015: "Institutional Theory and the Natural Environment: Research in (and on) the Anthropocene", in: *Organization and Environment,* 28,1: 8–31.

Hoffman, A.J.; Ehrenfeld, J.R., 2018: *Re-Engaging with Sustainability in the Anthropocene Era: An Institutional Approach,* Book Series: Elements in Organization Theory (Cambridge – New York: Cambridge University Press).

Hoffman, Andrew J.; Jennings, P. Devereaux, 2018a: "Free Accepted Article From Repository - Re-Engaging with Sustainability in the Anthropocene Era: An Institutional Approach", in: *Re-Engaging With Sustainability In The Anthropocene Era: An Institutional Approach*, Book Series: Elements in Organization Theory (Cambridge: Cambridge University Press): 1–93.

Hoffman, A.J., Jennings, P.D., 2018b, 2020, 2021: "Institutional-Political Scenarios for Anthropocene Society", in: *Business and Society*, 60,1: 57–94.

Hofman, C.A.; Rick, T.C., 2018: "Ancient Biological Invasions and Island Ecosystems: Tracking Translocations of Wild Plants and Animals", in: *Journal of Archaeological Research*, 26,1: 65–115.

Hofman, C.A.; Rick, T.C.; Fleischer, R.C.; Maldonado, J.E., 2015: "Conservation archaeogenomics: Ancient DNA and biodiversity in the Anthropocene", in: *Trends in Ecology and Evolution*, 30,9: 540–549.

Hofman, C.A.; Rick, T.C.; Maldonado, J.E.; (…); Vellanoweth, R.L.; Newsome, S.D., 2016: "Tracking the origins and diet of an endemic island canid (Urocyon littoralis) across 7300 years of human cultural and environmental change", in: *Quaternary Science Reviews*, 146: 147–160.

Holley, C., Mutongwizo, T., Shearing, C., 2020: "Conceptualizing Policing and Security: New Harmscapes, the Anthropocene, and Technology", in: *Annual Review of Criminology*, 3: 341–358.

Holley, C., Shearing, C., Harrington, C., Kennedy, A., Mutongwizo, T., 2018: Environmental security and the anthropocene: Law, criminology, and international relations", in: *Annual Review of Law and Social Science*, 14: 185–203.

Hoły-Łuczaj, M., Blok, V., 2019; "How to deal with hybrids in the anthropocene? Towards a philosophy of technology and environmental philosophy 2.0", in: *Environmental Values*, 28,3: 325–346.

Horn, E.; Bergthaller, H., 2019: *The Anthropocene: Key Issues for the Humanities* (London: Taylor and Francis).

Hornborg, A., 2015: "The Political Ecology of the Technocene", in: Hamilton, C.; Gemenne, F. (Eds.): *The Anthropocene and Global Environmental Crisis* (New York: Routledge).

Hornborg, A., 2016: "Post-Capitalist Ecologies: Energy, 'Value' and Fetishism in the Anthropocene", in: *Capitalism, Nature, Socialism*, 27,4: 61–76.

Hornborg, A., 2017a: "Artifacts have consequences, not agency: Toward a critical theory of global environmental history", in: *European Journal of Social Theory*, 20,1: 95–110.

Hornborg, A., 2017b: "Dithering while the planet burns: Anthropologists' approaches to the Anthropocene", in: *Reviews in Anthropology*, 46, 2–3: 61–77.

Houston, D., Hillier, J., MacCallum, D., Steele, W., Byrne, J., 2018: "Make kin, not cities! Multispecies entanglements and 'becoming-world' in planning theory", in: *Planning Theory*, 17,2: 190–212.

Howard-Williams, R., 2017: *A World Of Our Own: Climate Change Advocacy in the Anthropocene* (PhD dissertation, University of Pennsylvania).

Hudspeth, Thomas R., 2017: "Reimagining sustainability education to address Anthropocene challenges: envisioning, storytelling, community scenario planning", in: *Envisioning Futures for Environmental and Sustainability Education*: 407–418.

Huettmann, F. 2020: "The forgotten data: A rather short but deep story of museums and libraries in hkh and similar information sources in support of the global biodiversity information system (gbif.org) and model-predictions for improved conservation management", in: *Hindu Kush-Himalaya Watersheds Downhill: Landscape Ecology and Conservation Perspectives*: 497–520.

Huettmann, F.; Regmi, G.R., 2020: "Mountain landscapes and watersheds of the hindu kush-himalaya (HKH) and their biogeography: A descriptive overview and introduction for 18 nations in the anthropocene", in: *Hindu Kush-Himalaya Watersheds Downhill: Landscape Ecology and Conservation Perspectives*: 3–23.

Huijbens, E.H.; Gren, M., 2016: "Tourism and the anthropocene: An urgent emerging encounter", in: *Tourism and the anthropocene* (.London: Routledge): 1–13.

Ibáñez, C.; Alcaraz, C.; Caiola, N.; (...), Reyes; E., *Syvitski*, J.P.M., 2019: "Basin-scale land use impacts on world deltas: Human vs natural forcings", in: *Global and Planetary Change*, 173: 24–32.

Ingram, A., 2020: *Geopolitics and the Event: Rethinking Britain's Iraq War Through Art* (Oxford: John Wiley & Sons Inc.)

Ingram, M.; Ingram, H.; Lejano, R., 2015: "Environmental Action in the Anthropocene: The Power of Narrative Networks", in: *Journal of Environmental Policy & Planning*, 21,5: 492–503; DOI: https://doi.org/10.1080/1523908X.2015.1113513.

Irabien, M.J., *Cearreta*, A., Serrano, H., Villasante-Marcos, V., 2018: "Environmental regeneration processes in the Anthropocene: The Bilbao estuary case (northern Spain)", in: *Marine Pollution Bulletin*, 135: 977–987.

Irabien, M.J.; *Cearreta*, A.; Gómez-Arozamena, J.; García-Artola, A., 2020: "Holocene vs Anthropocene sedimentary records in a human-altered estuary: The Pasaia case (northern Spain)", in: *Marine Geology*: 429,106292.

Irwin, Ruth, 2020a: "Heidegger and Stiegler on failure and technology", in: *Educational Philosophy and Theory*, 52,2: 361–375.

Irwin, Ruth, 2020b: "Economics, ecology, and a new eco-social settlement informing education", in: *Educational Philosophy and Theory*, 52,8: 834–834.

Ison, R., 2016: "Governing in the Anthropocene: What Future Systems Thinking in Practice?", in: *Systems Research and Behavioral Science*, 33, 5: 595–613.

Ison, R., 2017: *Systems Practice: How to Act: In situations of uncertainty and complexity in a climate-change world* (London: Springer-Verlag).

Ison, R.; Shelley, Mo., 2016: "Governing in the Anthropocene: Contributions from Systems Thinking in Practice?", in: *Systems Research and Behavioral Science*, 33,5: 589–594.

Issberner, L.-R.; Léna, P., 2016: *Brazil in the Anthropocene: Conflicts Between Predatory Development and Environmental Policies* (Abingdon: Routledge).

Jagodzinski, J., 2018: *Interrogating the Anthropocene: Ecology, Aesthetics, Pedagogy, and the Future in Question* (Basingstoke: Palgrave).

Jaria-Manzano, J.; Borràs, S., 2019: *Research Handbook on Global Climate Constitutionalism* (Cheltenham, Glos.: Edward Elgar).

Jeanson, A.L.; Soroye, P.; Kadykalo, A.N.; (...); Bennett, J.R.; Cooke, S.J., 2020: "Twenty actions for a "good anthropocene"—perspectives from early-career conservation professionals", in: *Environmental Reviews*, 28,1: 99–108.

Jenkins, W., 2018: "The Mysterious Silence Of Mother Earth In Laudato Si'", in: *Journal of Religious Ethics*, 46,3: 441–462.

Jenkins, W., 2020: "Sacred Places and Planetary Stresses: Sanctuaries as Laboratories of Religious and Ecological Change", *Religions*, 11,5: Article 215.

Jennings, H., 2019a: "Encounters with the Wilderness: Unsettling Perspective in Margaret Atwood's The Journals of Susanna Moodie", in: *Tulsa Studies In Womens Literature*, 38,1: 131–152.

Jennings, H., 2019b: "Anthropocene Feminism, Companion Species, and the Madd Addam Trilogy", in: *Contemporary Womens Writing*, 13,1: 16–33.

Jennings, H., 2019c: "Anthropocene Storytelling: Extinction, D/Evolution, and Posthuman Ethics in Lidia Yuknavitch's The Book of Joan", in: *Lit-Literature Interpretation Theory*, 30,3: 191–210.

Jennings, P.D.; Hoffman, A.J., 2019: "Three Paradoxes of Climate Truth for the Anthropocene Social Scientist", in: *Organization and Environment*, Article in Press.

Jiménez-Aceituno, A.; Peterson, G.D.; Norström, A.V.; Wong, G.Y.; Downing, A.S., 2020: "Local lens for SDG implementation: lessons from bottom-up approaches in Africa", in: *Sustainability Science*, 15,3: 729–743.

Johnson, E.; Morehouse, H.; Dalby, S.; (...), Wakefield, S.; Yusoff, K., 2014: "After the Anthropocene: Politics and geographic inquiry for a new epoch", in: *Progress in Human Geography*, 38,3: 439–456.

Jonsson, F. A., 2012: "The Industrial Revolution in the Anthropocene", in: *Journal of Modern History*, 84,3: 679–696.

Jonsson, F. A., 2013: "A History of the Species?", in: *History and Theory*, 52,3: 462–472.

Jonsson, F. A., 2019: "Growth in the Anthropocene", in: *Scarcity in the Modern World: History, Politics, Society And Sustainability*, 1800-2075: 83–94.

Jorgensen, D., 2019: *Recovering Lost Species in the Modern Age: Histories of Longing and Belonging* (Cambridge, Mass.: MIT Press).

Kaczmarczyk, A., 2018: *Creation Hypothesis in the Anthropocene Epoch* (Basingstoke: Palgrave).

Kalantzakos, S., 2017a: *EU, US and China Tackling Climate Change: An Alliance for the Anthropocene* (London: Routledge).

Kalantzakos, S., 2017b: "US-China Rivalry trumps partnership in the Anthropocene", in: *EU, US and China Tackling Climate Change: An Alliance for the Anthropocene*, Book Series: Routledge Studies in Environmental Policy: 94–121;

Kalantzakos, S., 2017c: "What makes EU-China collaboration a better fit for the Anthropocene" in: *EU, US and China Tackling Climate Change: An Alliance for the Anthropocene*, Book Series: Routledge Studies in Environmental Policy: 122–144.

Kalantzakos, S., 2017d: "EU, US and China Tackling Climate Change: An Alliance for the Anthropocene", in: *EU, US and China Tackling Climate Change: An Alliance for the Anthropocene* (London: Routledge).

Kalantzakos, S., 2017e: "EU and China: An alliance for the Anthropocene Introduction", in: *EU, US and China Tackling Climate Change: An Alliance for the Anthropocene* (London: Routledge): 1–15.

Karlsson, R., 2016: "Three metaphors for sustainability in the anthropocene", in: *Anthropocene Review*, 3,1: 23–32.

Karlsson, R., 2017: "The Environmental Risks of Incomplete Globalisation", in: *Globalizations*, 14,4: 550–562.

Karlsson, R., 2018; "The high-energy planet", in: *Global Change, Peace and Security*, 30,1: 77–84.

Kashima, Y., 2016: "Culture and Psychology in the 21st Century: Conceptions of Culture and Person for Psychology Revisited", in: *Journal of Cross-Cultural Psychology*, 47,1: 4–20.

Kearnes, Matthew; Rickards, Lauren, 2017: "Earthly graves for environmental futures: Techno-burial practices", in: *Futures*, 92: 48–58.

Keim, B., 2017: *The Eye of the Sandpiper: stories from the living world* (Ithaca, NY: Cornell University Press).

Keller, L., 2016: "Twenty-First-Century Ecopoetry and the Scalar Challenges of the Anthropocene", in: *News From Poems: Essays on the 21st-Century American Poetry of Engagement*: 47–63.

Keller, L., 2017a: *Recomposing Ecopoetics: North American Poetry of the Self-Conscious Anthropocene* (Charlottesville, Va.: University of Virginia Press).

Keller, L., 2017b: "'In Deep Time into Deepsong' Writing the Scalar Challenges of the Anthropocene", in: *Recomposing Ecopoetics: North American Poetry of the Self-Conscious Anthropocene* , Under the Sign of Nature (Charlottesville, Va.: University of Virginia Press): 31–60.

Keller, L., 2017c: "Environmental Justice Poetry of the Self-Conscious Anthropocene", in: *Recomposing Ecopoetics: North American Poetry of the Self-Conscious Anthropocene*, Under the Sign of Nature (Charlottesville, Va.: University of Virginia Press): 208–238.

Keller, L., 2017d: "Coda Writing the Self-Conscious Anthropocene", in: *Recomposing Ecopoetics: North American Poetry of the Self-Conscious Anthropocene*, Under the Sign of Nature (Charlottesville, Va.: University of Virginia Press): 239–244.

Keller, R., 2008: *Wissenssoziologische Diskursanalyse. Grundlegung eines Forschungsprogramms* (Wiesbaden: VS Springer).

Keller, R., 2011: *Diskursforschung: Eine Einführung für SozialwissenschaftlerInnen* (Wiesbaden: VS Springer): 13–60.

Keller, R., 2012: *Das Interpretative Paradigma. Eine Einführung* (Wiesbaden: VS Springer).

Keller, R.; Hornidge, A.-K.; Schünemann, W.J. (Eds.), 2018: *The sociology of knowledge approach to discourse. Investigating the politics of knowledge and meaning-making* (London: Routledge).

Kelly, D., 2019: *Politics and the Anthropocene* (Cambridge: Polity Press).

Kelly, J.M.; Scarpino, P.V.; *Syvitski*, H.; Berry, J.; Meybeck, M., 2017: *Rivers of the Anthropocene* (Oakland: University of California Press).

Kettner, A.J.; Gomez, B.; *Syvitski*, J.P.M., 2008: "Will human catalysts or climate change have a greater impact on the sediment load of the W Waipaoa River in the 21st century?", in: *IAHS-AISH Publication*, 325: 425–431.

Keys, P.W.; Galaz, V.; Dyer, M.; (...); Nyström, M.; Cornell, S.E., 2019: "Anthropocene risk", in: *Nature Sustainability*, 2,8: 667–673.

Kim, R.E. 2015: "Global Environmental Governance, Technology and Politics: The Anthropocene Gap", in: *Transnational Environmental Law*, 4,1: 202–205.

Kim, R.E., 2016: "The Nexus between International Law and the Sustainable Development Goals", in: *Review of European Comparative & International Environmental Law*: 25,1: 15–26.

Kim, R.E.; Bosselmann, K., 2013: "International Environmental Law in the Anthropocene: Towards a Purposive System of Multilateral Environmental Agreements", in: *Transnational Environmental Law*, 2,2: 285–309.

Kim, R.E.; Bosselmann, K., 2015: "Operationalizing Sustainable Development: Ecological Integrity as a Grundnorm of International Law", in: *Review of European Comparative & International Environmental Law*, 24,2: 194–208.

Kim, Rakhyun E.; Kotze, Louis J., 2020: "Planetary boundaries at the intersection of Earth system law, science and governance: A state-of-the-art review", in: *Review of European Comparative & International Environmental Law* (Early access, Nov. 2020).

Kim, R.E.; Mackey, B., 2014: "International environmental law as a complex adaptive system", in: *International Environmental Agreements-Politics Law And Economics*,14,1: 5–24.

Kim, R.E.; van Asselt, H., 2016: "Global governance: Problem shifting in the Anthropocene and the limits of international law", in: *Research Handbook On International Law And Natural Resources*, Book Series: Research Handbooks in International Law (Cheltenham, Glos.: Edward Elgar): 473–495.

Klocker, N., Mbenna, P., Gibson, C., 2018: "From troublesome materials to fluid technologies: Making and playing with plastic-bag footballs", in: *Cultural Geographies*, 25,2: 301–318.

Korber, L.A.; MacKenzie, S.; Stenport, A.W., 2017: *Arctic environmental modernities: From the age of polar exploration to the era of the anthropocene* (London: Palgrave Macmillan).

Koselleck, Reinhard, 2006a: *Begriffsgeschichten. Studien zur Semantik und Pragmatik der politischen und sozialen Sprache* (Frankfurt am Main: Suhrkamp).

Koselleck, Reinhart, 1979: *Vergangene Zukunft: Zur Semantik geschichtlicher Zeiten* (Frankfurt/M.: Suhrkamp).

Koselleck, Reinhart, 1989: "Linguistic Change and the History of Events", in: *The Journal of Modern History*, 61: 649 –666.

Koselleck, Reinhart, 1994: "Some Reflections on the Temporal Structure of Conceptual Change", in: Melchung, Willem (Ed.): *Main Trends and Cultural History* (Amsterdam: Wyger Velen): 7–16.

Koselleck, Reinhart, 1996: "A Response to Comments on the Geschichtliche Grundbegriffe", in: Lehmann, Hartmut; Richter, Melvin (Ed.): *The Meaning of Historical Terms and Concepts*. Occational Paper 15 (Washington DC: German Historical Institute): 59–70.

Koselleck, Reinhart, 2000: *Zeitschichten: Studien zur Historik* (Frankfurt/M.: Suhrkamp).

Koselleck, Reinhart, 2002: *The Practice of Conceptual History: Timing History, Spacing Concepts* (Stanford: Stanford University Press); translation of: *Zeitschichten: Studien zur Historik* (Frankfurt am Main: Suhrkamp).

Koselleck, Reinhart, 2006b: *Begriffsgeschichten* (Frankfurt/M: Suhrkamp).

Kotzé, L.J. (Ed.), 2017b: *Environmental Law and Governance for the Anthropocene* (Portland, OR: Hart Publ.).

Kotzé, L.J., 2014: "Human rights and the environment in the anthropocene", in: *Anthropocene Review*, 1,3: 252–275.

Kotzé, L.J., 2015: "Human rights, the environment, and the global south", in: Kotzé, L.J.:(Ed.): *International Environmental Law and the Global South* (Cambridge – New York: Cambridge University Press): 171–191.

Kotzé, L.J., 2016a: "Constitutional Conversations in the Anthropocene: In Search of Environmental Jus Cogens Norms", in: *Netherlands Yearbook of International Law 2015: Jus Cogens: Quo Vadis?* (The Hague: T.M.C. Asser Press): 241–271.

Kotzé, L.J., 2016b: "Global Environmental Constitutionalism in the Anthropocene", in: *Global Environmental Constitutionalism In The Anthropocene* (Cumnor Hill, Oxford: Hart): 1–282.

Kotzé, L.J., 2016c: "Global Environmental Constitutionalism in the Anthropocene Introduction", in: *Global Environmental Constitutionalism in the Anthropocene* (Cumnor Hill, Oxford: Hart): 1–20.

Kotzé, L.J., 2016d: "Global Environmental Constitutionalism in the Anthropocene, Preface", in: *Global Environmental Constitutionalism In The Anthropocene* (Cumnor Hill, Oxford: Hart): II–+.

Kotzé, L.J., 2016e: "Law and the Anthropocene's Global Socio-Ecological Crisis", in: *Global Environmental Constitutionalism In The Anthropocene* (Cumnor Hill, Oxford: Hart): 21–42.

Kotzé, L.J., 2016f: "The Prospects of Environmental Constitutionalism in the Anthropocene", in: *Global Environmental Constitutionalism In The Anthropocene* (Cumnor Hill, Oxford: Hart): 176–200.

Kotzé, L.J., 2016g: "A Vision of Global Environmental Constitutionalism in the Anthropocene", in: *Global Environmental Constitutionalism In The Anthropocene* (Cumnor Hill, Oxford: Hart): 201–245.

Kotzé, L.J., 2016h: "Global Environmental Constitutionalism in the Anthropocene Conclusion", in: *Global Environmental Constitutionalism In The Anthropocene* (Cumnor Hill, Oxford: Hart): 246–247.

Kotzé, L.J., 2017a: "Preface: Discomforting Conversations In The Anthropocene", in: *Environmental Law and Governance for the Anthropocene* (Cumnor Hill, Oxford: Hart): VII–XV.

Kotzé, L.J., 2017b: "Global Environmental Constitutionalism in the Anthropocene", in: *Environmental Law and Governance for the Anthropocene* (Cumnor Hill, Oxford: Hart): 189–218.

Kotzé, L.J., 2018a: "Six Constitutional Elements for Implementing Environmental Constitutionalism in the Anthropocene", in: *Implementing Environmental Constitutionalism: Current Global Challenges*: 13–33.

Kotzé, L.J., 2018b: "The Sustainable Development Goals: an existential critique alongside three new-millennial analytical paradigms", in: *Sustainable Development Goals: Law, Theory and Implementation*: 41–65.

Kotzé, L.J., 2019a: "A global environmental constitution for the Anthropocene's climate crisis", in: *Research Handbook on Global Climate Constitutionalism* (): 50–74.

Kotzé, L.J., 2019b: "International Environmental Law and the Anthropocene's Energy Dilemma", in: *Environmental And Planning Law Journal*, 36,5: 437–458.

Kotzé, L.J., 2019c: "International environmental law's lack of normative ambition: An opportunity for the global pact for the environment?", in: *Journal for European Environmental and Planning Law*, 16,3: 213–236.

Kotzé, L.J., 2019d: "The anthropocene, earth system vulnerability and socio-ecological injustice in an age of human rights", in: *Journal of Human Rights and the Environment*, 10,1: 85–62.

Kotzé, L.J., 2019e: "Earth system law for the anthropocene", in: *Sustainability* (Switzerland), 11,23: 6796.

Kotzé, L.J., 2019f: "The Constitution of the Environmental Emergency", in: *Journal of Environmental Law*, 31,2: 367–382.

Kotzé, L.J., 2020: "Earth system law for the Anthropocene: rethinking environmental law alongside the Earth system metaphor", in: *Transnational Legal Theory*, 11,1–2: 75–104.

Kotzé, L.J.; Calzadilla, P.V., 2017: "Somewhere between Rhetoric and Reality: Environmental Constitutionalism and the Rights of Nature in Ecuador", in: *Transnational Environmental Law*, 6,3: 401–433.

Kotzé, L.J.; French, D, 2018: "A critique of the Global Pact for the environment: a stillborn initiative or the foundation for Lex Anthropocenae?", in: *International Environmental Agreements: Politics, Law and Economics*, 18,6: 811–838.

Kotzé, L.J.; Muzangaza, W., 2018: "Constitutional international environmental law for the Anthropocene?", in: *Review of European Comparative & International Environmental Law*, 27,3: 278–292.

Kraft, P.; Taylor, A.; Pacini-Ketchabaw, V., 2020: "Introduction to Symposium: childhood studies in the Anthropocene", in: *Discourse*, 41,3: 333–339.

Kumar, M.; Snow, D.D.; Honda, R. (Eds.), 2020: *Emerging Issues in the Water Environment during Anthropocene: A South East Asian Perspective* (Singapore: Springer Singapore).

Landwehr, Achim, 2009: *Historische Diskursanalyse* (Frankfurt/New York: Campus).

Lane, David C., 2016: "Till the Muddle in my Mind Have Cleared Awa': Can We Help Shape Policy Using Systems Modelling?", 59th Annual Conference of the International-Society-for-the-Systems-Sciences (ISSS), Berlin, Germany, 2-7 August 2015, in: *Systems Research and Behavioral Science,* 33,5: 633–650.

Lane, David C., 2019: "New truths begin as heresies: First thoughts on system dynamics and global modelling", 8th European System Dynamics Workshop (EuSDW) on Modelling Sustainability Pathways - Bridging Science, Policy and Society Location, Nova Univ Lisbon, Lisbon, Portugal, 31 May – 2 June 2017, in: *Systems Research and Behavioral Science*, 36,2: 233–243.

Lawrence, S.; Davies, P.; Turnbull, J., 2016: "The archaeology of Anthropocene rivers: water management and landscape change in 'Gold Rush' Australia ", in: *Antiquity*, 90, 353: 1348–1362.

Lawrence, S.; Davies, P.; Turnbull, J., 2017: "The Archaeology of Water on the Victorian Goldfields", in: *International Journal Of Historical Archaeology* , 21,1: 49–65.

Leach, M.; Reyers, B.; Bai, X.; (...); Stafford-Smith, M.; Subramanian, S.M., 2018: "Equity and sustainability in the anthropocene: A social-ecological systems perspective on their intertwined futures", in: *Global Sustainability*, 1: e13.

Leane, E.; McGee, J. (Eds.), 2019, 2020: *Anthropocene Antarctica: Perspectives from the Humanities, Law and Social Sciences* (London: Routledge).

Lecavalier, E.; Harrington, C., 2017 : "Entangling carbon lock-in: India's coal constituency", in: *Crime Law and Social Change*, 68,5: 529–546.

Leemans, R.; Rice, M.; Henderson-Sellers, A.; Noone, K., 2011: Research Agenda and Policy Input of the Earth Systems Science Partnership for Coping with Global Environmental Change" , in: Brauch, H.G.; Oswald Spring, Ú; Mesjasz, C., et al. (Eds.): *Coping with Global Environmental Change, Disasters and Security – Threats, Challenges, Vulnerabilities and Risks* (Berlin – Heidelberg – New York: Springer-Verlag): 1205–1220.

Levin, P.S.; Poe, M.R. (Eds.), 2017: *Conservation for the Anthropocene Ocean: Interdisciplinary Science in Support of Nature and People* (London: Academic Press Ltd. – Elsevier Science Ltd.).

Levit, G.S., 2001: *Biogeochemistry-Biosphere-Noosphere: The Growth of the Theoretical System of Vladimir Ivanovich Vernadsky* (Berlin: Verlag für Wissenschaft und Bildung).

Lewis, Simon L.; Maslin, Mark A., 2015: "Defining the Anthropocene", in: *Nature*, 519, 7542: 171–180.

Lewis, Simon L.; Maslin, Mark A., 2018: *The Human Planet: How we Created the Anthropocene* (New Haven: Yale University Press).

Lichtfouse, E.; Hamelin, M.; Navarrete, M.; et al., 2010: "Emerging agroscience", in: *Agronomy for Sustainable Development*, 30,1: 1–10.

Lichtfouse, E.; Hamelin, M.; Navarrete, M.; et al., 2011: "Emerging Agroscience", in: *Sustainable Agriculture*, 2: 3–14.

Logan, M., Khatun, F., 2019: "Socioecological learners as agentic: A posthumanist perspective", in: *Touchstones for Deterritorializing Socioecological Learning: The Anthropocene, Posthumanism and Common Worlds as Creative Milieux* (London: Palgrave Macmillan): 231–262.

Lövbrand, E.; Mobjörk, M.; Söder, R., 2020: "The Anthropocene and the geo-political imagination: Re-writing Earth as political space", in: *Earth System Governance*, 4 (June): 100051.

Lövbrand, E.; Beck, S.; Chilvers, J.; et al., 2015: "Who speaks for the future of Earth? How critical social science can extend the conversation on the Anthropocene", in: *Global Environmental Change-Human and Policy Dimensions*, 32: 211–218.

Lovelock, James E.; Appleyard, Bryan, 2019: *Novacene. The Coming Age of Hyperintelligence* (London: Penguin, Allen Lane).

Lubbe, W. D.; Kotzé, L. J., 2019: "Holistic Biodiversity Conservation In The Anthropocene: A Southern African Perspective", in: *African Journal of International and Comparative Law*, 27,1: 76–99.

Luke, R.W, 2018: "Reflections from a Damaged Planet: Adorno as Accompaniment to Environmentalism in the Anthropocene", in: *TELOS*, 183: 9–24.

Luke, R.W., 2015: "On the Politics of the Anthropocene", in: *TELOS*, 172: 139–162.

Luke, T.W., 2015: "Introduction: Political Critiques of the Anthropocene", in: *TELOS*, 172: 3–14.

Lynch, A.H.; Veland, S., 2018: *Urgency in the Anthropocene* (Cambridge, Mass.: MIT Press).

MacGregor, S., 2017: *Routledge Handbook of Gender and Environment* (London: Taylor and Francis).

Machin, Amanda, 2019a: "Agony and the Anthropos Democracy and Boundaries in the Anthropocene", in: *Nature + Culture*, 14,1: 1–6.

Machin, Amanda, 2019b: "Democracy and Agonism in the Anthropocene: The Challenges of Knowledge, Time and Boundary", in: *Environmental Values*, 28, 3: 347–365.

Macilenti, A., 2018: *Characterising the Anthropocene: Ecological Degradation in Italian Twenty-First Century Literary Writing* (Berlin, Bern, Bruxelles, New York, Oxford, Warszawa, Wien: Peter Lang AG).

Mackenzie, Fred T., et al., 2006; *Carbon in the Geobiosphere — Earth's Outer Shell* (Dordrecht: Springer Netherlands).

Mahaffy, P.G., 2014: "Telling Time: Chemistry Education in the Anthropocene Epoch", in: *Journal of Chemical Education*, 91,4: 463–465.

Mahaffy, P.G.; Holme, T.A.; Martin-Visscher, L.; et al., 2017: "Beyond 'Inert'Ideas to Teaching General Chemistry from Rich Contexts: Visualizing the Chemistry of Climate Change (VC3)", in: *Journal of Chemical Education*, 94, 8: 1027–1035.

Mahaffy, P.G.; Martin, B.E.; Kirchhoff, M.; et al., 2014: "Infusing Sustainability Science Literacy through Chemistry Education: Climate Science as a Rich Context for Learning Chemistry", in: *ACS Sustainable Chemistry & Engineering*, 2,11: 2488–2494.

Mahaffy, P.G.; Matlin, S.A.; Whalen, J.M.; et al., 2019: "Integrating the Molecular Basis of Sustainability into General Chemistry through Systems Thinking", in: *Journal of Chemical Education*, 96,12: 2730–2741.

Malm, A., Hornborg, A., 2014: "The geology of mankind? A critique of the anthropocene narrative", in: *Anthropocene Review*, 1,1: 62–69.

Malone, K., Logan, M., Siegel, L., Regalado, J., Wade-Leeuwen, B., 2020: "Shimmering with Deborah Rose: Posthuman theory-making with feminist ecophilosophers and social ecologists", in: *Australian Journal of Environmental Education*, 36,2: 129–145.

Malone, K.; Truong, S.; Gray, T., 2017: *Reimagining Sustainability in Precarious Times* (Singapore: Springer Science+Business Media Singapore).

Malone, N.; Wade, A.H.; Fuentes, A.;...; Remis, M.; Robinson, C.J., 2014: "Ethnoprimatology: Critical interdisciplinarity and multispecies approaches in anthropology", in: *Critique of Anthropology*, 34,1: 8–29.

Marcantonio, R.; Fuentes, A., 2020: "A clear past and a murky future: Life in the anthropocene on the Pampana river, Sierra Leone", in: *Land*, 9,3: 7.

Marchessault, J., 2017: *Ecstatic Worlds: Media, Utopias, Ecologies* (Cambridge, MA: MIT Press).

Marck, A.; Antero, J.; Berthelot, G.; et al., 2019: "Age-Related Upper Limits in Physical Performances", in: *Journals of Gerontology Series A - Biological Sciences and Medical Sciences*, 74,5 (May): 591–599.

Markl, H., 1986: *Natur als Kulturaufgabe. Über die Beziehung des Menschen zur lebendigen Natur* (Stuttgart: Deutsche Verlagsanstalt).

Marouby, C., 2018: *The Question of Limits: A Historical Perspective on the Environmental Crisis* (London: Taylor and Francis).

Marsh, G.P., 1864: *Man and Nature [reprinted in 1965 as]: The Earth as Modified by Human Action* (Cambridge, MA: Belknap Press).

Martin, L.J.; Quinn, J.E.; *Ellis*, E.C., (...); Shirey, P.D.; Wiederholt, R., 2014a, 2014b: "Conservation opportunities across the world's anthromes", in: *Diversity and Distributions*, 20,7: 745–755.

Mason, P.; Lang, T., 2017: *Sustainable Diets: How Ecological Nutrition Can Transform Consumption and the Food System* (London: Taylor and Francis).

Matthews, Daniel, 2019a: "From Global to Anthropocenic Assemblages: Re-Thinking Territory, Authority and Rights in the New Climatic Regime", in: *Modern Law Review*, 82,4: 665–691.

Matthews, Daniel, 2019b: "Law and Aesthetics in the Anthropocene: From the Rights of Nature to the Aesthesis of Obligations", in: *Law Culture and the Humanities*: Article UNSP 1743872119871830.

McBrien, J., 2016: "Accumulating Extinction: Planetary Catastrophism in the Necrocene", in: Moore, J. M. (Ed.): *Anthropocene or Capitalocene? Nature, History, and the Crisis of Capitalism* (Oakland, CA: PM Press): 78–115.

McCarthy, J.A.; Hilger, S.M.; Sullivan, H.I.; Saul, N., 2016: *The Early History of Embodied Cognition 1740-1920: The Lebenskraft-Debate and Radical Reality in German Science, Music, and Literature* (Leiden: Brill).

McEwan, C., 2018: *Postcolonialism, Decoloniality and Development, 2nd Edition* (London: Taylor and Francis).

McGregor, A., Houston, D., 2018: "Cattle in the Anthropocene: Four propositions", in: *Transactions of the Institute of British Geographers*, 43,1: 3–16.

McLean, J., 2019: *Changing Digital Geographies: Technologies, Environments and People*, (London: Palgrave Macmillan).

McNeill, J.R., 2015: "Energy, population, and environmental change since 1750: Entering the anthropocene", in: *The Cambridge World History*: 51–82.

McNeill, J.R., 2016 : "Introductory remarks: The anthropocene and the eighteenth century", in: *Eighteenth-Century Studies*, 49,2: 117–128.

McNeill, J.R., 2019: "The anthropocene and environmental history in the USA", in: *Historia Ambiental Latinoamericana y Caribena*, 9,1: 200–210.

MEA, 2005: *Ecosystems and Human Well-being: Synthesis Report* (Washington DC: Island Press).

Megoran, N., Dalby, S., 2018: "Geopolitics and Peace: A Century of Change in the Discipline of Geography", in: *Geopolitics*, 23,2: 251–276.

Melo Zurita, M.D.L., George Munro, P., Houston, D., 2018: "Un-earthing the Subterranean Anthropocene", in: *Area*, 50,3: 298–305.

Mendelsohn, E., 1977: "The Social Construction of Scientific Knowledge", in: Mendelsohn, E.; Weingart, P.; Whitley, R. (Eds.): *The Social Production of Scientific Knowledge. Sociology of the Sciences A Yearbook*, vol 1. (Dordrecht: Springer).

Merrill, E.M.; Giamarelos, S., 2019: "From the Pantheon to the Anthropocene: Introducing Resilience in Architectural History", in: *Architectural Histories*, 7,1: Article 7.

Meyer, W.B., 2016: *The progressive environmental prometheans: Left-wing heralds of a "good anthropocene"* (London: Palgrave Macmillan).

Moore, A., 2019: *Destination Anthropocene: Science and Tourism in the Bahamas* (Oakland: University of California Press).

Moore, J.M. (Ed.), 2016a: *Anthropocene or Capitalocene? Nature, History, and the Crisis of Capitalism* (Oakland, CA: PM Press).

Moore, J.W., 2016b: "Introduction: Anthropocene or Capitalocene? Nature, History, and the Crisis of Capitalism", in: Moore, Jason M. (Ed.): *Anthropocene or Capitalocene? Nature, History, and the Crisis of Capitalism* (Oakland, CA: PM Press): 1–13.

Moore, J.W., 2017: "The Capitalocene, Part I: on the nature and origins of our ecological crisis", in: *Journal of Peasant Studies*, 44,3: 594–630.

Moore, J.W., 2018: "The Capitalocene Part II: accumulation by appropriation and the centrality of unpaid work/energy", in: *Journal of Peasant Studies*, 45,2: 237–279.

Moser, K., 2018a: "Reviving the nuanced concept of mother earth in an era of non-sustainability: A serresian reading of Marcel Pagnol's L'EAU des collines", in: *Enthymema*, 21: 32–47.

Moser, K., 2018b: "The sensorial, biocentric philosophy of Michel Cerres and Michel Onfray: Rehabilitating the human body", in *Pacific Coast Philology*, 53,1: 95–110.

Moser, K., 2018c: "Decentring and rewriting the universal story of humanity: the cosmic historiography of J.M.G. Le Clézio and Michel Serres", in: *Green Letters*, 22,2: 129–147.

Moser, K., 2018d: "A biosemiotic reading of Michel Onfray's Cosmos: Rethinking the essence of communication from an ecocentric and scientific perspective", in: *Semiotica*, 225: 405–421.

Moser, K., 2019a: "Michel Onfray's decentered, ecocentric, atheistic philosophy: A user's guide for the anthropocene epoch?", in: *Worldviews: Environment, Culture, Religion*, 23,2: 113–131.

Moser, K., 2019b: "Jean-Marie Pelt and J.M.G. Le Clezio's Invitation to think and live otherwise in the anthropocene", in: *French Review*, 93,1: 63–78.

Moser, K., 2020: "A derridean interpretation of the biosemiosic dance of life in Yamen Manai's L'Amas Ardent", in: *Green Letters* (in press).

Mullan, B.; Haqq-Misra, J., 2019: "Population growth, energy use, and the implications for the search for extraterrestrial intelligence", in: *Futures*, 106: 4–17.

Müller, E.; Schmieder, F., 2016: *Begriffsgeschichte und historische Semantik* (Berlin: Suhrkamp).

Munns, D.P.D., 2017: *Engineering the Environment: Phytotrons and the Quest for Climate Control in the Cold War* (Pittsburgh, PA: University of Pittsburgh Press).

Nelson, N.; Pacini-Ketchabaw, V.; Nxumalo, F., 2018: "Rethinking Nature-Based Approaches in Early Childhood Education: Common Worlding Practices", in: *Journal of Childhood Studies*, 43,1: 4–14.

Neyrat, F.; Neyrat, D.S.B., 2018: *The Unconstructable Earth: An Ecology of Separation* (New York: Fordham University Press).

Nicholls, R.J.; Adger ,W.N.; Hutton C.W., et al. (Eds.), 2019: *Deltas in the Anthropocene* (Cham: Palgrave Macmillan).

Nicholson, S.; Jinnah, S.. (Eds.), 2016: *Earth System Governance: World Politics in the Anthropocene* (Cambridge, Mass.: MIT Press):

Noon, K.; Nobre, C.; Seitzinger, S., 2011: "The International Geosphere and Biosphere Programme's (IGBP) Scientific Research Agenda for Coping with Global Environmental Change", in: Brauch, H.G.; Oswald Spring, Ú; Mesjasz, C., et al. (Eds.): *Coping with Global Environmental Change, Disasters and Security – Threats, Challenges, Vulnerabilities and Risks* (Berlin – Heidelberg – New York: Springer-Verlag): 1249–1256.

Norgaard, Richard B., 2013: "The Econocene and the delta", in: *San Francisco Estuary and Watershed Science,* 11,3: 1–5.

Norman, B.; *Steffen*, W., 2018 : "Planning within planetary boundaries", in: *Sustainable Pathways For Our Cities And Regions: Planning Within Planetary Boundaries* (London: Routledge): 20–44.

Norström, A.V.; Dannenberg, A.; McCarney, G.; (...), Schultz, L.; Sjöstedt, M., 2014: "Three necessary conditions for establishing effective Sustainable Development Goals in the Anthropocene", in: *Ecology and Society*, 19,3: 8.

Notter, Harley, 1949: *Postwar Foreign Policy Preparation, 1939–1945* (Washington, D.C.: U.S. Government Printing Office).

Novak, J.; Cañas, A.J., 2008: "The Theory Underlying Concept Maps and How to Construct and Use Them", Technical Report IHMC CmapTools 2006-01, Rev 01-2008, Florida Institute for Human and Machine Cognition.

Noyer, J.-M., 2016: *Transformation of Collective Intelligences: Perspective of Transhumanism 2* Hoboken, N.J.: Wiley).

Nxumalo, F., 2018: "Stories for living on a damaged planet: Environmental education in a preschool classroom", in: *Journal of Early Childhood Research*, 16,2: 148–159.

Nxumalo, F.; Pacini-Ketchabaw, V., 2017: "'Staying with the trouble' in child-insect-educator common worlds", in: *Environmental Education Research*, 23,10: 1414–1426.

Nxumalo, F.; Ross, K.M., 2019: "Envisioning Black space in environmental education for young children", in: *Race Ethnicity And Education*, 22,4: 502–524.

Nyman, J., 2018: *The Energy Security Paradox: Rethinking Energy (In)security in the United States and China* (Oxford: Oxford University Press).

Oldfield, F., 2005: *Environmental change: Key issues and alternative perspectives* (Cambridge: Cambridge University Press).

Oldfield, F.; *Barnosky*, A.; Dearing, j., et al., 2014 "The Anthropocene Review: Its significance, implications and the rationale for a new transdisciplinary journal" in: *The Anthropocene Reviews*, 1,1 (April): 3–7.

Oldfield, F.; *Steffen*, W., 2014: "Anthropogenic climate change and the nature of Earth System science", in: The *Anthropocene Reviews*, 1,1 (April): 70–75.

Österblom, H.; Crona, B.I.; Folke, C.; Nyström, M.; Troell, M., 2017: "Marine Ecosystem Science on an Intertwined Planet", in: *Ecosystems*, 20,1: 54–61.

Oswald Spring, U., 2016: "Sustainability Transition in a Vulnerable River Basin in Mexico", in: Brauch, H.G.; Oswald Spring, Ú.; Grin, J.; Scheffran; J. (Eds.): *Handbook on Sustainability Transition And Sustainable Peace* (Cham: Springer International Publishing): 675–704.

Oswald Spring, U., 2017: "Development with Sustainable-Engendered Peace: A Challenge during the Anthropocene", Brauch, H.G.; Oswald Spring, Ú.; Grin, J.; Scheffran; J. (Eds.): *Handbook on Sustainability Transition And Sustainable Peace* (Cham: Springer International Publishing): 161–185.

Oswald Spring, Ú.; Brauch, H.G.; Scheffran, J., 2017: "Sustainability Transition with Sustainable Peace: Key Messages and Scientific Outlook", in: Brauch, H.G.; Oswald Spring, Ú.; Grin, J.; Scheffran; J. (Eds.): *Handbook on Sustainability Transition And Sustainable Peace* (Cham: Springer International Publishing): 887–927.

Oswald Spring, Ú.; Brauch, H.G.; Dalby, S., 2009: "Linking Anthropocene, HUGE and HESP: Fourth Phase of Environmental Security Research", in: Brauch, H.G.; Oswald Spring, Ú.; Grin, J.; Mesjasz, C.; Kameri-Mbote, P.; Behera, N.C.; Chourou, B.; Krummenacher, H. (Eds.), 2009: *Facing Global Environmental Change: Enivornmental, Human, Energy, Food, Health and Water Security Concepts* (Berlin—Heidelberg—New York: Springer-Verlag): (Heidelberg-Berlin: Springer).

Otlet, Paul, 1934: *Traité de Documentation: Le Livre sur le Livre, Théorie et Pratique* (Mons: Editiones Mundaneum).

Otter, C.; Bashford, A.; Brooke, J. L.; et al., 2018: "Roundtable: The Anthropocene in British History", in: *Journal of British Studies*, 57,3: 568–596.

Overeem, I.; Kettner, A.J.; *Syvitski*, J.P.M., 2013: "Impacts of Humans on River Fluxes and Morphology", in: *Treatise on Geomorphology* , 9: 828–842.

Palidda, S., 2016: *Governance of Security and Ignored Insecurities in Contemporary Europe* (London: Routledge).

Palsson, G.; Szerszynski, B.; Sörlin, S.; (...); Buendía, M.P.; Weehuizen, R., 2013: "Reconceptualizing the 'Anthropos' in the Anthropocene: Integrating the social sciences and humanities in global environmental change research", in: *Environmental Science and Policy*, 28: 3–13.

Parenti, C., 2016: "Environment-Making in the Capitalocene. Political Ecology of the State", in: Moore, J.M. (Ed.): *Anthropocene or Capitalocene? Nature, History, and the Crisis of Capitalism* (Oakland, CA: PM Press): 166–184.

Pasquero, C.; Poletto, M., 2020: "Bio-digital aesthetics as value system of post-Anthropocene architecture", in: *International Journal of Architectural Computing*, 18,2: 120–140, Article UNSP 1478077120922941.

Patel, R., 2013: "Misanthropocene?", in: *Earth Island Journal*, 28,1.

Pattberg, P., Weiland, S., 2020: "More Engagement of Political Science in the Anthropocene Debate" [Mehr Engagement der Politikwissenschaft in der Anthropozän-Debatte], in: *Politische Vierteljahresschrift*, 61,4: 659–670.

Pattberg, P., Widerberg, O., 2015: "Theorising global environmental governance: Key findings and future questions", in: *Millennium: Journal of International Studies*, 43,2: 684–705.

Pattberg, P., Zelli, F., 2016: *Environmental Politics and Governance in the Anthropocene: Institutions and legitimacy in a complex world* (Abingdon: Routledge): 1–254.

Pattberg, P.; Zelli, F., 2016b: "Global environmental governance in the Anthropocene: An introduction", in: *Environmental Politics and Governance in the Anthropocene: Institutions and legitimacy in a complex world* (Abingdon: Routledge): 1–12.

Pauly, D., 2010: *5 Easy Pieces: The Impact of Fisheries on Marine Systems* (Washington, D.C.: Island Press).

Pearce, Fred, 2007: *With speed and violence: why scientists fear tipping points in climate change* (Boston: Beacon Press).

Pearce, J., 2017: *Fundamentals for the Anthropocene* (Berlin: Walter De Gruyter).

Pennycook, A., 2017: *Posthumanist Applied Linguistics* (London: Taylor and Francis.

Pereira, J.; Preiser, R.; Biggs, R., 2018a: "Using futures methods to create transformative spaces: Visions of a good anthropocene in Southern Africa", in: *Ecology and Society*, 23,1: 19.

Pereira, J.C., 2017: "The limitations of IR theory regarding the environment: lessons from the anthropocene", in: *Revista Brasileira de Politica Internacional*, 60,1: e018.

Pereira, J.C., 2019: "Reducing Catastrophic Climate Risk by Revolutionizing the Amazon: Novel Pathways for Brazilian Diplomacy", in: Sequeira, T.; Reis, L. (Eds.): *Climate Change and Global Development: Market, Global Players and Empirical Evidence* (Cham: Springer International Publishing): 189–218.

Pereira, J.C.; Viola, E., 2018: "Catastrophic Climate Change and Forest Tipping Points: Blind Spots in International Politics and Policy", in: *Global Policy*, 9,4: 513–524.

Pereira, L.M.; Hichert, T.; Hamann, M.; Preiser, R.; Biggs, R., 2018b: "Using futures methods to create transformative spaces: Visions of a good anthropocene in Southern Africa", in: *Ecology and Society*, 23,1: 19.

Philippopoulos-Mihalopoulos, A.; Brooks, V., 2017: *Research Methods in Environmental Law: A Handbook* (Cheltenham, Glos.: Edward Elgar).

Polasky, S., Crépin, A.-S., Biggs, R.O., (...), Walker, B., Xepapadeas, A., 2020: "Corridors of clarity: Four principles to overcome uncertainty paralysis in the anthropocene", in: *BioScience*, 70,12: 1139–1144.

Pouw, N.; Gupta, J., 2017: "Inclusive development: a multi-disciplinary approach", in: *Current Opinion in Environmental Sustainability*, 24: 104–108.

Pritchard, Alan, 1969: "Statistical Bibliography or Bibliometrics?", in: *Journal of Documentation*, 25,4: 348–349.

Propen, A.D., 2018: *Visualizing Posthuman Conservation in the Age of the Anthropocene* (Columbus, OH: Ohio State University Press).

Pyhala, A.; Fernandez-Llamazares, A.; Lehvavirta, H.; et al., 2016: "Global environmental change: local perceptions, understandings, and explanations", *Ecology and Society*, 21,3: 25.

Raffnsoe, S., 2016a: "The Opening of a New Chapter in the World's History", in: *Philosophy of the Anthropocene: The Human Turn* (London: Palgrave Macmillan): 2–7.

Raffnsoe, S., 2016b: "A Prominent Role in a Landscape Lush with Mutual Mediation", in: *Philosophy of the Anthropocene: The Human Turn* (London: Palgrave Macmillan): 8–19.

Raffnsoe, S., 2016c: "The (Post)human Condition", in: *Philosophy of the Anthropocene: The Human Turn* (London: Palgrave Macmillan): 20–27.

Raffnsoe, S., 2016d: "The Turn within and of the Human", in: *Philosophy of the Anthropocene: The Human Turn*(London: Palgrave Macmillan): 29–34.

Raffnsoe, S., 2016e: "Philosophy of the Anthropocene: The Human Turn", in: *Philosophy of the Anthropocene: The Human Turn* (London: Palgrave Macmillan): CP1–75.

Ramsay, Christine, 2020: "Adrift in History, Who Is This One?: Art in the Critical Zone", in: *Ekphrasis-Images Cinema Theory Media*, 24,2: 167–192.

Randin, C.F.; Ashcroft, M.B.; Bolliger, J.; (...); Yoccoz, N.; Payne, D., 2020: "Monitoring biodiversity in the Anthropocene using remote sensing in species distribution models", in: *Remote Sensing of Environment*, 239: 111626.

Raudsepp-Hearne, C.; Peterson, G.D.; Bennett, E.M.; (...); Falardeau, M.; Aceituno, A.J., 2020: "Seeds of good anthropocenes: developing sustainability scenarios for Northern Europe", in: *Sustainability Science*, 15,2: 605–661.

Raworth, Kate, 2014: "Must the Anthropocene be a Manthropocene?", in: *The Guardian* (20 October).

Renn, J., 2020a: *Evolution of Knowledge: Rethinking Science for the Anthropocene* (Princeton - Oxford: Princeton University Press).

Renn, J., 2020b: "History of Science in the Anthropocene", in: *Evolution of Knowledge: Rethinking Science for the Anthropocene* (Princeton - Oxford: Princeton University Press): 3–22.

Renn, Jürgen, 2020c: "The Evolution Of Knowledge *Rethinking Science For The Anthropocene* The Story of This Book", in: *Evolution of Knowledge: Rethinking Science for the Anthropocene* (Princeton - Oxford: Princeton University Press): IX.

Renn, Jürgen, 2020d: "Knowledge for the Anthropocene", in: *Evolution of Knowledge: Rethinking Science for the Anthropocene* (Princeton - Oxford: Princeton University Press): 377–407.

Revkin, A.C., 1992: *Global Warming. Understanding the Forecast* (New York: Abbeville Press).

Reyers, B.; Folke, C.; Moore, M.-L.; Biggs, R.; Galaz, V., 2018: "Social-ecological systems insights for navigating the dynamics of the anthropocene", in: *Annual Review of Environment and Resources*, 43: 267–289.

Rick, T.; Ontiveros, M.Á.C.; Jerardino, A.; (...), Méndez, C.; Williams, A.N., 2020: "Human-environmental interactions in Mediterranean climate regions from the Pleistocene to the Anthropocene", in: *Anthropocene*, 31,100253.

Rick, T.C.; Braje, T.J.; Erlandson, J.M.; (...); Kirn, L.; McLaren-Dewey, L., 2018: "Horizon Scanning: Survey and Research Priorities for Cultural, Historical, and Paleobiological Resources of Santa Cruz Island, California", in: *Western North American Naturalist*, 78,4: 852–863.

Rick, T.C.; Kirch, P.V.; Erlandson, J.M.; Fitzpatrick, S.M., 2013: "Archeology, deep history, and the human transformation of island ecosystems", in: *Anthropocene*, 4: 33–45.

Rick, T.C.; Reeder-Myers, L.; Braje, T.J.; et al., 2020: "Human ecology, paleogeography, and biodiversity on California's small Islands", in: *Journal of Island & Coastal Archaeology*.

Rick, T.C.; Sandweiss, D.H., 2020: "Archaeology, climate, and global change in the Age of Humans", in: *Proceedings of the National Academy of Sciences of the United States of America*, 117,15: 8250–8253.

Rick, T.C.; Sillett, T.S.; Ghalambor, C.K.; (...); Still, C.; Morrison, S.A., 2014: "Ecological change on California's channel Islands from the pleistocene to the anthropocene", in: *BioScience*, 64,8: 680–692.

Rockström, J.; Falkenmark, M.; Allan, T.; (...), Turton, A., Varis, O., 2014: "The unfolding water drama in the Anthropocene: Towards a resilience-based perspective on water for global sustainability", in: *Ecohydrology*, 7,5: 1249–1261.

Rockström, J.; Williams, J.; Daily, G.; (...); Sibanda, L.; Smith, J.: 2017: "Sustainable intensification of agriculture for human prosperity and global sustainability", in: *Ambio*, 46,1: 4–17.

Rodhe, H.; Charlson, R.; Crawford, E., 1997: "Svante Arrhenius and the Greenhouse Effect", in: *Ambio*, 26,1 (February): 2–5; at: http://www.jstor.org/stable/4314542.

Rogers, K.G.; *Syvitski*, J.P.M.; Overeem, I.; Higgins, S.; Gilligan, J.M., 2013: "Farming practices and anthropogenic delta dynamics", in: *IAHS-AISH Proceedings and Reports*, 358: 133–142.

Rousseau, Ronald, 2014: "Library Science: Forgotten Founder of Bibliometrics", in: *Nature*, 510,7504: 218.

Rubin, F., 2020: A Discourse Analysis of Anthropocene in IHOPE Publications - Is There a Place for Archaeology? (Uppsala: Uppsala University, Uppsala University Publications, B.A. thesis).

Ruddiman, W.F., 2003: "The anthropogenic greenhouse era began thousands of years ago", in: *Climatic Change*, 61,3: 261–293; at: doi:https://doi.org/10.1023/B:CLIM.0000004577.17928.fa.

Ruddiman, W.F., 2005: *Plows, Plagues, and Petroleum: How Humans Took Control of Climate* (Princeton, N.J: Princeton University Press);

Ruddiman, W. F., 2016: "Late Holocene climate: natural and anthropogenic", in: *Review of Geophysics*, 54: 93–118.

Ruuska, T., Heikkurinen, P., Wilén, K., 2020: "Domination, power, supremacy: Confronting anthropolitics with ecological realism", in: *Sustainability* (Switzerland), 12,7: 2617.

Sachs, Jeffrey, 2007: *Survival in the Anthropocene* (London: BBC).

Sachs, Jeffrey, 2008: *Common wealth: economics for a crowded planet* (London: Penguin),

Sáez-Muñoz, M.; Ortiz, J.; Martorell, S.; Gómez-Arozamena, J.; *Cearreta*, A., 2020: "Sequential determination of uranium and plutonium in soil and sediment samples by borate salts fusion", in: *Journal of Radioanalytical and Nuclear Chemistry,* 323,3: 1167–1177.

Salottolo, D., 2019a: "The Extended Experience. Reflections on the Anthropocene", in: *S&F-Scienzaefilosofia It*, 21: 50–75.

Salottolo, D., 2019b : "Anthropocene or Capitalocene? Ecology- world scenarios in the era of planetary crisis, the Extended Experience. Reflections on the Anthropocene", in: *S&F-Scienzaefilosofia It*, 21: 323–328.

Salottolo, D., 2019c: "Considerations on the Concept of World and of Relationship to the World in the Age of Anthropocene: Essay on XXI Century Philosophy, in: *S&F-Scienzaefilosofia It*, 22: 200–250.

Salottolo, D., 2020: "How Long Can a Culture Persist Without The New? Tina and The Research Of A 'Cosmology' Worthy of the Anthropocene", in: *S&F-Scienzaefilosofia It*, 23: 350–385.

Sames, B.; *Wagreich*, M.; Conrad, C.P.; Iqbal, S., 2020: "Aquifer-eustasy as the main driver of short-term sea-level fluctuations during cretaceous hothouse climate phases", in: *Geological Society Special Publication*, 498,1: 9–38.

Sammonds, P R, et al., 2007: *Advances in earth science: from earthquakes to global warming* (London: Imperial College Press).

Sample, V.A; Bixler, R.P.; Miller, C. (Eds.), 2016: *Forest Conservation in the Anthropocene: Science, Policy, and Practice* (Boulder, CO.: Univ Press of Colorado).

Samways, M., 1999: "Translocating fauna to foreign lands: Here comes the Homogenocene", in: *Journal of Insect Conservation*, 3: 65–66.

Schaumann, C., Sullivan, H.I., 2011: "Introduction: Dirty nature: Grit, grime, and genre in the anthropocene", in: *Colloquia Germanica*, 44,2: 105–109.

Schaumann, C.; Sullivan, H.I., 2017a: *German Ecocriticism in the Anthropocene* (London: Palgrave Macmillan).

Schaumann, C.; Sullivan, H.I., 2017b: "German Ecocriticism in the Anthropocene Introduction", in: *German Ecocriticism in the Anthropocene* (London: Palgrave Macmillan): 7–21.

Scheffran, J., 2019: "The entwined Cold War roots of missile defense and climate geoengineering", in: *Bulletin of the Atomic Scientists*, 75,5: 222–228; DOI: https://doi.org/10.1080/00963402.2019.1654256.

Scheffran, J., 2020: "Weather, War, and Chaos: Richardson's Encounter with Molecules and Nations", in: Gleditsch, N.-P. (Ed.): *Lewis Fry Richardson: His Intellectual Legacy and Influence in the Social Sciences* (Cham: Springer International Publishing): 87–100.

Scheffran, J.; Brzoska, M.; Brauch, H.G.; Link, P.M.; Schilling, J. (Eds.), 2012: *Climate Change, Human Security and Violent Conflict: Challenges for Societal Stability* (Berlin—Heidelberg—New York: Springer-Verlag).

Schellnhuber, H.J., 2014a: "Evolution of the Atmosphere, Fire and the Anthropocene Climate Event Horizon, Foreword", in: Glikson, A.Y.: *Evolution of the Atmosphere, Fire and the Anthropocene Climate Event Horizon* (Dordrecht: Springer Netherlands): VII–VIII.

Schellnhuber, H.J., 2014b: "Evolution of the Atmosphere, Fire and the Anthropocene Climate Event Horizon, Preface", in: Glikson, A.Y.: *Evolution of the Atmosphere, Fire and the Anthropocene Climate Event Horizon* (Dordrecht: Springer Netherlands): IX–X.

Schellnhuber, H.J., 2014c: "A Flammable Biosphere", in: Glikson, A.Y.: *Evolution of the Atmosphere, Fire and the Anthropocene Climate Event Horizon* (Dordrecht: Springer Netherlands): 71–74.

Schellnhuber, H.J., 2014d: "A Fire Species", in: Glikson, A.Y.: *Evolution of the Atmosphere, Fire and the Anthropocene Climate Event Horizon* (Dordrecht: Springer Netherlands): 75–90.

Schellnhuber, H.J.; Cramer, W.; Nakícénovic, N.; Wigley, T.; Yohe, G. (Eds.), 2006: *Avoiding Dangerous Climate Change* (Cambridge: Cambridge University Press).

Schellnhuber, H.J.; Crutzen, P.J.; Clark, W.C.; Claussen, M.; Held, H. (Eds.), 2004: *Earth Systems Analysis for Sustainability* (Cambridge, Mass.: MIT Press).

Schellnhuber, H.J.; Wenzel, V. (Eds.). 1998: *Earth System Analysis – Integrating Science for Sustainability* (Berlin – Heidelberg: Springer-Verlag).

Schill, C.; Anderies, J.M.; Lindahl, T.; (...); Norberg, J.; Schlüter, M., 2019: "A more dynamic understanding of human behaviour for the Anthropocene", in: *Nature Sustainability*, 2,12: 1075–1082.

Schmidt, J.J., 2013: "Integrating Water Management in the Anthropocene", in: *Society and Natural Resources*, 26,1: 105–112.

Schmidt, J.J., 2017: "Social learning in the Anthropocene: Novel challenges, shadow networks, and ethical practices", in: *Journal of Environmental Management*, 193: 373–380.

Schmidt, J.J., 2020: "Settler geology: Earth's deep history and the governance of in situ oil spills in Alberta", in: *Political Geography*, 78: Article Number: 102132.

Schmidt, J.J.; Brown, P.G.; Orr, C.J., 2016: "Ethics in the Anthropocene: A research agenda", in: *Anthropocene Review*, 3,3: 188–200.

Schmidt, J.J.; Matthews, N., 2017a: "Environments", in: *Global Challenges in Water Governance: Environments, Economies, Societies*, Book Series: Global Challenges in Water Governance (London: Palgrave): 21–51.

Schmidt, J.J.; Matthews, N., 2017b: "Water Futures", in: *Global Challenges in Water Governance: Environments, Economies, Societies*, Book Series: Global Challenges in Water Governance.

Schwägerl, C., 2015: "A Concept with a Past", in: Möllers, N.; Schwägerl, C.; Trischler, H., (Eds.), 2014: *Welcome to the Anthropocene: The Earth in Our Hands* (München: Deutsches Museum): 128–129.

Schwartz, S.W., 2017: "Temperature and capital: Measuring the future with quantified heat", in: *Environment and Society: Advances in Research*, 8,1: 180–197.

Schwartz, S.W., 2019a: "Poetic metrics: Unfolding dissent in intensive space", in: *ACME*, 18,2: 264–284.

Schwartz, S.W., 2019b: "Measuring Vulnerability and Deferring Responsibility: Quantifying the Anthropocene", in: *Theory, Culture and Society*, 36,4: 73–93.

Scobie, M., 2019: *Global Environmental Governance and Small States: Architectures and Agency in the Caribbean* (Cheltenham, Glos.: Edward Elgar).

Seddon, A.W.R.; Mackay, A.W.; Baker, A.G.; (...); Williams, J.W.; Witkowski, A., 2014: "Looking forward through the past: Identification of 50 priority research questions in palaeoecology", in: *Journal of* Ecology, 102,1: 256–267.

Seitzinger, S.P.; Gaffney, O.; Brasseur, G.; (...); *Syvitski*, J.; Uematsu, M., 2015: "International Geosphere-Biosphere Programme and Earth system science: Three decades of co-evolution", in: *Anthropocene*, 12: 3–16.

Shaw, Wendy S.; Bonnett, Alastair, 2016: "Environmental crisis, narcissism and the work of grief", in: *Cultural Geographies*, 23,4: 565–579.

Sideris, L.H., 2013 : "Science as sacred myth? Ecospirituality in the anthropocene age", in: *Linking Ecology and Ethics for a Changing World: Values, Philosophy, and Action*: 147–162.

Sideris, L.H., 2015: "Science as sacred myth? Ecospirituality in the anthropocene age", in: *Journal for the Study of Religion, Nature and Culture*, 9,2; 136–153.

Sideris, L.H., 2016, 2017: "Biosphere, Noosphere, and the Anthropocene: Earth's Perilous Prospects in a Cosmic Context", *Journal for the Study of Religion Nature and Culture*, 11,4: 399–419.

Sideris, L.H., 2019: "The Last Biped Standing? Climate Change and Evolutionary Exceptionalism at the Smithsonian Hall of Human Origins", in: *Journal for the Study of Religion Nature and Culture*, 13,4: 455–478.

Sideris, L.H., 2020: "Grave Reminders: Grief and Vulnerability in the Anthropocene dagger", in: *Religions*, 11,6,293: 1–16.

Sideris, L.H., Whalen-Bridge, J., 2019: "Special Issue Introduction: Popular Culture, Religion, and the Anthropocene", in: *Journal for the Study of Religion, Nature and Culture*, 13,4: 409–413.

Simangan, D., 2019: "Situating the Asia Pacific in the age of the Anthropocene", in: *Australian Journal of International Affairs*, 73,6: 564–584.

Simangan, D., 2020a: "Literature review Where is the Anthropocene? IR in a new geological epoch", in: *International Affairs*, 96,1: 211–227.

Simangan, D., 2020b: "Where is the Asia Pacific in mainstream international relations scholarship on the Anthropocene?", in: *Pacific Review* (in press).

Simangan, Dahlia, 2020c: "Where is the Anthropocene? IR in a new geological epoch", in: *International Affairs*, 96,1 (January): 211–224.

Simmons, E.L., 2014: "Theology in the Anthropocene", in: *Dialog - A Journal of Theology*, 53,4: 271–273.

Simmons, E.L., 2018: "Vocation for planetary citizenship: Lutheran liberal arts education in the Anthropocene", in: *Dialog - A Journal of Theology*, 57,2: 99–106.

Simon, Z. B., 2015: "History manifested: making sense of unprecedented change", in: *European Review Of History-Revue Europeenne D Histoire*, 22,5: 819–834.

Simon, Z.B., 2017: "Why the Anthropocene has no history: Facing the unprecedented", in: *Anthropocene Review*, 4,3: 239–245.

Simon, Z. B., 2018a, 2018b: "(The impossibility of) acting upon a story that we can believe", in: *Rethinking History*, 22,1: 105–125.

Simon, Z. B., 2019: "Two Cultures of the Posthuman Future", in: *History and Theory*, 58,2: 171–184.

Simon, Z.B., 2020a: "The limits of Anthropocene narratives", in: *European Journal of Social Theory*, 23,2: 184–199.

Simon, Z.B., 2020b: "Utopia without us?", in: *Esbocos*, 27,46: 377–389.

Simpson, N.P.; Shearing, C.D.; Dupont, B., 2019a: "Climate gating: A case study of emerging responses to Anthropocene Risks", in: *Climate Risk Management*, 26,100196.

Simpson, N.P.; Simpson, K.J.; Shearing, C.D.; et al., 2019b: "Municipal finance and resilience lessons for urban infrastructure management: a case study from the Cape Town drought", in: *International Journal of Urban Sustainable Development*, 11,3: 257–276.

Sinaï, Agnès, 2005: *Penser la décroissance: politiques de l'anthropocène* (Paris: Presses de la Fondation nationale des sciences politiques).

Singer, B.; Giessen, L., 2017: "Towards a donut regime? Domestic actors, climatization, and the hollowing-out of the international forests regime in the Anthropocene ", in: *Forest Policy and Economics*, 79: 69–79.

Slaughter, R.A., 2012: "Welcome to the anthropocene", in: *Futures*, 44,2: 119–126.

Slaughter, R.A., 2018: "The IT revolution reassessed part three: Framing solutions", in: *Futures*, 100: 1–19.

Slaughter, R.A., 2020: "Farewell Alternative Futures?", in: *Futures*, 121: Art. No.: 102496.

Slaymaker, O., et al., 2006: *The cryosphere and global environmental change* (Blackwell).

Slesinger, Ryan, 2020: "John Steinbeck's To a God Unknown The Clearing Cycle and the Monterey Metaphysics of Ricketts, Steinbeck, and Campbell", in: *Steinbeck Review*, 17,2: 183–201.

Smith, J.K.A., 2008: *After modernity? Secularity, globalization, and the re-enchantment of the world* (Waco, TX: Baylor Press).

Smith, S., 2019: "Discursively constructed realities: exploring and extending the sociology of knowledge approach to discourse (SKAD)", *Lectures* [En ligne], *Les notes critiques*, online, 23 Mai 2019; at: http://journals.openedition.org/lectures/34700 (1 Nov. 2020).

Solli, B.; Burstrom, M.; Domanska, E.; et al., 2011: "Some Reflections on Heritage and Archaeology in the Anthropocene", in: *Norwegian Archaeological Review*, 44, 1: 40–88; Article: PII 938514980.

Solnick, S., 2017: *Poetry and the Anthropocene: Ecology, Biology and Technology in Contemporary British and Irish Poetry* (London: Routledge).

Southgate, E.W.B., 2019: *People and the Land through Time: Linking Ecology and History, Second Edition* (New Haven: Yale University Press).

Stager, C., 2011: "Deep Future. The Next 100,000 Years of Life on Earth", in: Thomas Dunne
    Books.
Steffen, W., 2004: "Group Report: Earth Systems Dynamics in the Anthropocene", in: Schellnhuber,
    H.J.; Crutzen; Clark, W.C.; et al. (Eds.): *Earth System Analysis for Sustainability* (Cambridge,
    MA - Berlin: MIT Press – Freie Universität Berlin): 313–340.
Steffen, W., 2006: "The anthropocene, global change and sleeping giants: Where on Earth are we
    going?", in: *Carbon Balance and Management*, 1,1: 3.
Steffen, W., 2010: "Observed trends in Earth System behavior", in: *Wiley Interdisciplinary Reviews:
    Climate Change*, 1,3: 428–449.
Steffen, W., 2013: "Commentary: Paul J. Crutzen and Eugene F. Stoermer", in: Robbin, L.; Sörlin,
    S.; Warde, P. (Eds.): *The Future of Nature. Documents of Global Change* (New Haven: Yale
    University Press): 486–490.
Steffen, W.; Andreae, M.O.; Bolin, B.; Cox, P.M.; *Crutzen*, P. J.; Cubasch, U.; Held, H.; Nakicenovic,
    N.; Talaue-McManus, L.; Turner II, B.L., 2004a: "Group Report: Earth Systems Dynamics in the
    Anthropocene", in: Schellnhuber, H.J.; Crutzen, P.J.; Clark, W.C.; Claussen, M.; Held, H. (Eds.):
    Earth System Analysis for Sustainability (Cambridge, MA; London: MIT Press): 313–340.
Steffen, W.; Andreae, M.O.; Cox, P.M.; *Crutzen*, P.J.; Cubasch, U.; Held, H.; Nakicenovic, N.;
    Talaue-McManus, L.; Turner II, B.L., 2004b: "Group Report: Earth System Dynamics in the
    Anthropocene", in: Schellnhuber, H.J; *Crutzen*, P.J.; Clark, W.C.; Claussen, M.; Held, H. (Eds.):
    Earth System Analysis for Sustainablility. Dahlem Workshop Report (Cambridge, CA: MIT
    Press): 313–340.
Steffen, W.; Sanderson, R.A.; Tyson, P.D.; Jäger, J.; Matson, P.A.; Moore III, B.; Oldfield, F.;
    Richardson, K.; Schellnhuber, H.J.; Turner, B.L.; Wasson, R.J., 2004c: *Global Change and the
    Earth System: A planet under pressure* (Berlin - Heidelberg: Springer-Verlag).
Steffen, W.; Broadgate, W.; Deutsch, L.; Gaffney, O.; Ludwig, C., 2015: "The trajectory of the
    anthropocene: The great acceleration", in: *Anthropocene Review*, 2,1: 81–98.
Steffen, W.; *Crutzen*, P.J.; *McNeill*, J.R., 2007: "The Anthropocene: Are Humans Now Over-
    whelming the Great Forces of Nature", in: *Ambio*, 36,8 (December): 614–621.
Steffen, W.; *Grinevald*, J.; *Crutzen*, P.; *McNeill*, J., 2011a: "The anthropocene: Conceptual and
    historical perspectives", in: *Philosophical Transactions of the Royal Society A: Mathematical,
    Physical and Engineering Science*s, 369,1938; 842–867.
Steffen, W., Jäger, J., Carson, D.J., Bradshaw, C. (Eds.), 2002: *Challenges of a Changing Earth
    - Proceedings of the Global Change Open Science Conference, Amsterdam, The Netherlands,
    10–13 July 2001* (Berlin Heidelberg: Springer–Verlag).
Steffen, W.; *Leinfelder*, R.; *Zalasiewicz*, J.; (...); Wolfe, A.P.; Schellnhuber, H.J., 2016: "Stratigraphic
    and Earth System approaches to defining the Anthropocene", in: *Earth's Future*, 4,8: 324–345.
Steffen, W.; Persson, Å.; Deutsch, L.; (...); Schellnhuber, H.J.; Svedin, U., 2011: "The anthropocene:
    From global change to planetary stewardship", in: *Ambio*, 40,7: 739–761.
Steffen, W.; Richardson, K.; Rockström, J. et al. 2020: "The emergence and evolution of Earth
    System Science", in: *Nat Rev Earth Environ*, **1**: 54–63; https://doi.org/10.1038/s43017-019-
    0005-6
Steffen, W.; Rockström, J.; Richardson, K.; Lenton, T. M.; Folke, C.; Liverman, D.; Summerhayes,
    C.P.; *Barnosky*, A.D.; Cornell, S.E.; Crucifix, M.; Donges, J.F.; Fetzer, I.; Lade, S.J.; Scheffer,
    M.; Winkelmann, R.; Schellnhuber, H.J., 2018: "Trajectories of the Earth System in the Anthro-
    pocene", in: *Proceedings of the National Academy of Sciences of the United States of America*,
    115,33: 8252–8259, DOI: https://doi.org/10.1073/pnas.1810141115.
Steffen, W.; Sanderson, A.; Tyson, P.D.; Jäger, J.; Matson, P.A.; Moore III., B.; Oldfield, F.;
    Richardson, K.; Schellnhuber, H.J.; Turner II, B.L.; Wasson, R.J., 2006: *Global Change and
    the Earth System. A Planet Under Pressure* (Berlin-Heidelberg-New York: Springer-Verlag).
Stephens, T., 2016: "Disasters, international environmental law and the Anthropocene", in: Breau,
    Susan, C.; Samuel, Katja L.H. (Eds.): *Research Handbook On Disasters And International Law*,
    Book Series: Research Handbooks in International Law (Cheltenham, Glos.: Edward Elgar):
    153–176.

Stephens, T., 2017: "Reimagining International Environmental Law in the Anthropocene", in: *Environmental Law and Governance for the Anthropocene*: 31–54.

Stephens, T., 2018: "Wishful thinking? The governance of climate change-related disasters in the Anthropocene", in: *Research Handbook on Climate Disaster Law: Barriers and Opportunities*, Book Series: Research Handbooks in Climate Law (Cheltenham, Glos.: Edward Elgar): 31–45.

Stephenson, M., 2018: *Energy and Climate Change: An Introduction to Geological Controls, Interventions and Mitigations* (Amsterdam: Elsevier).

Sterner, T.; Barbier, E.B.; Bateman, I.; (...); Pleijel, H.; Robinson, A., 2019: "Policy design for the Anthropocene", in: *Nature Sustainability*, 2,1: 14–21.

Stevens, L., 2019: "Anthroposcenic Performance and the Need For 'Deep Dramaturgy'", in: Performance Research, 24,8: 89–97.

Stevens, L.; Tait, P.; Varney, D., 2017a: "*Feminist Ecologies: Changing Environments in the Anthropocene* (London: Palgrave Macmillan).

Stevens, L.; Tait, P.; Varney, D., 2017b: "Introduction: 'Street-fighters and philosophers': Traversing ecofeminisms". In: *Feminist Ecologies: Changing Environments in the Anthropocene* (London: Palgrave Macmillan): 1–22.

Stone, K., 2018: "'Staying with the Trouble': Climates of Change in Biblical Studies", in: *Present and Future of Biblical Studies: Celebrating 25 Years of Brill's Biblical Interpretation* (Leiden: Brill): 161: 297–320.

Stone, K., 2019: "'All These Look to You': Reading Psalm 104 with Animals in the Anthropocene Epoch", in: *Interpretation - A Journal of Bible And Theology*, 73,3: 236–247.

Stoner, A.M., 2020: "Things are Getting Worse on Our Way to Catastrophe: Neoliberal Environmentalism, Repressive Desublimation, and the Autonomous Ecoconsumer", in: *Critical Sociology*, Article Number: 0896920520958099.

Stoner, A.M.; Melathopoulos, A., 2018:, "Stuck in the Anthropocene: The Problem of History, Theory, and Practice in Jason W. Moore and John Bellamy Foster's Eco-Marxism", in: *Interrogating the Anthropocene: Ecology, Aesthetics, Pedagogy, and the Future in Question* (London: Palgrave Macmillan): 105–132.

Stoppani, A., 1873: *Corsa di Geologia (*Milan: G. Bernadom – E.G. Brigola).

Subramanian, M., 2019: "Anthropocene now: influential panel votes to recognize Earth's new epoch", in: *Nature*, 19 May; at: doi: https://doi.org/10.1038/d41586-019-01641-5 (31 May 2019).

Sullivan, H.I., 2016a, 2016b: "Agency in the Anthropocene: Goethe, Radical Reality, and the New Materialisms", in: *Early History Of Embodied Cognition 1740–1920: The Lebenskraft-Debate And Radical Reality In German Science, Music, And Literature*, Internationale Forschungen zur Allgemeinen und Vergleichenden Literaturwissenschaft: 285–304.

Sullivan, H.I., 2016c: "The dark pastoral: Goethe and Atwood", in: *Green Letters*, 20,1: 47–59.

Sullivan, H.I., 2017a: "Goethe's colors: Revolutionary optics and the Anthropocene", in: *Eighteenth-Century Studies*, 51,1: 115–124.

Sullivan, H.I., 2017b: "The Dark Pastoral: A Trope for the Anthropocene", in: *German Ecocriticism in the Anthropocene, Literatures,Cultures and the Environment*: 25–44.

Sullivan, H.I., 2019: "Petro-texts, plants, and people in the Anthropocene: the dark green", in: *Green Letters*, 23,2: 152–167.

Sultana, F.; Loftus, A., 2019: *Water Politics: Governance, Justice and the Right to Water* (London: Taylor and Francis).

Summerhayes, C.P., Zalasiewicz, J., 2018: "Global warming and the Anthropocene", in: *Geology Today*, 34,5: 194–200.

Sundar, S.; Heino, J.; Roque, F. de Oliveira; et al., 2020: "Conservation of freshwater macroinvertebrate biodiversity in tropical regions", in: *Aquatic Conservation-Marine And Freshwater Ecosystems*, 30,6: 1238–1250.

Swanson, H.A., Bubandt, N., Tsing, A., 2015: "Less than one but more than many: Anthropocene as science fiction and scholarship-in-the-making", in: *Environment and Society: Advances in Research*, 6,1: 149–166.

Symons, J.; Karlsson, R., 2015: "Green political theory in a climate-changed world: between innovation and restraint", in: *Environmental Politics*, 24,2: 173–192.

Symons, J.; Karlsson, R., 2018: "Ecomodernist citizenship: rethinking political obligations in a climate-changed world", in: *Citizenship Studies*, 22,7: 685–704.

Syvitski, J.P.M., Kettner, A., 2011: "Sediment flux and the anthropocene", in: *Philosophical Transactions of the Royal Society A: Mathematical, Physical and Engineering Sciences*, 369,1938: 957–975.

Syvitski, J.P.M.; Kettner, A.J.; Overeem, I.; (...); Hannon, M.; Gilham, R., 2013: "Anthropocene metamorphosis of the Indus Delta and lower floodplain", in: *Anthropocene*, 3: 24–35.

Syvitski, J.P.M.; Saito, Y., 2007: "Morphodynamics of deltas under the influence of humans", in: *Global and Planetary Change*, 57,3–4: 261–282.

Szerszynski, B., 2015: "Commission on planetary ages decision CC87966424/49: The onomatophore of the Anthropocene", in: *The Anthropocene and the Global Environmental Crisis: Rethinking Modernity in a New Epoch* (London: Routledge): 177–183.

Szerszynski, B., 2017a: "The Anthropocene monument: On relating geological and human time", in: *European Journal of Social Theory*, 20,1: 111–131.

Szerszynski, B., 2017b: "Viewing the technosphere in an interplanetary light", in: *Anthropocene Review*, 4,2: 92–102.

Tait, P., 2018: "Replacing injured horses, cross-dressing and dust: modernist circus technologies in Asia", in: *Studies in Theatre and Performance*, 38.2: 149–164;

Tamm, M.; Simon, Z. B., 2020: "Historical Thinking and the Human: Introduction", in: *Journal of the Philosophy of History*, 14,3: 285–309.

Tarolli, P.; Cao, W.; Sofia, G.; Evans, D.: *Ellis*, E.C., 2019: "From features to fingerprints: A general diagnostic framework for anthropogenic geomorphology", in: *Progress in Physical Geography*, 43,1: 95–128.

Taylor, A., 2017a: "Beyond stewardship: common world pedagogies for the Anthropocene", in: *Environmental Education Research*, 23,10: 1448–1461.

Taylor, A., 2017b: "Romancing or re-configuring nature in the anthropocene? Towards common worlding pedagogies", in: *Reimagining Sustainability in Precarious Times*: 61–75.

Taylor, A., 2020: "Countering the conceits of the Anthropos: scaling down and researching with minor players", in: *Discourse*, 41,3: 340–358.

Taylor, A.; Pacini-Ketchabaw, V., 2015: "Learning with children, ants, and worms in the Anthropocene: towards a common world pedagogy of multispecies vulnerability", in: *Pedagogy, Culture and Society*, 23,4: 507–529.

Taylor, A.; Pacini-Ketchabaw, V., 2018: *The Common Worlds of Children and Animals: Relational Ethics for Entangled Lives* (London: Taylor and Francis).

Taylor, J.O., 2014: "Auras and Ice Cores: Atmospheric Archives and the Anthropocene", in: *Minnesota Review*, 83: 73–82.

Taylor, J.O., 2018a: "Anthropocene", in: *Victorian Literature and Culture*, 46,3–4: 573–577.

Taylor, J.O., 2018b: "The Novel After Nature, Nature After The Novel: Richard Jefferies's Anthropocene Romance", in: *Studies in the Novel*, 50,1: 108–133.

Teilhard de Chardin, P., 1959: *The Phenomenon of Man* (London: Collins).

Teilhard de Chardin, P., 1965: *The Appearance of Man* (New York: Harper & Row).

Terrington, R.L.; Silva, É.C.N.; *Waters*, C.N.; Smith, H.; Thorpe, S., 2018: "Quantifying anthropogenic modification of the shallow geosphere in central London, UK", in: *Geomorphology*, 319: 15–34.

Tobias, M.C; Morrison, J.G., 2016: *Anthrozoology: Embracing Co-Existence in the Anthropocene* 2016 (Cham: Springer Nature Switzerland).

Tobias, M.C; Morrison, J.G., 2018: *The Theoretical Individual: Imagination, Ethics and the Future of Humanity* (Cham: Springer International Publishing).

Torrescano-Valle, N.; Islebe, G.A.; Roy, P.D., 2019: *The Holocene and anthropocene environmental history of Mexico: A paleoecological approach on Mesoamerica* (Cham: Springer International Publishing).

Trischler, H. (Eds.), 2015: *Welcome to the Anthropocene: The Earth in Our Hands* (Munich: Deutsches Museum): 128–129.

Tritschler, H., 2016: "The Anthropocene: A Challenge for the History of Science, Technology, and the Environment", in: *NTM Zeitschrift für Geschichte der Wissenschaften, Technik und Medizin*, 24: 309–335; at: https://doi.org/10.1007/s00048-016-0146-3.

Tsing, A.L., Mathews, A.S., Bubandt, N., 2019: "Patchy Anthropocene: Landscape Structure, Multispecies History, and the Retooling of Anthropology: An Introduction to Supplement 20", in; *Current Anthropology*, 60, S20: S186–S197.

Tyszczuk, R., 2017: *Provisional Cities: Cautionary Tales for the Anthropocene* (London: Taylor and Francis).

Uhrqvist, O.; Lovbrand, E., 2014: "Rendering global change problematic: the constitutive effects of Earth System research in the IGBP and the IHDP", in: *Environmental Politics*, 23,2: 339–356.

Underdal, A.; Kim, R.E., 2017: "The Sustainable Development Goals and Multilateral Agreements", in: Kanie, N.; Biermann, F. (Eds.): *Governing Through Goals: Sustainable Development Goals As Governance Innovation* (Cambridge, MA: MIT Press): 241–258.

Usher, P.J., 2019: *Exterranean: Extraction in the Humanist Anthropocene* (New York: Fordham University Press).

Vaz, E., 2020: *Regional intelligence: Spatial analysis and anthropogenic regional challenges in the digital age* (Cham: Springer International Publishing).

Verburg, P.H.; Dearing, J.A.; Dyke, J.G.; et al., 2016: "Methods and approaches to modelling the Anthropocene", in: *Global Environmental Change - Human and Policy Dimensions*, 39: 328–340.

Vermeulen, P., 2015a: "Contemporary Literature and the End of the Novel: Creature, Affect, Form", in: *Contemporary Literature and the End of the Novel: Creature, Affect, Form*: 1–182.

Vermeulen, P., 2015b: "Don DeLillo's Point Omega, the Anthropocene, and the Scales of Literature", in: *Studia Neophilologica*, 87, Supplement: 1: 68–81.

Vermeulen, P., 2015c: *Contemporary Literature and the End of the Novel: Creature, Affect, Form* (Basingstoke: Palgrave Macmillan): 1–182.

Vermeulen, P., 2017a: "'The sea, not the ocean': Anthropocene fiction and the memory of (Non)human life", in: *Genre*, 50,2: 181–200.

Vermeulen, P., 2017b: "Future readers: narrating the human in the Anthropocene", in: *Textual Practice*, 31,5: 867–885.

Vermeulen, P., 2017c: "Creaturely Memory: Shakespeare, the Anthropocene and the New Nomos of the Earth", in: *Parallax*, 23,4: 384–397.

Vernadsky, V.I., 1929: *La Biosphere* (Paris: Librairie Felix Alcan, Nouvelle Collection scientifique), second revised and expanded edition 1926.

Vernadsky, V.I., 1945: "The Biosphere and the Noosphere", in: *Am. Sci.,* 33,1: 1–12.

Vernadsky, V.I., 1986: *The Biosphere* [reprinted] (Oracle AZ: Synergetic Press).

Vernadsky, V.I., 1998: *The Biosphere* (New York: Springer Science+Business Media).

Vidas, D., 2010: *Law, technology and science for Oceans in globalisation: IUU fishing, oil pollution, bioprospecting, outer continental shelf* (Leiden: Brill).

Vidas, D., 2011: "The anthropocene and the international law of the sea", in: *Philosophical Transactions of the Royal Society A: Mathematical, Physical and Engineering Sciences*, 369, 1938: 909–925.

Vidas, D., 2014: "Sea-level rise and international law: At the convergence of two epochs", in: *Climate Law*, 4,1–2: 70–84.

Vidas, D., 2017: "International Law at the Convergence of Two Epochs: Sea–Level Rise and the Law of the Sea for the Anthropocene", *Ocean Law And Policy: 20 Years Under UNCLOS*: 101–123.

Vidas, D.; Fauchald, O.K.; Jensen, T.M.W., 2015: "International law for the Anthropocene? Shifting perspectives in regulation of the oceans, environment and genetic resources", in: *Anthropocene*, 9,77: 1–13;

Vidas, D.; Schei, P.J., 2011: "The world ocean in globalisation: Challenges and responses for the anthropocene epoch", in: *The World Ocean in Globalisation: Climate Change, Sustainable Fisheries, Biodiversity, Shipping, Regional Issue*: 3–15.

Viola, E.; Franchini, M., 2017: *Brazil and Climate Change: Beyond the Amazon* (New York: Taylor and Francis).

Vitousek, P.M.; Mooney, H.A.; Lubchenco, J.; Melillo, J.M., 1997: "Human domination of Earth's ecosystems", in: *Science*, 277: 494–499.

Voeroesmarty, C.J.; Meybeck, M.; Pastore, C.L., 2015: "Impair-then-Repair: A Brief History & Global-Scale Hypothesis Regarding Human-Water Interactions in the Anthropocene", in: *Daedalus*, 144,3: 94–109.

Voeroesmarty, C.J.; Pahl-Wostl, C.; Bunn, S.E.; et al., 2013: "Global water, the anthropocene and the transformation of a science", in: *Current Opinion In Environmental Sustainability*, 5,6: 539–550.

Waddock, S., 2020: "Thinking Transformational System Change", in: *Journal of Change Management*, 20,3: 189–201.

Waddock, S.; Kuenkel, P., 2020: "What Gives Life to Large System Change?", in: *Organization & Environment*, 33,3: 342–358.

Wagreich, D., 2018: "Early mining and smelting lead anomalies in geological archives as potential stratigraphic markers for the base of an early Anthropocene", in: *The Anthropocene Review*, 5,2: 177–201.

Waitt, G.; Buchanan, I.; Duffy, M., 2020: "Lively cities made in sound: A study of the sonic sensibilities of listening and hearing in Wollongong, New South Wales", in: *Urban Studies*, 57,10: 2131–2146.

Waitt, G.; Harada, T., 2016: "Parenting, care and the family car [Education des enfants, soins et la voiture familiale]", in: *Social and Cultural* Geography, 17,8: 1079–1100.

Wakefield, S., 2018: "Infrastructures of liberal life: From modernity and progress to resilience and ruins", in: *Geography Compass*, 12,7: e12377.

Wakefield, S., 2020: "Urban resilience as critique: Problematizing infrastructure in post-Sandy New York City", in: *Political Geography*, 79, UNSP 102148.

Walther, B.A.; Larigauderie, A.; Loreau, M., 2011: "Diversitas: Biodiversity Science Integrating Research and Policy for Human Well-Being", in: Brauch, H.G.; Oswald Spring, Ú; Mesjasz, C., et al. (Eds.): *Coping with Global Environmental Change, Disasters and Security – Threats, Challenges, Vulnerabilities and Risks* (Berlin – Heidelberg – New York: Springer-Verlag): 1235–1248.

Walton, T.; Shaw, W.S., 2015: "Living with the Anthropocene blues", in: *Geoforum*, 60: 1–3.

Washington, C., 2019: *Romantic Revelations: Visions of Post-Apocalyptic Life and Hope in the Anthropocene* (Toronto: University of Toronto Press).

Waters, C.N.; Fairchild, I.J.; McCarthy, F.M.G.;...; *Zalasiewicz*, J.; *Williams*, M., 2018a: "How to date natural archives of the Anthropocene", in: *Geology Today*, 34,5: 182–187.

Waters, C.N.; *Syvitski*, J.P.M.; *Gałuszka*, A.;...; *Summerhayes*, C.; *Barnosky*, A., 2015: "Can nuclear weapons fallout mark the beginning of the Anthropocene Epoch", in: *Bulletin of the Atomic Scientists*, 71,3: 46–57.

Waters, C.N.; *Zalasiewicz*, J., 2017: "Concrete: The most abundant novel rock type of the anthropocene", in: *Encyclopedia of the Anthropocene*: 1–5: 75–85.

Waters, C.N.; *Zalasiewicz*, J.; *Summerhayes*, C.;...; *Oreskes*, N.; Wolfe, A.P., 2016: "The Anthropocene is functionally and stratigraphically distinct from the Holocene", in: *Science*, 351,6269: aad2622.

Waters, C.N.; *Zalasiewicz*, J.; *Summerhayes*, C.;...; *Poirier*, C.; *Edgeworth*, M., 2018b: "Global Boundary Stratotype Section and Point (GSSP) for the Anthropocene Series: Where and how to look for potential candidates", in: *Earth-Science Reviews*, 178: 379–429.

Waters, C.N.; *Zalasiewicz*, J.; *Williams*, M.;...; Ford, J.R.; Cooper, A.H., 2014a: "Evidence for a Stratigraphic Basis for the Anthropocene", in: *Springer Geology*, 3: 989–999.

Waters, C.N.; *Zalasiewicz*, J.A.; *Williams*, M., et al., 2014b:"A stratigraphical basis for the Anthropocene?", in: *Special Publication of the Geological Society* (London: Geological Society), Vol. 395: 1–21.

Waters, C.N.; *Zalasiewicz*, J.A.; *Williams*, M.; *Ellis*, M.A.; Snelling, A.M., 2014c: "A stratigraphical basis for the Anthropocene?", in: *Geological Society Special Publication* , 395,1: 1–21.

Weber, A., 2019: *Enlivenment: Toward a Poetics for the Anthropocene* (Cambridge, MA: MIT Press).

Westra, L.; Gray, J.; Gottwald, F.-T., 2017: *The Role of Integrity in the Governance of the Commons: Governance, Ecology, Law, Ethics* (Cham: Springer International Publishing).

Wheeler, S., 2017: "The corporation and the Anthropocene", in: Kotzé, Louis J. (Ed.): *Environmental Law and Governance for the Anthropocene* (Oxford – London – New York – New Delhi – Sydney: Hart): 289–308.

Willett, W.; Rockström, J.; Loken, B.;...; Nishtar, S.; Murray, C.J.L., 2019: "Food in the Anthropocene: the EAT–Lancet Commission on healthy diets from sustainable food systems", in: *The Lancet*, 393,10170: 447–492.

Williams, M.; Zalasiewicz, J.; Davies, N.;...; Goiran, J.-P.; Kane, S., 2014: "Humans as the third evolutionary stage of biosphere engineering of rivers", in: *Anthropocene, 7*: 57–63.

Williams, M.; Zalasiewicz, J.; Haff, P., et al., 2015a: "The Anthropocene biosphere", in: The *Anthropocene Review*, 2,3: 196–219.

Williams, M.; Zalasiewicz, J.; Haff, P.K.;...; Barnosky, A.D.; Ellis, E.C., 2015b: "The anthropocene biosphere", in: *Anthropocene Review*, 2,3: 196–219.

Williams, M.; Zalasiewicz, J.; Waters, C.N.;...; Wolfe, A.P.; Zhisheng, A., 2016: "The Anthropocene: A conspicuous stratigraphical signal of anthropogenic changes in production and consumption across the biosphere", in: *Earth's Future*, 4,3: 34–53.

Williams, M.; Zalasiewicz, J.; Waters, C., 2017: "The anthropocene: A geological perspective", in: Heikkurinen, P. (Ed.): *Sustainability and Peaceful Coexistence for the Anthropocene* (London: Routledge):16–30.

Williams, M.; Zalasiewicz, J.; Waters, C.;...; Barnosky, A.; Leinfelder, R., 2018: "The palaeontological record of the Anthropocene", in: *Geology Today*, 34,5: 188–193.

Wilson, E.O., 1992: *The Diversity of Life (London:* Allen Lane – Harmondsworth: The Penguin Press).

Wilson, E.O., 2002: *The Future of Life* (New York: Alfred A. Knopf).

Witmore, C., 2018: "The End of the 'Neolithic'? At the emergence of the Anthropocene", in: Pilaar Birch, S.E. (Ed.) *Multispecies Archaeology* (London – New York: Routledge): 26–46.

Witmore, C., 2019: "Hypanthropos: On Apprehending and Approaching That Which is in Excess of Monstrosity, with Special Consideration given to the Photography of Edward Burtynsky", in: *Journal of Contemporary Archaeology*, 6,1: 136–153.

Woods, D., 2014: "Scale Critique for the Anthropocene", in: *Minnesota Review*, 83: 133–142.

Wood, D., 2018a: *Deep Time, Dark Times: On Being Geologically Human* (New York, N.Y.: Fordham University Press).

Woods, D., 2018b: "Ecocriticism on the Edge: The Anthropocene as a Threshold Concept", in: *Configurations*, 26,4: 502–504.

Woods, D., 2019: "'Terraforming Earth' Climate And Recursivity", in: *Diacritics-A Review of Contemporary Criticism*, 47, 3: 6–29.

Wright, W., 2017: "Geophysical Agency in the Anthropocene: Engineering a Road and River to Rocky Mountain National Park", in: *Environmental History*, 22,4: 668–695.

Wright, W., 2018a: "Wildlife In The Anthropocene: Conservation After Nature", in: *Historical Studies in the Natural Sciences*, 48,1: 110–122.

Wright, W., 2018b: "The Great Acceleration: An Environmental History of the Anthropocene", in: *Historical Studies in the Natural Sciences*, 48,1: 110–122.

Wright, W., 2018c: "Defiant Earth: The Fate of Humans in the Anthropocene", in: *Historical Studies In The Natural Sciences*, 48,1: 110–122.

Wright, W., 2018d: "The New Ecology: Rethinking a Science for the Anthropocene", in: *Historical Studies in the Natural Sciences*, 48,1: 110–122.

Wright, C.; Nyberg, D.; Rickards, L.; et al., 2018: "Organizing in the Anthropocene", in: *Organization*, 25,4: 455–471.

Wright, L.D., Syvitski, J.P.M., Nichols, C.R., 2019a: "Coastal complexity and predictions of change", in: *Coastal Research Library*, 27: 3–23.

Wright, L.D., *Syvitski*, J.P.M., Nichols, C.R., 2019b: "Coastal systems in the Anthropocene", in: *Coastal Research Library*, 27: 85–99.

Yilamu, W., 2017: *Neoliberalism and Post–Soviet Transition: Kazakhstan and Uzbekistan* (London: Palgrave Macmillan).

Young, O.R., 2017: *Governing Complex Systems: Social Capital for the Anthropocene* (Cambridge, Mass.: MIT Press).

Young, O.R., 2013: "Sugaring off: enduring insights from long-term research on environmental governance", in: *International Environmental Agreements-Politics Law And Economics*, 13,1: 87–105.

Young, O.R.; Underdal, A.; Kanie, N.; et al., 2017: "Goal Setting in the Anthropocene: The Ultimate Challenge of Planetary Stewardship", in: *Governing Through Goals: Sustainable Development Goals As Governance Innovation* (Cambridge, MA: MIT Press): 53–74.

Yusoff, K. , 2013: "Geologic life: Prehistory, climate, futures in the Anthropocene", in: *Environment and Planning D: Society and Space*, 31,5: 779–795.

Yusoff, K., 2016: "Anthropogenesis: Origins and Endings in the Anthropocene", in: *Theory, Culture & Society*, 33,2: pp. 3–2.

Yusoff, K., 2017: "Geosocial Strata", in: *Theory, Culture and Society*, 34,2–3: 105–127.

Yusoff, K., 2018: "Politics of the Anthropocene: Formation of the Commons as a Geologic Process", in: *Antipode*, 50,1: 255–276.

Yusoff, K., 2020: "The Inhumanities", in: *Annals of the American Association of Geographers* (in press).

Zalasiewicz, J., 2008a: "Our Geological footprint", in: Wilkinson, C. (Ed.): *The Observer Book of the Earth* (London: Observer Books).

Zalasiewicz, J., 2008b: "Are we now living in the Anthropocene?", in: *Geological Society of America Today*, 18: 4–8.

Zalasiewicz, J., 2015: "Epochs. Disputed start dates for Anthropocene", in: *Nature*, 520: 436.

Zalasiewicz, J.; Williams, M., 2011: "The Anthropocene Ocean in its deep time context", in: *The World Ocean In Globalisation: Climate Change, Sustainable Fisheries, Biodiversity, Shipping, Regional Issue*: 19–35.

Zalasiewicz, J.; Williams, M., 2014 : "The anthropocene: A comparison with the Ordovician-Silurian boundary", in: *Rendiconti Lincei*, 25,1: 5–12.

Zalasiewicz, J.; Williams, M., 2016a: "Climate Change Through Earth's History"; in: *Climate Change: Observed Impacts on Planet Earth: Second Edition*. 3–17.

Zalasiewicz, J.; Waters, C., 2016b: "Geology and the Anthropocene", in: *Antiquity*, 90,350: 512–514.

Zalasiewicz, J.; Williams, M.; Smith, A., et al., 2008: "Are we now living in the Anthropocene?", in: *GSA Today*, 18,2 (February): 4–8.

Zalasiewicz, J.; Williams, M.; Steffen, W.; Crutzen, P., 2010a: "Response to 'the Anthropocene forces us to reconsider adaptationist models of human-environment interactions'", in: *Environmental Science and Technology*, 44,16: 600.

Zalasiewicz, J.; Williams, M.; Steffen, W.; Crutzen, P., 2010b, 2010c: "The New World of the Anthropocene", in: *Environmental Science & Technology*, 44.7: 2228–2231.

Zalasiewicz, J.; Williams, M.; Steffen, W.; Crutzen, P., 2010d: Response to "The Anthropocene forces us to reconsider adaptationist models of human-environment interactions", in: *Environ. Sci. Technol.*, 44,16, 6008, doi: https://doi.org/10.1021/es102062w.

Zalasiewicz, J.; Williams, M.; Fortey, R.;...; Oates, M.; Stone, P., 2011a: "Stratigraphy of the Anthropocene", in: *Philosophical Transactions of the Royal Society A: Mathematical, Physical and Engineering Sciences*, 369,1938: 1036–1055.

Zalasiewicz, J.; Williams, M.; Haywood, A.; Ellis, M., 2011b: "The anthropocene: A new epoch of geological time?", in: *Philosophical Transactions of the Royal Society A: Mathematical, Physical and Engineering Sciences*, 369,1938: 835–841.

Zalasiewicz, J.; Crutzen, P.J.; Steffen, W., 2012: "The anthropocene", in: Gradstein, F.M.; Ogg, J.G.; Schmitz, M.D.; Ogg, G.M. (Eds.): *The Geologic Time Scale 2012* (London: Elsevier): 1033–1040.

Zalasiewicz, J.; *Waters*, C.N.; *Williams*, M., 2014a: "Human bioturbation, and the subterranean landscape of the Anthropocene", in: *Anthropocene*, 6: 3–9.

Zalasiewicz, J.; *Williams*, M.; *Waters*, C.N, et al., 2014b: "The technofossil record of humans" in: *The Anthropocene Review*, 1,1: 34–43.

Zalasiewicz, J.; *Williams*, M.; *Waters*, C.N., 2014c: "Can an anthropocene series be defined and recognized?", in: *Geological Society Special Publication*, 395,1: 39–53.

Zalasiewicz, J.; *Waters*, C.; *Barnosky*, A., et al., 2015a: "Colonization of the Americas, 'Little Ice Age' climate, and bomb produced carbon: Their role in defining the Anthropocene", in: *The Anthropocene Review*, 2,1: 81–98.

Zalasiewicz, J.; *Waters*, C.N.; *Barnosky*, A.D.;...; *Williams*, M.; Wolfe, A.P., 2015b: "Colonization of the Americas, 'little ice age' climate, and bomb produced carbon: Their role in defining the anthropocene", in: *Anthropocene Review*, 2,2: 117–127.

Zalasiewicz, J.; *Waters*, C.N.; *Williams*, M.; *Barnosky*, A.D.; *Cearreta*, A.; *Crutzen*, P.; *Ellis*, E.; *Ellis*, M.A.; Fairchild, I.J.; *Grinevald*, J.; *Haff*, P.K.; *Hajdas*, I.; *Leinfelder*, R.; *McNeill*, J.; *Odada*, E.O.; *Poirier*, C.; *Richter*, D.; *Steffen*, W.; *Summerhayes*, C.; *Syvitski*, J.P.M.; *Vidas*, D.; *Wagreich*, M.; Wing, S.L.; Wolfe, A.P.; *Zhisheng*, A.; *Oreskes*, N., 2015c: "When did the Anthropocene begin? A mid-twentieth century boundary level is stratigraphically optimal", in: *Quaternary International*, 383 (5 October): 196–203.

Zalasiewicz, J.; *Waters*, C.N.; *Williams*, M.;...; *An*, Z., *Oreskes*, N., 2015d: "When did the Anthropocene begin? A mid-twentieth century boundary level is stratigraphically optimal", in: *Quaternary International*, 383: 196–220.

Zalasiewicz, J.; *Waters*, C.N.; *Williams*, M; et. al., 2015e: "When did the Anthropocene Begin? A Mid-twentieth Century Boundary Level Is Stratigraphically Optimal", in: *Quaternary International*, 183: 196–203;

Zalasiewicz, J.; *Waters*, C.N.; *Ivar do Sul*, J.A.;...; Wolfe, A.P.; Yonan, Y., 2016a, 2016b: "The geological cycle of plastics and their use as a stratigraphic indicator of the Anthropocene", in: *Anthropocene*, 13: 4–17.

Zalasiewicz, J., *Waters*, C.N., *Summerhayes*, C.P...., *Wagreich*, M., *Williams*, M., 2017a: "The Working Group on the Anthropocene: Summary of evidence and interim recommendations", in: *Anthropocene*, 19: 55–60.

Zalasiewicz, J.; *Steffen*, W.; *Leinfelder*, R.; *Williams*, M.; *Waters*, C., 2017b, 2017c: "Petrifying Earth Process: The Stratigraphic Imprint of Key Earth System Parameters in the Anthropocene", in: *Theory, Culture and Society*, 34,2–3: 83–104.

Zalasiewicz, J.; *Waters*, C.; *Head*, M.J., 2017d: "Anthropocene: Its stratigraphic basis", in: *Nature*, 541,637: 289.

Zalasiewicz, J.; *Waters*, C.; *Summerhayes*, C., et al., 2017e: "The Working Group on the Anthropocene: Summary of evidence and interim recommendations", in: *Anthropocene*, 19: 55–60.

Zalasiewicz, J.; *Waters*, C.; *Williams*, M., 2017f: "City-strata of the Anthropocene, [Les strates de la ville de l'Anthropocène]", in: *Annales*, 72,2: 329–351.

Zalasiewicz, J.; *Waters*, C.N.; *Summerhayes*, C.P.;...; *Wagreich*, M.; *Williams*, M., 2017g: "The Working Group on the Anthropocene: Summary of evidence and interim recommendations", in: *Anthropocene*, 19: 55–60.

Zalasiewicz, J.; *Waters*, C.N.; Wolfe, A.P.;...; *Wing*, S.; *Williams*, M., 2017i: "Making the case for a formal Anthropocene Epoch: An analysis of ongoing critiques", in: *Newsletters on Stratigraphy*, 50,2: 205–226.

Zalasiewicz, J.; *Williams*, M.; *Waters*, C.N.;...; Wing, S.; Wolfe, A.P., 2017j, 2017k: "Scale and diversity of the physical technosphere: A geological perspective", in: *Anthropocene Review*, 4,1: 9–22.

Zalasiewicz, J.; Waters, C.; Steffen, W., 2021: "Remembering the Extraordinary Scientist Paul Crutzen (1933–2021)", in: *Scientific American*; at: https://www.scientificamerican.com/article/remembering-the-extraordinary-scientist-paul-crutzen-1933-2021/.

Zalasiewicz, J.; *Waters*, C.; *Summerhayes*, C.; *Williams*, M., 2018a: "The Anthropocene", in: *Geology Today*, 34,5: 177-181.

Zalasiewicz, J.; *Waters*, C.; *Williams*, M.; Aldridge, D.C.; Wilkinson, I.P., 2018b: "The stratigraphical signature of the Anthropocene in England and its wider context", in: *Proceedings of the Geologists' Association*, 129,3: 482–491.

Zalasiewicz, J.; *Waters*, C.N.; *Williams*, M.; *Summerhayes*, C.P., 2018c: "The Anthropocene", in: *Geology Today*, 34,5 (Sept./October).

Zalasiewicz, J.; *Waters*, C.; *Williams*, M.; *Summerhayes*, C.; *Head*, M.; *Leinfelder*, R., 2019a: "A General Introduction to the Anthropocene", in: *Zalasiewicz*, J.; *Waters*, C.N.; *Williams*, M.; *Summerhayes*, C.P. (Eds.), 2019: *The Anthropocene as a Geological Time Unit. A Guide to the Scientific Evidence and Current Debate* (Cambridge: Cambridge University Press).

Zalasiewicz, J.; *Waters*, C.N.; *Head*, M.J.;...; *Barnosky*, A.D.; *Cearreta*, A., 2019b: "A formal Anthropocene is compatible with but distinct from its diachronous anthropogenic counterparts: a response to W.F. Ruddiman's 'three flaws in defining a formal Anthropocene'", in: *Progress in Physical Geography*, 43,3: 319–333.

Zalasiewicz, J.; *Waters*, C.N.; *Williams*, M.; *Summerhayes*, C.P. (Eds.), 2019c: *The Anthropocene as a Geological Time Unit. A Guide to the Scientific Evidence and Current Debate* (Cambridge: Cambridge University Press).

Zalasiewicz, J.; *Waters*, C.N.; *Williams*, M.; *Summerhayes*, C.P.; *Head*, M.J.; *Leinfelder*, R., 2019d: "A General Introduction to the Anthropocene", in: *Zalasiewicz*, J.; *Waters*, C.N.; *Williams*, M.; *Summerhayes*, C.P. (Eds.), 2019: *The Anthropocene as a Geological Time Unit. A Guide to the Scientific Evidence and Current Debate* (Cambridge: Cambridge University Press). 2–4.

Zeitschrift für Diskursforschung - Journal for Discourse Studies (ZfD) (Weinheim: Beltz Juvental).

Zelli, F.; Pattberg, P., 2016: "Conclusions: Complexity, responsibility and urgency in the anthropocene", in: Pattberg, P.; Zelli, F. (Eds): *Environmental Politics and Governance in the Anthropocene: Institutions and legitimacy in a complex world* (London – New York: Routledge): 231–242.

Zwier, J., Blok, V., 2017: "Saving earth: Encountering Heidegger's philosophy of technology in the anthropocene", in: *Techne: Research in Philosophy and Technology*, 21,2-3: 222–242.

Zwier, J., Blok, V., 2019: "Seeing Through the Fumes: Technology and Asymmetry in the Anthropocene", in: *Human Studies*, 42,4: 621–646.

Appendices

Paul J. Crutzen after receiving the Nobel Prize in Chemistry in 1995. *Source* Rolf Hofmann, MPIC.

S. Benner et al. (eds.), *Paul J. Crutzen and the Anthropocene: A New Epoch in Earth's History*, The Anthropocene: Politik–Economics–Society–Science 1,
https://doi.org/10.1007/978-3-030-82202-6

Appendix 1
Vita of Paul Josef Crutzen

Address (office) Born, 3 December 1933, Amsterdam, The Netherlands
Married, two children, deceased on 28 January 2021.

Max Planck Institute for Chemistry
Department of Atmospheric Chemistry, P.O. Box 3060
D-55020 Mainz/GERMANY
Telephone: +49 - (0) 6131 - 305-4640
Telefax: +49 - (0) 6131 - 305-4019

Education

1946–1951   High School:, St. Ignatius College, Amsterdam, The Netherlands
1951–1954   Civil Engineering, Amsterdam, The Netherlands
1959–1974   Academic Studies and Research Activities, Institute of Meteorology, University of Stockholm, Sweden
1963   M.Sc. (Filosofie Kandidat)
1968   Ph.D. (Filosofie Licentiat), Meteorology
            Title: *"Determination of parameters appearing in the 'dry' and the 'wet' photochemical theories for ozone in the stratosphere"*,
            Examiner: Prof. Dr. Bert Bolin, Stockholm.
1973   D.Sc. (Filosofie Doctor), Stockholm, Sweden
            Title: *"On the photochemistry of ozone in the stratosphere and troposphere and pollution of the stratosphere by high-flying aircraft"*,
            Promoters: Prof. Dr. John Houghton, FRS, Oxford, and Prof. Dr. R.P. Wayne, Oxford.
            (Ph.D. and D.Sc. degrees were given with the highest possible distinctions)

© The Editor(s) (if applicable) and The Author(s), under exclusive license   441
to Springer Nature Switzerland AG 2021
S. Benner et al. (eds.), *Paul J. Crutzen and the Anthropocene: A New Epoch in Earth's History*, The Anthropocene: Politik–Economics–Society–Science 1,
https://doi.org/10.1007/978-3-030-82202-6

Main Research Interest

Atmospheric chemistry and its role in biogeochemical cycles and climate; Climate-engineering; Climate and biofuels.

Employment

1954–1958	Bridge Construction Bureau of the City of Amsterdam, The Netherlands
1956–1958	Military Service (compulsory), The Netherlands
1958–1959	House Construction Bureau (HKB), Gävle, Sweden
1959–1974	Various computer consulting, teaching and research positions at the Department of Meteorology of the University of Stockholm, Sweden, Latest position: Research Associate Professor
1969–1971	Post-doctoral fellow of the European Space Research Organization at the Clarendon Laboratory of the University of Oxford, England
1969–1971	Visiting Fellow of St. Cross College, Oxford, England
1974–1977	Research Scientist in the Upper Atmosphere Project, National Center for Atmospheric Research (NCAR), Boulder, Colorado, USA (half time)
	Consultant at the Aeronomy Laboratory, Environmental Research ^Laboratories, National Oceanic and Atmospheric Administration (NOAA), Boulder, Colorado, USA (half time)
1976–1984	Adjunct professor at the Atmospheric Sciences Department, Colorado State University, Fort Collins, Colorado
1977–1980	Senior Scientist and Director of the Air Quality Division, National Center for Atmospheric Research (NCAR), Boulder, Colorado, USA
1980–2000	Director of the Atmospheric Chemistry Division, Max Planck Institute for Chemistry, Mainz, Germany
1987–1991	Professor (part-time) at the Department of Geophysical Sciences, University of Chicago, USA
1991–1992	Tage Erlander Professor of the Swedish Research Council at the University of Stockholm, Sweden
1992–2008	Distinguished Professor (part-time), Scripps Institution of Oceanography, University of California, San Diego, La Jolla, USA
1997–2000	Professor of Aeronomy (part-time), Utrecht University, at the faculty of Physics and Astronomy, The Netherlands
Since 1980	Scientific Member (for life) of the Max Planck Institute for Chemistry (Otto-Hahn-Institute), Mainz
Since 2000	Director Emeritus, Atmospheric Chemistry Department of the Max Planck Institute for Chemistry (Otto-Hahn-Institute), Mainz
Since 2000	Member of the Max-Planck-Society for the Advancement of Science
Since 2008	Emeritus Professor, Scripps Institution of Oceanography, University of California, San Diego, La Jolla, USA

Services

NASA Stratospheric Research Advisory Committee (1975–1977)

International Ozone Commission of IAMAS (International Association of Meteorology and Atmospheric Sciences) (1976–1984)

Atmospheric Sciences Advisory Committee, National Foundation, USA (1977–1979)

Committee of Atmospheric Sciences (CAS), National Academy of Sciences, USA (1978–1980)

Advisory Committee, High Altitude Pollution Programme, Federal Aviation Authority (FAA), USA (1978–1982)

Commission of Air Chemistry and Global Air Pollution (CACGP) of the International Association of Meteorology and Atmospheric Physics (IAMAP) (1979–1990)

Past Editor "Journal of Atmospheric Chemistry" (1983)

Special Inter-Ministerial Advisory Commission on Forest Damage in the Federal Republic of Germany (1983–1987)

Various research Advisory Committees of the German National Science Foundation (DFG) and Ministry of Research and Technology (BMFT) of the Federal Republic of Germany (since 1984)

Steering Committees of the SCOPE/ICSU effort to estimate the potential environmental consequences of a nuclear war (SCOPE/ENUWAR) (1984–1988)

Kuratorium (Board of Trustees) of the Max-Planck-Institut für Meteorologie, Hamburg (1984–2000)

Editorial Board "Climate Dynamics" (1985–2000)

Special Committee and Executive Committee, and Chairman of Coordinating Panel I of the International Geosphere–Biosphere Programme (IGBP) (1986–1990)

Executive Board of SCOPE (Scientific Committee on Problems of the Enviroment) of the International Council of Scientific Unions and Chairman of the National SCOPE Committee of the FRG (1986–1989)

Commission of the Parliament of the FRG for the "Protection of the Earth's Atmosphere" (Enquete-Kommission"Vorsorge zum Schutz der Erdatmosphäre") (1987–1990)

Steering Committee (Chairman) of the International Global Atmospheric Chemistry (IGAC) Programme, a Core Project of the IGBP (1987–1990); Vice-Chairman (1990–1996)

Kuratorium (Board of Trustees) (Chairman) of the Fraunhofer-Institut für atmosphärische Umweltforschung, Garmisch-Partenkirchen (1992–1994)

European IGAC Project Office (Chairman) (1992–1998)

Editorial Advisory Board "Issues in Environmental Science and Technology", British Royal Society of Chemistry, Britain (since 1993)

Reviewing Editor 'Science' (1993–1999)

Editorial Board "Mitigation and Adaptation Strategies for Global Change" (1993–2003)

Advisory Board of the Institute for Marine and Atmospheric Research, University Utrecht, The Netherlands (1993–1997)

STAP (Scientific and Technical Advisory Panel) Roster of Experts of the United Nations Environment Programme (1993–1998)

European Environmental Research Organisation (EERO) (1993–1999)

Advisory Council of the Volvo Environment Prize (1993)

Advisory Board Journal "Tellus B: Chemical and Physical Meteorology" (since 1993)

SPINOZA Prize Committee of Nederlandse Organisatie voor Wetenschappelijk Onderzoek (Dutch Organization for Scientific Research) (1994–1995)

Prix Lemaitre Committee, Belgium (1994–1997)

Steering Committee on Global Environmental Change of International Institute for Applied Systems Analysis (IIASA), Laxenburg, Austria (1995–1997)

Executive Board of Gesellschaft Deutscher Naturforscher und Ärzte (GDNÄ-German Society of Natural Scientists and Physicians) (1995–2000)

Scientific Advisory Group of the School of Environmental Sciences, University of East Anglia, Norwich, Britain (1995)

International Ozone Commission of IAMAS (International Association of Meteorology and Atmospheric Sciences) (1976–1984)

General Advisory Board of "Encyclopedia of Life Support Systems (EOLSS)" (1996–2002)

Editorial Board of "Earth and Planetary Science Letters" (1997/2001)

Board of Consulting Editors of "European Review" (Interdisciplinary Journal of the Academia Europaeae) (since 1997)

Editorial Advisory Board of "Current Topics in Meteorology" (since 1997)

International Advisory Board on the "Encyclopedia of Global Environment Change Project" (1997)

ESTA (European Science and Technology Assembly) of the European Union, Brussels (1997–2000)

Advisory Board of "The International Journal of Environmental Studies" (since 1997)

Editorial Board "Rendiconti Lincei" (Accademia Nazionale dei Lincei, Rome) (since 1997)

Deutscher Studienpreis, Jury Member (Körber Foundation, Germay (1998/1999)

Global Change Committee of the German Research Council and the Federal Ministry of Research and Technology, Germany (1998–1999)

Executive Advisory Board of the "Encyclopedia of Physical Science and Technology, 3rd edition" (1998–2003)

Editorial Board of "Ambio—A Journal of the Human Environment", Vol. 27 to 41 (1998–2012)

Scientific Committee for the International Geosphere-Biosphere Programme (SC-IGBP), Vice Chairman (1998–2003)

International Steering Committe of INDOEX (1999)

Scientific and Academic Advisory Committee, Weizmann Institute, Israel (1999–2010)

"Council on the Future", UNESCO, Paris (1999–2003)

Indian Ocean Experiment (INDOEX) Co-Chief Scientist (with Prof. V. Ramanathan, Scripps Institution of Oceanography) (1999)

Advisory Committee (Beirat), Jahrbuch Ökologie, Berlin (since 1999)

Editorial Advisory Board of 'ChemPhysChem' (2000)

Working Group on Establishing an Independent Advisory Body on European Research (2000–2002)

Advisory Body on Science and Technology in Europe (2000–2002)

Steering Committee of the Center for Atmospheric Sciences, University of California, Berkely (2000)

Steering Committee of the "Atmospheric Brown Clouds" programme, in collaboration with the United Nations Environmetal Programme (UNEP), La Jolla/Nairobi (2001)

Fifth Framework Programme Expert Advisory Group (EAG) on "Global change, climate and biodiversity", Eurpoean Commission, Brussels (2001–2003)

ABC (Atmospheric Brown Clouds) Science Team, Co-Chief scientist (2002–2005)

ABC (Atmospheric Brown Clouds) International Science Team (since 2002)

Advisory committee of the Institute: Urbanization, Emissions, and the Global Carbon Cycle, START, Washington DC (2002)

Board of the General Advisors (BGA) of the UNESCO-EOLSS (since 2002)

Scientific Council of the International Centre for Theoretical Physics (Abdus Salam), Trieste (2003–2005)

Advisory Board of the Atmospheric Chemistry and Physics, EGU Journal (since 2004)

Advisory Board of The National Society of High School Scholars, Atlanta (2005)

"Comité van Aanbeveling" (Committee of Recommendation) of the International Polar Year 2005, Nederlandse Organisatie voor Wetenschappelijk Onderzoek, Den Haag (2005)

Committee for the Gérard Mégie Prize, Académie des Sciences, France (2005)

Scientific Council of European Research Council, Brussels, Belgium (2005–2008)

International Advisory Board of the Mahatma Gandhi Center for Global Nonviolence, James Madison University, Harrisonburg, USA (2005–present)

Advsiory Board of the Ernst Strüngmann Forum, Frankfurt, Germany (since 2006)

Editorial Board "Anthropocene", Elsevier (2013–2016)

Memberships

Fellow of the American Geophysical Union (AGU) (since 1986), Member since 1973

Member of the Science Foundation (DFG) and Ministry of Research and Technology (BMFT) of the Federal Republic of Germany

Founder Member of The Academy of Europe (Academia Europaea), London (since 1988)

Corresponding Member of The Royal Netherlands Academy of Science (since 1990)

Member Leopoldina Nationale Akademie der Wissenschaften, Halle, Germany (1992–2014); now Honorary Member.

Member of the Royal Swedish Academy of Engineering Sciences (Kungl. IngenjörsVetenkaps Akademien IVA). (since 1992)

Member of the Royal Swedish Academy of Sciences (Kungl. Vetenskap-sakademien KVA) (since 1992)

Foreign Associate of the U.S. National Academy of Sciences, Washington (since 1994)

Member of the Institut Mondial des Sciences (World Institute of Science) Brussels, Belgium (since 1994)

Foreign Associate of the U.S. National Academy of Sciences, Washington (since 1994)

Titular Member of the European Academy of Arts, Sciences and Humanities, Paris (since 1996)

Member of the Pontifical Academy of Sciences (since 1996)

Member of Vereinigung Deutscher Wissenschaftler – VDW – (Association of German Scientists) (since 1997)

Foreign Member of Accademia Nazionale dei Lincei (Italian Academy of Sciences, Roma, Italy (since 1997)

Corresponding member of the Société Royale des Sciences de Liège (1997)

Member of the Board of Directors of the Mariolopoulos-Kanaguinis Foundation for Environmental Sciences, Athens (since 1997)

Member of the Deutsche Physikalische Gesellschaft (DPG), Bad Honnef, Germany (since 1998)

Member of the Fachverband Umweltphysik (Associaton of Environmental Physics) of the DPG (Deutsche Physikalische Gesellschaft) (since 1998)

Member of the Board of Governors, Weizmann Institute, Israel (1999–2012)

Foreign Member of the Russian Academy of Sciences (since 1999)

Member of the Council of the Pontifical Academy of Sciences (2001–2015)

Member of the Council of Chancellors of The Global Foundation of the Consejo Cultural Mundial, Mexico (since 2001)

Member of the Founders' Assembly, Foundation Lindau Nobel Laureates Meetings at Lake Constance, Lindau (since 2003)

Fellow of the Literary & Historical Society, University College of Dublin (since 2004)

Fellow of the American Association for the Advancement of Science (AAAS), Washington (since 2004)

Associate Fellow of TWAS (Third World Academy of Sciences), Trieste (since 2004)

Institute Scholar, International Institute for Applied Systems Analysis (IIASA), Laxenburg, Austria (since 2004)

Member of the ABC science team (2005 – present)

Fellow of the World Academy of Art and Science, San Francisco, USA (2005–2016). Emeritus Fellow (since 2016)

International Member of the American Philosophical Society, Class 1, USA (2007)

Member of the International Scientific Advisory Council of the CREF-Cyprus Institute, Cyprus (since 2008)

Founder Member of the Anthropocene Working Group (AWG) of the Subcommission on Quaternary Stratigraphy (part of the International Commission on Stratigraphy) (since 2009); (now Honorary member).

Honorary Doctoral/Professor Degrees

1986	Honorary Doctor of the York University, York, Canada
1992	Honorary Doctor of the Université Catholique de Louvain, Belgium
1993	Honorary Professor at the Johannes Gutenberg-University of Mainz
1994	Honorary Doctor of the University of East Anglia, Norwich, U.K.
1996	Honorary Doctor of the Aristotle University of Thessaloniki, Greece
1997	Honorary Doctor of the University of San José, Costa Rica
1997	Honorary Doctor of the Université de Liège, Belgium
1997	Honorary Doctor of the Tel Aviv University, Israel
1997	Honorary Doctor of the Oregon State University, Corvallis, USA
1997	Honorary Doctor of the Université de Bourgogne, Dijon, France
1997	Honorary Doctor of the University of Chile, Santiago, Chile
1998	Honorary Doctor of the University of Athens, Greece
2001	Honorary Doctor of the Democritus University of Thrace, Greece
2002	Honorary Doctor of the Nova Gorica Polytechnic, Slovenia (since 2006 Nova Gorica University)
2002	Honorary Doctor of the University of Hull, United Kingdom
2004	Honorary Professor of the College of Environmental Sciences, Peking University, China
2005	Honorary Doctor of the Université Joseph Fourier, Grenoble, France
2005	Honorary Professor of the Tongji University, Shanghai, China
2007	Honorary Doctor of the Environmental Engineering, Politecnico, Milan, Italy
2008	Honorary Professor at the Seoul National University, Seoul, South Korea
2010	Honorary Doctor of the University of Venice, Italy
2013	Honorary Doctor of the Maastricht University, Maastricht, The Netherlands

Special Distinctions, Awards and Honors

1975	Outstanding Publication Award, Environmental Research Laboratories, National Oceanic and Atmospheric Administration (NOAA), Boulder, Colorado, U.S.A
1977	Special Achievement Award, Environmental Research Laboratories, NOAA, Boulder, Colorado, U.S.A
1983	NASA Public Service Group Achievement Award to SME Project Operations and Control Center Team
1984	Rolex-Discover Scientist of the Year
1985	Centennial Medallion for participation in the Centennial Colloquium of the Georgia Institute of Technology

1985   "The Fourth Raymond R. Tucker Memorial Lecturer in Mechanical Engin-
        eering", Washington University, School of Engineering
1985   Recipient of the Leo Szilard Award for "Physics in the Public Interest" of
        the American Physical Society
1987   Lindsay Memorial Lecturer, Goddard Space Flight Center, National Aero-
        nautics and Space Administration
1989   Recipient of the Tyler Prize for Environmental Achievement
1990   Tracy and Ruth Storer Lecturer at the University of California, Davis, U.S.A
1991   Recipient of the Volvo Environmental Prize by the Royal Swedish Academy
        of Science, Stockholm, Sweden
1991   G.N. Lewis Lecturer at the University of California, Berkeley, USA
1993   Ida Beam Visiting Professor, The University of Iowa, USA
1994   Recipient of the Deutscher Umweltpreis of the Umweltstiftung (DBU)
        (German Environmental Prize of the Federal Foundation for the Environ-
        ment)
1994   Recipient of the Max-Planck Forschungspreis (Research Prize) (with M.
        Molina, USA)
1994   Raymond and Beverly Sackler Distinguished Lecturer in Geophysics and
        Planetary Sciences, Tel Aviv University, Israel
1994   Aristoteles Lecturer at the Aristotle University of Thessaloniki, Greece
1994   Group Achievement Award (to HALOE Science Data Validation, Data
        Processing, and Flight Operations Team) for outstanding contributions to
        the success of the HALOE/UARS satellite experiment by NASA Langley
        Research Center
1995   Recipient of the Global Ozone Award for "Outstanding Contribution to the
        Protection of the Ozone Layer" by UNEP (United Nations Environment
        Programme)
1995   "Commandeur in de Orde van de Nederlandse Leeuw" (knighted by the
        Queen of The Netherlands)
1995   Recipient of the Bundesverdienstkreuz mit Stern (Federal Merit Cross with
        Star Federal Republic of Germany)
1995   Recipient of the Nobel Prize in Chemistry (with M. Molina and F.S. Rowland,
        USA)
1996   Symons Memorial Lecturer of the Royal Meteorological Society of England,
        Imperial College, London
1996   Holder of the 'Ehrenring' (ring of honour) of the City of Mainz
1996   Name "Prof. Dr. P.J. Crutzenlaan" given to a street in De Bilt, The
        Netherlands
1996   Election to the Global 500 Roll of Honour of the United Nations Environment
        Programme (UNEP)
1996   Recipient of The Louis J. Battan Author's Award (with Dr. T.E. Graedel,
        USA) by the American Meteorological Society for "their book entitled
        Atmosphere, Climate and Change, an authoritative and beautifully illustrated
        introduction to the role of the atmosphere in global change"

1996   Recipient of The Minnie Rosen Award for "High Achievement in Service to Mankind" of Ross University, New York

1996   Recipient of the Médaille d'Or de la Ville de Grenoble, France

1997   Ceremonial Lecturer (Festvortrag) at the Annual General Assembly of the Max-Planck-Society, Bremen

1997   Erasmus Lecturer and Medalist of the Academia Europaeae

1997   Shipley Distinguished Lecturer, Clarkson University, Potsdam, USA

1997   "XXVII Krishnan Memorial Lecturer", Title: 30 Years of Progress in Atmospheric Chemistry, National Physical Laboratory, New Delhi, India

1998   45th Gilbert N. Lewis Lecturer at the University of California, Berkeley, USA

1998   "First Lecturer in the Thompson Lecture Series", Advanced Study Programme, National Center for Atmospheric Research, Boulder, Colorado, USA

1998   Ceremonial Lecturer (Festvortrag) at the Annual Assembly of the Deutsche Physikalische Gesellschaft (German Physical Society), Regensburg.

1998   Public lecturer at Michigan Technological University, Houghton, USA

1999   H. Julian Allan Award 1998, in recognition of the outstanding scientific paper (co-authored) for 1998 at NASA Ames

1999   Eyring Lecturer at the Arizona State University

       Worldwide most cited author in the Geosciences with 2911 citations from 110 publications during the decade 1991–2001, ISI (Institute for Scientific Information, Philadelphia, USA), issue November/December 2001

2000   Named among "Heroes for the Planet" by TIME Magazine, Special Issue April/May 2000, Earth Day 2000

2000   "Beatty Memorial" Lecturer at McGill University, Montreal, Canada

2000   Name 'Crutzen' given to asteroid n° 9679 on Nov. 11, 2000 (asteroid discovered by C.J. van Houten and I. van Houten-Groeneveld on September 24, 1960)

2000   The Third 'Rosenblith' Lecturer at the Massachusetts Institute of Technology: Atmospheric Chemistry in the 21st Century (Co-speaker M. Molina, F. Sherwood Rowland)

2001   Honorary Chairman "Climate Conference 2001", 20–24. August, Utrecht, The Netherlands.

2001   Honored by the Constantinos Karamanlis Institute for Democracy Athens, Greece, for outstanding contributions to Science and Society

2002   Worldwide most cited author in the Geosciences with 2911 citations from 110 publications during the decade 1991–2001, ISI (Institute for Scientific Information, Philadelphia, USA), issue November/December 2001

2003   Recipient of the Golden Medal (highest destination), The Academy of Athens Paul Crutzen Prize awarded to the best paper for participants in the International Young Scientists' Global Change START Conference, Trieste, Italy, November 16–19, 2003

2005   Recipient of the UNEP/WMO Vienna Convention Award

2005   Distinguished Lecturer in Science, The Hongkong University of Science and Technology School of Science

2006   Recipient of the Jawaharlal Nehru Birth Centenary Medal 2006, Indian
        National Science Academy, New Delhi, India
2007   Named among "Heroes of the Environment" by TIME Magazine
2008   Recipient of the Capo d'Orlando Award, Discepolo Foundation, Vico
        Equense, Italy
2008   Einstein Lecture Dahlem, Free University Berlin, Germany
2013   Recipient of the Landesverdienstorden of Rheinland-Pfalz (Order of Merit
        of the land Rhineland-Palatinate), Mainz, Germany
2018   Recipient of the Haagen-Smit Clean Air Award (California Air Resources
        Board), Sacramento, USA
2019   Recipient of the Lomonosov Gold Medal, Russian Academy of Sciences
        (RAS)

Foreign Honorary Member (IHM) of the American Academy of Arts and Sciences, Cambridge, USA (since 1986)

Honorary Member of the International Ozone Commission of IAMAS (International Association of Meteorology and Atmospheric Sciences) (since 1996)

Honorary Member of the International Ozone Commission (IO3C) (International Union of Geodesy and Geophysics) (since 1996)

Honorary Fellow of Physical Research Laboratory (PRL), Ahmedabad, India (since 1997)

Honorary Member of the American Meteorological Society (since 1997)

Honorary Member of the European Geophysical Society (EGS, since 1997)

Honorary Fellow of St. Cross College, Oxford, England (since 1998)

Honorary Member of the Committee of EUROSCIENCE (European Association for the Promotion of Science and Technology), Strasbourg (since 1999)

Honorary Member of the Swedish Meteorological Society (Svenska Meteorologiska Sällskapet) (since 2000)

Ambassador for the Environment of the European Commission for the Environment (since 2001)

Honorary Member of the The World Innovation Foundation (since 2002)

Honorary Member of the The International Raoul Wallenberg Foundation and the Angelo Roncalli International Committee, Jerusalem (since 2003)

Honorary Member of the International Polar Foundation, Brussels (since 2003)

Honorary Member of the European Geosciences Union (EGU) (since 2004)

Foreign Member of The British Royal Society (allowed to use the title Paul Josef Crutzen, ForMemRS), London, UK (since 2006)

Honorary Member of European Academy of Sciences and Arts (Academia Scientiarum et Artium Europaea), Salzburg, Austria (since 2007)

Honorary Fellow of the Institute of Green Professionals, Weston, Florida, USA. Entitled to use the identifiere "Hon. FIGP" (since 2007)

Honorary Member of the Commission on Atmospheric Chemistry and Global Pollution (iCACGP/CACGP) (since 2010)

Honorary Member of the „Naturforschende Gesellschaft zu Emden von 1814", Emden, Germany (since 2011)

Honorary Member of "Nationale Akademie der Wissenschaften Leopoldina", Halle, Germany (since 2014); (Member since 1992)

Honorary Member of the Institute for Earth System Preservation (IESP), Garching, Germany (since 2015)

Honorary Member of the Royal Netherlands Chemical Society (KNCV), The Hague, The Netherlands (since 2017)

Honorary Member of the Anthropocene Working Group (AWG) of the Subcommission on Quaternary Stratigraphy (part of the International Commission on Stratigraphy) (since 2019); (Founder Member 2009).

Paul J. Crutzen in his office at the MPIC in Mainz. *Source* MPIC, Mainz and Archives of the Max Planck Society, Berlin-Dahlem.

DOCTORAL STUDENTS AND POST-DOCTORAL CO-WORKERS

Ariya, Patricia (post-doc)	von Kuhlmann, Rolf
Bergamaschi, Peter	Krueger, Arlin
Berges, Markus	Kuhlbusch, Thomas
Biermann, Uta	Landgraf, Jochen
Bott, Andreas	Lawrence, Mark
Brühl, Christoph	Lelieveld, Jos
Burrows, John (post-doc)	Lobert, Jürgen
Chatfield, Robert	Müller, Rolf
de Liang, Chen	Peter, Thomas (post-doc)
Dentener, Frank	Pöschl, Ulrich (post-doc)
Dickerson, Russ (Don Stedman, main advisor)	Quesada, Jaime
Feichter, Johan (post-doc)	Reichenauer, Helga (Diplom = Masters degree)
Fishman, Jack	Röckmann, Thomas
von Glasow, Roland	Sander, Rolf
Gidel, Louis T. (post-doc)	Saueressig, Gerd
Grooß, Jens-Uwe	Schade, Gunnar
Harris, Geoff (post-doc)	Solomon, Susan
Hein, Ralf	Steil, Benedikt
Holzinger, Rupert (post-doc)	Valentin, Karen
Jöckel, Patrick	Valverde-Canossa, Jessica
Kanakidou, Maria (post-doc)	Vogt, Rainer (post-doc)
Kleiss, Bettina	Waibel, Andreas
Koop, Thomas	Zimmermann, Peter
Kouker, Wolfgang (with James Holton)	September 2019

Paul J. Crutzen speaking at the Symposium on the occasion of his 80th birthday in Mainz on 2 December 2013. *Source* Carsten Costard, MPIC, Mainz.

Appendix 2
Complete Bibliography of Paul J. Crutzen (1965–2020)

Books

1. Crutzen, P.J. and J. Hahn, 1985: Schwarzer Himmel. S. Fischer Verlag, 240 pp.
2. Pittock, A.B., T.P. Ackerman, P.J. Crutzen, M.C. MacCracken, C.S. Shapiro and R.P. Turco, 1986: Environmental Consequences of Nuclear War, SCOPE 28, Volume I: Physical and Atmospheric Effects, Wiley, Chichester, 359 pp; 2nd edition 1989.
3. Crutzen P.J. and M. Müller, 1989: Das Ende des blauen Planeten? C.H. Beck Verlag, 271 pp.
4. Graedel, T.E. and P.J. Crutzen, 1993: Atmospheric Change: An Earth System Perspective. W.H. Freeman, New York, 446 pp.
5. Graedel, T.E. and P.J. Crutzen, 1995: Atmosphere, Climate, and Change. W.H. Freeman, New York, 208 pp.
6. Graedel, T.E. and P.J. Crutzen, 1996: Atmosphäre im Wandel. Die empfindliche Lufthülle unseres Planeten. Spektrum Akademischer Verlag, Heidelberg, 221 pp.
7. Enquete Commission "Preventive Measures to Protect the Earth's Atmosphere", 1989: Interim Report: Protecting the Earth's Atmosphere: an International Challenge. Ed. Deutscher Bundestag, Referat Öffentlichkeitsarbeit, Bonn, 592 pp.
8. Enquete Commission "Preventive Measures to Protect the Earth's Atmosphere", 1990: Protecting the Tropical Forests: A High-Priority International Task. Ed. Deutscher Bundestag, Referat Öffentlichkeitsarbeit, Bonn, 968 pp.
9. Enquete Commission "Preventive Measures to Protect the Earth's Atmosphere", 1991: Protecting the Earth: A Status Report with Recommendations for a new Energy Policy. Ed. Deutscher Bundestag, Referat Öffentlichkeitsarbeit, Bonn, 2 Volumes.

© The Editor(s) (if applicable) and The Author(s), under exclusive license to Springer Nature Switzerland AG 2021
S. Benner et al. (eds.), *Paul J. Crutzen and the Anthropocene: A New Epoch in Earth's History*, The Anthropocene: Politik–Economics–Society–Science 1, https://doi.org/10.1007/978-3-030-82202-6

10. Crutzen, P.J., J.-C. Gerard and R. Zander (Eds.), 1989: Our Changing Atmosphere. Proceedings of the 28th Liège International Astrophysical Colloquium June 26–30, 1989, Université de Liège, Cointe-Ougree, Belgium, 534 pp.

11. Crutzen, P.J. and J.G. Goldammer, 1993: Fire in the Environment: The Ecological, Atmospheric, and Climatic Importance of Vegetation Fires. Dahlem Konferenz (15–20 March 1992, Berlin), ES13, Wiley, Chichester, 400 pp.

12. Graedel, T.E. and P.J. Crutzen, 1994: Chemie der Atmosphäre. Bedeutung für Klima und Umwelt, Spektrum Akademischer Verlag, Heidelberg, 511 pp.

13. Crutzen, P.J., 1996: Atmosphäre, Klima, Umwelt (Edited and with an introduction by P.J. Crutzen), Spektrum Akademischer Verlag, Heidelberg, 227 pp.

14. Crutzen, P. J., G. Komen, K. Verbeek and R. van Dorland, 2004: Veranderingen in het klimaat. Koninklijk Nederlands Meteorologisch Instituut, De Bilt, The Netherlands, 16 pp. (in Dutch) http://www.dbnl.org/basisbibliotheek/.

15. Earth System Analysis for Sustainablilty, edited by H. J. Schellnhuber, P. J. Crutzen, W. C. Clark, M. Claussen, and H. Held, Dahlem Workshop Reports, MIT Press, 2004.

16. Crutzen, P.J. and H.G. Brauch: Paul J. Crutzen: A Pioneer on Atmospheric Chemistry and Climate Change in the Anthropocene. Springer Briefs on Pioneers in Science and Practice – Nobel Laureates. Vol. 50, Springer, 2016, 248 pp.

(continued)

Crutzen P.J./M.. Müller, 1989: *Das Ende des blauen Planeten?* C.H. Beck Verlag

Crutzen, P.J./J. Hahn, 1985: *Schwarzer Himmel.* S. Fischer Verlag.

(continued)

Graedel, T.E./P.J. Crutzen, 1993:*Atmospheric Chan-ge: An Earth System Perspective.* W.H. Freeman

Crutzen, P.J./J.G. Goldammer, 1993: *Fire in the Environment: The Ecological, Atmospheric, and Climatic Importance of Vegetation Fires.* Wiley.

Autobiographical Papers

1. Crutzen, P.J., 1996: Mein Leben mit O_3, NO_x und anderen YZO_x-Verbindungen (Nobel-Vortrag). Angew. Chem., **108**, 1878–1898.
2. Crutzen, P.J., 1996: My life with O_3, NO_x, and other YZO_x compounds (Nobel Lecture). Angew. Chem. Int. Ed. Engl., **35**, 1758–1777.
3. Crutzen, P.J., 1997: Die Beobachtung atmosphärisch-chemischer Veränderungen: Ursachen und Folgen für Umwelt und Klima. (Festvortrag anläßlich der Hauptversammlung der Max-Planck-Gesellschaft in Bremen am 6. Juni 1997). In: Max-Planck-Gesellschft. Jahrbuch 1997. Ed. Generalverwaltung der Max-Planck-Gesellschaft München, Vandenhoek & Ruprecht, Göttingen, 1997, 51–71.
4. Crutzen, P.J., 1998: The BULLETIN Interviews. Professor Paul Josef Crutzen. WMO Bulletin, **47** (2), 3–15.
5. Crutzen, P. J., 1999: The nuclear winter. Proceedings of the international conference "The discovery of Polonium and Radium", Warsaw. Poland, 17–20 September 1998, Eds. J. Kornacki, R. Budzynski and J. Kotomycki, Warsawska Drukornia Nankowa PAN, 1998, 85–104.
6. Crutzen, P. J., 2004: How I became a scientist? The Abdus Salam Centre for Theoretical Physics, Trieste, Italy.
7. Crutzen, P. J., 2004: A late change to the programme - How an engineer became hooked on atmospheric chemistry. Nature, 429, 349.
8. Crutzen, P. J., 2005: Benvenuti nell'Antropocene! Arnoldo Modadori Editore S.p.A., Milano, Italy. (in Italian)

Journal articles (refereed)

1. Blankenship, J. R. and P. J. Crutzen, 1965: A photochemical model for the space-time variations of the oxygen allotropes in the 20 to 100 km layer. Tellus, 18, 160–175.
2. Crutzen, P. J., 1969: Determination of parameters appearing in the 'dry' and the 'wet' photochemical theories for ozone in the stratos. Tellus, **21**, 368–388.
3. Crutzen, P. J., 1969: Determination of parameters appearing in the oxygen-hydrogen atmosphere. Ann. Géophys., **25**, 275–279.
4. Crutzen, P. J., 1970: The influence of nitrogen oxides on the atmospheric ozone content. Q. J. R. Meteorol. Soc., **96**, 320–325.
5. Crutzen, P. J., 1970: Comments on "Absorption and emission by carbon dioxide in the mesosphere". Q. J. R. Meteorol. Soc., **96**, 767–769.
6. Crutzen, P. J., 1971: Energy conversions and mean vertical motions in the high latitude summer mesosphere and lower thermosphere. In Mesospheric Models and Related Experiments, G. Fiocco, (ed.), D. Reidel Publ. Co., Dordrecht, Holland, 78–88.
7. Crutzen, P. J., I. T. N. Jones and R. P. Wayne, 1971: Calculation of O2 ($1\Delta g$) in the atmosphere using new laboratory data. *J. Geophys. Res.,* **76**, 1490–1497.
8. Crutzen, P. J., 1971: Ozone production rates in an oxygen-hydrogen-nitrogen oxide atmosphere. *J. Geophys. Res.,* **76**, 7311–7327.

9.   Crutzen, P. J., 1972: SST's - a threat to the earth's ozone shield. Ambio, **1**, 41–51.

10.   Crutzen, P. J., 1973: A discussion of the chemistry of some minor constituents in the stratosphere and troposphere. Pure App. Geophys., **106–108**, 1385–1399.

11.   Crutzen, P. J., 1973: Gas-phase nitrogen and methane chemistry in the atmosphere. In: *Physics and Chemistry of Upper Atmospheres*, B.M. McCormac (ed.), Reidel, Dordrecht, Holland, 110–124.

12.   Crutzen, P. J., 1974: A review of upper atmospheric photochemistry. Can. J. Chem., **52**, 1569–1581.

13.   Crutzen, P. J., 1974: Estimates of possible future ozone reductions from continued use of fluorochloromethanes (CF_2Cl_2, $CFCl_3$). Geophys. Res. Lett., **1**, 205–208.

14.   Crutzen, P. J., 1974: Estimates of possible variations in total ozone due to natural causes and human activities. Ambio, **3**, 201–210.

15.   Crutzen, P. J., 1974: Photochemical reactions initiated by and influencing ozone in unpolluted tropospheric air. Tellus, **26**, 48–57.

16.   Cadle, R. D., P. J. Crutzen and D. H. Ehhalt, 1975: Heterogeneous chemical reactions in the stratosphere. *J. Geophys. Res.*, **80**, 3381–3385.

17.   Crutzen, P. J., I. S. A. Isaksen and G. C. Reid, 1975: Solar proton events: stratospheric sources of nitric oxide. Science, **189**, 457–459.

18.   Johnston, H. S., D. Garvin, M. L. Corrin, P. J. Crutzen, R. J. Cvetanovic, D. D. Davis, E. S. Domalski, E. E. Ferguson, R. F. Hampson, R. D. Hudson, L. J. Kieffer, H. I. Schiff, R. L. Taylor, D. D. Wagman and R. T. Watson, 1975: Chemistry in the stratosphere, Chapter 5, CIAP Monograph 1. The Natural Stratosphere of 1974, DOT-TST-75-51, U.S. Department of Transportation, Climate Impact Assessment Program.

19.   Schmeltekopf, A. L., P. D. Goldan, W. R. Henderson, W. J. Harrop, T. L. Thompson, F. C. Fehsenfeld, H. I. Schiff, P. J. Crutzen, I. S. A. Isaksen and E. E. Ferguson, 1975: Measurements of stratospheric $CFCl_3$, CF_2Cl_2, and N_2O. Geophys Res. Lett., **2**, 393–396.

20.   Zerefos, C. S. and P. J. Crutzen, 1975: Stratospheric thickness variations over the northern hemisphere and their possible relation to solar activity. *J. Geophys. Res.*, **80**, 5041–5043.

21.   Crutzen, P. J., 1976: Upper limits on atmospheric ozone reductions following increased application of fixed nitrogen to the soil. Geophys. Res. Lett., **3**, 169–172.

22.   Crutzen, P. J., 1976: The possible importance of CSO for the sulfate layer of the stratosphere. Geophys. Res. Lett., **3**, 73–76.

23.   Crutzen, P. J. and G. C. Reid, 1976: Comments on "Biotic extinctions by solar flares". Nature, **263**, 259.

24.   Fehsenfeld, F. C., P. J. Crutzen, A. L. Schmeltekopf, C. J. Howard, D. L. Albritton, E. E. Ferguson, J. A. Davidson and H. I. Schiff, 1976: Ion chemistry of chlorine compounds in the troposphere and stratosphere. *J. Geophys. Res.*, **81**, 4454–4460.

25.	Reid, G. C., I. S. A. Isaksen, T. E. Holzer and P. J. Crutzen, 1976: Influence of ancient solar proton events on the evolution of life. Nature, **259**, 177–179.

26.	Crutzen, P. J. and D. H. Ehhalt, 1977: Effects of nitrogen fertilizers and combustion on the stratospheric ozone layer. Ambio, **6**, 1–3, 112–117.

27.	Crutzen, P. J. and J. Fishman, 1977: Average concentrations of OH in the troposphere, and the budgets of CH_4, CO, H_2 and CH_3CCl_3. Geophys. Res. Lett., **4**, 321–324.

28.	Fishman, J. and P. J. Crutzen, 1977: A numerical study of tropospheric photochemistry using a one-dimensional model. *J. Geophys. Res.,* **82**, 5897–5906.

29.	Heath, D. F., A. J. Krueger and P. J. Crutzen, 1977: Solar proton event: influence on stratospheric ozone. Science, **197**, 886–889.

30.	Hidalgo, H. and P. J. Crutzen, 1977: The tropospheric and stratospheric composition perturbed by NO_x emissions of high altitude aircraft. *J. Geophys. Res.,* **82**, 5833–5866.

31.	Isaksen, I. S. A. and P. J. Crutzen, 1977: Uncertainties in calculated hydroxyl radical densities in the troposphere and stratosphere. Geophysica Norvegica, **31**, 4, 1–10.

32.	Isaksen, I. S. A., K. H. Midtboe, J. Sunde and P. J. Crutzen, 1977: A simplified method to include molecular scattering and reflection in calculations of photon fluxes and photo-dissociation rates. Geophysica Norvegica, **31**, 11–26.

33.	Schmeltekopf, A. L., D. L. Albritton, P. J. Crutzen, D. Goldan, W. J. Harrop, W. R. Henderson, J. R. McAfee, M. McFarland, H. I. Schiff, T. L. Thompson, D. J. Hofmann and N. T. Kjome, 1977: Stratospheric nitrous oxide altitude profiles at various latitudes. J. Atmos. Sci., **34**, 729–736.

34.	Crutzen, P. J. and C. J. Howard, 1978: The effect of the HO_2 + NO reaction rate constant on one-dimensional model calculations of stratospheric ozone perturbations. Pure Appl. Geophys., **116**, 487–510.

35.	Crutzen, P. J., I. S. A. Isaksen and J. R. McAfee, 1978: The impact of the chlorocarbon industry on the ozone layer. *J. Geophys. Res.,* **83**, 345–363.

36.	Fishman, J. and P. J. Crutzen, 1978: The origin of ozone in the troposphere. Nature, **274**, 855–858.

37.	Reid, G. C., J. R. McAfee and P. J. Crutzen, 1978: Effects of intense stratospheric ionization events. Nature, **257**, 489–492.

38.	Zimmerman, P. R., R. B. Chatfield, J. Fishman, P. J. Crutzen and P. L. Hanst, 1978: Estimates on the production of CO and H_2 from the oxidation of hydrocarbon emissions from vegetation. Geophys. Res. Lett., **5**, 679–682.

39.	Crutzen, P. J., 1979: The role of NO and NO_2 in the chemistry of the troposphere and stratosphere. Ann. Rev. Earth Planet. Sci., **7**, 443–472.

40.	Crutzen, P. J., 1979: Chlorofluoromethanes: threats to the ozone layer. Rev. Geophys. Space Phys., **17**, 1824–1832.

41.	Crutzen, P. J., L. E. Heidt, J. P. Krasnec, W. H. Pollock and W. Seiler, 1979: Biomass burning as a source of atmospheric gases CO, H_2, N_2O, NO, CH_3Cl and COS. Nature, **282**, 253–256.

42. Dickerson, R. R., D. H. Stedman, W. L. Chameides, P. J. Crutzen and J. Fishman, 1979: Actinometric measurements and theoretical calculations of $J(O_3)$, the rate of photolysis of ozone to $O(^1D)$. Geophys. Res. Lett., **6**, 833–836.

43. Fishman, J., V. Ramanathan, P. J. Crutzen and S. C. Liu, 1979: Tropospheric ozone and climate. Nature, **282**, 818–820.

44. Fishman, J., S. Solomon and P. J. Crutzen, 1979: Observational and theoretical evidence in support of a significant in situ photochemical source of tropospheric ozone. Tellus, **31**, 432–446.

45. Berg, W. W., P. J. Crutzen, F. E. Grahek, S. N. Gitlin and W. A. Sedlacek, 1980: First measurements of total chlorine and bromine in the lower stratosphere. Geophys. Res. Lett., **7**, 937–940

46. Crutzen, P. J. and S. Solomon, 1980: Response of mesospheric ozone to particle precipitation. Planet. Space Sci., **28**, 1147–1153.

47. Heidt, L. E., J. P. Krasnec, R. A. Lueb, W. H. Pollock, B. E. Henry and P. J. Crutzen, 1980: Latitudinal distributions of CO and CH_4 over the Pacific. *J. Geophys. Res.*, **85**, 7329–7336.

48. Seiler, W. and P. J. Crutzen, 1980: Estimates of gross and net fluxes of carbon between the biosphere and the atmosphere from biomass burning. Climatic Change, **2**, 207–247

49. Thomas, G. E., C. A. Barth, E. R. Hansen, C. W. Hord, G. M. Lawrence, G. H. Mount, G. J. Rottman, D. W. Rusch, A. I. Stewart, R. J. Thomas, J. London, P. L. Bailey, P. J. Crutzen, R. E. Dickinson, J. C. Gille, S.C. Liu, J. F. Noxon and C.B. Farmer, 1980: Scientific Objectives of the Solar Mesosphere Explorer Mission. Pure Appl. Geophys., **118**, 591–615.

50. Rodhe, H., P. J. Crutzen and A. Vanderpol, 1981: Formation of sulfuric acid in the atmosphere during long range transport. Tellus, **33**, 132–141.

51. Rusch, D. W., J. C. Gérard, S. Solomon, P. J. Crutzen and G. C. Reid, 1981: The effects of particle precipitation events on the neutral and ion chemistry of the middle atmosphere – I. Odd Nitrogen. Planet. Space Sci., **29**, 767–774.

52. Solomon, S. and P. J. Crutzen, 1981: Analysis of the August 1972 solar proton event including chlorine chemistry. *J. Geophys. Res.,* **86**, 1140–1146.

53. Solomon, S., D. W. Rusch, J. C. Gérard, G. C. Reid and P. J. Crutzen, 1981: The effect of particle precipitation events on the neutral and ion chemistry of the middle atmosphere - II. Odd Hydrogen. Planet. Space Sci., **29**, 885–892.

54. Baulch, D. L., R. A. Cox, P. J. Crutzen, R. F. Hampson Jr., J. A. Kerr and J. Troe, 1982: Evaluated kinetic and photochemical data for atmospheric chemistry: Supplement I. J. Phys. Chem. Ref. Data, **11**, 327–496.

55. Crutzen, P. J., 1982: The global distribution of hydroxyl. In: *Atmospheric Chemistry*, ed. E.D. Goldberg, pp. 313–328. Dahlem Konferenzen 1982. Berlin, Heidelberg, New York: Springer-Verlag.

56. Crutzen, P. J. and J. W. Birks, 1982: The atmosphere after a nuclear war: Twilight at Noon. Ambio, **2&3**, 114–125.

57. Hahn, J. and P. J. Crutzen, 1982: The role of fixed nitrogen in atmospheric photochemistry. Phil. Trans. R. Soc. Lond., **B 296**, 521–541.

58.   Solomon, S., P. J. Crutzen and R. G. Roble, 1982: Photochemical coupling between the thermosphere and the lower atmosphere. 1. Odd nitrogen from 50 to 120 km. *J. Geophys. Res.,* **87**, 7206–7220.

59.   Solomon, S., E. E. Ferguson, D. W. Fahey and P. J. Crutzen, 1982: On the chemistry of H_2O, H_2 and meteoritic ions in the mesosphere and lower thermosphere. Planet. Space Sci., **30**, 1117–1126.

60.   Solomon, S., G. C. Reid, R. G. Roble and P. J. Crutzen, 1982: Photochemical coupling between the thermosphere and the lower atmosphere. 2. D-region ion chemistry and the winter anomaly. *J. Geophys. Res.,* **87**, 7221–7227.

61.   Zimmerman, P. R., J. P. Greenberg, S. O. Wandiga and P. J. Crutzen, 1982. Termites: A Potentially Large Source of Atmospheric Methane, Carbon Dioxide, and Molecular Hydrogen. Science, **218**, 563–565.

62.   Bolin, B., P. J. Crutzen, P. M. Vitousek, R. G. Woodmansee, E. D. Goldberg and R. B. Cook, 1983: Interactions of biochemical cycles, in: B. Bolin and R.B. Cook, Eds: *The Major Biochemical Cycles and Their Interactions,* SCOPE 21, pp. 1–40, Wiley, Chichester.

63.   Crutzen, P. J., 1983: Atmospheric interactions - homogeneous gas reactions of C, N, and S containing compounds, in: B. Bolin and R.B. Cook, Eds.: *The Major Biochemical Cycles and Their Interactions,* SCOPE 21, pp. 67–114, Wiley, Chichester.

64.   Crutzen, P. J. and L. T. Gidel, 1983: A two-dimensional photochemical model of the atmosphere. 2. The tropospheric budgets of the anthropogenic chlorocarbons, CO, CH_4, CH_3Cl and the effect of various NO_x sources on tropospheric ozone. *J. Geophys. Res.,* **88**, 6641–6661.

65.   Crutzen, P. J. and U. Schmailzl, 1983: Chemical budgets of the stratosphere. Planet. Space. Sci., **31**, 1009–1032.

66.   Frederick, J. E., R. B. Abrams and P. J. Crutzen, 1983: The Delta Band Dissociation of Nitric Oxide: A Potential Mechanism for Coupling Thermospheric Variations to the Mesosphere and Stratosphere. *J. Geophys. Res.,* **88**, 3829–3835.

67.   Gidel, L. T., P. J. Crutzen and J. Fishman, 1983: A two-dimensional photochemical model of the atmosphere. 1: Chlorocarbon emissions and their effect on stratospheric ozone. *J. Geophys. Res.,* **88**, 6622–6640.

68.   Chatfield, R. B. and P. J. Crutzen, 1984: Sulfur Dioxide in Remote Oceanic Air: Cloud Transport of Reactive Precursors. *J. Geophys. Res.,* **89** (D5), 7111–7132.

69.   Crutzen, P. J., I. E. Galbally and C. Brühl, 1984: Atmospheric Effects from Postnuclear Fires. Climatic Change, **6**, 323–364.

70.   Crutzen, P. J., M. T. Coffey, A. C. Delany, J. Greenberg, P. Haagenson, L. Heidt, R. Heidt, L. Lueb, W. G. Mankin, W. Pollock, W. Seiler, A. Wartburg and P. Zimmerman, 1985: Observations of air composition in Brazil between the equator and 20°S during the dry season. Acta Amazonica, Manaus, **15**(1-2): 77–119.

71.   Crutzen, P. J., A. C. Delany, J. Greenberg, P. Haagenson, L. Heidt, R. Lueb, W. Pollock, W. Seiler, A. Wartburg and P. Zimmerman, 1985: Tropospheric

Chemical Composition Measurements in Brazil During the Dry Season. J. Atmos. Chem., **2**, 233–256.

72.   Crutzen, P. J., D. M. Whelpdale, D. Kley and L. A. Barrie, 1985: The cycling of sulfur and nitrogen in the remote atmosphere, in: *The Biogeochemical Cycling of Sulfur and Nitrogen in the Remote Atmosphere* (J. N. Galloway, R. J. Charlson, M. O. Andreae and H. Rodhe, Eds.), NATO ASI Series C 158, Reidel, Dordrecht, Holland, 203–212.

73.   Delany, A. C., P. Haagenson, S. Walters, A. F. Wartburg and P. J. Crutzen, 1985: Photochemically produced ozone in the emission of large scale tropical vegetation fires. *J. Geophys. Res.,* **90** (D1), 2425–2429.

74.   Crutzen, P. J. and F. Arnold, 1986: Nitric acid cloud formation in the cold Antarctic stratosphere: a major cause for the springtime "ozone hole". Nature, **324**, 651–655.

75.   Crutzen, P. J., I. Aselmann and W. Seiler, 1986: Methane production by domestic animals, wild ruminants, other herbivorous fauna, and humans. Tellus, **38B**, 271–284.

76.   Crutzen, P. J. and T. E. Graedel, 1986: The role of atmospheric chemistry in environment -development interactions, in: *Sustainable Development of the Environment* (W.C. Clark and R.E. Munn, Eds.), Cambridge University Press, 213–251.

77.   Pittock, A. B., T. P. Ackerman, P. J. Crutzen, J. C. MacCracken, C. S. Shapiro and R. P. Turco, 1986: Scenarios for a Nuclear Exchange. In: Environmental Consequences of Nuclear War Volume 1: Physical and Atmospheric Effects, SCOPE, John Wiley & Sons Ltd. (Eds.), 1986, 25–37.

78.   Pittock, A. B., T. P. Ackerman, P. J. Crutzen, J. C. MacCracken, C. S. Shapiro and R. P. Turco, 1986: Atmospheric Processes. In: Environmental Consequences of Nuclear War Volume 1: Physical and Atmospheric Effects, SCOPE, John Wiley & Sons Ltd. (Eds.), 1986, 105–147.

79.   Bingemer, H. G. and P. J. Crutzen, 1987: The production of methane from solid wastes. *J. Geophys. Res.,* **92** (D2), 2181–2187.

80.   Crutzen, P. J., 1987: Role of the tropics in the atmospheric chemistry, in: R. Dickinson, Ed: *Geophysiology of the Amazon*, Wiley, Chichester - New York, 107–131.

81.   Crutzen, P. J., 1987: Acid rain at the K/T boundary. Nature, **330**, 108–109.

82.   Barrie, L. A., J. W. Bottenheim, R. C. Schnell, P. J. Crutzen and R. A. Rasmussen, 1988: Ozone destruction and photochemical reactions at polar sunrise in the lower Arctic atmosphere. Nature, **334**, 138–141.

83.   Brühl, C. and P. J. Crutzen, 1988: Scenarios of possible changes in atmospheric temperatures and ozone concentrations due to man's activities as estimated with a one-dimensional coupled photochemical climate model. Climate Dynamics, **2**, 173–203.

84.   Crutzen, P. J., 1988: Tropospheric ozone: an Overview. In: *Tropospheric Ozone*, I.S.A. Isaksen Ed., Reidel, Dordrecht, 3–32.

85. Crutzen, P. J., 1988: Variability in Atmospheric-Chemical Systems, in: *Scales and Global Change*, SCOPE 35, T. Rosswall, R.G. Woodmansee and P.G. Risser, Eds., Wiley, Chichester, 81–108.

86. Crutzen, P. J., C. Brühl, U. Schmailzl and F. Arnold, 1988: Nitric acid haze formation in the lower stratosphere: a major contribution factor to the development of the Antarctic "ozone hole", in *Aerosols and Climate* (M.P. McCormick and P.V. Hobbs, Editors), A. Deepak Publ., Hampton, Virginia, USA, pp. 287–304.

87. Hao, W. M., D. Scharffe, E. Sanhueza and P. J. Crutzen, 1988: Production of N_2O, CH_4, and CO_2 from Soils in the Tropical Savanna During the Dry Season. J. Atmos. Chem., **7**, 93–105.

88. Horowitz, A., G. von Helden, W. Schneider, P. J. Crutzen and G. K. Moortgat, 1988: Ozone generation in the 214 nm photolysis of oxygen at 25 °C. J. Phys. Chem., **92**, 4956–4960.

89. Liu, S. C., R. A. Cox, P. J. Crutzen, D. H. Ehhalt, R. Guicherit, A. Hofzumahaus, D. Kley, S. A. Penkett, L. F. Phillips, D. Poppe and F. S. Rowland, 1988: Group Report: Oxidizing Capacity of the atmosphere, in: *The Changing Atmosphere*, F.S. Rowland and I.S.A. Isaksen, Eds., Wiley, Chichester, p. 219–232.

90. Wilson, S. R., P. J. Crutzen, G. Schuster, D. W. T. Griffith and G. Helas, 1988: Phosgene measurements in the upper troposphere and lower stratosphere. Nature, **334**, 689–691.

91. Aselmann, I. and P. J. Crutzen, 1989: Global distribution of natural freshwater wetlands and rice paddies, their net primary productivity, seasonality and possible methane emissions. J. Atmos. Chem., **8**, 307–358.

92. Brühl, C. and P. J. Crutzen, 1989: On the disproportionate role of tropospheric ozone as a filter against solar UV-B radiation. Geophys. Res. Lett., **16**, 703–706.

93. Crutzen, P. J. and C. Brühl, 1989: The impact of observed changes in atmospheric composition on global atmospheric chemistry and climate, in: *The Environmental Record in Glaciers and Ice Sheets*, Dahlem Konferenzen 1988, H. Oeschger and C.C. Langway, Eds., Wiley, Chichester, pp. 249–266.

94. Pearman, G. I., R. J. Charlson, T. Class, H. B. Clausen, P. J. Crutzen, T. Hughes, D. A. Peel, K. A. Rahn, J. Rudolph, U. Siegenthaler and D. S. Zardini, 1989: Group Report: What Anthropogenic Impacts are recorded in Glaciers? In: Dahlem Workshop Reports: *The Environmental Record in Glaciers and Ice Sheets*, Dahlem Konferenzen, H. Oeschger and C.C. Langway, Eds., Wiley, Chichester, pp. 269–286.

95. Robertson, R. P., M. O. Andreae, H. G. Bingemer, P. J. Crutzen, R. A. Delmas, J. H. Duizer, I. Fung, R. C. Harriss, M. Kanakidou, M. Keller, J. M. Melillo and G. A. Zavarzin, 1989: Group Report: Trace gas exchange and the chemical and physical climate: Critical interactions, In: Dahlem Workshop Reports: *Exchange of Trace Gases between Terrestrial Ecosystems and the Atmosphere*, Eds. M. O. Andreae and D. S. Schimel, Life Sciences Research Report 47, Wiley, Chichester, pp. 303–320.

96. Simon, F. G., J. P. Burrows, W. Schneider, G. K. Moortgat and P. J. Crutzen, 1989: Study of the reaction ClO + CH$_3$O$_2$ → Products at 300 K. J. Phys. Chem., **93**, 7807–7813.

97. Zimmermann, P. H., H. Feichter, H. K. Rath, P. J. Crutzen and W. Weiss, 1989: A global three-dimensional source-receptor model investigating Kr^{85}. Atmos. Environ. **23**, 25–35.

98. Brühl, C. and P. J. Crutzen, 1990: Ozone and climate changes in the light of the Montreal Protocol, a model study. Ambio, **19**, 293–301.

99. Chatfield, R. B. and P. J. Crutzen, 1990: Are there interactions of iodine and sulfur species in marine air photochemistry? *J. Geophys. Res.,* **95**, 22319–22341.

100. Crutzen, P. J. and M. O. Andreae, 1990: Biomass burning in the tropics: impact on atmospheric chemistry and biogeochemical cycles. Science, **250**, 1669–1678.

101. Feichter, J. and P. J. Crutzen, 1990: Parameterization of vertical tracer transport due to deep cumulus convection in a global transport model and its evaluation with 222 Radon measurements. Tellus, **42B**, 100–117.

102. Graedel, T. E. and P. J. Crutzen, 1990: Atmospheric trace constituents, in: *The Earth as Transformed by Human Action*, Eds. B.L. Turner II et al., Cambridge University Press, pp. 295–311.

103. Hao, W. M., M. H. Liu and P. J. Crutzen, 1990: Estimates of annual and regional releases of CO$_2$ and other trace gases to the atmosphere from fires in the tropics, based on the FAO statistics for the period 1975-1980, in: *Fire in the Tropical Biota*, Ecological Studies, 84, J. G. Goldammer, Ed., Springer-Verlag, Berlin, 440–462.

104. Lelieveld, J. and P. J. Crutzen, 1990: Influences of cloud and photochemical processes on tropospheric ozone. Nature, **343**, 227–233.

105. Lobert, J. M., D. H. Scharffe, W. M. Hao and P. J. Crutzen, 1990: Importance of biomass burning in the atmospheric budgets of nitrogen-containing gases. Nature, **346**, 552–554.

106. Sanhueza, E., W. M. Hao, D. Scharffe, L. Donoso and P. J. Crutzen, 1990: N$_2$O and NO emissions from soils of the northern part of the Guayana Shield, Venezuela. *J. Geophys. Res.,* **95** (D13), 22481–22488.

107. Scharffe, D., W. M. Hao, L. Donoso, P. J. Crutzen and E. Sanhueza, 1990: Soil fluxes and atmospheric concentrations of CO and CH$_4$ in the northern part of the Guayana Shield, Venezuela. *J. Geophys. Res.,* **95** (D13), 22475–22480.

108. Crutzen, P. J. and P. H. Zimmermann, 1991: The changing photochemistry of the troposphere. Tellus, **43 A/B**, 136–151.

109. Hao, W. M., D. Scharffe, J. M. Lobert and P. J. Crutzen, 1991: Emissions of N$_2$O from the burning of biomass in an experimental system. Geophys. Res. Lett., **18**, 999–1002.

110. Kanakidou, M., H. B. Singh, K. M. Valentin and P. J. Crutzen, 1991: A 2-D study of ethane and propane oxidation in the troposphere. *J. Geophys. Res.,* **96**, 15395–15413.

111. Kuhlbusch, A. T., J. M. Lobert, P. J. Crutzen and P. Warneck, 1991: Molecular nitrogen emissions from denitrification during biomass burning. Nature, **351**, 135–137.

112. Lelieveld, J. and P. J. Crutzen, 1991: The role of clouds in tropospheric photochemistry. J. Atmos. Chem., **12**, 229–267.

113. Lobert, J. M., D. H. Scharffe, W. M. Hao, T. A. Kuhlbusch, R. Seuwen, P. Warneck and P. J. Crutzen, 1991: Experimental evaluation of biomass burning emissions: Nitrogen and carbon containing compounds, in: *Global Biomass Burning: Atmospheric, Climatic and Biosphere Implications*, Ed. J. S. Levine, MIT Press, Cambridge, MA, pp. 122–125.

114. Peter, Th., C. Brühl and P. J. Crutzen, 1991: Increase in the PSC-formation probability caused by high-flying aircraft. Geophys. Res. Lett., **18**, 1465–1468.

115. Crutzen, P. J. and G. S. Golitsyn, 1992: Linkages between Global Warming, Ozone Depletion and Other Aspects of Global Environmental Change. In: *Confronting Climate Change*, I.M. Mintzer (Ed.), 15–32, Cambridge University Press.

116. Crutzen, P. J., R. Müller, Ch. Brühl and Th. Peter, 1992: On the potential importance of the gas phase reaction $CH_3O_2 + ClO \rightarrow ClOO + CH_3O$ and the hererogeneous reaction $HOCl + HCl \rightarrow H_2O + Cl_2$ in "ozone hole" chemistry. Geophys. Res. Lett., **19**, 1113–1116.

117. Kanakidou, M., P. J. Crutzen, P. H. Zimmermann and B. Bonsang, 1992: A 3-dimensional global study of the photochemistry of ethane and propane in the troposphere: Production and transport of organic nitrogen compounds, in: *Air Pollution Modeling and its Application IX*, H. van Dop and G. Kallos (Eds.), Plenum Press, New York, 415–426.

118. Langner, J., H. Rodhe, P. J. Crutzen and P. Zimmermann, 1992: Anthropogenic influence on the distribution of tropospheric sulphate aerosol. Nature, **359**, 712–715.

119. Lelieveld, J. and P. J. Crutzen, 1992: Indirect chemical effects of methane on climate warming. Nature, **355**, 339–342.

120. Luo, B. P., Th. Peter and P. J. Crutzen, 1992: Maximum supercooling of H_2SO_4 acid aerosol droplets. Ber. Bunsenges. Phys. Chem., **96**, 334–338.

121. Singh, H. B., D. O'Hara, D. Herlth, J. D. Bradshaw, S. T. Sandholm, G. L. Gregory, G. W. Sachse, D. R. Blake, P. J. Crutzen and M. Kanakidou, 1992: Atmospheric measurements of PAN and other organic nitrates at high latitudes: possible sources and sinks. *J. Geophys. Res.*, **97**, 16511–16522.

122. Singh, H. B., D. Herlth, K. Zahnle, D. O'Hara, J. Bradshaw, S. T. Sandholm, R. Talbot, P. J. Crutzen and M. Kanakidou, 1992: Relationship of PAN to active and total odd nitrogen at northern high latitudes: possible influence of reservoir species on NO_x and O_3. *J. Geophys. Res.*, **97**, 16523–16530.

123. Berges, M. G. M., R. M. Hofmann, D. Scharffe and P. J. Crutzen, 1993: Nitrous oxide emissions from motor vehicles in tunnels. *J. Geophys. Res.*, **98**, 18527–18531.

124. Crutzen, P. J. and C. Brühl, 1993: A model study of atmospheric temperatures and the concentrations of ozone, hydroxyl, and some other photochemically active gases during the glacial, the preindustrial holocene and the present. Geophys. Res. Lett., **20**, 1047–1050.

125. Crutzen, P. J. and G. R. Carmichael, 1993: Modeling the influence of fires on atmospheric chemistry, in: *Fire in the Environment: The Ecological, Atmospheric, and Climatic Importance of Vegetation Fires* (P. J. Crutzen and J. G. Goldammer, Eds.), op. cit., 90–105.

126. Dentener, F. and P. J. Crutzen, 1993: Reaction of N_2O_5 on tropospheric aerosols: Impact on the global distributions of NO_x, O_3 and OH. *J. Geophys. Res.*, **98**, 7149–7163.

127. Goldammer, J. G. and P. J. Crutzen, 1993: Fire in the Environment: Scientific Rationale and Summary of Results of the Dahlem Workshop, in: *Fire in the Environment: The Ecological, Atmospheric and Climatic Importance*, (P. J. Crutzen and J. G. Goldammer, Eds.), op. cit., 1–14.

128. Kanakidou, M. and P. J. Crutzen, 1993: Scale problems in global tropospheric chemistry modeling: Comparison of results obtained with a three-dimensional model, adopting longitudinally uniform and varying emissions of NO_x and NMHC. Chemosphere, **26**, 787–801.

129. Lelieveld, J., P. J. Crutzen and C. Brühl, 1993: Climate effects of atmospheric methane, Chemosphere, **26**, 739–768.

130. Müller, R. and P. J. Crutzen, 1993: A possible role of galactic cosmic rays in chlorine activation during polar night. *J. Geophys. Res.*, **98**, 20483–20490.

131. Peter, Th. and P. J. Crutzen, 1993: The role of stratospheric cloud particles in polar ozone depletion. An overview. J. Aerosol Sci., **24**, Suppl. 1, 119–120.

132. Russell III, J. M., A. F. Tuck, L. L. Gordley, J. H. Park, S. R. Drayson, J. E. Harries, R. J. Cicerone and P. J. Crutzen, 1993: HALOE Antarctic observations in the spring of 1991, Geophys. Res. Lett., **20**, 719–722.

133. Schupp, M., P. Bergamaschi, G. W. Harris and P. J. Crutzen, 1993: Development of a tunable diode laser absorption spectrometer for measurements of the $^{13}C/^{12}C$ ratio in methane. Chemosphere, **26**, 13–22.

134. Carslaw, K. S., B. P. Luo, S. L. Clegg, Th. Peter, P. Brimblecombe and P. J. Crutzen, 1994: Stratospheric aerosol growth and HNO_3 gas phase depletion from coupled HNO_3 and water uptake by liquid particles. Geophys. Res. Lett., **21**, 2479–2482.

135. Chen, J.-P. and P. J. Crutzen, 1994: Solute effects on the evaporation of ice particles, *J. Geophys. Res.*, **99**, 18847–18859.

136. Cox, R. A., A. R. MacKenzie, R. Müller, Th. Peter and P. J. Crutzen, 1994: Activation of stratospheric chlorine by reactions in liquid sulphuric acid. Geophys. Res. Lett., **21**, 1439–1442.

137. Crowley, J. N., F. Helleis, R. Müller, G. K. Moortgat and P. J. Crutzen, 1994: CH_3OCl: UV/visible absorption cross sections, J values and atmospheric significance, *J. Geophys. Res.*, **99**, 20683–20688.

138.   Crutzen, P. J., 1994: Global Tropospheric Chemistry, Proceedings of the NATO Advanced Study Institute on Low Temperature Chemistry of the Atmosphere, Maratea, Italy, August 29–September 11, 1993, NATO ASI Series I, 21 (Eds. G.K. Moortgat et al.), Springer, Heidelberg, 465–498.

139.   Crutzen, P. J., 1994: Global budgets for non-CO_2 greenhouse gases, Environmental Monitoring and Assessement, **31**, 1–15.

140.   Crutzen, P. J., J. Lelieveld and Ch. Brühl, 1994: Oxidation processes in the atmosphere and the role of human activities: Observations and model results, in: *Environmental Oxidants*, J.O. Nriagu and M.S. Simmons (Eds.), Vol. 28 in *"Advances in Environmental Science and Technology"*, Wiley, Chichester, 63–93.

141.   Dentener, F. J. and P. J. Crutzen, 1994: A three dimensional model of the global ammonia cycle, J. Atmos. Chem., **19**, 331–369.

142.   Deshler, T., Th. Peter, R. Müller and P. J. Crutzen, 1994: The lifetime of leewave-induced ice particles in the Arctic stratosphere: I. Balloonborne observations. Geophys. Res. Lett., **21**, 1327–1330.

143.   Lelieveld, J. and P. J. Crutzen, 1994: Role of deep cloud convection in the ozone budget of the troposphere, Science, **264**, 1759–1761.

144.   Luo, B. P., Th. Peter and P. J. Crutzen, 1994: Freezing of stratospheric aerosol droplets. Geophys. Res. Lett., **21**, 1447–1450.

145.   Luo, B. P., S. L. Clegg, Th. Peter, R. Müller and P. J. Crutzen, 1994: HCl solubility and liquid diffusion in aqueous sulfuric acid under stratospheric conditions, Geophys. Res. Lett., **21**, 49–52.

146.   Müller, R., Th. Peter, P. J. Crutzen, H. Oelhaf, G. Adrian, Th. v. Clarman, A. Wegner, U. Schmidt and D. Lary, 1994: Chlorine chemistry and the potential for ozone depletion in the Arctic stratosphere in the winter of 1991/92. Geophys. Res. Lett., **21**, 1427–1430.

147.   Peter, Th. and P. J. Crutzen, 1994: Modelling the chemistry and micro-physics of the could stratosphere. Proceedings of the NATO Advanced Study Institute on Low Temperature Chemistry of the Atmosphere, Maratea, Italy, August 29–September 11, 1993, NATO ASI Series I, 21 (Eds. G.K. Moortgat et al.), Springer, Heidelberg, 499–530.

148.   Peter, Th., P. J. Crutzen, R. Müller and T. Deshler, 1994: The lifetime of leewave- induced particles in the Artic stratosphere: II. Stabilization due to NAT-coating. Geophys. Res. Lett., **21**, 1331–1334.

149.   Sanhueza, E., L. Donoso, D. Scharffe and P. J. Crutzen, 1994: Carbon monoxide fluxes from natural, managed, or cultivated savannah grasslands, *J. Geophys. Res.,* **99**, 16421–16425.

150.   Sassen, K., Th. Peter, B. P. Luo and P. J. Crutzen, 1994: Volcanic Bishop's ring: Evidence for a sulfuric acid tetrahydrate particle aerosol, Appl. Optics., **33**, 4602–4606.

151.   Singh, H. B., D. O'Hara, D. Herlth, W. Sachse, D. R. Blake, J. D. Bradshaw, M. Kanakidou and P. J. Crutzen, 1994: Acetone in the atmosphere: Distribution, sources and sinks. *J. Geophys. Res.,* **99**, 1805–1819.

152.   Crutzen, P. J., 1995: On the role of CH_4 in atmospheric chemistry: Sources, sinks and possible reductions in anthropogenic sources. Ambio, **24**, 52–55.

153.   Crutzen, P. J., 1995: Introductory Lecture: Overview of tropospheric chemistry: Developments during the past quarter century and a look ahead. Faraday Discuss., **100**, 1–21.

154.   Crutzen, P. J., 1995: Ozone in the troposphere, in: Composition, Chemistry, and Climate of the Atmosphere. H.B. Singh (Ed.), Van Nostrand Reinhold Publ., New York, 349–393.

155.   Crutzen, P. J., 1995: The role of methane in atmospheric chemistry and climate, in: Ruminant Physiology: Digestion, Metabolism, Growth and Reproduction: Proceedings of the Eighth International Symposium on Ruminant Physiology, Eds. W. v. Engelhardt, S. Leonhard-Marek, G. Breves and D. Giesecke, Ferdinand Enke Verlag, Stuttgart, 291–315.

156.   Crutzen, P. J., J.-U. Grooß, C. Brühl, R. Müller, J. M. Russell III, 1995: A reevaluation of the ozone budget with HALOE UARS data: no evidence for the ozone deficit. Science, **268**, 705–708.

157.   Finkbeiner, M., J. N. Crowley, O. Horie, R. Müller, G. K. Moortgat and P. J. Crutzen, 1995: Reaction between HO_2 and ClO: Product formation between 210 and 300 K. J. Phys. Chem., **99**, 16264–16275.

158.   Kanakidou, M., F. J. Dentener and P. J. Crutzen, 1995: A global three-dimensional study of the fate of HCFCs and HFC-134a in the troposphere. *J. Geophys. Res.*, **100**, 18781–18801.

159.   Koop, T., U. M. Biermann, W. Raber, B. P. Luo, P. J. Crutzen and Th. Peter, 1995: Do stratospheric aerosol droplets freeze above the ice frost point? *J. Geophys. Res.*, **22**, 917–920.

160.   Kuhlbusch, T. A. J. and P. J. Crutzen, 1995: Toward a global estimate of black carbon in residues of vegetation fires representing a sink of atmospheric CO_2 and a source of O_2. Global Biogeochem. Cycles, **4**, 491–501.

161.   Meilinger, S. K., T. Koop, B. P. Luo, T. Huthwelker, K. S. Carslaw, P. J. Crutzen and Th. Peter, 1995: Size-dependent stratospheric droplet composition in lee wave temperature fluctuations and their potential role in PSC freezing. Geophys. Res. Lett., **22**, 3031–3034.

162.   Rodhe, H. and P. J. Crutzen, 1995: Climate and CCN, Nature, **375**, 111.

163.   Sander, R., J. Lelieveld and P. J. Crutzen, 1995: Modelling of the nighttime nitrogen and sulfur chemistry in size resolved droplets of an orographic cloud, J. Atmos. Chem., **20**, 89–116.

164.   Schade, G. W. and P. J. Crutzen, 1995: Emission of aliphatic amines from animal husbandry and their reactions: Potential source of N_2O and HCN. J. Atmos. Chem., **22**, 319–346.

165.   Singh, H. B., M. Kanakidou, P. J: Crutzen and D. J. Jacob, 1995: High concentrations and photochemical fate of oxygenated hydrocarbons in the global troposphere. Nature, **378**, 50–54.

166.   Vömel, H., S. J. Oltmans, D. Kley and P. J. Crutzen, 1995: New evidence for the stratospheric dehydration mechanism in the aquatorial Pacific. Geophys. Res. Lett., **22**, 3235–3238.

167. Wang, C. and P. J. Crutzen, 1995: Impact of a simulated severe local storm on the redistribution of sulfur dioxide. *J. Geophys. Res.,* **100**, 11357–11367.

168. Wang, C., P. J. Crutzen, V. Ramanathan, S. F. Williams, 1995: The role of a deep convective storm over the tropical Pacific Ocean in the redistribution of atmospheric chemical species. *J. Geophys. Res.,* **100**, 11509–11516.

169. Wayne, R. P., G. Poulet, P. Biggs, J. P. Burrows, R. A. Cox, P. J. Crutzen, G. D. Hayman, M. E. Jenkin, G. Le Bras, G. K. Moortgat, U. Platt, and R. N. Schindler, 1995: Halogen oxides: Radicals, sources and reservoirs in the laboratory and in the atmosphere. Atmos. Environ., **29**, 2677–2881 (special issue).

170. Bergamaschi, P., C. Brühl, C. A. M. Brenninkmeijer, G. Saueressig, J. N. Crowley, J. U. Grooß, H. Fischer and P. J. Crutzen, 1996: Implications of the large carbon kinetic isotope effect in the reaction $CH_4 + Cl$ for the $^{13}C/^{12}C$ ratio of stratospheric CH_4. Geophys. Res. Lett., **23**(17), 2227–2230.

171. Berges, M. G. M. and J. Crutzen, 1996: Estimates of global N_2O emissions from cattle, pig and chicken manure, including a discussion of CH_4 emissions. J. Atmos. Chem., **24**, 241–269.

172. Biermann, U. M., T. Presper, J. Mößinger, P. J. Crutzen and Th. Peter, 1996: The unsuitability of meteoritic and other nuclei for polar stratospheric cloud freezing, Geophys. Res. Lett., **23**, 1693–1696.

173. Brenninkmeijer, C. A. M., R. Müller, P. J. Crutzen, D. C. Lowe, M. R. Manning, R. J. Sparks and P. F. J. van Velthoven, 1996: A large ^{13}CO deficit in the lower Antarctic stratosphere due to "ozone hole" chemistry: Part I, observations. Geophys. Res. Lett., **23**(16), 2125–2128.

174. Brühl, C., S. R. Drayson, J. M. Russell III, P. J. Crutzen, J. M. McInerney, P. N. Purcell, H. Claude, H. Gernandt, T. J. McGee, I. S. McDermid and M. R. Gunson, 1996: Halogen Occultation Experiment ozone channel validation. *J. Geophys. Res.,* **101**, 10217–10240.

175. Chen, J.-P. and P. J. Crutzen, 1996: Reply. J. Geophys. Res., **101** (D17), 23037–23038.

176. Crutzen, P. J. and C. Brühl, 1996: Mass extinctions and supernova explosions. Proc. Natl. Acad. Sci. USA, **93**, 1582–1584.

177. Crutzen, P. J., G. S. Golitsyn, N. F. Elanskii, C. A. M. Brenninkmeijer, D. Scharffe, I. B. Belikov and A. S. Elokhov, 1996: Observations of minor impurities in the atmosphere over the Russian territory with the application of a railroad laboratory car. Trans. of the Russ. Acad. of Sciences/Earth Science Sections, **351**, 1289–1293 (Translated from Doklady Akademii Nauk, **350**, 819–823, 1996).

178. Dentener, F. J., G. R. Carmichael, Y. Zhang, J. Lelieveld, and P. J. Crutzen, 1996: Role of mineral aerosol as a reactive surface in the globel troposphere. *J. Geophys. Res.,* **101** (D17), 22869–22889.

179. Kley, D., P. J. Crutzen, H. G. J. Smit, H. Vömel, S. Oltmans, H. Grassl and V. Ramanathan, 1996: Observations of near-zero ozone concentrations over the convective Pacific: Effects on air chemistry. Science, **274**, 230–233.

180. Kuhlbusch, T. A. J., M. O. Andreae, H. Cachier, J. G. Goldammer, J.-P. Lacaux, R. Shea and P. J. Crutzen, 1996: Black carbon formation by savannah fires: Measurements and implications for the global carbeon cycle. *J. Geophys. Res.,* **101**, 23651–23665.

181. Müller, R., C. A. M. Brenninkmeijer and P. J. Crutzen, 1996: A large ^{13}CO deficit in the lower Antarctic stratosphere due to "ozone hole" chemistry: Part II, Modeling. Geophys. Res. Lett., **23**(16), 2129–2132.

182. Müller, R., P. J. Crutzen, J.-U. Grooß, C. Brühl, J. M. Russell III, A. F. Tuck, 1996: Chlorine activation and ozone depletion in the Arctic vortex: Observations by the Halogen Occultation Experiment on the Upper Atmosphere Research Satellite. *J. Geophys. Res.,* **101**, 12531–12554.

183. Sander, R. and P. J. Crutzen, 1996: Model study indicating halogen activation and ozone destruction in polluted air masses transported to the sea. *J. Geophys. Res.,* **101**D, 9121–9138.

184. Shorter, J. H., J. B. McManus, C. E. Kolb, E. J. Allewine, B. K. Lamb, B. W. Mosher, R. C. Harriss, U. Parchatka, H. Fischer, G. W. Harris, P. J. Crutzen and H.-J. Karbach, 1996: Methane emissions in urban aereas in Eastern Germany. J. Atmos. Chem., **24**, 121–140.

185. Singh, H. B., D. Herlth, R. Kolyer, L. Salas, J. D. Bradshaw, S. T. Sandholm, D. D. Davis, J. Crawford, Y. Kondo, M. Koike, R. Talbot, G. L. Gregory, G. W. Sachse, E. Browell, D. R. Blake, F. S. Rowland, R. Newell, J. Merrill, B. Heikes, S. C. Liu, P. J. Crutzen, M. Kanakidou, 1996: Reactive nitrogen and ozone over the western Pacific: Distribution, partitioning, and sources. *J. Geophys. Res.,* **101**, 1793–1808.

186. Vogt, R., P. J. Crutzen and R. Sander, 1996: A mechanism for halogen release form sea-salt aerosol in the remote marine boundary layer. Nature, **383**, 327–330.

187. Andreae, M. O. and P. J. Crutzen, 1997: Atmospheric Aerosols: Biogeochemical sources and role in atmospheric chemistry. Science, **276**, 1052–1058.

188. Fischer, H., A. E. Waibel, M. Welling, F. G. Wienhold, T. Zenker, P. J. Crutzen, F. Arnold, V. Bürger, J. Schneider, A. Bregman, J. Lelieveld and P. C. Siegmund, 1997: Observations of high concentration of total reactive nitrogen (NO_y) and nitric acid (HNO_3) in the lower Arctic stratosphere during the Stratosphere-Troposphere Experiment by Aircraft Measurements (STREAM) II campaign in February 1995. *J. Geophys. Res.,* **102**(D19), 23559–23571

189. Grooss, J.-U., R. B. Pierce, P. J. Crutzen, W. L. Grose and J. M. Russell III, 1997: Re-formation of chlorine reservoirs in southern hemisphere polar spring. *J. Geophys. Res.,* **102**, 13141–13152.

190. Hein, R. and P. J. Crutzen, 1997: An inverse modeling approach to investigate the global atmospheric methane cycle. Global Biogeochem. Cycles, **11**, 43–76.

191. Kley, D., H. G. J. Smit, H. Vömel, H. Grassl, V. Ramanathan, P. J. Crutzen, S. Williams, J. Meywerk and S. J. Oltmans, 1997: Tropospheric water-vapour

and ozone cross-sections in a zonal plane over the central equatorial Pacific Ocean. Q.J.R. Meteorol. Soc., **123**, 2009–2040.

192. Koop, T., B. P. Luo, U.-M. Biermann, P. J. Crutzen and Th. Peter, 1997: Freezing of $HNO_3/H_2SO_4/H_2O$ solutions at stratospheric temperatures: Nucleation statistics and experiments. J. Phys. Chem., **101**, 1117–1133.

193. Lelieveld, J., A. Bregman, F. Arnold, V. Bürger, P. J. Crutzen, H. Fischer, A. Waibel, P. Siegmund, and P. F. J. van Velthoven,, 1997: Chemical perturbation of the lowermost stratosphere through exchange with the troposphere. Geophys. Res. Lett., **24**, 603–606.

194. Müller, R., P. J. Crutzen, J.-U. Grooß, C. Brühl, J. M. Russell III, H. Gernandt, D. S. McKenna and A. F. Tuck, 1997: Severe chemical ozone loss in the Arctic during the winter of 1995-96. Nature, **389**, 709–712.

195. Müller, R., J.-U. Grooß, D. S. McKenna, P. J. Crutzen, C. Brühl, J. M. Russell III and A. F. Tuck, 1997: HALOE observations of the vertical structure of chemical ozone depletion in the Arctic vortex during winter and early spring 1996-1997. Geophys. Res. Lett., **24**, 2717–2720.

196. Pierce, R. B., J.-U. Grooss, W. L. Grose, J. M. Russell III, P. J. Crutzen, T. D. Fairlie and G. Lingenfelser, 1997: Photochemical calculations along air mass trajectories during ASHOE/MAESA. *J. Geophys. Res.,* **102**, 13153–13167.

197. Roehl, C. M., J. B. Burkholder, G. K. Moortgat, A. R. Ravishankara and P. J. Crutzen, 1997: Temperature dependence of UV absorption cross sections and atmospheric implications of several alkyl iodides. *J. Geophys. Res.,* **102**(D11), 12819–12829.

198. Sander, R., R. Vogt, G. W. Harris and P. J. Crutzen, 1997: Modeling the chemistry of ozone, halogen compounds, and hydrocarbons in the arctic troposphere during spring. Tellus, **49B**, 522–532.

199. Bergamaschi, P., C. A. M. Brenninkmeijer, M. Hahn, T. Röckmann, D. H. Scharffe, P. J. Crutzen, N. F. Elansky, I. B. Belikov, N. B. A. Trivett and D. E. J. Worthy, 1998: Isotope analysis based source identification for atmospheric CH_4 and CO sampled across Russia using the Trans-Siberian railroad. *J. Geophys. Res.,* **103**(D7), 8227–8235.

200. Biermann, U. M., J. N. Crowley, T. Huthwelker, G. K. Moortgat, P. J. Crutzen and Th. Peter, 1998: FTIR studies on lifetime prolongation of stratospheric ice particles due to NAT coating. Geophys. Res. Lett., **25**, 3939–3942.

201. Brühl, C., P. J. Crutzen and J.-U. Grooß, 1998: High-latitude, summertime NO_x activation and seasonal ozone decline in the lower stratosphere: Model calculations based on observations by HALOE on UARS. *J. Geophys. Res.,* **103**(D3), 3587–3597.

202. Crutzen, P. J., N. F. Elansky, M. Hahn, G. S. Golitsyn, C. A. M. Benninkmeijer, D. H. Scharffe, I. B. Belikov, M. Maiss, P. Bergamaschi, T. Röckmann, A. M. Grisenko and V. M. Sevostyanov, 1998: Trace gas measurements between Moscow and Vladivistok using the Trans-Siberian Railroad. J. Atmos. Chem., **29**, 179–194.

203. Dong, Y., D. Scharffe, J. M. Lobert, P. J. Crutzen and E. Sanhueza, 1998: Fluxes of CO_2, CH_4 and N_2O from a temperate forest soil: the effects of leaves and humus layers. Tellus, **50B**, 243–252.

204. Hegels, E., P. J. Crutzen, T. Klüpfel, D. Perner and J. P. Burrows, 1998: Global distribution of atmospheric bromine-monoxide from GOME on earth observing satellite ERS-2. Geophys. Res. Lett., **25**, 3127–3130.

205. Keene, W. C., R. Sander, A. A. P. Pszenny, R. Vogt, P. J. Crutzen and J. N. Galloway, 1998: Aerosol pH in the marine boundary layer: A review and model evaluation. J. Aerosol Sci., **29**, 339-356.

206. Landgraf, J. and P. J. Crutzen, 1998: An efficient method for online calculations of photolysis and heating rates. J. Atmos. Sci., **55**, 863–878.

207. Lawrence, M. G. and P. J. Crutzen, 1998: The impact of cloud particle gravitational settling on soluble trace gas distributions. Tellus, **50B**, 263–289.

208. Lelieveld, J., P. J. Crutzen and F. J. Dentener, 1998: Changing concentration, lifetime and climate forcing of atmospheric methane. Tellus, **50B**, 128–150.

209. Röckmann, Th., C. A. M. Brenninkmeijer, P. Neeb and P. J. Crutzen, 1998: Ozonolysis of nonmethane hydrocarbons as a source of the observed mass independent oxygen isotope enrichment in tropospheric CO. *J. Geophys. Res.,* **103**(D1), 1463–1470.

210. Röckmann, Th., C. A. M. Brenninkmeijer, G. Saueressig, P. Bergamaschi, J. N. Crowley, H. Fischer and P. J. Crutzen, 1998: Mass-independent oxygen isotope fractionation in atmospheric CO as a result of the reaction CO + OH. Science, **281**, 544–546.

211. Sanhueza, E. and P. J. Crutzen, 1998: Budgets of fixed nitrogen in the Orinoco Savannah region: Role of Pyrodenitrification. Global Biogeochem. Cycles, **12**, 653–666.

212. Sanhueza, E., Y. Dong, D. Scharffe, J. M. Lobert and P. J.Crutzen, 1998: Carbon monoxide uptake by temperature forest soils: the effects of leaves and humus layers. Tellus, **50B**, 51–58.

213. Steil, B. M. Dameris, Ch. Brühl, P. J. Crutzen, V. Grewe, M. Ponater and R. Sausen, 1998: Development of a chemistry module for GCMs: First results of a multiannual integration. Ann. Geophys., **16**, 205–228.

214. Brenninkmeijer, C. A. M., P. J. Crutzen, H. Fischer, H. Güsten, W. Hans, G. Heinrich, J. Heintzenberg, M. Hermann, T. Immelmann, D. Kersting, M. Maiss, M. Nolle, A. Pitscheider, H. Pohlkamp, D. Scharffe, K. Specht and A. Wiedensohler, 1999: Caribic – Civil aircraft for global measurement of trace gases and aerosols in the tropopause region. J. Atmos. Ocean. Technol., **16**, 1373–1383.

215. Brühl, Ch. and P. J. Crutzen, 1999: Reductions in the anthropogenic emissions of CO and their effect on CH_4. Chemosphere, **1**, 249–254.

216. Crutzen, P. J., M. Lawrence and U. Pöschl, 1999: On the background photochemistry of tropospheric ozone. Tellus, **51 A-B**, 123–146.

217. Crutzen, P. J., R. Fall, I. Galbally and W. Lindinger, 1999: Parameters for global ecosystem models. (Comment on "Effect of inter-annual climate variability on carbon storage in Amazonian ecosystems" (by Tian et al.). Nature, **399**, 535.

218. Dickerson, R. R., K. P. Rhoads, T. P. Carsey, S. J. Oltmans, J. P. Burrows and P. J. Crutzen, 1999: Ozone in the remote marine boundary layer: A possible role for halogens. *J. Geophys. Res.,* **104**(D17), 21, 385–21, 395.

219. Grooß, J. U., R. Müller, G. Becker, D. S. McKenna and P. J. Crutzen, 1999: The upper stratospheric ozone budget: An update of calculations based on HALOE data. J. Atmos. Chem., **34**, 171–183.

220. Holzinger, R., C. Warneke, A. Hansel, A. Jordan, W. Lindinger, D. Scharffe, G. Schade and P. J. Crutzen, 1999: Biomass burning as a source of formaldehyde, acetaldehyde, methanol, acetone, acetonitrile and hydrogen cyanide. Geophys. Res. Lett., **26**, 1161–1164.

221. Ingham, T., D. Bauer, R. Sander, P. J. Crutzen and J. N. Crowley, 1999: Kinetics and products of the reactions BrO + DMS and Br + DMS at 298 K. J. Phys. Chem., **103**, 7199–7209.

222. Kanakidou, M. and P. J. Crutzen, 1999: The photochemical source of carbon monoxide: Importance, uncertainties and feedbacks. Chemosphere, **1**, 91–109.

223. Lawrence, M. G., P. J. Crutzen and P. J. Rasch, 1999: Analysis of the CEPEX ozone data using a 3D chemistry-meteorology model. Q. J. R. Meteorol. Soc., **125**, 2987–3009.

224. Lawrence, M. G. and P. J. Crutzen, 1999: Influence of NO_x emissions from ships on tropospheric photochemistry and climate. Nature, **402**, 167–170.

225. Lawrence, M. G., P. J. Crutzen, P. J. Rasch, B. E. Eaton and N. M. Mahowald, 1999: A model for studies of tropospheric photochemistry: Description, global distributions and evaluation. *J. Geophys. Res.,* **104**(D21), 26245–26277.

226. Müller, R., J.-U. Grooß, D. S. McKenna, P. J. Crutzen, C. Brühl, J. M. Russell III, L. L. Gordley, J. P. Burrows and A. F. Tuck, 1999: Chemical ozone loss in the Arctic vortex in the winter 1995-96: HALOE measurements in conjunction with other observations. Ann. Geophysicae, **17**, 101–114.

227. Röckmann, T., C. A. M. Brenninkmeijer and P. J. Crutzen, 1999: Short-term variations in the $^{13}C/^{12}C$ ratio of CO as a measure of Cl activation during tropospheric ozone depletion events in the Arctic. *J. Geophys. Res.,* **104**(D21), 1691–1697.

228. Sander, R., Y. Rudich, R. von Glasow and P. J. Crutzen, 1999: The role of $BrNO_3$ in marine tropospheric chemistry: A model study. Geophys. Res. Letters, **26**, No.18, 2857–2860.

229. Sanhueza, E., P. J. Crutzen and E. Fernández, 1999: Production of boundary layer ozone from tropical American savannah biomass burning emissions. Atmos. Environ., **33**, 4969–4975.

230. Schade, G., R.-M. Hofmann and P. J. Crutzen, 1999: CO emissions from degrading plant matter. Part I: Measurements. Tellus, **51B**, 889–908.

231. Schade, G. and P. J. Crutzen, 1999: CO emissions from degrading plant matter. Part II: Estimate of a global source strength. Tellus, **51B**, 909–918.

232. Trautmann, T., I. Podgorny, J. Landgraf and P. J. Crutzen, 1999: Actinic fluxes and photodissociation coefficients in cloud fields embedded in realistic atmospheres. *J. Geophys. Res.,* **104**(D23), 30173–30192.

233. Vogt, R., R. Sander, R. von Glasow and P. J. Crutzen, 1999: Iodine Chemisty and its role in halogen activation and ozone loss in the marine boundary layer: A model study. J. Atmos. Chem., **32**, 375–395.

234. Waibel, A. E., H. Fischer, F. G. Wienhold, P. C. Siegmund, B. Lee, J. Ström, J. Lelieveld and P. J. Crutzen, 1999: Highly elevated carbon monoxide concentrations in the upper troposphere and lowermost stratosphere at northern midlatitudes during the STREAM II summer campaign in 1994. Chemosphere, **1**, 233–248.

235. Waibel, A. E., Th. Peter, K. S. Carslaw, H. Oelhaf, G. Wetzel, P. J. Crutzen, U. Pöschl, A. Tsias, E. Reimer, and H. Fischer, 1999: Arctic ozone loss due to denitrification. Science, **283**, 2064–2069.

236. Warneke, C., T. Karl, H. Judmaier, A. Hansel, A. Jordan, W. Lindinger and P. J. Crutzen, 1999: Acetone, methanol, and other partially oxidized volatile organic emissions from dead plant matter by abiological processes: Significance for atmospheric HO_x chemistry. Global Biogeochem. Cycles, **13**, 9–17.

237. Ariya, P. A., R. Sander and P. J. Crutzen, 2000: Significance of HO_x and peroxides production due to alkene ozonolysis during fall and winter: A modeling study. *J. Geophys. Res.,* **105** (D14), 17721–17738.

238. Bergamaschi, P., R. Hein, M. Heimann and P. J. Crutzen, 2000: Inverse modeling of the global CO cycle: 1. Inversion of CO mixing ratios. *J. Geophys. Res.,* **105** (D2), 1909–1927.

239. Bergamaschi, P., R. Hein, C. A. M. Brenninkmeijer and P. J. Crutzen, 2000: Inverse modeling of the global CO cycle: 2. Inversion of $^{13}C/^{12}C$ and $^{18}O/^{16}O$ isotope ratios. *J. Geophys. Res.,* **105** (D2), 1929–1945.

240. Brühl, Ch. and P. J. Crutzen, 2000: NO_x-catalyzed ozone destruction and NO_x activation at mid to high latitudes as the main cause of the spring to fall ozone decline in the Northern Hemisphere. *J. Geophys. Res.,* **105** (D10), 12163–12168.

241. Brühl, Ch., U. Pöschl, P. J. Crutzen and B. Steil, 2000: Acetone and PAN in the upper troposphere: impact on ozone production from aircraft emissions. Atmos. Environ., **34**, 3931–3938.

242. Crutzen, P. J. and M. G. Lawrence, 2000: The impact of precipitation scavenging on the transport of trace gases: A 3-dimensional model sensitivity study. J. Atmos. Chem., **37**, 81–112.

243. Crutzen, P. J. and V. Ramanathan, 2000: The Ascent of Atmospheric Sciences. Science, **290**, 299–304.

244. Crutzen, P. J., J. Williams, U. Pöschl, P. Hoor, H. Fischer, C. Warneke, R. Holzinger, A. Hansel, W. Lindinger, B. Scheeren and J. Lelieveld, 2000: High spatial and temporal resolution measurements of primary organics and their

oxidation products over the tropical forests of Surinam. Atmos. Environ., **34**, No. 8, 1161–1165.

245. Holzinger, R., L. Sandoval-Soto, S. Rottenberger, P. J. Crutzen, and J. Kesselmeier, 2000: Emissions of volatile organic compounds from *Quercus ilex* L. measured by Proton Transfer Reaction Mass Spectrometry under different environmental conditions. *J. Geophys. Res.,* **105** (D16), 20573-20579.

246. Kanakidou, M., K. Tsigaridis, F. J. Dentener, and P. J. Crutzen, 2000: Human-activity-enhanced formation of organic aerosols by biogenic hydrocarbon oxidation. *J. Geophys. Res.,* **105** (D7), 9243–9254.

247. Law, K. S., P.-H. Plantevin, V. Thouret, A. Marenco, W. A. H. Asman, M. Lawrence, P. J. Crutzen, J.-F. Muller, D. A. Hauglustaine, and M. Kanakidou, 2000: Comparison between global chemistry transport model results and Measurement of Ozone and Water Vapor by Airbus In-Service Aircraft (MOZAIC) data. *J. Geophys. Res.,* **105** (D1), 1503–1525.

248. Pöschl, U., R. von Kuhlmann, N. Poisson and P. J. Crutzen, 2000: Development and intercomparison of condensed isoprene oxidation mechanisms for global atmospheric modeling. J. Atmos. Chem., **37**, 29–52.

249. Pöschl, U., M. G. Lawrence, R. von Kuhlmann, and P. J. Crutzen, 2000: Comment on "Methane photooxidation in the atmosphere: Contrast between two methods of analysis" by Harold Johnston and Douglas Kinnison. *J. Geophys. Res.,* **105** (D1), 1431–1433.

250. Poisson, N., M. Kanakidou and P. J. Crutzen, 2000: Impact of non-methane hydrocarbons on tropospheric chemistry and the oxidizing power of the global troposphere: 3-dimensional modelling results. J. Atmos. Chem., **36**, 157–230.

251. Röckmann, T., C. A. M. Brenninkmeijer, M. Wollenhaupt, J. N. Crowley and P. J. Crutzen, 2000: Measurement of the isotopic fractionation of $^{15}N^{14}N^{16}O$, $^{14}N^{15}N^{16}O$ and $^{14}N^{14}N^{18}O$ in the UV photolysis of nitrous oxide. Geophys. Res. Lett., **27** (No. 9), 1399–1402.

252. Sander, R. and P. J. Crutzen, 2000: Comment on "A chemical aqueous phase radical mechanism for tropospheric chemistry" by Herrmann et al. Chemosphere, **41**, 631–632.

253. Williams, J., H. Fischer, G. W. Harris, P. J. Crutzen, P. Hoor, A. Hansel, R. Holzinger, C. Warnecke, W. Lindinger, B. Scheeren and J. Lelieveld, 2000: Variability-lifetime relationship for organic trace gases: A novel aid to compound identification and estimation of HO concentrations. *J. Geophys. Res.,* **105** (D16), 20473–20486.

254. Heintzenberg, A. Wiedensohler, H. Güsten, G. Heinrich, H. Fischer, J. W. M. Cuijpers and P. F. J. van Velthoven, 2000: Identification of extratropical two-way troposphere-stratosphere mixing based on CARIBIC measurements of O_3, CO, and ultrafine particles. *J. Geophys. Res.,* **105** (D1), 1527–1535.

255. Crutzen, P. J. and Ch. Brühl, 2001: Catalysis by NO_x as the Main Cause of the Spring to Fall Stratospheric Ozone Decline in the Northern Hemisphere. J. Phys. Chem., **105**, 1579–1582.

256. Crutzen, P. J. and J. Lelieveld, 2001: Human Impacts on Atmospheric Chemistry. Ann. Rev. Earth Planet. Sci., **29**, 17–45.

257. Crutzen, P. J. and V. Ramanathan, 2001: Foreword, INDOEX special issue. *J. Geophys. Res.,* **106** (D22), 28369–28370.

258. Jöckel, P., R. v. Kuhlmann, M. G. Lawrence, B. Steil, C. A. M. Brenninkmeijer, P. J. Crutzen, P. J. Rasch and B. Eaton, 2001: On a fundamental problem in implementing flux-form advection schemes for tracer transport in 3-dimensional general circulation and chemistry transport models. Q. J. R. Meteorol. Soc., **127**, 1035–1052.

259. Karl, T., R. Fall, P. J. Crutzen, A. Jordan and W. Lindinger, 2001: High concentrations of reactive biogenic VOCs at a high altitude site in late autumn. Geophys. Res. Lett., **28**, 507–510.

260. Karl, T., P. J. Crutzen, M. Mandl, M. Staudinger, A. Guenther, A. Jordan, R. Fall and W. Lindinger, 2001: Variability-lifetime relationship of VOCs observed at the Sonnblick Observatory 1999-estimation of HO-densities. Atmos. Environ., **35**, 5287–5300.

261. Lelieveld, J., P. J. Crutzen, V. Ramanathan, M. O. Andreae, C. A. M. Brenninkmeijer, T. Campos, G. R. Cass, R. R. Dickerson, H. Fischer, J. A. de Gouw, A. Hansel, A. Jefferson, D. Kley, A. T. J. de Laat, S. Lal, M. G. Lawrence, J. M. Lobert, O. L. Mayol-Bracero, A. P. Mitra, T. Novakov, S. J. Oltmans, K. A. Prather, T. Reiner, H. Rodhe, H. A. Scheeren, D. Sikka, J. Williams, 2001: The Indian Ocean Experiment: Widespread Air Pollution from South to Southeast Asia. Science, **291**, 1031–1036.

262. Pöschl, U., J. Williams, P. Hoor, H. Fischer, P. J. Crutzen, C. Warneke, R. Holzinger, A. Hansel, A. Jordan, W. Lindinger, H. A. Scheeren, W. Peters and J. Lelieveld, 2001: High acetone concentrations throughout the 0-12 km altitude range over the tropical rainforest in Surinam. J. Atmos. Chem., **38**, 115–132.

263. Quesada, J., D. Grossmann, E. Fernández, J. Romero, E. Sanhueza, G. Moortgat and P. J. Crutzen, 2001: Ground Based Gas Phase Measurements in Surinam during the LBA-Claire 98 Experiment. J. Atmos. Chem., **39**, 15–36.

264. Ramanathan, V., P. J. Crutzen, J. T. Kiehl and D. Rosenfeld, 2001: Aerosols, Climate, and the Hydrological Cycle. Science, **294**, 2119–2124.

265. Ramanathan, V., P. J. Crutzen, J. Lelieveld, A. P. Mitra, D. Althausen, J. Anderson, M. O. Andreae, W. Cantrell, G. R. Cass, C. E. Chung, A. D. Clarke, J. A. Coakley, W. D. Collins, W. C. Conant, F. Dulac, J. Heintzenberg, A. J. Heymsfield, B. Holben, S. Howell, J. Hudson, A. Jayaraman, J. T. Kiehl, T. N. Krishnamurti, D. Lubin, G. McFarquhar, T. Novakov, J. A. Ogren, I. A. Podgorny, K. Prather, K. Priestley, J. M. Prospero, P. K. Quinn, K. Rajeev, P. Rasch, S. Rupert, R. Sadourny, S. K. Satheesh, G. E. Shaw, P. Sheridan and F. P. J. Valero, 2001: Indian Ocean Experiment: An integrated analysis of the climate forcing and effects of the great Indo-Asian haze. *J. Geophys. Res.,* **106** (D22), 28371–28398.

266. Röckmann, T., J. Kaiser, J. N. Crowley, C. A. M. Brenninkmeijer and P. J. Crutzen, 2001: The origin of the anomalous or "mass-independent" oxygen

isotope fractionation in tropospheric N_2O. Geophys. Res. Lett., **28** No. 3, 503–506.

267.	Röckmann, T., J. Kaiser, C. A. M. Brenninkmeijer, J. N. Crowley, R. Borchers, W. A. Brand and P. J. Crutzen, 2001: Isotopic enrichment of nitrous oxide ($^{15}N^{14}NO$, $^{14}N^{15}NO$, $^{14}N^{14}N^{18}O$) in the stratosphere and in the laboratory. *J. Geophys. Res.,* **106** (D10), 10403–10410.

268.	Röckmann, T., J. Kaiser, C. A. M. Brenninkmeijer, W. A. Brand, R. Borchers, J. N. Crowley, M. Wollenhaupt and P. J. Crutzen, 2001: The position dependent ^{15}N enrichment of nitrous oxide in the stratosphere. Isotopes Environ. Health Stud., **37**, 91–95.

269.	Warneke, C., R. Holzinger, A. Hansel, A. Jordan, W. Lindinger, U. Pöschl, J. Williams, P. Hoor, H. Fischer, P. J. Crutzen, H. A. Scheeren and J. Lelieveld, 2001: Isoprene and its oxidation products methyl vinyl ketone, methacrolein, and isoprene related peroxides measured online over the tropical rain forest of Surinam in March 1998. J. Atmos. Chem., **38**, 167–185.

270.	Williams, J., H. Fischer, P. Hoor, U. Pöschl, P. J. Crutzen, M. O. Andreae and J. Lelieveld, 2001: The influence of the tropical rainforest on atmospheric CO and CO_2 as measured by aircraft over Surinam, South America. Chemosphere, **3**, 157–170.

271.	Williams, J., U. Pöschl, P. J. Crutzen, A. Hansel, R. Holzinger, C. Warneke, W. Lindinger and J. Lelieveld, 2001: An atmospheric chemistry interpretation of mass scans obtained from a proton transfer mass spectrometer flown over the tropical rainforest of Surinam. J. Atmos. Chem., **38**, 133–166.

272.	Crutzen, P. J. and V. Ramanathan, 2002: The Ascent of Atmospheric Sciences. In: *Science – Pathways of Discovery*, I. Amato (Ed.), John Wiley & Sons, New York, 2002, 175–188.

273.	Gabriel, R., R. von Glasow, R. Sander, M. O. Andreae and P. J. Crutzen, 2002: Bromide content of sea-salt aerosol particles collected over the Indian Ocean during INDOEX 1999 *J. Geophys. Res.,* **107D**, 8032, doi:10.1029/2001JD001133.

274.	von Glasow, R., R. Sander, A. Bott and P. J. Crutzen, 2002: Modeling halogen chemistry in the marine boundary layer. 1. Cloud-free MBL *J. Geophys. Res.,* **107D**, 4341, doi:10.1029/2001JD000942.

275.	von Glasow, R., R. Sander, A. Bott and P. J. Crutzen, 2002: Modeling halogen chemistry in the marine boundary layer. 2. Interactions with sulfur and the cloud-covered MBL *J. Geophys. Res.,* **107D**, 4323, doi:10.1029/2001JD000943.

276.	Holzinger, R., E. Sanhueza, R. von Kuhlmann, B. Kleiss, L. Donoso and P. J. Crutzen, 2002: Diurnal cycles and seasonal variation of isoprene and its oxidation products in the tropical savanna atmosphere. Global Biogeochem. Cycles, **16**, No. 4, 1074, https://doi.org/10.1029/2001GB001421.

277.	Lelieveld, J., H. Berresheim, S. Borrmann, P. J. Crutzen, F. J. Dentener, H. Fischer, J. Feichter, P. F. Flatau, J. Heland, R. Holzinger, R. Korrmann, M. G. Lawrence, Z. Levin, K. M. Markowicz, N. Mihalopoulos, A. Minikin, V. Ramanathan, M. de Reus, G. J. Roelofs, H. A. Scheeren, J. Sciare, H.

Schlager, M. Schultz, P. Siegmund, B. Steil, E. G. Stephanou, P. Stier, M. Traub, C. Warneke, J. Williams and H. Ziereis, 2002: Global Air Pollution Crossroads over the Mediterranean. Science, **298**, 794–799.

278. Markowicz, K. M., P. J. Flatau, M. V. Ramana and P. J. Crutzen, 2002: Absorbing mediterranean aerosols lead to a large reduction in the solar radiation at the surface. Geophys. Res. Lett., **29**, https://doi.org/10.1029/2002GL015767.

279. Mühle, J., A. Zahn, C. A. M. Brenninkmeijer, V. Gros and P. J. Crutzen, 2002: Air mass classifictation during the INDOEX R/V Ronald Brown cruise using measurements of nonmethane hydrocarbons, CH_4, CO_2, CO, ^{14}CO, and $\delta^{18}O(CO)$. *J. Geophys. Res.*, **107**, 10.1029/2001JD000730.

280. Oberlander, E. A., C. A. M. Brenninkmeijer, P. J. Crutzen, N. F. Elansky, G. S. Golitsyn, I. G. Granberg, D. H. Scharffe, R. Hofmann, I. B. Belikov, H. G. Paretzke and P. F. J. van Velthoven, 2002: Trace gas measurements along the Trans-Siberian railroad: The TROICA 5 expedition. *J. Geophys. Res.*, **107**, 10.1029/2001DJ000953.

281. Ramanathan, V., P. J. Crutzen, A. P. Mitra and D. Sikka, 2002: The Indian Ocean Experiment and the Asian Brown Cloud. Current Science, **83**, 947–955.

282. Wagner, V., R. von Glasow, H. Fischer and P. J. Crutzen, 2002: Are CH_2O measurements in the marine boundary layer suitable for testing the current understanding of CH_4 photooxidation? A model study. *J. Geophys. Res.*, **107** (D3), 10.1029/2001JD000722.

283. Williams, J., H. Fischer, S. Wong, P. J. Crutzen, M. P. Scheele and J. Lelieveld, 2002: Near equatorial CO and O_3 profiles over the Indian Ocean during the winter monsoon: High O_3 levels in the middle troposphere and interhemispheric exchange. *J. Geophys. Res.*, **107**, 10.1029/2001JD001126.

284. Wisthaler, A., A. Hansel, R. R. Dickerson and P. J. Crutzen, 2002: Organic trace gas measurements by PTR-MS during INDOEX 1999. *J. Geophys. Res.*, **107** (D19), https://doi.org/10.1029/2001JD000576.

285. Wuebbles, D. J., G. P. Brasseur, H. Rodhe, L. A. Barrie, P. J. Crutzen, R. J. Delmas, D. J. Jacob, C. Kolb, A. Pszenny, W. Steffen and R. F. Weiss, 2002, Changes in the Chemical Composition of the Atmosphere and Potential Impacts. In: Atmospheric Chemistry in a Changing World, G. P. Brasseur, R. G. Prinn and A. A. P. Pszenny (Eds.), Springer Verlag, 1–17.

286. Zahn, A., C. A. M. Brenninkmeijer, W. A. H. Asman, P. J. Crutzen, G. Heinrich, H. Fischer, J. W. M. Cuijpers and P. F. J. van Velthoven, 2002: Budgets of O_3 and CO in the upper troposphere: CARIBIC passenger aircraft results 1997–2001. *J. Geophys. Res.*, **107**, 10.1029/2001JD001529.

287. Zahn, A., C. A. M. Brenninkmeijer, P. J. Crutzen, D. D. Parrish, D. Sueper, G. Heinrich, H. Güsten, H. Fischer, M. Hermann and J. Heintzenberg, 2002: Electrical discharge source for tropospheric „ozone-rich transients". *J. Geophys. Res.*, **107**, 10.1029/2001JD002345.

288. Asman, W. A. H., M. G. Lawrence, C. A. M. Brenninkmeijer, P. J. Crutzen, J. W. M. Cuijpers and P. Nédélec, 2003: Rarity of upper-tropospheric low

O_3 concentration events during MOZAIC flights. Atmos. Chem. Phys., **3**, 1541–1549.

289. Christian, T. J., B. Kleiss, R. J. Yokelson, R. Holzinger, P. J. Crutzen, W. M. Hao, B. H. Saharjo and D. E. Ward, 2003: Comprehensive laboratory measurements of biomass-burning emissions: 1. Emissions from Indonesian, African, and other fuels. *J. Geophys. Res.*, **108**, 10.1029/JD003704.

290. Crutzen, P. J. and J. Lelieveld, 2003: Comment on the paper by C. G. Roberts et al. "Cloud condensation nuclei in the Amazon Basin: "Marine" conditions over a continent?". Geophys. Res. Lett. **30**, 10.1029/2002GL015206.

291. Crutzen, P. J. and V. Ramanathan, 2003: The Parasol Effect on Climate. Science, **302**, 1679–1680.

292. Crutzen, P. J. and W. Steffen, 2003: How long have you been in the Anthropocene Era? An Editorial Comment, Climatic Change, **61**, 251–257.

293. von Glasow, R. and P. J. Crutzen, 2003: Tropospheric Halogen Chemistry. In: Treatise on Geochemistry, H. D. Holland, K. K. Turekian and R. F. Keeling (Eds.), Elsevier Pergamon, 21–64.

294. von Glasow, R., M. G. Lawrence, R. Sander and P. J. Crutzen, 2003: Modeling the chemical effects of ship exhaust in the cloud-free marine boundary layer. Atmos. Chem. Phys., **3**, 233–250.

295. Guazzotti, S. A., D. T. Suess, K. R. Coffee, P. K. Quinn, T. S. Bates, A. Wisthaler, A. Hansel, W. P. Ball, R. R. Dickerson, C. Neusüß, P. J. Crutzen and K. A. Prather, 2003: Characterization of carbonaceous aerosols outflow from India and Arabia: Biomass/biofuel burning and fossil fuel combustion. *J. Geophys. Res.*, **108**, 10.1029/2002JD003277.

296. Jöckel, P., C. A. M. Brenninkmeijer and P. J. Crutzen, 2003: A discussion on the determination of atmospheric OH and its trends. Atmos. Chem. Phys., **3**, 107–118.

297. Kaiser, J., C. A. M. Brenninkmeijer, T. Röckmann and P. J. Crutzen, 2003: Wavelength dependence of isotope fractionation in N_2O photolysis, Atmos. Chem. Phys., **3**, 303–313.

298. von Kuhlmann, R., M. G. Lawrence, P. J. Crutzen, and P. J. Rasch, 2003: A model for studies of tropospheric ozone and non-methane hydrocarbons: Model description and ozone results, *J. Geophys. Res.*, **108**, https://doi.org/10.1029/2002JD002893.

299. von Kuhlmann, R., M. G. Lawrence, P. J. Crutzen and P. J. Rasch, 2003: A model for studies of tropospheric ozone and nonmethane hydrocarbons: Model evaluation of ozone-related species. *J. Geophys. Res.*, **108**, https://doi.org/10.1029/2002JD003348.

300. Lawrence, M. G., P. J. Rasch, R. von Kuhlmann, J. Williams, H. Fischer, M. de Reus, J. Lelieveld, P. J. Crutzen, M. Schultz, P. Stier, H. Huntrieser, J. Helans, A. Stohl, C. Forster, H. Elbern, H. Jakobs and R. R. Dickerson, 2003: Global chemical weather forecasts for field campaign planning: predictions and observations of large-scale features during MINOS, CONTRACE, and INDOEX. Atmos. Chem. Phys., **3**, 267–289.

301. Ramanathan, V. and P. J. Crutzen, 2003: New Directions: Atmospheric Brown "Clouds". Atmos. Environ., **37**, 4033–4035.

302. Sander, R., W. C. Keene, A. A. P. Pszenny, R. Arimoto, G. P. Ayers, E. Baboukas, J. M. Cainey, P. J. Crutzen, R. A. Duce, G. Hönninger, B. J. Huebert, W. Maenhaut, N. Mihalopoulos, V. C. Turekian, and R. Van Dingenen, 2003: Inorganic bromine in the marine boundary layer: a critical review. Atmos. Chem. Phys., **3**, 1301–1336.

303. Steil, B., C. Brühl, E. Manzini, P. J. Crutzen, J. Lelieveld, P. J. Rasch, E. Roeckner and K. Krüger, 2003: A new interactive chemistry-climate model: 1. Present-day climatology and interannual variability of the middle atmosphere using the model and 9 years of HALOE/UARS data. *J. Geophys. Res.,* **108**, https://doi.org/10.1029/2002JD002971.

304. van Aalst, M. K., M. M. P. van den Broek, A. Bregman, C. Brühl, B. Steil, G. C. Toon, S. Garcelon, G. M. Hansford, R. L. Jones, T. D. Gardiner, G. J. Roelofs, J. Lelieveld and P. J. Crutzen, 2004: Trace gas transport in the 1999/2000 Arctic winter: comparison of nudged GCM runs with observations. Atmos. Chem. Phys., **4**, 81–93.

305. Christian, T. J., B. Kleiss, R. J. Yokelson, R. Holzinger, P. J. Crutzen, W. M. Hao, T. Shirai and D. R. Blake, 2004: Comprehensive laboratory measurements of biomass-burning emissions: 2. First intercomparison of open-path FTIR, PTR-MS, GC-MS/FID/ECD. *J. Geophys. Res.,* **109**, https://doi.org/10.1029/2003JD003874.

306. Clark W. C., P. J. Crutzen and H. J. Schellnhuber, 2004: Science for Global Sustainability. In: Earth System Analysis for Sustainability. Dahlem Workshop Report. H. J. Schellnhuber, P. J. Crutzen, W. C. Clark, M. Claussen and H. Held (Eds.), pp. 1-28, MIT Press, Cambridge, USA.

307. Crutzen, P. J., 2004: New Directions: The growing urban heat and pollution "island" effect - impact on chemistry and climate. Atmos. Env., **38**, 3539–3540.

308. Crutzen, P. J. and V. Ramanathan, 2004: Atmospheric Chemistry and Climate in the Anthropocene. Where are we Heading? In: Earth System Analysis for Sustainability. Dahlem Workshop Report. H. J. Schellnhuber, P. J. Crutzen, W. C. Clark, M. Claussen and H. Held (Eds.), pp. 265–292, MIT Press, Cambridge, USA.

309. Hurst, D.F., P.A. Romashkin, J.W. Elkins, E.A. Oberlander, N.F. Elansky, I.B. Belikov, I.G. Granberg, G.S. Golitsyn, A.M. Grisenko, C.A.M. Brenninkmeijer and P.J. Crutzen, 2004: Emissions of ozone-depleting substances in Russia during 2001. *J. Geophys. Res.,* 109, https://doi.org/10.1029/2004JD004633.

310. Gabrielli, P., C. Barbante, J. M. C. Plane, A. Varga, S. Hong, G. Cozzi, V. Gaspari, F. A. M. Planchon, W. Cairns, C. Ferrari, P. Crutzen, P. Cescon and C. F. Boutron, 2004: Meteoric smoke fallout over the Holocene epoch revealed by iridium and platiunum in Greenland ice. Nature, **432**, 1011–1014.

311. von Glasow, R. and P. J. Crutzen, 2004: Model study of multiphase DMS oxidation with a focus on halogens. Atmos. Chem. Phys., **4**, 589–608.

312. von Glasow, R., R. von Kuhlmann, M. G. Lawrence, U. Platt and P. J. Crutzen, 2004: Impact of reactive bromine chemistry in the troposphere. Atmos. Chem. Phys., **4**, 2481–2497.

313. von Kuhlmann, R., M. G. Lawrence, U. Pöschl and P. J. Crutzen, 2004: Sensitivities in global scale modeling of isoprene. Atmos. Chem. Phys., **4**, 1–17.

314. Peters, W., M. C. Krol, J. P. F. Fortuin, H. M. Kelder, A. M. Thompson, C. R. Becker, J. Lelieveld and P. J. Crutzen, 2004: Tropospheric ozone over a tropical Atlantic station in the Northern Hemisphere: Paramaribo, Surinam (6N, 55W). Tellus B, **56**(1), 21–34.

315. Pszenny, A. P., J. Moldanová, W. C. Keene, R. Sander, J. R. Maben, M. Martinez, P. J. Crutzen, D. Perner and R. G. Prinn, 2004: Halogen Cycling and Aerosol pH in the Hawaiian Marine Boundary Layer. Atmos. Chem. Phys., **4**, 147–168.

316. Richter, A., V. Eyring, J. P. Burrows, H. Bovensmann, A. Lauer, B. Sierk and P. J. Crutzen, 2004: Satellite measurementsof NO2 from international shipping emissions. Geophys. Res. Lett., **31**, https://doi.org/10.1029/2004GL020822.

317. Sander, R., P. J. Crutzen and R. von Glasow, 2004: Comment on „Reactions at Interfaces As a Source of Sulfate Formation in Sea-Salt Particles" (II). Science, **303**, 628.

318. Sanhueza, E., R. Holzinger, B. Kleiss, L. Donoso and P. J. Crutzen, 2004: New insights in the global cycle of acetonitrile: release from the ocean and dry deposition in the tropical savanna of Venezuela, Atmos. Chem. Phys., **4**, 275–280.

319. Steffen, W., M. O. Andreae, B. Bolin, P. M. Cox, P. J. Crutzen, U. Cubasch, H. Held, N. Nakićenović, R. J. Scholes, L. Talaue-McManus and B. L. Turner II, 2004: Abrupt Changes: The Achilles' Heels of the Earth System. Environment, **46**, 8–20 (also published as IIASA rapport RR-04-006, June 2004).

320. Steffen, W., M. O. Andreae, P. M. Cox, P. J. Crutzen, U. Cubasch, H. Held, N. Nakicenovic, L. Talaue-McManus and B. L. Turner II, 2004: Group Report: Earth System Dynamics in the Anthropocene. In: Earth System Analysis for Sustainablility. Dahlem Workshop Report. H. J. Schellnhuber, P. J. Crutzen, W. C. Clark, M. Claussen and H. Held (Eds.), pp. 313–340, MIT Press, Cambridge, USA.

321. Vrekoussis, M., M. Kanakidou, N. Mihalopoulos, P. J. Crutzen, J. Lelieveld, D. Perner, H. Berresheim and E. Baboukas, 2004: Role of NO_3 radicals in oxidation processes in the eastern Mediterranean troposphere during the MINOS campaign. Atmos. Chem. Phys., **4**, 169–182.

322. Wallström, M., B. Bolin, P. J. Crutzen and W. Steffen, 2004: The Earth's Life-support System is in Peril. Global Change NewsLetter, **57**, 22–23.

323. Brenninkmeijer, C. A. M., F. Slemr, C. Koeppel, D. S. Scharffe, M. Pupek, J. Lelieveld, P. Crutzen, A. Zahn, D. Sprung, H. Fischer, M. Hermann, M. Reichelt, J. Heintzenberg, H. Schlager, H. Ziereis, U. Schumann, B. Dix, U. Platt, R. Ebinghaus, B. Martinsson, P. Ciais, D. Flippi, M. Leuenberger, D.

Oram, S. Penkett, P. van Velthoven and A. Waibel, 2005: Analyzing Atmospheric Trace Gases and Aerosols Using Passenger Aircraft. EOS, **86**(8), 77, 82, 83.

324.   Fishman, J., J.K. Creilson, A.E. Wozniak and P.J. Crutzen, 2005: Interannual variability of stratospheric and tropospheric ozone determined from satellite measurements. *J. Geophys. Res.*, **110**, https://doi.org/10.1029/200 5JD005868.

325.   Gabrielli, P., J. M. C. Plane, C. F. Boutron, S. Hong, G. Cozzi, P. Cescon, C. Ferrari, P. J. Crutzen, J. R. Petit, V. Y. Lipenkov and C. Barbante, 2006: A climatic control on the accretion of meteoric and super-chondritic iridium-platinum to the Antarctic ice cap. Earth and Planetary Science Letters, **250**, 459-469.

326.   Holzinger, R., J. Williams, G. Salisbury, T. Klüpfel, M. de Reus, M. Traub, P. J. Crutzen and J. Lelieveld, 2005: Oxygenated compounds in aged biomass burning plumes over the Eastern Mediterranean: evidence for strong secondary production of methanol and acetone. Atmos. Chem. Phys. **5**, 39–46.

327.   Schellnhuber, H. J., P. J. Crutzen, W. C. Clark and J. Hunt, 2005: Earth System Analysis for Sustainability. Environment, **47**, 11–25.

328.   Crutzen, P. J., 2006: Albedo Enhancement by Stratospheric Sulfur Injections: A Contribution to Resolve a Policy Dilemma? Climatic Change, https://doi.org/10.1007/s10584-006-9101-y (online published).

329.   Keene, W. C., J. M. Lobert, P. J. Crutzen, J. R. Maben, D. H. Scharffe, T. Landmann, C. Hély and C. Brain, 2006: Emissions of major gaseous and particulate species during experimental burns of southern African biomass. J. Geophy. Res., **111**, https://doi.org/10.1029/2005JD006319.

330.   Vrekoussis, M., E. Liakakou, N. Mihalopoulos, M. Kanakidou, P.J. Crutzen and J. Lelieveld, 2006: Formation of HNO_3 and NO_3 in the anthropogenically-influenced eastern Mediterranean marine boundary layer. Geopyhs. Res. Lett., **33**, https://doi.org/10.1029/2005GL025069.

331.   Birks, J. W., P. J. Crutzen and R. G. Roble, 2007: Frequent Ozone Depletion Resulting from Impacts of Asteroids and Comets. In: Comet/Asteroid Impacts and Human Society, P. Bobrowsky and H. Rickman (Eds.), pp. 225–245, https://doi.org/10.1007/978-3-540-32711-0, Springer Verlag Berlin Heidelberg.

332.   Brenninkmeijer, C. A. M., P. Crutzen, F. Boumard, T. Dauer, B. Dix, R. Ebinghaus, D. Filippi, H. Fischer, H. Franke, U. Frieß, J. Heintzenberg, F. Helleis, M. Hermann, H. H. Kock, C. Koeppel, J. Lelieveld, M. Leuenberger, B. G. Martinsson, S. Miemczyk, H. P. Moret, H. N. Nguyen, P. Nyfeler, D. Oram, D. O'Sullivan, S. Penkett, U. Platt, M. Pupek, M. Ramonet, B. Randa, M. Reichelt, T. S. Rhee, J. Rohwer, K. Rosenfeld, D. Scharffe, H. Schlager, U. Schumann, F. Slemr, D. Sprung, P. Stock, R. Thaler, F. Valentino, P. van Velthoven, A. Waibel, A. Wandel, K. Waschitschek, A. Wiedensohler, I. Xueref-Remy, A. Zahn, U. Zech and H. Ziereis, 2007: Civil Aircraft for the regular investigation of the atmosphere based on an instrumented container: The new CARIBIC system. Atmos. Chem. Phys., **7**, 4953–4976.

333. von Glasow, R. and P. J. Crutzen, 2007: Tropospheric Halogen Chemistry. In: Treatise on Geochemistry Update1, Holland H. D. and Turekian K. K. (eds.), Vol. 4.02, 1–67.

334. Lelieveld, J. C. Brühl, P. Jöckel, B. Steil, P.J. Crutzen, H. Fischer, M.A. Giorgetta, P. Hoor, M.G. Lawrence, R. Sausen and H. Tost, 2007: Stratospheric dryness: model simulations and satellite observations. Atmos. Chem. Phys., **7**, 1313–1332.

335. Steffen, W., P. J. Crutzen and J. R. McNeill, 2007: The Anthropocene: Are Humans Now Overwhelming the Great Forces of Nature? Ambio, **36**, 614–621.

336. Vreskoussis, M., N. Mihalopoulos, E. Gerasopoulos, M. Kanakidou, P.J. Crutzen and J. Lelieveld, 2007: Two-years of NO_3 radical observations in the boundary layer over the Eastern Mediterranean. Atmos. Chem. Phys., **7**, 315–327.

337. Crutzen, P. J. and M. Oppenheimer, 2008: Learning about ozone depletion. Climatic Change, **89**, 143–154, https://doi.org/10.1007/s10584-008-9400-6.

338. Crutzen, P. J., A. R. Mosier, K. A. Smith and W. Winiwarter, 2008 N_2O release from agro-biofuel production negates global warming reduction by replacing fossil fuels. Atmos, Chem. Phys., **8**, 389–395.

339. O'Neill, B. C., P. Crutzen, A. Grübler, M. H. Duong, K. Keller, C. Kolstad, J. Koomey, A. Lange, M. Obersteiner, M. Oppenheimer, W. Pepper, W. Sanderson, M. Schlesinger, N. Treich, A. Ulph, M. Webster and C. Wilson, 2008: Learning and climate change. Climate Policy, **6**, 585–589.

340. Rasch, P. J., P. J. Crutzen and D. B. Coleman, 2008: Exploring the geoengineering of climate using stratospheric sulfate aerosols: The role of particle size. Geophys. Res. Lett., **35**, https://doi.org/10.1029/2007GL032179.

341. Mosier A. R., P. J. Crutzen, K. A. Smith and W. Winiwarter, 2009: Nitrous oxide's impact on net greenhouse gas savings from biofuels: life-cycle analysis comparison. Int. J. Biotechnology, **11**, 60–74.

342. Rockström, J., W. Steffen, K. Noone, A. Persson, F. S. Chapin, E. F. Lambin, T. M. Lenton, M. Scheffer, C. Folke, H. J. Schellnhuber, B. Nykvist, C. A. de Wit, T. Hughes, S. van der Leeuw, H. Rodhe, S. Sörlin, P. K. Snyder, R. Costanza, U. Svedin, M. Falkenmark, L. Karlberg, R. W. Corell, V. J. Fabry, J. Hansen, B. Walker, D. Liverman, K. Richardson, P. Crutzen and J. Foley, 2009: A safe operating space for humanity. Nature, **461**, 472–475.

343. Rockström, J., W. Steffen, K. Noone, A. Persson, F. S. Chapin, E. F. Lambin, T. M. Lenton, M. Scheffer, C. Folke, H. J. Schellnhuber, B. Nykvist, C. A. de Wit, T. Hughes, S. van der Leeuw, H. Rodhe, S. Sörlin, P. K. Snyder, R. Costanza, U. Svedin, M. Falkenmark, L. Karlberg, R. W. Corell, V. J. Fabry, J. Hansen, B. Walker, D. Liverman, K. Richardson, P. Crutzen and J. Foley, 2009: Planetary Boundaries: Exploring the Safe Operating Space for Humanity. Ecology and Society, **14(2)**, art. 32.

344. Williams, J. and P. J. Crutzen, 2010: Nitrous oxide from aquaculture. Nature Geoscience, **3**, 143.

345. Gleick, P.H. et al., 2010: Climate Change and the Integrity of Science. Science
      Letters, **328**, 689–690.
346. Zalasiewicz, J., M. Williams, W. Steffen and P. Crutzen, 2010: The New
      World of the Anthropocene. Environ. Sci. Technol., **44**, 2228–2231.
347. Zalasiewicz J., M. Williams, W. Steffen and P. Crutzen, 2010: Response to
      „The Anthropocene forces us to reconsider adaptationist models of human-
      environment interactions". Environ. Sci. Technol., **44 (16)**, 6008, https://doi.
      org/10.1021/es102062w.
348. Pierazzo E., R. R. Garcia, D. E. Kinnison, D. R. Marsh, J. Lee-Taylor and P.
      J. Crutzen, 2010: Ozone perturbation from medium-size asteroid impacts in
      the ocean. Earth and Planetary Science Letters, **299**, 263–272.
349. Zander, R., P. Duchatelet, E. Mahieu, P. Demoulin, G. Roland, C. Servais, J. V.
      Auwera, A. Perrin, C. P. Rinsland and P. J. Crutzen, 2010: Formic acid above
      the Jungfraujoch during 1985–2007: observed variability, seasonality, but no
      long-term backround evolution. Atmos. Chem. Phys., **10**, 10047–10065.
350. Brühl, C., J. Lelieveld, P.J. Crutzen and H. Tost: the role of carbonyl sulphide
      as a source of stratospheric sulphate aerosol and its impact on climate. Atmos.
      Chem. Phys., 12, 1239-1253 (2012)
351. Crutzen, P.J.: Sherry Rowland: Ozone and advocacy. Nature Geosci., **5**, 311
      (2012).
352. De Fries, R. S., E. C. Ellis, F. Stuart Chapin III, P. A. Matson, B. L. Turner
      II, A. Agrawal, P. J. Crutzen, C. Field, P. Gleick, P. M. Kareiva, E. Lambin,
      D. Liverman, E. Ostrom, P. A. Sanchez and J. Syvitski, 2012: Planetary
      Opportunities: A Social Contract for Global Change Science to Contribute
      to a Sustainable Future. BioScience, **62**, 603–606.
353. Smith, K. A., A. R. Mosier, P. J. Crutzen and W. Winiwarter, 2012: The role
      of N_2O derived from crop-based biofuels, and from agriculture in general, in
      Earth's climate. Phil. Trans. R. Soc. B, **367**, 1169–1174.
354. Reay, D. S., E. A. Davidson, K. A. Smith, P. Smith, J. M. Melillo, F. Dentener
      and P. J. Crutzen, 2012: Global agriculture and nitrous oxide emissions.
      Nature Climate Change, **2**, 410–416.
355. Williams, J. and P.J. Crutzen: Perspectives on our planet in the Atmosphere.
      Environmental Chemistry, **10**, 269–280 http://dx.doi.org/10.1071/EN13061
      (2013).
356. Elshorbany, Y.F., P. Crutzen, B. Steil, A. Pozzer, H. Tost and J. Lelieveld:
      Global and regional impacts of HONO on the chemical composition of clouds
      and aerosols. Atmos. Chem. Phys., **14**, 1167–1184 (2014).
357. Foley, S.F., D. Gronenborn, M.O. Andreae, J.W. Kadereit, J. Esper, D.
      Scholz, U. Pöschl, D.E. Jacob, B.R. Schöne, R. Schreg, A. Vött, D. Jordan, J.
      Lelieveld, C.G. Weller, K.W. Alt, S. Gaudzinski-Windheuser, K.-C. Bruhn,
      H. Tost, F. Sirocko and P.J. Crutzen: (2013) The Palaeoanthropocene – The
      beginnings of anthropogenic environmental change. Anthropocene, **3**, 83–88
      (2014).

358. Crutzen, P.J: The Anthropocene: When Humankind overrides Nature. In: "Contributions towards a sustainable world – in dialogue with Klaus Töpfer", F. Schmidt, N. Nuttall (Eds.), oekom Verlag München, 2014, 21–27.

359. Lenhart, K., B. Weber, W. Elbert, J. Steinkamp, T. Clough, P. Crutzen, U. Pöschl and F. Keppler: Nitrous oxide and methane emissions from cryptogamic covers. Glob. Change Biol., **21,** 3889–3900, https://onlinelibrary.wiley.com/doi/abs/10.1111/gcb.12995 (2015).

360. Weber, B., D. Wu, A. Tamm, N. Ruckteschler, E. Rodriguez-Caballero, J. Steinkamp, H. Meusel, W. Elbert, T. Behrend, M. Sörgel, Y. Cheng, P.J. Crutzen, H. Su and U. Pöschl. Biological soil crusts accelerate the nitrogen cycle through large NO and NONO emissions in drylands. Proceedings of the National Academy of Sciences of the United States of America, **112,** 15384–15389, https://doi.org/10.1073/pnas.1515818112 (2015).

361. Zalasiewicz, J., C. N. Water, M. Williams, A. D. Barnosky, A. Cearreta, P. Crutzen, E. Ellis, M. A. Ellis, I. J. Fairchild, J. Grinevald, P. K. Haff, I. Hajdas, R. Leinfelder, J. McNeill, E.O. Odada, C. Poirier, D. Richter, W. Steffen, C. Summerhayes, J. P.M. Syvitski, D. Vidas, M. Wagreich, S. L. Wing, A. P. Wolfe, A. Zhisheng and N. Oreskes: When did the Anthropocene begin? A mid-twentieth century boundary level is stratigraphically optimal. Quat. Int., **383**, 196–203 (2015).

362. Steffen, W., R. Leinfelder, J. Zalasiewicz, C.N. Waters, M. Williams, C. Summerhayes, A.D. Barnosky, A. Cearreta, P.J. Crutzen, M. Edgeworth, E.C. Ellis, I.J. Fairchild, A. Galuszka, J. Grinevald, A. Haywood, J. Ivar do Sul, C. Jeandel, J.R. McNeill, E. Odada, N. Oreskes, A. Revkin, D. deB.Richter, J. Syvitsky, D. Vidas, M. Wagreich, S.L. Wing, A.P. Wolfe and H.J. Schellnhuber: Stratigraphic and Earth System approaches in defining the Anthropocenc. Earth's Future, **8**, 324–345 (2016).

363. Lawrence Mark G. and P.J. Crutzen: Was breaking the taboo on research on climate engineering via albedo modification amoral hazard, or a moral imperative? Earth's Future, **5**, 136–143, https://doi.org/10.1002/2016EF000463 (2017).

364. Zalasiewicz, J., C.N. Waters, C.P. Summerhayes, A.P. Wolfe, A.D. Barnosky, A. Cearreta, P. Crutzen, E. Ellis, I.J. Fairchild, A. Gałuszka, P. Haff, I. Hajdas, M. J. Head, J. A. Ivar do Sul, C. Jeandel, R. Leinfelder, J. R. McNeill, C. Neal, E. Odada, N. Oreskes, W. Steffen, J. Syvitski, D. Vidas, M. Wagreich and M. Williams: The Working Group on the Anthropocene: Summary of evidence and interim recommendations. Anthropocene, **19**, 55–60 (2017).

365. Rodríguez-Caballero, E., J. Belnap, B. Büdel, P. Crutzen, M.O. Andreae, U. Pöschl and B. Weber: Dryland photoautotrophic soil surface communities endangered by global change. Nature Geoscience, **11**, 185–189, https://pubs.er.usgs.gov/publication/70197641 (2018).

366. Lawrence, M.G. and P.J. Crutzen: The Evolution of Climate Engineering Research. In: Geoengineering our Climate? Ethics, Politics and Governance. Eds. J.J. Blackstock and S. Low, Routledge, Abingdon, 2019, 89–94.

367. Duprey, N.N., T.X. Wang, T. Kim, J.D. Cybulski, H.B. Vonhof, P.J. Crutzen, G.H. Haug, D.M. Sigman, A. Martinez-Garcia and D.M. Baker: Megacity development and the demise of coastal coral communities: Evidence from coral skeleton δ15N records in the Pearl River estuary. Glob Change Biol. 2020, 26, 1338–1353, (2020).

Other Publications (unrefereed)

A1.   Crutzen, P. J., 1969: Koldioxiden och klimatet (Carbon dioxide and climate), Forskning och Framsteg, **5**, 7–9.

A2.   Crutzen, P. J., 1971: On some photochemical and meteorological factors determining the distribution of ozone in the stratosphere: Effects on contamination by NO_x emitted from aircraft, Technical Report UDC 551.510.4, Institute of Meteorology, University of Stockholm.

A3.   Crutzen, P. J., 1972: Gas-phase nitrogen and methane chemistry in the atmosphere, Report AP-10, Institute of Meteorology, University of Stockholm, 20pp.

A4.   Crutzen, P. J., 1972: The photochemistry of the stratosphere with special attention given to the effects of NOx emitted by supersonic aircraft, First Conference on CIAP, United States Department of Transportation, 80–88.

A5.   Crutzen, P. J., 1972: Liten risk för klimatändring (Small risk for climatic change; in Swedish). *Forskning och Framsteg*, **2**, 27.

A6.   Crutzen, P. J., 1974: Artificial increases of the stratospheric nitrogen oxide content and possible consequences for the atmopsheric ozone, Technical Report UDC 551.510.4:546.2, Institute of Meteorology, University of Stockholm.

A7.   Crutzen, P. J., 1974: Väderforskning med matematik (Weather research with mathematics; in Swedish). *Forskning och Framsteg*, **6**, 22–23, 26.

A8.   Crutzen, P. J., 1975: Physical and chemical processes which control the production, destruction and distribution of ozone and some other chemically active minor constituents. *GARP Publications Series 16*, World Meterological Organization, Geneva, Switzerland.

A9.   Crutzen, P. J., 1975: A two-dimensional photochemical model of the atmosphere below 55 km. In: *Estimates of natural and man-caused ozone perturbations due to NO_x. Proceedings of 4th CIAP Conference*, U.S. Department of Transportation, Cambridge, DOT-TSC-OST-75-38, T.M. Hard and A.J. Brodrick (eds.), 264–279.

A10.   Crutzen, P. J., 1976: Ozonhöljet tunnas ut: Begränsa spray-gaserna (The ozone shield is thinning: limit the use of aerosol propellants; in Swedish). *Forskning och Framsteg*, **5**, 29–35.

A11.   Heath, D. F., A. J. Krueger and P. J. Crutzen, 1976: Influence of a solar proton event on stratospheric ozone. Goddard Space Flight Center, Greenbelt, Maryland.

A12.   Crutzen, P. J., 1977: The Stratosphere-Meosphere. *Solar Output and Its Variations*, O.R. White (ed.), Colorado Associated University Press, Boulder, Colorado, 13–16.

A13.   Crutzen, P. J., J. Fishman, L. T. Gidel and R. B. Chatfield, 1978: Numerical investigations of the photochemical and transport processes which affect halocarbons and ozone in the atmosphere. *Annual Summary of Research*, Dept. of Atmospheric Science, Colorado State Univ., Fort Collins, CO.

A14.   Fishman, J. and P. J. Crutzen, 1978: The distribution of the hydroxyl radical in the troposphere. Atmos. Sci. Paper 284 (Dept. of Atmos. Sci., Colorado State University, Fort. Collins, CO).

A15.   Crutzen, P. J. L. T. Gidel and J. Fishman, 1979: Numerical investigations of the photochemical and transport processes which affect ozone and other trace constituents in the atmosphere. *Annual Summary of Research*, Dept. of Atmospheric Science, Colorado State University, Fort Collins, CO.

A16.   Fishman, J. and P. J. Crutzen, 1979: A preliminary estimate of stratospheric ozone depletion by the release of chlorocarbon chemicals as calculated by a two-dimensional photochemical model of the atmosphere to 55 km. First Quarterly Progress Report EPA Grant R804921-03

A17.   Crutzen, P. J., 1981: Atmospheric chemical processes of the oxides of nitrogen including nitrous oxide. In: *Denitrification, Nitrification and Atmospheric Nitrous Oxide*. Ed. C.C. Delwiche, John Wiley and Sons, New York 1981, 17–44.

A18.   Birks, J. W. and P. J. Crutzen, 1983: Atmospheric effects of a nuclear war, Chemistry in Britain, **19**, 927–930.

A19.   Galbally, I. E., P. J. Crutzen and H. Rodhe, 1983: Some changes in the Atmosphere over Australia that may occur due to a Nuclear War, pp. 161–185 in *"Australia and Nuclear War"*, (Ed. M.A. Denborough), Croom Helm Ltd., Canberrra, Australia, 270 pp.

A20.   Brühl, C. and P. J. Crutzen, 1984: A radiative convective model to study the sensitivity of climate and chemical composition to a variety of human activities, Proceedings of a working party meeting, Brussels, 18th May 1984, Ed. A. Ghazi, CEC, pp. 84–94.

A21.   Crutzen, P. J., 1985: The Global Environment After Nuclear War. Environment, **27**, 6–11.

A22.   Crutzen, P. J., 1985: Global Aspects of Atmospheric Chemistry: Natural and Anthropogenic Influences. presented at the III. Bi-National Colloquium of the Alexander von Humboldt Foundation (Bonn, FRG) at Northern University, Evanston, Illinois, USA, September 17–20, 1985.

A23.   Crutzen, P. J. and M. O. Andreae, 1985: Atmospheric Chemistry, in T.F. Malone and J.G. Roederer, Eds., *Global Change*, Cambridge University Press, Cambridge, 75–113.

A24.   Crutzen, P. J. and I. E. Galbally, 1985: Atmospheric conditions after a nuclear war. In: *Chemical Events in the Atmosphere and Their Impact on the Environment*. (G. B. Marini-Bettolo, Editor), Pontificiae Academiae Scientiarum Scripta Varia, Città del Vaticano, pp. 457–502.

A25. Crutzen, P. J. and J. Hahn, 1985: Atmosphärische Auswirkungen eines Atomkrieges. Physik in unserer Zeit, **16**, 4–15.

A26. Klose, W., H. Butin. P. J. Crutzen, F. Führ, H. Greim, W. Haber, K. Hahlbrock, A. Hüttermann. W. Klein, W. Klug, H. U. Moosmayer, W. Obländer, B. Prinz, K. E. Rehfuess and O. Rentz, 1985: Forschungs-beirat Waldschäden/Luftverunreinigungen der Bundesregierung und der Länder, Zwischenbericht Dezember 1984, in *Bericht über den Stand der Erkenntnisse zur Ursache.*

A27. Crutzen, P. J., 1986: Globale Aspekte der amtosphärischen Chemie: Natürliche und anthropogene Einflüsse, Vorträge Rheinisch-Westfälische Akademie der Wissenschaften, S. 41-72, Westdeutscher Verlag GmbH, Opladen.

A28. Klose, W., H. Butin and P. J. Crutzen, 1986: u.a. Forschungsbeirat Wald-schäden/Luftverunreinigungen der Bundesregierung und der Länder, 2. Bericht, 229 S.

A29. Crutzen, P. J., 1987: Recent depletions of ozone with emphasis on the polar "ozone hole". Källa, **28**, (Stockholm; in Swedish).

A30. Crutzen, P. J., 1987: Climatic Effects of Nuclear War, Annex 2 in *Effects of Nuclear War on Health and Health Services*, Report A40/11 of the World Health Organization to the 40th World Health Assembly, 18 March 1987; WHO, Geneva.

A31. Crutzen, P. J., 1987: Ozonloch und Spurengase - Menschliche Einflüsse auf Klima und Chemie der Atmosphäre, Max-Planck-Gesellschaft, Jahrbuch 1987, München, S. 27–40.

A32. Darmstadter, J., L. W. Ayres, R. U. Ayres, W. C. Clark, R. P. Crosson, P. J. Crutzen, T. E. Graedel, R. McGill, J. F. Richards and J. A. Torr, 1987: Impacts of World Development on Selected Characteristics of the Atmosphere: An Integrative Approach, Oak Ridge National Laboratory, 2 Volumes, ORNL/Sub/86-22033/1/V2, Oak Ridge, Tennessee 37931, USA.

A33. Crutzen, P. J., 1988: Das Ozonloch – Menschliche Einflüsse auf die Chemie der globalen Atmosphäre. In: Gewerkschaftliche Monatshefte 12'88 "Der blaue Planet in der Krise", 731–745.

A34. Brühl, C. and P. J. Crutzen, 1989: The potential role of odd hydrogen in the ozone hole photochemistry, in *Our Changing Atmosphere* (P. J. Crutzen, J.-C. Gerard and R. Zander, Editors), Université de Liège, Institut d'Astrophysique, B-4200 Cointe-Ougree, Belgium, pp. 171–177.

A35. Crutzen, P. J., W. M. Hao, M. H. Liu, J. M. Lobert and D. Scharffe, 1989: Emissions of CO_2 and other trace gases to the atmosphere from fires in the tropics. In: P. J. Crutzen, J.-C. Gerard and R. Zander (Eds.): *Our Changing Atmosphere*, Proceedings of the 28th Liège International Astrophysical Colloqium, Université de Liège, Belgium, 449–471.

A36. Graedel, T. E. and P. J. Crutzen, 1989: The Changing Atmosphere. *Scientific American*, **160**, 58–68 (in deutsch: Veränderungen der Atmosphäre. *Spektrum der Wissenschaften*, **11**, 58–68).

A37.   Lelieveld, J., P. J. Crutzen and H. Rodhe, 1989: Zonal average cloud characteristics for global atmospheric chemistry modelling. Report CM-76, UDC 551.510.4, Glomac 89/1. International Meteorological Institute in Stockholm, University of Stockholm, 54 pp.

A38.   Crutzen, P. J., 1990: Auswirkungen menschlicher Aktivitäten auf die Erdatmosphäre: Was zu forschen, was zu tun? DLR-Nachrichten, Heft **59**, 5–13.

A39.   Crutzen, P. J., 1990: Comments on George Reid's "Quo Vadimus" contribution "Climate". In: Quo Vadimus. Geophysics for the next generation, Geophysical Monograph 60, IUGG Vol 10, Eds. G.D. Garland and John R. Apel, American Geophysical Union, Washington, USA, p. 47.

A40.   Crutzen, P. J., 1990: Global changes in tropospheric chemistry, Proceedings of Summer School on Remote Sensing and the Earth's Environment, Alpbach, Austria, 26 July - 4 August 1989, pp. 105–113.

A41.   Crutzen, P. J., C. Brühl, 1990: The potential role of HO_x and ClO_x interactions in the ozone hole photochemistry, in *Dynamics, Transport and Photochemistry in the Middle Atmosphere of the Southern Hemisphere*, Ed. A. O'Neil, Kluwer, Dordrecht, 203–212.

A42.   Crutzen, P. J. and C. Brühl, 1990: The atmospheric chemical effects of aircraft operations. In: *Air Traffic and the Environment - Background, Tendencies and Potential Global Atmospheric Effects.* Proceedings of a DLR International Colloqium Bonn, Germany, November 15/16, 1990, Ed. Schumann, Springer-Verlag, Heidelberg 1990, pp. 96–106.

A43.   Horowitz, A., G. von Helden, W. Schneider, F. G. Simon, P. J. Crutzen and G. K. Moortgat, 1990: Oxygen photolysis at 214 nm and 25 °C, Proceedings of the Quadrennial Ozone Symposium, Göttingen 8-13 August 1988, Eds. R. D. Boikov and P. Fabian, Deepak Publ. Co., pp. 690–693.

A44.   Brühl, Ch., P. J. Crutzen, E. F. Danielsen, H. Graßl, H.-D. Hollweg and D. Kley, 1991: Umweltverträglichkeitsstudie für das Raumtransportsystem SÄNGER, Teil 1 Unterstufe, Ed. Max-Planck-Institut für Meteorologie Hamburg, 142 pp.

A45.   Crutzen, P. J., 1991: Methane's sinks and sources. Nature, **350**, pp. 380–382.

A46.   Lelieveld, J. and P. J. Crutzen, 1991: Climate discussion and fossil fuels. Oil Gas - European Magazine, **4**, 11–15.

A47.   Crutzen, P. J., 1992: Ozone depletion: Ultraviolet on the increase. Nature, **356**, 104–105.

A48.   Crutzen, P. J., 1992: Menschliche Einflüsse auf das Klima und die Chemie der globalen Atmosphäre, in: *Stadtwerke der Zukunft* - ASEW-Fachtagung Kassel, 1991, Ed. ASEW, Köln, Ponte Press, Bochum, 7–27.

A49.   Graedel, T. E. and P. J. Crutzen, 1992: Ensemble assessments of atmospheric emissions and impacts, in: *Energy and the Environment in the 21st Century*, pp 1–24, Energy Laboratory, Massachusetts Institute of Technology, Cambridge.

A50.   Sander, R., J. Lelieveld and P. J. Crutzen, 1992: Model calculations of the nighttime aqueous phase oxidation of S(IV) in an orographic cloud. Proceedings of Joint CEC/EUROTRAC Workshop and LACTOZ-HALIPP Working Group on Chemical Mechanisms Describing Tropospheric Processes, Leuven, Belgium, September 23–25, 1992, Air Pollution Research Report 45, Ed. J. Peeters, E. Guyot SA, Brussels, 285–290.

A51. Brühl, Ch., P. J. Crutzen, H. Graßl and D. Kley, 1993: The impact of the spacecraft system Sänger on the composition of the middle atmosphere, in: *AIAA Fourth International Aerospace Planes Conference*, Orlando/Florida, 1–4 December 1992, American Institute of Aeronautics and Astronautics, Washington DC, 1–9.

A52. Crutzen, P. J., 1993: Die Beobachtung atmosphärisch-chemischer Veränderungen: Ursachen und Folgen für Umwelt und Klima. In: *Klima: Vorträge im Wintersemester 1992/93*, Sammelband der Vorträge des Studium Generale der Ruprecht-Karls-Universität Heidelberg (Ed.), Heidelberger Verlagsanstalt, 31–48.

A53. Kanakidou, M., F. J. Dentener and P. J. Crutzen, 1993: A global three-dimensional study of the degradation of HCFC's and HFC-134a in the troposphere. Proceedings of STEP-HALOCSIDE/AFEAS Workshop on Kinetics and Mechanisms for the Reactions of Halogenated Organic Compounds in the Troposphere, Dublin, Ireland, March 23-25, 1993, Campus Printing Unit, University College Dublin, 113–129.

A54. Grooß, J. U., Th. Peter, C. Brühl and P. J. Crutzen, 1994: The influence of high flying aircraft on polar heterogeneous chemistry, Proceedings of an International Scientific Colloquium on Impact of Emissions from Aircraft and Spacecraft upon the Atmosphere, Köln, Germany, April 18–20, 1994, DLR-Mitteilung 94-06, U. Schumann and D. Wurzel (Eds.), 229–234.

A55. Lelieveld, J. and P. J. Crutzen, 1994: Emissionen klimawirksamer Spurengase durch die Nutzung von Öl und Erdgas. Energiewirtschaftliche Tagesfragen, **7**, 435–440.

A56. Kanakidou, M., P. J. Crutzen and P.H. Zimmermann, 1994: Estimates of the changes in tropospheric chemistry which result from human activity and their dependence on NOx emissions and model resolution, Proceedings of the Quadrennial Ozone Symposium, June 4-13, 1992, Charlottsville, Virginia, U.S., NASA Conference Publication 3266, 66–69

A57. Müller, R. and P. J. Crutzen, 1994: On the relevance of the methane oxidation cycle to "ozone hole chemistry", Proceedings of the Quadrennial Ozone Symposium, June 4–13, 1992, Charlottsville, Virginia, U.S., NASA Conference Publication 3266, 298–301.

A58. Peter, Th. and P. J. Crutzen, 1994: Das Ozonloch: Wie kam es dazu und was sollten wir daraus lernen? In: *Der Mensch im Strahlungsfeld der Sonne*. Konstanz und Wandel in Natur und Gesellschaft. Ed. C. Fröhlich, Forum Davos, Wissenschaftliches Studienzentrum, Davos, 31–44.

A59. Steil, B., C. Brühl, P. J. Crutzen, M. Dameris, M. Ponater, R. Sausen, E. Roeckner, U. Schlese and G. J. Roelofs, 1994: A chemistry model for use in comprehensive climate models, Proceedings of an International Scientific Colloquium on Impact of Emissions from Aircraft and Spacecraft upon the

Atmosphere, Köln, Germany, April 18-20, 1994, DLR-Mitteilung 94-06, U. Schumann and D. Wurzel (Eds.), 235–240.

A60.   Andreae, M. O., W. R. Cofer III, P. J. Crutzen, P. V. Hobbs, J. M. Hollander, T. Kuhlbusch, R. Novakov, J. E. Penner, 1995: Climate impacts of carbonaceous and other non-sulfate aerosols: A proposed study. Lawrence Berkely Laboratory Document - PUB-5411.

A61.   Crutzen, P. J., 1995: On the role of ozone in atmospheric chemistry. In: *"The Chemistry of the Atmosphere. Oxidants and Oxidation in the Earth's Atmosphere"*, Proceedings of the 7th BOC Priestley Conference, Lewisburg, Pennsylvania, U.S.A., June 25–27, 1994, Ed. A.R. Bandy, The Royal Society of Chemistry, Cambridge, UK, 3–22.

A62.   Crowley, J. N., P. Campuzano-Jost, S. A. Carl, P. J. Crutzen, M. Finkbeiner, F. Helleis, O.Horie, G. K. Moortgat, R. Müller and C. Roehl, 1996: Laboratory investigations of the production and loss of hypochlorite in the stratosphere. Proceedings of the 3rd European Workshop on "Polar Stratospheric Ozone", Schliersee, Germany, September 18–22, 1995, Air Pollution Research Report 56, J.A Pyle, N.R.P. Harris and G.T. Amanatidis (Eds.), European Commission, Luxembourg, 1996, 679–683.

A63.   Crutzen, P. J., 1996: Das stratosphärische Ozonloch: Eine durch menschliche Aktivitäten erzeugte chemische Instabilität in der Atmosphäre. Festvortrag und Festansprache anlässlich der Verleihung des Internationalen Rheinlandpreises für Umweltschutz 1996 in Köln, TÜV Rheinland, 17–29.

A64.   Vogt, R. and P. J. Crutzen, 1996: Modelling of halogen chemistry in the remote marine boundary layer. Proceedings of the EUROTRAC Symposium '96, Garmisch-Partenkirchen, Germany, 25-29 March 1996 on "Transport and Transformation of Pollutants in the Troposphere, Vol. 1, Clouds, Aerosols, Modelling and Photo-oxidants", P. M. Borrell, T. Cvitas, K. Kelly and W. Seiler (Eds.), Computational Mechanics Publications, Southampton, 1996, 445–449.

A65.   Crutzen, P. J., 1997: Entdeckung des Ozonlochs - Wissen und Vision. GIT Labor-Fachzeitschrift, **41**, 110–112.

A66.   Crutzen, P. J., 1997: Mesospheric mysteries. Science, **277**, 1951–1952.

A67.   Crutzen, P. J., 1997: Problems in global atmospheric chemistry. In: *"The oxidizing capacity of the troposphere"*, Proceedings of the 7th European symposium on 'Physico-chemical behaviour of atmospheric pollutants', Venice, Italy, 2–4 Ocotber 1996, B. Larsen, B. Versino and G. Angeletti (Eds.), European Commission, Brussels, 1997, 1–13.

A68.   Crutzen, P. J., 1997: Die Beobachtung atmosphärisch-chemischer Veränderungen: Ursachen und Folgen für Umwelt und Klima. (Festvortrag anläßlich der Hauptversammlung der Max-Planck-Gesellschaft in Bremen am 6. Juni 1997). In: Max-Planck-Gesellschft. Jahrbuch 1997. Ed. Generalverwaltung der Max-Planck-Gesellschaft München, Vandenhoek & Ruprecht, Göttingen, 1997, 51–71.

A69.   Crutzen, P. J. and P. Braesicke, 1997: Rule-of-Thumb: Converting potential temperature to altitude in the stratosphere. EOS, **78**, 410.

A70.   Crutzen, P. J. and M. Lawrence, 1997: Ozone clouds over the Atlantic. Nature, **388**, 625–626.

A71.   Lelieveld, J., P. J. Crutzen, D. J. Jacob and A. M. Thompson, 1997: Modeling of biomass burning influences on tropospheric ozone. In: *Fire in the Southern African Savannas*. Eds. B.W. van Wilgen, M. O. Andreae, J. G. Goldammer and J. A. Lindsay, Witwatersrand University Press, 1997, 217–238.

A72.   Perner, D., T. Klüpfel, E. Hegels, P. J. Crutzen and J. P. Burrows, 1997: First results on tropospheric observations by the global ozone monitoring experiment, GOME, on ERS 2. Proceedings of the 3rd. ERS Symposium on "Space at the service of our Environment", Florence, Italy, March 17–21, 1997, ESA SP-414, 3 Vols., 647–652.

A73.   Steil, B., M. Dameris, Ch. Brühl, P. J. Crutzen, V. Grewe, M. Ponater and R. Sausen, 1997: Development of a chemistry module for GCMs: First results of a multi-annual integration.DLR, Institut für Physik der Atmosphäre, Report No. **74**, U. Schumann (Ed.), DLR-Oberpfaffenhofen, 50 pp.

A74.   Crutzen, P. J., 1998: What is happening to our precious air? The dramatic role of trace components in atmospheric chemistry. Science Spectra, **14**, 22–31.

A75.   Crutzen, P. J., 1998: ERASMUS Lecture: Changing atmospheric chemistry: Causes and consequences for environment and climate. European Review, **6** (1), 7–23.

A76.   Crutzen, P. J., 1998: How the atmosphere keeps itself clean and how this is affected by human activities, Pure & Appl. Chem., **70** (7), 1319–1326 (IUPAC Symposium on "Degradation Processes in the Environment", 24-28 May 1998, Dubrovnik, Croatia).

A77.   Hegels, E., H. Harder, T. Klüpfel, P. J. Crutzen and D. Perner, 1998: On the global distribution of the halogen oxides from observations by GOME (Global Ozone Monitoring Experiment) on ERS-2. Proceedings of the 4th European Symposium "Polar stratosperic ozone 1997", Schliersee, Germany, September 22–26, 1997, Air Pollution Research Report 66, N. R. P. Harris, I. Kilbane-Dawe and G.T. Amanatidis (Eds.), European Commission, Luxembourg, 1998, 343–346.

A78.   Hegels, E., P. J. Crutzen, T. Klüpfel, D. Perner, J. P. Burrows, A. Ladstätter-Weißenmayer, M. Eisinger, J. Callies, A. Hahne, K. Chance, U. Platt and W. Balzer, 1998: Satellite measurements of halogen oxides by the Global Ozone Monitoring Experiment, GOME, on ERS2: Distribution of BrO and comparison with groundbased observations. Proceedings of the Quadrennial Ozone Symposium on "Atmospheric Ozone", l'Aquila, Italy, September 12–21, 1996, R. D. Bojkov and G. Visconti (Eds.), Edigrafital S.p.A., S. Atto, Italy, 1998, 293–296.

A79.   Müller, R., J.-U. Grooß, D. S. McKenna, P. J. Crutzen, C. Brühl, J. M. Russell III, and A. F. Tuck, 1998: HALOE observations of ozone depletion in the arctic vortex during winter and early spring 1996–1997. Proceedings of the 4th European Symposium "Polar stratosperic ozone 1997", Schliersee, Germany, September 22–26, 1997, Air Pollution Research Report 66, N.R.P. Harris, I. Kilbane-Dawe and G. T. Amanatidis (Eds.), European Commission, Luxembourg, 1998, 301–304.

A80. Müller, R., P. J. Crutzen, J.-U. Grooß, C. Brühl, H. Gernandt, J. M. Russell III, A. F. Tuck, 1998: Chlorine activation and ozone depletion in the Arctic stratospheric vortex during the first five winters of HALOE observations on the UARS. Proceedings of the Quadrennial Ozone Symposium on "Atmospheric Ozone", l'Aquila, Italy, September 12–21, 1996, R.D. Bojkov and G. Visconti (Eds.), Edigrafital S.p.A., S. Atto, Italy, 1998, 225–228.

A81. Waibel, A., H. Fischer, M. Welling, F.G. Wienhold, Th. Peter, K.S. Carslaw, Ch. Brühl, J.-U. Grooß and P. J. Crutzen, 1998: Nitrification and denitrification of the Arctic stratosphere during winter 1994-1995 due to ice particle sedimentation. Proceedings of the Quadrennial Ozone Symposium on "Atmospheric Ozone", l'Aquila, Italy, September 12–21, 1996, R.D. Bojkov and G. Visconti (Eds.), Edigrafital S.p.A., S. Atto, Italy, 1998, 233–236.

A82. Crutzen, P. J., 1999: Die Beobachtung atmosphärisch-chemischer Veränderungen: Ursachen und Folgen für Umwelt und Klima. Leopoldina, **44**, 351-368 (Jahrbuch 1998 der Deutschen Akademie der Naturforscher Leopoldina, Halle/Saale).

A83. Crutzen, P. J., 1999: Verschmutzung und Selbstreinigung der Atmosphäre. Naturw. Rdsch., **52**, 1–5.

A84. Crutzen, P. J., 1999: Das stratosphärische Ozonloch: Durch menschliche Aktivitäten erzeugte chemische Instabilität in der Atmosphäre. In: Mainz wie es lebt und denkt. Ein Mainzer Mosaik, Vereinigung der Freunde des Lions Club Mainz-Schönborn e.V. (Herausgeber), pp.181–189.

A85. Crutzen, P. J., 1999: Global problems of atmospheric chemistry – The story of man's impact on atmospheric ozone. In: *Atmospheric Environmental Research*, D. Möller (Ed.), Springer-Verlag, Heidelberg, 1999, 3–30.

A86. Crutzen, P. J., 1999: An Essay in Atmospheric Chemistry and Global Change. G. P. Brasseur, J. J. Orlando and G. S. Tyndall (Eds.), Oxford University Press, New York – Oxford, pp. 486.

A87. Ladstätter-Weißenmayer, A., J. P. Burrows, P. J. Crutzen and A. Richter, 1999: Biomass burning and its influence on the troposphere, in Atmospheric Measurements from Space. ESA WPP-**161**, 1369–1374.

A88. Lelieveld J., V. Ramanathan and P. J. Crutzen, 1999: The global effects of Asian haze. IEEE Spectrum, December 1999, 50–54.

A89. Crutzen, P. J., 2000: Dowsing the human volcano. Nature, **407**, 674–675.

A90. Crutzen, P. J., 2000: Developments in tropospheric chemistry. In: *Chemistry and radiation changes in the ozone layer*, C. S. Zerefos et al. (Eds.), Kluwer Academic Publishers, The Netherlands, 2000, 1–12.

A91.   Crutzen, P. J., 2000: The Changing Chemistry of the Atmosphere. In: *Climate Impact Research: Why, How and When.* Joint International Symposium, Berlin, October 28-29, 1997, B. Parthier and D. Simon (Eds.), Akademie Verlag, Berlin, 2000, 47–68.

A92.   Crutzen, P. J. and E. F. Stoermer, 2000: The "Anthropocene". IGBP Newsletter, **41**, 17–18.

A93.   Steil, B., C. Brühl, P. J. Crutzen and E. Manzini, 2000: A MA-GCM with interactive chemistry (MAECHAM4-CHEM), simulations for present, past and future. Proceedings Quadrennial Ozone Symposium, Sapporo 2000, NASDA, 221–222.

A94.   Crutzen, P. J., 2001: The role of tropical atmospheric chemistry in global change research: The need for research in the tropics and subtropics. Proceedings of the Preparatory Session 12–14 November 1999 and the Jubilee Plenary Session 10–13 November 2000 on "Science and the Future of Mankind: Science for Man and Man for Science", Vatican City, Pontifical Academia Scientiarum, Vatican City, 2001, 110–114.

A95a.   Crutzen, P. J., 2001: Was ist Luft? Süddeutsche Zeitung Magazin, **40**, 20–23.

A95b.   Crutzen, P. J., 2001: Was ist Luft? In: *Kinder fragen, Nobelpreisträger antworten*, B. Stiekel (Ed.), Wilhelm Heyne Verlag, München, 2001, 129–136.

A95c.   Crutzen, P. J., 2003: What is Air? In: Nobel Book of Answers. B. Stiekel (Ed.), Atheneum Books for Young Readers, pp. 166–191.

A96.   Crutzen, P.J., 2001: The Antarctic Ozone Hole, a Human-Caused Chemical Instability in the Stratosphere. What Should We Learn from It? In: Geosphere – Biosphere Interactions and Climate, L. O. Bengtsson and C. U. Hammer (Eds.), Cambridge University Press, 2001, 1–11.

A97.   Crutzen, P. J., R. Sander and R. Vogt, 2001: The Influence of Aerosols on the Photochemistry of the Atmosphere. In: *Dynamics and Chemistry of Hydrometers.* Final Report of the Collaborative Research Centre 233 "Dynamik und Chemie der Hydrometeore", R. Jaenicke (Ed.), Wiley-VCH Verlag GmbH, Weinheim, 2001, 130–147.

A98.   Crutzen, P. J., E. Oberlander, W. Peters and A. Römpp, 2001: Overview of atmospheric chemistry. In: *Chemical, Physical and Biogenic Processes.* Notes from the 3rd COACh International School, G. Moortgat (Ed.), Max-Planck-Institut für Chemie, Mainz, 2001, 7–17.

A99.   Sander, R. and P. J. Crutzen, 2001: Bodennahes Ozonloch in der Arktis. Spektrum der Wissenschaft, Jan. 2001, 12–13.

A100.   Keene, W. C., J. M. Lobert, J. R. Maben, D. Scharffe and P. J. Crutzen, 2001: Emissions of volatile inorganic halogens, carboxylic acids, ammonia and sulfur dioxide from experimental burns of southern African biofuels, Eos Trans. AGU, **82(47)**, F112.

A101.   Crutzen, P. J., 2002: Geology of mankind – The Anthropocene. Nature, **415**, 23.

A102.   Crutzen, P. J., 2002: The importance of tropical atmospheric chemistry in global change research. In: *Human Development and the Environment*, H.

Ginkel, B. Barrett, J. Court and J. Velasquez (Eds.), The United Nations University, 2002, 213–219.

A103. Crutzen, P. J., 2002: Eine schillernde Hypothese…mit der sich's leben läßt und ohne die man als Forscher auskommt. GAIA, **1/2002**, 19–21.

A104. Crutzen, P. J., 2002: A Critical Analysis of the Gaia Hypothesis as a Model for Climate/Biosphere Interactions. GAIA, **2/2002**, 96–103.

A105. Crutzen, P. J., 2002: The effects of industrial and agricultural practices on atmospheric chemistry and climate during the anthropocene. J. Environ. Sci. Health, **37**, 423–424.

A106. Crutzen, P. J., 2002: Atmospheric Chemistry in the "Anthropocene". In: Challenges of a Changing Earth. Proceedings of the Global Change Open Science Conference, Amsterdam, The Netherlands, 10-13 July 2001. W. Steffen, J. Jäger, D. J. Carson and C. Bradshaw (Eds.), Springer Verlag, 2002, 45–48.

A107. Crutzen, P. J., 2002: The "anthropocene". In: ERCA, Vol. 5, From the Impacts of Human Activities on our Climate and Environment to the Mysteries of Titan. C Boutron (Ed.), EDP Sciences, 2002, 1–5.

A108. Levin, Z., Y. Rudich, P. J. Crutzen, M. Andreae and A. Bott, 2002: The role of cloud processing and of organic matter on the formation of soluble layers on mineral dust particles and the impact on clouds characteristics. Final Report to GIF, April 2002.

A109. Oberlander, E. A., C. A. M. Brenninkmeijer, P. J. Crutzen, J. Lelieveld and N. F. Elansky, 2002: Why Not Take the Train? Trans-Siberian Atmospheric Chemistry Observations across Central and East Asia. EOS, **83**, 509/515/516

A110. Crutzen, P. J., 2003: Schutz der Ozonschicht – ein Beipsiel gelungener Umweltpolitik. In: Jahrbuch Ökologie 2004. G. Altner, H. Leitschuh-Fecht, G. Michelsen, U. E. Simonis and E. U. von Weizsäcker (Eds.), Verlag C.H. Beck, 2003, 132–145.

A111. Crutzen, P. J., 2003: High flyer. NewScientist, 5 July 2003, 44–47.

A112. Crutzen, P. J., 2003: Het antropoceen: op de drempel naar de toekomst. In : Systeem Aarde, cahiers bio-wetenschappen en maatschappij, D. W. van Bekkum, H. N. A. Priem, G. J. van der Zwaan (Eds.), 60-64.

A113. Crutzen, P. J., 2004: Anti-Gaia. In: Global Change and the Earth System. Springer Verlag, Berlin, 72.

A114. Crutzen, P. J., 2004: The Ozone Hole. In: Global Change and the Earth System. Springer Verlag, Berlin, 236–237.

A115. Crutzen, P. J., 2004: Keynote address 3. In: Partnerships for Sustainable Development addressing the WEHAB agenda. TERI – Delhi Sustainable Development Summit 2004, R. K. Pachauri (Ed.), TERI Press, New Delhi, 229–234.

A116. Wallström, M., B. Bolin, P. J. Crutzen and W. Steffen, 2004: The Earth's life-support system is in peril. International Herald Tribune, January 20, 2004.

A117.   Crutzen, P. J., 2005: Das stratosphärische Ozonloch: eine durch Menschen verursachte chemische Instabilität in der Atmosphäre. Was können wir daraus lernen? In: Vermächtnis und Vision der Wissenschaft, Zeitdiagnosen 7, Karl Acham (Ed.), Passagen Verlag, Wien (Austria), 31–38.

A118.   Crutzen, P. J., 2006: The "Anthropocene". In: Earth System Science in the Anthropocene – Emerging Issues and Problems. E. Ehlers, T. Krafft (Eds.), Springer, Heidelberg, 13–18.

A119.   Crutzen, P. J.: Impact of China's air pollution. Frontier in Ecology and the Environment, 4, p. 340, 2006.

A120.   Brenninkmeijer, C., F. Slemr, T. Schuck, D. Scharffe, C. Koeppel, M. Pupek, P. Jöckel, J. Lelieveld, P. Crutzen, T. S. Rhee, M. Hermann, A. Weigelt, M. Reichelt, J. Heintzenberg, A. Zahn, D. Sprung, H. Fischer, H. Ziereis, H. Schlager, U. Schumann, B. Dix, U. Friess, U. Platt, R. Ebinghaus, B. Martinsson, H. N. Nguyen, D. Oram, D. O'Sullivan, S. Penkett, P. van Velthoven, T. Röckmann, G. Pieterse, S. Assonov, M. Ramonet, I. Xueref-Remy, P. Ciais, S. Reimann, M. Vollmer, M. Leuenberger and F. L. Valentino, 2007: The CARIBIC aircraft system for detailed, long-term, global-scale measurement of trace gases and aerosol in a changing atmosphere. IG Activities, 37, 2–9.

Graedel, T.E./ P.J. Crutzen, 1996: *Atmosphäre im Wandel. Die empfindliche Lufthülle unseres Planeten.* (Heidelberg; Spektrum Akademischer Verlag).

Graedel, T.E./ P.J. Crutzen, 199x: Atmosphere, Climate and Change (New York: W.H. Freeman.,).

(continued)

(continued)

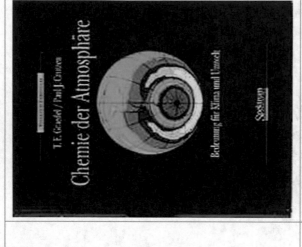

Crutzen, P.J., 1996: *Atmosphäre, Klima, Umwelt* (Edited and with an introduction by P.J. Crutzen), (Heidelberg: Spektrum Akademischer Verlag).

Graedel, T.E. and P.J. Crutzen, 1994: *Chemie der Atmosphäre. Bedeutung für Klima und Umwelt*, (Heidelberg: Spektrum Akademischer Verlag).

A121.   Crutzen, P.J.: Atmospheric chemistry and climate in the Anthropocene. In: Making Peace with the Earth – What Future for the Human Species and the Planet? J. Bindé (Ed.), Berghahn Books, UNESCO Publishing, 113–120. 2007.

A122.   Winiwarter, W., P. Crutzen, A. Mosier and K. Smith: N_2O in Treibhausgasbilanzen von Biotreibstoffen – eine globale Perspektive. In: Verband Deutscher Landwirtschaftlicher Untersuchungs- und Forschungsanstalten, Kongressband 2008 Jena, Erhöhte Biomassenachfrage – eine neue Herausforderung für die Landwirtschaft. VDLUFA Schriftenreihe Bd. 64, pp. 75–82, VDLUFA-Verlag, Darmstadt, 2008.

A123.   Crutzen, P. J., A. Mosier, K. Smith and W. Winiwarter: Atmospheric N_2O Releases from Biofuel Production Systems: A Major Factor Against "CO_2 Emission Savings": A Global View. In: Twenty Years of Ozone Decline – Proceedings of the Symposium for the 20th Anniversary of the Montreal Protocol. C. Zerefos, G. Contopoulos and G. Skalkeas (Eds.), pp. 67–70, Springer-Verlag, 2009.

A124.   Elansky, N. F., I. B. Belikov, E. V. Berezina, C. A. M. Brenninkmeijer, N. N. Buklikova, P. J. Crutzen, S. N. Elansky, J. V. Elkins, A. S. Elokhov, G. S. Golitsyn, G. I. Gorchakov, I. G. Granberg, A. M. Grisenko, R. Holzinger, D. F. Hurst, A. I. Igaev, A. A. Kozlova, V. M. Kopeikin, S. Kuokka, O. V. Lavrova, L. V. Lisitsyna, K. B. Moeseenko, E. A. Oberlander, Yu. I. Obvintsev, L. A. Obvintseva, N. V. Pankratova, O. V. Postylyakov, E. Putz, P. A. Romashkin, A. N. Safronov, K. P. Shenfeld, A. I. Skorokhod, R. A. Shumsky, O.A.Tarasova, J. C. Turnbull, E. Vartiainen, L. Weissflog and K. V. Zhernikov: Atmospheric composition observations over the Northern Eurasia using the mobile laboratory (Troica experiments), Obukhov Institute of Atmospheric Physics, Russian Academy of Sciences, International Science and Technology Center, European Union, pp. 1–75, 2009.

A125.   Crutzen, P.J.: Cooling the Earth's Surface by Stratospheric Sulphur Injections, WZB, 2010.

A126a.   Ramanathan, V., M. Agrawal, H. Akimoto, M. Aufhammer, S. Devotta, L. Emberson, S.I. Hasnain, M. Iyngararasan, A. Jayaraman, M. Lawrance, T. Nakajima, T. Oki, H. Rodhe, M. Ruchirawat, S.K. Tan, J. Vincent, J.Y. Wang, D. Yang, Y.H. Zhang, H. Autrup, L. Barregard, P. Bonasoni, M. Brauer, B. Brunekreef, G. Carmichael, C.E. Chung, J. Dahe, Y. Feng, S. Fuzzi, T. Gordon, A.K. Gosain, N. Htun, J. Kim, S. Mourato, L. Naeher, P. Navasumrit, B. Ostro, T. Panwar, M.R. Rahman, M.V. Ramana, M. Rupakheti, D. Settachan, A. K. Singh, G. St. Helen, P. V. Tan, P.H. Viet, J. Yinlong, S.C. Yoon, W.-C. Chang, X. Wang, J. Zelikoff and A. Zhu: Atmospheric Brown Clouds. Regional Assessment Report with focus on Asia, UNEP, 2008.

A126b.   Ramanathan, V., M. Agrawal, H. Akimoto, M. Aufhammer, S. Devotta, L. Emberson, S.I. Hasnain, M. Iyngararasan, A. Jayaraman, M. Lawrance, T. Nakajima, T. Oki, H. Rodhe, M. Ruchirawat, S.K. Tan, J. Vincent, J.Y. Wang, D. Yang, Y.H. Zhang, H. Autrup, L. Barregard, P. Bonasoni, M.

Brauer, B. Brunekreef, G. Carmichael, C.E. Chung, J. Dahe, Y. Feng, S. Fuzzi, T. Gordon, A.K. Gosain, N. Htun, J. Kim, S. Mourato, L. Naeher, P. Navasumrit, B. Ostro, T. Panwar, M.R. Rahman, M.V. Ramana, M. Rupakheti, D. Settachan, A. K. Singh, G. St. Helen, P. V. Tan, P.H. Viet, J. Yinlong, S.C. Yoon, W.-C. Chang, X. Wang, J. Zelikoff and A. Zhu: Atmospheric Brown Clouds. Regional Assessment Report with focus on Asia. Summary, UNEP, 2008.

A127.   Crutzen, P. J., 2010: Erdabkühlung durch Sulfatinjektionen in die Stratosphäre. In: Jahrbuch Ökologie 2010 (Die Klima-Manipulateure). G. Altner, H. Leitschuh, G. Michelsen, U. E. Simonis and E. U. von Weizsäcker (Eds.), S. Hirzel Verlag Stuttgart, 2010, 33–39.

A128.   Smith, K., P. J. Crutzen, A. Mosier and W. Winiwarter, 2010: The Global N_2O Budget: A Reassessment. In: Nitrous Oxide and Climate Change. K. Smith (Ed.), Earthscan, May 2010.

A129.   Crutzen, P. J., 2010: Anthropocene man. Nature, 467, S10.

A130.   Ajai, L. Bengtsson, D. Breashears, P. J. Crutzen, S. Fuzzi, W. Haeberli, W. W. Immerzeel, G. Kaser, C. Kennel, A. Kulkarni, R. Pachauri, T. H. Painter, J. Rabassa, V. Ramanathan, A. Robock, C. Rubbia, L. Russell, M. Sánchez Sorondo, H. J. Schellnhuber, S. Sorooshian, T. F. Stocker, L. G. Thompson, O. B. Toon, D. Zaelke and J. Mittelstraß, 2011: Fate of Mountain Glaciers in the Anthropocene. A Report by the Working Group on 2-4 April, Commissioned by the Pontifical Academy of Sciences, Vatican City, 15 pp. (2011).

A131.   Crutzen, P.J. and C. Schwägerl: Living in the Anthropocene: Toward a New Global Ethos. Yale Environment 360, Online-publication of the Yale School of Forestry & Environmental Studies, Yale University, New Haven (Connecticut), 24 January 2011.

A132.   Crutzen, P.J., 2011: Die Geologie der Menschheit. In: Das Raumschiff Erde hat keinen Notausgang. Crutzen, Paul; Davis, Mike; Mastrandrea, Michael D., Suhrkamp Verlag, edition unseld.

A133.   Crutzen, P.J., 2011: The Anthropocene: Geology by Mankind. In: Coping with Global Environmental Change, Disasters and Security. Threats, Challenges, Vulnerabilities and Risks, Brauch, Hans Günter; Oswald Spring, Úrsula; Mesjasz, Czeslaw; Grin, John; Kameri_Mbote, Patricia; Chourou, Béchir; Dunay, Pál; Birkmann, Jörn (Eds.), Springer-Verlag, Berlin-Heidelberg-New York, 3–4.

A134.   Crutzen, P. J., 2012: Ozone and advocacy (Sherry Rowland), Nature Geoscience, 5, 311.

A135.   Crutzen, P.J., 2012: Climate, Atmospheric Chemistry and Biogenic Processes in the Anthropocene. In: „100 Jahre Kaiser-Wilhelm-/Max-Planck-Institut für Chemie (Otto-Hahn-Institut): Facetten seiner Geschichte", H. Kant und C. Reinhardt (Eds.), Veröffentlichungen aus dem Archiv der Max-Planck-Gesellschaft, Band 22, Berlin, 241–249.

A136.   Crutzen, P., G. Lax and C. Reinhardt: Paul Crutzen on the Ozone Hole, Nitrogen Oxides, and the Nobel Prize. Angewandte Chemie - International Edition, **52**, 48–50. https://doi.org/10.1002/anie.201208700 (2013).

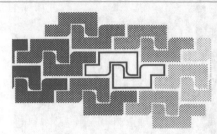

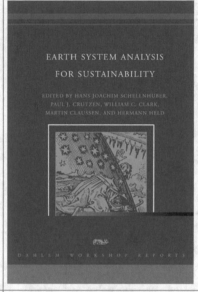

Crutzen, Paul J., Ramanathan, Veerabhadran (Eds.), 1996: *Clouds, Chemistry and Climate,* Nato ASI Subseries I	H. J. Schellnhuber, P. J. Crutzen, W. C. Clark, M. Claussen, H. Held, 2004:, *Earth System Analysis for Sustainablilty,* Dahlem Workshop Reports, MIT Press.

(continued)

(continued)

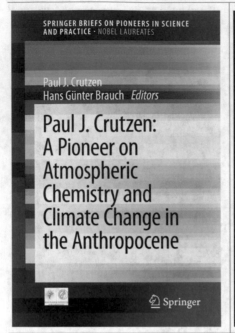

| Paul J. Crutzen; Hans Günter Brauch (Eds.): Paul J. Crutzen: A Pioneer on Atmospheric Chemistry and Climate Change in the Anthropocene (Cham: Springer International Publishing, 2016). | Paul J. Crutzen, Michael Müller (Eds.): Das Anthropozän - Schlüsseltexte des Nobelpreisträgers für das neue Erdzeitalter. Mit Einführungen u.a. von Hans J. Schellnhuber und Klaus Töpfer (Munich oekom, 2019) |

Appendix 3
Facsimile of P.J. Crutzen's First Powerpoint on the Anthropocene

---

Atmospheric Chemistry, Biosphere and Climate in the Anthropocene
Paul J. Crutzen, MPI for Chemistry, Mainz, Germany

I first want to introduce myself. Here I am in this picture in the lap of my grandmother almost 80 years ago. I have changed a lot, but so has the whole world.

(continued)

S. Benner et al. (eds.), *Paul J. Crutzen and the Anthropocene: A New Epoch in Earth's History*, The Anthropocene: Politik–Economics–Society–Science 1, https://doi.org/10.1007/978-3-030-82202-6

(continued)

- **During the past 3 centuries human population has increased tenfold to 7000 million**
- **Cattle population increased to 1400 million (one cow/family)**
- **There are currently some 20 billion (20,000 million) of farm animals worldwide**
- **Urbanisation grew more than tenfold in the past century; almost half of the people live in cities and megacities**
- **Industrial output increased 40 times during the past century; energy use 16 times**
- **Almost 80% of the Land surface has been transformed by human action**
- **Annual releases of C from fossil fuel burning: $6x10^{15}$g and 10^{15}g from deforestation**

- **Fish catch increased 40 times**
- **The release of SO_2(115Tg/year) by coal and oil burning is at least twice the sum of all natural emissions; over land the increase has been 7 fold, causing acid rain, health effects, poor visibility, and climate changes due to sulfate aerosols**
S.J. Smith at al, Atmos. Chem. Phys, 11, 1101-1116, 2011
- **Several climatically important "greenhouse gases" have substantially increased in the atmosphere, e.g. CO_2 by 40%, CH_4 by more than 100%**
- **Releases of NO to the atmosphere from N fertilizer use. Lightning, fossil fuel and biomass burning, 240Tg/year, is 2 times larger than its natural inputs, causing regional high surface ozone levels**
Galloway and Cowling, Ambio, 1999
- **Water use increased 9 fold during the past century to 600 m^3 per capita /year; 65% for irrigation, 25% industry, ~10% households**

Industrial and agricultural activities have grown dramatically especially since the second world war, the so-called "great acceleration". While the main activity has been in the developed world, the developing countries are followed rapidly, especially in Asia.
Steffen W., P. J. Crutzen and J.R. Mc Neill. Ambio, 36 (8), 2007

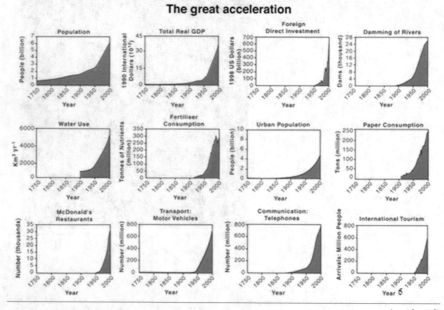

The great acceleration

(continued)

(continued)

- **Humanity is also responsible for the presence of many toxic substances in the environment and even some which are not toxic at all, but which have, nevertheless, led to the ozone hole.**
- **Among the „greenhouse gases" are also the almost inert CFCs (chlorofluorocarbons) gases. However, their photochemical breakdown in the stratosphere gives rise to highly reactive chlorine and bromine gases (radicals), which destroy ozone by catalytic reactions. As a consequence UY-B radiation at the earth surface from the sun increases, leading for instance to enhanced risk of skin cancer.**

 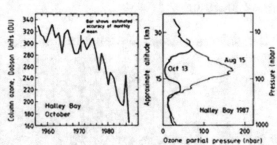

Fortunately, the CFC gases are no longer produced, but it will take 50 years or more to heal the ozone hole.

Figure on left hand side: loss of total ozone in polar regions. Most ozone loss occurs during spring time in the lower stratosphere between altitudes of 12 and 25 km and at temperatures below about -80 °C.

Fortunately, the CFC gases are no longer produced, but it will take 50 years or more to heal the ozone hole.

Figure on left hand side: loss of total ozone in polar regions. Most ozone loss occurs during spring time in the lower stratosphere between altitudes of 12 and 25 km and at temperatures below about –80 °C.

(continued)

(continued)

Since the beginning of the 19th Century, by its own growing activities, Mankind opened a new geological epoch: the Anthropocene.
We are clearly affecting climate and can deliberately do so by geo-egineering.
Humans are now overwhelming the great forces of nature.

Since the beginning of the 19th Century, by its own growing activities, Mankind opened a new geological epoch: the Anthropocene.

We are clearly affecting climate and can deliberately do so by geo-egineering.

Humans are now overwhelming the great forces of nature.

Wastes: Only 20-30% of N fertilizer is taken up by plants.
About 30% of food produced is wasted.
Future agriculture: Loss of agricultural soil through erosions is a serious problem.
Even worse is the loss of phosphorous. Some studies indicate dangerous depletion in agricultural regions (tropics).
Mankind is the only species that produces weapons of mass destruction (nuclear, chemical, biological).
Mankind will remain a major environmental force for many millennia. A daunting task lies ahead for scientists and engineers to guide society towards environmentally sustainable management during the era of the Anthropocene. This will require appropriate human behaviour at all scales.

(continued)

(continued)

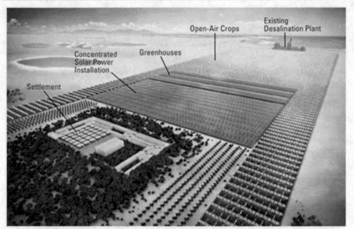

Electricity and fresh water from unproductive desert.

Electricity and fresh water from unproductive desert.

Appendix 4
Abstracts of Anthropocene Texts

For the following texts the editors failed to get a free permission from the original publishers to include these texts in this anthology of texts on the Anthropocene, a term Nobel Laureate Prof. Dr. Paul J. Crutzen had coined in February 2000 during a conference of the International Geophysical Biological Programme (IGBP). The editors decided therefore to include only the abstracts of these texts authored and co-authored by Paul J. Crutzen and his co-authors under the provision of 'fair use'[1] permitted under the copyright law. According to Guidelines of Wiley on 'fair use' no permissions are required for citations from books for a total of 300 words.[2] However, there is no agreement among major publishers whether for the republication of abstracts a written permission is needed.

---

[1] *Fair use* is a doctrine in the law of the United States that permits limited use of copyrighted material without having to first acquire permission from the copyright holder. Fair use is one of the *limitations to copyright* intended to balance the interests of copyright holders with the public interest in the wider distribution and use of creative works by allowing as a defence to copyright infringement claims certain limited uses that might otherwise be considered infringement. While about 40 countries adopted a similar 'fair use' policy, several others referred to a 'fair dealing'; see at: https://en.wikipedia.org/wiki/Fair_use (20 October 2019).

[2] There is no consensus among major British publishers whether a republication of abstracts requires a permission. According to Wiley: "*Abstracts specifically created for abstract publications:* If you have written an abstract of a previously published article for publication in another journal (e.g. a reviews-type journal), we do require you to sign an Agreement. *Reproduction of existing abstracts:* If you wish to include verbatim abstracts from previously published articles, you do not need written permission from the publisher of those articles, although the source should be cited. Abstracts are covered by copyright and are not in the public domain but there is an exception in UK law which permits the copying and publication of scientific and technical abstracts accompanying published periodical articles." See at: https://authorservices.wiley.com/author-resources/Journal-Authors/licensing/licensing-info-faqs.html (20 October 2019).

S. Benner et al. (eds.), *Paul J. Crutzen and the Anthropocene: A New Epoch in Earth's History*, The Anthropocene: Politik–Economics–Society–Science 1,
https://doi.org/10.1007/978-3-030-82202-6

1. **Paul J. Crutzen, 2002: "The Effects of Industrial and Agricultural Practices on Atmospheric Chemistry and Climate During the Anthropocene", in:** *J. Environ. Sci. Health*, **37, 423–424; see at:** https://www.tandfonline.com/doi/abs/10.1081/ESE-120003224.

Abstract

The impact of human activities on the environment has grown so much that there are good reasons to call the past 2–3 centuries the "Anthropocene" epoch. The impact is also clearly discernible in rising concentrations of several radioactively and chemically important gases and particles since preindustrial times. This is due to direct increases in primary emissions of carbon dioxide and monoxide, methane, nitrogen oxides, and chlorofluorocarbon gases (CFCs) and indirectly through chemical transformations, which affect especially the concentrations of ozone and particulate matter with repercussions for climate and the biosphere. The precipitous loss of ozone in the lower stratosphere at heights between 12 and 22km during the late winter/spring period over Antarctica (the so-called "ozone hole") has been the most dramatic example of a human-caused change in the global atmospheric chemistry system. It occurred unexpectedly in a region of the atmosphere the farthest removed from the regions where almost all emissions of the CFCs took place. The loss of ozone has a substantial effect on the penetration of biologically harmful, solar ultraviolet radiation to Earth's surface. Ozone in the troposphere (lower atmosphere), making up only about 10% of all ozone in the atmosphere too has been strongly affected by human activities, but opposite to the stratosphere. Increased industrial and agricultural emissions of nitrogen oxides, carbon monoxide, and methane lead to growing ozone concentrations, which have been observed at a number of stations around the world. The increase in ozone can affect human health, as well as agricultural and plant productivity. Furthermore, as ozone is a greenhouse gas, the observed changes in ozone will influence climate.

2. **Nebojsa Nakicenovic, Liana Talaue-McManus and Billie. L. Turner II, 2004: "Earth System Dynamics in the Anthropocene: Group Report", in: H. J. Schellnhuber, P. J. Crutzen, W. C. Clark, M. Claussen and H. Held (Eds.):** *Earth System Analysis for Sustainablility. Dahlem Workshop Report*(Cambridge: MIT Press): **313–340, see at:** https://mitpress.mit.edu/books/earth-system-analysis-sustainability.

Summary

Demonstrates how understanding the intertwined evolution of the Earth's geosphere and biosphere can contribute to the achievement of global sustainability.

Earth System Analysis for Sustainability uses an integrated systems approach to provide a panoramic view of planetary dynamics since the inception of life some four billion years ago and to identify principles for responsible management of the global environment in the future. Perceiving our planet as a single entity with hypercomplex, often unpredictable behavior, the authors use Earth system analysis to study global changes past and future. They explore the question of whether the unprecedented human-originated changes transforming the ecosphere today will end a 10,000-year period of climate stability.

The book presents the complete story of the inseparably intertwined evolution of life and matter on Earth, focusing on four major topics: long-term geosphere-biosphere interaction and the possibility of using extrasolar planets to test various geophysical hypotheses; the Quaternary Earth System's modes of operation; current planetary dynamics under human pressure; and transition to global sustainability. Written by leading figures in the disciplines of geology, climatology, evolution, biogeochemistry, macroeconomics, and institutions theory, *Earth System Analysis for Sustainability* analyzes the driving forces behind global change and uses this knowledge to propose principles to propose principles for global stewardship.

3. **Paul J. Crutzen, 2010: "Anthropocene Man", in:** *Nature,* **467: S10 (no abstract).** https://doi.org/10.1038/467S10a. **This text is accessible** https://www.nature.com/articles/467S10a.

How important is an interdisciplinary approach in addressing urgent scientific questions, and how can we foster such collaborations?

Interdisciplinary research has always played a major role in my career. In my earliest papers I proposed that NO and NO_2 could act as catalysts to destroy ozone in the stratosphere or to produce ozone in the troposphere, yet I had no formal education in chemistry. In fact, I was trained as civil engineer; I shifted to working as a computer specialist at the University of Stockholm to run meteorological models and, while helping to develop a program for classical ozone chemistry, got hooked on atmospheric chemistry. I wrote the 'NOx papers' when I was 36 years old.

Are scientists under-represented in politics? And do established scientists, especially Nobel laureates, have a duty to become active in politics and science policy?

There is, in fact, a dearth of scientists and scientifically educated representatives in politics, which is a worthwhile subject for analysis. I am only aware of three national leaders who have had scientific backgrounds: Margaret Thatcher, Angela Merkel and Gro Harlem Brundtland of Norway; interestingly, all women.

I don't think that Nobel laureates have a special duty to become more active in politics, except if their research could lead to problems for society. In the first place, they should remain scientists. However, the general public expects Nobelists to have an above average interest in social issues and the increased participation of scientists here should be applauded, as is the case under President Obama.

To establish closer contact between politicians and scientists, the German parliament has created commissions consisting of equal numbers of politicians and scientists to study and highlight problems of general importance, such as climate change and stratospheric ozone depletion. The advice of these commissions has been very successful in the political process.

How can the public be convinced of the importance of fundamental research with no applications in sight?

By pointing to the examples of great scientists like Einstein, Newton, Darwin, Faraday and others. In my own, much more modest, case, when I wrote the article about the potential effects of NOx on ozone, I had not the slightest idea about its relevance.

Things changed in the fall of 1970 when I read a report of a major international conference in the US, which mentioned NOx emissions but denied their importance. Independently, Harold Johnston and I drew attention to the problem of NOx emissions from supersonic aircraft destroying stratospheric ozone. The same happened in the case of chlorofluorocarbon emissions: the societal relevance was initially not recognized.

You must have experienced a lull at some point in your research career. What kept you going?

"After my first research successes I lived in a constant fear of running out of problems to work on: it would have been terrible for a well-funded research director."

After my first research successes I lived in a constant fear of running out of problems to work on: it would have been terrible for a well-funded research director. That did not happen. The main reason is that my research field is highly multi-faceted and interdisciplinary. There is room for many projects.

What advice would you give all young researchers who are starting their research life so as to become a good scientist?

Stay away from big science in which the resulting papers have many authors. And choose a professor who gives you a lot of freedom. I was lucky to find Bert Bolin.

Aside from as a Nobel laureate, how do you want to be remembered?

As the person, who significantly increased knowledge about the processes that determine the distribution of ozone in the atmosphere.

And as the scientist who coined the term 'Anthropocene': A new geologic epoch dominated by human activities, actually first published in *Nature*.

And as one of the scientists who drew attention to the potentially devastating climatic consequences of a nuclear war, the so-called 'nuclear winter'. More people would die of the indirect consequences of mass starvation and disease than would be killed by the nuclear bombs.

4.   **Jan Zalasiewicz; Mark Williams, M.; Will Steffen; Paul Crutzen, 2010: "Response to 'The Anthropocene Forces us to Reconsider Adaptationist Models of Human-Environment Interactions", in:** *Environ. Sci. Technol.,* **44, 16: 6008,** https://doi.org/10.1021/es102062w. **This text is accessible at:** https://pubs.acs.org/doi/pdf/10.1021/es102062w **and** https://academictree.org/chemistry/publications.php?pid=52216.

Abstract (not available)

5. **Jan Zalasiewicz, Mark Williams, Will Steffen and Paul J. Crutzen, 2010; "The New World of the Anthropocene", in:** *Environ. Sci. Technol.*, **44, 7 (25 February): 2228-2231;** https://doi.org/10.1021/es903118j;

The Anthropocene, following the lost world of the Holocene, holds challenges for both science and society. The notion that humankind has changed the world is not new. Over a century ago, terms such as the Anthropozoic (1), Psychozoic (2), and Noosphere (3) were conceived to denote the idea of humans as a new global forcing agent.

These ideas received short shrift in the geological community (4), seeming absurd when set aside the vastness (newly realized, also) of geological time. Moreover, the scarring of the landscape associated with industrialization may appear as transformation, but the vicissitudes of the geological pasts meteorite strikes, extraordinary volcanic outbursts, colliding continents, and disappearing oceans seemed of an epic scale beyond the largest factories and most populous cities.

So when one of us (P.C.) proposed the new term "Anthropocene" for this concept a decade ago (5), why did it not also become a discarded footnote in the history of geological ideas? It helps that the term is vivid, as much for the public as for scientists. More importantly, it was coined at a time of dawning realization that human activity was indeed changing the Earth on a scale comparable with some of the major events of the ancient past. Some of these changes are now seen as permanent, even on a geological time-scale.

Hence, the term Anthropocene quickly began to be used by practicing scientists (6) to denote the current interval of time, one dominated by human activity. The term, though, was (and currently remains) informal and not precisely defined. However, in 2008, the Stratigraphy Commission of the Geological Society of London decided, by a large majority, that there was merit in considering the possible formalization of this term: that is, that it might eventually join the Cambrian, Jurassic, Pleistocene, and other such units on the Geological Time Scale (7).

Note the careful wording. This was not the same as formalizing the term (this Commission does not have that power). Nevertheless, it was a clear signal that a body of independent geologists (each chosen for their technical expertise in the discipline of stratigraphy) thought that the case should be examined further.

The first (of many) formal steps are now being taken. An Anthropocene Working Group has been initiated, as part of the Subcommission on Quaternary Stratigraphy (the body that deals with formal units of the current Ice Ages). That is itself part of the International Commission on Stratigraphy, in turn answerable to the International Union of Geological Sciences. All of these bodies will have to be convinced that the case to formally include the Anthropocene in the Geological Time Scale is overwhelming, and, if so, to agree on a formulation of it that will be widely acceptable. The work involved will take several years to accomplish, and the outcome is not certain. The Geological Time Scale is held dear by geologists (because it is fundamental to their work), and it is not amended lightly.

In this article, therefore, we outline the scale of human modification of the Earth on which the concept of the Anthropocene rests, describe the means by which geologicaltime units are established, and discuss the particular problems and implications of discussing the Anthropocene as a formal geological time term.

6. **Will Steffen, Jacques Grinevald, Paul J. Crutzen and John R. McNeill, 2011: "The Anthropocene: Conceptual and Historical Perspectives", in:** *Phil. Trans. R. Soc. A***369: 835–841;** https://doi.org/10.1098/rsta.2010.0327**; and this text at:** https://royalsocietypublishing.org/doi/full/10.1098/rsta.2010. 0327.

The human imprint on the global environment has now become so large and active that it rivals some of the great forces of Nature in its impact on the functioning of the Earth system. Although global-scale human influence on the environment has been recognized since the 1800s, the term *Anthropocene*, introduced about a decade ago, has only recently become widely, but informally, used in the global change research community. However, the term has yet to be accepted formally as a new geological epoch or era in Earth history. In this paper, we put forward the case for formally recognizing the Anthropocene as a new epoch in Earth history, arguing that the advent of the Industrial Revolution around 1800 provides a logical start date for the new epoch. We then explore recent trends in the evolution of the Anthropocene as humanity proceeds into the twenty-first century, focusing on the profound changes to our relationship with the rest of the living world and on early attempts and proposals for managing our relationship with the large geophysical cycles that drive the Earth's climate system.

7. **Jan Zalasiewicz; Mark Williams; A. Haywood; M. Ellis, 2011: "The Anthropocene: A New Epoch of Geological Time?", in:** *Philosophical Transactions of The Royal Society A-Mathematical Physical And Engineering Sciences,***369, 1938: 835–841 (Mar 13 2011). See at:** https://royals ocietypublishing.org/doi/10.1098/rsta.2010.0339 **where the text is available for a free download.**

Abstract

1. Introduction
2. What characterizes the Anthropocene?
3. Dealing with geological time
4. Examining the Anthropocene

Footnotes

Abstract

Anthropogenic changes to the Earth's climate, land, oceans and biosphere are now so great and so rapid that the concept of a new geological epoch defined by the action of humans, the Anthropocene, is widely and seriously debated. Questions of the scale, magnitude and significance of this environmental change, particularly in the context of the Earth's geological history, provide the basis for this Theme Issue. The Anthropocene, on current evidence, seems to show global change consistent with the suggestion that an epoch-scale boundary has been crossed within the last two centuries.

8. **Will Steffen, Asa Persson, Lisa Deutsch, Jan Zalasiewicz, Mark Williams, Katherine Richardson, Carole Crumley, Paul Crutzen, Carl Folke, Line Gordon, Mario Molina, Veerabhadran Ramanathan, Johan Rockström, Marten Scheffer, Hans Joachim Schellnhuber, Uno Svedin, 2011: "The Anthropocene: From Global Change to Planetary Stewardship", in: Ambio; 40,7 (Nov): 739–761; 2011 Oct 12;** https://doi.org/10.1007/s13280-011-0185-x; **see at:** https://europepmc.org/article/med/22338713.

Abstract

Over the past century, the total material wealth of humanity has been enhanced. However, in the twenty- first century, we face scarcity in critical resources, the degradation of ecosystem services, and the erosion of the planet's capability to absorb our wastes. Equity issues remain stubbornly difficult to solve. This situation is novel in its speed, its global scale and its threat to the resilience of the Earth System. The advent of the Anthropene, the time interval in which human activities now rival global geophysical processes, suggests that we need to fundamentally alter our relationship with the planet we inhabit. Many approaches could be adopted, ranging from geoengineering solutions that purposefully manipulate parts of the Earth System to becoming active stewards of our own life support system. The Anthropocene is a reminder that the Holocene, during which complex human societies have developed, has been a stable, accommodating environment and is the only state of the Earth System that we know for sure can support contemporary society. The need to achieve effective planetary stewardship is urgent. As we go further into the Anthropocene, we risk driving the Earth System onto a trajectory toward more hostile states from which we cannot easily return.

9. **Jan Zalasiewicz, Paul J. Crutzen and Will Steffen, 2012: "The Anthropocene", in: Gradstein, F.M.; Ogg, J.G.; Schmitz, M.D.; Ogg, Gabi M. (Eds.): Geologic Time Scale, vol. 1 & 2 (Amsterdam: Elsevier): 1033–1040; vol. 2 (London Geological Society, Elsevier): see at:** https://www.elsevier.com/books/the-geologic-time-scale-2012/gradstein/978-0-444-59425-9.

Abstract on this chapter

The Anthropocene is a currently informal term to signify a contemporary time interval in which surface geological processes are dominated by human activities, now being studied by an ICS working group as regards potential formalization within the Geological Time Scale. Its developing stratigraphic signature includes components that are lithostratigraphic, biostratigraphic, and chemostratigraphic; and these vary from being approximately synchronous to strongly diachronous. Formalization will depend upon both scientific justification and utility to working scientists, and upon the choice of an effective boundary, whether by GSSP or GSSA.

Blurb About this Book

The Geologic Time Scale 2012, winner of a 2012 PROSE Award Honorable Mention for Best Multi-volume Reference in Science from the Association of American Publishers, is the framework for deciphering the history of our planet Earth. The authors have been at the forefront of chronostratigraphic research and initiatives to create an international geologic time scale for many years, and the charts in this book present the most up-to-date, international standard, as ratified by the International Commission on Stratigraphy and the International Union of Geological Sciences. This 2012 geologic time scale is an enhanced, improved and expanded version of the GTS2004, including chapters on planetary scales, the Cryogenian-Ediacaran periods/systems, a prehistory scale of human development, a survey of sequence stratigraphy, and an extensive compilation of stable-isotope chemostratigraphy.

This book is an essential reference for all geoscientists, including researchers, students, and petroleum and mining professionals. The presentation is non-technical and illustrated with numerous colour charts, maps and photographs. The book also includes a detachable wall chart of the complete time scale for use as a handy reference in the office, laboratory or field.

Key Features

- The most detailed international geologic time scale available that contextualizes information in one single reference for quick desktop access
- Gives insights in the construction, strengths, and limitations of the geological time scale that greatly enhances its function and its utility
- Aids understanding by combining with the mathematical and statistical methods to scaled composites of global succession of events

Meets the needs of a range of users at various points in the workflow (researchers extracting linear time from rock records, students recognizing the geologic stage by their content).

10. **Klaus Töpfer, 2012": "On the Way to the Anthropocene. Consequences of Scientific Research, Societal Understanding and Political Responsibility", in: Frauke Kraas, Dietrich Soyez, Carsten Butsch, Franziska Krachten, Holger Kretschmer (Eds.):** *IGC Cologne 2012. Down to Earth. Documenting the 32nd International*, **Geographical Congress in Cologne 26-30 August 2012; see at:** https://www.researchgate.net/publication/80802677_ Frauke_Kraas_Dietrich_Soyez_Carsten_Butsch_Franziska_Krachten_Hol ger_Kretschmer_eds_2015_IGC_Cologne_2012_Down_to_Earth_Documen ting_the_32nd_International_Geographical_Congress_in_Cologne_26-30_ August**: 36-40; see at:** https://docplayer.net/60878856-Igc-cologne-down-to-earth.html.

11. **Klaus Töpfer (Potsdam), 2013: "Nachhaltigkeit im Anthropozän", in:** *Nova Acta Leopoldina*, **NF 117,398: 31-40, see at:** https://www.leopoldina. org/uploads/tx_leopublication/Probekapitel_NAL398.pdf.

Human alterations of nature and the effects of these actions have reached a new dimension in quantity and quality. This caused intense discussions regarding the coining of the term "anthropocene" as a new geological era. Not only the physical, or natural, environment bears the marks of the acceleration of environmental change. The acceleration of decision making processes which too often results in the allocation of risks in the future is also a great challenge for social and finance systems, and altogether the foundations of life and therefore also for peaceful development.

The Anthropocene, however, is not only about the remodelling of the earth but also about a readjustment of the man-environment-relationship in the sense of taking the responsibility for human actions toward nature. Man is part of nature and cannot survive without it.

Sustainable decision making and action, in the sense of a sustainability which integrates time as a central aspect in decision making processes, are one answer to the tremendous global changes and the resulting risks. Middle and long-term effects of today's actions have to be considered in the moment of decision making, as explicitly stated by VON CARLOWITZ 300 years ago. Sustainability today needs to be understood as thinking and acting responsibly and anticipatory with regard to today's and tomorrow's, local and global effects.

12. **Jonathan Williams and Paul J. Crutzen, 2013: "Perspectives on our Planet in the Anthropocene", in:** *Environ. Chem.*, **10: 269–280 (CSIRO Publishing);** http://dx.doi.org/10.1071/EN130; **see at:** https://www.publish. csiro.au/en/ExportCitation/EN13061 **and** https://www.researchgate.net/pub lication/264194902_Perspectives_on_our_planet_in_the_Anthropocene.

Environmental context. The term Anthropocene has been proposed as a name for the present geological epoch in recognition of the recent rise of humans to being a geophysical force of planetary importance. This paper provides an overview of humanity's global impact in terms of population, energy and food demands, climate, air and ocean pollution, biodiversity and erosion, before giving a perspective on our collective future in the Anthropocene.

Abstract

Within the last 70 years (an average person's lifetime), the human population has more than tripled. Our energy, food and space demands as well as the associated waste products have affected the Earth to such an extent that humanity may be considered a geophysical force in its own right. As a result it has been proposed to name the current epoch the 'Anthropocene'. Here we draw on a broad range of references to provide an overview of these changes in terms of population, energy and food demands, climate, air and ocean pollution, biodiversity and erosion. The challenges for the future in the Anthropocene are highlighted. We hope that in the future, the 'Anthropocene' will not only be characterised by continued human plundering of the Earth's resources and dumping of excessive amounts of waste products in the environment, but also by vastly improved technology and management, wise use of the Earth's resources, control of the human and domestic animal population, and overall careful manipulation and restoration of the natural environment. This paper is the first in a series of annual invited papers commemorating Professor Sherwood (Sherry) Rowland, Nobel laureate and founding Board Member of Environmental Chemistry.

13.   **Stephen F. Foley, Detlef Gronenborn, Meinrat O. Andreae, Joachim W. Kadereit, Jan Esper, Denis Scholz, Ulrich Pöschl, Dorrit E. Jacob, Bernd R. Schöne, Rainer Schreg, Andreas Vött, David Jordan, Jos Lelieveld, Christine G. Weller, Kurt W. Alt, Sabine Gaudzinski-Windheuser, Kai-Christian Bruhn, Holger Tost, Frank Sirocko and Paul J. Crutzen, 2013: "The Palaeoanthropocene – The Beginnings of Anthropogenic Environmental Change", in: *Anthropocene*, 3: 83-88, see at:** https://researchers.mq.edu.au/en/publications/the-palaeoan-thropo cene-the-beginnings-of-anthropogenic-environment.

Abstract

As efforts to recognize the Anthropocene as a new epoch of geological time are mounting, the controversial debate about the time of its beginning continues. Here, we suggest the term Palaeoanthropocene for the period between the first, barely recognizable, anthropogenic environmental changes and the industrial revolution when anthropogenically induced changes of climate, land use and biodiversity began to increase very rapidly. The concept of the Palaeoanthropocene recognizes that humans are an integral part of the Earth system rather than merely an external forcing factor. The delineation of the beginning of the Palaeoanthropocene will require an increase in the understanding and precision of palaeoclimate indicators, the recognition of archaeological sites as environmental archives, and interlinking palaeoclimate, palaeoenvironmental changes and human development with changes in the distribution of Quaternary plant and animal species and socio-economic models of population subsistence and demise.

14.   **S. Schäfer; H. Stelzer; A. Maas; M.G. Lawrence, 2014: "Earth's future in the Anthropocene: Technological interventions between piecemeal and utopian social engineering". in: *Earths Future*, 2, 4 (April): 239–243; see at:** https://publications.iass-potsdam.de/rest/items/item_336638_16/com ponent/file_652894/content.

An extensive discussion in the academic and policy communities is developing around the possibility of climate engineering through stratospheric aerosol injection (SAI). In this contribution, we develop a perspective on this issue in the context of the wider setting of societal development in the Anthropocene. We draw on Karl Popper's concepts of piecemeal and utopian social engineering to examine how different visions of societal development relate to SAI. Based on this reflection, we argue that the debate on SAI is fueled not only by the inequitable distribution of its effects and potential atmospheric and climatic side effects, as disconcerting as some of these effects and side effects may be, but also, and perhaps primarily, by its apparent privileging of the status quo and incremental change over a more immediate and radical change in societal organization. Although differing ideological orientations might thus help explain the intensity of parts of the debate, the understanding from which they follow, in which societal development is deduced from postulated technological characteristics and assumptions about a technology's use, hides from view a more subtle understanding of the relationship between technology and politics.

15. **C.N. Waters; J.A. Zalasiewicz; M. Williams; M.A. Ellis; A.M. Snelling, 2014: "A Stratigraphical Basis for the Anthropocene?", in:** *Stratigraphical Basis for the Anthropocene*, **Book Series (London: Geological Society), Special Publication, 395: 1–21.**

Abstract

- Status as epoch or age
- Absolute and relative dating techniques
- Definition of a boundary stratotype or numerical age
- Summary and conclusions
- Acknowledgments
- Appendix
- References

Abstract

Recognition of intimate feedback mechanisms linking changes across the atmosphere, biosphere, geosphere and hydrosphere demonstrates the pervasive nature of humankind's influence, perhaps to the point that we have fashioned a new geological epoch, the Anthropocene. To what extent will these changes be evident as long-lasting signatures in the geological record? To establish the Anthropocene as a formal chronostratigraphical unit it is necessary to consider a spectrum of indicators of anthropogenically induced environmental change, and to determine how these show as stratigraphic signals that can be used to characterize an Anthropocene unit and to recognize its base. It is important to consider these signals against a context of Holocene and earlier stratigraphic patterns. Here we review the parameters used by stratigraphers to identify chronostratigraphical units and how these could apply to the definition of the Anthropocene. The onset of the range of signatures is diachronous, although many show maximum signatures which post-date 1945, leading to the suggestion that this date may be a suitable age for the start of the Anthropocene.

16. **Paul J. Crutzen; Stanisław Wacławek, 2014: "Atmospheric Chemistry and Climate in the Anthropocene", in:** *Chem. Didact. Ecol. Metrol.,* **19, 1–2: 9-28;** https://doi.org/10.1515/cdem-2014-0001; **at:** https://www.researchgate.net/publication/276511675_Atmospheric_Chemistry_and_Cli mate_in_the_Anthropocene_Chemia_Atmosferyczna_I_Klimat_W_Antropo cenie.

Abstract

Humankind actions are exerting increasing effect on the environment on all scales, in a lot of ways overcoming natural processes. During the last 100 years human population went up from little more than one to six billion and economic activity increased nearly ten times between 1950 and the present time. In the last few decades of the twentieth century, anthropogenic chlorofluorocarbon release have led to a dramatic decrease in levels of stratospheric ozone, creating ozone hole over the Antarctic, as a result UV-B radiation from the sun increased, leading for example to enhanced risk of skin cancer. Releasing more of a greenhouse gases by mankind, such as CO_2, CH_4, NO_x to the atmosphere increases the greenhouse effect. Even if emission increase has held back, atmospheric greenhouse gas concentrations would continue to raise and remain high for hundreds of years, thus warming Earth's climate. Warming temperatures contribute to sea level growth by melting mountain glaciers and ice caps, because of these portions of the Greenland and Antarctic ice sheets melt or flow into the ocean. Ice loss from the Greenland and Antarctic ice sheets could contribute an additional 19–58 cm of sea level rise, hinge on how the ice sheets react. Taking into account these and many other major and still growing footprints of human activities on earth and atmosphere without any doubt we can conclude that we are living in new geological epoch named by P. Crutzen and E. Stoermer in 2000 – "Anthropocene". For the benefit of our children and their future, we must do more to struggle climate changes that have had occurred gradually over the last century.

Keywords: greenhouse gases, greenhouse effect, climate changes, ozone hole, Anthropocene.

17. **J. Zalasiewicz; M. Williams; C.N. Waters, 2014: "Can an Anthropocene Series be Defined and Recognized?" in:** *Stratigraphical Basis For The Anthropocene Book Series*(**London: Geological Society, Special Publica-tion Volume: 395. 39–53: This text is accessible at:** https://doi.org/10.1144/SP395.16.

Abstract

- Anthropocene boundary level
- Chronostratigraphy, scale-dependence and the Anthropocene
- Components of an Anthropocene Series
- Terrestrial settings
- Marine settings
- Duration of the Anthropocene: the long-term perspective
- Discussion

- Conclusions
- Acknowledgments
- References

Abstract

We consider the Anthropocene as a physical, chronostratigraphic unit across terrestrial and marine sedimentary facies, from both a present and a far future perspective, provisionally using an approximately 1950 CE base that approximates with the 'Great Acceleration', worldwide sedimentary incorporation of A-bomb-derived radionuclides and light nitrogen isotopes linked to the growth in fertilizer use, and other markers. More or less effective recognition of such a unit today (with annual/decadal resolution) is facies-dependent and variably compromised by the disturbance of stratigraphic superposition that commonly occurs at geologically brief temporal scales, and that particularly affects soils, deep marine deposits and the pre-1950 parts of current urban areas. The Anthropocene, thus, more than any other geological time unit, is locally affected by such blurring of its chronostratigraphic boundary with Holocene strata. Nevertheless, clearly separable representatives of an Anthropocene Series may be found in lakes, land ice, certain river/delta systems, in the widespread dredged parts of shallow-marine systems on continental shelves and slopes, and in those parts of deep-water systems where human-rafted debris is common. From a far future perspective, the boundary is likely to appear geologically instantaneous and stratigraphically significant.

18. **Will Steffen, Katherine Richardson, Johan Rockström, Sarah E. Cornell, Ingo Fetzer, Elena M. Bennett, Reinette Biggs, Stephen R. Carpenter, Wim de Vries, Cynthia A. de Wit, Carl Folke, Dieter Gerten, Jens Heinke, Georgina M. Mace, Linn M. Persson, Veerabhadran Ramanathan, Belinda Reyers, Sverker Sörlin, 2015: "Planetary boundaries: Guiding human development on a changing planet", in: *Science*, 347, 6223 (13 February),** https://doi.org/10.1126/science.1259855; **see at:** https://science.sciencemag.org/content/347/6223/1259855/tab-figures-data.

The planetary boundaries framework defines a safe operating space for humanity based on the intrinsic biophysical processes that regulate the stability of the Earth system. Here, we revise and update the planetary boundary framework, with a focus on the underpinning biophysical science, based on targeted input from expert research communities and on more general scientific advances over the past 5 years. Several of the boundaries now have a two-tier approach, reflecting the importance of cross-scale interactions and the regional-level heterogeneity of the processes that underpin the boundaries. Two core boundaries—climate change and biosphere integrity—have been identified, each of which has the potential on its own to drive the Earth system into a new state should they be substantially and persistently transgressed.

19. **Simon L. Lewis and Mark A. Maslin, 2015: "Defining the Anthropocene",** in: *Nature*, **519: 171–180; at:** https://doi.org/10.1038/nature14258.

Abstract

Time is divided by geologists according to marked shifts in Earth's state. Recent global environmental changes suggest that Earth may have entered a new human-dominated geological epoch, the Anthropocene. Here we review the historical genesis of the idea and assess anthropogenic signatures in the geological record against the formal requirements for the recognition of a new epoch. The evidence suggests that of the various proposed dates two do appear to conform to the criteria to mark the beginning of the Anthropocene: 1610 and 1964. The formal establishment of an Anthropocene Epoch would mark a fundamental change in the relationship between humans and the Earth system.

20.   **Will Steffen, Wendy Broadgate, Lisa Deutsch, Owen Gaffney, Cornelia Ludwig, 2015; "The Trajectory of the Anthropocene: The Great Acceleration", in:** *The Anthropocene Review*, **2,1: 81–98; at:** https://journals.sag epub.com/doi/10.1177/2053019614564785 .

Abstract

The 'Great Acceleration' graphs, originally published in 2004 to show socio-economic and Earth System trends from 1750 to 2000, have now been updated to 2010. In the graphs of socio-economic trends, where the data permit, the activity of the wealthy (OECD) countries, those countries with emerging economies, and the rest of the world have now been differentiated. The dominant feature of the socio-economic trends is that the economic activity of the human enterprise continues to grow at a rapid rate. However, the differentiated graphs clearly show that strong equity issues are masked by considering global aggregates only. Most of the population growth since 1950 has been in the non-OECD world but the world's economy (GDP), and hence consumption, is still strongly dominated by the OECD world. The Earth System indicators, in general, continued their long-term, post-industrial rise, although a few, such as atmospheric methane concentration and stratospheric ozone loss, showed a slowing or apparent stabilisation over the past decade. The post-1950 acceleration in the Earth System indicators remains clear. Only beyond the mid-20th century is there clear evidence for fundamental shifts in the state and functioning of the Earth System that are beyond the range of variability of the Holocene and driven by human activities. Thus, of all the candidates for a start date for the Anthropocene, the beginning of the Great Acceleration is by far the most convincing from an Earth System science perspective.

21.   **Mark Williams, Jan Zalasiewicz, P.K. Haff, Christian Schwägerl, Anthony D. Barnosky and Erle C. Ellis, 2015: "The Anthropocene Biosphere", in:** *Anthropocene Review,* **2: 196–219; see at:** https://journals.sagepub.com/doi/abs/10.1177/2053019615591020.

The geological record preserves evidence for two fundamental stages in the evolution of Earth's biosphere, a microbial stage from ~3.5 to 0.65 Ga, and a metazoan stage evident by c. 650 Ma. We suggest that the modern biosphere differs significantly from these previous stages and shows early signs of a new, third stage of biosphere evolution characterised by: (1) global homogenisation of flora and fauna; (2) a single species *(Homo sapiens)* commandeering 25–40% of net primary production and also mining fossil net primary production (fossil fuels) to break through the photosynthetic energy barrier; (3) human-directed evolution of other species; and (4) increasing interaction of the biosphere with the technosphere (the global emergent system that includes humans, technological artefacts, and associated social and technological networks). These unique features of today's biosphere may herald a new era in the planet's history that could persist over geological timescales.

22.   **J. Lelieveld, J. S. Evans, M. Fnais, D. Giannadaki and A. Pozzer, 2015: "The contribution of outdoor air pollution sources to premature mortality on a global scale", in:** *Nature*, **525: 367–371; see** https://www.nature.com/articles/nature15371.

Assessment of the global burden of disease is based on epidemiological cohort studies that connect premature mortality to a wide range of causes, including the long-term health impacts of ozone and fine particulate matter with a diameter smaller than 2.5 μm ($PM_{2.5}$). It has proved difficult to quantify premature mortality related to air pollution, notably in regions where air quality is not monitored, and also because the toxicity of particles from various sources may vary. Here we use a global atmospheric chemistry model to investigate the link between premature mortality and seven emission source categories in urban and rural environments. In accord with the global burden of disease for 2010, we calculate that outdoor air pollution, mostly by PM2.5, leads to 3.3 (95 percent confidence interval 1.61–4.81) million premature deaths per year worldwide, predominantly in Asia. We primarily assume that all particles are equally toxic[5], but also include a sensitivity study that accounts for differential toxicity. We find that emissions from residential energy use such as heating and cooking, prevalent in India and China, have the largest impact on premature mortality globally, being even more dominant if carbonaceous particles are assumed to be most toxic. Whereas in much of the USA and in a few other countries emissions from traffic and power generation are important, in eastern USA, Europe, Russia and East Asia agricultural emissions make the largest relative contribution to PM2.5, with the estimate of overall health impact depending on assumptions regarding particle toxicity. Model projections based on a business-as-usual emission scenario indicate that the contribution of outdoor air pollution to premature mortality could double by 2050.

23.   **T. Suni, A. Guenther, H.C. Hansson, M. Kulmala, M.O. Andreae, A. Arneth[e], P. Artaxo, E. Blyth, M. Brus, L. Ganzeveld, P. Kabat, N. de. Noblet-Ducoudre, M. Reichstein, A. Reissell, D. Rosenfeld, S. Seneviratne, 2015: "The significance of land-atmosphereinteractions in the Earth system— iLEAPS achievements and perspectives", in:** *Anthropocene*, **12: 69–84; see at:** https://www.sciencedirect.com/science/article/pii/S2213305415300254.

The integrated land ecosystem-atmosphere processes study (iLEAPS) is an international research project focussing on the fundamental processes that link land-atmosphere exchange, climate, the water cycle, and tropospheric chemistry. The project, iLEAPS, was established 2004 within the International Geosphere-Biosphere Programme (IGBP). During its first decade, iLEAPS has proven to be a vital proJect, well equipped to build a community to address the challenges involved in understanding the complex Earth system: multidisciplinary, integrative approaches for both observations and modeling. The iLEAPS community has made major advances in process understanding, land-surface modeling, and observation techniques and networks. The modes of iLEAPS operation include elucidating specific iLEAPS scientific questions through networks of process studies, field campaigns, modeling, long-term integrated field studies, international interdisciplinary mega-campaigns, synthesis studies, databases, as well as conferences on specific scientific questions and synthesis meetings. Another essential component of iLEAPS is knowledge transfer and it also encourages community- and policy-related outreach activities associated with the regional integrative projects. As a result of its first decade of work, iLEAPS is now setting the agenda for its next phase (2014–2024) under the new international initiative, future Earth. Human influence has always been an important part of land-atmosphere science but in order to respond to the new challenges of global sustainability, closer ties with social science and economics groups will be necessary to produce realistic estimates of land use and anthropogenic emissions by analysing future population increase, migration patterns, food production allocation, land management practices, energy production, industrial development, and urbanization.

24. **Ulrich Pöschl and Manabu Shiraiwa, 2015: "Multiphase Chemistry at the Atmosphere-Biosphere Interface Influencing Climate and Public Health in the Anthropocene", in:** *Chem. Rev.*, **115, 10: 4440–4475 (Washington, D.C.; ACS Publishing); see at:** https://pubs.acs.org/doi/10.1021/cr500487s.

Multiphase chemistry plays a vital role in the Earth system, climate, and health. Chemical reactions, mass transport, and phase transitions between gases, liquids, and solids are essential for the interaction and coevolution of life and climate. Knowledge of the mechanisms and kinetics of these processes is also required to address societally relevant questions of global environmental change and public health in the Anthropocene, that is, in the present era of globally pervasive and steeply increasing human influence on planet Earth. In this work, we review the current scientific understanding and recent advances in the investigation of short-lived health- and climate-relevant air contaminants (SHCC) and their multiphase chemical interactions at the atmosphere–biosphere interface, including human lungs and skin, plant leaves, cryptogamic covers, soil, and aquatic surfaces. After an overview of different groups of SHCC, we address the chemical interactions of reactive oxygen species and reactive nitrogen species (ROS, RNS), primary biological and secondary organic aerosols (PBA, SOA), as well as carbonaceous combustion aerosols (CCA) including soot, black/elemental carbon, polycyclic aromatic hydrocarbons, and related compounds (PAH, PAC). ROS and RNS interact strongly with other SHCC and are central to both atmospheric and physiological processes and their coupling through the atmosphere–biosphere interface, for example, in the formation and aging of biogenic and combustion aerosols as well as in inflammatory and allergic immune responses triggered by air pollution. Deposition of atmospheric ROS/RNS and aerosols can damage biological tissues, modify surface microbiomes, and induce oxidative stress through Fenton-like reactions and immune responses. The chemical mechanisms and kinetics are not yet fully elucidated, but the available evidence suggests that multiphase processes are crucial for the assessment, prediction, and handling of air quality, climate, and public health. Caution should be taken to avoid that human activities shaping the Anthropocene create a hazardous or pathogenic atmosphere overloaded with allergenic, corrosive, toxic, or infectious contaminants. [no abstract, first paragraph]

25. **Jan Zalasiewicz, Colin N. Waters, Mark Williams, Anthony D. Barnosky, Alejandro Cearreta, Paul J. Crutzen, Erle Ellis, Michael A. Ellis, Ian J. Fairchild, Jacques Grinevald, Peter K. Haff, Irka Hajdas, Reinhold Leinfelder, John McNeill, Eric O. Odada, Clement Poirier, Daniel Richter, Will Steffen, Colin Summerhayes, James P.M. Syvitski, Davor Vidas, Michael Wagreich, Scott L. Wing, Alexander P. Wolfe, An Zhisheng and Naomi Oreskes, 2015: "When Did the AnthropoceneBegin? A Mid-twentieth Century Boundary Level is Stratigraphically Optimal", in:** *Quaternary International,***383: 196–203; see at:** https://hal.archives-ouv ertes.fr/hal-01446643**and** http://ib.berkeley.edu/labs/barnosky/Zalasiewicz% 20et%20al%20QI2015.pdf.

Abstract

We evaluate the boundary of the Anthropocene geological time interval as an epoch, since it is useful to have a consistent temporal definition for this increasingly used unit, whether the presently informal term is eventually formalized or not. Of the three main levels suggested – an 'early Anthropocene' level some thousands of years ago; the beginning of the Industrial Revolution at ~1800 CE (Common Era); and the 'Great Acceleration' of the mid-twentieth century – current evidence suggests that the last of these has the most pronounced and globally synchronous signal. A boundary at this time need not have a Global Boundary Stratotype Section and Point (GSSP or 'golden spike') but can be defined by a Global Standard Stratigraphic Age (GSSA), i.e. a point in time of the human calendar.We propose an appropriate boundary level here to be the time of the world's first nuclear bomb explosion, on July 16th 1945 at Alamogordo, New Mexico; additional bombs were detonated at the average rate of one every 9.6 days until 1988 with attendant worldwide fallout easily identifiable in the chemostratigraphic record. Hence, Anthropocene deposits would be those that may include the globally distributed primary artificial radionuclide signal, while also being recognized using a wide range of other stratigraphic criteria. This suggestion for the Holocenee Anthropocene boundary may ultimately be superseded, as the Anthropocene is only in its early phases, but it should remain practical and effective for use by at least the current generation of scientists.

26.   C.N. Waters; J. Zalasiewicz; C. Summerhayes; A.D. Barnosky; C. Poirier;
       A. Galuszka; A. Cearreta; M. Edgeworth; E.C. Ellis; M. Ellis; C. Jeandel;
       R. Leinfelder; J.R. McNeill; D.D. Richter; W. Steffen; J. Syvitski; D. Vidas;
       M. Wagreich; M. Williams;, Z.S. An; J. Grinevald; E. Odada; N. Oreskes;
       A.P. Wolfe, 2016: "The Anthropoceneis functionally and stratigraphically
       distinct from the Holocene", in: *Science*, 351, 6269: 137; see at: https://sci
       ence.sciencemag.org/content/351/6269/aad2622,

Abstract

Human activity is leaving a pervasive and persistent signature on Earth. Vigorous
debate continues about whether this warrants recognition as a new geologic time unit
known as the Anthropocene. We review anthropogenic markers of functional changes
in the Earth system through the stratigraphic record. The appearance of manufac-
tured materials in sediments, including aluminum, plastics, and concrete, coincides
with global spikes in fallout radionuclides and particulates from fossil fuel combustion. Carbon, nitrogen, and phosphorus cycles have been substantially modified over
the past century. Rates of sea-level rise and the extent of human perturbation of the
climate system exceed Late Holocene changes. Biotic changes include species invasions worldwide and accelerating rates of extinction. These combined signals render
the Anthropocene stratigraphically distinct from the Holocene and earlier epochs.

27.   M. Williams; J. Zalasiewicz; C.N. Waters; M. Edgeworth; C. Bennett;
       A.D. Barnosky; E.C. Ellis; M.A. Ellis; A. Cearreta; P.K. Haff; J.A.I.
       do Sul; R. Leinfelder; J.R. McNeill; E. Odada; N. Oreskes; A. Revkin;
       D.D. Richter; W. Steffen; C. Summerhayes; J.P. Syvitski; D. Vidas; M.
       Wagreich; S.L. Wing; A.P. Wolfe; Z.S. An, 2016: "The Anthropocene: a
       conspicuous stratigraphical signal of anthropogenic changes in produc-
       tion and consumption across the biosphere", in: *Earths Future*, 4,3: 34-53;
       see at: https://doi.org/10.1002/2015EF000339; see at: https://agupubs.online
       library.wiley.com/doi/full/10.1002/2015EF000339.

Abstract

Biospheric relationships between production and consumption of biomass have been
resilient to changes in the Earth system over billions of years. This relationship
has increased in its complexity, from localized ecosystems predicated on anaerobic
microbial production and consumption to a global biosphere founded on primary
production from oxygenic photoautotrophs, through the evolution of Eukarya, metazoans, and the complexly networked ecosystems of microbes, animals, fungi, and
plants that characterize the Phanerozoic Eon (the last ~541 million years of Earth
history). At present, one species, *Homo sapiens,* is refashioning this relationship
between consumption and production in the biosphere with unknown consequences.
This has left a distinctive stratigraphy of the production and consumption of biomass,
of natural resources, and of produced goods. This can be traced through stone tool
technologies and geochemical signals, later unfolding into a diachronous signal

of technofossils and human bioturbation across the planet, leading to stratigraphically almost isochronous signals developing by the mid-20th century. These latter signals may provide an invaluable resource for informing and constraining a formal Anthropocene chronostratigraphy, but are perhaps yet more important as tracers of a biosphere state that is characterized by a geologically unprecedented pattern of global energy flow that is now pervasively influenced and mediated by humans, and which is necessary for maintaining the complexity of modern human societies.

28. **Clive Hamilton, 2016: "The Anthropoceneas rupture",** *The Anthropocene Review*, **3,2; see at:** https://journals.sagepub.com/doi/10.1177/205301961663 4741.

I argue that Earth System science – a recent paradigm shift in the earth and life sciences (Hamilton C and Grinevald J (2015) Was the Anthropocene anticipated? *The Anthropocene Review* 2(1): 59–72) – named the Anthropocene as the very recent rupture in Earth history arising from the impact of human activity on the Earth System as a whole. Many have mistakenly treated the new concept of the Earth System as if it were equivalent to 'the landscape', 'ecosystems' or 'the environment'. The new paradigm of Earth System science is erroneously understood as no more than a variation or development of established ecological sciences. Various attempts to invent new starting dates for the new epoch are based on these misconceptions, as are a number of arguments deployed to reject the Anthropocene altogether. In this context I consider the early Anthropocene hypothesis, three readings of the Anthropocene as instances of ecosystem change, and the notion of the 'good Anthropocene'. Using this frame I also assess the arguments of those who do not accept the idea of the new epoch. I defend the view that disciplines other than Earth System science distort the idea of the Anthropocene when they read it through their own lenses.

29. **E.S. Brondizio; K. O'Brien; X.M. Bai; F. Biermann; W. Steffen; F. Berkhout; C. Cudennec; M.C. Lemos; A. Wolfe; J. Palma-Oliveira; C.T.A. Chen, 2016: "Re-conceptualizing the Anthropocene: A call for collaboration", in:** *Global Environmental Change-Human And Policy Dimensions*, **39 (July): 318–327; See at:** https://hal.archives-ouvertes.fr/hal-01377890.

Since it was first proposed in 2000, the concept of the Anthropocene has evolved in breadth and diversely. The concept encapsulates the new and unprecedented planetary-scale changes resulting from societal transformations and has brought to the fore the social drivers of global change. The concept has revealed tensions between generalized interpretations of humanity's contribution to global change, and interpretations that are historically, politicallyand culturally situated. It motivates deep ethical questions about the politics and economics of global change, including diverse interpretations of past causes and future possibilities. As such, more than other concepts, the Anthropocene concept has brought front-and-center epistemological divides between and within the natural and social sciences, and the humanities. It has also brought new opportunities for collaboration. Here we explore the potential and challenges of the concept to encourage integrative understandings of global change and sustainability. Based on bibliometric analysis and literature review, we discuss the now wide acceptance of the term, its interpretive flexibility, the emerging narratives as well as the debates the concept has inspired. We argue that without truly collaborative and integrative research, many of the critical exchanges around the concept are likely to perpetuate fragmented research agendas and to reinforce disciplinary boundaries. This means appreciating the strengths and limitations of different knowledge domains, approaches and perspectives, with the concept of the Anthropocene serving as a bridge, which we encourage researchers and others to cross. This calls for institutional arrangements that facilitate collaborative research, training, and action, yet also depends on more robust and sustained funding for such activities. To illustrate, we briefly discuss three overarching global change problems where novel types of collaborative research could make a difference: (1) Emergent properties of socioecological systems; (2) Urbanization and resource nexus; and (3) Systemic risks and tipping points. Creative tensions around the Anthropocene concept can help the research community to move toward new conceptual syntheses and integrative action-oriented approaches that are needed to producing useful knowledge commensurable with the challenges of global change and sustainability

30. **Xuemei Bai, Sander van der Leeuw, Karen O'Brien, Frans Berkhout, Frank Biermann, Eduardo S. Brondizio, Christophe Cudennec, John Dearing, Anantha Duraiappah, Marion Glaser, Andrew Revkin, Will Steffen, James Syvitski, 2016. "Plausible and desirable futures in the Anthropocene: A new research agenda", in:** *Global Environmental Change*, **39 (July): 351–362;** https://doi.org/10.1016/j.gloenvcha. 2015.09.017; **see at:** https://asu.pure.elsevier.com/en/publications/plausible-and-desirable-futures-in-the-anthropocene-a-new-researc.

While the concept of the Anthropocene reflects the past and present nature, scale and magnitude of human impacts on the Earth System, its true significance lies in how it can be used to guide attitudes, choices, policies and actions that influence the future. Yet, to date much of the research on the Anthropocene has focused on interpreting past and present changes, while saying little about the future. Likewise, many futures studies have been insufficiently rooted in an understanding of past changes, in particular the long-term co-evolution of bio-physical and human systems. The Anthropocene perspective is one that encapsulates a world of intertwined drivers, complex dynamic structures, emergent phenomena and unintended consequences, manifest across different scales and within interlinked biophysical constraints and social conditions. In this paper we discuss the changing role of science and the theoretical, methodological and analytical challenges in considering futures of the Anthropocene. We resent three broad groups of research questions on: (1) societal goals for the future; (2) major trends and dynamics that might favor or hinder them; (3) and factors that might propel or impede transformations towards desirable futures. Tackling these questions requires the development of novel approaches integrating natural and social sciences as well as the humanities beyond what is current today. We present three examples, one from each group of questions, illustrating how science might contribute to the identification of desirable and plausible futures and pave the way for transformations towards them. We argue that it is time for debates on the sustainability of the Anthropocene to focus on opportunities for realizing desirable and plausible futures.

31. **Kathrin Reinmuth-Selzle, Christopher J. Kampf, Kurt Lucas, Naama Lang-Yona, Janine Fröhlich-Nowoisky, Manabu Shiraiwa, Pascale S. J. Lakey, Senchao Lai, Fobang Liu, Anna T. Kunert, Kira Ziegler, Fangxia Shen, Rossella Sgarbanti, Bettina Weber, Iris Bellinghausen, Joachim Saloga, Michael G. Weller, Albert Duschl, Detlef Schuppan, and Ulrich Pöschl, 2017: "Air pollution and climate change effects on allergies in the Anthropocene", in: *Environmental Science & Technology*, 51,8: 4119–4141, see at:** https://doi.org/10.1021/acs.est.6b04908.

Abstract

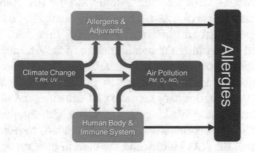

Air pollution and climate change are potential drivers for the increasing burden of allergic diseases. The molecular mechanisms by which air pollutants and climate parameters may influence allergic diseases, however, are complex and elusive. This article provides an overview of physical, chemical and biological interactions between air pollution, climate change, allergens, adjuvants and the immune system, addressing how these interactions may promote the development of allergies. We reviewed and synthesized key findings from atmospheric, climate, and biomedical research. The current state of knowledge, open questions, and future research perspectives are outlined and discussed. The Anthropocene, as the present era of globally pervasive anthropogenic influence on planet Earth and, thus, on the human environment, is characterized by a strong increase of carbon dioxide, ozone, nitrogen oxides, and combustion- or traffic-related particulate matter in the atmosphere. These environmental factors can enhance the abundance and induce chemical modifications of allergens, increase oxidative stress in the human body, and skew the immune system toward allergic reactions. In particular, air pollutants can act as adjuvants and alter the immunogenicity of allergenic proteins, while climate change affects the atmospheric abundance and human exposure to bioaerosols and aeroallergens. To fully understand and effectively mitigate the adverse effects of air pollution and climate change on allergic diseases, several challenges remain to be resolved. Among these are the identification and quantification of immunochemical reaction pathways involving allergens and adjuvants under relevant environmental and physiological conditions.

32. **Jan Zalasiewicz, Colin N. Waters, Colin P. Summerhayes, Alexander P. Wolfe, Anthony D. Barnosky, Alejandro Cearreta, Paul Crutzen, Erle Ellis, Ian J. Fairchild, Agnieszka Galuszka, Peter Haff, Irka Hajdas, Martin J. Head, Juliana A Ivar do Sul, Catherine Jeandel, Reinhold Leinfelder, John R. McNeill, Cath Neal, Eric Odada, Naomi Oreskes, Will Steffen, James Syvitski, Davor Vidas, Michael Wagreich, Mark Williams, 2017: "The Working Group on the Anthropocene: Summary of Evidence and Interim Recommendations", in: *Anthropocene*, 19: 55-60; see at:** https://www.sciencedirect.com/science/article/abs/pii/S22133054 17300097; https://www.researchgate.net/publication/319613362_The_Wor king_Group_on_the_Anthropocene_Summary_of_evidence_and_interim_r ecommendations.

Since 2009, the Working Group on the 'Anthropocene' (or, commonly, AWG for Anthropocene Working Group), has been critically analysing the case for formalization of this proposed but still informal geological time unit. The study to date has mainly involved establishing the overall nature of the Anthropocene as a potential chronostratigraphic/geochronologic unit, and exploring the stratigraphic proxies, including several that are novel in geology, that might be applied to its characterization and definition. A preliminary summary of evidence and interim recommendations was presented by the Working Group at the 35th International Geological Congress in Cape Town, South Africa, in August 2016, together with results of voting by members of the AWG indicating the current balance of opinion on major questions surrounding the Anthropocene. The majority opinion within the AWG holds the Anthropocene to be stratigraphically real, and recommends formalization at epoch/series rank based on a mid-20th century boundary. Work is proceeding towards a formal proposal based upon selection of an appropriate Global boundary Stratotype Section and Point (GSSP), as well as auxiliary stratotypes. Among the array of proxies that might be used as a primary marker, anthropogenic radionuclides associated with nuclear arms testing are the most promising; potential secondary markers include plastic, carbon in a wide variety of sedimentary bodies, both marine and non-marine.

Appendix 5

Websites on the Anthropocene with Hyperlinks (Compiled by H.G. Brauch based on google.com, 10 October 2020).

© The Editor(s) (if applicable) and The Author(s), under exclusive license   531
to Springer Nature Switzerland AG 2021
S. Benner et al. (eds.), *Paul J. Crutzen and the Anthropocene: A New Epoch in Earth's History*, The Anthropocene: Politik–Economics–Society–Science 1,
https://doi.org/10.1007/978-3-030-82202-6

No.	Name of the Website	Hyperlink	Purpose	Audience	Discipline	Language	Country	Source	Funding	Comments
1	Welcome to the Anthropocene	http://www.anthropocene.info/	Educational	Broad		English	Sweden	Globaïa	Resilience centre	Slides, films, figures, graphics
2	Anthropocene: Age of Mankind, VPRO document 2017	https://www.youtube.com/watch?v=AW138ZTKioM&vl=en	Educational			English	Netherlands			Film youtube
3	Anthropocene: (Newsletter)	https://www.anthropocenemagazine.org/	Educational	Science focused		English	USA	Weekly science dispatch	Future Earth	Reporting (environment)
4	Anthropocene: History of an Idea	https://www.carbonbrief.org/anthropocene-journey-to-new-geological-epoch	Educational			English	UK	Carbon brief	European Climate Foun-dation	Reporting (environment)
5	Subcommission on Quaternary Stratigraphy Anthropocene Working Group (AWG)	http://quaternary.stratigraphy.org/working-groups/anthropocene/	Scientific	Natural sciences, geology	Geology	English	UK	International union of geological sciences (IUGS)		Documentation of decisions
6	The Anthropocene	https://www.journals.elsevier.com/anthropocene	Scientific			English		Elsevier	Elsevier	Peer-reviewed scien-tific journal
7	The Anthropocene at the House of Cultures (HKW)	https://www.hkw.de/en/programm/themen/das_anthropozaen_am_hkw/das_anthropozaen_am_hkw_start.php	Educational			English	Germany		Berlin, House of Cultures	
8	The Anthropocene Symposium	https://www.mpic.de/3537874/The_Anthropocene	Scientific			English	Germany	MPIC Mainz	MPG	
9	Anthropocene	https://www.ted.com/topics/anthropocene				English				Videos
10	Welcome to the Anthropocene	https://www.carsoncenter.uni-muenchen.de/events_conf_seminars/exhibitions/anthropocene/index.html	Educational (exhibition)			English	Germany	LMU Rachel Carson Center		Catalogue
11	What is the Anthropocene? Why does it matter?	https://www.weforum.org/agenda/2016/08/what-is-the-anthropocene-and-why-does-it-matter/	Educational			English	Switzer-land	World economic forum Davos		Conference (background)

(continued)

(continued)

No.	Name of the Website	Hyperlink	Purpose	Audience	Discipline	Language	Country	Source	Funding	Comments
12	The Age of Humans: Evolutionary Perspectives on the Anthropocene	https://humanorigins.si.edu/research/age-humans-evolutionary-perspectives-anthropocene	Educational	Broad	Multiple	English	USA			Smithsonian Museum, Washington, D.C.
13	Anthropocene	http://www.igbp.net/globalchange/anthropocene.4.1b8ae20512db692f2a680009238.html	Scientific	Scholars	Global environmental change (multidiscip.)	English	Sweden	IGBP		
14	Elementa –science of the Anthropocene	https://www.elementascience.org/	Scientific			English	USA			Journal
15	Knowledge in and of the Anthropocene	https://www.mpiwg-berlin.mpg.de/project/knowledge-anthropocene.	Scientific	Scholars	History of science	English	Germany	MPIWG	MPG	Research department Anthropocene Lectures
16	What is the Anthropocene?	https://www.nationaltrust.org.uk/features/what-is-the-anthropocene	Scientific			English	UK	National Trust		
17	Edward Burtynsky: The Anthropocene Project	https://www.edwardburtynsky.com/projects/the-anthropocene-project	Educational		Photography, arts	English				Photography, Art
18	The Anthropocene	https://monoskop.org/Anthropocene	Scientific			English		Monoskop (Wiki)		
19	Anthropocene: The Human Epoch	https://theanthropocene.org/film/	Educational	The Film Project	Film	English			A cinematic meditation on humanity's massive reengineering of the planet, *Anthropocene: The Human Epoch* is a … documentary film from the the team of **Jennifer Baichwal, Nicholas de Pencier and Edward Burtynsky.**	Berlin film festival 2018
20	Anthropocene Design	https://www.dezeen.com/tag/anthropocene/	Educational			English				Design
21	Anthropocene September 28, 2018 – January 6, 2019	https://ago.ca/exhibitions/anthropocene	Educational			English	Canada	Ottawa		Art exhibition

(continued)

(continued)

No.	Name of the Website	Hyperlink	Purpose	Audience	Discipline	Language	Country	Source	Funding	Comments
22	Welcome to the Anthropocene: Population matters	https://populationmatters.org/campaigns/anthropocene	Educational		Demography	English	UK	2020 Population matters		
23	Anthropocene, a film by Steve Bradshaw	http://www.visualanthropology.net/anthropocene-a-film-by-steve-bradshaw/	Educational			English				Movie
24	BZPB Dosssier Anthropozän	https://www.bpb.de/gesellschaft/umwelt/anthropozaen/	Educational			German	Germany			
25	Anthropozän – Das Zeit-alter des Menschen: Drei- teilige Dokumenta tions-reihe mit D. Steffens	https://www.zdf.de/dokumenta tion/terra-x/anthropozaen-das-zeitalter-des-menschen-trailer-100.html	Educational			German	Germany	ZDF – Second German TV station		three films
26	Anthropozän	https://www.sueddeutsche.de/thema/Anthropoz%C3%A4n	Educational			German	Germany	Süddeutsche Zeitung		
27	Qu'est-ce que l'Anthropocène?	https://www.sciencesetavenir.fr/fondamental/geologie/question-de-la-semaine-c-est-quoi-l-ant hropocene_153518	Educational			French	France	La Documentation française		
28	Notion en débat: Anthropocène	http://geoconfluences.ens-lyon.fr/informations-scientifiques/a-la-une/notion-a-la-une/anthro pocene	Educational	Scientific	Geography	French	France	Ressources de géogra-phie pour les enseig-nants		
29	Anthropocène ; L'épisté-mologie flottante du concept d'Anthropocène	https://parcs.hypotheses.org/ant hropocene	Educational			French	France			
30	Avis de tempête sur le climat ? (2/4); L'anthropocène : par-delà nature et culture	https://www.franceculture.fr/emi ssions/les-nouveaux-chemins-de-la-connaissance/avis-de-tem pete-sur-le-climat-24-l-anthro pocene	Educational			French	France			
31	¿Sabes qué es el Antropoceno?	https://www.ecointeligencia.com/2014/11/antropoceno/	Educational			Spanish				

(continued)

(continued)

No.	Name of the Website	Hyperlink	Purpose	Audience	Discipline	Language	Country	Source	Funding	Comments
32	O que é o antropoceno, a época em que os humanos tomam controle do planeta	https://epoca.oglobo.globo.com/colunas-e-blogs/blog-do-pla neta/noticia/2015/12/o-que-e-o-antropoceno-epoca-em-que-os-humanos-tomam-controle-do-planeta.html	Educational			Portuguese				
33	La Tierra en la era del Antropoceno	https://www.lavanguardia.com/vida/junior-report/20190919/47447907776/antropoceno-nueva-ra-impacto-humano.html	Educational			Spanish (Catalan)				

Appendix 6
Anthropocene Videos with Hyperlinks

Videos with lectures by and interviews with Paul J. Crutzen speaking on the Anthropocene are available on the Internet, as well as documentary films and educational material is. A selection was compiled by Hans Günter Brauch on 21 January 2021 based on the search engine of google.com.

Neither the compiler nor the coeditors are responsible for the quality of the claims in these links. Some of these links may disappear while others may be made available. For more recent hyperlinks the readers should search with google.com and other search engines for: "Paul J. Crutzen, Anthropocene, videos". As Youtube does not seem to permit any longer the use of the URL links provided by Google.com, the readers of the Ebook are asked to copy the hyperlinks into the browser of their computer and then click manually to connect to the Videos on Youtube.

© The Editor(s) (if applicable) and The Author(s), under exclusive license to Springer Nature Switzerland AG 2021
S. Benner et al. (eds.), *Paul J. Crutzen and the Anthropocene: A New Epoch in Earth's History*, The Anthropocene: Politik–Economics–Society–Science 1, https://doi.org/10.1007/978-3-030-82202-6

No.	Website Name	hyperlink	country	Language	year
I.	**Lectures by and interviews with Paul J. Crutzen**				
1	Paul Crutzen: The Impor-tant Role of the Tropics in the Self-Cleaning Capacity of the Atmosphere, lecture Lindau Nobel Laureates Forum	https://www.mediatheque.lindau-nobel.org/videos/36459/1998-the-important-role-of-the-tropics-in-the-self-cleaning-capacity-of-the-atmosphere/meeting-1998	Netherlands, MPIC in Mainz, Germany	English	1998
2	Interview with Paul Crutzen at Lindau Nobel Laureates Forum	https://www.nobelprize.org/mediaplayer/?id=734	English	2000	
3	Paul J. Crutzen's keynote: speech "Air pollution in Asia and its impact on regional and global climate"	http://peace-foundation.net.7host.com/video/thailand/Paul%5FCrutzen/ and https://www.youtube.com/watch?v=vCIWBIXFMAE	Chulalong-korn Univ. in Bangkok (Thailand) 2.12.2003	English	2003
4	Paul J. Crutzen's keynote "The Antarctic ozone hole - A manmade chemical instability of the strato-sphere - What should we learn from it?"	https://www.youtube.com/watch?v=Ay9nvcv2YDY https://www.youtube.com/watch?v=vCIWBIXFMAE	Asian Insti-tute of Technology 3.12.2003	English	2003
5	Prof. Paul J. Crutzen at AIT facilitated by the International Peace Foundation, part 2	https://www.youtube.com/watch?v=iaQBdPBItjY	Asian Institute of Technology 3.12.2003	English	2003
6	Paul J. Crutzen at CMU facilitated by the International Peace Foundation	https://www.youtube.com/watch?v=BacSUk118rs	Lecture in Chiang Mai, Thailand	English	2003
7	Paul J. Crutzen, Interview by Harry Kroto on his research work on ozone.	http://www.vega.org.uk/video/programme/111	English	2005	
8	Paul J. Crutzen, lecture Lindau Nobel Laureates Forum	http://www.vega.org.uk/video/programme/272	English	2006	
9	Interview with. Paul J. Crutzen	https://www.youtube.com/watch?v=77FUaqXdSYA	Univers. of Cambridge, UK,	English	2010
10	Premio Capo d'Orlando 2010 - Vico Equense (third part)	https://www.youtube.com/watch?v=YvCA9zQNlqM		English Italian	2010

(continued)

(continued)

No.	Website Name	hyperlink	country	Language	year
11	Crutzen: Atmospheric Science in the human dominated era of the Anthropocene	https://www.youtube.com/watch?v=vd7utmeP1qo	Politechni-cal Univers Nova Goria Slovenia	Slove-nian English	2011
12	Paul Crutzen: lecture, Lindau Nobel Laureates Forum	https://www.mediatheque.lindau-nobel.org/videos/31236/atmospheric-chemistry-and-climate-in-the-anthropocene-2012/laureate-crutzen	Netherlands MPIC Mainz Germany	English	2012
13	Paul Crutzen zur Klima-schädlichkeit von Biokraft-stoffen und Biomasse	https://www.youtube.com/watch?v=iRZWVCQx9S0		German	2015

(continued)

(continued)

No.	Website Name	hyperlink	country	Language	year
14	Interview with Paul J. Crutzen	https://www.youtube.com/watch?v=IS-6M6fypho			2016
15	Paul Crutzen: brief remark	https://www.youtube.com/watch?v=CWk5VQqadCE		German	2017
II.	**Crutzen Symposium on the occasion of his 80th Birthday, 2 December 2013**				

(continued)

(continued)

No.	Website Name	hyperlink	country	Language	year

These presentations can be listened to on the website of the Max Planck Institute for Chemistry (MPIC) English programme is at: http://www.mpic.de/fileadmin/user_upload/images_presse/Images_Pls/Crutzen_Symposium/Program_Anthropocene_Symposium_SB11_small. The pre-sentations can be approached at: https://www.youtube.com/watch?v=g0HuKpbMREU. See also: https://www.mpic.de/3537874/The_Anthropocene

| 1 | Hartmut Grassl: Shaping Germany's Role in Ozone and Climate Policy. The Push by P. Crutzen | https://www.youtube.com/watch?v=HD2BjfgGhgQ | Germany | English | 2013 |

(continued)

(continued)

(continued)

No.	Website Name	hyperlink	country	Language	year
Prof- Hartmut Grassl, Hamburg, Germany			Prof. Mario Molina, Mexico/USA		
2	Mario J. Molina: Climate ChangeScience and Policy	https://www.youtube.com/watch?v=gOf5FFqe0Y4	Mexico	English	2013
3	Susan Solomon: Ozone Depletion: An Enduring Challenge	https://www.youtube.com/watch?v=Dbgz2XI-8YI	USA (MIT)	English	2013

(continued)

No.	Website Name	hyperlink	country	Language	year
Prof. Susan Solomon (USA)					
4	Veerabhadran Ramanathan: The Two worlds in the Anthropocene: A new Approach for Climate Change Mitigation	https://www.youtube.com/watch?v=LWh0ea6mA48	USA	English	2013
5	Jack Fishman: Troposphe-ric Ozone in the Anthropo-cene: Are We Creating a Toxic Atmosphere?	https://www.youtube.com/watch?v=n_gw5gKJtGM	USA	English	2013
6	John P. Burrows: Li-ving in and Observing the Anthropocenefrom Space	https://www.youtube.com/watch?v=gkdWTTi7M2E	Univ.Bremen Germany	English	2013
Prof. Veerabhadran Rama-nathan					

(continued)

(continued)

No.	Website Name	hyperlink	country	Language	year
	Jack Fishman (USA)		John P. Burrows (UK)		
7	Klaus Töpfer: The AnthropoceneSustainability in a World of 9 Billion People	https://www.youtube.com/watch?v=yzxL9bz76R0	Germany	English	2013

(continued)

(continued)

No.	Website Name	hyperlink	country	Language	year
Prof. Klaus Töpfer (Germany)			Prof. Meinrat O. Andreae (Germany)		
8	Meinrat O. Andreae: 400,000,036 Years of Biomass Burning	https://www.youtube.com/watch?v=nmd9vA-w-Aq4	Germany	English	2013
9	Summary	https://www.youtube.com/watch?v=g0HuKpbMREU	Germany	English	2013

(continued)

(continued)

No.	Website Name	hyperlink	country	Language	year
Mrs. Tertu Crutzen; Paul J. Crutzen, Doris Ahnen			K. Töpfer, P.J. Crutzen, S. Solomon, J. Lelieveld		
III	Videos with members of the Anthropocene Working Group				
1	Jan Zalasiewicz, Univ. of Leicester, UK, chairman of AWG (interview The Anthropocene- with Jan Zalasiewicz and Christian Schwägerl	https://www.youtube.com/watch?v=xP9P2i5jx-4	UK Royal Institution (RI)	English	2015
2	Will Steffen: Affinity, Intercultural Foundation	http://www.igbp.net/globalchange/anthropocene.4.1b8ae20512db692f2a680009238.html	Canberra, Au-stralia		20.10.2010
3	Will Steffen, Opening the Anthropocene project at the House of Cultures of the World (HKW) Berlin	https://www.hkw.de/de/app/mediathek/video/22385	MunichGerman Museum	English	2015
4	Will Steffen: Stocholm Resilience Centre	https://www.youtube.com/watch?v=lgxFKi3cis4	Sweden	English	31.10.2016
5	Will Steffen - The Anthropocene: Challenges of the Human Age	https://www.youtube.com/watch?v=RitK73xnB0M	English		10.9. 2018

(continued)

(continued)

No.	Website Name	hyperlink	country	Language	year
6	NCAR Explorer Series: Will Steffen (ANU), Paul Crutzen Lecture: The Anthropocene: Where on Earth are We Going?	https://www2.acom.ucar.edu/annual-paul-crutzen-lecture and https://ncar.ucar.edu/what-we-offer/education-outreach/public/ncar-explorer-series/anthropocene	Austra-lia	English	20.6. 2019
7	Erle Ellis, Univ. of Maryland, AWG (Economist, Interview)	https://www.youtube.com/watch?v=XLCa1njCK0E	BoulderCO, USA	English	2011
8	James P. M. Syvitski (AWG) Human Impacts and Their Consequences	https://www.youtube.com/watch?v=yot-Z7zldzo	USA	English	2017
9	Prof. Gupta: Human Impacts and Their Consequences	https://www.youtube.com/watch?v=6_Bht2Zwd9s	Netherlands	English	2014
10	Earth Systems Sycience: Johan Rockström: Beyond the Anthropocene	https://www.youtube.com/watch?v=V9ETiSaxyfk	Sweden	English	2017
IV	Videos from the Social Sciences on the Anthropocene (Selected Lectures)				
1	**Anthropology:** Donna Haraway - Staying with the Trouble: Making Kin in the Chthulucene	https://www.youtube.com/watch?v=GrYA7sMQaBQ&feature=emb_rel_pause	San Francisco	English	2017
2	**History:** Dipesh Chakra-barty: HKW: The Anthropocene Project. An Opening	https://www.youtube.com/watch?v=svgqLPFpaOg&list=PLrKjatlXyX_4MFxPp4qL3xtiEhvkPI_vo&index=4	Berlin HKW	English	2013
3	**History:** John Green: the Anthropocene	https://www.youtube.com/watch?v=3WpaLt_Blr4&feature=emb_rel_pause		English	2014

(continued)

(continued)

(continued)

No.	Website Name	hyperlink	country	Language	year
4	**Philosophy:** Welcome to the Anthropocene, Debate with philosophers Peter Sloterdijk, Bernard Stiegler	https://www.youtube.com/watch?v=ETHOqqKluC4&list=PLrKjatlXyX_4MFxPp4qL3xtiEhvkPI_vo&index=6	Radboud Univ. Netherlands	English	11.1. 2016
5	**Philosophy:** Graham Harman: Morton's Hyperobjects and the Anthropocene	https://www.youtube.com/watch?v=Id4FF7lO2wU		English	2017
6	Graham Harman; Anthropocene Ontology	https://www.youtube.com/watch?v=cR1A4ILPmjE&t=408s		English	2017
7	**Politics:** The Anthropocene - Consequences for Parlia-mentary Democracy? Klaus Töpfer in Augsburg	https://www.uni-augsburg.de/en/forschung/einrichtungen/institute/jfz/veranstaltungen/visiting-professorship/klaus_toepfer/	Augsburg	German	7.5. 2019
8	**Politics:** Climate Enginee-ring Klaus Töpfer in Augsburg	https://www.youtube.com/watch?v=xnk3A0SB-so&feature=emb_title	Augsburg	German	23.10.2019
9	**Politics:** Die Krise des Multilateralismus in einer Welt mit über 9 Milliarden Menschen Klaus Töpfer in Augsburg	https://www.uni-augsburg.de/en/forschung/einrichtungen/institute/jfz/veranstaltungen/visiting-professorship/klaus_toepfer/	Augsburg	German	18.6. 2019
10	**Politics:** Ernst Ulrich von Weizsäcker: Klimaschutz, Naturschutz und eine neue Aufklärung	https://www.youtube.com/watch?v=w6VhNlkZex0	Heidelberg	German	10.7. 2019
11	**Sociology:** Bruno Latour AnthropoceneLecture	https://www.hkw.de/de/app/mediathek/video/63344	France	English	4.5. 2018
12	**Sociology:** Conversation between Bruno Latour and Hans Joachim Schellnhuber	https://www.youtube.com/watch?v=Z-n_44M2nLw	Harvard Univ.	English	2018

(continued)

No.	Website Name	hyperlink	country	Language	year
13	**Theology**: Anthropology as Cosmic Diplomacy: Toward an Ecological Ethics for the Anthropocene (Harvard Divinity School)	https://www.youtube.com/watch?v=87yJKnVSd0k&feature=emb_rel_pause	USA	English	2017
14	**Mathematics/climate change**: Thomas Stocker: Anthropocene: The closing doors of climate targets	https://www.youtube.com/watch?v=Vn3_hviRfdU	Imperial College London	English	06.12.2013
V	**Berlin Lecture Series organised by the House of Cultures of the World (HKW), the Max-Plank Institute of the History of Science (MPIWG) and the Institute of Advanced Sustainability Studies (IASS) in Potsdam**				
1	HKW, Berlin: Anthropo-cene Lecture Series	https://www.hkw.de/en/programm/projekte/2017/anthropocene_lectures/anthropocene_lectures_start.php	Berlin	English	2017/ 2018
2	HWK, Berlin: Anthropocene Mediathek Project	https://www.hkw.de/en/app/mediathek/project/133043-anthropocene-lectures	Berlin	English	
3	HKW: Anthropocene Lecture – McKenzie Wark: Bogdanov, Platonov, Haraway & Robinson: An unexpected canon for a theory for the Anthropocene	https://www.anthropocene-curriculum.org/contribution/anthropocene-lecture-mckenzie-wark	Berlin	English	18.5 .2017

(continued)

(continued)

No.	Website Name	hyperlink	country	Language	year
4	HKW: John Mc Neill (environmental historian)	https://www.anthropocene-curriculum.org/contribution/anthropocene-lecture-john-mcneill	Berlin	English	21.9. 2017
5	HKW: Julia Adeney Thomas: The historians' task in the age of the Anthropocene: Finding hope in Japan?	https://www.anthropocene-curriculum.org/contribution/anthropocene-lecture-julia-adeney-thoma s	Berlin	English	12.10. 2017
6	HKW: Karen Liftin: The Body Politic	https://www.anthropocene-curriculum.org/contribution/anthropocene-lecture-karen-litfin	Berlin	English	19. 12. 2017
7	HKW: Anthropocene Lecture – Anna Lowenhaupt Tsing	https://www.anthropocene-curriculum.org/contribution/anthropocene-lecture-anna-lowenhaupt-tsing	Berlin	English	27.3. 2018
8	HKW: Philippe Descola Is the Anthropocene soluble in ontological pluralism	https://www.anthropocene-curriculum.org/contribution/anthropocene-lecture-philippe-descola	Berlin	English	25.5. 2018
9	HWK: Prasannan Parthasarathi: Nature and the Writing of History	https://www.anthropocene-curriculum.org/contribution/anthropocene-lecture-prasannan-parthasarathi	Berlin	English	1.6. 2018
10	HKW: Sheila Jasanoff: The Human Imprint: Nature, Time, and Law in the Anthropocene	https://www.mpiwg-berlin.mpg.de/video/human-imprint-nature-time-and-law-anthropocene	MPIWG, HDK, IASS	English	25.6 2018
VI	**Videos and Documentary Films on the Anthropocene**				
1	Globaia: Welcome to the Anthropocene	https://globaia.org/anthropocene		English	2009-2021
2	Osborne, Mike,/Miles Traer "Generation Anthropocene is Upon Us"	http://www.environmentandsociety.org/mml/generation-anthropocene-upon-us		English	2013
3	Steve Bradshaw: Movie: anthropocene	https://filmsfortheearth.org/en/films/anthropocene		English	2018
4	Christian Schwägerl on The Anthropocene	https://www.youtube.com/watch?v=og4qPXg1veY		English	2015

(continued)

(continued)

No.	Website Name	hyperlink	country	Language	year
5	Anthropocene: The new age of humans	https://topdocumentaryfilms.com/anthropocene/		English	2016
6	The Age of Mankind (VPRO)	https://topdocumentaryfilms.com/age-mankind/	Netherlands	English	2017
7	Biosphere	https://topdocumentaryfilms.com/biosphere/		English	2011
8	Hot Planet	https://www.bbc.co.uk/programmes/b00jf6md	BBC	English	2009
9	The Age of Man – Docu-mentary about the Anthropocene	https://www.youtube.com/watch?v=wJAbHssn0GI		English	8.6. 2019

(continued)

(continued)

No.	Website Name	hyperlink	country	Language	year
10	The risk of sustainability	https://www.youtube.com/watch?v=bjrPtlem30g		English	20.08.2019
11	Dipesh Chakrabarty, Clima- te Change as Epochal Con-sciousness, Yale, Tanner Lectures on Human Values	https://www.youtube.com/watch?v=CEPTyrQGgdI	Univ. of Chi-cago, Yale	English	18.2. 2015
12	Dipesh Chakrabarty: De-centering the Human? Gaia	https://www.youtube.com/watch?v=r_8w2LDgPWM		English	19.2. 2015
13	Dipesh Chakrabarty: The Human Condition in the Anthropocene, Roundtable	https://whc.yale.edu/videos/human-condition-anthropoc ene-roundtable-discussion		English	20.2. 2015
VII	**Educational videos**				
1	Making Sense of Climate Change (Smithsonian Environmeantal. Research Center, Washington, DC). Bert Drake				2017
	1: History and Physical Science of Global Warming	https://serc.si.edu/making-sense-of-climate-change	USA	English	31.1. 2017
	2: Global Warming, Rising Seas and Extreme Weather				7.2.17
	3: Beginning of the Age of Humans; People & Climate				14.2.17
	4.CO$_2$, Plants and Food				21.1.17
	5.Controlling CO$_2$				28.2.17
	6. Moving Forward: Con-fronting Denial and the Truth About Uncertainty				7.3.17
2	Anthropocene on Hold (House of Cultures of the World, Berlin)				
	Humanity's Epoch: ANTHROPOCENE	https://www.youtube.com/watch?v=mfDm7rM9_-8&fea ture=emb_rel_end	Berlin	German	2013
	Welcome to the Anthropocene	https://www.youtube.com/watch?v=_JRu4RVuuto&fea ture=emb_rel_end			2012

(continued)

(continued)

No.	Website Name	hyperlink	country	Language	year
	Exhibition with Georg Baselitz	https://www.youtube.com/watch?v=-E243mAL9Nw&feature=emb_rel_end			2014
3	Polyeco Contemporary Art Initiative aims at raising environmental awareness	https://www.youtube.com/channel/UCKThYXergdup6X4g6ndTC2w		English	2017
		https://www.youtube.com/channel/UCKThYXergdup6X4g6ndTC2w			
		https://www.youtube.com/watch?v=DvfNZ_xXjyQ&list=PL3PQrka5So1ge0S6t0djEul3c5djRTHVm			

(continued)

(continued)

No.	Website Name	hyperlink	country	Language	year
4	Anthropocene on Hold: 20 artists address the impact of a global pandemic	https://www.youtube.com/watch?v=mfDm7rM9_-8 https://www.youtube.com/watch?v=l5eg9T_uq4U https://www.pcai.gr/news/anthropocene-on-hold/			2020/2021
5	AFES-PRESS Brainstorming on 31 May 2017 in Mosbach, Germany 'Politik' in and for the Anthropocene: Obstacles, Challenges, Opportunities and Tasks for the Social Sciences in the 21stCentury with Paul J. C rutzen as a guest of honour	http://www.afes-press-books.de/html/Brainstorming_Mosbach.html Overview Password anthropocene2017 		English	31.5. 2017
6	**Construction, circulation et *praxisde l'Anthropocène* : défis épistémologiques et enjeux normatifs du processus de catégorisation**	https://parcs.hypotheses.org/932	France	French	4.11. 2018

(continued)

(continued)

No.	Website Name	hyperlink	country	Language	year
7	L'anthropocene : Un Concept Socioscientifique Pour Une Education Au Developpement Durable?	https://parcs.hypotheses.org/lanthropocene-un-concept-socioscientifique-pour-une-education-au-developpe ment-durable			
8	Global Politics in the Anthropocene (Macquarie University)	https://www.coursera.org/lecture/analysing-complexity/global-politics-in-the-anthropocene-KfSxK	Austra-lia	English	
9	Prof. Yadvinder Malhi on the emerging concept of the Anthropocene.	https://www.youtube.com/watch?v=rLM1MEhvZJA	Oxford Univ. UK	English	2017
10	Reading the *Anthropocene* and the Technosphere from *Africa*. (HKW)	https://www.anthropocene-curriculum.org/project/cam pus-2016/seminar-whose-reading-the-anthropocene-and-the-technosphere-from-africa	MIT	English	19.4.2016

(continued)

(continued)

No.	Website Name	hyperlink	country	Language	year
11	Prof Oonsie Biggs, Lecture Tackling the Anthropocene Challenge	https://www.youtube.com/watch?v=DyFDT4I60OY	Steelenbosch, S. Africa	English	2019
12	Jess Franco, Johns Hopkins Univ.: Eating in the Anthropocene - Food Systems	https://www.youtube.com/watch%3Fv%3DyXwh6xz HT14	USA	English	2020
13	Prof. Jan Zalasiewicz: Big History Anthropocene	https://www.youtube.com/watch?v=dLLRCaRIqRM	UK, Australia	English	2016
14	Urbanism and the Anthropocene (ETH Zurich)	http://ooze.eu.com/en/lectures/sessions_on_territory_urbanism_and_the_anthropocene_eth_zurich_116/ove rlay/	Switzerland	English	2019
15	Historians David Christian and John R. McNeill, *The Anthropocene: Are We There Yet?*	https://www.loc.gov/item/webcast-4829/	Washington, DC, LC USA	English	2009

(continued)

(continued)

No.	Website Name	hyperlink	country	Language	year
16	Bruno Latour thinks about the Anthropocene (Six Gifford Lectures *Facing Gaia: A new enquiry into Natural Religion,*	https://www.cigionline.org/events/fate-humans-new-geological-age-beginning-think-about-anthropocene	Edinburgh, UK	English	2012
17	Clive Hamilton: The Fate of Humans in a New Geolo-gical Age: Beginning to Think About the Anthropocene (CIGI)	https://www.cigionline.org/events/fate-humans-new-geological-age-beginning-think-about-anthropocene	CanadaOntariaWater-loo	English	2012-2013
18	ATC Lecture—Kim Stanley Robinson's "The Good Anthropocene: Terraforming Earth"	https://www.youtube.com/watch?v=VmCx4qcehqc	USA	English	15.3. 2015
19	The Future of Nature: Conservation in the Anthropocene with Emma Marris UC San Diogo	https://www.youtube.com/watch?v=eZnAompGPVE	USA	English	16.05.2019
20	Media Fossils and the *Anthropocene:* A production of an archaeological future - Uploaded by Birkbeck, University of London	https://www.youtube.com/watch?v=hz4I09a8HiA&list=PLpANdMEGSnFd4yKUBCbP2V49JhO-GRBww&index=32&t=0s	Birkbeck, Univ. of Lon-don, UK	English	11.6. 2020
21	Dr. Joshua Tewksbury: Living in the Anthropocene: Science, Sustainability and Society	https://vtechworks.lib.vt.edu/handle/10919/70902		Univ. of Vir-ginia, USA	English

(continued)

(continued)

No.	Website Name	hyperlink	country	Language	year
22	Nigel Thrift "Cities in the Anthropocene",	https://www.youtube.com/watch?v=ePA3e0NuqGs	UK	English	26.2. 2015
23	Climate Change and the Anthropocene, École Polytechnique & Columbia University lecture	https://www.youtube.com/watch?v=ukaVLYYQZXY	France USA	English	
24	Postcolonial Anthropocene: Artists Lectures: Sammy Baloji, Jennifer Allora, Guillermo Calzadilla	https://www.youtube.com/watch?v=hN6LFOSHzHY	USA Dart-mouth College	English	3.6. 2020

About the Nobel Prize in Chemistry in 1995

The Royal Swedish Academy of Sciences has decided to award the 1995 Nobel Prize in Chemistry to Professor Paul Crutzen, Max Planck Institute for Chemistry, Mainz, Germany (Dutch citizen), Professor Mario Molina, Department of Earth, Atmospheric and Planetary Sciences and Department of Chemistry, MIT, Cambridge, MA, USA and Professor F. Sherwood Rowland, Department of Chemistry, University of California, Irvine, CA, USA *for their work in atmospheric chemistry, particularly concerning the formation and decomposition of ozone.*

Paul J. Crutzen
Prize share: 1/3

Mario J. Molina
Prize share: 1/3

F. Sherwood Rowland
Prize share: 1/3

- **Paul Crutzen** was born in 1933 in Amsterdam. Dutch citizen. Doctor's degree in meteorology, Stockholm University, 1973. Member of the Royal Swedish Academy of Sciences, the Royal Swedish Academy of Engineering Sciences and Academia Europaea. Professor **Paul Crutzen,** Max Planck Institute for Chemistry, P.O. Box 3060 D-55020 Mainz, Germany. He passed away in 2021.

- **Mario Molina** was born in 1943 in Mexico City, Mexico. PhD in physical chemistry, University of California, Berkeley. Member of the US National Academy of Sciences. Professor **Mario Molina,** Department of Earth, Atmospheric and Planetary Sciences MIT 54 – 1312, Cambridge MA 02139, USA. He died in 2020.
- **F. Sherwood Rowland** was born in Delaware, Ohio, USA, 1927 and deceased in 2012. Doctor's degree in chemistry, University of Chicago, 1952. Member of the American Academy of Arts and Sciences and of the US National Academy of Sciences, where he was Foreign Secretary.

Source "Press Release: The 1995 Nobel Prize in Chemistry". *Nobelprize.org.* Nobel Media AB 2014. Web. 3 Jan 2015. http://www.nobelprize.org/nobel_pri zes/chemistry/laureates/1995/press.html and at: http://www.nobelprize.org/nobel_ prizes/chemistry/laureates/1995/illpres/reading.html and for the Award Ceremony Speech by Professor Ingmar Grenthe of the Royal Swedish Academy of Sciences is at: http://www.nobelprize.org/nobel_prizes/chemistry/laureates/1995/presentation-speech.html.

Interview with *Paul Crutzen* by Astrid Gräslund at the meeting of Nobel Laureates in Lindau, Germany, June 2000. Paul Crutzen talks about family background, early education and interest in natural science; his work in the Institute of Meteorology in Stockholm (5:02); his discovery (6:55); the ozone layer (15:42); the Greenhouse Effect (19:10); ozone holes (23:43); and the consequences of a 'Nuclear winter' (27:09). See a Video of the Interview, 34 min., at: https://www.nobelprize.org/pri zes/chemistry/1995/crutzen/interview/.

Prof. Paul J. Crutzen on the eve of his 80th birthday on 2 December 2013 at the symposium The Anthropocene organised in his honour by the Max Planck Institute for Chemistry in Mainz. Photo by Carsten Costard who granted permission.

About the Max Planck Institute for Chemistry

Aiming at an integral scientific understanding of chemical processes in the Earth System from molecular to global scales.

Current research at the Max Planck Institute for Chemistry in Mainz aims at an integral understanding of chemical processes in the Earth system, particularly in the atmosphere and biosphere. Investigations address a wide range of interactions between air, water, soil, life and climate in the course of Earth history up to today´s human-driven epoch, the **Anthropocene**.

The Max Planck Institute for Chemistry is one of the two oldest institutes of the Max Planck Society. It was founded in 1912 as the Kaiser Wilhelm Institute for Chemistry in Berlin, and it was relocated to Mainz in 1949. Particularly well-known scientists in the Institute´s history are the Nobel laureates Richard Willstätter, Otto Hahn, and Paul Crutzen. In honor of the former director and president of the Max Planck Society, the Max Planck Institute for Chemistry also carries the epithet Otto Hahn Institute. Further information about the institute history is available **here**.

Currently, the institute employs some 300 staff in four departments (Atmospheric Chemistry, Climate Geochemistry, Multiphase Chemistry, and Particle Chemistry) and additional research groups. Scientists conduct laboratory experiments, collect samples and record measurement data during field campaigns utilizing airplanes, ships, and vehicles. The practical work is complemented with mathematical models that simulate chemical, physical, and biological processes from molecular to global scales. One of the major goals is to find out how air pollution, including reactive trace gases and aerosols, affect the atmosphere, biosphere, climate, and public health.

A description of current research topics is given in the Institute Reports https://www.mpic.de/3916170/scientific-report.

On the Major Author Paul J. Crutzen

The research of Paul J. Crutzen has been mainly concerned with the role of chemistry in climate and biogeochemistry, and in particular the photochemistry of ozone in the stratosphere and troposphere. In 1970 he hypothesised that natural ozone production by the action of solar ultraviolet radiation on molecular oxygen (O_2) is mainly balanced by destruction processes, involving NO and NO_2 as catalysts. These catalysts in turn result from the oxidation of N_2O, a product of microbiological nitrogen conversion in soils and waters. He and Prof. Harold Johnston of the University of California, Berkeley, pointed out that NO emissions from large fleets of supersonic aircraft could cause substantial ozone losses in the stratosphere.

In the years 1972–1974 Crutzen proposed that NO and NO_2 could catalyse ozone production in the background troposphere by reactions occurring in the CO and CH_4 oxidation chains. Additional photochemical reactions leading to ozone loss were likewise identified. These gross ozone production and destruction terms are each substantially larger than the downward flux of ozone from the stratosphere, which until then had been considered the main source of tropospheric ozone.

In 1979–1980 Crutzen and co-workers drew attention to the great importance of the tropics in atmospheric chemistry. In particular, some measurement campaigns in Brazil clearly showed that biomass burning in the tropics was a major source of air pollutants, on a par with, or larger than, industrial pollution in the developed world.

n 1982 Crutzen, together with Prof. John Birks of the University of Colorado, drew attention to the risk of darkness and strong cooling at the earth surface as a consequence of heavy smoke production by extensive fires in a nuclear war ('nuclear winter'). This study and additional studies by R. Turco, B. Toon, T. Ackerman, J. Pollack and C. Sagan and by the Scientific Committee on Problems of the Environment (SCOPE), to which Crutzen contributed, showed that more people could die from the indirect consequences of a nuclear war than by the direct impacts of the nuclear explosions.

In 1986, together with Dr. F. Arnold of the Max Planck Institute of Nuclear Physics in Heidelberg, Crutzen showed that nitric acid and water vapour could co-condense in the stratosphere at higher temperatures than required for water ice formation, providing a significant part of a chain of events leading to rapid ozone depletion at high latitudes during late winter and spring (the so-called Antarctic 'ozone hole').

S. Benner et al. (eds.), *Paul J. Crutzen and the Anthropocene: A New Epoch in Earth's History*, The Anthropocene: Politik–Economics–Society–Science 1, https://doi.org/10.1007/978-3-030-82202-6

His most recent research is concerned with the role of clouds in atmospheric chemistry as well as photochemical reactions taking place in the marine boundary layer, involving catalysis by halogen gases produced by marine organisms. Also, his current research deals with the chemical and climatic effects of the heavy air pollution which is found over Asia and other regions in the developing world: the so-called ABC (Atmospheric Brown Clouds) phenomenon.

Address: Prof. Paul J. Crutzen, via Ms. Astrid Kaltenbach, Max Planck Institute for Chemistry, Otto Hahn Institute, Hahn-Meitner-Weg 1, 55128 Mainz, Germany.
Email: astrid.kaltenbach@mpic.de
Website: http://www.mpic.de/index.php?id=31&type=0

On the Co-Editors

Photo: © Thomas Hartmann

Susanne Benner is head of communications at the Max Planck Institute for Chemistry since 2011. She has more than 20 years of experience working in science and industry communications. Susanne studied biology at the RWTH Aachen University in Germany. During her Ph.D. time in organic chemistry at the Swiss Federal University Zurich (ETHZ) she was working as editor and coordinator of the Swiss University Newspaper "Synthese". Afterwards, she has received a sound education in science communica-tion at the Technoseum – a technology museum in Mannheim, Germany. From 1997 until 2004 she has been working as manager for public affairs at the Max Planck Institute for Plant Breeding Research in Cologne, Germany and afterwards as head of Communications at BASF Plant Science.

Address: Dr. Susanne Benner, Max-Planck-Institut für Chemie (Otto-Hahn-Institut), Hahn-Meitner-Weg 1, 55128 Mainz, Germany.

Email: susanne.benner@mpic.de

Website: https://www.mpic.de/person/49889

Gregor Lax is a research scholar at the Max Planck Institute for the History of Science. His research is directed at science policy stuctures and discourses, as well as institutional and epistemic developments, especially in climate and earth system sciences. After his doctorate from the University of Bielefeld in 2014, he worked on the history of atmospheric chemistry and biogeochemistry at the Max Planck Institute for Chemistry and supervised the evaluation of large research infrastructures as a science-administrator for the German Council for Science and Humanities (Wissenschaftsrat).

His current research addresses the *History of Climate and Earth System Science (ESS) in Germany,* focusing on integrative research approaches, which play an active role on the epistemical, conceptual and science-policy level. The ESS adresses a couple of highly relevant topics, including climate change, nuclear winter, CFC's, and the ozone layer as well as current debates on the anthropocene and geo-engineering which achieved a broad societal attention. The project focuses on a large cluster of Max Planck Institutes with regard to the development of the ESS, the internal and external negotiations of these institutes and their role in national and international scientific and political discourses.

Address: Dr. Gregor Lax, Max-Planck-Institute for the History of Science, Boltzmannstraße 22, 14195 Berlin, Germany.
Email: glax@mpiwg-berlin.mpg.de and gregor.lax@gmx.de
Website: http://gmpg.mpiwg-berlin.mpg.de/en/staff/research-scholars/dr-gregor-lax

Jos Lelieveld was born on July 25, 1955 in The Hague. Study of natural sciences Leiden Univ. (1984), research associate at Geosens B.V. (1984-1987), research scientist at the Max Planck Institute for Chemistry (1987-1993), PhD in Physics and Astronomy Utrecht Univ. (1990), Professor of Atmospheric Physics and Chemistry Univ. of Wageningen and Utrecht (1993-2000), Director and Scientific Member at the Max Planck Institute for Chemistry (since 2000), Professor in Atmospheric Physics, University of Mainz.

- Extended visits at Stockholm University (1991) and the University of California, San Diego (1992).
- Professor in Air Quality at Wageningen University in 1993-1995.Professor in Atmospheric Physics and Chemistry at Utrecht University 1996–2000.
- Director of the international research school COACh (Co-operation on Oceanic, Atmospheric and Climate Change studies) 1997–2000.Director of the Atmospheric Chemistry Department since 2000.
- Professor in Atmospheric Physics at Mainz University, Spokesperson of the Paul Crutzen Graduate School.
- Part-time professor at the Cyprus Institute, Nicosia.

Address: Prof. Dr. Jos Lelieveld, Max-Planck-Institut für Chemie (Otto-Hahn-Institut), Hahn-Meitner-Weg 1, 55128 Mainz, Germany.
Email: jos.lelieveld@mpic.de
Website: https://www.mpic.de/3783327/profile-jos-lelieveld

Ulrich Pöschl is director of the Multiphase Chemistry Department at the Max Planck Institute for Chemistry and professor at the Johannes Gutenberg University in Mainz, Germany. He has studied chemistry at the Technical University of Graz, Austria, and he has worked as a postdoctoral fellow, research scientist, group leader, and university lecturer at the Massachusetts Institute of Technology, Departments of Chemistry and of Earth, Atmospheric, and Planetary Sciences; at the Max Planck Institute for Chemistry, Atmospheric Chemistry and Biogeochemistry Departments; and at the Technical University of Munich, Institute of Hydrochemistry. His current scientific research and teaching are focused on the effects of multiphase processes in the Earth system, climate, life & public health.

Pöschl is actively engaged in the promotion open science, and he is the initiator of interactive open access publishing with public peer review and interactive discussion (multi-stage open peer review) as established with the international scientific journal *Atmospheric Chemistry and Physics* (ACP, http://www.atmospheric-chemistry-and-physics.net) and the *European Geosciences Union* (EGU, http://www.egu.eu). He is also initiator and co-chair of global open access initiative OA2020 (https://oa2020.org/).

Address: Prof. Dr. Ulrich Pöschl, Max-Planck-Institut für Chemie (Otto-Hahn-Institut), Hahn-Meitner-Weg 1, 55128 Mainz, Germany.
Email: u.poschl@mpic.de
Website: https://www.mpic.de/3785120/profile-poeschl

Hans Günter Brauch (Germany), Dr. phil. habil, was, until 2012, Adj. Prof. (Privatdozent) at the Faculty of Political and Social Sciences, Free University of Berlin; since 1987 he has been chairman of *Peace Research and European Security Studies* (AFES-PRESS). In 2020 he set up the Hans Günter Brauch Foundation on Peace and Ecology in the Anthropocene (HGBS), of which he is the managing board chairman.

He is editor of: *Hexagon Book Series on Human and Environmental Security and Peace (HESP)*; *Springer Briefs on Environment, Security, Development and Peace* (ESDP); *Springer Briefs on Pioneers in Science and Practice (PSP)*; *Pioneers in Arts, Humanities, Science, Engineering, Practice (PAHSEP)*; and of *The Anthropocene: Politik – Economics – Society – Science (APESS)* with Springer International Publishing,

He was a guest professor of international relations at the universities of Frankfurt am Main, Leipzig, Greifswald, and Erfurt, a research associate at Heidelberg and Stuttgart, and a research fellow at Harvard and Stanford Universities. In the autumn and winter of 2013/2014 he was a guest professor at Chulalongkorn University in Bangkok. He has been a member of the IPRA Council (1996–2000 and 2016-2020) and of its Executive Committee (2016–2018).

He has published on security, armaments, climate, energy, migration, and Mediterranean issues in English and German, and his works have been translated into Chinese, Spanish, Greek, French, Danish, Finnish, Russian, Japanese, Portuguese, Serbo-Croatian, and Turkish. Recent books in English: (co-edited with Scheffran, Brzoska, Link, Schilling, 2012): *Climate Change, Human Security and Violent Conflict: Challenges for Societal Stability*; (co-edited with Oswald Spring, Grin, Scheffran, 2016): *Handbook on Sustainability Transition and Sustainable Peace;* (co-edited with Oswald Spring, Bennett and Serrano Oswald, 2016): *Addressing Global Environmental Challenges from a Peace Ecology Perspective*; (co-edited with Oswald Spring, Bennett and Serrano

Oswald, 2016): *Regional Ecological Challenges for Peace in Africa, the Middle East, Latin America and Asia Pacific*; co-edited with Oswald Spring, Collins and Serrano Oswald, 2018: *Climate Change, Disasters, Sustainability Transition and Peace in the Anthropocene*; (co-edited with Oswald Spring): *Decolonising Conflicts, Security, Peace, Gender, Environment and Development in the Anthropocene*, 2021.

Address: PD Dr. Hans Günter Brauch, Alte Bergsteige 47, 74821 Mosbach, Germany.
Email: brauch@afes-press.de
Websites: http://www.afes-press.de; http://www.afes-press-books.de/; http://www.hgb-stiftung.de and http://www.hgb-stiftung.org.

About the Affiliations of the Contributors

Ajai [11], Dr. Ajai, Group director, MPsG, Earth, Ocean, Atmosphere, Planetary Sciences & Applications Area, Space Applications Centre (IsrO), Ahmedabad (India).

Alt, Kurt W. [16], head, Centre for the History of Nature and Culture, Danube Private University, Austria; was affiliated with Institute of Anthropology, University of Mainz, 55099 Mainz, Germany; E-mail: kurt.alt@dp-uni.ac.at.

Andreae, Meinrat O. [8, 16] was Director of the Biogeochemistry Department from 1987 until 2017 at the Max Planck Institute for Chemistry, Hahn-Meitner-Weg 1, P.O. Box 3060, 55020 Mainz, Germany. Website: https://www.mpic.de/4245984/profile-meinrat-o-andreae.

Barnosky, Anthony D. [17] worked with *the* Dept. of Integrative Biology, Museum of Paleontology, Museum of Vertebrate Zoology, University of California, Berkeley, CA 94720, USA; Jasper Ridge Biological Preserve, Stanford University, Stanford, California, USA.

Bengtsson, Lennart [11], Prof., director, Earth sciences, ISSI (International space science Institute), Bern (Switzerland).

Brauch, Hans Günter [22], political science and modern history, Adj. Prof. (ret.), PD Dr. phil. habil., Otto-Suhr Institute, Free University of Berlin, editor of five English language book series with Springer Nature, Chairman of Peace Research and European Security Studies (AFES-PRESS), Mosbach, Germany. E-mail: hg.brauch@onlinehome.de.

Breashears, David [11], David Breashears Arcturus Motion Pictures, Boston, MA (USA).

Bruhn, Kai-Christian [16] worked with Geocycles Research Centre, University of Mainz, 55099 Mainz, Germany; and Institute for Spatial Information and Surveying Technology (i3Mainz), Mainz University of Applied Sciences, Lucy-Hillebrand Strasse 2, 55128 Mainz, Germany; E-mail: kai-christian.bruhn@hs-mainz.de.

© The Editor(s) (if applicable) and The Author(s), under exclusive license   571
to Springer Nature Switzerland AG 2021
S. Benner et al. (eds.), *Paul J. Crutzen and the Anthropocene: A New Epoch in Earth's History*, The Anthropocene: Politik–Economics–Society–Science 1,
https://doi.org/10.1007/978-3-030-82202-6

Cearreta, Alejandro [17] worked with Departamento de Estratígrafa y Paleon-tología, Facultad de Ciencia y Tecnologia, Universidad del País Vasco UPV/EHU, Apartado 644,48080 Bilbao, Spain.

Cox, Peter M. [8], Hadley Centre for Climate Prediction and Research. Met Office, Fitz Roy Road, Exeter, Devon. EX1 3PB, UK.

Crumley, Carole [13] is Professor of Anthropology at the University of North Carolina, Chapel Hill (emer.), Senior Social Scientist at the Stockholm Resilience Centre, Stockholm University (Sweden), and Research Director of the IHOPE project. She holds a doctorate in anthropology from the University of Wisconsin, Madison (USA), an M.A. in archaeology from the University of Calgary (Alberta, Canada), and a B.A. in anthropology from the University of Michigan, Ann Arbor (USA).

Crutzen, Paul J. [2, 3, 4, 5, 6, 7, 8, 9, 10, 11, 12, 13, 14, 15, 16, 17, 18, 20], chemistry, Prof. Dr. hc mult., Nobel Laureate in Chemistry (1995), Max Planck Institute for Chemistry, Mainz, Germany. He was a former Director of the Atmospheric Chemistry Division of the Max Planck Institute for Chemistry in Mainz, Germany where he was Director of the Atmospheric Chemistry Department from 1978 until 2000. He also was (part-time) Professor at the Scripps Institute of Oceanograpy, University of California, La Jolla, USA. His research has been mainly concerned with the role of chemistry in climate and biogeochemistry, and in particular the photochemistry of ozone in the stratosphere and troposphere. Crutzen's work has also drawn attention to the great importance of the tropics in atmospheric chemistry. In addition, his research has shown the risk of darkness and strong cooling at the earth surface as a consequence of heavy smoke production by extensive fires in a nuclear war ("nuclear winter"). His research over the past 1-2 decades is concerned with the role of clouds in atmospheric chemistry as well as photochemical reactions taking place in marine air. More recently, he has focused on the climatic effects of bio-fuel production, in particular the emissions of N_2O derived from nitrogen fertilizers. E-mail: astrid.kaltenbach@mpic.de.

Cubasch, Ulrich [8], Chair of 'Interactions of Earth's Climate System'; since 2011 director, Meteorologisches Institut der Freien Universität Berlin, Carl-Heinrich-Becker-Weg 6-10, 12165 Berlin; website: E-mail: cubasch@zedat.fu-berlin.de.

Deutsch, Lisa [13] is Senior Lecturer, Director of Studies and Programme Director Sustainable Enterprising at the Stockholm Resilience Centre. Her research examines the couplings between the ecological effects of globalisation of food production systems and national policy and economic accounts. She particularly focuses on the implications of trade for freshwater and coastal ecosystem func-tioning relative to intensive livestock and aquaculture production systems. Her work contributes to the development of a set of complementary tools that can be used in economic accounting at national and international scales that address ecosystem support and performance. She was an Adjunct Fellow of the Australian National University within the Fenner School of Environment & Society. Her most recent publication: Deutsch, L., Troell, M., Limburg, L. and Huitric, M. 2011. Global trade of fisheries products-implications for marine ecosystems and their services.

In Kollner, T, editor. *Ecosystem Services and Global Trade of Natural Resources: Ecology, Economics and Policies* (London: Routledge).

Edgeworth, Matt [17] is affiliated with the School of Archaeology and Ancient History, University of Leicester, Leicester, UK.

Ellis, Erle C. [17] worked with Department of Geography and Environmental Systems, University of Maryland Baltimore County, Baltimore, MD 21250, USA; E-mail: ece@umbc.edu.

Ellis, Michael A., British Geological Survey, Keyworth, Nottingham NG12 5GG, UK; E-mail: mich3@bgs.ac.uk.

Esper, Jan [16], head of unit, Department of Geography, University of Mainz, 55099 Mainz, Germany, E-mail: esper@uni-mainz.de.

Fairchild, Ian J. [17] worked with School of Geography, Earth & Environmental Sciences, University of Birmingham, Birmingham, UK.

Foley, Stephen F. [16] has worked with the Geocycles Research Centre, University of Mainz, 55099 Mainz, Germany; *and the ARC* Centre of Excellence for Core to Crust Fluid Systems and Department of Earth Sciences, Macquarie University, North Ryde, New South Wales 2109, Australia; E-mail: stephen.foley@mq.edu.au.

Folke, Carl [13] is Professor in natural resource management, Science Director of the Stockholm Resilience Centre at Stockholm University and Director of the Beijer Institute of Ecological Economics of the Royal Swedish Academy of Sciences. His research is on the role that living systems at different scales play in social and economic development and how to govern and manage for resilience in integrated social-ecological systems.

Fuzzi, Sandro [11], Institute of Atmospheric Sciences and Climate of CNR, Bologna (Italy).

Galuszka, Agnieszka [17] works with Geochemistry and the Environment Division, Institute of Chemistry, Jan Kochanowski University, Kielce, Poland.

Gaudzinski-Windheuser, Sabine [16] Professor, Institute of Prehistory and Early History, University of Mainz, 55099 Mainz, Germany; and Monrepos Archaeologisches Forschungszentrum und Museum für Menschliche Verhaltensevolution, Schloss Monrepos, 56567 Neuwied, Germany; E-mail: gaudzinski@rgzm.de.

Gordon, Line [13] is a researcher at Stockholm Resilience Centre. She holds a PhD in Natural Resources Management, from Department of Systems Ecology at Stockholm University. She works with interactions among freshwater resources, agricultural production and ecosystem services with a particular focus on resilience, development and global change.

Grinevald, Jacques [17] works with Institut de Hautes Études Internationales et du Développement, Geneva, Switzerland.

Gronenborn, Detlef [16] Römisch-Germanisches Zentralmuseum, Ernst-Ludwig-Platz 2, 55116 Mainz, Germany; Institute of Prehistory and Early History, University of Mainz, 55099 Mainz, Germany; Geocycles Research Centre, University of Mainz, 55099 Mainz, Germany; E-mail: gronenborn@rgzm.de.

Haeberli, Wilfried [11], University of Zürich, Physical Geography division, Department of Geography, Zürich (Switzerland).

Haywood, Alan [17] is affiliated with School of Earth and Environment, University of Leeds, Leeds, UK; E-mail: A.M.Haywood@leeds.ac.uk.

Held, Hermann [8], Prof. Dr., Chair Sustainability & Global Change, Research Unit Sustainability & Global Change, Hamburg University; Website: https://www.fnu.uni-hamburg.de/en/staff/held.html; E-mail: hermann.held@uni-hamburg.de.

Immerzeel, Walter W. [11], Utrecht University, Utrecht (The Netherlands), Future Water, Wageningen (The Netherlands).

Ivar do Sul, Juliana [17] is affiliated with the Institute of Oceanography, Federal University of Rio Grande, Rio Grande, Brazil.

Jacob, Dorrit E. [16] *ARC* Centre of Excellence for Core to Crust Fluid Systems and Department of Earth Sciences, Macquarie University, North Ryde, New South Wales 2109, Australia; and Institute for Geosciences, University of Mainz, 55099 Mainz, Germany; E-mail: dorrit.jacob@mq.edu.au.

Jeandel, Catherine [17] is affiliated with Laboratoire d'Etudes en Géophysique et Océanographie Spatiales (CNRS, Centre National d'Etudes Spatiales, Institut de Recherche pour le Développement, Toulouse, France.

Jordan, David [16] worked with Geocycles Research Centre, University of Mainz, 55099 Mainz, Germany; and Institute for Geosciences, University of Mainz, 55099 Mainz, Germany.

Kadereit, Joachim W. [16] Institute of Botany and Botanical Garden, University of Mainz, 55099 Mainz, Germany; E-mail: kadereit@uni-mainz.de.

Kaser, Georg [11], University of Innsbruck, Institute of Geography, Innsbruck (Austria).

Kennel, Charles F. [11], Distinguished Professor of Atmospheric Science, Emeritus Senior Advisor, Sustainability Solutions Institute, UCSd, La Jolla, CA (USA).

Kulkarni, Anil [11], Distinguished Visiting Scientist, Divecha Center for Climate Change, Centre for Atmospheric & Oceanic Sciences, Indian Institute of Science, Bangalore (India).

Lal, Rattan [14] †, was affiliated with Carbon Management and Sequestration Center, The Ohio State University, 2021 Coffey Road, 210 Kottman Hall 422b, Columbus, OH, USA; E-mail: lal.1@osu.edu.

Lawrence, M. [18], Prof. Dr., Institute for Advanced Sustainability Studies, Potsdam, Germany; E-mail: mark.lawrence@iass-potsdam.de.

Lax, Gregor [1], history, Dr., MA, Research Fellow, Max Planck Institute for the History of Science, Boltzmannstraße 22, 14195 Berlin, E-mail: glax@mpiwg@mpiwg-berlin.mpg.de.

Leinfelder, Reinhold [17] is affiliated with Department of Geological Sciences, Freie Universität Berlin, Berlin, Germany; E-mail: reinhold.leinfelder@fu-berlin.de.

Lelieveld, Jos [16], biology, chemistry, Prof. Dr. Max Planck Institute for Chemistry, Mainz, Hahn-Meitner-Weg 1, P.O. Box 3060, 55020 Mainz, Germany and Geocycles Research Centre, University of Mainz, Germany.

Lorenz, Klaus [14] works with Global Soil Forum, IASS Institute for Advanced Sustainability Studies e.V., Berliner Strasse 130, 14467 Potsdam, Germany; E-mail: klaus.lorenz@iass-potsdam.de.

McNeill, John R. [9, 17], history, Professor of History and University Professor at Georgetown University. His research interests lie chiefly in the environmental history of the Mediterranean world, the tropical Atlantic world, and Pacific islands. His most recent books, both global in scope, are *Something New Under the Sun: An Environmental History of the Twentieth-Century World* and *The Human Web,* co-authored with William H. McNeill. E-mail: mcneillj@georgetown.edu.

Mittelstraß, Jürgen [11], Philosophy, University of Konstanz; E-mail: Juergen.Mittelstrass@uni-konstanz.de.

Molina, Mario [13], †, chemistry, Prof. Dr. h.c. mult., Nobel Laureate in Chemistry (1995), holds a Chemical Engineer degree (1965) from UNAM (Mexico), and a Ph.D. in Physical Chemistry (1972) from the University of California, Berkeley. He was the President of the Mario Molina Center in Mexico City, Professor at the University of California, San Diego (UCSD), and was formerly an Institute Professor at the Massachusetts Institute of Technology (MIT). He serves on the U.S. President's Committee of Advisors in Science and Technology. In the 1970s he drew attention to the threat to the ozone layer from industrial chlorofluorocarbon (CFC) gases that were being used as propellants in spray cans, refrigerants, solvents, etc. More recently, he has been involved with the chemistry of air pollution of the lower atmosphere, and with the science and policy of climate change. He is a member of the U.S. National Academy of Sciences and the Institute of Medicine, and of the Pontifical Academy of Sciences of the Vatican. He has received more than thirty honorary degrees, as well as numerous awards for his scientific work including the Tyler Prize in 1983, the UNEP-Sasakawa Award in 1999, and the 1995 Nobel Prize in Chemistry.

Nakicenovic, Nebosja [8] is an energy economist who was Deputy Director General of the International Institute for Applied Systems Analysis (IIASA) in Laxenburg, Austria and former Full Professor of Energy Economics at the Vienna University of Technology, Austria.

Odada, Eric [17] is affiliated with Department of Geology, University of Nairobi, Nairobi, Kenya.

Oreskes, Naomi [17] works with Department of the History of Science, Harvard University, Cambridge, Massachusetts, USA.

Pachauri, Rajendra [11], †, engineer, economist, was chairman of the International Panel on Climate Change, New Delhi, India.

Painter, Thomas H. [11], Jet Propulsion Laboratory, California Institute of Technology Pasadena, CA (USA).

Persson, Åsa [13], is a Researcher at the Stockholm Resilience Centre and Research Fellow at the Stockholm Environment Institute, Sweden. With a background in human geography, she now specialises in global environmental governance, in particular governance of and financial mechanisms for climate adaptation.

Pontifical Academy of Sciences [19], Rome, The Holy Sea, The Vatican.

Pöschl, Ulrich [16], biology, chemistry, Prof. Dr., Director, Max Planck Institute for Chemistry, Mainz, Germany.

Rabassa, Jorge [11], Centro Austral de Investigaciones Cientificas(CAdIC) Ushuaia, Tierra del Fuego (Argentina).

Ramanathan, Veerabhadran [7, 11, 13] is a Distinguished Professor of Atmospheric and Climate Sciences at the Scripps Institution of Oceanography, University of California, San Diego. In the 1970s, he discovered the greenhouse effect of CFCs and numerous other manmade trace gases and forecasted in 1980, along with R. Madden that the global warming would be detectable by the year 2000. He, along with Paul Crutzen, led an international team that first discovered the widespread Atmospheric Brown Clouds (ABCs). He showed that ABCs led to large scale dimming, decreased monsoon rainfall and rice harvest in India and played a dominant role in melting of the Himalayan glaciers. His team developed unmanned aerial vehicles with miniaturised instruments to measure black carbon in soot over S Asia and to track pollution from Beijing during the Olympics. He has estimated that reduction of black carbon can reduce global warming significantly and is following this up with a climate mitigation Project Surya which will reduce soot emissions from bio-fuel cooking in rural India. He chaired a National Academy report that calls for a major restructuring of the Climate Change Science Program and it was received favourably by the Obama administration. *Address:* Scripps Institution of Oceanography, University of California, San Diego, CA, USA.

Renn, Jürgen, Preface; Prof. Dr., Director, Max Planck Institute for Hisotry of Sciecne, Berlin; he is honorary professor for History of Science at both the Humboldt-Universität zu Berlin and the Freie Universität Berlin. He has taught at Boston University, at the ETH in Zurich, and at the University of Tel Aviv. He is a member of the Leopoldina as well as of further national and international scientific and editorial boards. From 2017 to 2019, Jürgen Renn served as Director and Scientific Member at the Max Planck Institute for the History of Science (since 1994). His most recent publication is *The Evolution of Knowledge: Rethinking Science for the Anthropocene* (2020, Princeton University Press).

Revkin, Andrew [17] works with Dyson College Institute for Sustainability and the Environment, Pace University, Pleasantville, New York, USA.

Richardson, Katherine [13] is a professor of biological oceanography and Leader of the Sustainability Science Center at the University of Copenhagen. She chaired the Danish Commission on Climate Change Policy that presented a plan for removing fossil fuels from the country's energy system.

Richter, Daniel de B. [17] works with Nicholas School of the Environment, Duke University, Durham, North Carolina, USA.

Robock, Alan [11], Editor, Reviews of Geophysics; Director, Meteorology Undergraduate Program, Department of Environmental Sciences, Rutgers University, New Brunswick, NJ (USA).

Rockström, Johan [13] is Director of the Potsdam Institute for Climate Impact Research and Professor in Earth System Science at the University of Potsdam. Before he was a Professor in natural resource management at Stockholm University, Executive Director of the Stockholm Environment Institute and the Stockholm Resilience Centre. He is an internationally recognised systems researcher on global sustainability issues, and a leading scientist on integrated land and water resource management, with a particular focus on resilience and development. In addition Rockström is active as a consultant for several governments and business networks, as an advisor for sustainable development issues at international meetings including the World Economic Forum, the *United Nations Sustainable Development Solutions Network* (SDSN) and the *United Nations Framework Convention on Climate Change Conferences* (UNFCCC). Professor Rockström chairs the advisory board for the EAT Foundation and the Earth League and has been appointed as chair of the Earth Commission. He has more than 25 years experience from applied water research in tropical regions, and has more than 150 research publications in fields of agricultural water management, watershed hydrology, global water resources, eco-hydrology, resilience and global sustainability. He chaired the visioning process on global environmental change of ICSU, the *International Council for Science.* He was awarded the title "Swede of the Year" in 2009 for his work on bridging science on climate change to policy and society. Email: presse@pik-potsdam.de.

Rubbia, Carlo [11], Scientific Adviser of CIEMAT (Spain)

Russell, Lynn M. [11], Scripps Institution of Oceanography, University of California San Diego, CA (USA).

Sánchez Sorondo, Marcelo [11], H.E. Msgr., Chancellor, The Pontifical Academy of Sciences (Vatican City).

Scheffer, Marten [13], Department of Environmental Sciences, Wageningen University, Wageningen, The Netherlands, Email: Marten.Scheffer@wur.NL.

Schellnhuber, Hans Joachim [11, 13, 17], was the Director of *Potsdam Institute for Climate Impact Research* (PIK) since he founded the institute in 1991 until 2018. He was Professor for Theoretical Physics at Potsdam University, External Professor at the Santa Fe Institute and Chairman of the *German Advisory Council on Global Change* (WBGU). He is a member of numerous national and international panels for scientific strategies and policy advice on environment and development matters and elected member of the Max Planck Society, the German National Academy (Leopoldina), the US National Academy of Sciences, the Leibniz-Sozietät, the Geological Society of London, and the International Research Society Sigma Xi. He authored and co-authored about 250 articles and more than 50 books in the fields of condensed matter physics, complex systems dynamics, climate change research, Earth System analysis, and sustainability science.

Scholz, Denis [16] Institute for Geosciences, University of Mainz, 55099 Mainz, Germany. Email: scholzd@uni-mainz.de.

Schöne, Bernd R. [16] Institute for Geosciences, University of Mainz, 55099 Mainz, Germany; Email: schoeneb@uni-mainz.de.

Schreg, Rainer [16] Institute for Archaeology, University Bamberg, 96047 Bamberg, Germany; Email: rainer.schreg@uni-bamberg.de.

Schwägerl, Christian [12], a Berlin-based journalist who writes for *GEO* magazine, the German newspaper *Frankfurter Allgemeine*, and other media outlets. He is co-founder and chair of Riff Reporter, a freelance cooperative, and author of *The Anthropocene: The Human Era and How it Shapes Our Planet.*

Sirocko, Frank [16] Head Climate and Sediment, Institute for Geosciences, University of Mainz, 55099 Mainz, Germany; Email: sirocko@uni-mainz.de.

Sorooshian, Soroosh [11], UCI Distinguished Professor and Director, Center for Hydrometeorology and Remote Sensing (CHRS), University of California Irvine, CA (USA).

Steffen, Will [6, 8, 9, 13, 17], physics, global environmental change, was Director of the Fenner School of Environment and Society, Australian National University, Canberra. From 1998 to mid-2004, he served as Executive Director of the International Geosphere-Biosphere Programme, based in Stockholm, Sweden. His research interests span a broad range within the field of Earth System Science, with a special emphasis on terrestrial ecosystem interactions with global change, the global carbon cycle, incorporation of human processes in Earth System modelling and analysis, and sustainability and the Earth System. Fenner School of Environment and Society, The Australian National University, Acton, Australia; Stockholm Resilience Centre, Stockholm University, Stockholm, Sweden; E-mail: will.steffen@anu.edu.au.

Stocker, Thomas F. [11], University of Bern, Climate and Environmental Physics, Bern (Switzerland).

Stoermer, Eugene F. [2], (1934-2012), †, was a leading researcher in diatoms, with a special emphasis on freshwater species of the North American Great Lakes. He was a professor of biology at the University of Michigan School of Natural Resources and Environment.

Summerhayes, Colin [17] is affiliated with Scott Polar Research Institute, Cambridge University, Cambridge, UK.

Svedin, Uno [13] is a senior research fellow at the Stockholm Resilience Centre, Stockholm, Sweden. Email: uno.svedin@su.se.

Syvitski, James [17] works with Department of Geological Sciences, University of Colorado-Boulder, Boulder, Colorado, USA.

Talaue-McManus, Liana [8] was Scientist of Marine Affairs, University of Miami, USA. At the global scale, she chaired the Scientific Steering Committee of the Land-Ocean Interactions in the Coastal Zone Project from 2002 to 2004; was Co-Chair of the Coastal Panel of the Integrated Global Observing System from 2003 to 2007. She is lead author in developing assessment methods to examine the socioeconomic impacts of policy interventions by the Global Environment Facility in transboundary waters. She is a founding member of the GEO (Group on Earth Observations) Coastal Zone Community of Practice.

Thompson, Lonnie G., [11], The Ohio State University, Byrd Polar Research Center School of Earth Sciences, Columbus, OH (USA).

Toon, Owen B. [11], University of Colorado, Department of Atmospheric and Oceanic Sciences, Boulder, CO (USA).

Töpfer, Klaus [14] was director of IASS Institute for Advanced Sustainability Studies e.V., Berliner StraBe 130, 14467 Potsdam, Germany; E-mail: klaus.toepfer@iass-potsdam.de.

Tost, Holger [16] Institute for Atmospheric Physics, Johannes Gutenberg University Mainz, J.-J.-Becherweg 21, 55128 Mainz, Germany; Email: tost@uni-mainz.de.

Turner II, Billie [8] Regents Professor, School of Sustainability, College of Global Futures, School of Geographical Sciences and Urban Planning, Arizona State University; *E-mail*: Billie.L.Turner@asu.edu.

Vidas, Davor [17] works with Marine Affairs and Law of the Sea Programme, The Fridtjof Nansen Institute, Lysaker, Norway; Email: Davor.Vidas@fni.no.

Vött, Andreas [16] Natural Hazard Research and Geoarchaeology, Department of Geography, University of Mainz, 55099 Mainz, Germany; and Institute for Physics of the Atmosphere, University of Mainz, 55099 Mainz, Germany; Email: voett@uni-mainz.de.

Wagreich, Michael [17] works with Department of Paleobiology, National Museum of Natural History, Smithsonian Institution, Washington, District of Columbia, USA

Waters, Colin N. [17] was affiliated with British Geological Survey, Keyworth, Nottingham NG12 5GG, UK.

Weller, Christine G. [16] worked with Geocycles Research Centre, University of Mainz, 55099 Mainz, Germany; and Institute for Geosciences, University of Mainz, 55099 Mainz, Germany.

Williams, Mark [13, 17] worked with Department of Geology, University of Leicester, University Road, Leicester LEI 7RH, UK; Email: mri@le.ac.uk.

Wing, Scott L. [17] works with Department of Paleobiology, National Museum of Natural History, Smithsonian Institution, Washington, District of Columbia, USA.

Wolfe, Alexander P. [17] works with Department of Biological Sciences, University of Alberta, Edmonton, Canada.

Zaelke, Durwood [11], President, IGSD, Director, INECE Secretariat, Washington, DC (USA).

Zalasiewicz, Jan [13, 17] was a professor and Senior Lecturer at the Department of Geology of the University of Leicester, and formerly was a field geologist and biostratigrapher at the British Geological Survey. He was a former Chair of the Stratigraphy Commission of the Geological Society of London, Chair of the Anthropocene Working Group of the International Commission on Stratigraphy, vice-Chair of the International Subcommission on Stratigraphic Classification and Secretary of the Subcommission on Quaternary Stratigraphy. Broadly interested in Earth evolution and the deep time context of current anthropogenic change, his book *The Earth After Us* explores the geological record of humankind, while the *Planet in a Pebble* explains palaeoenvironmental analysis and *The Goldilocks Planet* (with Mark Williams) is a geological history of Earth's climate. Department of Geology, University of Leicester, University Road, Leicester LEI 7RH, UK; Email: jaz1@le.ac.uk.

Index

A

Abrupt changes, 76, 80–84, 86, 89, 91, 93, 97, 168, 195, 219

Abrupt changes in the Earth system, 81, 82, 96, 97

Abrupt changes in the physical Earth system, 81

Absorbing particulate pollution, 134

Achilles' Heels in the Earth System, 80

Acidification, 29, 40, 53, 66, 117, 168, 358

Acid rain, 47, 52, 53, 115, 124

Adaptation, 115, 137–139, 160, 162, 177, 211, 254

Adapt to the climatic changes, 133

AERONET network of ground-based radiometers, 58

Aeronian Age of the Silurian Period, 220

Aerosol, 4, 8, 13, 34, 37, 47, 50, 54, 56–58, 60, 61, 63–65, 76–78, 80, 87, 90, 96, 97, 117, 124, 125, 135, 137, 139, 158, 163, 164, 180, 181, 187, 211, 253, 254, 256, 257, 261, 262, 518, 524, 525

Aerosol optical depth, 60

Agriculture, 20, 24, 29, 37, 39, 40, 48, 52, 66, 79, 91, 106, 107, 109, 116, 124, 126, 143, 149, 150, 152, 155, 157, 158, 178–181, 183, 204, 234, 236, 268, 276, 299, 310, 311, 315, 319, 327, 328, 332, 349

Air pollution, 37, 47, 54, 57, 60, 65, 126, 137, 138, 198, 257, 268–270, 523, 525, 529, 538

Ajai, 129

Albedo enhancement, 128

Albedo modification, 253–255, 257–262

Alliance for sustainability, 269

Alliances among government, industry, and universities, 114

Alt, Kurt W., 203

Amazon, 79, 168, 240, 290, 307, 308, 320, 326, 328, 334, 344, 346–350, 367, 377

American way of life, 142, 297

Amsterdam Conference, 352, 353, 355

Amsterdam declaration on Earth system science, 354

Andreae, Meinrat O., 57, 58, 75, 117, 203, 211, 293, 518, 524, 545

Antarctic ozone hole, 8, 86, 259, 262, 538

Anthrocene, 218, 303

Anthropocene, 1–3, 5, 12, 13, 19, 23, 24, 28, 33, 39, 41, 47, 48, 52, 66, 67, 75, 76, 78, 85, 87, 88, 97, 103, 104–106, 108–111, 114–117, 119, 123, 130, 132, 134, 135, 139, 141–145, 149, 155, 156, 158, 160, 163, 166, 167, 169, 170, 175, 180, 181, 193, 196, 198, 200, 203–207, 209, 217, 218, 221, 226, 230, 232–238, 240, 241, 267, 275–277, 281–283, 286, 287, 289–315, 317–326, 328–369, 373–390, 395, 445, 447, 451, 454, 503, 509–530, 532–534, 537, 539, 541, 543, 544, 546–552, 543, 554–558, 561, 563, 569, 570

Anthropocene: a concept in the humanities, 344

Anthropocene and environmental humanities, 281

Anthropocene: an emerging concept in the social sciences, 334

Anthropocene: a new geological epoch, 134

Anthropocene as a context, turning point and challenge, 305

Anthropocene as a geological time unit, 305

Anthropocene as a new epoch of geological time, 203, 204, 518

Anthropocene as a new geological epoch, 298

Anthropocene: a theme in development studies, policy and politics, 338

Anthropocene: a theme in environmental politics, 338

Anthropocene: a theme in gender studies, 340

Anthropocene: a theme in peace studies, 339

Anthropocene: a theme in security studies, 340

Anthropocene concept in anthropology, 335

Anthropocene concept in archaeology, 344

Anthropocene concept in biology and in biosciences, 332

Anthropocene concept in economics, 342

Anthropocene concept in education, 343

Anthropocene concept in geology and in earth systems science, 328

Anthropocene concept in history, 344

Anthropocene concept in human geography, 334

Anthropocene concept in philosophy, 345

Anthropocene concept in political science, 336

Anthropocene concept in psychology, 342

Anthropocene concept in sociology, 341

Anthropocene concept in the natural sciences, 321, 326, 363

Anthropocene concept in theology and religion, 346

Anthropocene concept in the social sciences, 289

Anthropocene from a complex systems perspective, 167

Anthropocene: from global change to planetary stewardship, 145

Anthropocene: from hunter-gatherers to a global geophysical force, 104

Anthropocene geopolitics, 303

Anthropocene has no past, 282

Anthropocene: impact on international relations, 336

Anthropocene in linguistics and in literature, 347

Anthropocene in medicine and health, 349

Anthropocene in national and international law, 346

Anthropocene in technology and engineering, 349

Anthropocene in the performing arts, 348

Anthropocene: politik–economics–society–science, 1, 19, 23, 27, 33, 39, 47, 75, 103, 123, 129, 141, 145, 175, 193, 203, 217, 253, 267, 275, 281, 289

Anthropocene project, 2, 533, 546

Anthropocene review, 330, 522, 527

Anthropocene system, 105

Anthropocene–the age of humans, 275

Anthropocene, the Age of Men, 142

Anthropocene videos with hyperlinks, 537

Anthropocene Working Group (AWG), 1, 282, 290, 291, 293, 297–299, 301, 304–308, 314, 322, 325, 329–338, 343, 344, 346, 351–353, 357–361, 363, 364, 367–369, 376, 378, 387, 390, 395, 447, 451, 530, 532, 546, 547

Anthropogenic climate change, 29, 67, 164, 254, 292, 295

Anthropogenic emission of aerosol particles, 117

Anthropogenic emissions, 20, 23, 30, 36, 54, 67, 87, 105, 117, 130, 140, 180, 184, 240, 255, 524

Anthropogenic emissions of greenhouse gases and other pollutants, 130, 140

Anthropogenic emissions of particulate matter, 36

Anthropogenic risk to the climate system, 134

Anthropos, 2, 283, 302, 305

Anthropozoic era, 19, 23, 27, 105, 302

Anthropozoikum, 302

Applied sciences, 290, 316, 324, 327, 357, 363, 367, 368, 379

Architecture in the anthropocene, 281

Arctic warming, 63, 137

Arias-Maldonado, Manuel, 326, 336, 338, 345, 375

Arrhenius, Svante, 48, 150, 287, 296

Artificial Intelligence (AI), 283

Artificial photosynthesis, 143

Artificial releases of CO_2, 31, 117

Art in the anthropocene, 281

Arts & Humanities Citation Index (A&HCI), 311

Atmosphere, 4–7, 9, 12, 13, 20, 29–31, 34–37, 39, 40, 41, 47, 48, 50–53, 56, 57, 61–63, 66, 67, 76–78, 80, 81, 83, 84, 93, 96, 98, 104, 109, 110, 115, 117, 118, 123–126, 128, 132, 133, 137, 138, 143, 150, 152, 155, 156, 158, 163, 168, 169, 175, 176, 178–181, 183, 184, 186, 187, 198, 207, 219, 230, 233, 237, 239, 254, 259, 262, 269, 296, 302, 329, 331, 379, 448, 453, 510, 512, 519, 520, 523–525, 529, 538, 563

Atmosphere after a nuclear war, 9

Atmospheric chemistry, 1, 3, 4, 8, 33, 47, 123, 175, 179, 193, 354, 441–443, 445, 449, 450, 454, 510, 520, 563, 568, 569

Atmospheric chemistry and climate in the anthropocene, 47, 123, 175, 520

Atmospheric chemistry in the 'Anthropocene', 33, 354

Atmospheric chemistry models and GCMs, 88

Atmospheric concentration of CO_2, 117, 163, 184, 227, 235

Avoiding 'dangerous anthropogenic interference', 136

Avoid removal of carbon sinks, 133

B

Barnosky, Anthony D., 217

Beautiful Anthropocene, 277

Beginning of the Anthropocene, 107, 109, 200, 204, 205, 211, 296, 522

Beginning of the Nuclear Era, 293

Bengtsson, Lennart, 129

Bibliometric analysis, 289, 291, 308, 328, 350, 352, 367, 379, 527

Bibliometric approach, 290, 306

Bibliometric or scientometric analysis, 307

Biermann, Frank, 318, 373, 528

Bifurcation, 81, 82, 168

Biggs, Reinette, 373, 521

Bio-adaptive technologies, 143

Biodiversity change, 155, 158, 231

Biodiversity in Earth system dynamics, 158

Biofuels, 113, 116, 185, 442

Biomass burning, 34, 37, 47, 48, 51–54, 57, 58, 60, 65, 124, 180, 184, 545, 565

Biosphere, 19, 48, 50, 52, 79, 84, 90, 91, 93–95, 108, 124, 161, 168, 169, 176–179, 218, 219, 223, 224, 228, 230, 231, 234, 237–241, 302, 329, 356, 361, 510, 511, 514, 519, 521, 522, 524, 526, 551, 563

Biosphere-Climate Interaction—The Earth System, 228

Biostratigraphic unit, 221

Black carbon, 37, 50, 57, 60, 65, 136, 137, 179, 198, 256, 269

Black carbon management, 137

Black carbon soot, tropospheric ozone, 137

Bolin, Bert, 4, 47, 353, 512

Book Citation Index – Science (BKCI-S), 311

Book Citation Index – Social Sciences & Humanities (BKCI-SSH), 311

Boundaries for critical Earth system processes, 170

Braje, Todd J., 218, 319, 335, 344, 370, 376

Brauch, Hans Günter, 289

Braudel, Fernand, 296, 306

Breashears, David, 129

Brown clouds' of black carbon, 136

Bruhn, Kai-Christian, 203

Brühl, Christoph, 7, 8

Burning of fossil fuels and solid biomass, 268

Burning of large amounts of biomass, 36

Business as usual, 95, 297

C

Cambrian adaptive radiation, 223

Cambrian explosion, 223

Cambrian fauna, 223

Capitalocene, 304, 305

Carbon capture and storage, 143, 163, 185

Carbon dioxide removal, 186

Carboniferous, 107, 226

Catastrophism, 379

Cattle population, 19, 23, 28, 40, 66, 68, 123

Cause of the great acceleration, 204

Cearreta, Alejandro, 217

Cenozoic, 221, 225, 226, 241

CFC case, 62

Challenges of peak oil, peak phosphorus, 147

Chandler, David, 318, 325, 326, 337, 339–341, 345, 346, 348, 370, 374

Changes in societal values, 117

Changes in the biosphere, 177, 299

Changes in the structure and functioning of the Earth System, 150

Changes to the climate system, 117

Change techno-managerial strategies, 92
Changing global environment, 115
Changing pattern of CO_2 emissions, 155
Changing the climate, 134, 142
Chemical composition of the Atmosphere, 33, 47, 48, 50
Chemistry of the atmosphere, 83, 123
Chlorofluorocarbon, 7, 20, 24, 30, 34, 41, 47, 67, 69, 93, 115, 124, 175, 180, 184, 187, 195, 510, 512, 520
Chlorofluoromethane, 7
Chronostratigraphic boundary, 221, 521
Chronostratigraphic classification, 220
Chronostratigraphic record, 223
Chronostratigraphic unit, 220, 221, 300, 304, 521
Chthulucene, 304, 547
Circular economies, 275
City populations, 114
Clarivate analytics, 291, 308, 310, 314, 350
Clean-up of air pollution, 117
Climate and biosphere models, 91
Climate and clean air coalition, 257
Climate, atmospheric chemistry and biogenic processes in the anthropocene, 193
Climate Change, 29, 62, 110, 116, 125, 128, 135, 136, 181, 183, 260, 267, 272, 275, 295, 353, 454, 542, 543, 552, 558, 568, 569, 575
Climate change 2001
    the scientific basis, 125, 128
Climate change, air pollution and health workshop, 267
Climate change and air pollution, 267, 268
Climate-cooling and climate-warming pollutants, 257
Climate engineering, 175, 186, 253–256, 258, 259, 261, 262, 518
Climate engineering via albedo modification, 253
Climate impact assessment program, 6
Climate intervention, 254
Climate in the anthropocene, 47, 123, 175, 180, 520
Climate of renewed prosperity, 114
Climate policy is to stabilise greenhouse gas emissions, 136
Climate sensitivity, 76–78, 96, 97
Climate warming, 24, 33, 47, 48, 52, 64, 65, 126, 165
Cloud and hydrological cycle feedbacks provide major challenges, 47

Cloud darkening observations, 63
CMIP5 simulations, 260
CO_2 concentration, 41, 43, 49, 65, 103, 109, 111, 114, 118, 128, 143, 150, 154, 158, 180, 183, 227, 234, 239
CO_2 sequestration, 118, 185
Cold War, 9, 11, 12, 295, 357
Colebrook, Claire, 319, 345, 348, 371, 375
Collapse of coastal eco¬systems, 85
Collapse of modern, globalised society, 116
Columbian Exchange, 293, 294
Compensation of greenhouse warming by aerosols and clouds, 64
Complete Earth system models, 89, 90
Complex Earth system, 34, 71, 170, 524
Complexity in the chemistry of the Atmosphere, 83
Complex physical/chemical/biological climate system, 33
Concentration of carbon dioxide, 132, 134, 200
Concentrations of ozone and hydroxyl radicals, 51
Concept history approach, 291, 306
Concept of a planetary-scale social-ecological-geophysical system, 148
Concept of Earth system goods and services, 152
Concept of social-ecological systems, 148
Concept of the anthropocene, 149, 166, 207, 241, 303, 308, 326, 527, 528, 555
Concept of the palaeoanthropocene, 203, 518
Concept or conceptual/concept history, 306
Conference Proceedings Citation Index – Science (CPCI-S), 311
Conference Proceedings Citation Index – Social Science & Humanities (CPCI-SSH), 311
Consequences of land-use change, 54
Consequences on the global life support system, 147
Conservation management, 142
Conserving (ecosystem-biosphere), 95
Consumption of plant-based diets, 269
Contemporary Earth system science, 219
Contrasting trajectories for the anthropocene, 238
Control of human and domestic animal population, 31, 116
Conversion of mangrove forests to shrimp farms, 150

Cooke, Steven J., 370, 376
Cooling pollutants, 257
Cooling the Earth, 254
Cornucopian, 303
Correlation between the extinction events and human migration patterns, 106
Countries and languages, 291, 316, 382
Coupled model intercomparison project, 260
Cox, Peter M., 75
Cretaceous, 204, 226, 234
Critical climate and radiative forcings on mountain glaciers, 138
Crumley, Carole, 145
Crutzen, Paul J., 19, 23, 33, 39, 47, 75, 103, 123, 129, 141, 145, 175, 193, 203, 217, 253, 275
Crutzen, Paul J., anthropocene texts, abstracts, 510
Crutzen, Paul J., bibliography, 453
Crutzen, Paul J., facsimile, 503
Crutzen, Paul J., lectures and interviews, 537
Crutzen, Paul J., Scientific and Human Legacy, 378
Crutzen, Paul J., vita, 441
Cryospheric models, 116
Cubasch, Ulrich, 75
Cuernavaca, 1, 290, 292, 293, 350, 352, 366
Current chemical reactions, 311
Current knowledge base on abrupt changes, 86
Current state-of-the-art in climate projection, 87
Current trajectory of human societies, 241

D
Dahlem Workshop, 293, 352, 353, 355, 356, 510
Dalby, Simon, 303, 318, 374
Dangerous anthropogenic interference with the climate system, 136
Dangerous levels of climate change, 117
Decarbonisation, 239, 269
Decarbonise the energy system, 269
Declaration of the health of people, health of planet and our responsibility, 267
Decreases in stratospheric ozone, 60
Deep sea drilling project, 230
Defining the anthropocene, 217, 232, 234, 305, 329, 522
Deforestation, 29, 33, 36, 40, 54, 67, 107, 109, 133, 137, 152, 168, 180, 204, 267, 269

Degradation of ecosystem services, 145, 515
Dematerialisation, 116, 117
Denitrification, 84
Desert and soil dust, 57
Deutsch, Lisa, 145
Develop geoengineering capabilities, 143
Direct cooling effect of anthropogenic sulphate particles, 63
Disasters of world wars and depression, 114
Disruptions to the global economic system, 115
Diversitas, 231, 353, 354
Drivers of climate change, 87
Droughts caused by climate change, 268
Durable 'bio-economy', 143
During the Pleistocene, 226
Dynamic global vegetation models, 91

E
Early anthropocene debate, 321
Early Holocene, 107, 209, 228
Earth's albedo, 117, 257
Earth's critical zone network, 231
Earth's future, 331
Earth's glaciers are retreating, 135
Earth's metabolism, 142
Earth's stratospheric ozone layer, 115
Earth as our life support system, 156
Earth has now left its natural geological epoch, 104
Earth into a new geological era, 123
Earth is currently operating in a no-analogue state, 41
Earth is warming, 110, 135, 146
Earth system, 1, 2, 5, 34, 35, 41–44, 47, 63, 71, 75, 76, 78–97, 103–105, 107–109, 112, 114–117, 132, 134, 138, 145, 147–150, 152, 154, 155, 158, 159, 162–170, 175–177, 179, 187, 203, 206, 217–226, 228–241, 261, 292–296, 298–300, 303–305, 324, 329, 331, 351–357, 361, 366, 370, 371–373, 377, 379, 380, 385, 393, 521, 522, 526, 525–528
Earth System Analysis (ESA), 235, 352, 353, 355, 356, 510, 511
Earth system analysis for sustainability, 352, 353, 356, 510, 511
Earth system anthropocene, 234, 235
Earth system behaviour, 175–177, 187
Earth system changes, 87, 88, 94, 116, 220, 221, 233, 237

Earth system definition, 235
Earth system definition of the anthropocene, 235
Earth system dynamics in the anthropocene, 75, 510
Earth system dynamics in the late quaternary, 167
Earth system functioning, 79, 84, 86, 97, 103, 105, 107, 112, 149, 162, 222
Earth system geography, 78, 97
Earth system goods and services, 147, 148, 152, 158
Earth system in the anthropocene, 75, 78, 96, 168, 237, 355
Earth system modelling, 90
Earth system Models of Intermediate Complexity (EMICs), 231
Earth system science, 1, 2, 5, 96, 217–222, 226, 231, 232, 234, 237, 241, 261, 300, 304, 324, 351–356, 361, 364, 366, 368, 372, 377, 393, 522, 527
Earth system science community, 218, 232, 304, 352, 366
Earth system science in the anthropocene, 324, 355
Earth system scientist, 292, 321, 351
Earth Systems Science Partnership (ESSP), 353, 359
Earth system stewardship, 169
Earth system's architecture, xv
Ecological complexity, 84
Ecological footprint, 159, 160
Ecological system theory, 211
Ecomodernists, 303
Ecosphere in the anthropocene, 68
Ecosystem effects of eutrophication, 84
Ecosystem services, 91, 118, 146, 147, 158, 160, 187
Edgeworth, Matt, 217
Effects of human activities on climate, 123
Ehlers, Eckhart, 302, 321, 323, 324, 354, 356
Ellis, Erle C., 217
Ellis, Michael A., 525
El Niño, 54, 79, 219
Elsevier, 291, 308, 312, 314, 318, 350, 377, 378, 515, 532
Emerging Sources Citation Index (ESCI), 311
Emissions of greenhouse gases in the atmosphere, 128
End-Permian extinction, 219
End-Pleistocene climate change, 228
Energy conservation, 116

Energy production, 29, 31, 176, 179, 184, 185, 524
Energy use, 24, 28, 29, 31, 40, 67, 68, 108, 116, 123, 184, 185, 195, 294, 523
Engineering, 13, 21, 24, 31, 68, 92, 117, 139, 163, 164, 177, 198, 200, 231, 259, 261, 290, 311, 442, 518
Environmental Consequences of Nuclear War (ENUWAR), 10
Eocene, 226, 233, 234
Epistemic community, 290, 291, 306, 359
Epoch to the geological time, 281
Era of humankind, 123
Erlandson, Jon M., 319, 325, 335, 372, 373
Esper, Jan, 203
European Green Deal, 305
Eutrophication of surface waters, 29, 40, 66
Evolution of the biosphere, 223, 230
Evolution of the climate system, 223, 224
Exploitation of fisheries, 150
Exponential character of the great acceleration, 114
Extreme events, 65, 97, 146, 162, 209, 210, 355

F
Fairchild, Ian J., 217
Feedback processes, 33, 132, 168, 224, 228
Fish catch, 28, 68, 124, 176
Foley, Stephen F., 203
Folke, Carl, 145
Food production is not threatened, 136
Food supplies, 209
Foreign direct investment, 112, 150
Forward-looking preventive action, 379
Fossil fuel, 1, 20, 24, 29, 33, 34, 37, 40, 62, 66, 94, 103, 107–110, 113, 114, 116, 118, 124, 132, 133, 137, 141, 146, 147, 150, 155, 159, 163, 175, 176, 179, 181, 183–185, 187, 200, 204, 205, 207, 232, 233, 239, 240, 268, 269, 294, 296, 523, 526
Future climate change, 77, 78, 207
Future Earth, 231, 297, 353–355, 524, 532
Future trajectory of the anthropocene, 237
Fuzzi, Sandro, 129

G
Gaia hypothesis, 219, 304, 355, 356
Galaz, Victor, 337–339, 341, 372, 375
Galuszka, Agnieszka, 217, 329, 526, 530
Garfield, Eugene, 307, 310, 311

Gaudzinski-Windheuser, Sabine, 203, 518
General Circulation Models (GCMs), 78, 81, 82, 88, 89, 231
Genetic Engineering (GE), 262, 283
Geoanthropology, xvi
Geo¬sequestration, 117
Geoengineering [geo-engineering], 12, 13, 118, 139, 143, 145, 164, 175, 186, 187, 201, 254–256, 258, 259, 261, 296, 297, 303, 307, 379, 515, 568
Geological epoch of the holocene, xxi
Geoengineering model intercomparison project, 256
Geological time, 2, 48, 177, 207, 220, 226, 281, 282, 296–300, 329, 358, 513–516, 521, 525, 530
Geologic time scale, 218, 222, 232, 328, 351, 515, 516
Geology of mankind, 23, 354
GeoMIP, 256, 260–262
Geosciences, 218, 219, 328, 369
Geosphere-biosphere coevolution on Earth, 230
German Advisory Council on Global Change (WBGU), 239, 355
German Committee Future Earth (DKN), 355
German National Committee on Global Change Research (NKGCF), 355
Gibson-Graham, J. Katherine, 335, 342, 371, 374
Glacier fragmentation, 135
Glacier measurements, 138
Glacier retreat, 132
Glaciers, 116, 130, 132–140, 147, 175, 182, 187, 224, 230, 275, 287, 520
Glaser, Marion, 376, 528
Global average radiative (cooling), 37
Global biogeochemical cycles, 177, 219
Global boundary Stratotype Section and Point (GSSP), 221, 298, 299, 329, 330, 357, 516, 525, 530
Global capacity¬ building initiative, 133
Global change and Earth system science communities, 352
Global change and the earth system, 104, 354, 366
Global climate models, 61
Global economy and material consumption, 150
Global endurable risk, 283
Global environmental change, 93, 166, 291, 295, 297, 303, 304, 321, 323, 324, 328, 330, 352, 355, 357, 359, 361, 362, 366, 378, 379, 522, 524, 527, 528
Global governance and stewardship, 164
Globalisation, 95, 96, 104, 110, 112, 148, 169
Global mean radiative forcing of the climate system, 64, 125, 198
Global scale of environmental degradation in the anthropocene, 161
Global Standard Stratigraphic Age (GSSA), 221, 329, 516, 525
Global sustainability, 510, 511, 524
Global terrestrial nitrogen budget, 113
Global warming, 13, 51, 62, 63, 65, 77, 103, 115, 128, 130, 133, 137, 140, 186, 253, 256, 257, 267, 287, 299, 379, 552
Gordon, Line, 145, 515
Governance system, 373
Grear, Anna, 340, 346, 347, 372, 376
Great acceleration, 103, 110, 112–115, 117, 118, 141, 150, 153, 155, 161, 193, 195, 196, 204, 211, 233, 235, 237, 281, 293–297, 299, 356, 377, 521, 522, 525
Great depression, 114
Greenhouse-effect, 5
Greenhouse gas, 1, 12, 20, 29–31, 33, 37, 40, 47, 48, 51–53, 63, 65, 68, 76–78, 80, 83, 87, 90, 96, 97, 110, 117, 118, 124–126, 128, 130, 132–138, 140, 146, 151, 152, 155, 156, 163, 164, 168, 169, 175, 179, 183, 184, 187, 193, 197–200, 209, 219, 226–228, 233, 238–240, 257–259, 262, 268, 287, 294, 296, 510, 520
Greenhouse warming, 82, 117, 126, 137, 258
Green revolution, 93, 95
Grinevald, Jacques, 217
Gronenborn, Detlef, 203
Growing impacts upon the Earth system, 114
Growth in human population, 28, 66, 95
Growth of 'greenhouse gases', 63
Growth of democratic political systems, 115
Gupta, Joyeeta, 336, 338, 339, 342, 343, 348, 373

H
Haber-Bosch, 40, 66, 109, 113, 197
Haeberli, Wilfried, 129
Hamilton, Clive, 374, 527, 557

Haus der Kulturen der Welt, 2
Haywood, Alan, 217
Hazards framing, 258
Heikkurinen, Pasi, 342, 371, 375, 388
Held, Hermann, 75
Historical relationship between stratigraphy
        and Earth system science, 218
History of concepts, 306
Holistic understanding of our evolving
        planet, 218
Holocene, 1, 2, 19, 23, 27, 82, 104–107, 109,
        111, 114, 141, 145, 149, 150, 156–
        158, 164, 166–170, 181, 204, 205,
        217, 218, 221, 227, 230, 232–241,
        275, 282, 292, 293, 295, 296, 298,
        299, 306, 329, 330, 351, 353, 356–
        358, 363, 513, 515, 519, 521, 522,
        526
Homo erectus, 106
Homogenocene, 304
Homo sapiens, 31, 33, 48, 149, 158, 170,
        228, 233, 241, 268, 283, 523, 526
Horowitz, Abraham, 281
Hot spots, 78, 79, 82
Human appropriation, 116, 124, 240
Human causation, 150, 152
Human-caused changes in the composition
        of the air, 133
Human Development Index (HDI), 159, 160
Human-driven changes to Earth system
        functioning, 75
Human-earth nexus, xv
Human-environment problems, 95
Human impacts on Earth system, 41, 162
Human imprint on the Earth system, 109,
        114
Human-inclusive Earth system, 148
Human-induced stresses, 68, 114
Human influence on global climate, 29, 40,
        67
Human influence on the Earth system, 115
Humanities, 2, 91, 114, 115, 118, 124, 134,
        139, 142, 144–147, 149, 150, 155,
        157, 159, 160, 162, 164, 169, 170,
        218, 219, 262, 268, 275, 276, 283,
        286, 289–291, 302–305, 310–312,
        316, 318, 321, 322, 324, 327, 328,
        330, 332, 344, 349–352, 356, 357,
        359, 363, 366–368, 376–379, 514,
        515, 517, 521, 527, 528, 552, 568,
        569
Humankind, 31, 47, 48, 65–68, 81, 88, 93,
        94, 103, 105, 106, 108, 114, 123, 128,
        193, 206, 207, 295, 297, 302, 303,
        329, 379, 380, 513, 519, 520
Human-made chemicals in the reduction of
        stratospheric ozone, 151
Human-nature interactions in the anthro-
        pocene, 281
Human Planet, 294
Human politics and governance in the
        anthropocene, 281
Human population, 1, 19, 23, 28, 35, 39, 40,
        48, 66, 85, 109, 113, 123, 134, 142,
        146, 150, 163, 177, 187, 195, 197,
        204, 209, 302, 517, 520
Human response to the biodiversity decline,
        155
Humans as a mere biological species, xv
Human subsistence and migration, 208
Human systems with natural ecosystems
        embedded, 142
Hydrosphere, 230, 329, 519
Hyper-connectivity of the human enterprise,
        148

I
IGBP, 1, 19, 231, 290–293, 296, 297, 302,
        350–352, 354, 359, 361, 366, 443,
        533
IGBP/Earth system science community, 351
Immerzeel, Walter W., 129
Impacts of climate change, 89, 130, 133, 135,
        138, 140, 254
Impacts of global change on human—envi-
        ronment, 76
Impacts of particles on human health and
        climate, 65
Implications and Risks of Novel Options to
        Limit Climate Change (IMPLICC),
        260
Improved technology and management, 31,
        116, 518
Increasing competition for land, 146
Increasing global population, 128
Increasing loss of biodiversity, 151
Increasing sea level rise, 135
Index Chemicus, 311
Indirect aerosol effect, 50, 65
INDOEX [Indian Ocean Experiment], 37,
        56, 57
Industrial emissions of SO_2 and NO, 53
Industrial era, 108, 109, 150, 294, 297
Industrial fixation of nitrogen from the atmo-
        sphere, 150

Industrialisation, 103, 108, 109, 112, 133, 143, 176
Industrial output, 28, 68, 123, 195
Industrial revolution, 42, 93, 110, 180, 200, 203–205, 233, 234, 236, 293, 294, 357, 514, 518, 525
Institute for scientific information, 307, 310
Integrated Assessment Models (IAMs), 88, 92, 231
Integrated History, 104, 118
Integrated mitigation strategy, 137
Integrated socioeconomic-biophysical models, 91
Integrating stratigraphic and earth system approaches, 232
Interdisciplinary work on human-environment systems, 115
Intergovernmental panel on climate change, 24, 110, 116, 117, 135, 201, 231
Intermediate range nuclear forces, 11
International commission on stratigraphy, 1, 2, 141, 149, 218, 220, 297–299, 358, 447, 451, 513, 516
International Council for Scientific Unions, 5
International environmental policies, 115
International geosphere biosphere programme, 5
International global warming targets, 133
International human dimension programme, 353, 354
International multidisciplinary scientific network, 291
International tourism, 150
International Union of Biological Science, 354
International Union of Geological Sciences, 1, 218, 297, 300, 358, 516, 532
Interpretation, 10, 77, 220, 239, 241, 289, 300, 304, 306, 350, 352, 355, 377, 527
Intertropical Convergence Zone, 37
IPAT, 150
IPCC, 39, 40, 50–52, 57, 63, 64, 66, 67, 76–78, 82, 87, 88, 116, 128, 139, 146, 151, 152, 162, 181, 182, 184, 198, 231, 239, 254, 256, 260, 270, 377
Ivar do Sul, Juliana, 217

J
Jacob, Dorrit E., 203
Jeandel, Catherine, 217

Jordan, David, 203

K
Kadereit, Joachim W., 203
Kaser, Georg, 129
Keeling's Mauna Loa CO_2 growth curve, 48
Kennel, Charles F., 129
Kim, Rakhyun E., 319, 337, 338, 346, 370, 375
Kotzé, Louis J., 325, 326, 337, 346, 347, 375, 387
Kulkarni, Anil, 129
Kyoto Protocol, 138, 296

L
Lal, Rattan, 175
Land-clearing for agriculture, 149
Land surface, 20, 24, 29, 40, 66, 123, 134, 149, 178, 195, 230, 302
Large-scale Biosphere-Atmosphere Experiment in Amazonia, 79
Largest environmental cause of premature death, 257
Late Pleistocene, 106, 156, 177, 211
Late quaternary, 86, 109, 156, 169, 217, 230, 238–241
Law, 286, 289, 294, 310, 315, 316, 324, 327, 332, 334, 346, 347, 352, 357, 359, 361–363, 366–370, 372, 375, 376, 378, 379, 393, 550
Lawrence, M., 253
Lax, Gregor, 1
Leaders of sustainable development, 269
Leinfelder, Reinhold, 217
Lelieveld, Jos, 203
Light-absorbing and light-scattering particles, 56
Linear stability theory, 82
Linguistics and related literatures, 291
Literature, 262, 290, 291, 302, 304, 305, 307–311, 314, 315, 317, 320–322, 327, 344, 347, 348, 350–352, 361, 366, 367, 370, 371, 373, 375, 376, 378, 527
Lithosphere, 224, 230
Lithostratigraphic unit, 220, 221, 329, 330, 515
Little Ice Age, 132, 156, 294
Living in the anthropocene, 141, 142, 275, 337, 357, 557
Long-term sustainability, 166
Lorenz, Klaus, 175

Loss of glaciers, ice, and snow, 132
Lovelock, James, 219, 355

M
Manhattan District Project, 295
Mankind's future in the anthropocene, 284
Man-made ecosystems, 142
Manthropocene, 304
Marine ecosystems of the phanerozoic, 223
Market-oriented economic system, 115
Markl, H., 302
Marsh, G.P., 19, 27, 42, 105, 302
Marxism and the anthropocene, 281
Mauna Loa CO_2 records of C.D. Keeling, 62
Max Planck Institute for Chemistry, 3, 395, 541
Max Planck Institute for the History of Science, 1, 568
McNeill, John R., 103, 217
Medicine, 285, 290, 291, 311, 312, 315, 316, 321, 327, 328, 349, 357, 367, 368, 378
Meghalayan Stage/Age, 299
Melting mountain glaciers and snows, 132, 520
Mesozoic, 221, 225, 230, 241
Meteorological Institute of Stockholm University, 4
Methane, 12, 23, 24, 30, 35, 39, 51–53, 66, 83, 107, 123, 124, 126, 133, 137, 144, 164, 165, 175, 176, 198–200, 226, 227, 269
Microbialisation of coastal systems, 84
Mid-Piacenzian Warm Period, 226, 227
Milankovich orbital parameters, 39
Millennium Ecosystem Assessment, 146, 293
Miocene, 211, 233
Mitigate, 115, 133, 134, 139, 176, 183–185, 187, 529
Mitigating, 117, 183, 255
Mitigation, 85, 116, 117, 137–139, 185, 187, 254, 258, 261, 268, 269, 297, 543
Mittelstraß, Jürgen, 129
Modelling studies, 259, 260
Models and observations, 87, 97
Models of biodiversity, 92
Modern environmentalism, 115
Molina, Mario, 7, 9, 61, 145, 221, 379, 448, 449, 560
Monitoring of the Earth system, 90
Montreal Protocol, 9, 12, 53, 94, 138, 184

Moral hazard issue, 255
Moratorium on implementation, 258
Mountain glaciers in the anthropocene, 129
Mount Pinatubo, 117, 187
Music in the anthropocene, 281
Myxocene, 304

N
Nakicenovic, Nebosja, 75
National Academy of Sciences, 257, 322, 373, 560
National and international law, 291, 321, 346
National Center for Atmospheric Research, 4, 8, 128
NATO Double Track Decision, 10
Natural carbon cycle dynamics, 107
Natural feedback processes in the climate system, 132
Natural Sciences, 218, 290, 291, 297, 305, 306, 316, 321, 322, 324–326, 331, 332, 337, 351, 352, 356, 357, 359, 363, 366–368, 376, 378, 532, 568
Nature, 2, 12, 31, 34, 41, 81, 86, 93, 97, 104–106, 126, 139, 142–144, 148, 159, 161, 170, 195, 196, 204, 209, 217, 220, 224, 235–237, 239, 240, 258, 275, 287, 294, 302, 304, 305, 329, 331, 333, 337, 355, 378, 379, 517, 519, 528, 530, 534
Nature of human-environment interactions, 161
Necrocene, 304
Neolithic or agricultural revolution, 293, 294
Neoproterozoic and Cambrian sedimentary strata, 223
New geological epoch, 5, 13, 33, 134, 141, 149, 175, 196, 200, 218, 298, 303, 329, 351, 356, 377, 514, 519, 520
New global ethos, 141
New phase of Earth history, 337
New planetary sphere, xv
New political, economic, security and ideological order, 293
New regime of international institutions, 114
New technologies, 95, 96, 114, 141, 206, 285, 295
Nitrate aerosol, 58
Nitrogen emissions in the environment, 124
Nitrogen fertilizers, 24, 29, 40, 52, 66, 126
Nitrous dioxide, 124
Nitrous oxide, 4, 12, 30, 84, 109, 126, 175, 176, 198, 199

No-analogue situation, 93
Nobel Prize in chemistry, 378, 448, 559, 560
Noble goal of sustainable development, 275
Noösphere, 19, 23, 27, 30, 105
Novacene, 304
Nuclear energy, 185
Nuclear era, 295, 297, 377
Nuclear fallout, 293
Nuclear fission, 116, 295
Nuclear Winter, 1, 5, 9–12, 379, 512, 560, 565, 568
Numbers of motor vehicles, 150

O

Ocean acidification, 163, 164, 204, 239, 255, 299
Odada, Eric, 217
Orbis Spike, 294
Ordovician-Silurian boundary, 222
Ordovician to Devonian stratigraphic records, 223
Oreskes, Naomi, 217
Oxidation efficiency, 35, 36
Ozone, 3–9, 12, 20, 24, 29, 30, 33–36, 40, 41, 43, 47–54, 60–62, 67, 69–71, 81, 83, 86, 94, 115, 123, 124, 126, 133, 137, 147, 148, 154, 164, 165, 175, 176, 180, 187, 193–195, 200, 201, 259, 262, 269, 378, 441, 510–512, 520, 522, 523, 529, 538, 559, 560, 565, 568
Ozone depletion, 34, 48, 69, 70, 193, 511, 565
Ozone hole, 7–9, 12, 20, 24, 30, 34, 41, 48, 60–62, 67, 69, 70, 81, 83, 94, 123, 124, 126, 147, 180, 193, 195, 201, 510, 520, 560, 565
Ozone layer, 5–9, 20, 30, 41, 61, 67, 124, 138, 201, 378, 448, 560, 568

P

Pachauri, Rajendra, 129
Pacific decadal oscillation, 219
Painter, Thomas H., 129
Palaeoanthropocene, 203, 205–209, 211, 518
Palaeoclimate reconstructions, 210, 212
Palaeoenvironmental Sciences, 211
Paleocene-Eocene thermal maximum, 226, 227, 234, 237
Pandemic, 24, 85, 86, 163
Paris climate agreement, 269

Past Global Changes (PAGES), 230, 238, 239
Pattberg, Philipp, 304, 336, 337, 339, 342, 371, 374
Peace ecology, 379, 380, 569
Peace Research and European Security Studies (AFES-PRESS), 289, 387, 388, 569
Peak oil, 146, 147
People and the planet, 146, 148, 275
Performing arts, 290, 291, 315, 321, 327, 344, 348, 367, 378
Persson, Åsa, 145
Phanerozoic Eon, 241, 526
Phanerozoic marine diversity curve, 224
Phase of the 'anthropocene', 44
Philosophy of the anthropocene, 281
Photochemical smog, 34, 36, 47
Photosynthesis, 31, 50, 108, 116, 117
Photovoltaic, 116
Physical climate system, 79, 81
Planetary boundaries, 161, 164, 355, 356, 521
Planetary environment, 152
Planet's capability to absorb our wastes, 145, 515
Planetary stewardship, 145, 152, 159, 165, 169, 170, 515
Plasticene, 304
Pleistocene, 149, 158, 162, 204, 205, 221, 226, 228, 233, 234, 357, 513
Pliocene, 226, 227, 233, 234
Polar stratospheric clouds, 8, 193, 194
Policy dilemma framing, 256
Political geoecology, 380
Political time of longue durée, 296
Pontifical Academy of Sciences, 267
Pope Francis's encyclical Laudato Si, 269
Population growth, 87, 88, 112, 114, 118, 155, 160, 240, 299, 522
Pöschl, Ulrich, 203
Positive and negative feedbacks, 33, 34, 226
Potential shortage of the mineral phosphorus, 146
Potsdam institute on climate change impact research, 353
Pre-Anthropocene events, 106
Precautionary principle, 95
Pre-industrial anthropos, xv
Primary triggers for ice ages and interglacials, 132
Proactive, 115, 116
Protect people and the planet, 275

Provisioning goods, 147, 148
Provisioning (resources), 95, 147, 148, 153, 160
Provisioning services, 152, 160

Q
Quaternary, 1, 156, 205, 210, 211, 228, 297, 329, 390, 447, 451, 511, 513
Quaternary glacial-interglacial cycles, 226
Quaternary period, 228, 330
Quaternary plant and animal species, 203, 518

R
Rabassa, Jorge, 129
Radiation management, 254
Radiogenic isotope systems, 209
Ramanathan, Veerabhadran, 47, 129, 145
Receding glaciers require urgent responses, 131
Reduce hazardous air pollutants, 269
Reduce the emission of greenhouse gases, 128
Reduce the human modification of the global environment, 116
Reduce worldwide carbon dioxide emissions, 133
Reducing climate change, 258
Reduction in CFC emissions, 94
Reductions in emissions, 126
Reflecting particles, 256
Reflectivity of the Earth, 253
Regional Palaeoclimate, 209
Regional surface climate forcing, 56
Regulating services, 148, 153, 158, 160
Renewable energy, 62, 110, 133, 179, 184, 185, 275
Renn, Jürgen, 293–296, 302, 303, 305, 344, 345, 372, 395
Research challenges, 96
Research on geoengineering, 258
Resilience, 76, 145, 158, 160, 167, 168, 170, 353, 355, 361, 368, 375, 515, 546
Resilience Centre in Stockholm, 361
Resource depletion, 92, 94, 133
Responsibility of the scientific community, 259
Restoration of the natural environment, 31, 116, 518
Review of global debate on the Anthropocene, 281
Revkin, Andrew, 217

Richardson, Katherine, 145
Richter, Daniel de B., 217
Rise of sulphur dioxide, 124
Rising temperatures, 276
Risk, 1, 65, 89, 116, 134, 137–139, 145, 146, 163, 175, 176, 180, 185, 201, 209, 256, 259–262, 267, 269, 283, 284, 305, 354, 355, 515, 517, 520, 528, 565
Robock, Alan, 129
Rockström, Johan, 145
Rowland, Frank Sherwood, 7, 9, 61, 559, 560
Rubbia, Carlo, 129
Ruddiman, William F., 39, 41, 107, 150, 180, 204, 205, 209, 232, 233, 293, 294
Russell, Lynn M., 129

S
Sánchez Sorondo, Marcelo, 129
SCAD, 306
Scheffer, Marten, 145
Schellnhuber, Hans Joachim, 130, 145, 217
Scholz, Denis, 203
Schöne, Bernd R., 203
Schreg, Rainer, 203
Schwägerl, Christian, 141
Science Citation Index, 307, 311
Science for sustainable development, 93
Scientific Committee on Problems of the Environment (SCOPE), 10, 354, 565
Scientific history, 289
Scientific knowledge community, 357, 359, 363, 368, 376, 378
Scientific lead authors, 368, 374, 378
Scientific revolutions, 294, 379
Scopus, 290, 291, 307, 308, 310, 312–326, 328, 331, 332, 334–351, 359–364, 367–378, 380–384, 386, 389
Scripps institution of oceanography, 62, 442
SDI programme, 11
Sea level rise [sea-level rise], 29, 41, 62, 65, 67, 116, 132, 133, 135, 146, 182, 198, 227, 240, 257, 276, 299, 520, 526
Sea salt, 58, 60
Secondary consequences of eutrophication, 84
Second stage of the anthropocene, 112, 150
Sediment of global industrialisation, 296
Self-reliance and preservation of local identities, 128
Severity of global change, 117
Shifting precipitation patterns, 276

Shifting societal values, 117
Shifts in rainfall patterns, 146
Short-lived climate-forcing pollutants, 257
Single Scattering Albedo (SSA), 57
Sirocko, Frank, 203
Sixth great extinction, 110, 238
Sixth mass extinction event, 175, 177, 195, 241
Social sciences, 2, 218, 290, 291, 297, 303, 305, 306, 310–312, 315, 321, 322, 324, 327, 328, 331, 332, 334, 336, 342, 350, 352, 356, 357, 359, 361, 363, 366–368, 370, 376–380, 524, 527, 528, 547, 554, 569
Social Sciences Citation Index (SSCI), 311
Societies collapse, 160
Sociological approaches of discourse analysis, 290
Sociotechnical transformations, xiv
Solar forcing in the late quaternary, 107
Solar radiation, 8, 31, 34, 37, 47, 48, 50, 54, 57, 60, 62, 63, 65, 116, 117, 125, 126, 175, 187, 198, 219, 228, 231, 240
Solar radiation management, 163, 186, 254
Solar thermal, 116
Sorooshian, Soroosh, 129
Species extinction, 20, 24, 29, 135, 177, 209, 211, 267, 302
Speed of environmental change, 162
Spheroidal carbonaceous particles, 232
SRM geoengineering, 261
Stability of the cryosphere, 116
Stage 3 of the anthropocene, 114
Start of the Anthropocene, 39, 204, 293, 294, 329, 519
Steffen, Will, 39, 75, 103, 145, 217
Stewardship of the Earth system, 44, 155
Stewards of the Earth system, 114, 149
Stochastic resonance, 82, 83, 97
Stocker, Thomas F., 129, 204, 549
Stockholm resilence centre, 353
Stockholm University, 4, 353, 559, 568
Stoermer, Eugene F., 19
Stoppani, Antonio, 19, 23, 27, 105, 302
Stratigraphic and Earth system approaches, 217, 232
Stratigraphic anthropocene, 232, 233, 235
Stratigraphic evidence from artificial radionuclides, 296
Stratigraphic handbook of the International Commission on Stratigraphy, 220
Stratigraphic Holocene Epoch, 235
Stratigraphic perspectives, 241

Stratigraphy, 1, 2, 149, 217–220, 232, 241, 281, 297, 299, 329, 330, 358, 390, 447, 451, 513, 516, 526
Stratospheric aerosol injections, 253, 261, 518
Stratospheric aerosol particle injections, 256, 262
Stratospheric ozone chemistry, 29, 34, 35, 40, 67
Stratospheric 'ozone hole', 520
Strengthen carbon sinks by reforestation, 133
Subcommission on quaternary stratigraphy, 298, 299, 358, 359, 532
Summerhayes, Colin, 217
Super Sonic transports, 5
Supersonic transports and stratospheric ozone, 5
Sustainability of Earth's life support system, 114
Sustainability transition, 291, 305, 379
Sustainable development goals, 217, 238, 269
Svedin, Uno, 145
Svenning, Jens-Christian, 325, 333, 345, 369, 373
Systems of models and observations, 87
Syvitski, Jaia, 334, 362, 393, 526
Syvitski, James, 217

T
Taboo toward research, 253
Talaue-McManus, Liana, 75
Techno-fix approach, 256
Technological fix, 62, 92, 93, 95, 96, 98, 303
Technological opportunities, 31, 116
Technological options, 92
Technological substitution, 92, 96, 98
Technologies to recycle substances, 143
Technology, 21, 30, 31, 44, 48, 62, 65, 68, 90, 91, 93, 94, 96, 114, 116, 117, 142, 143, 150, 177, 256, 260, 261, 270, 271, 284–286, 291, 303, 309–312, 315, 321, 322, 324, 327, 349, 367, 368, 377, 378, 443–445, 447, 449, 450, 518, 519, 529, 538, 569
Technology and engineering, 291, 321, 349
Technosphere, 232, 523, 555
Teilhard de Chardin, Pierre, 19, 23, 27, 105, 200, 302
Temperature climbs toward 2°C above pre-industrial, 146

The good, the bad and the ugly, 303

Thermohaline Circulation (THC), 79, 81, 82

Third stage of the Anthropocene, 115

Thompson, Lonnie G., 129

Threat of nuclear war, 134

Timescale of human history, 221

Tipping elements analysis of Lenton et al., 168

Toon, Owen B., 129

Töpfer, Klaus, 175

Tost, Holger, 203

Toward planetary stewardship, 159

Trace gases, 4, 176, 187, 211

Traditional air pollution, 60

Transformed mobility, 112

Transition from a green to an arid Sahara, 82

Transition of our globalising society, 117

Transition to a safe anthropocene, 275

Tropical deforestation, 36, 150

Tropics and subtropics in atmospheric chemistry, 35, 36

Tropospheric NO_2, O_3, and CH_2O measured with the DOAS, 55

TTAPS Study, 10

Turner II, Billie, 75

Turning point, 294–296, 305, 320, 326, 333, 356, 379

Twilight at noon, 9

U

Uncertainties in the projections, 89

Uncontrollable environmental change, 116

Uncontrolled scientific and technological progress, 284

Understanding planetary dynamics, 156

United Nations Charter, 295

United Nations Educational, Scientific and Cultural Organization, 354

United Nations Environment Programme, 353

United States, 7, 11, 53, 114, 185, 186, 284, 292, 509

Unmitigated anthropocene trajectory, 240

Unravelling Earth system evolution, 223

Urban environments, 113, 150

Urbanisation, 1, 20, 28, 104, 114, 123, 125, 169, 177, 195, 302

US role in world politics, 295

Utility of chronostratigraphy for Earth system science, 221

V

Variations of the Earth's surface temperature, 125

Vernadsky, Vladimir I., 19, 23, 27, 42, 48, 105, 200, 218, 219, 302, 355

Vidas, Davor, 217

Vitousek, 20, 29, 40, 66, 178, 302

Volcanic emissions, 58

Von Humboldt, Alexander, 143, 218, 355

Vött, Andreas, 203

W

Wagreich, Michael, 217

Warming of the atmosphere by greenhouse gases, 124

Warming of the Earth is unequivocal, 135

Warming by the 'greenhouse gases', 34

Warning to humanity and a call for fast action, 134

Water and food shortages for many vulnerable peoples, 135

Waters, Colin N., 217

Water use, 28, 68, 91, 123, 195

We are taking control of Nature's realm, 141

Web of Science, 255, 290, 291, 307, 308, 310–323, 325–328, 331, 332, 334–351, 359–369, 373–378, 380–389

Web of Science Core Collection, 310, 311, 314, 376

Websites on the anthropocene, 320, 531

We fundamentally change the biology and the geology of the planet, 142

We humans are becoming the dominant force for change on Earth, 141

Welcome to the anthropocene, 3, 532, 534, 548, 550, 552

Weller, Christine G., 203

Widespread biodiversity loss, 158

Williams, Mark, 145, 217

Wind power, 108, 116

Wing, Scott L., 217

Wise use of Earth's resources, 116

Wolfe, Alexander P., 217

World Catalogue, 290, 291, 308, 309, 313, 314, 320, 322, 325, 326, 328, 334, 344, 346–351, 359, 367

World Climate Research Programme, 231, 352, 353

World economy, 28, 68, 86, 112–114

World Meteorological Organisation, 7

World of Science, 351

World War I, 114

World War II, 84, 114, 161, 293–296, 357

Y
Younger Dryas event, 81
Yusoff, Kathryn, 334, 335, 345, 370, 374

Z
Zaelke, Durwood, 129
Zalasiewicz, Jan, 145, 217

Printed in the United States
by Baker & Taylor Publisher Services